Algebra and Trigonometry

Third Edition

Margaret L. Lial/Charles D. Miller
American River College

Scott, Foresman and Company
Glenview, Illinois

Dallas, Tex.   Oakland, N.J.
Palo Alto, Cal.   Tucker, Ga.   London, England

Cover—*Angle of Incidence* by McCrystle Wood. Collection of Federated Department Stores, Inc., Cincinnati, Ohio.

Library of Congress Cataloging in Publication Data

Lial, Margaret L.
  Algebra and trigonometry.

  Includes index.
  1. Algebra. 2. Trigonometry. I. Miller,
Charles David, 1942–   .II. Title.
QA154.2.L49 1983   512'.13   82-24998
ISBN 0-673-15794-6

Certain portions of this book were previously published in *Fundamentals of College Algebra*, Copyright © 1982 Scott, Foresman and Company, and *Trigonometry, 2nd ed.*, Copyright © 1981 Scott, Foresman and Company.

1 2 3 4 5 RRC 86 85 84 83 82

Preface

The third edition of ALGEBRA AND TRIGONOMETRY provides a comprehensive and mathematically sound treatment of those topics in algebra and trigonometry needed for further study in mathematics. Explanations are precise yet written for student understanding. Students who use this text will be well prepared for success in such courses as calculus, statistics, finite mathematics, etc. High school algebra or a college course in intermediate algebra is the prerequisite assumed for the course.

HIGHLIGHTS OF THE TEXT

Emphasis on Functions. This text features a thorough treatment of functions and of the techniques of graphing in general. Linear, polynomial, and quadratic functions, conic sections, exponential and logarithmic functions, and the trigonometric functions are covered in depth. General techniques of graphing are presented and new graphs are obtained from familiar ones by investigating symmetry and translations.

Examples. The text provides an abundance of examples. All major points are developed with detailed, step-by-step examples.

Pedagogically Oriented Format. The book was designed to facilitate studying. Definitions and rules are emphasized and highlighted. Examples are easy to follow.

Applications. A systematic approach to problem solving, based on the work of George Polya, is introduced early in the text. The number of problems involving applications has been increased to over 500. This edition includes applications from the fields of business, engineering, computer science, physics, chemistry, biology, astronomy, navigation, demography, and political science.

Calculators. The use of calculators has been integrated into the text and numerous calculator exercises, identified by the symbol ⌨, have been included. Brief discussions of the difficulties that can arise when using calculators are also presented throughout the text.

EXERCISES **Expanded Exercise Sets.** This edition contains over 5700 graded drill exercises. Answers to the odd-numbered exercises are located at the back of the text.

Review Exercises. An extensive set of Review Exercises is included at the end of each chapter. There are over 1000 Review Exercises in this edition.

Problems from Calculus Textbooks. Most of the Review Exercise sets end with a selection of problems from standard calculus texts. These exercises help students see the importance of this course for their future work in calculus.

Cumulative Review Exercises. A set of Cumulative Review Exercises is included at the end of Chapter 2 and can be used by the instructor to decide which portions of Chapter 1, Fundamentals of Algebra, and Chapter 2, Equations and Inequalities, need be covered.

CONTENT FEATURES **Thorough Algebra Review in Chapter 1.** This chapter has been revised to be as complete as possible. The inclusion of a large number of examples and exercises makes the review material virtually self-teaching. More applied problems have also been included. In addition to Chapter 1, much of Chapter 2 (on equation solving) may also be considered review for many students. The Cumulative Review Exercises at the end of Chapter 2 will be of help in deciding what needs to be reviewed.

Complex Numbers. A new introductory section on complex numbers has been included (Section 1.9, pp. 51–57).

Polar Equations and Rotation of Axes. Coverage of topics in analytic geometry has been extended with the addition of a section on polar equations (Section 8.7, pp. 426–30), and an appendix on Rotation of Axes (pp. 577–82).

Conic Sections. General properties of conic sections are thoroughly discussed. This material, however, is not a prerequisite for any other topic presented.

Polynomial and Rational Functions. These topics are introduced early in the book through a discussion of graphing, and zeros of polynomials are thoroughly discussed in Chapter 10.

Partial Fractions. A full section in the chapter on Zeros of Polynomials is devoted to this topic.

Basics of Probability. A full section on probability theory is preceded by a section on permutation and combination.

SUPPLEMENTS The STUDY GUIDE features additional examples for each section, as well as solutions to selected text exercises, and a chapter test with answers for each chapter.

The INSTRUCTOR'S MANUAL contains a diagnostic pretest, answers to even-numbered exercises, a test bank, a section on linear programming, and answers to the pretest and test bank items.

Acknowledgments

In particular, we would like to express our gratitude to the following individuals who reviewed the entire manuscript: Chris Cosner, Texas A & M University; Milton Cox, Miami University; John Kuisti, Michigan Technological University; Doris Nice, University of Wisconsin–Parkside; Rosanne Proga, Boston University; James Rauff, College of Lake County; Charles Sinclair, Portland State University; Betty Travis, University of Texas–San Antonio; and Stewart Venit, California State University, Los Angeles.

We would also like to thank those who reviewed portions of the material used in this text and offered many useful suggestions: James E. Arnold, Jr., University of Wisconsin at Milwaukee; Romae J. Cormier, Northern Illinois University; Dale Ewen, Parkland Community College; William A. Ferguson, University of Illinois; Joan Golladay, Santa Fe Community College; James Hall, University of Wisconsin; Louis Hoelzle, Bucks County Community College; Eldon Miller, University of Mississippi; Chester Miracle, University of Minnesota; Ruth Murray, College of Du Page; and Anthony Peressini, University of Illinois.

Our appreciation also goes to the staff at Scott, Foresman and Company who did an excellent job in working with us toward publication.

Contents

1 Fundamentals of Algebra

Why study algebra? For many students, the answer is, "It's required." Today algebra is required in a great variety of fields, ranging from accounting to zoology. The reason it is required in all these fields is that algebra is *useful*. Equations, graphs, and functions occur again and again in many different areas. We study all these topics in detail in this book. For example, in this text we use algebra to predict population growth, to determine the path of objects orbiting in space, and to investigate the costs versus the benefits of removing pollutants from a substance. To begin, we review the basics of algebra in this first chapter.

1.1 The Real Numbers

Numbers are the foundation of mathematics. The most common numbers in mathematics are the **real numbers,** which are all the numbers that can be written as a decimal, either repeating, such as

$$\frac{1}{3} = .33333 \ldots, \quad \frac{3}{4} = .75000 \ldots, \quad \text{or} \quad 2\frac{4}{7} = 2.571428571428 \ldots,$$

or nonrepeating, such as

$$\sqrt{2} = 1.4142135 \ldots \quad \text{or} \quad \pi = 3.14159 \ldots.$$

There are four fundamental operations on real numbers: addition, subtraction, multiplication, and division. Addition is indicated with the symbol $+$. Subtraction is written with the symbol $-$, as in $8 - 2 = 6$. Multiplication is written in a variety of ways. All the symbols 2×8, $2 \cdot 8$, $2(8)$, and $(2)(8)$ represent the product of 2 and 8, or 16. When writing products involving variables (letters used to represent numbers), no operation symbols may be necessary: $2x$ represents the product of 2 and x, while xy indicates the product of x and y. Division of the real numbers a and b is written $a \div b$, or more commonly a/b.

1

The set of real numbers, together with the relation of equality and the operations of addition and multiplication, forms the **real number system.** The key properties of the real numbers are given in the box below, where a, b, and c are letters used to represent any real numbers. (In fact, for the rest of this book we shall assume that the first few letters of the alphabet represent real numbers unless otherwise stated.)

Properties of the Real Numbers

closure properties	$a + b$ is a real number ab is a real number
commutative properties	$a + b = b + a$ $ab = ba$
associative properties	$(a + b) + c = a + (b + c)$ $(ab)c = a(bc)$
identity properties	There exists a unique real number 0 such that $$a + 0 = a \quad \text{and} \quad 0 + a = a.$$ There exists a unique real number 1 such that $$a \cdot 1 = a \quad \text{and} \quad 1 \cdot a = a.$$
inverse properties	There exists a unique real number $-a$ such that $$a + (-a) = 0 \text{ and } (-a) + a = 0.$$ If $a \neq 0$, there exists a unique real number $1/a$ such that $$a \cdot \frac{1}{a} = 1 \text{ and } \frac{1}{a} \cdot a = 1.$$
distributive property	$a(b + c) = ab + ac$

EXAMPLE 1   ◻   By the closure properties, the sum or product of two real numbers is a real number. Thus,

(a) $4 + 5$ is a real number, 9.

(b) $-9(-4)$ is a real number, 36.

(c) $8 \cdot 0$ is a real number, 0.

(d) $\sqrt{5} + \sqrt{3}$ is a real number.

(e) $\sqrt{2} \cdot \sqrt{11}$ is a real number.   ◻

EXAMPLE 2   ◻   The following statements are examples of the commutative property. Notice that the order of the numbers changes from one side of the equals sign to the other.

(a) $6 + x = x + 6$

(b) $(6 + x) + 9 = (x + 6) + 9$

(c) $(6 + x) + 9 = 9 + (6 + x)$

(d) $5 \cdot (9 \cdot 8) = (9 \cdot 8) \cdot 5$

(e) $5 \cdot (9 \cdot 8) = 5 \cdot (8 \cdot 9)$ ◻

The associative properties are used when we have three numbers to add or multiply. For example, the associative property for addition says that to find the sum

$$a + b + c$$

of the real numbers a, b, and c, we can associate a and b, adding as follows

$$(a + b) + c;$$

or we may associate b and c,

$$a + (b + c).$$

By the associative property, either method gives the same result.

EXAMPLE 3 ☐ The following statements are examples of the associative properties. Here the order of the numbers does not change, but the placement of the parentheses does change.

(a) $4 + (9 + 8) = (4 + 9) + 8$

(b) $3(9x) = (3 \cdot 9)x$

(c) $(\sqrt{3} + \sqrt{7}) + 2\sqrt{6} = \sqrt{3} + (\sqrt{7} + 2\sqrt{6})$ ☐

The identity properties show the special properties of the numbers 0 and 1. If we find the sum of 0 and any real number a, we get the number a for an answer. Thus, 0 preserves the identity of a real number under addition. For this reason, 0 is the **identity element for addition.** In the same way, 1 preserves the identity of a real number under multiplication. Thus, 1 is the **identity element for multiplication.**

EXAMPLE 4 ☐ By the identity properties,

(a) $-8 + 0 = -8$

(b) $(-9)1 = -9$ ☐

According to the addition inverse property, if we start with any real number a, we will be able to find a real number, written $-a$, such that the sum of these two numbers is 0, or $a + (-a) = 0$. The number $-a$ is called the **additive inverse** or **negative** of a.

Don't confuse the *negative of a number* with a *negative number*. Since a is a variable, it can represent a positive or a negative number (as well as zero). The negative of a, written $-a$, can also be either a negative or a positive number (or zero). Don't make the common mistake of thinking that $-a$ *must* be a negative number. For example if a is -3, then $-a$ is $-(-3)$ or 3.

For each real number a (except 0), there is a real number $1/a$ such that the product of a and $1/a$ is 1, or

$$a \cdot \frac{1}{a} = 1 \quad (a \neq 0).$$

The symbol $1/a$ is often written a^{-1}, so that, by definition,

$$a^{-1} = \frac{1}{a}.$$

The number $1/a$ or a^{-1} is called the **multiplicative inverse** or **reciprocal** of the number a. Every real number except 0 has a reciprocal.

EXAMPLE 5   By the inverse properties,

(a) $9 + (-9) = 0$

(b) $-15 + 15 = 0$

(c) $6 \cdot \frac{1}{6} = 1$ $\left(\text{so that } 6^{-1} = \frac{1}{6}\right)$

(d) $-8 \cdot \left(\frac{1}{-8}\right) = 1$

(e) $\frac{1}{\sqrt{5}} \cdot \sqrt{5} = 1.$

(f) There is no real number x such that $0 \cdot x = 1$, so that 0 has no inverse for multiplication. ☐

One of the most important properties of the real numbers, and the only one that involves both addition and multiplication, is the distributive property. The next example shows how this property is applied.

EXAMPLE 6   By the distributive property,

(a) $9(6 + 4) = 9 \cdot 6 + 9 \cdot 4$ (b) $3(x + y) = 3x + 3y$

(c) $-\sqrt{5}(m + 2) = -m\sqrt{5} - 2\sqrt{5}$ (Here the radical is written second, so we won't confuse, for example, $-\sqrt{5}m$ and $-\sqrt{5m}$.) ☐

The distributive property can be extended to include more than two numbers in the sum, as follows.

$$a(b + c + d + e + \cdots + n) = ab + ac + ad + ae + \cdots + an$$

This form is called the **extended distributive property.**

Another key property is called the **substitution property.**

Substitution Property

If $a = b$, then a may replace b in any expression without affecting the truth or falsity of the statement.

Many further properties of the real numbers can be proven directly from those given above. For example, the following two important properties are used in the next chapter in solving equations.

Addition and Multiplication Properties

If $a = b$, then $a + c = b + c$, and $ac = bc$.

In words, these properties say that the same number may be added or multiplied on both sides of an equation.

The next box shows two special properties of the number 0.

Properties of Zero

$a \cdot 0 = 0$

$ab = 0$ if and only if $a = 0$ or $b = 0$

The second property above contains the phrase "if and only if." By definition,

If and Only If

p if and only if q

means

if p, then q and   if q, then p.

Finally, we list some properties of negatives.

Properties of Negatives

$$-(-a) = a$$
$$-a(b) = -(ab)$$
$$a(-b) = -(ab)$$
$$(-a)(-b) = ab$$

The proofs of many of these properties are included in Exercises 61–66 below. Example 7 shows a general style of proof that can be used for many of these properties.

EXAMPLE 7 ☐   If a is a real number, prove that

$$-(-a) = a.$$

We shall prove that both $-(-a)$ and a are additive inverses of $-a$. Since a real number has only one additive inverse, this will show that $-(-a)$ and a are equal. First,

$$-a + [-(-a)] = 0$$

since $-(-a)$ is the additive inverse of $-a$. Also,

$$-a + a = 0,$$

since a is the additive inverse of $-a$. Since both $-(-a)$ and a are additive inverses of $-a$, and since $-a$ has only one additive inverse, we must have

$$-(-a) = a. \quad \blacksquare \quad \square$$

The Properties of Real Numbers above apply to addition or multiplication. Two other common operations for the real numbers are subtraction and division. These two operations are defined in terms of the operations of addition and multiplication, respectively.

Subtraction is defined by saying that the difference of the numbers a and b, written $a - b$, is

Subtraction

$$a - b = a + (-b).$$

That is, to subtract b from a, add a and the additive inverse of b.

EXAMPLE 8

(a) $7 - 2 = 7 + (-2) = 5$

(b) $-8 - (-3) = -8 + (+3) = -5$

(c) $6 - (-15) = 6 + (+15) = 21$

Division of a real number a by a nonzero real number b is defined as follows.

Division

$$\frac{a}{b} = a \cdot \frac{1}{b} = ab^{-1}$$

That is, to divide a by b, multiply a by the reciprocal of b.

EXAMPLE 9

(a) $\frac{6}{3} = 6 \cdot \frac{1}{3} = 2$ (b) $\frac{-8}{4} = -8 \cdot \frac{1}{4} = -2$

(c) $\frac{-9}{0}$ is meaningless since there is no reciprocal for 0; also, $\frac{0}{0}$ is meaningless.

The properties of quotients are listed in the following box. (All denominators represent nonzero real numbers.)

Properties of Quotients

$$\frac{a}{b} = \frac{c}{d} \text{ if and only if } ad = bc \qquad \frac{a}{b} + \frac{c}{d} = \frac{ad + bc}{bd}$$

$$\frac{ac}{bc} = \frac{a}{b} \qquad\qquad\qquad \frac{a}{b} - \frac{c}{d} = \frac{ad - bc}{bd}$$

$$\frac{a}{-b} = \frac{-a}{b} = -\frac{a}{b} \qquad\qquad \frac{a}{b} \cdot \frac{c}{d} = \frac{ac}{bd}$$

$$\frac{-a}{-b} = \frac{a}{b} \qquad\qquad\qquad \frac{a}{b} \div \frac{c}{d} = \frac{a}{b} \cdot \frac{d}{c}$$

To avoid possible ambiguity when working problems, use the following **order of operations,** which has been generally agreed upon. (By the way, this order of operations is used by computers and many calculators.)

Order of Operations

1. Find any indicated powers.

2. Do any work indicated inside parentheses or brackets.

3. Perform any multiplications or divisions, in order, from left to right.

4. Do any additions or subtractions, in order, from left to right.

5. If the problem involves a fraction bar $\left(\text{the } - \text{ in } \dfrac{a}{b}\right)$, treat the numerator and the denominator separately.

EXAMPLE 10   □   Use the order of operations given above to simplify the following.

(a) $(6 \div 3) + 2 \cdot 4 = 2 + 2 \cdot 4 = 2 + 8 = 10$

(b) $\dfrac{-9(-3) + (-5)}{2(-8) - 5(3)} = \dfrac{27 + (-5)}{-16 - 15} = \dfrac{22}{-31} = -\dfrac{22}{31}$

(c) $4 \cdot 2^3 = 4 \cdot 8 = 32$ □

So far, we have discussed only real numbers. However, there are several sub-sets* of the set of real numbers that come up so often that they are given special names. Some of the sets in the following box are written with **set-builder notation;** for example, with this notation, $\{x \mid x \text{ has property } P\}$† represents the set of all elements having some specified property P.

Subsets of the Real Numbers

Natural numbers	$\{1, 2, 3, 4, \ldots\}$
Whole numbers	$\{0, 1, 2, 3, 4, \ldots\}$
Integers	$\{\ldots, -3, -2, -1, 0, 1, 2, 3, \ldots\}$
Rational numbers	$\left\{\dfrac{p}{q} \mid p \text{ and } q \text{ are integers, } q \neq 0\right\}$, or
	$\{x \mid x \text{ has a repeating or terminating decimal expansion}\}$
Irrational numbers	$\{x \mid x \text{ has a nonrepeating, nonterminating decimal expansion}\}$ or $\{x \mid x \text{ is a real number that is not rational}\}$

*Set A is a subset of set B if and only if every element of set A is also an element of set B.

†$\{x \mid x \text{ has property } P\}$ is read "The set of all x, such that x has property P."

EXAMPLE   ☐   Let set $A = \{-8, -6, -3/4, 0, 3/8, 1/2, 1, \sqrt{2}, \sqrt{5}, 6, \sqrt{-1}\}$. List the
11   elements from set A that belong to each of the following sets.

(a) The *natural numbers* in set A are 1 and 6.

(b) The *whole numbers* are 0, 1, and 6.

(c) The *integers* are -8, -6, 0, 1, and 6.

(d) The *rational numbers* are -8, -6, $-3/4$, 0, 3/8, 1/2, 1, and 6.

(e) The *irrational numbers* are $\sqrt{2}$ and $\sqrt{5}$. (In further mathematics courses you will
see that these numbers do not have repeating or terminating decimal expansions.)

(f) All elements of A are *real numbers* except $\sqrt{-1}$. Square roots of negative
numbers are discussed later in this chapter.   ☐

The relationships among the various sets of numbers are shown in Figure 1.1.
Note that all the numbers shown are real numbers.

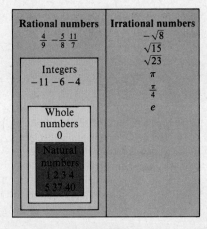

Figure 1.1

*Identify the properties which are illustrated in each
of the following. Some will require more than one
property. Assume all variables represent real numbers.*

1. $8 \cdot 9 = 9 \cdot 8$

2. $3 + (-3) = 0$

3. $3 + (-3) = (-3) + 3$

4. $0 + (-7) = (-7) + 0$

5. $-7 + 0 = -7$

6. $8 + (12 + 6) = (8 + 12) + 6$

7. $[9(-3)] \cdot 2 = 9[(-3) \cdot 2]$

8. $8(m + 4) = 8m + 8 \cdot 4$

9. If x is a real number, then $x + 2$ is a real number.

10. $(7 - y) + 0 = 7 - y$

11. $8(4 + 2) = (2 + 4)8$

12. $x \cdot \dfrac{1}{x} + x \cdot \dfrac{1}{x} = x\left(\dfrac{1}{x} + \dfrac{1}{x}\right)$ (if $x \neq 0$)

For Exercises 13–22, tell if the statement is true or false. If it is false, tell why.

13. By the identity property, $8 \cdot 0 = 0$.

14. By the identity property, $8 + 1 = 9$.

15. The sum of two whole numbers is always a whole number.

16. The quotient of two whole numbers is always a whole number.

17. The quotient of two rational numbers is always a rational number.

18. The difference of two real numbers is always a real number.

19. The set $\{0, 1\}$ is closed with respect to subtraction.

20. The set $\{0, 1\}$ is closed with respect to multiplication.

21. The set $\{1, -1\}$ is closed with respect to division.

22. The set of irrational numbers contains an identity element for multiplication.

Is each of the following sets closed with respect to the indicated operations?

23. rationals, subtraction

24. rationals, division

25. irrationals, addition

26. reals, subtraction

27. irrationals, multiplication

28. irrationals, division

29. $\{0\}$, addition

30. $\{0\}$, multiplication

31. $\{1\}$, addition

32. $\{1\}$, multiplication

Simplify each of the following expressions using the order of operations given in the text.

33. $9 \div 3 \cdot 4 \cdot 2$ 34. $18 \cdot 3 \div 9 \div 2$

35. $8 + 7 \cdot 2 + (-5)$

36. $-9 + 6 \cdot 5 + (-8)$

37. $(-9 + 4 \cdot 3)(-7)$

38. $-15(-8 - 4 \div 2)$

39. $\dfrac{-8 + (-4)(-6) \div 12}{4 - (-3)}$

40. $\dfrac{15 \div 5 \cdot 4 \div 6 - 8}{-6 - (-5) - 8 \div 2}$

41. $\dfrac{17 \div (3 \cdot 5 + 2) \div 8}{-6 \cdot 5 - 3 - 3(-11)}$

42. $\dfrac{-12(-3) + (-8)(-5) + (-6)}{17(-3) + 4 \cdot 8 + (-7)(-2) - (-5)}$

43. $\dfrac{-9.23(5.87) + 6.993}{1.225(-8.601) - 148(.0723)}$

44. $\dfrac{189.4(3.221) - 9.447(-8.772)}{4.889[3.177 - 8.291(3.427)]}$

For Exercises 45–56, choose all words from the following list which apply.

(a) natural number   (b) whole number

(c) integer   (d) rational number

(e) irrational number   (f) real number

(g) meaningless

45. 12   46. 0

47. -9 48. 3/4

49. $-5/9$ 50. $\sqrt{8}$

51. $-\sqrt{2}$ 52. $\sqrt{25}$

53. $-\sqrt{36}$ 54. 8/0

55. $-9/(0 - 0)$ 56. $0/(3 + 3)$

57. We know that $a(b + c) = ab + ac$. This property is the distributive property of multiplication over addition. Is there a distributive property of addition over multiplication? That is, does

$$a + (b \cdot c) = (a + b)(a + c)$$

for all real numbers a, b, and c? To find out, try various sample values of a, b, and c.

Give a reason for each step in the following:

58. Prove: $a - a = 0$.

$a - a = a + (-a)$

$a + (-a) = 0$

$a - a = 0$

59. Prove: $(a + b)c = ac + bc$.

$(a + b)c = c(a + b)$

$c(a + b) = ca + cb$

$(a + b)c = ca + cb$

$ca + cb = ac + bc$

$(a + b)c = ac + bc$

60. Prove: $(a + b) + (-a) = b$.

$(a + b) + (-a) = a + [b + (-a)]$

$a + [b + (-a)] = a + [-a + b]$

$a + [-a + b] = [a + (-a)] + b$

$[a + (-a)] + b = 0 + b$

$0 + b = b$

$(a + b) + (-a) = b$

Prove each of the following statements about real numbers a, b, and c.

61. $-a(b) = -(ab)$ **62.** $(-a)(-b) = ab$

63. $a \cdot 0 = 0$

64. If $a = b$, then $a + c = b + c$.

65. If $a = b$, then $ac = bc$.

66. If $ab = 0$, then $a = 0$ or $b = 0$.

67. $ab(a^{-1} + b^{-1}) = b + a$

68. $(a + b)(a^{-1} + b^{-1}) = 2 + ba^{-1} + ab^{-1}$

1.2   The Number Line and Absolute Value

In this section we study the order relationship between real numbers. (For example, we decide which of two given real numbers is smaller.) Such work is often easier if we use a **number line**, a geometric representation of the set of real numbers. To each real number, we can associate a point on a line, and to each point on a line we can associate a real number. One way to set up this correspondence between points and numbers is as follows: Draw a line and choose any point on the line to represent 0. (See Figure 1.2.) Then choose any point to the right of 0 and label it 1. The distance from 0 to 1 sets up a unit measure that can be used to locate other points to the right of 1, which are labeled 2, 3, 4, 5, and so on, and points to the left of 0, labeled -1, -2, -3, -4, and so on.

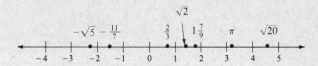

Figure 1.2

Points representing rational numbers, such as 2/3, 16/9, $-11/7$, and so on, can be found by dividing the intervals between integers. For example, 16/9 (or 1 7/9) can be found by dividing the interval from 1 to 2 into 9 equal parts (using methods given in geometry), and then choosing the correct point. Numbers such as $\sqrt{2}$, $\sqrt{3}$, $\sqrt{5}$, and so on, can be located by other geometric constructions. A point corresponding to other irrational numbers can be found to any desired degree of accuracy by using decimal approximations for the numbers.

Figure 1.2 shows a number line with the points corresponding to several different numbers marked on the line. A number that corresponds to a particular point on a line is called the **coordinate** of the point. For example, the leftmost marked point in Figure 1.2 has coordinate -4. The correspondence between points on a line and the real numbers is called a **coordinate system** for the line. (From now on, the phrase "the point on a number line with coordinate a" will be abbreviated as "the point with coordinate a," or simply "the point a.")

EXAMPLE 1   Locate the elements of the set $\{-2/3, 0, \sqrt{2}, \sqrt{5}, \pi, 4\}$ on a number line.

The number π is irrational, with $\pi \approx 3.14159$ ($\approx$ means "approximately equal to"). From the square root table in the back on this book or from a calculator, we can find that $\sqrt{2} \approx 1.414$ and $\sqrt{5} \approx 2.236$. Using this information, we get the points on the number line shown in Figure 1.3.

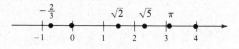

Figure 1.3

Suppose a and b are two real numbers. If the difference $a - b$ is positive, then a **is greater than** b, written $a > b$. If the difference $a - b$ is negative, then a **is less than** b, written $a < b$.

These algebraic statements can be given a geometric interpretation. If $a - b$ is positive, so that $a > b$, then a would be to the *right* of b on a number line. Also, if $a < b$, then a would have to be to the *left* of b. We can summarize both the algebraic and geometric statements.

Statement	Algebraic form	Geometric form
$a > b$	$a - b$ is positive	a is to the right of b
$a < b$	$a - b$ is negative	a is to the left of b

EXAMPLE 2   (a) In Figure 1.3 above, $-2/3$ is to the left of $\sqrt{2}$, so that $-2/3 < \sqrt{2}$.

Also, $\sqrt{2}$ is to the right of $-2/3$, giving $\sqrt{2} > -2/3$.

(b) The difference $-3 - (-8)$ is positive, showing that $-3 > -8$.

The difference $-8 - (-3)$ is negative, so that $-8 < -3$.

The following variations on $<$ and $>$ are often used.

Symbol	Meaning
$\leq$	is less than or equal to
$\geq$	is greater than or equal to
$\not<$	is not less than
$\not>$	is not greater than

Statements involving these symbols, as well as $<$ and $>$, are called **inequalities.**

EXAMPLE 3   (a) $8 \leq 10$ (since $8 < 10$)   (d) $-8 \not> -2$ (since $-8 < -2$)
(b) $8 \leq 8$ (since $8 = 8$)   (e) $4 \not< 2$ (since $4 > 2$) □
(c) $-9 \geq -14$ (since $-9 > -14$)

The expression $a < b < c$ says that b is **between** a and c, since $a < b < c$ means $a < b$ and $b < c$. Also, $a \leq b \leq c$ means $a \leq b$ and $b \leq c$.

The distance on the number line from a number to 0 is called the **absolute value** of that number. The absolute value of the number a is written $|a|$. For example, the distance on the number line from 9 to 0 is 9, as is the distance from -9 to 0. (See Figure 1.4.) Therefore,

$$|9| = 9 \text{ and } |-9| = 9.$$

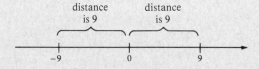

distance is 9   distance is 9

-9 0 9

Figure 1.4

EXAMPLE 4   (a) $|-4| = 4$ (d) $-|8| = -(8) = -8$
(b) $|2\pi| = 2\pi$ (e) $-|-2| = -(2) = -2$ ▨
(c) $\left|-\dfrac{5}{8}\right| = \dfrac{5}{8}$

The definition of the absolute value of the real number a can be stated as follows

Absolute Value

$$|a| = \begin{cases} a \text{ if } a \geq 0 \\ -a \text{ if } a < 0 \end{cases}$$

The second part of this definition requires some thought. If a is a negative number, that is, if $a < 0$, then $-a$ is positive. Thus, for a *negative a*,

$$|a| = -a.$$

For example, if $a = -5$, then $|a| = |-5| = -(-5) = 5$.

EXAMPLE 5 □   Write each of the following without absolute value bars.

(a) $|-8 + 2|$ Since $-8 + 2 = -6$, $|-8 + 2| = |-6| = 6$.

(b) $|\sqrt{5} - 2|$ Since $\sqrt{5} > 2$ (see Figure 1.3), $\sqrt{5} - 2 > 0$, and $|\sqrt{5} - 2| = \sqrt{5} - 2$.

(c) $|\pi - 4|$ We know $\pi < 4$, so $\pi - 4 < 0$, and therefore

$$|\pi - 4| = -(\pi - 4) = -\pi + 4 = 4 - \pi.$$

(d) $|m - 2|$ if $m < 2$ If $m < 2$, then $m - 2 < 0$, so $|m - 2| = -(m - 2) = 2 - m$. □

The definition of absolute value can be used to prove the following properties of absolute value. (See the exercises below.)

Properties of Absolute Value

$$|a| \geq 0$$

$$|-a| = |a|$$

$$|a - b| = |b - a|$$

$$|a| \cdot |b| = |ab|$$

$$\left|\frac{a}{b}\right| = \frac{|a|}{|b|} \quad (b \neq 0)$$

$$-|a| \leq a \leq |a|$$

$$|a + b| \leq |a| + |b| \quad \text{(the triangle inequality)}$$

EXAMPLE 6 □   Prove $|a| \geq 0$.

If $a \geq 0$, then by the definition of absolute value, $|a| = a$, which is assumed to be greater than or equal to 0. Thus, if $a \geq 0$, then $|a| \geq 0$.

If $a < 0$, then $|a| = -a$, which is positive. If either $a \geq 0$ or $a < 0$, we have $|a| \geq 0$, so that $|a| \geq 0$ for every real number a. □

The number line of Figure 1.5 shows the point A, with coordinate -3, and the point B, with coordinate 5. The distance between points A and B is 8 units, which can be found by subtracting the smaller coordinate from the larger. If $d(A, B)$ represents the distance between points A and B, then

$$d(A, B) = 5 - (-3) = 8.$$

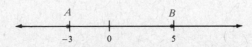

Figure 1.5

To avoid worrying about which coordinate is smaller, use absolute value as in the following definition.

Distance

> Suppose points A and B have coordinates a and b respectively. The distance between A and B, written $d(A, B)$, is
>
> $$d(A, B) = |a - b|.$$

EXAMPLE 7   ☐   Let points A, B, C, D, and E have coordinates as shown on the number line of Figure 1.6. Find the indicated distances.

(a) $d(B, E)$

Since B has coordinate -1 and E has coordinate 5,

$$d(B, E) = d(-1, 5) = |-1 - 5| = 6.$$

(b) $d(D, A) = \left|2\frac{1}{2} - (-3)\right| = 5\frac{1}{2}$

(c) $d(B, C) = |-1 - 0| = 1$

(d) $d(E, E) = |5 - 5| = 0$ ☐

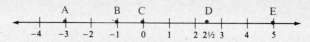

Figure 1.6

Since distance on a number line is defined in terms of absolute value, many of the properties of absolute value can be rewritten in terms of distance. Some of these properties are included in the following box. Here A and B are any two points on a number line, with A having coordinate a.

Properties of Distance

$$d(A, B) = d(B, A)$$

$$d(A, B) \geq 0$$

$$d(A, 0) = |a|$$

$$d(A, A) = 0$$

EXERCISES 1.2

Write the following numbers in numerical order, from smallest to largest.

1. $-9, -2, 3, -4, 8$

2. $7, -6, 0, -2, -3$

3. $|-8|, -|9|, -|-6|$

4. $-|-9|, -|7|, -|-2|$

5. $\sqrt{8}, -4, -\sqrt{3}, -2, -5, \sqrt{6}, 3$

6. $\sqrt{2}, -1, 4, 3, \sqrt{8}, -\sqrt{6}, \sqrt{7}$

7. $3/4, \sqrt{2}, 7/5, 8/5, 22/15$

8. $-9/8, -3, -\sqrt{3}, -\sqrt{5}, -9/5, -8/5$

Let $x = -4$ and $y = 2$. Evaluate each of the following.

9. $|2x|$ 10. $|-3y|$

11. $|x - y|$ 12. $|2x + 5y|$

13. $|3x + 4y|$ 14. $|-5y + x|$

15. $|-4x + y| - |y|$ 16. $|-8y + x| - |x|$

Write each of the following without absolute value bars.

17. $|-6|$ 18. $-|-2|$

19. $-|-8| + |-2|$ 20. $3 - |-4|$

21. $|8 - \sqrt{50}|$ 22. $|2 - \sqrt{3}|$

23. $|\sqrt{7} - 5|$ 24. $|\sqrt{2} - 3|$

25. $|\pi - 3|$ 26. $|\pi - 5|$

27. $|x - 4|$, if $x > 4$

28. $|y - 3|$, if $y < 3$

29. $|2k - 8|$, if $k < 4$

30. $|3r - 15|$, if $r > 5$

31. $|7m - 56|$, if $m < 4$

32. $|2k - 7|$, if $k > 4$

33. $|-8 - 4m|$, if $m > -2$

34. $|6 - 5r|$, if $r < -2$

35. $|x - y|$, if $x < y$ 36. $|x - y|$, if $x > y$

37. $|3 + x^2|$ 38. $|x^2 + 4|$

In the following exercises, the coordinates of four points are given. Find (a) $d(A, B)$; (b) $d(B, C)$; (c) $d(B, D)$; (d) $d(D, A)$; (e) $d(A, B) + d(B, C)$.

39. $A, -4; B, -3; C, -2; D, 10$

40. $A, -8; B, -7; C, 11; D, 5$

41. $A, -3; B, -5; C, -12; D, -3$

42. $A, 0; B, 6; C, 9; D, -1$

Under what conditions are the following true?

43. $|x| = |y|$

44. $|x + y| = |x| + |y|$

45. $|x + y| = |x| - |y|$

46. $|x + y| = ||x + y||$

Prove each of the following properties.

47. $|-a| = |a|$

48. $|a - b| = |b - a|$

49. $|a| \cdot |b| = |ab|$

50. $\left|\dfrac{a}{b}\right| = \dfrac{|a|}{|b|}$, if $b \neq 0$

51. $-|a| \le a \le |a|$

52. $|a + b| \le |a| + |b|$

53. $d(A, B) = d(B, A)$ 54. $d(A, B) \ge 0$

55. $d(A, 0) = |a|$ 56. $d(A, A) = 0$

Evaluate each of the following if x is a nonzero real number.

57. $\dfrac{|x|}{x}$ 58. $\left|\dfrac{|x|}{x}\right|$

1.3   Integer Exponents

Exponents are used to write repeated products. If n is any positive integer and a is any real number, then the symbol a^n is defined as follows.

Definition of a^n

$a^n = a \cdot a \cdot a \cdots a$, where a appears n times.

Here n is called the **exponent,** and a is called the **base.** (Read a^n as "a to the nth.") For example,

$$(-6)^2 = (-6)(-6) = 36, \qquad 4^3 = 4 \cdot 4 \cdot 4 = 64, \qquad \text{and}$$

$$\left(\frac{2}{3}\right)^4 = \frac{2}{3} \cdot \frac{2}{3} \cdot \frac{2}{3} \cdot \frac{2}{3} = \frac{16}{81}.$$

One common error with exponents occurs with expressions such as $4 \cdot 3^2$. The exponent of 2 applies only to the base 3, so that

$$4 \cdot 3^2 = 4 \cdot 3 \cdot 3 = 36.$$

On the other hand,

$$(4 \cdot 3)^2 = (4 \cdot 3)(4 \cdot 3) = 12 \cdot 12 = 144.$$

Work with exponents can be simplified by using various **properties of exponents.** By definition, a^m (where m is a positive integer and a is a real number) means a appears as a factor m times. In the same way, a^n (where n is a positive integer) means that a appears as a factor n times. Thus, in the product $a^m \cdot a^n$, a appears as a factor $m + n$ times. Therefore, for any positive integers m and n and any real number a,

$$a^m \cdot a^n = a^{m+n}.$$

Similar arguments can be given for the other properties of exponents given in the following box.

Properties of Exponents

For any positive integers m and n and for any real numbers a and b,

(a) $a^m \cdot a^n = a^{m+n}$

(b) $\dfrac{a^m}{a^n} = \begin{cases} a^{m-n} \text{ if } m > n \\ 1 \text{ if } m = n \qquad (a \neq 0) \\ \dfrac{1}{a^{n-m}} \text{ if } m < n \end{cases}$

(c) $(a^m)^n = a^{mn}$

(d) $(ab)^m = a^m b^m$

(e) $\left(\dfrac{a}{b}\right)^m = \dfrac{a^m}{b^m} \quad (b \neq 0)$

Complete proofs of the various properties included in the box require the method of mathematical induction, discussed in the final chapter of this book. Properties (a) and (d) can be generalized further; for example,

$$a^m \cdot a^n \cdot a^p = a^{m+n+p} \qquad \text{and} \qquad (abc)^n = a^n \cdot b^n \cdot c^n$$

and so on.

EXAMPLE 1   ☐   Each of the following examples is justified by one or more of the properties of exponents and by the properties of real numbers.

(a) $(4x^3)(-3x^5) = (4)(-3)x^3 \cdot x^5 = -12x^{3+5} = -12x^8$

(b) $\dfrac{m^4 p^2}{m^3 p^5} = \dfrac{m^4}{m^3} \cdot \dfrac{p^2}{p^5} = m \cdot \dfrac{1}{p^3} = \dfrac{m}{p^3}$ $(m \neq 0,\ p \neq 0)$

(c) $\left(\dfrac{2x^5}{7}\right)^3 = \dfrac{2^3(x^5)^3}{7^3} = \dfrac{8x^{15}}{343}$

(d) $(2z^{2r})(z^{r+1}) = 2z^{2r+(r+1)} = 2z^{3r+1}$ (r is a positive integer)

(e) $\dfrac{k^{r+1}}{k^r} = k^{(r+1)-r} = k$ (r is a positive integer and k is nonzero)   ▢

By part (a) of the properties of exponents, $a^m \cdot a^n = a^{m+n}$. Suppose we try to extend this property so that it is valid for an exponent of 0. If $n = 0$, then

$$a^m \cdot a^n = a^{m+n} \text{ becomes } a^m \cdot a^0 = a^{m+0} = a^m.$$

The only way that $a^m \cdot a^0$ can equal a^m is if $a^0 = 1$. Therefore, to be consistent with past work, we make the following definition. If a is any nonzero real number, then

Zero Exponent $a^0 = 1.$

The symbol 0^0 is not defined.

EXAMPLE 2   ▢   (a) $3^0 = 1$ (d) $(-\sqrt{11})^0 = 1$

(b) $(-4)^0 = 1$ (e) $-4^0 = -1$

(c) $\left(\dfrac{1}{7}\right)^0 = 1$ (f) $-(-4)^0 = -1$ ▢

In Section 1.1 we defined the symbol a^{-1} as

$$a^{-1} = \frac{1}{a}$$

for any nonzero real number a. Now we wish to extend this definition so as to give meaning to an expression of the form a^{-n}, where n is any positive integer. As with zero exponents, we wish any definition we give for a^{-n} to be consistent with the properties of exponents given earlier. For property (c) from above to be valid, we must have

$$a^{-n} = a^{(-1)n} = (a^{-1})^n.$$

Since $a^{-1} = \dfrac{1}{a}$, we have

$$a^{-n} = (a^{-1})^n = \underbrace{\frac{1}{a} \cdot \frac{1}{a} \cdot \frac{1}{a} \cdot \ldots \frac{1}{a}}_{n \text{ factors}} = \frac{1}{a^n}.$$

Based on this, for our past properties of exponents to remain valid, we must define a^{-n} as $1/a^n$. Thus, for all integers n and all nonzero real numbers a,

Negative Exponent

$$a^{-n} = \frac{1}{a^n}.$$

EXAMPLE 3

(a) $4^{-2} = \dfrac{1}{4^2} = \dfrac{1}{16}$

(b) $2^{-3} = \dfrac{1}{2^3} = \dfrac{1}{8}$

(c) $\left(\dfrac{1}{3}\right)^{-2} = \dfrac{1}{\left(\dfrac{1}{3}\right)^2} = \dfrac{1}{\dfrac{1}{9}} = 9$

(d) $\left(\dfrac{2}{3}\right)^{-3} = \dfrac{1}{\left(\dfrac{2}{3}\right)^3} = \dfrac{1}{\dfrac{8}{27}} = \dfrac{27}{8}$

(e) $-4^{-2} = -\dfrac{1}{4^2} = -\dfrac{1}{16}$

Parts (c) and (d) of Example 3 involve work with fractions of a type that can lead to errors. For a shortcut useful with such fractions, first note that $(a/b)(b/a) = 1$, so that $(a/b)^{-1} = b/a$. Use this fact, along with properties of exponents, to get

$$\left(\frac{a}{b}\right)^{-n} = \left[\left(\frac{a}{b}\right)^{-1}\right]^n = \left(\frac{b}{a}\right)^n.$$

In summary,

$$\left(\frac{a}{b}\right)^{-n} = \left(\frac{b}{a}\right)^n.$$

EXAMPLE 4

(a) $\left(\dfrac{1}{3}\right)^{-2} = \left(\dfrac{3}{1}\right)^2 = 9$ (b) $\left(\dfrac{2}{3}\right)^{-3} = \left(\dfrac{3}{2}\right)^3 = \dfrac{27}{8}$

It can be shown that all the properties of exponents given above are valid for *all* integer exponents, and not just positive integer exponents. In particular, property (b) from above can now be simplified to just one case:

$$\frac{a^m}{a^n} = a^{m-n}.$$

We shall now summarize all our properties and definitions for exponents. All the statements in the box are true for any integers m and n, and any real numbers a and b (we assume no denominators are zero).

Summary for Exponents

(a) $a^m \cdot a^n = a^{m+n}$ (f) $a^0 = 1$

(b) $\dfrac{a^m}{a^n} = a^{m-n}$ (g) $a^{-n} = \dfrac{1}{a^n}$

(c) $(a^m)^n = a^{mn}$

(d) $(ab)^m = a^m b^m$ (h) $\left(\dfrac{a}{b}\right)^{-n} = \left(\dfrac{b}{a}\right)^n$

(e) $\left(\dfrac{a}{b}\right)^m = \dfrac{a^m}{b^m}$

EXAMPLE 5

(a) $2^{-4} \cdot 2^5 = 2^{-4+5} = 2^1 = 2$

(b) $3x^{-2}(4^{-1}x^{-5})^2 = 3x^{-2}(4^{-2}x^{-10})$

$$= 3 \cdot 4^{-2} \cdot x^{-2+(-10)}$$

$$= 3 \cdot 4^{-2} \cdot x^{-12}$$

$$= \frac{3}{16x^{12}}$$

(c) $\dfrac{5m^{-3}}{10m^{-5}} = \dfrac{5}{10}m^{-3-(-5)} = \dfrac{1}{2}m^2$, or $\dfrac{m^2}{2}$

EXAMPLE 6   Simplify each of the following expressions by writing without negative exponents.

(a) $\dfrac{12p^3 q^{-1}}{8p^{-2} q}$

Use the properties of exponents, working through the following sequence of steps.

$$\frac{12p^3 q^{-1}}{8p^{-2} q} = \frac{12}{8} \cdot \frac{p^3}{p^{-2}} \cdot \frac{q^{-1}}{q}$$

$$= \frac{3}{2} \cdot p^{3-(-2)} q^{-1-1} \quad (q = q^1)$$

$$= \frac{3}{2} p^5 q^{-2} = \frac{3p^5}{2q^2}$$

(b) $\dfrac{(3x^2)^{-1}(3x^5)^{-2}}{(3^{-1}x^{-2})^2} = \dfrac{3^{-1}x^{-2}3^{-2}x^{-10}}{3^{-2}x^{-4}}$

$$= \frac{3^{-1+(-2)}x^{-2+(-10)}}{3^{-2}x^{-4}}$$

$$= \frac{3^{-3}x^{-12}}{3^{-2}x^{-4}}$$

$$= 3^{-3-(-2)}x^{-12-(-4)} = 3^{-1}x^{-8} = \frac{1}{3x^8}$$

Scientific work often involves numbers that are very large or very small. To avoid working with many zeros, scientists usually express such numbers in **scientific notation.**

Scientific Notation

In *scientific notation,* a number is written as the product of a number whose absolute value is at least 1, but less than 10, and a power of 10.

For example,

$$125,000 = 1.25 \times 10^5,$$
$$7900 = 7.9 \times 10^3,$$
$$.008 = 8 \times 10^{-3},$$
$$.00000563 = 5.63 \times 10^{-6}.$$

One advantage of scientific notation is that some calculations can be simplified by using the properties of exponents. This process is shown in the next example.

EXAMPLE 7   ☐   Find each of the following.

(a) $(28,000,000)(.000012)$

Writing each number in scientific notation, and using the properties of exponents, we get

$$(28,000,000)(.000012) = (2.8 \times 10^7)(1.2 \times 10^{-5})$$
$$= (2.8 \times 1.2) \times 10^{7+(-5)}$$
$$= 3.36 \times 10^2 = 336.$$

(b) $\dfrac{.041}{.000082} = \dfrac{4.1 \times 10^{-2}}{8.2 \times 10^{-5}}$

$$= \frac{4.1}{8.2} \times 10^{-2-(-5)} = .5 \times 10^3 = 500 \quad \square$$

Some calculators have the capability of displaying scientific notation. For example, such a calculator might display the number 1.46×10^{12} as

1.46   12.

A common mistake when working with calculators is to quote the displayed answer to more accuracy than is warranted by the original data. This often happens because almost all numbers encountered in real-world problems are approximations, and the result of a calculation is no more accurate than the least accurate number involved in it.

For example, if we measure a wall to the nearest meter and say that it is 18 meters long, then we are really saying that the wall has a length in the range of 17.5 meters to 18.5 meters. If we measure the wall more accurately, and say that it is 18.3 meters long, then we know that its length is really in the range 18.25 meters to 18.35 meters. A measurement of 18.00 meters would indicate that the wall's length is in the range 17.995 to 18.005 meters. The measurement 18 meters is said to have 2 significant digits of accuracy; 18.3 has 3 significant digits and 18.00 has 4.

Significant Digits

In general, to find the number of **significant digits** in a measurement, count from left to right, starting at the first nonzero digit and counting to the right to the last nonzero digit.

EXAMPLE 8   ☐   The following chart shows some numbers, the number of significant digits in each number, and the range represented by each number.

Number	Number of significant digits	Range represented by number
29.6	3	29.55 to 29.65
1.39	3	1.385 to 1.395
.000096	2	.0000955 to .0000965
.03	1	.025 to .035
100.2	4	100.15 to 100.25 ☐

There is one possible place for trouble when finding significant digits. We know that the measurement 19.00 meters is a measurement to the nearest hundredth meter. What about the measurement 93,000 meters? Does it represent a measurement to the nearest meter? the nearest ten meters? hundred meters? thousand meters? We cannot tell by the way the number is written. To get around this problem, we write the number in scientific notation. Depending on what we know about the accuracy of the measurement, we could write 93,000 using scientific notation as follows.

Measurement to the nearest . . .	Scientific notation	Number of significant digits
meter	9.3000×10^4	5
ten meters	9.300×10^4	4
hundred meters	9.30×10^4	3
thousand meters	9.3×10^4	2

When calculating with approximate data, use rules given in the following box.

Calculation with Significant Digits

For *adding* and *subtracting,* add or subtract normally; then round the answer so that the last digit you keep is in the rightmost column in which all the numbers have significant digits.

When *multiplying* or *dividing,* round the answer to the *least* number of significant digits found in any of the given numbers.

For *powers* and *roots,* round the answer so that it has the same number of significant digits as the number whose power or root you are finding.

**EXAMPLE
9**

☐ (a) $23.1 + 17.92 + 19.004 = 60.024$

Since 23.1 has just three significant digits, round the answer to 60.0.

(b) $(2.9328)(30.531)(12.82) = 1148$

The answer was rounded to four significant digits, the *least* number of significant digits found in the three numbers.

(c) $\sqrt{284.8} = 16.88$

Note that the answer has the same number of significant digits as the root. ☐

Calculator Errors

A calculator can store only so many digits in its memory. Because of this, numbers which have more digits than can be stored must be rounded. For example, 1/3 is not stored as the exact fraction 1/3, but rather as a decimal, perhaps .3333333333333. Since this rounded form of 1/3 is used, errors can occur in calculations. To see how this happens, use a calculator to divide 1 by 3, and then multiply the result by 3. You should get 1 (exactly), but many machines produce

$$(1 \div 3) \times 3 = \left(\frac{1}{3}\right) \times 3 = .9999999999.$$

Some machines round this result to 1; however, the machine does not treat the number internally as 1. To see this, subtract 1 from the result above; you should get 0 but probably won't. On one expensive Texas Instruments machine,

$$(1 \div 3) \times 3 - 1 = -1 \times 10^{-12}.$$

Another calculator error results when numbers of greatly different size are used in addition. For example, although

$$10^9 + 10^{-5} - 10^9 = 10^{-5},$$

most calculators will give

$$10^9 + 10^{-5} - 10^9 = 0.$$

These calculator errors seldom occur in realistic problems, but if they do occur you should know the basis of the errors.

**EXERCISES
1.3**

Simplify each of the following. Write all answers without exponents.

1. 15^3

2. 21^4

3. 2^{-3}

4. 3^{-2}

5. $4^{-1} + 3^{-3}$

6. $5^{-2} - 2^{-1}$

7. $(-4)^{-3}$

8. $(-5)^{-2}$

9. $\left(\frac{1}{2}\right)^{-3}$

10. $\left(\frac{2}{7}\right)^{-2}$

🖩 11. $(9.864)^{-3}$

12. $(14.259)^{-2}$

Simplify each of the following by writing each with only positive exponents. Assume that all variables represent nonzero real numbers.

13. $(3m)^2(-2m)^3$

14. $(-2a^4b^7)(3b^2)$

15. $(6^{-2})(6^{-5})(6^3)$

16. $(4^8)(4^{-10})(4^2)$

17. $(5^3)(5^{-5})(5^{-2})$

18. $(m^{10})(m^{-4})(m^{-5})$

19. $\dfrac{(3^5)(3^{-2})}{3^{-4}}$

20. $\dfrac{(2^2)(2^{-5})}{2^7}$

21. $\dfrac{(d^{-1})(d^{-2})}{(d^8)(d^{-3})}$

22. $\dfrac{(t^5)(t^{-3})}{(t^4)(t^{-7})}$

23. $\dfrac{2^{-1}x^3y^{-3}}{xy^{-2}}$

24. $\dfrac{5^{-2}m^2y^{-2}}{5^2m^{-1}y^{-2}}$

25. $\dfrac{(4+s)^3(4+s)^{-2}}{(4+s)^{-5}}$

26. $\dfrac{(m+n)^4(m+n)^{-2}}{(m+n)^{-5}}$

27. $[(m^2n)^{-1}]^4$

28. $(x^5t^{-2})^{-2}$

29. $\left(\dfrac{a^{-1}}{b^2}\right)^{-3}$

30. $\left(\dfrac{2c^2}{d^3}\right)^{-2}$

31. $(-2x^{-3}y^2)^{-2}$

32. $(-3p^2q^{-2}s^{-1})^7$

33. $\dfrac{(3^{-1}m^{-2}n^2)^{-2}}{(mn)^{-1}}$

34. $\dfrac{(-4x^3y^{-2})^{-2}}{(4x^5y^4)^{-1}}$

35. $\dfrac{(r^2s^3t^4)^{-2}(r^3s^2t)^{-1}}{(rst)^{-3}(r^2st^2)^3}$

36. $\dfrac{(ab^2c)^{-3}(a^2bc)^{-2}}{(abc^2)^4}$

37. $\dfrac{[k^2(p+q)^4]^{-1}}{k^{-4}(p+q)^3}$

38. $\dfrac{t^4(x+y)^4t^{-8}}{(x+y)^{-2}t^{-5}}$

Evaluate each of the following, assuming that $a = 2$ and $b = -3$.

39. $a^{-1} + b^{-1}$

40. $b^{-2} - a$

41. $\dfrac{2b^{-1} - 3a^{-1}}{a + b^2}$

42. $\dfrac{3a^2 - b^2}{b^{-3} + 2a^{-1}}$

43. $\left(\dfrac{a}{3}\right)^{-1} + \left(\dfrac{b}{2}\right)^{-2}$

44. $\left(\dfrac{2b}{5}\right)^2 - 3\left(\dfrac{a-1}{4}\right)$

Express each of the following numbers in scientific notation.

45. 69,300

46. 5000

47. 6,000,000,000

48. .0001

49. .00792

50. .054

Express each of the following numbers without exponents.

51. 8.2×10^5

52. 3.7×10^3

53. 1.7×10^8

54. 3.61×10^{-2}

55. 6.15×10^{-3}

56. 9.3×10^{-6}

Use scientific notation to evaluate each of the following.

57. $(4600)(.00092)$

58. $(.87)(.0004)$

59. $(.00002)(.00009)$

60. $(.0004)(.0000015)$

61. $\dfrac{.000034}{.017}$

62. $\dfrac{.0000042}{.0006}$

63. $\dfrac{28}{.0004}$

64. $\dfrac{50}{.0000025}$

🖩 Use a calculator to evaluate each of the following. Write all answers in scientific notation rounded to three significant digits.

65. $(1.66 \times 10^4)(2.93 \times 10^3)$

66. $(6.92 \times 10^7)(8.14 \times 10^5)$

67. $(4.08 \times 10^{-9})(3.172 \times 10^4)$

68. $(9.113 \times 10^{12})(8.72 \times 10^{-11})$

69. $\dfrac{(-4.389 \times 10^4)(2.421 \times 10^{-2})}{1.76 \times 10^{-9}}$

70. $\dfrac{(1.392 \times 10^{10})(5.746 \times 10^{-8})}{4.93 \times 10^{-15}}$

71. $\dfrac{-3.9801 \times 10^{-6}}{(7.4993 \times 10^{-8})(2.117 \times 10^{-4})}$

72. $\dfrac{-5.421 \times 10^{-7}}{(4.2803 \times 10^{-4})(9.982 \times 10^{-8})}$

Find each of the following products or quotients. Assume all variables appearing as exponents represent integers.

73. $(2k^m)(k^{1-m})$

74. $(y^m)(y^{1+m})$

75. $\dfrac{x^{2r}}{x^r}$

76. $\dfrac{4a^{m+2}}{2a^{m-1}}$

77. $\dfrac{(b^2)^y}{(2b^y)^3}$

78. $\left(\dfrac{3p^a}{2m^b}\right)^c$

79. $\dfrac{(2m^n)^2(-4m^{2+n})}{8m^{4n}}$

80. $\dfrac{(3-y)^k(3-y)^{2k}}{(3-y)^2}$

1.4 Polynomials

A variable is a letter used to represent an element from a given set. Unless otherwise specified, in this book variables will represent real numbers. An **algebraic expression** is the result of performing the basic operations of addition, subtraction, multiplication, division (except by 0), or extraction of roots, on any collection of variables and numbers. The simplest algebraic expression is a polynomial, which we discuss in this section.

Before we define a polynomial, we must define *term:* a **term** is the product of a real number and one or more variables raised to powers. The real number is called the **numerical coefficient,** or just the **coefficient.** For example, -3 is the coefficient in $-3m^4$, while -1 is the coefficient in $-p^2$.

A **polynomial** is defined as a finite sum of terms, with only nonnegative integer exponents permitted on the variables. If the terms of a polynomial contain only the variable x, then the polynomial is called a **polynomial in x.** (Polynomials in other variables are defined similarly.)

In symbols, a **polynomial in x** is an expression of the form

Polynomial in x

$$a_n x^n + a_{n-1} x^{n-1} + \cdots + a_1 x + a_0,$$

where a_n (read "a-sub-n"), $a_{n-1}, \cdots, a_1$, and a_0 are real numbers and n is a nonnegative integer.

If $a_n \neq 0$, then n is the **degree** of the polynomial, and a_n is called the **leading coefficient.** By this definition,

$$3x^4 - 5x^2 + 2$$

is a polynomial of degree 4 with leading coefficient 3.

A nonzero constant is said to have degree 0, but no degree is assigned to the real number 0. If all the coefficients of a polynomial are 0, the polynomial is called the **zero polynomial.**

A polynomial can have more than one variable. A term containing more than one variable is said to have degree equal to the sum of all the exponents appearing on the variables in the term. For example, $-3x^4 y^3 z^5$ is of degree 12 because $4 + 3 + 5 = 12$.

The degree of a polynomial in more than one variable is the highest degree of any term appearing in the polynomial. Thus, the polynomial

$$2x^4 y^3 - 3x^5 y + x^6 y^2$$

is of degree 8 because of the $x^6 y^2$ term.

A polynomial containing exactly three terms is called a **trinomial**; one containing exactly two terms is a **binomial**; and a single-term polynomial is called a **monomial.** Thus, $7x^9 - \sqrt{2}x^4 + 1$ is a trinomial of degree 9.

Since the variables used in polynomials represent real numbers, a polynomial represents a real number. This means that all the properties of the real numbers mentioned at the beginning of this chapter hold for polynomials. In particular, the distributive property holds, so that

$$3m^5 - 7m^5 = (3 - 7)m^5 = -4m^5.$$

To *add* polynomials of one variable, add coefficients of the same powers; to *subtract* polynomials, subtract coefficients of the same powers. (Notice how the distributive property was used to suggest these definitions.)

EXAMPLE 1 ☐ Add or subtract, as indicated.

(a) $(2y^4 - 3y^2 + y) + (4y^4 + 7y^2 + 6y) = (2 + 4)y^4 + (-3 + 7)y^2$
$$+ (1 + 6)y$$
$$= 6y^4 + 4y^2 + 7y$$

(b) $(6r^4 + 2r^2) + (3r^3 + 9r) = 6r^4 + 3r^3 + 2r^2 + 9r$

(c) $(-3m^3 - 8m^2 + 4) - (m^3 + 7m^2 - 3) = (-3 - 1)m^3 + (-8 - 7)m^2$
$$+ [4 - (-3)]$$
$$= -4m^3 - 15m^2 + 7 \quad ☐$$

The distributive property, together with the properties of exponents, can also be used to find the product of two polynomials. For example, to find the product of $3x - 4$ and $2x^2 - 3x + 5$, treat $3x - 4$ as a single expression and use the distributive property as follows.

$$(3x - 4)(2x^2 - 3x + 5) = (3x - 4)(2x^2) - (3x - 4)(3x)$$
$$+ (3x - 4)(5)$$

Now use the distributive property three separate times on the right of the equals sign to get

$$(3x - 4)(2x^2 - 3x + 5) = (3x)(2x^2) - 4(2x^2) - (3x)(3x) - (-4)(3x)$$
$$+ (3x)5 - 4(5)$$
$$= 6x^3 - 8x^2 - 9x^2 + 12x + 15x - 20$$
$$= 6x^3 - 17x^2 + 27x - 20.$$

It is sometimes more convenient to write such a product as

$$
\begin{array}{r}
2x^2 - 3x + 5 \\
3x - 4 \\
\hline
-8x^2 + 12x - 20 \\
6x^3 - 9x^2 + 15x \\
\hline
6x^3 - 17x^2 + 27x - 20.
\end{array}
$$

EXAMPLE 2 ☐ Find each product.

(a) $(6m + 1)(4m - 3) = (6m)(4m) - (6m)(3) + 1(4m) - 1(3)$

$$= 24m^2 - 14m - 3$$

(b) $(2k^n - 5)(k^n + 3) = 2k^{2n} + 6k^n - 5k^n - 15$

$$= 2k^{2n} + k^n - 15$$

(c) $(2x + 7)(2x - 7) = 4x^2 - 14x + 14x - 49$

$$= 4x^2 - 49 \quad ☐$$

In Examples 2(a) and 2(b), the product of two binomials was a trinomial, while in Example 2(c) the product of two binomials was a binomial. The product of two binomials of the form $x + y$ and $x - y$ is a binomial. We multiply $x + y$ and $x - y$ to find that

Difference of Two Squares

$$(x + y)(x - y) = x^2 - y^2.$$

This product is called the **difference of two squares**.

EXAMPLE 3 ☐ Find each product

(a) $(3p + 11)(3p - 11)$

Using the pattern in the box, replace x with $3p$ and y with 11. This gives

$$(3p + 11)(3p - 11) = (3p)^2 - 11^2 = 9p^2 - 121.$$

(b) $(5m^3 - 3)(5m^3 + 3) = (5m^3)^2 - 3^2$

$$= 25m^6 - 9$$

(c) $(9k - 11r^3)(9k + 11r^3) = (9k)^2 - (11r^3)^2 = 81k^2 - 121r^6 \quad ☐$

The **square of a binomial** is another special product. If we multiply $x + y$ and $x + y$—or $(x + y)^2$—as well as finding $(x - y)^2$, we get the results shown in the following box.

Square of a Binomial

$$(x + y)^2 = x^2 + 2xy + y^2$$
$$(x - y)^2 = x^2 - 2xy + y^2$$

EXAMPLE 4 ☐ (a) $(2m + 5)^2 = (2m)^2 + 2(2m)(5) + (5)^2$

$$= 4m^2 + 20m + 25$$

(b) $(3z - 7y^4)^2 = (3z)^2 - 2(3z)(7y^4) + (7y^4)^2$

$$= 9z^2 - 42zy^4 + 49y^8 \quad ☐$$

Two last special products show how to find the cube of a binomial; for all real numbers x and y,

Cube of a Binomial

$$(x + y)^3 = x^3 + 3x^2y + 3xy^2 + y^3$$
$$(x - y)^3 = x^3 - 3x^2y + 3xy^2 - y^3.$$

EXAMPLE 5 ☐ Find each product

(a) $(m + 4n)^3$

Replace x with m and y with $4n$ in the pattern for $(x + y)^3$ to get

$$(m + 4n)^3 = m^3 + 3m^2(4n) + 3m(4n)^2 + (4n)^3$$
$$= m^3 + 12m^2n + 48mn^2 + 64n^3.$$

(b) $(5k - 2z^5)^3 = (5k)^3 - 3(5k)^2(2z^5) + 3(5k)(2z^5)^2 - (2z^5)^3$

$$= 125k^3 - 150k^2z^5 + 60kz^{10} - 8z^{15}. \quad ☐$$

It is useful to memorize and be able to apply these special products. (Both the square of a binomial and the cube of a binomial are special cases of the binomial theorem, discussed in the last chapter of this book.)

To divide a polynomial by a monomial, divide each term of the polynomial by the monomial. The result is called the **quotient** of the two polynomials.

EXAMPLE 6 ☐ Find each quotient

(a) $\dfrac{2m^5 - 6m^3}{2m^3} = \dfrac{2m^5}{2m^3} - \dfrac{6m^3}{2m^3} = m^2 - 3$

The polynomial $m^2 - 3$ is called the quotient of $2m^5 - 6m^3$ and $2m^3$.

(b) $\dfrac{3y^6x^3 - 6y^3x^6 + 8y^5x}{3y^3x^3} = \dfrac{3y^6x^3}{3y^3x^3} - \dfrac{6y^3x^6}{3y^3x^3} + \dfrac{8y^5x}{3y^3x^3}$

$$= y^3 - 2x^3 + \dfrac{8y^2}{3x^2} \quad ☐$$

In part (b) of this example, the quotient is not a polynomial.

─────────────────────────────────────

EXERCISES 1.4

Perform the indicated operations.

1. $(3x^2 - 4x + 5) + (-2x^2 + 3x - 2)$

2. $(4m^3 - 3m^2 + 5) + (-3m^3 - m^2 + 5)$

3. $(x^2 + 4x) + (3x^3 - 4x^2 + 2x + 2)$

4. $(r^5 - r^3 + r) + (3r^5 - 4r^4 + r^3 + 2r)$

5. $(12y^2 - 8y + 6) - (3y^2 - 4y + 2)$

6. $(8p^2 - 5p) - (3p^2 - 2p + 4)$

7. $(p^3 - 4p^2 + p) - (3p^2 + 2p + 7)$

8. $(4y^2 - 2y + 7) - (6y - 9)$

9. $(3a^2 - 2a) + (4a^2 + 3a + 1) - (a^2 + 2)$

10. $(6m^4 - 3m^2 + m) - (2m^3 + 5m^2 + 4m) + (m^2 - m)$

11. $(5b^2 - 4b + 3) - (2b^2 + b) - (3b + 4)$

12. $-(8x^3 + x - 3) + (2x^3 + x^2) - (4x^2 + 3x - 1)$

13. $-3(4q^2 - 3q + 2) + 2(-q^2 + q - 4)$

14. $2(3r^2 + 4r + 2) - 3(-r^2 + 4r - 5)$

15. $(4r - 1)(7r + 2)$

16. $(5m - 6)(3m + 4)$

17. $(6p + 5q)(3p - 7q)$

18. $(2z + y)(3z - 4y)$

19. $\left(3x - \dfrac{2}{3}\right)\left(5x + \dfrac{1}{3}\right)$

20. $\left(2m - \dfrac{1}{4}\right)\left(3m + \dfrac{1}{2}\right)$

21. $\left(\dfrac{2}{5}y + \dfrac{1}{8}z\right)\left(\dfrac{3}{5}y + \dfrac{1}{2}z\right)$

22. $\left(\dfrac{3}{4}r - \dfrac{2}{3}s\right)\left(\dfrac{5}{4}r + \dfrac{1}{3}s\right)$

23. $(5r + 2)(5r - 2)$　　**24.** $(6z + 5)(6z - 5)$

25. $(4x + 3y)(4x - 3y)$

26. $(7m + 2n)(7m - 2n)$

27. $(6k - 3)^2$　　　　**28.** $(3p + 5)^2$

29. $(4m + 2n)^2$　　　**30.** $(a - 6b)^2$

31. $(2z - 1)^3$　　　　**32.** $(3m + 2)^3$

33. $4x^2(3x^3 + 2x^2 - 5x + 1)$

34. $2b^3(b^2 - 4b + 3)$

35. $5m(3m^3 - 2m^2 + m - 1)$

36. $4y^3(y^3 + 2y^2 - 6y + 3)$

37. $(2z - 1)(-z^2 + 3z - 4)$

38. $(x - 1)(x^2 - 1)$

39. $(3p - 1)(9p^2 + 3p + 1)$

40. $(2p - 1)(3p^2 - 4p + 5)$

41. $(2m + 1)(4m^2 - 2m + 1)$

42. $(k + 2)(12k^3 - 3k^2 + k + 1)$

43. $(m - n + k)(m + 2n - 3k)$

44. $(r - 3s + t)(2r - s + t)$

45. $(a - b + 2c)^2$　　**46.** $(k - y + 3m)^2$

47. $(x^{-1} - 2)^2$　　　**48.** $(1 + 2y^{-1})^2$

49. $(3m^{-1} - 2n^{-1})(4m^{-1} + n^{-1})$

50. $(-k^{-1} + 3q^{-1})(4k^{-1} - 3q^{-1})$

Perform each of the following divisions. Assume that all variables appearing in denominators represent nonzero real numbers.

51. $\dfrac{4m^3 - 8m^2 + 16m}{2m}$

52. $\dfrac{30k^5 - 12k^3 + 18k^2}{6k^2}$

53. $\dfrac{15x^4 + 30x^3 + 12x^2 - 9}{3x}$

54. $\dfrac{16a^6 + 24a^5 - 48a^4 + 12a}{8a^2}$

55. $\dfrac{25x^2y^4 - 15x^3y^3 + 40x^4y^2}{5x^2y^2}$

56. $\dfrac{-8r^3s - 12r^2s^2 + 20rs^3}{4rs}$

In each of the following, find the coefficient of x^3 without finding the entire product.

57. $(x^2 + 4x)(-3x^2 + 4x - 1)$

58. $(4x^3 - 2x + 5)(x^3 + 2)$

59. $(1 + x^2)(1 + x)$　　**60.** $(3 - x)(2 - x^2)$

61. $x^2(4 - 3x)^2$

62. $-4x^2(2 - x)(2 + x)$

Find each of the following products. Assume all variables used as exponents represent integers.

63. $(k^m + 2)(k^m - 2)$　　**64.** $(y^x - 4)(y^x + 4)$

65. $(b^r + 3)(b^r - 2)$　　**66.** $(q^y + 4)(q^y + 3)$

67. $(3p^x + 1)(p^x - 2)$　　**68.** $(2^a + 5)(2^a + 3)$

69. $(m^x - 2)^2$　　　　**70.** $(z^r + 5)^2$

71. $(3k^a - 2)^3$　　　　**72.** $(r^x - 4)^3$

1.5 Factoring

The process of finding the polynomials whose product equals a given polynomial is called **factoring.** Each polynomial in the product is called a **factor.** For example, since $x(x - 8) = x^2 - 8x$, both x and $x - 8$ are factors of $x^2 - 8x$.

The first step in factoring a polynomial is to look for any **common factors;** that is, factors that are factors of each term of the given polynomial.

EXAMPLE 1 ☐ Factor a common factor from each of the following polynomials.

(a) $6x^2y^3 + 9xy^4 + 18y^5$

Each term of this polynomial can be written with a factor of $3y^3$, so that $3y^3$ is a common factor. Use the distributive property to get

$$6x^2y^3 + 9xy^4 + 18y^5 = (3y^3)(2x^2) + (3y^3)(3xy) + (3y^3)(6y^2)$$
$$= 3y^3(2x^2 + 3xy + 6y^2).$$

(b) $6x^2t + 8xt + 12t = 2t(3x^2 + 4x + 6)$ ☐

Each of the special patterns of multiplication from the previous section can be used in reverse to get a pattern for factoring. One of the most common of these is listed in the following box.

Difference of Two Squares

$$x^2 - y^2 = (x + y)(x - y)$$

EXAMPLE 2 ☐ Factor each of the following polynomials.

(a) $4m^2 - 9$

First, recognize that $4m^2 - 9$ is the difference of two squares, since $4m^2 = (2m)^2$ and $9 = 3^2$. We can use the pattern for the difference of two squares if we let $2m$ replace x and 3 replace y. Doing this, we have

$$4m^2 - 9 = (2m)^2 - 3^2$$
$$= (2m + 3)(2m - 3).$$

(b) $144r^2 - 25s^2 = (12r + 5s)(12r - 5s)$

(c) $256k^4 - 625m^4$

Here we must use the difference of two squares pattern twice, as follows:

$$256k^4 - 625m^4 = (16k^2)^2 - (25m^2)^2$$
$$= (16k^2 + 25m^2)(16k^2 - 25m^2)$$
$$= (16k^2 + 25m^2)(4k + 5m)(4k - 5m)$$

(d) $(a + 2b)^2 - 4c^2 = (a + 2b)^2 - (2c)^2$

$$= [(a + 2b) + 2c][(a + 2b) - 2c]$$
$$= (a + 2b + 2c)(a + 2b - 2c) \quad ☐$$

In this chapter, we assume that a polynomial with only integer coefficients will be factored so that all factors have only integer coefficients. This assumption is sometimes summarized by saying that we permit only **factoring over the integers.** If only factoring over the integers is permitted, then the polynomial $x^2 - 5$, for example, cannot be factored. While it is true that

$$x^2 - 5 = (x + \sqrt{5})(x - \sqrt{5}),$$

the two factors $x + \sqrt{5}$ and $x - \sqrt{5}$ have noninteger coefficients.

A polynomial is **prime** or **irreducible** if it cannot be written as the product of two polynomials of positive degree. The polynomial $x^2 - 5$ is prime (over the integers).

Now we need to look at methods of factoring a trinomial of degree 2, such as $kx^2 + mx + n$, where k, m, and n are integers. Any factorization will be of the form $(ax + b)(cx + d)$ with a, b, c, and d integers. Multiplying out the product $(ax + b)(cx + d)$ gives

$$(ax + b)(cx + d) = acx^2 + (ad + bc)x + bd,$$

which equals $kx^2 + mx + n$ only if

$$ac = k, \qquad ad + bc = m, \qquad \text{and} \qquad bd = n. \hspace{2cm} (*)$$

Thus, to factor a trinomial $kx^2 + mx + n$, we must find four integers a, b, c, and d satisfying the conditions given in $(*)$. If no such numbers exist, the trinomial is prime.

EXAMPLE 3 ☐ Factor each of the following polynomials.

(a) $6p^2 - 7p - 5$

To find integers a, b, c, and d so that

$$6p^2 - 7p - 5 = (ap + b)(cp + d),$$

we use the results given in $(*)$ above and try to find integers a, b, c, and d such that $ac = 6$, $ad + bc = -7$, and $bd = -5$. To find these integers, we try various possibilities. Since $ac = 6$, we might let $a = 2$ and $c = 3$. Since $bd = -5$, we might let $b = -5$ and $d = 1$, giving

$$(2p - 5)(3p + 1) = 6p^2 - 13p - 5. \qquad \text{(incorrect)}$$

We need to make another attempt; we'll try

$$(3p - 5)(2p + 1) = 6p^2 - 7p - 5. \qquad \text{(correct)}$$

Thus, $6p^2 - 7p - 5$ factors as $(3p - 5)(2p + 1)$.

(b) $4x^3 + 6x^2r - 10xr^2$

There is a common factor of $2x$.

$$4x^3 + 6x^2r - 10xr^2 = 2x(2x^2 + 3xr - 5r^2).$$

To factor $2x^2 + 3xr - 5r^2$, we need the factors of $2x^2$ and of $-5r^2$ that will yield the correct middle term of $3xr$. By inspection, we end up with

$$4x^3 + 6x^2r - 10xr^2 = 2x(2x + 5r)(x - r).$$

(c) $r^2 + 6r + 7$

It is not possible to find integers a, b, c, and d so that

$$r^2 + 6r + 7 = (ar + b)(cr + d);$$

therefore, $r^2 + 6r + 7$ is a prime polynomial. ☐

Two other special types of factoring are listed in the next box.

Difference and Sum of Cubes

$x^3 - y^3 = (x - y)(x^2 + xy + y^2)$	difference of two cubes
$x^3 + y^3 = (x + y)(x^2 - xy + y^2)$	sum of two cubes

EXAMPLE 4   ☐   Factor each polynomial.

(a) $m^3 - 64n^3$

Since $64n^3 = (4n)^3$, we have a difference of two cubes. To factor, use the first pattern in the box above, replacing x with m and y with $4n$, to get

$$m^3 - 64n^3 = m^3 - (4n)^3$$
$$= (m - 4n)[m^2 + m(4n) + (4n)^2]$$
$$= (m - 4n)(m^2 + 4mn + 16n^2).$$

(b) $8q^6 + 125p^9$

Write $8q^6$ as $(2q^2)^3$ and $125p^9$ as $(5p^3)^3$, showing that the given polynomial is a sum of two cubes. Factor it as

$$8q^6 + 125p^9 = (2q^2)^3 + (5p^3)^3$$
$$= (2q^2 + 5p^3)[(2q^2)^2 - (2q^2)(5p^3) + (5p^3)^2]$$
$$= (2q^2 + 5p^3)(4q^4 - 10q^2p^3 + 25p^6).$$

(c) $(2a - 1)^3 + 8$

Use the pattern for the sum of two cubes, with $x = 2a - 1$ and $y = 2$. Doing so gives

$$(2a - 1)^3 + 8 = [(2a - 1) + 2][(2a - 1)^2 - (2a - 1)2 + 2^2]$$
$$= (2a - 1 + 2)(4a^2 - 4a + 1 - 4a + 2 + 4)$$
$$= (2a + 1)(4a^2 - 8a + 7). \ ☐$$

When a polynomial has more than three terms, it can often be factored by a method called **factoring by grouping.** For example, to factor

$$ax + ay + 6x + 6y,$$

collect the terms into two groups,

$$ax + ay + 6x + 6y = (ax + ay) + (6x + 6y),$$

and then factor each group, getting

$$ax + ay + 6x + 6y = a(x + y) + 6(x + y).$$

The quantity $(x + y)$ is now a common factor, which can be factored out, producing

$$ax + ay + 6x + 6y = (x + y)(a + 6).$$

EXAMPLE ☐ Factor $x^2 + 4x + 4 - y^2$ by grouping.
5   Here, it will be advantageous to group the first three terms, getting

$$x^2 + 4x + 4 - y^2 = (x^2 + 4x + 4) - y^2$$
$$= (x + 2)^2 - y^2.$$

Now factor on the right as a difference of two squares; this yields

$$x^2 + 4x + 4 - y^2 = (x + 2 + y)(x + 2 - y). \quad ☐$$

**EXERCISES
1.5**

Factor as completely as possible. Assume all variables appearing as exponents represent integers.

1. $12mn - 8m$ **2.** $3pq - 18pqr$

3. $12r^3 + 6r^2 - 3r$

4. $-5k^2g - 25kg - 30kg^2$

5. $6px^2 - 8px^3 - 12px$

6. $9m^2n^3 - 18m^3n^2 + 27m^2n^4$

7. $2(a + b) + 4m(a + b)$

8. $4(y - 2)^2 + 3(y - 2)$

9. $x^2 - 11x + 24$ **10.** $y^2 - 2y - 35$

11. $4p^2 + 3p - 1$ **12.** $6x^2 + 7x - 3$

13. $2z^2 + 7z - 30$

14. $4m^2 + m - 3$

15. $12r^2 + 24r - 15$

16. $12p^2 + p - 20$

17. $18r^2 - 3rs - 10s^2$

18. $12m^2 + 16mn - 35n^2$

19. $15x^2 - 14xy - 8y^2$

20. $12t^2 - 25tv + 12v^2$

21. $9x^2 - 6x^3 + x^4$ **22.** $30a + am - am^2$

23. $2r^2z - rz - 3z$

24. $3m^4 + 7m^3 + 2m^2$

25. $4m^2 - 25$ **26.** $25a^2 - 16$

27. $144r^2 - 81s^2$ **28.** $81m^2 - 16n^2$

29. $121p^4 - 9q^4$ **30.** $4z^4 - w^8$

31. $16x^4 - y^4$ **32.** $81q^4 - 256m^4$

33. $p^8 - 1$ **34.** $y^{16} - 1$

35. $8m^3 - 27n^3$ **36.** $125x^3 - 1$

37. $64 - x^6$ **38.** $m^6 - 1$

39. $4x + 4y + mx + my$

40. $x^2 + xy + 5x + 5y$

41. $q^2 + 6q + 9 - p^2$

42. $4b^2 + 4bc + c^2 - 16$

43. $a^2 + 2ab + b^2 - x^2 - 2xy - y^2$

44. $d^2 - 10d + 25 - c^2 + 4c - 4$

45. $x^2 - (x - y)^2$

46. $16m^2 - 25(m - n)^2$

47. $8y^3 + 27(x - y)^3$

48. $64p^6 - 27(p^2 - q)^3$

49. $(x + y)^2 + 2(x + y)z - 15z^2$

50. $(m + n)^2 + 3(m + n)p - 10p^2$

51. $6(a + b)^2 + (a + b)c - 40c^2$

52. $12(g + h)^2 - 5(g + h)j - 25j^2$

53. $(p + q)^2 - (p - q)^2$

54. $(p - q)^2 - (p + q)^2$

55. $m^{2n} - 16$ **56.** $p^{4n} - 49$

57. $x^{3n} - y^{6n}$ **58.** $a^{4p} - b^{12p}$

59. $2x^{2n} - 23x^ny^n - 39y^{2n}$

60. $3a^{2x} + 7a^xb^x + 2b^{2x}$

Factor the coefficient of smallest absolute value from each of the following.

61. $9.44x^3 + 48.144x^2 - 30.208x + 37.76$

62. $12.915m^2 + 2.05m - 23.575m^3 + 20.5$

63. $64.616z^4 - 23.64z^2 + 7.88z + 114.26$

64. $1.47y^5 - 8.526y^3 + 69.384y^2 - 36.75$

Factor the variable of smallest exponent, together with any numerical common factor, from each of the following expressions. (For example, factor $9x^{-2} - 6x^{-3}$ as $3x^{-3}(3x - 2)$.)

65. $p^{-4} - p^{-2}$

66. $m^{-1} + 3m^{-5}$

67. $12k^{-3} + 4k^{-2} - 8k^{-1}$

68. $6a^{-5} - 10a^{-4} + 18a^{-2}$

69. $100p^{-6} - 50p^{-2} + 75p^2$

70. $32y^{-3} + 48y - 64y^2$

1.6 Fractional Expressions

An expression that is the quotient of two algebraic expressions (with denominator not 0) is called a **fractional expression.** The most common fractional expressions are those that are the quotients of two polynomials; these are called **rational expressions.** Since fractional expressions involve quotients, it is important to keep track of the values of the variables that satisfy the necessity that no denominator be 0. The set of such values is the **domain** of the expression.

Domain

> The domain of a fractional expression is the set of all real numbers for which the expression is defined.

For example, -2 is not in the domain of the rational expression

$$\frac{x + 6}{x + 2}$$

since -2 (when substituted for x) makes the denominator equal 0. The domain of this rational expression can be written $\{x | x \neq -2\}$. In a similar way, the domain of

$$\frac{(x + 6)(x + 4)}{(x + 2)(x + 4)}$$

can be written $\{x | x \neq -2 \text{ and } x \neq -4\}$.

Just as the fraction 6/8 is reduced to 3/4, it is also necessary to reduce rational expressions. Since the variables in a rational expression represent real numbers, we can use the properties of real numbers in Section 1.1. To reduce a rational expression to lowest terms, we use the rule

$$\frac{ac}{bc} = \frac{a}{b}, \quad \text{if } c \neq 0,$$

as shown in Example 1.

EXAMPLE 1   ☐   Reduce to lowest terms.

(a) $\dfrac{2p^2 + 7p - 4}{5p^2 + 20p}$

Factor in both numerator and denominator to get

$$\frac{2p^2 + 7p - 4}{5p^2 + 20p} = \frac{(2p - 1)(p + 4)}{5p(p + 4)}.$$

By the property mentioned above, this becomes

$$\frac{2p^2 + 7p - 4}{5p^2 + 20p} = \frac{2p - 1}{5p}.$$

The domain of the original expression is $\{p \mid p \neq 0 \text{ and } p \neq -4\}$; so our result is valid only for values of p other than 0 and -4. We shall always assume such restrictions when reducing rational expressions.

(b) $\dfrac{6 - 3k}{k^2 - 4}$

Factor to get $\dfrac{6 - 3k}{k^2 - 4} = \dfrac{3(2 - k)}{(k + 2)(k - 2)}.$

The factors $2 - k$ and $k - 2$ have exactly opposite signs. Because of this, multiply numerator and denominator by -1, as follows.

$$\frac{6 - 3k}{k^2 - 4} = \frac{3(2 - k)(-1)}{(k + 2)(k - 2)(-1)}$$

Since $(k - 2)(-1) = -k + 2$, or $2 - k$, the denominator becomes

$$\frac{6 - 3k}{k^2 - 4} = \frac{3(2 - k)(-1)}{(k + 2)(2 - k)},$$

finally giving $\dfrac{6 - 3k}{k^2 - 4} = \dfrac{-3}{k + 2}.$

By reducing in an alternate way, we might get the equivalent result

$$\frac{3}{-k - 2}. \quad \square$$

To multiply or divide rational expressions, we again use properties and definitions from Section 1.1.

Multiplication and Division

$$\frac{a}{b} \cdot \frac{c}{d} = \frac{ac}{bd},$$

$$\frac{a}{b} \div \frac{c}{d} = \frac{a}{b} \cdot \frac{d}{c}$$

EXAMPLE 2 Multiply or divide, as indicated.

(a) $\dfrac{3m^2 - 2m - 8}{3m^2 + 14m + 8} \cdot \dfrac{3m + 2}{3m + 4} = \dfrac{(m - 2)(3m + 4)}{(m + 4)(3m + 2)} \cdot \dfrac{3m + 2}{3m + 4}$

$$= \frac{(m - 2)(3m + 4)(3m + 2)}{(m + 4)(3m + 2)(3m + 4)} = \frac{m - 2}{m + 4}$$

(b) $\dfrac{3p^2 + 11p - 4}{24p^3 - 8p^2} \div \dfrac{9p + 36}{24p^4 - 36p^3} = \dfrac{(p + 4)(3p - 1)}{8p^2(3p - 1)} \div \dfrac{9(p + 4)}{12p^3(2p - 3)}$

$$= \frac{(p + 4)(3p - 1)(12p^3)(2p - 3)}{8p^2(3p - 1)(9)(p + 4)}$$

$$= \frac{12p^3(2p - 3)}{9 \cdot 8p^2} = \frac{p(2p - 3)}{6} \quad \square$$

To add or subtract two rational expressions, we again use the appropriate properties from Section 1.1.

Addition and Subtraction

$$\frac{a}{b} + \frac{c}{d} = \frac{ad + bc}{bd},$$

$$\frac{a}{b} - \frac{c}{d} = \frac{ad - bc}{bd}$$

In practice, rational expressions are normally added or subtracted after rewriting all the rational expressions so that they have the same denominator. This common denominator is found with the steps given in the following box.

Common Denominator

1. Write each denominator as a product of prime factors.

2. Form a product of all the different prime factors. Each factor should have as exponent the *highest* exponent which appears on that factor.

EXAMPLE
3

☐   Add or subtract, as indicated.

(a) $\dfrac{5}{9x^2} + \dfrac{1}{6x}$

Write each denominator as a product of prime factors, as follows.

$$9x^2 = 3^2 \cdot x^2 \quad \text{and} \quad 6x = 2^1 \cdot 3^1 x^1$$

To get the common denominator, form the product of all the prime factors, with each factor having the highest exponent which appears on it. Here the highest exponent on 2 is 1, while both 3 and x have a highest exponent of 2; therefore the common denominator is

$$2^1 \cdot 3^2 \cdot x^2 = 18x^2.$$

Now write both of the given rational expressions with this denominator, giving

$$\frac{5}{9x^2} + \frac{1}{6x} = \frac{5 \cdot 2}{9x^2 \cdot 2} + \frac{1 \cdot 3x}{6x \cdot 3x} = \frac{10}{18x^2} + \frac{3x}{18x^2} = \frac{10 + 3x}{18x^2}.$$

(b) $\dfrac{y + 2}{y^2 - y} - \dfrac{3y}{2y^2 - 4y + 2}$

Factor each denominator, giving

$$\frac{y + 2}{y^2 - y} - \frac{3y}{2y^2 - 4y + 2} = \frac{y + 2}{y(y - 1)} - \frac{3y}{2(y - 1)^2}.$$

The common denominator, by the method above, is $2y(y - 1)^2$. Write each rational expression with this denominator, as follows.

$$\frac{y + 2}{y(y - 1)} - \frac{3y}{2(y - 1)^2} = \frac{2(y - 1)(y + 2)}{2y(y - 1)^2} - \frac{y \cdot 3y}{2y(y - 1)^2}$$

$$= \frac{2(y^2 + y - 2)}{2y(y - 1)^2} - \frac{3y^2}{2y(y - 1)^2}$$

$$= \frac{2y^2 + 2y - 4 - 3y^2}{2y(y - 1)^2} = \frac{-y^2 + 2y - 4}{2y(y - 1)^2} \quad ☐$$

In the remainder of this section we look at methods for simplifying expressions which are not quotients of polynomials.

EXAMPLE
4

☐   Simplify $\dfrac{\dfrac{a}{a + 1} + \dfrac{1}{a}}{\dfrac{1}{a} + \dfrac{1}{a + 1}}$.

First simplify numerator and denominator separately, as follows.

$$\frac{\dfrac{a}{a + 1} + \dfrac{1}{a}}{\dfrac{1}{a} + \dfrac{1}{a + 1}} = \frac{\dfrac{a^2 + 1(a + 1)}{a(a + 1)}}{\dfrac{1(a + 1) + 1(a)}{a(a + 1)}}$$

$$= \frac{\dfrac{a^2 + a + 1}{a(a + 1)}}{\dfrac{2a + 1}{a(a + 1)}}$$

$$= \frac{a^2 + a + 1}{a(a + 1)} \cdot \frac{a(a + 1)}{2a + 1}$$

$$= \frac{a^2 + a + 1}{2a + 1}$$

We could also simplify the given expression by multiplying both numerator and denominator by the common denominator of all the fractions, in this case $a(a + 1)$. Doing so gives

$$\frac{\dfrac{a}{a + 1} + \dfrac{1}{a}}{\dfrac{1}{a} + \dfrac{1}{a + 1}} = \frac{\left(\dfrac{a}{a + 1} + \dfrac{1}{a}\right) a(a + 1)}{\left(\dfrac{1}{a} + \dfrac{1}{a + 1}\right) a(a + 1)}$$

$$= \frac{a^2 + (a + 1)}{(a + 1) + a} = \frac{a^2 + a + 1}{2a + 1}. \quad \square$$

The next example shows how negative exponents can lead to rational expressions.

EXAMPLE 5 $\square$ Simplify $\dfrac{(x + y)^{-1}}{x^{-1} + y^{-1}}$. Write the result with only positive exponents.

Use the definition of negative integer exponent to get

$$\frac{(x + y)^{-1}}{x^{-1} + y^{-1}} = \frac{\dfrac{1}{x + y}}{\dfrac{1}{x} + \dfrac{1}{y}} = \frac{\dfrac{1}{x + y}}{\dfrac{y + x}{xy}} = \frac{1}{x + y} \cdot \frac{xy}{x + y} = \frac{xy}{(x + y)^2}. \quad \square$$

EXERCISES 1.6

Give the domain of each of the following.

1. $\dfrac{2x}{5x - 3}$

2. $\dfrac{6x}{2x - 8}$

3. $\dfrac{-8}{x^2 + 1}$

4. $\dfrac{3x}{2x^2 + 3}$

Simplify each of the following.

5. $\dfrac{6p^2 - 3p}{4p^2 - 1}$

6. $\dfrac{24y - 56y^2}{12y^4}$

7. $\dfrac{6k^2 + 7k + 2}{8k^2 - 2k - 3}$

8. $\dfrac{12r^2 + r - 1}{4r^2 + 23r - 6}$

9. $\dfrac{8 - 10m - 3m^2}{15m^3 - 10m^2}$ **10.** $\dfrac{36z^3 - 4z^4}{z^2 - 81}$

11. $\dfrac{a^2 - a - 6}{a + 2} \div \dfrac{a - 3}{a + 2}$

12. $\dfrac{m^2 + 11m + 30}{m + 6} \div \dfrac{m + 5}{m + 6}$

13. $\dfrac{p^2 - p - 12}{p^2 - 2p - 15} \cdot \dfrac{p^2 - 9p + 20}{p^2 - 8p + 16}$

14. $\dfrac{x^2 + 2x - 15}{x^2 + 11x + 30} \cdot \dfrac{x^2 + 2x - 24}{x^2 - 8x + 15}$

15. $\left(1 + \dfrac{1}{x}\right)\left(1 - \dfrac{1}{x}\right)$

16. $\left(3 + \dfrac{2}{y}\right)\left(3 - \dfrac{2}{y}\right)$

17. $\dfrac{x^3 + y^3}{x^2 - y^2} \cdot \dfrac{x + y}{x^2 - xy + y^2}$

18. $\dfrac{8y^3 - 125}{4y^2 - 20y + 25} \cdot \dfrac{2y - 5}{y}$

19. $\dfrac{x^3 + y^3}{x^3 - y^3} \cdot \dfrac{x^2 - y^2}{x^2 + 2xy + y^2}$

20. $\dfrac{x^2 - y^2}{(x - y)^2} \cdot \dfrac{x^2 - xy + y^2}{x^2 - 2xy + y^2} \div \dfrac{x^3 + y^3}{(x - y)^4}$

21. $\dfrac{3}{5x} - \dfrac{1}{10x^2}$ **22.** $\dfrac{9}{4p^3} + \dfrac{5}{6p^2}$

23. $\dfrac{1}{y} + \dfrac{1}{y + 1}$

24. $\dfrac{2}{3x - 1} + \dfrac{1}{4(x - 1)}$

25. $\dfrac{2}{a + b} - \dfrac{1}{2(a + b)}$

26. $\dfrac{3}{m} - \dfrac{1}{m - 1}$

27. $\dfrac{m + 1}{m - 1} + \dfrac{m - 1}{m + 1}$ **28.** $\dfrac{2}{x - 1} + \dfrac{1}{1 - x}$

29. $\dfrac{3}{a - 2} - \dfrac{1}{2 - a}$ **30.** $\dfrac{q}{p - q} - \dfrac{q}{q - p}$

31. $\dfrac{2}{k + 3} + \dfrac{1}{(k + 3)^2} - \dfrac{3}{(k + 3)^3}$

32. $\dfrac{1}{y + 1} - \dfrac{2}{(y + 1)^2} - \dfrac{4}{(y + 1)^3}$

33. $\dfrac{r}{r^2 + 5r + 6} + \dfrac{5r}{r^2 + 2r - 3}$

34. $\dfrac{5}{t^2 - 1} - \dfrac{8}{t^2 + 2t + 1}$

35. $\dfrac{1}{a^2 - 5a + 6} - \dfrac{1}{a^2 - 4}$

36. $\dfrac{-3}{m^2 - m - 2} - \dfrac{1}{m^2 + 3m + 2}$

37. $\left(\dfrac{3}{p - 1} - \dfrac{2}{p + 1}\right)\left(\dfrac{p - 1}{p}\right)$

38. $\left(\dfrac{y}{y^2 - 1} - \dfrac{y}{y^2 - 2y + 1}\right)\left(\dfrac{y - 1}{y + 1}\right)$

39. $\dfrac{3a}{a^2 + 5a - 6} - \dfrac{2a}{a^2 + 7a + 6}$

40. $\dfrac{2k}{k^2 + 4k + 3} + \dfrac{3k}{k^2 + 5k + 6}$

41. $\dfrac{4x - 1}{x^2 + 3x - 10} + \dfrac{2x + 3}{x^2 + 4x - 5}$

42. $\dfrac{3y + 5}{y^2 - 9y + 20} + \dfrac{2y - 7}{y^2 - 2y - 8}$

43. $\dfrac{\dfrac{1}{x + h} - \dfrac{1}{x}}{h}$

44. $\dfrac{1}{h}\left(\dfrac{1}{(x + h)^2 + 9} - \dfrac{1}{x^2 + 9}\right)$

Perform all indicated operations. Write all answers with positive integer exponents. Assume that all variables represent nonzero real numbers.

45. $\dfrac{a^{-1} + b^{-1}}{(ab)^{-1}}$ **46.** $\dfrac{p^{-1} - q^{-1}}{(pq)^{-1}}$

47. $\dfrac{r^{-1} + q^{-1}}{r^{-1} - q^{-1}} \cdot \dfrac{r - q}{r + q}$

48. $\dfrac{xy^{-1} + yx^{-1}}{x^2 + y^2}$

49. $(a + b)^{-1}(a^{-1} + b^{-1})$

50. $(m^{-1} + n^{-1})^{-1}$

51. $(x - 9y^{-1})[(x - 3y^{-1})(x + 3y^{-1})]^{-1}$

52. $(m + n)^{-1}(m^{-2} - n^{-2})^{-1}$

Perform the indicated operations.

53. $\dfrac{1 + \dfrac{1}{x}}{1 - \dfrac{1}{x}}$

54. $\dfrac{2 - \dfrac{2}{y}}{2 + \dfrac{2}{y}}$

57. $1 + \dfrac{1}{1 + \dfrac{1}{x}}$

58. $1 - \dfrac{1}{1 - \dfrac{1}{x}}$

55. $\dfrac{1 + \dfrac{1}{1 - b}}{1 - \dfrac{1}{1 + b}}$

56. $m - \dfrac{m}{m + \dfrac{1}{2}}$

1.7 Radicals

The positive number whose square is 144 is 12; this is written

$$\sqrt{144} = 12.$$

Also,

$$\sqrt{400} = 20, \qquad \sqrt{0.0001} = .01, \qquad \text{and} \quad \sqrt{\frac{25}{16}} = \frac{5}{4}.$$

Generalizing, we can define $\sqrt{a}$ as follows for positive real numbers a and b.

Square Root

$$\sqrt{a} = b \text{ if and only if } a = b^2$$

Under these conditions, b is the **square root** of a.

If $a > 0$ and if $a = b^2$, then it is also true that $a = (-b)^2$. Thus, there are always two real numbers, one positive and one negative, whose square equals a given positive real number. However, it is important to remember that the symbol

$\sqrt{a}$ is reserved for the *positive* number whose square is a.

The symbol $-\sqrt{a}$ is used for the negative square root. We can define the **cube root** for a number in a similar manner to that used for square roots. For any real numbers a and b,

Cube Root

$$\sqrt[3]{a} = b \qquad \text{if and only if} \qquad a = b^3.$$

There is a fundamental difference between the definitions of square root and cube root. We defined $\sqrt{a}$ only for *positive* a, while $\sqrt[3]{a}$ is defined for *any* a.

EXAMPLE 1 □

(a) $\sqrt{64} = 8$ since $8 > 0$ and $8^2 = 64$

(b) $-\sqrt{100} = -10$

(c) $\sqrt{-4}$ is not a real number, since there is no real number whose square is -4

(d) $\sqrt[3]{8} = 2$ since $2^3 = 8$

(e) $\sqrt[3]{-1000} = -10$ since $(-10)^3 = -1000$ □

If $a > 0$, then both $\sqrt{a}$ and $\sqrt[3]{a}$ are positive. If $a < 0$, then $\sqrt{a}$ is not a real number, while $\sqrt[3]{a} < 0$. Generalizing, if a and b are nonnegative real numbers and n is a positive integer, or if both a and b are negative and n is an odd positive integer, then

nth Root

$$\sqrt[n]{a} = b \qquad \text{if and only if} \qquad a = b^n.$$

We call $\sqrt[n]{a}$ the **principal nth root** of a (often abbreviated as just the **nth root of** a); n is the **index** of the **radical expression** $\sqrt[n]{a}$. The number a is the **radicand,** and $\sqrt[n]{}$ is a **radical.** We abbreviate $\sqrt[2]{a}$ as just $\sqrt{a}$.

The following box summarizes the conditions necessary for $\sqrt[n]{a}$ to exist.

If n is a positive integer, and a is a real number, then $\sqrt[n]{a}$ is defined as follows.

n is	$a > 0$	$a < 0$	$a = 0$
even	$\sqrt[n]{a}$ is the positive real number such that $(\sqrt[n]{a})^n = a$	$\sqrt[n]{a}$ is not a real number	$\sqrt[n]{a}$ is 0
odd	$\sqrt[n]{a}$ is the real number such that $(\sqrt[n]{a})^n = a$		$\sqrt[n]{a}$ is 0

EXAMPLE 2 □

(a) $\sqrt[5]{32} = 2$ since $2^5 = 32$

(b) $\sqrt[4]{256} = 4$, since $4^4 = 256$

(c) $\sqrt[7]{-128} = -2$, since $(-2)^7 = -128$

(d) $\sqrt[6]{-64}$ is not a real number   □

By the definition of $\sqrt[n]{a}$, for any positive integer n, if $\sqrt[n]{a}$ exists, we have

$$(\sqrt[n]{a})^n = a.$$

If a is positive, or if a is negative and n is an odd positive integer, then

$$\sqrt[n]{a^n} = a.$$

Notice that because of the conditions just given, we *cannot* simply write $\sqrt{x^2} = x$. For example, if $x = -5$,

$$\sqrt{x^2} = \sqrt{(-5)^2} = \sqrt{25} = 5 \neq x.$$

To take care of the fact that a negative value of x can produce a positive result for the square root, we must use the following rule, which involves absolute value. For any real number a,

$$\sqrt{a^2} = |a|.$$

This statement is certainly true if a is positive. If a is negative, then $-a$ is positive. Since $(-a)^2 = a^2$, we have

$$\sqrt{(-a)^2} = -a.$$

Since $|a| = -a$ if a is negative, $\sqrt{a^2} = |a|$ for *all* real numbers a. For example,

$$\sqrt{(-9)^2} = |-9| = 9, \quad \text{and} \quad \sqrt{13^2} = |13| = 13.$$

To avoid difficulties when working with variable radicands, we will assume that all variables in radicands represent only nonnegative real numbers.

Three key rules for working with radicals are given in the box below. These rules are valid for all real numbers a and b and positive integers m and n for which the indicated roots exist.

Rules for Radicals

$$\sqrt[n]{a} \cdot \sqrt[n]{b} = \sqrt[n]{ab}$$

$$\sqrt[n]{\frac{a}{b}} = \frac{\sqrt[n]{a}}{\sqrt[n]{b}} \quad (b \neq 0)$$

$$\sqrt[m]{\sqrt[n]{a}} = \sqrt[mn]{a}$$

We shall prove only the first of these rules, with the other parts left for the exercises. To prove the first rule, let $x = \sqrt[n]{a}$ and let $y = \sqrt[n]{b}$. Then, by the definition of nth root, $x^n = a$ and $y^n = b$. Hence, $x^n \cdot y^n = ab$. However, by our previous work with exponents, $x^n \cdot y^n = (xy)^n$, giving

$$(xy)^n = ab.$$

Thus, xy is an nth root of ab, or

$$xy = \sqrt[n]{ab}.$$

Substituting the original values of x and y gives

$$\sqrt[n]{a} \cdot \sqrt[n]{b} = \sqrt[n]{ab}.$$

EXAMPLE 3   Use the rules for radicals to simplify each of the following:

(a) $\sqrt{6} \cdot \sqrt{54} = \sqrt{6 \cdot 54} = \sqrt{324} = 18$

(b) $\sqrt{\dfrac{7}{64}} = \dfrac{\sqrt{7}}{\sqrt{64}} = \dfrac{\sqrt{7}}{8}$

(c) $\sqrt[3]{\dfrac{27}{5}} = \dfrac{\sqrt[3]{27}}{\sqrt[3]{5}} = \dfrac{3}{\sqrt[3]{5}}$

(d) $\sqrt[7]{\sqrt[3]{2}} = \sqrt[21]{2}$

(e) $\sqrt[4]{\sqrt{3}} = \sqrt[8]{3}$. $\square$

The various rules for radicals can be used to simplify certain expressions. For example, since $300 = 100 \cdot 3$,

$$\sqrt{300} = \sqrt{100 \cdot 3} = \sqrt{100} \cdot \sqrt{3} = 10\sqrt{3}.$$

In general, an expression containing a radical is **simplified** when the conditions in the following box are satisfied.

Simplified Radicals

1. All possible factors have been removed from under the radical sign.

2. The index on the radical is as small as possible.

3. All radicals are removed from any denominators (a process called **rationalizing** the denominator).

4. All indicated operations have been performed (if possible).

EXAMPLE 4   Simplify each of the following. Assume that all variables represent nonnegative real numbers.

(a) $\sqrt{175} = \sqrt{25 \cdot 7} = \sqrt{25} \cdot \sqrt{7} = 5\sqrt{7}$

(b) $\sqrt[3]{81x^5y^7z^6} = \sqrt[3]{27 \cdot 3 \cdot x^3 \cdot x^2 \cdot y^6 \cdot y \cdot z^6}$

$$= \sqrt[3]{(27x^3y^6z^6)(3x^2y)} = 3xy^2z^2\sqrt[3]{3x^2y}$$

(c) $\sqrt{98x^3y} + 3x\sqrt{32xy}$

First remove all perfect square factors from under the radical. Then use the distributive property, as follows.

$$\sqrt{98x^3y} + 3x\sqrt{32xy} = \sqrt{49 \cdot 2 \cdot x^2 \cdot x \cdot y} + 3x\sqrt{16 \cdot 2 \cdot x \cdot y}$$
$$= 7x\sqrt{2xy} + (3x)(4)\sqrt{2xy}$$
$$= 7x\sqrt{2xy} + 12x\sqrt{2xy} = 19x\sqrt{2xy}$$

(d) $\sqrt[3]{64m^4n^5} - \sqrt[3]{-27m^{10}n^{14}} = \sqrt[3]{(64m^3n^3)(mn^2)} - \sqrt[3]{(-27m^9n^{12})(mn^2)}$

$$= 4mn\sqrt[3]{mn^2} - (-3)m^3n^4\sqrt[3]{mn^2}$$
$$= 4mn\sqrt[3]{mn^2} + 3m^3n^4\sqrt[3]{mn^2}$$
$$= (4 + 3m^2n^3)mn\sqrt[3]{mn^2} \quad \blacksquare$$

Multiplying radical expressions is much like multiplying polynomials.

EXAMPLE 5 $\square$ $(\sqrt{2} + 3)(\sqrt{8} - 5) = \sqrt{2}(\sqrt{8}) - \sqrt{2}(5) + 3\sqrt{8} - 3(5)$

$$= \sqrt{16} - 5\sqrt{2} + 3(2\sqrt{2}) - 15$$
$$= 4 - 5\sqrt{2} + 6\sqrt{2} - 15$$
$$= -11 + \sqrt{2} \quad \blacksquare$$

The next example shows how to rationalize the denominator in an expression containing radicals.

EXAMPLE 6 $\square$ Simplify each of the following expressions.

(a) $\dfrac{4}{\sqrt{3}}$

To rationalize the denominator, multiply by $\sqrt{3}/\sqrt{3}$ (or 1) so that the denominator of the product is a rational number. Work as follows.

$$\frac{4}{\sqrt{3}} \cdot \frac{\sqrt{3}}{\sqrt{3}} = \frac{4\sqrt{3}}{3}$$

(b) $\sqrt[4]{\dfrac{3}{5}}$

Start by using the fact that the radical of a quotient can be written as the quotient of radicals.

$$\sqrt[4]{\frac{3}{5}} = \frac{\sqrt[4]{3}}{\sqrt[4]{5}}.$$

To rationalize the denominator, multiply numerator and denominator by $\sqrt[4]{5^3}$. We use this number so that the denominator will be a rational number. This multiplication gives

$$\frac{\sqrt[4]{3}}{\sqrt[4]{5}} = \frac{\sqrt[4]{3} \cdot \sqrt[4]{5^3}}{\sqrt[4]{5} \cdot \sqrt[4]{5^3}} = \frac{\sqrt[4]{3 \cdot 5^3}}{\sqrt[4]{5^4}} = \frac{\sqrt[4]{375}}{5}. \quad \square$$

EXAMPLE 7 $\square$ Rationalize the denominator of $\dfrac{1}{1 - \sqrt{2}}$.

First, note that

$$(1 - \sqrt{2})(1 + \sqrt{2}) = 1 - 2 = -1.$$

Thus, the best approach here is to multiply both numerator and denominator by the number $1 + \sqrt{2}$. Doing so gives

$$\frac{1}{1 - \sqrt{2}} = \frac{1(1 + \sqrt{2})}{(1 - \sqrt{2})(1 + \sqrt{2})} = \frac{1 + \sqrt{2}}{-1} = -1 - \sqrt{2}. \quad \square$$

EXERCISES 1.7

Simplify each of the following. Assume that all variables represent nonnegative real numbers, and that no denominators are zero.

1. $\sqrt{50}$
2. $\sqrt{12}$
3. $\sqrt[3]{250}$
4. $\sqrt[3]{128}$
5. $-\sqrt{\dfrac{9}{5}}$
6. $\sqrt{\dfrac{3}{8}}$
7. $\sqrt{5} + \sqrt{45}$
8. $\sqrt{6} + \sqrt{54}$
9. $4\sqrt{3} - 5\sqrt{12} + 3\sqrt{75}$
10. $2\sqrt{5} - 3\sqrt{20} + 2\sqrt{45}$
11. $3\sqrt{28} - 5\sqrt{63} + \sqrt{112}$
12. $-\sqrt{12} + 2\sqrt{27} + 6\sqrt{48}$
13. $-\sqrt[3]{\dfrac{3}{2}}$
14. $-\sqrt[3]{\dfrac{4}{5}}$
15. $\sqrt[4]{\dfrac{3}{2}}$
16. $\sqrt[4]{\dfrac{32}{81}}$
17. $3\sqrt[3]{16} - 4\sqrt[3]{2}$
18. $\sqrt[3]{2} - \sqrt[3]{16} + 2\sqrt[3]{54}$

19. $2\sqrt[3]{3} + 4\sqrt[3]{24} - \sqrt[3]{81}$
20. $\sqrt[3]{32} - 5\sqrt[3]{4} + 2\sqrt[3]{108}$
21. $\dfrac{1}{\sqrt{3}} - \dfrac{2}{\sqrt{12}} + 2\sqrt{3}$
22. $\dfrac{1}{\sqrt{2}} + \dfrac{3}{\sqrt{8}} + \dfrac{3}{\sqrt{32}}$
23. $\dfrac{5}{\sqrt[3]{2}} - \dfrac{2}{\sqrt[3]{16}} + \dfrac{1}{\sqrt[3]{54}}$
24. $\dfrac{-4}{\sqrt[3]{3}} + \dfrac{1}{\sqrt[3]{24}} - \dfrac{2}{\sqrt[3]{81}}$
25. $\sqrt{2x^3y^2z^4}$
26. $\sqrt{98r^3s^4t^{10}}$
27. $\sqrt[3]{16z^5x^8y^4}$
28. $\sqrt[4]{x^8y^6z^{10}}$
29. $\sqrt{a^3b^5} - 2\sqrt{a^7b^3} + \sqrt{a^3b^9}$
30. $\sqrt{p^7q^3} - \sqrt{p^5q^9} + \sqrt{p^9q}$
31. $(\sqrt{2} + 3)(\sqrt{2} - 3)$
32. $(\sqrt{5} + \sqrt{2})(\sqrt{5} - \sqrt{2})$
33. $(\sqrt[3]{11} - 1)(\sqrt[3]{11^2} + \sqrt[3]{11} + 1)$
34. $(\sqrt[3]{7} + 3)(\sqrt[3]{7^2} - 3\sqrt[3]{7} + 9)$

35. $(\sqrt{3} + \sqrt{8})^2$

36. $(\sqrt{2} - 1)^2$

37. $(3\sqrt{2} + \sqrt{3})(2\sqrt{3} - \sqrt{2})$

38. $(4\sqrt{5} - 1)(3\sqrt{5} + 2)$

39. $(2\sqrt[3]{3} + 1)(\sqrt[3]{3} - 4)$

40. $(\sqrt[3]{4} + 3)(5\sqrt[3]{4} + 1)$

41. $\sqrt{\dfrac{2}{3x}}$

42. $\sqrt{\dfrac{5}{3p}}$

43. $\sqrt{\dfrac{x^5y^3}{z^2}}$

44. $\sqrt{\dfrac{g^3h^5}{r^3}}$

45. $\sqrt[3]{\dfrac{8}{x^2}}$

46. $\sqrt[3]{\dfrac{9}{16p^4}}$

47. $-\sqrt[3]{\dfrac{k^5m^3r^2}{r^8}}$

48. $-\sqrt[3]{\dfrac{9x^5y^6}{z^5w^2}}$

49. $\sqrt[4]{\dfrac{g^3h^5}{9r^6}}$

50. $\sqrt[4]{\dfrac{32x^6}{y^5}}$

51. $\dfrac{\sqrt[3]{mn} \cdot \sqrt[3]{m^2}}{\sqrt[3]{n^2}}$

52. $\dfrac{\sqrt[3]{8m^2n^3} \cdot \sqrt[3]{2m^2}}{\sqrt[3]{32m^4n^3}}$

53. $\dfrac{\sqrt[4]{32x^5y} \cdot \sqrt[4]{2xy^4}}{\sqrt[4]{4x^3y^2}}$

54. $\dfrac{\sqrt[4]{rs^2t^3} \cdot \sqrt[4]{r^3s^2t}}{\sqrt[4]{r^2t^3}}$

55. $\sqrt[3]{\sqrt{\sqrt[3]{4}}}$

56. $\sqrt[4]{\sqrt{\sqrt[3]{2}}}$

57. $\sqrt[6]{\sqrt[3]{\sqrt{x}}}$

58. $\sqrt[8]{\sqrt{\sqrt[4]{y}}}$

59. $\dfrac{3}{1 - \sqrt{2}}$

60. $\dfrac{2}{1 + \sqrt{5}}$

61. $\dfrac{\sqrt{3}}{4 + \sqrt{3}}$

62. $\dfrac{2\sqrt{7}}{3 - \sqrt{7}}$

63. $\dfrac{4}{2 - \sqrt{y}}$

64. $\dfrac{-1}{\sqrt{k} - 2}$

65. $\dfrac{1}{\sqrt{m} - \sqrt{p}}$

66. $\dfrac{\sqrt{z}}{1 - \sqrt{z}}$

67. $\dfrac{y}{\sqrt{x + y + z}}$

68. $\dfrac{p}{p + q - \sqrt{r}}$

69. $\dfrac{2}{3 + \sqrt{1 + k}}$

70. $\dfrac{-5}{1 - \sqrt{3 - p}}$

71. $\dfrac{m}{\sqrt{p}} + \dfrac{p}{\sqrt{m}}$

72. $\dfrac{a}{\sqrt{b}} - \dfrac{b}{\sqrt{a}}$

73. $\dfrac{\sqrt{x} + \sqrt{x + 1}}{\sqrt{x} - \sqrt{x + 1}}$

74. $\dfrac{\sqrt{p} + \sqrt{p^2 - 1}}{\sqrt{p} - \sqrt{p^2 - 1}}$

75. $\dfrac{5}{\sqrt[3]{a} + \sqrt[3]{b}}$

76. $\dfrac{1}{\sqrt[3]{m} - \sqrt[3]{n}}$

In advanced mathematics it is sometimes useful to write a radical expression with a rational numerator. The procedure is similar to rationalizing the denominator. Rationalize the numerator of each of the following.

77. $\dfrac{\sqrt{3}}{2}$

78. $\dfrac{\sqrt{6}}{5}$

79. $\dfrac{1 + \sqrt{2}}{2}$

80. $\dfrac{1 - \sqrt{3}}{3}$

81. $\dfrac{\sqrt{x}}{1 + \sqrt{x}}$

82. $\dfrac{\sqrt{p}}{1 - \sqrt{p}}$

83. $\dfrac{\sqrt{x} + \sqrt{x + 1}}{\sqrt{x} - \sqrt{x + 1}}$

84. $\dfrac{\sqrt{p} + \sqrt{p^2 - 1}}{\sqrt{p} - \sqrt{p^2 - 1}}$

Evaluate each of the following. Write all answers in scientific notation with three significant digits. For fourth roots, use the square root key twice. 🖩

85. $\sqrt{2.876 \times 10^7}$

86. $\sqrt{5.432 \times 10^9}$

87. $\sqrt[4]{3.87 \times 10^{-4}}$

88. $\sqrt[4]{5.913 \times 10^{-8}}$

89. $\dfrac{2.04 \times 10^{-3}}{\sqrt{5.97 \times 10^{-5}}}$

90. $\dfrac{3.86 \times 10^{-5}}{\sqrt{4.82 \times 10^{-5}}}$

91. $\sqrt{(4.721)^2 + (8.963)^2 - 2(4.721)(8.963)(.0468)}$

92. $\sqrt{(157.3)^2 + (184.7)^2 - 2(157.3)(184.7)(.9082)}$

Prove each of the following (for all real numbers a and b and for any positive integers m and n for which all the following are real numbers).

93. $\dfrac{\sqrt[n]{a}}{\sqrt[n]{b}} = \sqrt[n]{\dfrac{a}{b}} \quad (b \neq 0)$

94. $\sqrt[m]{\sqrt[n]{a}} = \sqrt[mn]{a}$

95. Use a calculator to find an approximate value for $\sqrt{5 + 2\sqrt{6}}$.

96. Show that $\sqrt{5 + 2\sqrt{6}} = \sqrt{2} + \sqrt{3}$.

1.8 Rational Exponents

We have now defined a^n for all integer values of n. In this section we wish to extend this definition to give meaning to a^n for *rational* (and not just integer) values of the exponent.

To start, let us define $a^{1/n}$ for positive integers n. Any definition of $a^{1/n}$ that we give should be consistent with past rules of exponents. In particular, we want the rule $(b^m)^n = b^{mn}$ to still be valid. If we replace b^m with $a^{1/n}$, then

$$(a^{1/n})^n = a^{n/n} = a^1 = a.$$

Thus, $a^{1/n}$ must be an nth root of a, or, as a definition,

$a^{1/n}$

If a is a real number, n is a positive integer, and $\sqrt[n]{a}$ is a real number, then

$$a^{1/n} = \sqrt[n]{a}.$$

EXAMPLE 1

(a) $16^{1/2} = \sqrt{16} = 4$

(b) $(-32)^{1/5} = \sqrt[5]{-32} = -2$

What about rational exponents in general? We want to define $a^{m/n}$ so that all the rules for integer exponents still hold. For the rule $(b^m)^n = b^{mn}$ to hold, we must have $(a^{1/n})^m = a^{m/n}$. Therefore, $a^{m/n}$ is defined as follows.

$a^{m/n}$

For all integers m, all positive integers n, and all real numbers a for which $\sqrt[n]{a}$ is a real number,

$$a^{m/n} = (\sqrt[n]{a})^m.$$

EXAMPLE 2

(a) $125^{2/3} = (\sqrt[3]{125})^2 = 5^2 = 25$

(b) $x^{5/2} = (\sqrt{x})^5 = x^2\sqrt{x}$ $(x \ge 0)$

(c) $(16m)^{3/4} = (\sqrt[4]{16m})^3 = (2\sqrt[4]{m})^3 = 8(\sqrt[4]{m})^3$ $(m \ge 0)$

We shall now show that $(a^{1/n})^m = (a^m)^{1/n}$, so that $a^{m/n}$ is also equal to $\sqrt[n]{a^m}$. To prove that $(a^{1/n})^m = (a^m)^{1/n}$, we shall use some of the rules for radicals and exponents developed in this chapter. Start with $(a^{1/n})^m$, and raise it to the nth power, to get

$$[(a^{1/n})^m]^n.$$

Now use properties of exponents, getting

$$[(a^{1/n})^m]^n = (a^{1/n})^{mn} = [(a^{1/n})^n]^m = a^m.$$

Because of this result, $(a^{1/n})^m$ must be an nth root of a^m, so that

$$(a^{1/n})^m = (a^m)^{1/n}.$$

Since $a^{m/n} = (a^{1/n})^m$ by definition, then we must also have

$$a^{m/n} = (a^m)^{1/n}.$$

Using radicals,

$$a^{m/n} = (\sqrt[n]{a})^m \quad \text{or} \quad a^{m/n} = \sqrt[n]{a^m}.$$

Thus, we have two ways to evaluate $a^{m/n}$: we can find $\sqrt[n]{a}$ and raise the result to the mth power, or we can find the nth root of a^m. In practice, it is usually easier to find $(\sqrt[n]{a})^m$. For example, $27^{4/3}$ can be evaluated in either of two ways:

$$27^{4/3} = (27^{1/3})^4 = 3^4 = 81$$
$$27^{4/3} = (27^4)^{1/3} = 531,441^{1/3} = 81.$$

The form $(27^{1/3})^4$ is easier to evaluate.

It can be shown that all the earlier results concerning integer exponents also apply to rational exponents. We list these rules in the following box.

Rules for Exponents

Let r and s be rational numbers. The results below are valid for all real numbers a and b for which all indicated powers exist.

$$a^r \cdot a^s = a^{r+s} \qquad\qquad \left(\frac{a}{b}\right)^r = \frac{a^r}{b^r}$$

$$\frac{a^r}{a^s} = a^{r-s} \qquad\qquad (a^r)^s = a^{rs}$$

$$(ab)^r = a^r \cdot b^r \qquad\qquad a^{-r} = \frac{1}{a^r}$$

EXAMPLE 3 Evaluate each of the following. Assume that all variables represent positive real numbers.

(a) $81^{5/4} \cdot 4^{-3/2} = (81^{1/4})^5 \cdot \dfrac{1}{(4^{1/2})^3} = 3^5 \cdot \dfrac{1}{2^3} = \dfrac{243}{8}$

(b) $6y^{2/3} \cdot 2y^{1/2} = 12y^{2/3+1/2} = 12y^{7/6}$

(c) $\left(\dfrac{3m^{5/6}}{y^{3/4}}\right)^2 \cdot \left(\dfrac{8y^3}{m^6}\right)^{2/3} = \dfrac{9m^{5/3}}{y^{3/2}} \cdot \dfrac{4y^2}{m^4} = 36m^{5/3-4}y^{2-3/2} = \dfrac{36y^{1/2}}{m^{7/3}}$

Rational exponents can be used to simplify expressions with radicals. Convert the radicals into expressions with rational exponents, simplify, and then convert back to radical form. The next example shows how this process works.

EXAMPLE 4 ☐ Evaluate each of the following. Assume that all variables represent positive real numbers.

(a) $\dfrac{\sqrt[4]{9}}{\sqrt{3}} = \dfrac{9^{1/4}}{3^{1/2}} = \dfrac{(3^2)^{1/4}}{3^{1/2}} = \dfrac{3^{1/2}}{3^{1/2}} = 1$

(b) $\dfrac{\sqrt[3]{m^8}}{\sqrt{m^5}} = \dfrac{m^{8/3}}{m^{5/2}} = m^{8/3 - 5/2} = m^{1/6} = \sqrt[6]{m}$

(c) $\sqrt{a^3 b^5}\ \sqrt[3]{a^4 b^2} = (a^3 b^5)^{1/2}(a^4 b^2)^{1/3}$

$= (a^{3/2} b^{5/2})(a^{4/3} b^{2/3})$

$= (a^{3/2 + 4/3})(b^{5/2 + 2/3})$

$= a^{17/6} b^{19/6} = a^2 \cdot a^{5/6} \cdot b^3 \cdot b^{1/6} = a^2 b^3 \sqrt[6]{a^5 b}$ ☐

The next examples show how to factor with rational exponents and how to simplify fractional expressions involving rational exponents.

EXAMPLE 5 ☐ Factor out the lowest power of the variable. Assume all variables represent positive real numbers.

(a) $4m^{1/2} + 3m^{3/2} = m^{1/2}(4 + 3m)$

To check this result, multiply $m^{1/2}$ and $4 + 3m$.

(b) $y^{-1/3} + y^{2/3} = y^{-1/3}(1 + y)$ ☐

EXAMPLE 6 ☐ Simplify $\dfrac{1}{(m-1)^{1/2}} + 3(m-1)^{1/2}$.

The common denominator is $(m-1)^{1/2}$. Work as follows.

$$\frac{1}{(m-1)^{1/2}} + 3(m-1)^{1/2} = \frac{1}{(m-1)^{1/2}} + \frac{3(m-1)^{1/2}(m-1)^{1/2}}{(m-1)^{1/2}}$$

$$= \frac{1}{(m-1)^{1/2}} + \frac{3(m-1)}{(m-1)^{1/2}}$$

$$= \frac{1 + 3(m-1)}{(m-1)^{1/2}}$$

$$= \frac{3m - 2}{(m-1)^{1/2}}$$ ☐

EXERCISES 1.8

Simplify each of the following. Assume all variables represent nonnegative real numbers.

1. $4^{1/2}$

2. $25^{1/2}$

3. $8^{2/3}$

4. $81^{3/4}$

5. $27^{-2/3}$

6. $32^{-4/5}$

7. $\left(\dfrac{4}{9}\right)^{-3/2}$

8. $\left(\dfrac{1}{8}\right)^{-5/3}$

9. $(16p^4)^{1/2}$

10. $(36r^6)^{1/2}$

11. $(27x^6)^{2/3}$

12. $(64a^{12})^{5/6}$

Write each of the following using rational exponents instead of radical signs. Assume that all variables represent positive real numbers.

13. $\sqrt[3]{x^2}$

14. $\sqrt[3]{y^4}$

15. $\sqrt[3]{z^5}$

16. $p\sqrt[3]{p^4}$

17. $y^3\sqrt[4]{y^6}$

18. $p^2\sqrt[3]{\sqrt[4]{p^8}}$

19. $m^{-2/3}\sqrt[4]{m^6}$

20. $y^{-3/2}\sqrt[4]{\sqrt{y^6}}$

Rewrite each of the following, using only positive exponents. Assume that all variables represent positive real numbers and that variables used as exponents represent rational numbers.

21. $(m^{2/3})(m^{5/3})$

22. $(x^{4/5})(x^{2/5})$

23. $(1 + n)^{1/2}(1 + n)^{3/4}$

24. $(m + 7)^{-1/6}(m + 7)^{-2/3}$

25. $(2y^{3/4}z)(3y^{1/4}z^{-1/3})$

26. $(4a^{-1/2}b^{2/3})(a^{3/2}b^{-1/3})$

27. $(r^{5/7}s^{3/5})(r^{-1/2}s^{-1/4})$

28. $(t^{3/2}w^{3/5})(t^{-1/3}w^{-1/5})$

29. $\dfrac{a^{4/3} \cdot b^{1/2}}{a^{2/3} \cdot b^{-3/2}}$

30. $\dfrac{x^{1/3} \cdot y^{2/3} \cdot z^{1/4}}{x^{5/3} \cdot y^{-1/3} \cdot z^{3/4}}$

31. $\dfrac{k^{-3/5} \cdot h^{-1/3} \cdot t^{2/5}}{k^{-1/5} \cdot h^{-2/3} \cdot t^{1/5}}$

32. $\dfrac{m^{7/3} \cdot n^{-2/5} \cdot p^{3/8}}{m^{-2/3} \cdot n^{3/5} \cdot p^{-5/8}}$

33. $\dfrac{k^{3/2} \cdot k^{-1/2}}{k^{1/4} \cdot k^{3/4}}$

34. $\dfrac{m^{2/5} \cdot m^{3/5} \cdot m^{-4/5}}{m^{1/5} \cdot m^{-6/5}}$

35. $\dfrac{x^{-2/3} \cdot x^{4/3}}{x^{1/2} \cdot x^{-3/4}}$

36. $\dfrac{-4a^{1/2} \cdot a^{2/3}}{a^{-5/6}}$

37. $\dfrac{8y^{2/3}y^{-1}}{2^{-1}y^{3/4} \cdot y^{-1/6}}$

38. $\dfrac{9 \cdot k^{1/3} \cdot k^{-1/2} \cdot k^{-1/6}}{k^{-2/3}}$

39. $\left(\dfrac{x^4y^3z}{16x^{-6}yz^5}\right)^{-1/2}$

40. $\left(\dfrac{p^3r^9}{27p^{-3}r^{-6}}\right)^{-1/3}$

41. $(r^{3/p})^{2p}(r^{1/p})^{p^2}$

42. $(m^{2/x})^{x/3}(m^{x/4})^{2/x}$

43. $\dfrac{m^{1-a} \cdot m^a}{m^{-1/2}}$

44. $\dfrac{(y^{3-b})(y^{2b-1})}{y^{1/2}}$

45. $\dfrac{(p^{1/n})(1/m)}{p^{-m/n}}$

46. $\dfrac{(q^{2r/3})(q^r)^{-1/3}}{(q^{4/3})^{1/r}}$

47. $(r^{1/2} - r^{-1/2})^2$

48. $(p^{1/2} - p^{-1/2})(p^{1/2} + p^{-1/2})$

49. $[(x^{1/2} - x^{-1/2})^2 + 4]^{1/2}$

50. $[(x^{1/2} + x^{-1/2})^2 - 4]^{1/2}$

Write each of the following as a single radical. Assume all variables represent nonnegative real numbers.

51. $\sqrt{r^3} \cdot \sqrt[3]{r}$

52. $\sqrt[6]{y^3} \cdot \sqrt[4]{y^3}$

53. $\sqrt[10]{p^8} \cdot \sqrt{p}$

54. $\sqrt[8]{y^6} \cdot \sqrt[3]{y}$

55. $\sqrt[4]{p^2q^8} \cdot \sqrt[3]{p^5}$

56. $\sqrt[6]{q^4r^{18}} \cdot \sqrt[3]{r^5}$

57. $\sqrt{m^2n} \cdot \sqrt[3]{m^4n^2}$

58. $\sqrt[4]{y^2r^3} \cdot \sqrt[3]{y^6}$

59. $\sqrt[5]{m^3n^6} \cdot \sqrt[3]{m^4n^5}$

60. $\sqrt[6]{r^8s^5} \cdot \sqrt[3]{r^3s^7}$

Factor, using the given common factor. Assume all variables represent positive real numbers.

61. $4k^{7/4} + k^{3/4}$; $k^{3/4}$

62. $y^{9/2} - 3y^{5/2}$; $y^{5/2}$

63. $9z^{-1/2} + 2z^{1/2}$; $z^{-1/2}$

64. $3m^{2/3} - 4m^{-1/3}$; $m^{-1/3}$

65. $p^{-3/4} - 2p^{-1/4}$; $p^{-1/4}$

66. $6r^{-2/3} - 5r^{-5/3}$; $r^{-5/3}$

67. $(p + 4)^{-3/2} + (p + 4)^{-1/2} + (p + 4)^{1/2}$;
$(p + 4)^{-3/2}$

68. $(3r + 1)^{-2/3} + (3r + 1)^{1/3} + (3r + 1)^{4/3}$;
$(3r + 1)^{-2/3}$

Simplify each of the following. Assume that all variables represent positive real numbers.

69. $\dfrac{2}{p^{1/2}} + 5p^{1/2}$

70. $\dfrac{3}{r^{2/3}} + r^{1/3}$

71. $\dfrac{(p + 1)^{1/2} - p(\frac{1}{2})(p + 1)^{-1/2}}{p + 1}$

72. $\dfrac{(r - 2)^{2/3} - r(\frac{2}{3})(r - 2)^{-1/3}}{(r - 2)^{4/3}}$

73. $\dfrac{3(2x^2 + 5)^{1/3} - x(2x^2 + 5)^{-2/3}(4x)}{(2x^2 + 5)^{2/3}}$

74. $\dfrac{-(m^3 + m)^{2/3} + m(\frac{2}{3})(m^3 + m)^{-1/3}(3m^2 + 1)}{(m^3 + m)^{4/3}}$

One important application of mathematics to business and management concerns supply and demand. Usually, as the price of an item increases, the supply increases and the demand decreases. By studying past records of supply and demand at different prices, economists can construct an equation which describes (approximately) supply and demand for a given item. The next two exercises show examples of this.

75. The price of a certain type of solar heater is approximated by p, where

$$p = 2x^{1/2} + 3x^{2/3}$$

and x is the number of units supplied. Find the price when the supply is 64 units.

76. The demand for a certain commodity and the price are related by the equation

$$p = 1000 - 200x^{-2/3} \qquad (x > 0),$$

where x is the number of units of the product demanded. Find the price when the demand is 27.

In our system of government, the president is elected by the electoral college and not by individual voters. Because of this, smaller states have a greater voice in the selection of a president than they would otherwise. Two political scientists have studied the problems of compaigning for president under the current system and have concluded that candidates should allot their money according to the formula:

$$\begin{array}{c} \text{amount} \\ \text{for} \\ \text{large} \\ \text{state} \end{array} = \left(\frac{E_{large}}{E_{small}}\right)^{3/2} \times \begin{array}{c} \text{amount} \\ \text{for} \\ \text{small} \\ \text{state} \end{array}$$

Here E_{large} represents the electoral vote of the large state, and E_{small} represents the electoral vote of the small state. Find the amount that should be spent in each of the following larger states if $1,000,000 is spent in the small state and the following statements are true.

77. The large state has 48 electoral votes, and the small state has 3.

78. The large state has 36 electoral votes, and the small state has 4.

79. Six votes in a small state; 28 in a large.

80. Nine votes in a small state; 32 in a large.

A Delta Airlines map gives a formula for calculating the visible distance from a jet plane to the horizon. On a clear day, this distance is approximated by

$$D = 1.22x^{1/2},$$

where x is altitude in feet, and D is distance to the horizon in miles. Find D for an altitude of

81. 5000 feet   **82.** 10,000 feet

83. 30,000 feet   **84.** 40,000 feet.

The Galápagos Islands are a chain of islands ranging in size from 2 to 2249 square miles. A biologist has shown that the number of different land-plant species on an island in this chain is related to the size of the island by

$$S = 28.6A^{0.32},$$

where A is the area of an island in square miles and S is the number of different plant species on that island. Estimate S (rounding to the nearest whole number) for islands of area

85. 1 square mile   **86.** 25 square miles

87. 300 square miles   **88.** 2000 square miles.

1.9   Complex Numbers

So far in this book, we have dealt only with real numbers. However, the set of real numbers is not large enough to include all the numbers that we need in algebra. For example, using real numbers alone, we cannot find a number whose square is -1. To find such a number, it is necessary to extend the real number system.

To do this, we introduce numbers of the form $a + bi$, where a and b are real numbers, and i is defined by

Definition of *i*

$$i^2 = -1.$$

These new numbers $a + bi$ are called **complex numbers.** Each real number is a complex number, since a real number a may be thought of as the complex number $a + 0i$. A complex number of the form $0 + bi$, where b is nonzero, is called an **imaginary number.** Both the set of real numbers and the set of imaginary numbers are subsets of the set of complex numbers. (See Figure 1.7, which is an extension of Figure 1.1 on page 8.) A complex number written in the form $a + bi$ or $a + ib$ is in **standard form.** (The form $a + ib$ is used to simplify certain symbols such as $i\sqrt{5}$, since $\sqrt{5}\,i$ could be too easily mistaken for $\sqrt{5i}$.)

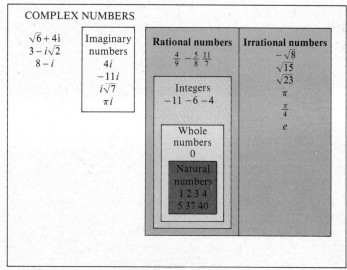

Real numbers are shaded.

Figure 1.7

EXAMPLE 1   ☐   (a) $6 + 2i$ is a complex number but not an imaginary number or real number.

(b) -2 is a complex number and a real number but not an imaginary number.

(c) $-5i$ is a complex number and an imaginary number but not a real number. ☐

EXAMPLE 2   ☐   The list below shows several numbers, along with the standard form of the number.

Number	Standard form
$6i$	$0 + 6i$
-9	$-9 + 0i$
0	$0 + 0i$
$9 - i$	$9 - i$
$i - 1$	$-1 + i$ ☐

Equality for complex numbers is defined as follows: for real numbers a, b, c, and d,

Equality

$$a + bi = c + di \quad \text{if and only if} \quad a = c \text{ and } b = d.$$

EXAMPLE 3   ☐   Solve $2 + mi = k + 3i$ for real numbers m and k.

By the definition of equality, $2 + mi = k + 3i$ if and only if $2 = k$ and $m = 3$.   ☐

The **sum** of two complex numbers $a + bi$ and $c + di$ is defined as

Sum

$$(a + bi) + (c + di) = (a + c) + (b + d)i.$$

EXAMPLE 4   ☐   (a) $(3 - 4i) + (-2 + 6i) = [3 + (-2)] + [-4 + 6]i$
$$= 1 + 2i$$

(b) $(-9 + 7i) + (3 - 15i) = -6 - 8i$ ☐

Since $(a + bi) + (0 + 0i) = a + bi$ for all complex numbers $a + bi$, the number $0 + 0i$ is called the **additive identity** for complex numbers. The sum of $a + bi$ and $-a - bi$ is $0 + 0i$, so that the number $-a - bi$ is called the **negative** or **additive inverse** of $a + bi$.

Using this definition of additive inverse, **subtraction** of the complex numbers $a + bi$ and $c + di$ is defined as follows:

$$(a + bi) - (c + di) = (a + bi) + (-c - di)$$

Rearranging to put the result in standard form we obtain,

Subtraction

$$(a + bi) - (c + di) = (a - c) + (b - d)i.$$

EXAMPLE 5 ☐ Subtract as indicated.

(a) $(-4 + 3i) - (6 - 7i) = (-4 + 3i) + (-6 + 7i)$
$$= -10 + 10i$$

(b) $(12 - 5i) - (8 - 3i) = 4 - 2i$ ☐

The product of two complex numbers can be found by multiplying as if the numbers were binomials and using the fact that $i^2 = -1$, as follows.

$$(a + bi)(c + di) = ac + adi + bic + bidi$$
$$= ac + adi + bci + bdi^2$$
$$= ac + (ad + bc)i + bd(-1)$$
$$(a + bi)(c + di) = (ac - bd) + (ad + bc)i.$$

Thus, the **product** of the complex numbers $a + bi$ and $c + di$ is defined as

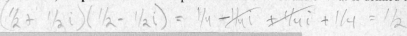

$$\left(\tfrac{1}{2} + \tfrac{1}{2}i\right)\left(\tfrac{1}{2} - \tfrac{1}{2}i\right) = \tfrac{1}{4} - \tfrac{1}{4}i + \tfrac{1}{4}i + \tfrac{1}{4} = \tfrac{1}{2}$$

Product

$$(a + bi)(c + di) = (ac - bd) + (ad + bc)i.$$

This definition is hard to remember. To find a given product, it is better to just multiply as we did with binomials. The next example shows this.

EXAMPLE 6 ☐ Find each of the following products.

(a) $(2 - 3i)(3 + 4i) = 2(3) - 3i(3) + 2(4i) - 3i(4i)$
$$= 6 - 9i + 8i - 12i^2$$
$$= 6 - i - 12(-1)$$
$$= 18 - i$$

(b) $(5 - 4i)(7 - 2i) = 5(7) - 4i(7) + 5(-2i) - 4i(-2i)$
$$= 35 - 28i - 10i + 8i^2$$
$$= 35 - 38i + 8(-1)$$
$$= 27 - 38i$$

(c) i^{15}

We know that $i^2 = -1$. Therefore, $i^3 = i^2 \cdot i = (-1) \cdot i = -i$. Furthermore, $i^4 = i^2 \cdot i^2 = (-1) \cdot (-1) = 1$. By writing a number such as i^{15} as a product involving a power of i^4, we can quickly simplify it. By properties of exponents,

$$i^{15} = i^{12} \cdot i^3 = (i^4)^3 \cdot i^3 = (1)^3(-i) = -i.$$ ☐

Using methods similar to those of part (c) of this last example, we can construct a table of **powers of** *i*:

Powers of i

$$i^1 = i \qquad i^5 = i \qquad i^9 = i$$
$$i^2 = -1 \qquad i^6 = -1 \qquad i^{10} = -1$$
$$i^3 = -i \qquad i^7 = -i \qquad i^{11} = -i$$
$$i^4 = 1 \qquad i^8 = 1 \qquad i^{12} = 1,$$

and so on.

Since

$$(a + bi)(1 + 0i) = a \cdot 1 + a \cdot 0i + bi \cdot 1 + bi \cdot 0i$$
$$= a + bi,$$

$1 + 0i$ is called the **multiplicative identity** for complex numbers. To find the multiplicative inverse of $a + bi$, when $a \neq 0$ or $b \neq 0$, we start with $1/(a + bi)$. We want $1/(a + bi)$ to be written in standard form for complex numbers. Since

$$(a + bi)(a - bi) = a^2 - (b^2 i^2) = a^2 - b^2 (-1) = a^2 + b^2,$$

multiply numerator and denominator of $1/(a + bi)$ by $a - bi$ to get

$$\frac{1}{a + bi} \cdot \frac{a - bi}{a - bi} = \frac{1(a - bi)}{(a + bi)(a - bi)}$$

$$= \frac{a - bi}{a^2 + b^2}$$

$$\frac{1}{a + bi} = \frac{a}{a^2 + b^2} - \frac{b}{a^2 + b^2} i.$$

Thus, the multiplicative inverse of $a + bi$, in standard form, is

Multiplicative Inverse

$$\frac{a}{a^2 + b^2} - \frac{b}{a^2 + b^2} i.$$

The number $a - bi$ is called the **conjugate** of $a + bi$.

EXAMPLE 7   ☐   Find the multiplicative inverse of $2 - 3i$.

We could use the formula given above to find this inverse. However, this formula is hard to remember, so it is best to go back to the basics and multiply numerator and denominator of $1/(2 - 3i)$ by the conjugate of $2 - 3i$ which is $2 + 3i$. Work as follows.

$$\frac{1}{2 - 3i} = \frac{1 \cdot (2 + 3i)}{(2 - 3i)(2 + 3i)} = \frac{2 + 3i}{4 - 9i^2}$$

$$= \frac{2 + 3i}{13}$$

$$\frac{1}{2 - 3i} = \frac{2}{13} + \frac{3}{13} i.$$

To check that this is correct, multiply $2 - 3i$ by its inverse to see that the product is 1. ☐

The conjugate of the divisor is used to find the **quotient** of two complex numbers, as shown in the next example.

EXAMPLE 8   ☐   (a) $\dfrac{3 + 2i}{5 - i}$

Multiply the numerator and denominator by the conjugate of $5 - i$.

$$\frac{3 + 2i}{5 - i} = \frac{(3 + 2i)(5 + i)}{(5 - i)(5 + i)}$$

$$= \frac{15 + 3i + 10i + 2i^2}{25 - i^2}$$

$$= \frac{13 + 13i}{26}$$

$$= \frac{1}{2} + \frac{1}{2} i$$

To check this answer, show that

$$(5 - i)\left(\frac{1}{2} + \frac{1}{2} i\right) = 3 + 2i.$$

(b) $\dfrac{3}{i} = \dfrac{3(-i)}{i(-i)}$ $-i$ is the conjugate of i

$$= \frac{-3i}{-i^2}$$

$$= \frac{-3i}{1}$$ $-i^2 = -(-1) = 1$

$$= -3i$$ $0 - 3i$ in standard form   ☐

When solving certain equations in the next chapter, it is common to encounter expressions such as $\sqrt{-a}$, for a positive real number a. We shall define $\sqrt{-a}$ as

$$\sqrt{-a} = i\sqrt{a}, \qquad \text{if } a > 0.$$

EXAMPLE 9
   □   (a) $\sqrt{-16} = i\sqrt{16} = 4i$

         (b) $\sqrt{-70} = i\sqrt{70}$ □

Products or quotients with square roots of negative numbers may be simplified using the fact that $\sqrt{-a} = i\sqrt{a}$, for positive numbers a. The next example shows how to do this.

EXAMPLE 10
   □   (a) $\sqrt{-7} \cdot \sqrt{-7} = i\sqrt{7} \cdot i\sqrt{7}$

$$= i^2 \cdot (\sqrt{7})^2 = (-1) \cdot 7 = -7$$

         (b) $\sqrt{-6} \cdot \sqrt{-10} = i\sqrt{6} \cdot i\sqrt{10}$

$$= i^2 \cdot \sqrt{6 \cdot 10} = -1 \cdot 2\sqrt{15} = -2\sqrt{15}$$

         (c) $\dfrac{\sqrt{-20}}{\sqrt{-2}} = \dfrac{i\sqrt{20}}{i\sqrt{2}} = \sqrt{10}$

         (d) $\dfrac{\sqrt{-48}}{\sqrt{24}} = \dfrac{i\sqrt{48}}{\sqrt{24}} = i\sqrt{2}$ □

When working with negative radicands, it is very important to use the definition $\sqrt{-a} = i\sqrt{a}$ before using any of the other rules for radicals. In particular, the rule $\sqrt{c} \cdot \sqrt{d} = \sqrt{cd}$ is valid only when c and d are *not* both negative.

EXERCISES 1.9

Add or subtract as indicated.

1. $(3 + 2i) + (4 - 3i)$
2. $(4 - i) + (2 + 5i)$
3. $(6 - 4i) - (3 + 2i)$
4. $(5 - 2i) - (5 + 3i)$
5. $(2 - 5i) - (3 + 4i) - (-2 + i)$
6. $(-4 - i) - (2 + 3i) + (-4 + 5i)$
7. $-i - 2 - (3 - 4i) - (5 - 2i)$
8. $3 - (4 - i) - 4i + (-2 + 5i)$

Multiply as indicated.

9. $(2 + i)(3 - 2i)$
10. $(-2 + 3i)(4 - 2i)$
11. $(5 + 2i)(5 - 3i)$ 12. $(-8 + i)(4 - 2i)$

13. $(-3 + 2i)^2$ 14. $(2 + i)^2$
15. $(3 + i)(-3 - i)$ 16. $(-5 - i)(5 + i)$
17. $(2 + 3i)(2 - 3i)$ 18. $(6 - 4i)(6 + 4i)$
19. $(2 - 5i)(2 + 5i)$
20. $(-3 + i)(-3 - i)$
21. $(\sqrt{6} + i)(\sqrt{6} - i)$
22. $(\sqrt{2} - 4i)(\sqrt{2} - 4i)$
23. $i(3 - 4i)(3 + 4i)$ 24. $i(2 + 7i)(2 - 7i)$
25. $(3 + 5i)^3$ 26. $(2 - 3i)^3$
27. $(4 + i)^4$ 28. $(1 - i)^4$

Simplify each of the following.

29. $\sqrt{-9}$ 30. $\sqrt{-25}$
31. $-\sqrt{-400}$ 32. $-\sqrt{-225}$

33. $\sqrt{-7}$ **34.** $\sqrt{-21}$

35. $\sqrt{-5} \cdot \sqrt{-5}$ **36.** $\sqrt{-20} \cdot \sqrt{-20}$

37. $\sqrt{-8} \cdot \sqrt{-2}$ **38.** $\sqrt{-27} \cdot \sqrt{-3}$

39. $\dfrac{\sqrt{-40}}{\sqrt{-10}}$ **40.** $\dfrac{\sqrt{-190}}{\sqrt{-19}}$

41. $\dfrac{\sqrt{-6} \cdot \sqrt{-2}}{\sqrt{3}}$ **42.** $\dfrac{\sqrt{-12} \cdot \sqrt{-6}}{\sqrt{8}}$

Find each of the following powers of i.

43. i^5 **44.** i^6

45. i^8 **46.** i^9

47. i^{11} **48.** i^{12}

49. i^{20} **50.** i^{25}

51. i^{43} **52.** i^{57}

53. $1/i^9$ **54.** $1/i^{12}$

Divide as indicated.

55. $\dfrac{1+i}{1-i}$ **56.** $\dfrac{2-i}{2+i}$

57. $\dfrac{4-3i}{4+3i}$ **58.** $\dfrac{5+6i}{5-6i}$

59. $\dfrac{4+i}{6+2i}$ **60.** $\dfrac{3-2i}{5+3i}$

61. $\dfrac{5-2i}{6-i}$ **62.** $\dfrac{3-4i}{2-5i}$

Perform the indicated operations and write your answers in standard form.

63. $\dfrac{2+i}{3-i} \cdot \dfrac{5+2i}{1+i}$ **64.** $\dfrac{1-i}{2+i} \cdot \dfrac{4+3i}{1+i}$

65. $\dfrac{6+2i}{5-i} \cdot \dfrac{1-3i}{2+6i}$ **66.** $\dfrac{5-3i}{1+2i} \cdot \dfrac{2-4i}{1+i}$

67. $\dfrac{5-i}{3+i} + \dfrac{2+7i}{3+i}$ **68.** $\dfrac{4-3i}{2+5i} + \dfrac{8-i}{2+5i}$

69. $\dfrac{6+2i}{1+3i} + \dfrac{2-i}{1-3i}$ **70.** $\dfrac{4-i}{3+4i} - \dfrac{3+2i}{3-4i}$

71. $\dfrac{6+3i}{1-i} - \dfrac{2-i}{4+i}$ **72.** $\dfrac{2-3i}{2+i} + \dfrac{6+i}{3+5i}$

Use the definition of equality for complex numbers to solve the following equations for real numbers a and b.

73. $a + bi = 23 + 5i$

74. $a + bi = -2 + 4i$

75. $a + bi = 18 - 3i$

76. $2 + bi = a - 4i$

77. $a + 3i = 5 + 3bi + 2a$

78. $4a - 2bi + 7 = 3i + 3a + 5$

79. $i(2b + 6) - 3 = 4(bi + a)$

80. $3i + 2(a - 1) = 4 + 2i(b + 3)$

Prove that the complex numbers $z_1 = a + bi$, $z_2 = c + di$, and $z_3 = e + fi$ satisfy each of the following properties.

81. commutative property for addition:
$z_1 + z_2 = z_2 + z_1$

82. commutative property for multiplication:
$z_1 z_2 = z_2 z_1$

83. associative property for addition:
$(z_1 + z_2) + z_3 = z_1 + (z_2 + z_3)$

84. associative property for multiplication:
$(z_1 z_2) z_3 = z_1 (z_2 z_3)$

85. distributive property: $z_1(z_2 + z_3) = z_1 z_2 + z_1 z_3$

86. closure property of addition: $z_1 + z_2$ is a complex number

87. closure property of multiplication: $z_1 z_2$ is a complex number

Evaluate $8z - z^2$ by replacing z with the indicated complex number.

88. $2 + i$ **89.** $4 - 3i$ **90.** $-6i$

Find all complex numbers $a + bi$ such that the square $(a + bi)^2$ is

91. real   **92.** imaginary

93. Show that $\dfrac{\sqrt{2}}{2} + \dfrac{\sqrt{2}}{2} i$ is a square root of i.

94. Show that $\dfrac{\sqrt{3}}{2} - \dfrac{1}{2} i$ is a cube root of i.

CHAPTER SUMMARY
1

Key Words

real numbers	greater than	common factor
identity element	inequalities	prime polynomial
additive inverse	absolute value	rational expressions
negative	exponent	complex fraction
multiplicative inverse	scientific notation	square root
reciprocal	algebraic expression	radical
subtraction	term	radicand
order of operations	coefficient	nth root
subset	polynomial	index
natural numbers	degree of a polynomial	principal nth root
whole numbers	trinomial	radical expression
integers	monomial	simplified
rational numbers	difference of two squares	rationalizing
irrational numbers	square of a binomial	conjugate
number line	cube of a binomial	complex number
coordinate	quotient	imaginary number
less than	factoring	standard form

CHAPTER 1   REVIEW EXERCISES

Identify the property which tells why the following are true.

1. $6 \cdot 4 = 4 \cdot 6$

2. $8(5 + 9) = (5 + 9)8$

3. $3 + (-3) = 0$

4. If $x \neq 0$, then $x \cdot \dfrac{1}{x} = 1$

5. $4 \cdot 6 + 4 \cdot 12 = 4(6 + 12)$

6. $3(4 \cdot 2) = (3 \cdot 4)2$

7. Is the set of rational numbers closed for division?

8. Is the set of irrationals closed for division?

Simplify each of the following.

9. $(4 + 2 \cdot 8) \div 3$

10. $-7 + (-6)(-2) - 8$

11. $\dfrac{-9 + (2)(-8) + 7}{3(-4) - (-1 - 3)}$

12. $\dfrac{4 - (-8)(-2) - 7}{(-2)(-9) - 4(-4)}$

Let set $K = \{-12, -6, -9/10, -\sqrt{7}, 0, 1/8, \pi/4, 6, \sqrt{11}\}$. List the elements of K that are

13. natural numbers

14. whole numbers

15. integers

16. rational numbers

17. irrational numbers

18. real numbers.

For Exercises 19–27, choose all words from the following list which apply.

(a) natural numbers

(b) whole numbers

(c) integers

(d) rational numbers

(e) irrational numbers

(f) real numbers

(g) meaningless

19. 22

20. 0

21. $\sqrt{36}$

22. $-\sqrt{25}$

23. -1

24. 5/8

25. $\sqrt{15}$

26. $-6/0$

27. $3\pi/4$

Write the following numbers in numerical order, from smallest to largest.

28. $\dfrac{5}{6}, \dfrac{1}{2}, -\dfrac{2}{3}, -\dfrac{5}{4}, -\dfrac{3}{8}$

29. $|6 - 4|, -|-2|, |8 + 1|, -|3 - (-2)|$

30. $\sqrt{7}, -\sqrt{8}, -|\sqrt{16}|, |-\sqrt{12}|$

Write without absolute value bars.

31. $|3 - \sqrt{7}|$

32. $|\sqrt{8} - 3|$

33. $|m - 3|$, if $m > 3$

34. $|-6 - x^2|$

In each of the following exercises, the coordinates of three points are given. Find (a) $d(A, B)$ (b) $d(A, B) + d(B, C)$.

35. $A, -2; B, -1; C, 10$

36. $A, -8; B, -12; C, -15$

Under what conditions are the following statements true?

37. $|x| = x$

38. $|a + b| = -a - b$

39. $|x + y| = -|x| - |y|$

40. $|x| \le 0$

41. $d(A, B) = 0$

42. $d(A, B) + d(B, C) = d(A, C)$

Simplify each of the following.

43. $(-6x^2 - 4x + 11) + (-2x^2 - 11x + 5)$

44. $(3x^3 - 9x^2 - 5) - (-4x^3 + 6x^2)$
$+ (2x^3 - 9)$

45. $(8k - 7)(3k + 2)$

46. $(3r - 2)(r^2 + 4r - 8)$

47. $(x + 2y - z)^2$

48. $(r^z - 8)(r^z + 8)$

49. $\dfrac{2x^5t^2 \cdot 4x^3t}{16x^2t^3}$

50. $\dfrac{(2k^2)^2(3k^3)}{(4k^4)^2}$

Factor as completely as possible.

51. $7z^2 - 9z^3 + z$

52. $12p^5 - 8p^4 + 20p^3$

53. $6m^2 - 13m - 5$

54. $15a^2 + 7ab - 2b^2$

55. $30m^5 - 35m^4n - 25m^3n^2$

56. $2x^2p^3 - 8xp^4 + 6p^5$

57. $169y^4 - 1$

58. $49m^8 - 9n^2$

59. $(x - 1)^2 - 4$

60. $a^3 - 27b^3$

61. $r^9 - 8(r^3 - 1)^3$

62. $(2p - 1)^3 + 27p^3$

63. $3(m - n) + 4k(m - n)$

64. $6(z - 4)^2 + 9(z - 4)^3$

65. $2bx - b + 6x - 3$

66. $3x(x - 2) + 9x^2(x - 2) - 12x^3(x - 2)$

67. $y^{2k} - 9$

68. $z^{6n} - 100$

Perform the indicated operation.

69. $\dfrac{3m - 9}{8m} \cdot \dfrac{16m + 24}{15}$

70. $\dfrac{5x^2y}{x + y} \cdot \dfrac{3x + 3y}{30xy^2}$

71. $\dfrac{x^2 + x - 2}{x^2 + 5x + 6} \div \dfrac{x^2 + 3x - 4}{x^2 + 4x + 3}$

72. $\dfrac{27m^3 - n^3}{3m - n} \div \dfrac{9m^2 + 3mn + n^2}{9m^2 - n^2}$

73. $\dfrac{p^2 - 36q^2}{(p - 6q)^2} \cdot \dfrac{p^2 - 5pq - 6q^2}{p^2 - 6pq + 36q^2} \div \dfrac{5p}{p^3 + 216q^3}$

74. $\dfrac{3r^3 - 9r^2}{r^2 - 9} \div \dfrac{8r^3}{r + 3}$

75. $\dfrac{1}{4y} + \dfrac{8}{5y}$

76. $\dfrac{m}{4 - m} + \dfrac{3m}{m - 4}$

77. $\left(1 - \dfrac{3}{p}\right)\left(1 + \dfrac{3}{p}\right)$

78. $\dfrac{3}{x^2 - 4x + 3} - \dfrac{2}{x^2 - 1}$

79. $\dfrac{\dfrac{1}{p} + \dfrac{1}{q}}{1 - \dfrac{1}{pq}}$

80. $\left(\dfrac{1}{(x + h)^2 + 16} - \dfrac{1}{x^2 + 16}\right) \div h$

81. $\dfrac{3 + \dfrac{2m}{m^2 - 4}}{\dfrac{5}{m - 2}}$

82. $\dfrac{x^{-1} - 2y^{-1}}{y^{-1} - x^{-1}}$

Simplify each of the following. Assume all variables represent positive real numbers.

83. $\dfrac{(p^4)(p^{-2})}{p^{-6}}$

84. $(-6x^2y^{-3}z^2)^{-2}$

85. $\dfrac{6^{-1}r^3s^{-2}}{6r^4s^{-3}}$

86. $\dfrac{(3m^{-2})^{-2}(m^2n^{-4})^3}{9m^{-3}n^{-5}}$

87. $\dfrac{(2x^{-3})^2(3x^2)^{-2}}{6(x^2y^3)}$

88. $\dfrac{k^{2+p} \cdot k^{-4p}}{k^{6p}}$ (p is a rational number)

Simplify. Assume that all variables represent positive real numbers.

89. $\sqrt{200}$

90. $\sqrt[4]{1250}$

91. $\sqrt{\dfrac{7}{3r}}$

92. $\sqrt{\dfrac{27y^8}{m^3}}$

93. $-\sqrt[3]{\dfrac{r^6m^5}{z^2}}$

94. $\sqrt[4]{\sqrt[3]{m}}$

95. $(\sqrt[3]{2} + 4)(\sqrt[3]{2^2} - 4\sqrt[3]{2} + 16)$

96. $\dfrac{\sqrt[4]{8p^2q^5} \cdot \sqrt[4]{2p^3q}}{\sqrt[4]{p^5q^2}}$

97. $\sqrt{18m^3} - 3m\sqrt{32m} + 5\sqrt{m^3}$

98. $\sqrt{75y^5} - y^2\sqrt{108y} + 2y\sqrt{27y^3}$

99. $\dfrac{3}{\sqrt{5}} - \dfrac{2}{\sqrt{45}} + \dfrac{6}{\sqrt{80}}$

100. $\dfrac{-12}{\sqrt[3]{4}}$

101. $\dfrac{z}{\sqrt{z - 1}}$

102. $\dfrac{6}{3 - \sqrt{2}}$

103. $\dfrac{1}{\sqrt{5} + 2}$

104. $\dfrac{\sqrt{x} - \sqrt{x - 2}}{\sqrt{x} + \sqrt{x - 2}}$

Simplify each of the following. Assume that all variables represent positive real numbers, and variables used as exponents are rational numbers.

105. $36^{-3/2}$

106. $(125m^6)^{-2/3}$

107. $(8r^{3/4}s^{2/3})(2r^{3/2}s^{5/3})$

108. $(7r^{1/2})(2r^{3/4})(-r^{1/6})$

109. $\dfrac{p^{-3/4} \cdot p^{5/4} \cdot p^{-1/4}}{p \cdot p^{3/4}}$

110. $\left(\dfrac{y^6x^3z^{-2}}{16x^5z^4}\right)^{-1/2}$

111. $\dfrac{m^{2+p} \cdot m^{-2}}{m^{3p}}$

112. $\dfrac{z^{-p+1} \cdot z^{-8p}}{z^{-9p}}$

Find each product. Assume that all variables represent positive real numbers.

113. $2z^{1/3}(5z^2 - 2)$

114. $-m^{3/4}(8m^{1/2} + 4m^{-3/2})$

115. $(p + p^{1/2})(3p - 5)$

116. $(m^{1/2} - 4m^{-1/2})^2$

Write in standard form.

117. $(1 - i) - (3 + 4i) + 2i$

118. $(2 - 5i) + (9 - 10i) - 3$

119. $(6 - 5i) + (2 + 7i) - (3 - 2i)$

120. $(4 - 2i) - (6 + 5i) - (3 - i)$

121. $(3 + 5i)(8 - i)$　　**122.** $(4 - i)(5 + 2i)$

123. $(2 + 6i)^2$　　**124.** $(6 - 3i)^2$

125. $(1 - i)^3$　　**126.** $(2 + i)^3$

127. i^{17}　　**128.** i^{52}

129. $\dfrac{6 + 2i}{3 - i}$　　**130.** $\dfrac{2 - 5i}{1 + i}$

131. $\dfrac{2 + i}{1 - 5i} \cdot \dfrac{1 + i}{3 - i}$ **132.** $\dfrac{4 + 3i}{1 - i} \cdot \dfrac{2 - 3i}{2 + i}$

133. $\dfrac{8 - i}{2 + i} + \dfrac{3 + 2i}{4i}$

134. $\dfrac{6 + 3i}{1 + i} + \dfrac{1 - i}{2 + 2i}$

135. $\sqrt{-12}$ **136.** $\sqrt{-18}$

Simplify each of the following.

137. $(|a| + a)(|a| - a)$

138. $[2a^4 + (b^2 - a^2)^2 - (a^2 - b^2)^2]^{1/2} \times$
     $[a^6 - (-a^2)^3]^{1/2}$

Suppose a value of the number a has been given. Let $b = a + 1$ and $c = \sqrt{a + b}$. Show that

139. $b(a - c)^2 + a(b + c)^2 = c^2(ab + c^2)$

140. $b(a - c)^2 + ab(1 + c)^2 = b^2(c^2 + a)$

141. If $x + |x| = x - |x|$, what is x?

142. Prove that if $(x + 1)^2 = (|x| + 1)^2$,
     then $x \geq 0$.

Suppose $0 < a < 1$ and $a + b = 1$. Are the following less than, equal to, or greater than 1?

143. $a^2 + b^2$ **144.** $\sqrt{a} + \sqrt{b}$

What are the order relations among 1, a, a^2, $\sqrt{a}$, $1/a$, b, b^2, $\sqrt{b}$, and $1/b$ if

145. $\sqrt{b} < a < b$

146. $b < a < \sqrt{b}$

147. If $a > 0$, show that $a + 1/a \geq 2$.

148. If $a > 0$, show that
     $4a + 1 > (\sqrt{a} + \sqrt{a + 1})^2 < 4a + 2.$

Exercises 137–148: Reproduced from *Calculus*, 2nd edition, by Leonard Gillman and Robert H. McDowell, by permission of W. W. Norton & Company, Inc. Copyright © 1978, 1973 by W. W. Norton & Company, Inc.

2 Equations and Inequalities

For many people, the study of algebra is really the study of equations. Many applications of mathematics require the solution of one or more equations. The study of inequalities has also become important as more and more applications—especially in fields such as business—utilize inequalities.

An **equation** is a statement that two expressions are equal. Examples of equations include

$$x + 2 = 9, \qquad 11y = 5y + 6y, \qquad x^2 - 2x - 1 = 0,$$

and so on. In this chapter we discuss the solution of several different kinds of equations and inequalities.

2.1 Linear Equations

To **solve** an equation means to find the number or numbers that make the equation a true statement. A number that is a solution of an equation is said to **satisfy** the equation. The set of all solutions for an equation makes up its **solution set.** An equation which is satisfied by every number which is a meaningful replacement for the variable is called an **identity.** Examples of identities are

$$3x + 4x = 7x, \qquad \text{and} \qquad x^2 - 3x + 2 = (x - 2)(x - 1).$$

Equations that are satisfied by some numbers, but not satisfied by others, are called **conditional equations.** Examples of conditional equations are

$$2m + 3 = 7, \qquad \text{and} \qquad \frac{5r}{r - 1} = 7.$$

To verify that an equation is an identity, show that the two sides are algebraically equivalent for the same set of values of x.

EXAMPLE 1 ☐ Decide if the following equations are identities or conditional equations.

(a) $9p^2 - 25 = (3p + 5)(3p - 5)$

Since the product of $3p + 5$ and $3p - 5$ is $9p^2 - 25$, the given equation is true for *every* value of p and thus is an identity.

(b) $\dfrac{(x - 3)(x + 2)}{(x - 3)} = x + 2$

On the left side of the equation, 3 cannot be used as a replacement for x. For this reason, the equation is an identity only for $x \neq 3$.

(c) $5y - 4 = 11$

If $y = 3$, we can replace y with 3 to get

$$5 \cdot 3 - 4 = 11$$
$$11 = 11,$$

a true statement. On the other hand, if $y = 4$, we get

$$5 \cdot 4 - 4 = 11$$
$$16 = 11,$$

a false statement. The equation $5y - 4 = 11$ is true for some values of y, but not all, and thus is a conditional equation. (By the way, the word *some* in mathematics means "at least one." We can therefore say that $5y - 4 = 11$ is true for *some* replacements of y, even though it turns out to be true only for $y = 3$.) ☐

Any two equations with the same solution set are called **equivalent equations.** For example $x + 1 = 5$ and $6x + 3 = 27$ are equivalent equations since they both have the same solution set, $\{4\}$. On the other hand, the equations

$$x + 1 = 5 \quad \text{and} \quad (x - 4)(x + 2) = 0$$

are not equivalent. The number 4 is a solution of both equations, but the equation $(x - 4)(x + 2) = 0$ also has -2 as a solution.

One way to solve an equation is to rewrite it as successively simpler equivalent equations. These simpler equivalent equations are derived using the properties of Chapter 1 that allow us to add the same number to each side of an equation or to multiply each side of an equation by the same nonzero number. We can expand those properties to include not just numbers that are added or multiplied but also algebraic expressions.

In this section, we see how to solve equations that are equivalent to linear equations.

Linear Equation

A *linear equation* is an equation of the form

$$ax + b = 0,$$

where a and b are real numbers, with $a \neq 0$.

The equation $ax + b = 0$ can be solved by writing the following sequence of equivalent equations:

$$ax + b = 0 \qquad \text{given equation}$$
$$ax + b + (-b) = 0 + (-b) \qquad \text{add } -b \text{ on each side}$$
$$ax = -b \qquad \text{inverse and identity properties}$$
$$\frac{1}{a} \cdot ax = \frac{1}{a} \cdot (-b) \qquad \text{multiply by } 1/a \text{ on each side}$$
$$x = -\frac{b}{a} \qquad \text{inverse and identity properties}$$

At which point in this solution did we use the fact that $a \neq 0$?

We have now shown that $-b/a$ is the only possible solution of the equation $ax + b = 0$. To show that $-b/a$ is indeed a solution, replace x with $-b/a$, getting

$$a\left(-\frac{b}{a}\right) + b = 0$$
$$-b + b = 0$$
$$0 = 0,$$

a true statement.

Solution of a Linear Equation

The linear equation $ax + b = 0$ has exactly one solution, $-\dfrac{b}{a}$.

It is not really necessary to remember this result. To find the solution for a given linear equation, go through the steps necessary to produce a sequence of simpler equivalent equations.

EXAMPLE 2   Solve $3(2x - 4) = 7 - (x + 5)$.

By using the distributive property and then collecting like terms, we can get a sequence of simpler equivalent equations as follows:

$$3(2x - 4) = 7 - (x + 5)$$
$$6x - 12 = 7 - x - 5$$
$$6x - 12 = 2 - x.$$

Now adding the same expressions to each side of the equation gives

$$x + 6x - 12 = x + 2 - x$$
$$7x - 12 = 2$$
$$12 + 7x - 12 = 12 + 2$$
$$7x = 14.$$

Finally, multiplying each side by the same number, $\frac{1}{7}$, produces

$$\frac{1}{7} \cdot 7x = \frac{1}{7} \cdot 14$$

$$x = 2.$$

To check, replace x with 2 in the original equation, getting

$$3(2x - 4) = 7 - (x + 5) \qquad \text{original equation}$$
$$3(2 \cdot 2 - 4) = 7 - (2 + 5) \qquad \text{let } x = 2$$
$$3(4 - 4) = 7 - (7)$$
$$0 = 0. \qquad\qquad\qquad \text{true}$$

Since replacing x with 2 results in a true statement, 2 is the solution of the given equation. The solution set is therefore {2}. ◻

EXAMPLE 3 ◻ Solve $\dfrac{3p - 1}{3} - \dfrac{2p}{p - 1} = p$.

First obtain a simpler equivalent equation by multiplying both sides by their common denominator, $3(p - 1)$, where we must assume $p \neq 1$. (Why?) Doing this gives

$$3(p - 1)\left(\frac{3p - 1}{3}\right) - 3(p - 1)\left(\frac{2p}{p - 1}\right) = 3(p - 1)p$$
$$(p - 1)(3p - 1) - 3(2p) = 3p(p - 1)$$
$$3p^2 - 4p + 1 - 6p = 3p^2 - 3p.$$

We can obtain an even simpler equivalent equation by combining terms and adding $-3p^2$ to both sides, producing

$$-10p + 1 = -3p.$$

We now add $10p$ to both sides, to get

$$1 = 7p.$$

Finally, multiplying both sides by 1/7 gives

$$1/7 = p.$$

Check that the solution of the given equation is 1/7 to verify that the solution set is {1/7}. Note that our restriction $p \neq 1$ does not affect the solution set here, since 1/7 $\neq$ 1. ◻

EXAMPLE 4 ◻ Solve $\dfrac{x}{x - 2} = \dfrac{2}{x - 2} + 2$.

Multiplying both sides of the equation by $x - 2$, assuming that $x - 2 \neq 0$ (or $x \neq 2$), gives

$$x = 2 + 2(x - 2)$$
$$x = 2 + 2x - 4$$
$$x = 2.$$

We had to assume $x - 2 \neq 0$ in order to multiply both sides of the equation by $x -$ 2. Our proposed solution of 2, however, makes $x - 2 = 0$. Because of this, the given equation has no solution. (To see that 2 is not a solution, substitute 2 for x in the given equation.) The solution set is $\emptyset$, the set containing no elements. ($\emptyset$ is called the **empty set.**)   ☐

Sometimes we need to solve an equation with more than one letter in which one letter represents a variable and the others represent constants. The next example shows how this works.

EXAMPLE 5   ☐   Solve the equation $3(2x - 5a) + 4b = 4x - 2$ for x.

Using the distributive property gives

$$6x - 15a + 4b = 4x - 2.$$

Arranging all terms with x on one side of the equation and all terms without x on the other side yields

$$6x - 4x = 15a - 4b - 2$$
$$2x = 15a - 4b - 2$$
$$x = \frac{15a - 4b - 2}{2}. \quad ☐$$

EXERCISES 2.1

Decide which of the following pairs of equations are equivalent.

1. $3x - 5 = 7$
$-6x + 10 = 14$

2. $-x = 2x + 3$
$-3x = 3$

3. $\dfrac{3x}{x - 1} = \dfrac{2}{x - 1}$
$3x = 2$

4. $\dfrac{x + 1}{12} = \dfrac{5}{12}$
$x + 1 = 5$

5. $\dfrac{x}{x - 2} = \dfrac{2}{x - 2}$
$x = 2$

6. $\dfrac{x + 3}{x + 1} = \dfrac{2}{x + 1}$
$x = -1$

7. $x = 4$
$x^2 = 16$

8. $z^2 = 9$
$z = 3$

Decide whether each of the following equations is an identity or a conditional equation.

9. $x^2 + 5x = x(x + 5)$

10. $3y + 4 = 5(y - 2)$

11. $2(x - 7) = 5x + 3 - x$

12. $2x - 4 = 2(x - 2)$

13. $\dfrac{m + 3}{m} = 1 + \dfrac{3}{m}$ **14.** $\dfrac{p}{2 - p} = \dfrac{2}{p} - 1$

15. $4q^2 - 25 = (2q + 5)(2q - 5)$

16. $3(k + 2) - 5(k + 2) = -2k - 4$

Solve each of the following equations.

17. $4x - 1 = 15$

18. $-3y + 2 = 5$

19. $.2m - .5 = .1m + .7$

20. $.01p + 3.1 = 2.03p - 2.96$

21. $\dfrac{5}{6}k - 2k + \dfrac{1}{3} = \dfrac{2}{3}$

22. $\dfrac{3}{4} + \dfrac{1}{5}r - \dfrac{1}{2} = \dfrac{4}{5}r$

23. $2[m - (4 + 2m) + 3] = 2m + 2$

24. $4[2p - (3 - p) + 5] = -7p - 2$

25. $\dfrac{3x - 2}{7} = \dfrac{x + 2}{5}$

26. $\dfrac{2p + 5}{5} = \dfrac{p + 2}{3}$

27. $\dfrac{3k - 1}{4} = \dfrac{5k + 2}{8}$

28. $\dfrac{9x - 1}{6} = \dfrac{2x + 7}{3}$

29. $\dfrac{1}{4p} + \dfrac{2}{p} = 3$

30. $\dfrac{2}{t} + 6 = \dfrac{5}{2t}$

31. $\dfrac{m}{2} - \dfrac{1}{m} = \dfrac{6m + 5}{12}$

32. $\dfrac{-3k}{2} + \dfrac{9k - 5}{6} = \dfrac{11k + 8}{k}$

33. $\dfrac{2r}{r - 1} = 5 + \dfrac{2}{r - 1}$

34. $\dfrac{3x}{x + 2} = \dfrac{1}{x + 2} - 4$

35. $\dfrac{5}{2a + 3} + \dfrac{1}{a - 6} = 0$

36. $\dfrac{2}{x + 1} = \dfrac{3}{2x - 5}$

37. $\dfrac{4}{x - 3} - \dfrac{8}{2x + 5} + \dfrac{3}{x - 3} = 0$

38. $\dfrac{5}{2p + 3} - \dfrac{3}{p - 2} = \dfrac{4}{2p + 3}$

39. $\dfrac{2p}{p - 2} = 3 + \dfrac{4}{p - 2}$

40. $\dfrac{5k}{k + 4} = 3 - \dfrac{20}{k + 4}$

41. $2(m + 1)(m - 1) = (2m + 3)(m - 2)$

42. $(2y - 1)(3y + 2) = 6(y + 2)^2$

Solve each of the following equations for x.

43. $2(x - a) + b = 3x + a$

44. $5x - (2a + c) = a(x + 1)$

45. $ax + b = 3(x - a)$

46. $4a - ax = 3b + bx$

47. $\dfrac{x}{a - 1} = ax + 3$

48. $\dfrac{2a}{x - 1} = a - b$

49. $a^2x + 3x = 2a^2$

50. $ax + b^2 = bx - a^2$

51. Find the error in the following.

$$x^2 + 2x - 15 = x^2 - 3x$$
$$(x + 5)(x - 3) = x(x - 3)$$
$$x + 5 = x$$
$$5 = 0$$

Find a value of k so that the following equations have the solution set {2}.

52. $9x - 7 = k$

53. $-5x + 11x - 2 = k + 4$

54. $\dfrac{8}{k + x} = 1$

55. $\sqrt{x + k} = 0$

56. $\sqrt{3x - 2k} = 4$

Solve each of the following equations. Round solutions to the nearest hundredth. 🖩

57. $9.06x + 3.59(8x - 5) = 12.07x + .5612$

58. $-5.74(3.1 - 2.7p) = 1.09p + 5.2588$

59. $\dfrac{2.5x - 7.8}{3.2} + \dfrac{1.2x + 11.5}{5.8} = 6$

60. $\dfrac{4.19x + 2.42}{.05} - \dfrac{5.03x - 9.74}{.02} = 1$

61. $\dfrac{2.63r - 8.99}{1.25} - \dfrac{3.90r - 1.77}{2.45} = r$

62. $\dfrac{8.19m + 2.55}{4.34} - \dfrac{8.17m - 9.94}{1.04} = 4m$

2.2   Formulas and Applications

Mathematics is an important problem-solving tool. Many times the solution of a problem depends on the use of a formula which expresses a relationship among several variables. For example, the formula

$$A = \frac{24\,f}{b(p + 1)} \tag{1}$$

gives the approximate true annual interest rate for a consumer loan paid off with monthly payments. Here f is the finance charge on the loan, p is the number of payments, and b is the original amount of the loan.

Suppose we need to find the number of payments, p, when the other quantities are known. To do this, solve the equation for p by treating p as the variable and the other letters as constants. Begin by multiplying both sides of formula (1) by $p + 1$. This gives

$$(p + 1)A = (p + 1)\frac{24f}{b(p + 1)}$$

$$(p + 1)A = \frac{24f}{b}.$$

Multiplying both sides by $1/A$ produces

$$\frac{1}{A}(p + 1)A = \frac{1}{A} \cdot \frac{24f}{b}$$

$$p + 1 = \frac{24f}{Ab}.$$

(Here we must assume $A \neq 0$. Why is this a very safe assumption?)
Finally, add -1 to both sides to get

$$p = \frac{24f}{Ab} - 1.$$

EXAMPLE 1   □   Solve $J\left(\dfrac{x}{k} + a\right) = x$ for x.

We want to get all terms with x on one side of the equation and all terms without x on the other. To do this, first use the distributive property, as follows.

$$J\left(\frac{x}{k}\right) + Ja = x$$

Now multiply both sides by k (assuming $k \neq 0$) to get

$$kJ\left(\frac{x}{k}\right) + kJa = kx$$

$$Jx + kJA = kx.$$

Then add $-Jx$ to both sides, yielding

$$kJA = kx - Jx$$
$$kJa = x(k - J).$$

If we assume $k \neq J$, we can multiply both sides by $1/(k - J)$ to find x. Doing this gives

$$x = \frac{kJa}{k - J}. \quad \square$$

For many students, learning how to apply mathematical skills to applications or word problems is the most challenging task they face. When trying to solve word problems, you should find the following steps helpful. They are from *How to Solve It*, by George Polya, published by Princeton University Press. This book is a modern classic among mathematical books.

First.
You have to
understand **the**
problem.

Understanding the problem
What is the unknown? What are the data? What is the condition? Is it possible to satisfy the condition? Is the condition sufficient to determine the unknown? Or is it insufficient? Or redundant? Or contradictory?
Draw a figure. Introduce suitable notation.
Separate the various parts of the conditions. Can you write them down?

Second.
Find the
connection
between the data
and the unknown.

Devising a plan
Have you seen it before? Or have you seen the same problem in a slightly different form?
Do you know a related problem? Do you know a theorem that could be useful?
Look at the unknown! And try to think of a familiar problem having the same or a similar unknown.

Third.
Carry out **your**
plan.

Carrying out the plan
Carrying out your plan of the solution, *check each step*. Can you see clearly that the step is correct? Can you prove that it is correct?

Fourth.
Examine **the**
solution obtained.

Looking back
Can you *check the result?* Can you check the argument? Can you derive the result differently? Can you see it at a glance? Can you use the result, or the method, for some other problem?

Notice how each of these steps is used in the following examples.

EXAMPLE 2 ☐ If the length of a side of a square is increased by 3 cm, the perimeter of the new square is 40 cm more than twice the length of a side of the original square. Find the dimensions of the original square.

First, what should the variable represent? We want to find the length of a side of the original square, so let the variable represent that.

x = length of side of the original square.

Now draw a figure using the given information, as in Figure 2.1.

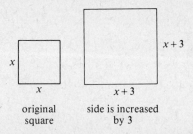

x

x

original
square

$x + 3$

$x + 3$

side is increased
by 3

Figure 2.1

The length of a side of the new square is 3 cm more than the length of a side of the old square. Write a variable expression for that.

$x + 3$ = length of side of the new square.

Now write a variable expression for the perimeter of the new square. Since the perimeter of a square is 4 times the length of a side,

$4(x + 3)$ = perimeter of the new square.

We are now ready to use the given information of the problem to write an equation. The perimeter of the new square is 40 cm more than twice the length of a side of the original square, so the equation is

the new perimeter	is	40	more than	twice the side of the original square
$4(x + 3)$	=	40	+	$2x$.

Now solve the equation as follows:

$$4(x + 3) = 40 + 2x$$
$$4x + 12 = 40 + 2x$$
$$2x = 28$$
$$x = 14.$$

This result should be checked by using the wording of the original problem. The length of a side of the new square would be $14 + 3 = 17$ cm; its perimeter would be

4(17) = 68 cm. Twice the length of a side of the original square is 2(14) = 28 cm. Since 40 + 28 = 68, the solution satisfies the problem. ▢

EXAMPLE 3 ▢ Chuck travels 80 km in the same time that Mary travels 180 km. Mary travels 50 km per hour faster than Chuck. Find the rate of each person.

Let x represent Chuck's rate. Since Mary traveled 50 km per hour faster, her rate is

$$x + 50 = \text{rate for Mary.}$$

To solve constant velocity problems of this kind, we need the formula $d = rt$, where d is distance traveled in t hours at a constant rate r. We can use a chart to organize the information given in the problem and the variable expressions that we need to write.

For Chuck, $d = 80$ and $r = x$. From the formula $d = rt$, we have $t = d/r$, so that, for Chuck, $t = 80/x$. For Mary, $d = 180$, $r = x + 50$, so that $t = 180/(x + 50)$. We use these facts to complete the following chart.

	d	r	t
Chuck	80	x	$\dfrac{80}{x}$
Mary	180	$x + 50$	$\dfrac{180}{x + 50}$

Since they both traveled for the same time, we have the equation

$$\frac{80}{x} = \frac{180}{x + 50}.$$

To solve this equation, multiply both sides by $x(x + 50)$, getting

$$x(x + 50) \cdot \frac{80}{x} = x(x + 50) \cdot \frac{180}{x + 50}$$

$$80(x + 50) = 180x$$

$$80x + 4000 = 180x$$

$$4000 = 100x$$

$$40 = x.$$

Since x represents Chuck's rate, we see that Chuck went 40 km per hour. Mary's rate is thus $x + 50$, or $40 + 50 = 90$ km per hour. ▢

EXAMPLE 4 ▢ Rick and Joann are refinishing an antique table. Working alone, Rick would need 8 days to finish the table, though Joann would require 10. If they work together, how long will it take them to complete the project?

Let x represent the number of days it will take Rick and Joann working together to finish the table. Then, in one day, they will together complete $1/x$ of the job.

In one day, Rick will do 1/8 of the task, while Joann will do 1/10. The amount of the job done by each in one day must equal the amount done working together in one day, or

$$\frac{1}{8} + \frac{1}{10} = \frac{1}{x}.$$

Multiplying both sides of this equation by $40x$ gives

$$40x\left(\frac{1}{8} + \frac{1}{10}\right) = 40x \cdot \frac{1}{x}$$

$$5x + 4x = 40$$

$$9x = 40$$

$$x = 40/9 \quad \left(\text{or } 4\frac{4}{9}\right).$$

Hence, working together they can finish the table in $4\frac{4}{9}$ days.   □

EXAMPLE 5   □   A homeowner opens a valve to fill her swimming pool. The pool would normally fill in 10 hours. However, after 3 hours, a mistake is made and the drain is also opened. The drain can empty a full pool in 15 hours. How long does it now take to fill the pool?

Let us first find the time it would take to fill the pool if the inlet pipe and the drain were open *from the beginning*. Let t represent this time. In one hour, the inlet will fill 1/10 of the pool, while the outlet would empty 1/15 of the pool. With both open, $1/t$ of the pool would be filled in one hour, so that

$$\frac{1}{10} - \frac{1}{15} = \frac{1}{t}.$$

(Why did we use a $-$ sign?) To solve this equation, start by multiplying both sides by $30t$, getting

$$30t\left(\frac{1}{10} - \frac{1}{15}\right) = 30t\left(\frac{1}{t}\right)$$

$$30t\left(\frac{1}{10}\right) - 30t\left(\frac{1}{15}\right) = 30$$

$$3t - 2t = 30$$

$$t = 30.$$

It would take 30 hours to fill the pool if both the inlet and drain were open from the beginning. However, we were told in the problem that only the inlet was open for the initial 3 hours. In 3 hours, the inlet would fill 3/10 of the pool, leaving 7/10 unfilled. When the drain was opened, 7/10 of the pool remained to be filled; this would take

$$\frac{7}{10}(30) = 21 \text{ hours.}$$

Altogether it takes $3 + 21 = 24$ hours to fill the pool.   □

EXAMPLE 6 ☐ One day a chemist needed a 20% solution of potassium permanganate. She had a 15% solution on hand as well as a 30% solution. How many liters of the 15% solution should she add to 3 liters of the 30% solution to get her 20% solution?

Let x be the number of liters of the 15% solution to be added. Arrange the information of the problem in a table.

strength	liters of solution	liters of pure potassium permanganate
15%	x	$.15x$
30%	3	$.30(3)$
20%	$3 + x$	$.20(3 + x)$

Since the number of liters of potassium permanganate in the 15% solution plus the number of liters in the 30% solution must equal the number of liters in the final 20% solution, we have

$$\underset{\text{liters in 15\%}}{.15x} + \underset{\text{liters in 30\%}}{.30(3)} = \underset{\text{liters in 20\%}}{.20(3 + x)}.$$

Solve this equation, as follows:

$$.15x + .90 = .60 + .20x$$
$$.30 = .05x$$
$$6 = x$$

Thus, 6 liters of the 15% solution should be mixed with 3 liters of the 30% solution. The result will be $6 + 3 = 9$ liters of 20% solution. ☐

EXAMPLE 7 ☐ Richard Jones receives a $14,000 bonus from his company. He invests part of the money in 6% tax-free bonds, and the remainder at 15%. He earns $1335 per year in interest from the investments. Find the amount he has invested at each rate.

Let x represent the amount Jones invests at 6%, so that $14,000 - x$ is the amount invested at 15%. Since interest is given by the product of principal, rate, and time ($i = prt$), we have for one year

$$\text{interest at 6\%} = x \cdot 6\% \cdot 1 = .06x$$
$$\text{interest at 15\%} = (14,000 - x) \cdot 15\% \cdot 1 = .15(14,000 - x).$$

Since the total interest is $1335, we have

$$.06x + .15(14,000 - x) = 1335.$$

Solve this equation, getting

$$.06x + .15(14,000) - .15x = 1335$$
$$-.09x + 2100 = 1335$$
$$-.09x = -765$$
$$x = 8500$$

Thus, $8500 was invested at 6%, and $14,000 - \$8500 = \$5500$ at 15%. ☐

EXERCISES 2.2

Solve each of the following for the variable indicated. Assume all denominators are nonzero.

1. $PV = k$ for V

2. $F = whA$ for h

3. $V = lwh$ for l

4. $i = prt$ for p

5. $V = V_0 + gt$ for g

6. $s = s_0 + gt^2 + k$ for g

7. $s = \dfrac{1}{2} gt^2$ for g

8. $A = \dfrac{1}{2}(B + b)h$ for h

9. $A = \dfrac{1}{2}(B + b)h$ for B

10. $C = \dfrac{5}{9}(F - 32)$ for F

11. $S = 2\pi(r_1 + r_2)h$ for r_1

12. $A = P\left(1 + \dfrac{i}{m}\right)$ for m

13. $g = \dfrac{4\pi^2 l}{t^2}$ for l

14. $P = \dfrac{E^2 R}{r + R}$ for R

15. $S = 2\pi rh + 2\pi r^2$ for h

16. $u = f \cdot \dfrac{k(k + 1)}{n(n + 1)}$ for f

17. $A = \dfrac{24f}{b(p + 1)}$ for f

18. $A = \dfrac{24f}{b(p + 1)}$ for b

19. $\dfrac{1}{R} = \dfrac{1}{r_1} + \dfrac{1}{r_2}$ for R

20. $m = \dfrac{Ft}{v_1 - v_2}$ for v_2

Solve each of the following that have solutions.

21. A triangle has a perimeter of 30 cm. Two sides of the triangle are both twice as long as the shortest side. Find the length of the short side.

22. The length of a rectangle is three inches less than twice the width. The perimeter is 54 cm. Find the width.

23. A clock-radio is on sale for $49. If the sale price is 15% less than the regular price, what was the regular price?

24. A shopkeeper prices his items 20% over their wholesale price. If a lamp is marked $74, what was its wholesale price?

25. Ms. Prullage invests $20,000 received from an insurance settlement in two ways, some at 13% and some at 16%. Altogether, she makes $2840 per year interest. How much is invested at each rate?

26. Bruce Hallett received $52,000 profit from the sale of some land. He invested part at $7\frac{1}{2}\%$ interest and the rest at $9\frac{1}{2}\%$ interest. He earned a total of $4520 interest per year. How much did he invest at each rate?

27. John Raton won $100,000 in a state lottery. He first paid income tax of 40% on the winnings. Of the rest, he invested some at $8\frac{1}{2}\%$ and some at 16%, making $5550 interest per year. How much is invested at each rate?

28. Mary Collins earned $48,000 from royalties on her cookbook. She paid a 40% income tax on these royalties. The balance was invested in two ways, at $7\frac{1}{2}\%$ and at $10\frac{1}{2}\%$. The investments produced $2550 interest income per year. Find the amount invested at each rate.

29. Susan Futterman bought two plots of land for a total of $120,000. On the first plot, she made a profit of 15%. On the second, she lost 10%. Her total profit was $5500. How much did she pay for each piece of land?

30. Suppose $10,000 is invested at 6%. How much additional money must be invested at 8% to produce a yield on the entire amount invested of 7.2%?

31. Mario Nunez earns take-home pay of $198 a week. If his deductions for retirement, union dues, medical plan, and so on amount to 26% of his wages, what is his weekly pay before deductions?

32. Janet gives 10% of her net income to the church. This amounts to $80 a month. In addition, her paycheck deductions are 24% of her gross monthly income. What is her gross monthly income?

33. A bank pays 7% interest on passbook accounts and 10% interest on long-term deposits. Suppose a depositor divides $20,000 among the two types of deposits. Find the amount deposited at each rate if the total annual income from interest is $2500.

34. An investor wishes to sell a piece of property for $125,000. She wishes the money to be paid off in two ways—a short-term note at 12% and a long-term note at 10%. Find the amount of each note if the total annual interest income is $2000.

35. In planning his retirement, Mr. Wangler deposits some money at 12% with twice as much deposited at 10%. The total deposit is $60,000. Find the amount deposited at each rate if the total annual interest income is $6400.

36. A church building fund has invested $75,000 in two ways: some at 9% and four times as much at 12%. Find the amount invested at each rate if the total annual income from interest is $8550.

37. A pharmacist wishes to strengthen a mixture which is 10% alcohol to one which is 30% alcohol. How much pure alcohol should be added to 7 liters of the 10% mixture?

38. A student needs some 10% hydrochloric acid for a chemistry experiment. How much 5% acid should be mixed with 60 ml of 20% acid to get a 10% solution?

Exercises 39 and 40 depend on the idea of the octane rating of gasoline, a measure of its antiknock qualities. In one measure of octane, a standard fuel is made with only two ingredients: heptane and isooctane. For this fuel, the octane rating is the percent of isooctane. An actual gasoline blend is then compared to a standard fuel. For example, a gasoline with an octane rating of 98 has the same antiknock properties as a standard fuel that is 98% isooctane.

39. How many liters of 94-octane gasoline should be mixed with 200 liters of 99-octane gasoline to get a mixture that is 97-octane?

40. A service station has 92-octane and 98-octane gasoline. How many liters of each should be mixed to provide 12 liters of 96-octane gasoline needed for chemical research?

41. The Old Time Goodies Store sells mixed nuts. Cashews sell for $4 per quarter kilogram, hazelnuts for $3 per quarter kilogram, and peanuts for $1 per quarter kilogram. How many kilograms of peanuts should be added to 10 kilograms of cashews and 8 kilograms of hazelnuts to make a mixture which will sell for $2.50 per quarter kilogram?

42. The Old Time Goodies Store also sells candy. They want to prepare 200 kilograms of a mixture for a special Halloween promotion to sell at $4.84 per kilogram. How much $4-per-kilogram candy should be mixed with $5.20-per-kilogram candy for the required mix?

43. On a vacation trip, Toni Tyson averaged 50 mph traveling from Amarillo to Flagstaff. Returning by a different route that covered the same number of miles, she averaged 55 mph. What is the distance between the two cities if her total traveling time was 32 hours?

44. Paisley left by plane to visit her mother in Hartford, 420 kilometers away. Fifteen minutes later, her mother left to meet her at the airport. She drove the 20 kilometers to the airport at 40 kph and arrived just as the plane taxied in. What was the speed of the plane?

45. Russ and Janet are running in the Apple Hill Fun Run. Russ runs at 7 mph, Janet at 5 mph. If they start at the same time, how long will it be before they are 1/2 mile apart?

46. If the run in Exercise 45 has a staggered start with Janet starting first and Russ starting 10 minutes later, how long will it be before he catches up with her?

47. Joann took 20 minutes to drive her boat upstream to water-ski at her favorite spot. Coming back later in the day, at the same boat speed, took her 15 minutes. If the current in that part of the river is 5 kph, what was her boat speed?

48. Joe traveled against the wind in a small plane at 180 mph for 3 hours. The return trip at the same speed took 2.8 hours. What was the wind speed?

49. Mark can clean the house in 9 hours and Wendy in 6. How long will it take them if they work together to clean it?

50. Helen can paint a room in 5 hours. Jay can paint the same room in 4 hours. (He does a sloppier job.) How long will it take them to paint the room together?

51. Two chemical plants are polluting a river. If plant A produces a predetermined maximum amount of pollution twice as fast as plant B, and together they produce the maximum pollution in 26 hours, how long will it take plant B alone?

52. A sewage treatment plant has two inlet pipes to its settling pond. One can fill the pond in 10 hours, the other in 12 hours. If the first pipe is open for 5 hours, and then the second pipe is opened, how long will it take to fill the pond?

53. An inlet pipe can fill Dominic's pool in 5 hours, while an outlet pipe can empty it in 8 hours. In his haste to watch television, Dominic left both pipes open. How long would it then take to fill the pool?

54. Suppose Dominic discovered his error (see Exercise 53) after an hour-long program. If he then closed the outlet pipe, how much longer would be needed to fill the pool?

55. A sink will fill in 1/2 hour if both faucets are on and the drain is closed. The sink will empty in 2/3 hour. If both faucets are on and the drain is open, how long will it take before the sink overflows?

56. A cashier has some $5 bills and some $10 bills. The total number of bills is n, and the total value of the money is v. Find the number of each kind of bill that the cashier has.

57. Let $0 < a < b < c < 100$. How many liters of $a\%$ solution should be mixed with m liters of $c\%$ solution to make a $b\%$ mixture?

58. One way that biologists estimate the number of individuals of a species in an area is as follows. First, 100 animals of the species are caught and marked. A period of time is permitted to elapse, and then b animals are caught. If c of these ($c \leq b$) are marked, find the number of individuals in the area.

59. Suppose B dollars are invested, some at $m\%$ and the rest at $n\%$. If a total of I dollars in interest is earned per year, find the amount invested at each rate.

2.3 Quadratic Equations

An equation of the form $ax + b = 0$ is a linear equation, while an equation with an x^2 term is called **quadratic.** That is,

Quadratic Equation

$ax^2 + bx + c = 0$, where $a \neq 0$, is a **quadratic equation.**

(Why is the restriction $a \neq 0$ necessary?) A quadratic equation written in the form $ax^2 + bx + c = 0$ is said to be in **standard form.** Any equation that can be written in this standard form is equivalent to a quadratic equation.

The simplest method of solving a quadratic equation, but one that is not always easily applied, is by factoring. This method depends on the following property mentioned in Chapter 1.

If a and b are real numbers, with $ab = 0$, then $a = 0$ or $b = 0$ or both.

EXAMPLE 1

☐ Solve $6r^2 + 7r = 3$.

First write the equation in standard form, yielding

$$6r^2 + 7r - 3 = 0.$$

Now factor $6r^2 + 7r - 3$ to get

$$(3r - 1)(2r + 3) = 0.$$

By the property above, $(3r - 1)(2r + 3)$ can equal 0 only if

$$3r - 1 = 0 \quad \text{or} \quad 2r + 3 = 0.$$

Solve each of these linear equations separately to find that the solutions of the original equation are $1/3$ and $-3/2$. Check these solutions by substituting them back in the original equation. The solution set is $\{1/3, -3/2\}$.   ☐

We can use factoring to solve a quadratic equation of the form $x^2 = k$ by writing the following sequence of equivalent equations:

$$x^2 = k$$
$$x^2 - k = 0$$
$$(x - \sqrt{k})(x + \sqrt{k}) = 0$$
$$x - \sqrt{k} = 0 \quad \text{or} \quad x + \sqrt{k} = 0$$
$$x = \sqrt{k} \quad \text{or} \quad x = -\sqrt{k}.$$

This proves the following result, sometimes called the **square root property.**

Square Root Property

The solution set of $x^2 = k$ is $\{\sqrt{k}, -\sqrt{k}\}$.

The solution set is often abbreviated $\{\pm\sqrt{k}\}$. Both solutions are real if $k > 0$ and imaginary if $k < 0$. (If $k = 0$, there is exactly one real solution.)

EXAMPLE
2

☐ Solve each equation.

(a) $z^2 = 17$

This equation has solution set $\{\pm\sqrt{17}\}$.

(b) $m^2 = -25$

Since $\sqrt{-25} = 5i$, the solution set of $m^2 = -25$ is $\{\pm 5i\}$.

(c) $(y - 4)^2 = 12$

Use a generalization of the square root property, working as follows.

$$(y - 4)^2 = 12$$
$$y - 4 = \pm\sqrt{12}$$
$$y = 4 \pm \sqrt{12}$$
$$y = 4 \pm 2\sqrt{3}$$

The solution set is $\{4 \pm 2\sqrt{3}\}$. ☐

As suggested by Example 2(c), any quadratic equation can be solved using the square root property if it can be written in the form $(x + n)^2 = k$ for suitable numbers n and k. The next example shows how to write a quadratic equation in this form.

EXAMPLE
3

☐ Solve $9z^2 - 12z - 1 = 0$.

We need to find constants n and k so that the given equation can be written in the form $(z + n)^2 = k$. If we expand $(z + n)^2$, we get $z^2 + 2zn + n^2$, where 1 is the coefficient of z^2. To get a leading coefficient of 1 in our given equation, multiply both sides by $1/9$, giving

$$z^2 - \frac{4}{3}z - \frac{1}{9} = 0.$$

We need to replace $-1/9$ by a number that will make the left side a perfect square. To find this number, first add $1/9$ on both sides, getting,

$$z^2 - \frac{4}{3}z = \frac{1}{9}.$$

The term $-4z/3$ is twice the product of z and n, the number we need to make the left side a perfect square. That is, n satisfies the equation

$$2zn = -\frac{4}{3}z$$
$$n = -\frac{2}{3}.$$

If $n = -2/3$, then $n^2 = 4/9$, which we add to both sides, getting

$$z^2 - \frac{4}{3}z + \frac{4}{9} = \frac{1}{9} + \frac{4}{9}.$$

Factoring on the left yields

$$\left(z - \frac{2}{3}\right)^2 = \frac{5}{9}.$$

Now use the square root property and rationalize denominators to get

$$z - \frac{2}{3} = \pm \sqrt{\frac{5}{9}}$$

$$z - \frac{2}{3} = \pm \frac{\sqrt{5}}{3}$$

$$z = \frac{2}{3} \pm \frac{\sqrt{5}}{3}.$$

These two solutions can be written as

$$\frac{2 \pm \sqrt{5}}{3},$$

with solution set $\{(2 \pm \sqrt{5})/3\}$. ☐

This process of changing $9z^2 - 12z - 1 = 0$ into the equivalent equation $(z - 2/3)^2 = 5/9$—as in the example above—is called **completing the square.**

**Completing
the Square**

> To solve $ax^2 + bx + c = 0$ (with $a \neq 0$) by completing the square:
> 1. If $a \neq 1$, multiply both sides by $1/a$. Then rewrite the equation so that the constant is alone on one side of the equals sign.
> 2. Square half the coefficient of x, and add the square to both sides.
> 3. Factor, and use the square root property.

We shall now use the method of completing the square on the general quadratic equation,

$$ax^2 + bx + c = 0 \qquad (a \neq 0),$$

In order to convert it to one whose solution can be found by the square root property. We can thus find a general formula for solving any quadratic equation. First multiply both sides by $1/a$ (here we use the fact that $a \neq 0$) to get

$$x^2 + \frac{b}{a}x + \frac{c}{a} = 0.$$

Then, adding $-c/a$ to both sides gives

$$x^2 + \frac{b}{a}x = -\frac{c}{a}.$$

Now take half of b/a, square the result, and add the square to both sides, producing

$$x^2 + \frac{b}{a}x + \frac{b^2}{4a^2} = \frac{b^2}{4a^2} - \frac{c}{a}.$$

We can now write the expression on the left-hand side as the square of a binomial, while the expression on the right can be simplified. Doing this yields

$$\left(x + \frac{b}{2a}\right)^2 = \frac{b^2 - 4ac}{4a^2}.$$

This last statement leads to

$$x + \frac{b}{2a} = \sqrt{\frac{b^2 - 4ac}{4a^2}} \quad \text{or} \quad x + \frac{b}{2a} = -\sqrt{\frac{b^2 - 4ac}{4a^2}}.$$

Since $4a^2 = (2a)^2$, we get

$$x + \frac{b}{2a} = \frac{\sqrt{b^2 - 4ac}}{|2a|} \quad \text{or} \quad x + \frac{b}{2a} = \frac{-\sqrt{b^2 - 4ac}}{|2a|}.$$

If $a > 0$, then $|2a| = 2a$, giving

$$x = \frac{-b + \sqrt{b^2 - 4ac}}{2a} \quad \text{or} \quad x = \frac{-b - \sqrt{b^2 - 4ac}}{2a}. \qquad (*)$$

If $a < 0$, then $|2a| = -2a$, giving the same two solutions as in $(*)$, except in reversed order. In either case, the solutions can be written as

$$x = \frac{-b + \sqrt{b^2 - 4ac}}{2a} \quad \text{or} \quad x = \frac{-b - \sqrt{b^2 - 4ac}}{2a}.$$

A more compact form of this result, called the **quadratic formula,** is given in the next box.

Quadratic Formula	The solution set of the quadratic equation $ax^2 + bx + c = 0$ $(a \neq 0)$ is $$\left\{\frac{-b \pm \sqrt{b^2 - 4ac}}{2a}\right\}.$$

EXAMPLE 4   Solve $x^2 - 4x + 1 = 0$.

Here $a = 1$, $b = -4$, and $c = 1$. Substitute these values into the quadratic formula, producing

$$x = \frac{-b \pm \sqrt{b^2 - 4ac}}{2a}$$

$$= \frac{-(-4) \pm \sqrt{(-4)^2 - 4(1)(1)}}{2(1)}$$

$$= \frac{4 \pm \sqrt{16 - 4}}{2}$$

$$x = \frac{4 \pm 2\sqrt{3}}{2} = \frac{2(2 \pm \sqrt{3})}{2} = 2 \pm \sqrt{3}.$$

The solution set is $\{2 + \sqrt{3}, 2 - \sqrt{3}\}$, abbreviated $\{2 \pm \sqrt{3}\}$. ☐

EXAMPLE 5 ☐ Solve $2y^2 = y - 4$.

To find the values of a, b, and c, first rewrite the equation as $2y^2 - y + 4 = 0$. Then $a = 2$, $b = -1$, and $c = 4$. By the quadratic formula,

$$y = \frac{-(-1) \pm \sqrt{(-1)^2 - 4(2)(4)}}{2(2)}$$

$$= \frac{1 \pm \sqrt{1 - 32}}{4}$$

$$y = \frac{1 \pm \sqrt{-31}}{4} = \frac{1 \pm i\sqrt{31}}{4}.$$

The solutions are complex numbers; the solution set is $\{(1 \pm i\sqrt{31})/4\}$. ☐

The quantity under the radical in the quadratic formula, $b^2 - 4ac$, is called the **discriminant.** From the sign of the discriminant, when a, b, and c are *real numbers,* we can determine the nature of the solutions, real or complex (that is, not real), and the number of solutions, one or two. The discriminant indicates the following information about the solutions.

Discriminant

Discriminant, $b^2 - 4ac$	Number of solutions	Kind of solutions
positive	two	real
zero	one	real
negative	two	complex

EXAMPLE 6 ☐ Find a value of k so that the equation

$$16p^2 + kp + 25 = 0$$

has exactly one solution.

A quadratic equation with real coefficients will have exactly one solution if the discriminant is zero. Here, $a = 16$, $b = k$, and $c = 25$, giving the discriminant

$$b^2 - 4ac = k^2 - 4(16)(25) = k^2 - 1600.$$

The discriminant is 0 if

$$k^2 - 1600 = 0$$

or

$$k^2 = 1600,$$

from which $k = \pm 40$. ☐

The next example illustrates the use of the quadratic formula when one or more of the coefficients of the quadratic equation are imaginary numbers.

EXAMPLE 7   ☐   Solve $im^2 + 5m - 3i = 0$.

Here, $a = i$, $b = 5$, and $c = -3i$, with

$$m = \frac{-5 \pm \sqrt{5^2 - 4(i)(-3i)}}{2i}$$

$$= \frac{-5 \pm \sqrt{25 + 12i^2}}{2i}$$

$$= \frac{-5 \pm \sqrt{25 - 12}}{2i}$$

$$m = \frac{-5 \pm \sqrt{13}}{2i}.$$

Simplify by multiplying numerator and denominator by $-i$ (the same result would be obtained if we used i). This gives

$$m = \frac{(-5 \pm \sqrt{13})(-i)}{2i(-i)}.$$

In the denominator, $2i(-i) = -2i^2 = -2(-1) = 2$, so that the final result is

$$m = \frac{5i \pm i\sqrt{13}}{2}.$$

The solution set is $\{(5i \pm i\sqrt{13})/2\}$.   ☐

EXAMPLE 8   ☐   Michael wants to make an exposed gravel border of uniform width around a rectangular pool in his garden. The pool is 10 feet by 6 feet. He has enough material to cover 36 square feet. How wide should the border be?

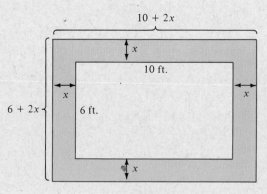

Figure 2.2

A sketch of the pool with border is shown in Figure 2.2. Let x represent the width of the border. Then the width of the large rectangle is $6 + 2x$ and its length is $10 + 2x$. The area of the large rectangle is $(6 + 2x)(10 + 2x)$. The area of the pool is $6 \cdot 10 = 60$. The area of the border is found by subtracting the area of the pool from the area of the large rectangle. The difference should be 36 square feet, giving the equation

$$(6 + 2x)(10 + 2x) - 60 = 36.$$

To solve this equation go through the following sequence of steps, getting

$$60 + 32x + 4x^2 - 60 = 36$$
$$4x^2 + 32x - 36 = 0$$
$$x^2 + 8x - 9 = 0$$
$$(x + 9)(x - 1) = 0.$$

The solutions are -9 and 1. We cannot use -9 as the width of the border, so the solution is to make the border one foot wide. ▢

EXERCISES
2.3

Solve the following equations by factoring or by using the square root property.

1. $p^2 = 16$
2. $k^2 = 25$
3. $x^2 = 27$
4. $y^2 = 24$
5. $(m - 3)^2 = 5$
6. $(p + 2)^2 = 7$
7. $(3k - 1)^2 = 12$
8. $(4t + 1)^2 = 20$
9. $p^2 - 5p + 6 = 0$
10. $q^2 + 2q - 8 = 0$
11. $6z^2 - 5z - 50 = 0$
12. $6r^2 + 7r = 3$
13. $8k^2 + 14k + 3 = 0$
14. $18r^2 - 9r - 2 = 0$

Solve the following equations by completing the square.

15. $p^2 - 8p + 15 = 0$
16. $m^2 + 5m = 6$
17. $x^2 - 2x - 4 = 0$
18. $r^2 + 8r + 13 = 0$
19. $2p^2 + 2p + 1 = 0$
20. $9z^2 - 12z + 8 = 0$

Solve the following equations.

21. $m^2 - m - 1 = 0$
22. $y^2 - 3y - 2 = 0$
23. $2s^2 + 2s = 3$
24. $t^2 - t = 3$
25. $x^2 - 6x + 7 = 0$
26. $11p^2 - 7p + 1 = 0$
27. $n^2 + 4 = 3n$
28. $9p^2 = 25 + 30p$
29. $2m^2 = m - 1$
30. $4z^2 - 12z + 11 = 0$
31. $x^2 = 2x - 5$
32. $3k^2 + 2 = k$
33. $4 - \dfrac{11}{x} - \dfrac{3}{x^2} = 0$
34. $3 - \dfrac{4}{p} = \dfrac{2}{p^2}$
35. $2 - \dfrac{5}{r} + \dfrac{3}{r^2} = 0$
36. $2 - \dfrac{5}{k} + \dfrac{2}{k^2} = 0$

Use a calculator to give the solutions of the following equations to the nearest thousandth.

37. $5n^2 - 4 = 3n$ **38.** $9p^2 - 30p = 25$

39. $2m^2 = m + 2$

40. $4n^2 + 12n - 9 = 0$

41. $x^2 - 2 = 3x$ **42.** $3k^2 - 2 = 4k$

Use the quadratic formula to solve the following equations.

43. $m^2 - \sqrt{2}m - 1 = 0$

44. $z^2 - \sqrt{3}z - 2 = 0$

45. $\sqrt{2}p^2 - 3p + \sqrt{2} = 0$

46. $-\sqrt{6}k^2 - 2k + \sqrt{6} = 0$

47. $x^2 + ix + 1 = 0$

48. $3im^2 - 2m + i = 0$

49. $ip^2 - 2p + i = 0$

50. $2ix^2 - 3x + 3i = 0$

51. $(1 + i)x^2 - x + (1 - i) = 0$

52. $(2 + i)r^2 - 3r + (2 - i) = 0$

Identify the values of a, b, and c for each of the following; then evaluate the discriminant $b^2 - 4ac$ and use it to predict the type of solutions. Do not solve the equations.

53. $x^2 + 8x + 16 = 0$

54. $x^2 - 5x + 4 = 0$

55. $3m^2 - 5m + 2 = 0$

56. $8y^2 = 14y - 3$

57. $4p^2 = 6p + 3$

58. $2r^2 - 4r + 1 = 0$

59. $9k^2 + 11k + 4 = 0$

60. $3z^2 = 4z - 5$

Solve each of the following for the indicated variable. Assume all denominators are nonzero.

61. $s = \dfrac{1}{2}gt^2$ for t

62. $A = \pi r^2$ for r

63. $L = \dfrac{d^4k}{h^2}$ for h

64. $F = \dfrac{kMv^2}{r}$ for v

65. $s = s_0 + gt^2 + k$ for t

66. $g = \dfrac{4\pi^2}{t^2}$ for t

67. $P = \dfrac{E^2R}{(r + R)^2}$ for R

68. $S = 2\pi rh + 2\pi r^2$ for r

Solve the following problems.

69. Find two consecutive even integers whose product is 288.

70. The sum of the squares of two consecutive integers is 481. Find the integers.

71. Two integers have a sum of 10. The sum of the squares of the integers is 148. Find the integers.

72. A shopping center has a rectangular area of 40,000 square yards enclosed on three sides for a parking lot. The length is 200 yards more than twice the width. What are the dimensions of the lot?

73. An ecology center wants to set up an experimental garden. It has 300 meters of fencing to enclose a rectangular area of 5000 square meters. Find the dimensions of the rectangle.

74. Alfredo went into a frame-it-yourself shop. He wanted a frame 3 centimeters longer than wide. The frame he chose extends 1.5 centimeters beyond the picture on each side. Find the outside dimensions of the frame if the area of the unframed picture is 70 square centimeters.

75. Joan wants to buy a rug for a room that is 12 feet by 15 feet. She wants to leave a uniform strip of floor around the rug. She can afford 108 square feet of carpeting. What dimensions should the rug have?

76. Max can clean the garage in 9 hours less time than his brother Paul. Working together, they can do the job in 20 hours. How long would it take each one to do the job alone?

77. Dolores drives 10 mph faster than Steve. Both start at the same time for Atlanta from Chattanooga, a distance of about 100 miles. It takes Steve 1/3 of an hour longer than Dolores to make the trip. What is Steve's average speed?

78. Amy walks 1 mph faster than her friend Lisa. In a walk for charity, both walked the full distance of 24 miles. Lisa took 2 hours longer than Amy. What was Lisa's average speed?

Find the solution for each of the following. Round to the nearest hundredth.

79. One leg of a right triangle is 4 cm longer than the other leg. The hypotenuse (longest side) is 10 cm longer than the shorter leg. Find each of the three sides of the triangle.

80. A rectangle has a diagonal of 16 m. The width of the rectangle is 3.2 cm less than the length. Find the length and width of the rectangle.

Find all values for k for which the following equations have exactly one solution.

81. $9x^2 + kx + 4 = 0$

82. $25m^2 - 10m + k = 0$

83. $y^2 + 11y + k = 0$

84. $z^2 - 5z + k = 0$

85. $kr^2 + (2k + 6)r + 16 = 0$

86. $ky^2 + 2(k + 4)y + 25 = 0$

Let r_1 and r_2 be the solutions of the quadratic equation $ax^2 + bx + c = 0$. Show that

87. $r_1 + r_2 = -\dfrac{b}{a}$

88. $r_1 r_2 = \dfrac{c}{a}$

89. Let m, n, and k be real numbers. Find the solutions of the equation $(x - m)(x - n) = k^2$ and show that they are real.

90. Suppose one solution of the equation $km^2 + 10m = 8$ is -4. Find the value of k, and the other solution.

For each of the following equations, (a) solve for x in terms of y, (b) solve for y in terms of x.

91. $4x^2 - 2xy + 3y^2 = 2$

92. $3y^2 + 4xy - 9x^2 = -1$

2.4 Equations Reducible to Quadratics

The equation $12m^4 - 11m^2 + 2 = 0$ is not a quadratic equation. However, if we make the substitution

$$x = m^2, \qquad \text{so that } x^2 = m^4,$$

then the given equation becomes

$$12x^2 - 11x + 2 = 0,$$

which is a quadratic equation. We can then solve this quadratic equation to find x, and then from the fact that $x = m^2$, we can find the values of m, which are the solutions to our original equation.

An equation, such as $12m^4 - 11m^2 + 2 = 0$, is said to be **quadratic in form** if it can be written as

$$au^2 + bu + c = 0,$$

where $a \neq 0$ and u is some algebraic expression.

As a further example,

$$6p^{-2} + p^{-1} = 2$$

is quadratic in form, since substituting u for p^{-1} and rearranging terms permits the equation to be written as

$$6u^2 + u - 2 = 0.$$

The next few examples show how to solve equations that are quadratic in form.

EXAMPLE 1 ☐ Solve $12m^4 - 11m^2 + 2 = 0$.

As mentioned above, this equation is quadratic in form. By making the substitution $x = m^2$, the equation becomes

$$12x^2 - 11x + 2 = 0,$$

which can be solved by factoring, as follows.

$$(3x - 2)(4x - 1) = 0$$
$$x = 2/3 \quad \text{or} \quad x = 1/4$$

Now we know that $x = 2/3$ or $x = 1/4$. To find m, use the fact that $x = m^2$ and replace x with m^2, getting

$$m^2 = \frac{2}{3} \quad \text{or} \quad m^2 = \frac{1}{4}$$

$$m = \pm\sqrt{\frac{2}{3}} \qquad m = \pm\sqrt{\frac{1}{4}}$$

$$m = \pm\frac{\sqrt{2}}{\sqrt{3}}$$

$$m = \frac{\pm\sqrt{6}}{3} \quad \text{or} \quad m = \pm\frac{1}{2}.$$

These four solutions of the given equation $12m^4 - 11m^2 + 2 = 0$ make up the solution set $\{\sqrt{6}/3, -\sqrt{6}/3, 1/2, -1/2\}$, abbreviated $\{\pm\sqrt{6}/3, \pm 1/2\}$. ☐

EXAMPLE 2 ☐ Solve $6p^{-2} + p^{-1} = 2$.

Let $u = p^{-1}$ and rearrange terms to get

$$6u^2 + u - 2 = 0.$$

Factor on the left, and then place each factor equal to 0, giving

$$(3u + 2)(2u - 1) = 0$$
$$3u + 2 = 0 \quad \text{or} \quad 2u - 1 = 0$$
$$u = -\frac{2}{3} \qquad u = \frac{1}{2}.$$

Since $u = p^{-1}$, we have

$$p^{-1} = -\frac{2}{3} \quad \text{or} \quad p^{-1} = \frac{1}{2},$$

from which

$$p = -\frac{3}{2} \quad \text{or} \quad p = 2.$$

The solution set of $6p^{-2} + p^{-1} = 2$ is thus $\{-3/2, 2\}$. ☐

EXAMPLE 3 ☐ Solve $4p^4 - 16p^2 + 13 = 0$.
Let $u = p^2$ to get

$$4u^2 - 16u + 13 = 0.$$

To solve this equation, we can use the quadratic formula with $a = 4$, $b = -16$, and $c = 13$. Substituting these values into the quadratic formula gives

$$u = \frac{-(-16) \pm \sqrt{(-16)^2 - 4(4)(13)}}{2(4)}$$

$$u = \frac{16 \pm \sqrt{48}}{8} = \frac{16 \pm 4\sqrt{3}}{8} = \frac{4 \pm \sqrt{3}}{2}.$$

Since $p^2 = u$, we have

$$p^2 = \frac{4 + \sqrt{3}}{2} \quad \text{or} \quad p^2 = \frac{4 - \sqrt{3}}{2}.$$

Finally, $\quad p = \pm \sqrt{\dfrac{4 + \sqrt{3}}{2}} \quad \text{or} \quad p = \pm \sqrt{\dfrac{4 - \sqrt{3}}{2}}.$

The solution set of $4p^4 - 16p^2 + 13 = 0$ is thus $\left\{ \sqrt{\dfrac{4 + \sqrt{3}}{2}}, \; -\sqrt{\dfrac{4 + \sqrt{3}}{2}}, \right.$
$\left. \sqrt{\dfrac{4 - \sqrt{3}}{2}}, \; -\sqrt{\dfrac{4 - \sqrt{3}}{2}} \right\}.$ ☐

To solve equations containing radicals or rational exponents, such as $x = \sqrt{15 - 2x}$, or $(x + 1)^{1/2} = x$, we use the following rule.

> If P and Q are algebraic expressions, then every solution of the equation $P = Q$ is also a solution of the equation $(P)^n = (Q)^n$, for any positive integer n.

Be very careful when using this rule. It does *not* say that the equations $P = Q$ and $(P)^n = (Q)^n$ are equivalent; it says only that each solution of the original equation $P = Q$ is also a solution of the new equation $(P)^n = (Q)^n$. However, the new equation may have *more* solutions than the original equation. For example, the solution set of the equation $x = -2$ is $\{-2\}$. If we square both sides of the equation $x = -2$, we get the new equation $x^2 = 4$, which has solution set $\{-2, 2\}$. Since the solution sets are not equal, the equations are not equivalent. Because of this, it is essential to check all proposed solutions back in the original equation.

EXAMPLE 4   ☐   Solve $x = \sqrt{15 - 2x}$.

The equation $x = \sqrt{15 - 2x}$ can be solved by squaring both sides as follows:

$$x^2 = (\sqrt{15 - 2x})^2$$

$$x^2 = 15 - 2x$$

$$x^2 + 2x - 15 = 0$$

$$(x + 5)(x - 3) = 0$$

$$x = -5 \quad \text{or} \quad x = 3$$

Now it is necessary to check the proposed solutions in the original equation,

$$x = \sqrt{15 - 2x}.$$

If $x = -5$, does $x = \sqrt{15 - 2x}$?   If $x = 3$, does $x = \sqrt{15 - 2x}$?

$$-5 = \sqrt{15 + 10} \qquad\qquad\qquad 3 = \sqrt{15 - 6}$$

$$-5 = 5 \quad \text{(false)} \qquad\qquad\quad 3 = 3 \quad \text{(true)}$$

As this check shows, only 3 is a solution, giving the solution set $\{3\}$.   ☐

EXAMPLE 5   ☐   Solve $\sqrt{2x + 3} - \sqrt{x + 1} = 1$.

Write the equation as

$$\sqrt{2x + 3} = 1 + \sqrt{x + 1}.$$

Now we need to square both sides. Be very careful when squaring on the right side of this equation. Recall that $(a + b)^2 = a^2 + 2ab + b^2$; replace a with 1 and b with $\sqrt{x + 1}$ to get the next equation, the result of squaring both sides of $\sqrt{2x + 3} = 1 + \sqrt{x + 1}$.

$$2x + 3 = 1 + 2\sqrt{x + 1} + x + 1$$

$$x + 1 = 2\sqrt{x + 1}.$$

One side of the equation still contains a radical, and to eliminate it, we need to square both sides again. This gives

$$x^2 + 2x + 1 = 4(x + 1)$$

$$x^2 - 2x - 3 = 0$$

$$(x - 3)(x + 1) = 0$$

$$x = 3 \quad \text{or} \quad x = -1.$$

Check these proposed solutions in the original equation.

$$\text{If } x = 3, \text{ does } \sqrt{2x + 3} - \sqrt{x + 1} = 1?$$
$$\sqrt{9} - \sqrt{4} = 1$$
$$3 - 2 = 1 \quad \text{(true)}$$
$$\text{If } x = -1, \text{ does } \sqrt{2x + 3} - \sqrt{x + 1} = 1?$$
$$\sqrt{1} - \sqrt{0} = 1$$
$$1 - 0 = 1 \quad \text{(true)}$$

Here both the proposed solutions 3 and -1 are solutions of the original equation, giving $\{3, -1\}$ as the solution set. ☐

EXAMPLE 6 ☐ Solve $(5x^2 - 6)^{1/4} = x$.

Since the equation involves a fourth root, begin by raising both sides to the fourth power. This gives

$$[(5x^2 - 6)^{1/4}]^4 = x^4$$
$$5x^2 - 6 = x^4$$
$$x^4 - 5x^2 + 6 = 0.$$

Now substitute y for x^2, yielding

$$y^2 - 5y + 6 = 0$$
$$(y - 3)(y - 2) = 0$$
$$y = 3 \quad \text{or} \quad y = 2.$$

Since $y = x^2$, we have

$$x^2 = 3 \quad \text{or} \quad x^2 = 2$$
$$x = \pm\sqrt{3} \quad \text{or} \quad x = \pm\sqrt{2}.$$

By checking the four proposed solutions, $\sqrt{3}, -\sqrt{3}, \sqrt{2}$, and $-\sqrt{2}$, in the original equation, we find that only $\sqrt{3}$ and $\sqrt{2}$ are solutions, so that the solution set is $\{\sqrt{3}, \sqrt{2}\}$. ☐

EXERCISES 2.4

Solve the following equations.

1. $m^4 - 13m^2 + 36 = 0$

2. $p^4 - 17p^2 + 16 = 0$

3. $2r^4 - 7r^2 + 5 = 0$

4. $4x^4 - 8x^2 + 3 = 0$

5. $(g - 2)^2 - 6(g - 2) + 8 = 0$

6. $(p + 2)^2 - 2(p + 2) - 15 = 0$

7. $6(k + 2)^4 - 11(k + 2)^2 + 4 = 0$

8. $8(m - 4)^4 - 10(m - 4)^2 + 3 = 0$

9. $7p^{-2} + 19p^{-1} = 6$

10. $5k^{-2} - 43k^{-1} = 18$

11. $(r - 1)^{2/3} + (r - 1)^{1/3} = 12$

12. $(y + 3)^{2/3} - 2(y + 3)^{1/3} - 3 = 0$

13. $m^{1/3} = m^{2/3}$

14. $a^{-1} = a^{-2}$

15. $3 + \dfrac{5}{p^2 + 1} = \dfrac{2}{(p^2 + 1)^2}$

16. $6 = \dfrac{7}{2y - 3} + \dfrac{3}{(2y - 3)^2}$

17. $\sqrt{2m + 1} = 2\sqrt{m}$

18. $3\sqrt{p} = \sqrt{8p + 16}$

19. $\sqrt{3z + 7} = 3z + 5$

20. $\sqrt{4r + 13} = 2r - 1$

21. $\sqrt{4k + 5} - 2 = 2k - 7$

22. $\sqrt{6m + 7} - 1 = m + 1$

23. $\sqrt{4x} - x + 3 = 0$

24. $\sqrt{2t} - t + 4 = 0$

25. $\sqrt{y} = \sqrt{y - 5} + 1$

26. $\sqrt{2m} = \sqrt{m + 7} - 1$

27. $\sqrt{r + 5} - 2 = \sqrt{r - 1}$

28. $\sqrt{m + 7} + 3 = \sqrt{m - 4}$

29. $\sqrt{y + 2} = \sqrt{2y + 5} - 1$

30. $\sqrt{4x + 1} = \sqrt{x - 1} + 2$

31. $\sqrt{2\sqrt{7x + 2}} = \sqrt{3x + 2}$

32. $\sqrt{3\sqrt{2m + 3}} = \sqrt{5m - 6}$

33. $\sqrt{x + 2} = \sqrt{4 + 7\sqrt{x}}$

34. $3 - \sqrt{x} = \sqrt{2\sqrt{x} - 3}$

35. $(2r + 5)^{1/3} = (6r - 1)^{1/3}$

36. $(3m + 7)^{1/3} = (4m + 2)^{1/3}$

37. $\sqrt[4]{q - 15} = 2$ **38.** $\sqrt[4]{3x + 1} = 1$

39. $\sqrt[4]{y^2 + 2y} = \sqrt[4]{3}$ **40.** $\sqrt[4]{k^2 + 6k} = 2$

41. $(2r - 1)^{2/3} = r^{1/3}$ **42.** $(z - 3)^{2/5} = (4z)^{1/5}$

43. $k^{2/3} = 2k^{1/3}$ **44.** $3m^{3/4} = m^{1/2}$

Find all solutions for each of the following, rounded to the nearest thousandth.

45. $6p^4 - 41p^2 + 63 = 0$

46. $20z^4 - 67z^2 + 56 = 0$

47. $3k^2 - \sqrt{4k^2 + 3} = 0$

48. $2r^2 - \sqrt{r^2 + 1} = 0$

Solve the following equations for the specified variable. Assume all denominators are nonzero.

49. $d = k\sqrt{h}$ for h

50. $v = \dfrac{k}{\sqrt{d}}$ for d

51. $P = 2\sqrt{\dfrac{L}{g}}$ for L

52. $c = \sqrt{a^2 + b^2}$ for a

53. $x^{2/3} + y^{2/3} = a^{2/3}$ for y

54. $m^{3/4} + n^{3/4} = 1$ for m

2.5   Inequalities

An equation says that two expressions are equal, while an **inequality** says that one expression is greater than, greater than or equal to, less than, or less than or equal to, another. As with equations, a value of the variable for which the inequality is true is a **solution** of the inequality; similarly, the set of all such solutions is called the **solution set** of the inequality. Two inequalities with the same solution set are **equivalent**. Inequalities are solved with the following properties of real numbers:

Properties of Inequalities

For real numbers a, b, and c,

(a) if $a < b$, then $a + c < b + c$

(b) if $a < b$, and if $c > 0$, then $ac < bc$

(c) if $a < b$, and if $c < 0$, then $ac > bc$.

Similar properties are valid if $<$ is replaced with $>$, $\leq$, or $\geq$. Pay careful attention to part (c): if both sides of an inequality are multiplied by a negative number, the direction of the inequality symbol must be reversed. For example, if we start with the true statement $-3 < 5$, and multiply both sides by the *positive* number 2, we get

$$-3 \cdot 2 < 5 \cdot 2$$
$$-6 < 10,$$

still a true statement. On the other hand, if we start with $-3 < 5$ and multiply both sides by the *negative* number -2, we get a true result only if we reverse the direction of the inequality symbol:

$$-3(-2) > 5(-2)$$
$$6 > -10$$

To prove the properties of inequalities in the box above, first recall from Chapter 1 that $a < b$ means that $b - a$ is positive. The proofs also depend on the fact that the sum or product of two positive numbers is positive.

To prove part (a), use the fact that $a < b$ to see that $b - a$ is positive. Rewrite $b - a$ as

$$b - a = (b + c) - (a + c).$$

Since $b - a$ is positive, $(b + c) - (a + c)$ must also be positive, so that

$$a + c < b + c.$$

To prove part (b), again use the assumption $a < b$ to see that $b - a$ is positive. Since c is assumed positive, the product $(b - a)c$ is positive. By the distributive property,

$$(b - a)c = bc - ac,$$

and $bc - ac$ must be positive, giving

$$ac < bc.$$

For part (c), again we are given $a < b$, so that $b - a$ is positive. We are also told that $c < 0$. Since c is less than 0, we must have $0 - c$, or $-c$, as positive. Thus,

$$(b - a)(-c) = -bc + ac = ac - bc$$

is positive. If $ac - bc$ is positive, we have

$$bc < ac.$$

Similar proofs could be given if we replaced $<$ with $>$, $\leq$, or $\geq$.

EXAMPLE ☐ Solve the inequality $-3x + 5 > -7$.
    1   Use the properties of inequalities. First, adding -5 on both sides gives

$$-3x + 5 + (-5) > -7 + (-5)$$
$$-3x > -12.$$

Now multiply both sides by $-1/3$. Since $-1/3 < 0$, we must reverse the direction of the inequality symbol, to get

$$-\frac{1}{3}(-3x) < -\frac{1}{3}(-12)$$

$$x < 4.$$

The original inequality is thus satisfied by any real number less than 4. The solution set can be written $\{x | x < 4\}$. A graph of the solution set is shown in Figure 2.3, where the parenthesis is used to show that 4 itself does not belong to the solution set. ☐

There is a shortcut way of writing a set such as $\{x | x < 4\}$. This set can be written in **interval notation** as $(-\infty, 4)$. The symbol ∞ is not a real number—it merely tells us that the interval includes all real numbers less than 4. Examples of other sets written in interval notation are shown below. Square brackets are used to show that the given number *is* part of the graph. Whenever two real numbers a and b are used to write an interval, it is always assumed that $a < b$.

Type of interval	Set	Interval notation	Graph
open interval	$\{x |\ a < x\}$	$(a, +\infty)$	
	$\{x |\ a < x < b\}$	(a, b)	
	$\{x |\ x < b\}$	$(-\infty, b)$	
half-open interval	$\{x |\ a \leq x\}$	$[a, +\infty)$	
	$\{x |\ a < x \leq b\}$	$(a, b]$	
	$\{x |\ a \leq x < b\}$	$[a, b)$	
	$\{x |\ x \leq b\}$	$(-\infty, b]$	
closed interval	$\{x | a \leq x \leq b\}$	$[a, b]$	

Figure 2.3

Figure 2.4

EXAMPLE 2 ☐   Solve $4 - 3y \leq 7 + 2y$. Write the solution in interval notation and graph the solution on a number line.

Write the following series of equivalent inequalities.

$$4 - 3y \leq 7 + 2y$$
$$-4 - 2y + 4 - 3y \leq -4 - 2y + 7 + 2y$$
$$-5y \leq 3$$
$$(-1/5)(-5y) \geq (-1/5)(3)$$
$$y \geq -3/5$$

In interval notation, the solution set is $[-3/5, +\infty)$. See Figure 2.4 for the graph. ☐

From now on, we shall write the solutions of all inequalities with interval notation.

EXAMPLE 3 ☐   Solve $-2 < 5 + 3m < 20$.

We are to find all values of m so that $5 + 3m$ is between -2 and 20. This can be true if and only if m satisfies *both* of the inequalities

$$-2 < 5 + 3m \qquad \text{and} \qquad 5 + 3m < 20.$$

Solve each of these separately. If we begin with $-2 < 5 + 3m$, we have

$$-2 < 5 + 3m$$
$$-7 < 3m$$
$$-\frac{7}{3} < m.$$

Now solve $5 + 3m < 20$, getting

$$5 + 3m < 20$$
$$3m < 15$$
$$m < 5.$$

Thus, our given inequality is true only when

$$-\frac{7}{3} < m \qquad \text{and} \qquad m < 5,$$

or $\qquad -\frac{7}{3} < m < 5.$

The solution set, written in interval notation as $(-7/3, 5)$, is graphed in Figure 2.5.

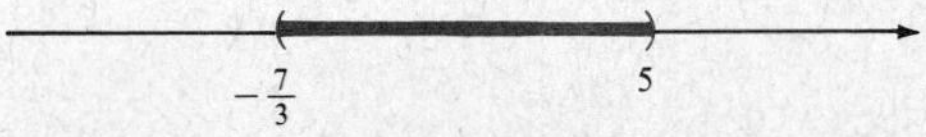

Figure 2.5

The inequality $-2 < 5 + 3m < 20$ can be solved more quickly as follows.

$$-2 < 5 + 3m < 20$$
$$-2 + (-5) < 5 + 3m + (-5) < 20 + (-5)$$
$$-7 < 3m < 15$$
$$-\frac{7}{3} < m < 5 \quad \square$$

Quadratic Inequality

A **quadratic inequality** is an inequality that can be written in the form

$$ax^2 + bx + c < 0,$$

for real numbers $a \neq 0$, b, and c. (The symbol $<$ can be replaced with $>$, $\leq$, or $\geq$.)

EXAMPLE 4

$\square$ Solve the quadratic inequality $x^2 - x - 12 < 0$.
Start by writing the inequality in factored form as

$$(x - 4)(x + 3) < 0.$$

To solve the inequality, find all values of x so that the product $(x - 4)(x + 3)$ is negative (<0). The only way that this product can be negative is if the two factors are opposite in sign. This happens if either

$$x - 4 < 0 \quad \text{and} \quad x + 3 > 0.$$

or if $x - 4 > 0 \quad$ and $\quad x + 3 < 0.$

Since the factor $x - 4$ is 0 when $x = 4$, and the factor $x + 3$ is 0 when $x = -3$, the sign of the product $(x - 4)(x + 3)$ can change only at 4 or -3, with each of the intervals $(-\infty, -3)$, $(-3, 4)$, or $(4, +\infty)$ either part of the solution or not. (See Figure 2.6.) To decide whether a given interval is part of the solution, sketch a **sign graph**, as in Figure 2.7. First decide on the sign of the factor $x - 4$ in each region. The factor $x - 4$ is positive if

$$x - 4 > 0$$

or $\qquad x > 4,$

and negative if $x < 4$, while $x + 3$ is positive if $x > -3$ and negative if $x < -3$. These results are sketched in Figure 2.7.

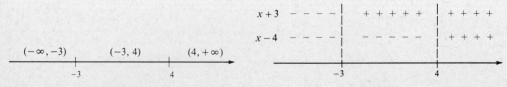

Figure 2.6   Figure 2.7

Now consider the sign of the product of the two factors in each region. As the sign graph shows, both factors are negative in the interval $(-\infty, -3)$; therefore their product is positive in that interval. For the interval $(-3, 4)$, one factor is positive, while the other is negative, giving a negative product. In the last region, $(4, +\infty)$, both factors are positive so their product is positive. The polynomial $x^2 - x - 12$ is negative (what the original inequality calls for) when the product of its factors is negative, that is, for the interval $(-3, 4)$. The graph of this solution set is shown in Figure 2.8. ☐

Figure 2.8

EXAMPLE 5 ☐ Solve the inequality $2x^2 + 5x \geq 12$.

Start with $\qquad\qquad\qquad 2x^2 + 5x - 12 \geq 0$.

Factoring gives $\qquad\qquad (2x - 3)(x + 4) \geq 0$.

Solving $2x - 3 = 0$ and $x + 4 = 0$ shows that the product can change its sign only at $3/2$ or -4. These two points divide the number line into the three regions shown in the sign graph of Figure 2.9. Since both factors are negative in the first interval, their product, $2x^2 + 5x - 12$, is positive there. In the second interval, the factors have opposite signs, and therefore their product is negative. Both factors are positive in the third interval with their product also positive there. Thus, the polynomial $2x^2 + 5x - 12$ is positive or zero in the interval $(-\infty, -4]$ and also in the interval $[3/2, +\infty)$. Since both of these intervals belong to the solution set, the result can be written as the *union** of the two intervals, or

$$(-\infty, -4] \cup [3/2, +\infty).$$

The graph of the solution set is shown in Figure 2.10. ☐

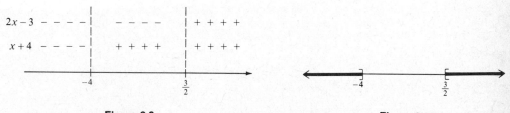

Figure 2.9 $\qquad\qquad\qquad\qquad\qquad\qquad$ **Figure 2.10**

The inequalities in the rest of this section are not quadratic inequalities but they can still be solved by the method of sign graphs presented above.

*The **union** of sets A and B, written $A \cup B$, is defined as $A \cup B = \{x | x$ is an element of A or x is an element of $B\}$.

EXAMPLE
6

☐   Solve the inequality $\dfrac{5}{x + 4} \geq 1$.

We could begin by multiplying both sides of the inequality by $x + 4$, but we would then have to consider whether $x + 4$ is positive or negative. Instead, subtract 1 from both sides of the inequality, getting

$$\frac{5}{x + 4} - 1 \geq 0.$$

Writing the left side as a single fraction yields

$$\frac{5 - (x + 4)}{x + 4} \geq 0$$

or $$\frac{1 - x}{x + 4} \geq 0.$$

The quotient can change sign only where

$$1 - x = 0 \qquad \text{or} \qquad x + 4 = 0$$
$$x = 1 \qquad \text{or} \qquad x = -4.$$

Use these two numbers to make a sign graph as before. This time consider the sign of the *quotient* of the two quantities rather than their product. See Figure 2.11.

We know that the quotient of two numbers is positive if both numbers are positive or if both numbers are negative. On the other hand, the quotient is negative if the two numbers have opposite signs. From the sign graph, since we want a positive quotient, we see that the interval $(-4, 1)$ is part of the solution. When we have a quotient, we must consider the endpoints separately to make sure that no denominator is 0. Here -4 leads to a zero denominator while 1 satisfies the given inequality. Thus, the solution is the interval $(-4, 1]$.   ☐

EXAMPLE
7

☐   Solve $\dfrac{2x - 1}{3x + 4} < 5$.

Begin by subtracting 5 on both sides and combining the terms on the left into a single fraction. This process gives

$$\frac{2x - 1}{3x + 4} < 5$$

$$\frac{2x - 1}{3x + 4} - 5 < 0$$

$$\frac{2x - 1 - 5(3x + 4)}{3x + 4} < 0$$

$$\frac{-13x - 21}{3x + 4} < 0.$$

To draw a sign graph, we must first solve the equations

$$-13x - 21 = 0 \qquad \text{and} \qquad 3x + 4 = 0,$$

getting the solutions

$$x = -\frac{21}{13} \quad \text{and} \quad x = -\frac{4}{3}.$$

Use the values $-21/13$ and $-4/3$ to divide the number line into three intervals. Now complete a sign graph and find the intervals where the quotient is negative. See Figure 2.12.

From the sign graph, we see that values of x in the two intervals $(-\infty, -21/13)$ and $(-4/3, +\infty)$ make the quotient negative, as required. Neither endpoint satisfies the strict inequality, so the solutions set is written $(-\infty, -21/13) \cup (-4/3, +\infty)$. ☐

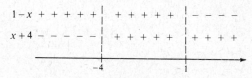

Figure 2.11

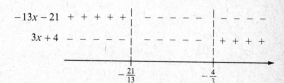

Figure 2.12

EXERCISES
2.5

Write each of the following in interval notation. Graph each interval.

1. $-1 < x < 4$

2. $x \geq -3$

3. $x < 0$

4. $8 > x > 3$

5. $2 > x \geq 1$

6. $-4 \geq x > -5$

7. $-9 > x$

8. $6 \leq x$

Using the variable x, write each of the following intervals as an inequality.

9. $(-4, 3)$

10. $[2, 7)$

11. $(-\infty, -1]$

12. $(3, +\infty)$

13.

14.

15.

16.

Solve the following inequalities. Write the solutions in interval notation.

17. $2x + 1 \leq 9$

18. $3y - 2 \leq 10$

19. $-3p - 2 \leq 1$

20. $-5r + 3 \geq -2$

21. $2(m + 5) - 3m + 1 \geq 5$

22. $6m - (2m + 3) \geq 4m - 5$

23. $8k - 3k + 2 < 2(k + 7)$

24. $2 - 4x + 5(x - 1) < -6(x - 2)$

25. $\dfrac{4x + 7}{-3} \leq 2x + 5$

26. $\dfrac{2z - 5}{-8} \leq 1 - z$

27. $2 \le y + 1 \le 5$ 28. $-3 \le 2t \le 6$

29. $-10 > 3r + 2 > -16$

30. $4 > 6a + 5 > -1$

31. $-3 \le \dfrac{x - 4}{-5} < 4$ 32. $1 < \dfrac{4m - 5}{-2} < 9$

33. $y^2 - 10y + 25 < 25$

34. $m^2 + 6m + 9 < 9$

35. $x^2 - x \le 6$

36. $r^2 + r < 12$

37. $2k^2 - 9k > -4$

38. $3n^2 < -10 - 13n$

39. $x^2 + 5x - 2 < 0$

40. $4x^2 + 3x + 1 \le 0$

41. $x^3 - 4x \le 0$ 42. $r^3 - 9r \ge 0$

43. $4m^3 + 7m^2 - 2m > 0$

44. $6p^3 - 11p^2 + 3p > 0$

45. $\dfrac{2}{p - 1} \le 0$ 46. $\dfrac{3}{8 - k} > 0$

47. $\dfrac{5c + 2}{c} > 3$ 48. $\dfrac{6 - 8m}{m} > -5$

49. $\dfrac{3}{x - 6} \le 2$ 50. $\dfrac{1}{k - 2} < \dfrac{1}{3}$

51. $\dfrac{1}{m - 1} < \dfrac{5}{4}$ 52. $\dfrac{6}{5 - 3x} \le 2$

53. $\dfrac{a + 2}{3 + 2a} \le 5$ 54. $\dfrac{x + 1}{x + 2} \ge 3$

55. $\dfrac{2}{x^2 + 1} \ge 0$ 56. $\dfrac{1}{3 + p^2} < 0$

57. $\dfrac{2x - 3}{x^2 + 1} \ge 0$ 58. $\dfrac{9x - 8}{4x^2 + 25} < 0$

59. $(x - 2)^2(x - 4) < 0$

60. $(x + 1)^2(x - 3) \ge 0$

61. $\dfrac{(3x - 5)^2}{(2x - 5)^3} > 0$ 62. $\dfrac{(5x - 3)^3}{(8x - 25)^2} \le 0$

63. $\dfrac{(2x - 3)(3x + 8)}{(x - 6)^3} \ge 0$

64. $\dfrac{(9x - 11)(2x + 7)}{(3x - 8)^3} > 0$

Use the discriminant to find the values of k where the following equations have real solutions.

65. $x^2 - kx + 8 = 0$ 66. $x^2 + kx - 5 = 0$

67. $x^2 + kx + 2k = 0$

68. $kx^2 + 4x + k = 0$

A product will break-even or produce a profit only if the revenue from selling the product at least equals the cost of producing it. Find all intervals when the following products will at least break-even.

69. The cost to produce x units of wire is $C = 50x + 5000$, while the revenue is $R = 60x$.

70. The cost to produce x units of squash is $C = 100x + 6000$, while the revenue is $R = 500x$.

71. $C = 70x + 500; R = 60x$.

72. $C = 1000x + 5000; R = 900x$.

73. The commodity market is very unstable; money can be made or lost quickly when investing in soybeans, wheat, and so on. Suppose that an investor kept track of her total profit, P, at time t, measured in months, after she began investing, and found that

$$P = 4t^2 - 29t + 30.$$

Find the time intervals where she has been ahead. (Hint: $t > 0$ in this case.)

74. Suppose the velocity of an object is given by

$$v = 2t^2 - 5t - 12,$$

where t is time in seconds. (Here, t can be positive or negative.) Find the intervals where the velocity is negative.

75. An analyst has found that his company's profits, in hundred thousands of dollars, are given by

$$P = 3x^2 - 35x + 50,$$

where x is the amount, in hundreds, spent on advertising. For what values of x does the company make a profit?

76. The manager of a large apartment complex has found that the amount of profit he makes is given by

$$P = -x^2 + 250x - 15,000,$$

where x is the number of units rented. For what values of x does the complex produce a profit?

77. The formula for converting from Celsius to Fahrenheit temperature is F = 9C/5 + 32. What temperature range in °F corresponds to 0° to 30°C?

78. A projectile is fired from ground level. After t seconds its height above the ground is $220t - 16t^2$ feet. For what time period is the projectile at least 624 feet above the ground?

79. If $a > b > 0$, show that $1/a < 1/b$.

80. If $a > b$, is it always true that $1/a < 1/b$?

81. Suppose $a > b > 0$. Show that $a^2 > b^2$.

82. Suppose $a > b > 0$. Show that $(\sqrt{b} - \sqrt{a})^2 > 0$ and that $a < \sqrt{ab} < \dfrac{a + b}{2}$.

83. Let $b > 0$. When is $b^2 > b$?

84. If $a < b$ and $c < d$, show that $a + c < b + d$.

85. Solve $\dfrac{(m - 1)(m - 2)}{(m - 3)(m - 4)} \geq 0$.

86. Solve $\dfrac{(a + 2)(a + 3)}{(a + 4)(a - 5)} \leq 0$.

2.6 Absolute Value Equations and Inequalities

In this section we study methods of solving equations and inequalities involving absolute value. Recall from Chapter 1 that the absolute value of a number a, written $|a|$, gives the distance from a to 0 on a number line. By this definition, we can solve the absolute value equation $|x| = 3$ by finding all real numbers at a distance of 3 units from 0. As shown in the graph of Figure 2.13, there are two numbers satisfying this condition, 3 and -3. Thus, the solution set of the equation $|x| = 3$ is $\{3, -3\}$.

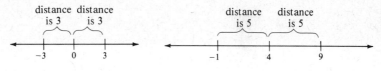

Figure 2.13   **Figure 2.14**

EXAMPLE 1   Solve $|p - 4| = 5$.

The expression $|p - 4|$ represents the distance between p and 4. Thus, to solve the equation $|p - 4| = 5$, we must find all real numbers that are 5 units from 4. As shown in Figure 2.14, these numbers are -1 and 9 so that the solution set is $\{-1, 9\}$. ☐

The definition of absolute value could be used to prove the following property of absolute value:

If b is positive, then

$|a| = b$ if and only if $a = b$ or $a = -b$.

EXAMPLE
2

☐ Solve $|4m - 3| = |m + 6|$.

By a generalization of the property just given, this equation will be true if either

$$4m - 3 = m + 6 \quad \text{or} \quad 4m - 3 = -(m + 6).$$

Solve each of these equations separately. Starting with $4m - 3 = m + 6$, we have

$$4m - 3 = m + 6$$
$$3m = 9$$
$$m = 3.$$

If $4m - 3 = -(m + 6)$, we get

$$4m - 3 = -(m + 6)$$
$$4m - 3 = -m - 6$$
$$5m = -3$$
$$m = -\frac{3}{5}.$$

The solution set of $|4m - 3| = |m + 6|$ is thus $\{3, -3/5\}$.   ☐

In the remainder of this section we shall solve inequalities involving absolute value.

EXAMPLE
3

☐ Solve $|x| < 5$.

Since absolute value gives the distance from a number to 0, the inequality $|x| < 5$ will be satisfied by all real numbers whose distance from 0 is less than 5. As shown in Figure 2.15, the solution includes all numbers from -5 to 5, or $-5 < x < 5$. In interval notation, the solution is written as the open interval $(-5, 5)$. A graph of the solution set is shown in Figure 2.15.   ☐

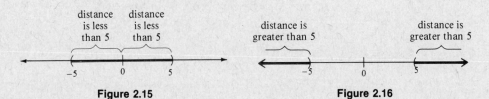

Figure 2.15   Figure 2.16

In a similar way, the solution of $|x| > 5$ is made up of all real numbers whose distance from 0 is greater than 5. This includes those numbers greater than 5 or those less than -5, or

$$x < -5 \quad \text{or} \quad x > 5.$$

In interval notation, the solution is written $(-\infty, -5) \cup (5, +\infty)$. A graph of the solution set is shown in Figure 2.16.

Using the definition of absolute value, we could prove the following additional properties of absolute value.

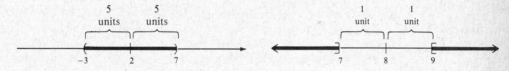

If b is a positive number,

(a) $|a| < b$ if and only if $-b < a < b$;

(b) $|a| > b$ if and only if $a < -b$ or $a > b$.

EXAMPLE 4   Solve $|x - 2| < 5$.

To solve this inequality, we need to find all real numbers whose distance from 2 is less than 5. As shown in Figure 2.17, the solution set is the interval $(-3, 7)$. We can also find the solution using property (a) above. Let $a = x - 2$ and $b = 5$, so that $|x - 2| < 5$ if and only if

$$-5 < x - 2 < 5.$$

Adding 2 to each portion of this inequality produces

$$-3 < x < 7,$$

again giving the interval solution $(-3, 7)$.   ◻

Figure 2.17

Figure 2.18

EXAMPLE 5   Solve $|x - 8| \geq 1$.

We need to find all numbers whose distance from 8 is greater than or equal to 1. As shown in Figure 2.18, the solution set is $(-\infty, 7] \cup [9, +\infty)$. To find the solution using property (b) above, let $a = x - 8$ and $b = 1$ so that $|x - 8| \geq 1$ if and only if

$$x - 8 \leq -1 \quad \text{or} \quad x - 8 \geq 1.$$

Solve each inequality separately to get the same solution set, $(-\infty, 7] \cup [9, +\infty)$, as mentioned above.   ◻

EXAMPLE 6   Solve $|2 - 7m| - 1 > 4$.

In order to use the properties of absolute value given above, first add 1 to both sides; this gives

$$|2 - 7m| > 5.$$

Now use property (b) given above. By this property, $|2 - 7m| > 5$ if and only if

$$2 - 7m < -5 \quad \text{or} \quad 2 - 7m > 5.$$

Solve each of these inequalities separately to get the solution set $(-\infty, -3/7) \cup (1, +\infty)$.   ◻

EXAMPLE ☐ Solve $|2 - 5x| \geq -4$.
7

The absolute value of a number is always nonnegative. Thus $|2 - 5x| \geq -4$ is always true, and the solution set includes all real numbers. In interval notation, the solution set is $(-\infty, +\infty)$. ☐

Now let us see how to solve absolute value inequalities with variable denominators. The following examples show a method for solving this kind of inequality.

EXAMPLE ☐ Solve $\left|\dfrac{3k + 1}{k - 1}\right| < 2$.
8

By property (a) of absolute values, this inequality is satisfied if and only if

$$-2 < \frac{3k + 1}{k - 1} < 2.$$

As we have seen, this last inequality is equivalent to

$$-2 < \frac{3k + 1}{k - 1} \quad \text{and} \quad \frac{3k + 1}{k - 1} < 2. \tag{*}$$

Now we must solve each of these inequalities separately. Start by rewriting the inequality on the left; this produces the following sequence of equivalent inequalities.

$$\frac{3k + 1}{k - 1} > -2$$

$$\frac{3k + 1}{k - 1} + 2 > 0$$

$$\frac{3k + 1 + 2(k - 1)}{k - 1} > 0$$

$$\frac{5k - 1}{k - 1} > 0$$

By the methods presented in the previous section, the solution set of this inequality is

$$(-\infty, 1/5) \cup (1, +\infty).$$

Solving the second inequality from (*) above gives

$$\frac{3k + 1}{k - 1} < 2$$

$$\frac{3k + 1}{k - 1} - 2 < 0$$

$$\frac{3k + 1 - 2(k - 1)}{k - 1} < 0$$

$$\frac{k + 3}{k - 1} < 0.$$

Again using the methods of the previous section, the solution set here is the interval $(-3, 1)$.

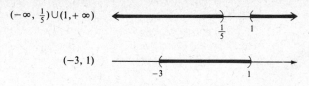

Figure 2.19

In (*) above, we said that our given inequality is true if and only if *both* the separate inequalities are true. To find where both are true, draw graphs of the separate solution sets, as in Figure 2.19.

By inspecting these graphs, we can see that the only values that make both inequalities true at the same time are those in the interval $(-3, 1/5)$. So this interval, shown in Figure 2.20, is the solution set of our original inequality.   □

Figure 2.20

EXAMPLE 9   □   Solve $\left| \dfrac{y}{2y + 1} \right| > 3.$

By property (b) above, this inequality will be true if and only if

$$\frac{y}{2y + 1} < -3 \quad \text{or} \quad \frac{y}{2y + 1} > 3.$$

Again, we solve each inequality separately. Starting with the one on the left, we get

$$\frac{y}{2y + 1} + 3 < 0$$

$$\frac{y + 3(2y + 1)}{2y + 1} < 0$$

$$\frac{7y + 3}{2y + 1} < 0.$$

The solution set here can be shown to equal $(-1/2, -3/7)$. Now solve the inequality on the right above, getting

$$\frac{y}{2y + 1} > 3$$

$$\frac{y}{2y + 1} - 3 > 0$$

$$\frac{y - 3(2y + 1)}{2y + 1} > 0$$

$$\frac{-5y - 3}{2y + 1} > 0$$

The solution set here is $(-3/5, -1/2)$. In the first step of this example, we said that the given inequality is true if either of our separate inequalities is true. Thus, the solution of the given inequality is the union of the solution sets of the separate inequalities. As shown in Figure 2.21, this union is

$$(-3/5, -1/2) \cup (-1/2, -3/7).$$

This solution set is graphed in Figure 2.22. ☐

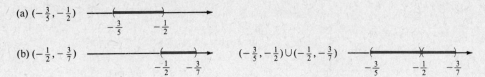

Figure 2.21   Figure 2.22

EXERCISES 2.6

Solve each of the following equations.

1. $|3m - 1| = 2$
2. $|4p + 2| = 5$
3. $|5 - 3x| = 3$
4. $|-2 + 5a| = 3$
5. $\left|\dfrac{a - 4}{2}\right| = 5$
6. $\left|\dfrac{2}{m + 2}\right| = 7$
7. $\left|\dfrac{5}{r - 3}\right| = 10$
8. $\left|\dfrac{2h - 1}{5h}\right| = 4$
9. $|2k - 3| = |5k + 4|$
10. $|p + 1| = |3p - 1|$
11. $|4 - 3y| = |7 + 2y|$
12. $|2 + 5a| = |4 - 6a|$
13. $|x + 2| = |x - 1|$
14. $|y - 5| = |y + 3|$

Solve each of the following inequalities. Write the solutions in interval notation.

15. $|x| \le 3$
16. $|y| \le 10$
17. $|m| > 1$
18. $|z| > 3$
19. $|x| - 3 \le 7$
20. $|r| + 3 \le 10$

21. $|3m - 2| > 4$
22. $|2x + 5| < 3$
23. $|5 - 3y| \le 1$
24. $|6 - 4x| > 10$
25. $|4z + 6| \ge 7$
26. $|8b + 5| \ge 7$
27. $|2m - 5| + 2 > 3$
28. $|4 - 3k| - 3 > 2$
29. $|5x + 1/2| - 2 < 5$
30. $|x + 2/3| + 1 < 4$
31. $\left|\dfrac{2x + 3}{x}\right| < 1$
32. $\left|\dfrac{7 - 4y}{y}\right| < 3$
33. $\left|\dfrac{6 + 2y}{y - 5}\right| > 2$
34. $\left|\dfrac{3 - 3p}{p + 4}\right| > 5$
35. $\left|\dfrac{2}{q - 2}\right| \le 4$
36. $\left|\dfrac{5}{t - 1}\right| \le 3$
37. $\left|\dfrac{x - 1}{x - 2}\right| \ge 3$
38. $\left|\dfrac{2r + 1}{r}\right| \ge 5$
39. $\left|\dfrac{3z - 5}{2z}\right| > 4$
40. $\left|\dfrac{m + 4}{3m}\right| \ge 2$
41. $\left|\dfrac{6y - 5/2}{2y - 1}\right| > 1$
42. $\left|\dfrac{2k + 1}{3 - k}\right| < \dfrac{5}{2}$

Write each of the following, using absolute value statements.

43. x is within 4 units of 2

44. m is no more than 8 units from 9

45. z is no less than 2 units from 12

46. p is at least 5 units from 9

47. k is 6 units from 1

48. r is 5 units from 3

49. If x is within .0004 units of 2, then y is within .00001 units of 7.

50. y is within 10^{-6} units of 10 whenever x is within 2×10^{-4} units of 5.

51. If $|x - 2| < 3$, find the values of m and n such that $m < 3x + 5 < n$.

52. If $|x + 8| < 16$, find the values of p and q so that $p < 2x - 1 < q$.

53. Solve $|m + 2| \le |4m - 1|$ by first dividing each side by $|4m - 1|$.

54. Solve $|3x - 1| < 2|2x + 1|$.

CHAPTER SUMMARY
2

Key Words

equation	linear equation	discriminant
solve	quadratic equation	quadratic in form
satisfy	standard form	inequality
solution set	square root property	quadratic inequality
identity	completing the square	sign graph
conditional equations	quadratic formula	union
equivalent equations		

CHAPTER REVIEW EXERCISES
2

Solve each of the following equations.

1. $2m + 7 = 3m + 1$

2. $4k - 2(k - 1) = 12$

3. $5y - 2(y + 4) = 3(2y + 1)$

4. $\dfrac{x - 3}{2} = \dfrac{2x + 1}{3}$

5. $\dfrac{p}{2} - \dfrac{3p}{4} = 8 + \dfrac{p}{3}$

6. $\dfrac{2r}{5} - \dfrac{r - 3}{10} = \dfrac{3r}{5}$

7. $\dfrac{2z}{5} - \dfrac{4z - 3}{10} = \dfrac{1 - z}{10}$

8. $\dfrac{p}{p + 2} - \dfrac{3}{4} = \dfrac{2}{p + 2}$

9. $(x - 3)(2x + 1) = 2(x + 2)(x - 4)$

10. $(3k + 1)^2 = 6k^2 + 3(k - 1)^2$

Solve for x.

11. $3(x + 2b) + a = 2x - 6$

12. $9x - 11(k + p) = x(a - 1)$

13. $\dfrac{x}{m - 2} = kx - 3$ **14.** $r^2x - 5x = 3r^2$

Solve each of the following for the indicated variable.

15. $2a + ay = 4y - 4a$ for y

16. $\dfrac{3m}{m - x} = 2m + x$ for x

17. $F = \dfrac{9}{5}C + 32$ for C

18. $A = P + Pi$ for P

19. $A = I\left(1 - \dfrac{j}{n}\right)$ for j

20. $A = \dfrac{24f}{b(p + 1)}$ for f

21. $\dfrac{1}{k} = \dfrac{1}{r_1} + \dfrac{1}{r_2}$ for r_1

22. $m = \dfrac{Ft}{\sqrt{I} - \sqrt{2}}$ for t

23. $V = \pi r^2 L$ for L

24. $P(r + R)^2 = E^2 R$ for P

25. $\dfrac{xy^2 - 5xy + 4}{3x} = 2p$ for x

26. $\dfrac{zx^2 - 5x + z}{z + 1} = 9$ for z

Solve each of the following problems.

27. A stereo is on sale for 15% off. The sale price is $425. What was the original price?

28. To make a special mix for Valentine's Day, the owner of a candy store wants to combine chocolate hearts which sell for $5 per pound with candy kisses which sell for $3.50 per pound. How many pounds of each should be used to get 30 pounds of a mix which can be sold for $4.50 per pound?

29. Two people are stuffing envelopes for a political campaign. Working together, they can stuff 5000 envelopes in 4 hours. If the first person worked alone, it would take 6 hours to stuff the envelopes. How long would it take the second person, working alone, to stuff the envelopes?

30. Maria can ride her bike to the university library in 20 minutes. The trip home, which is all uphill, takes her half an hour. If her rate is 8

mph slower on the return trip, how far does she live from the library?

Solve each equation.

31. $(b + 7)^2 = 5$

32. $(3y - 2)^2 = 8$

33. $2a^2 + a - 15 = 0$

34. $12x^2 = 8x - 1$

35. $2q^2 - 11q = 21$

36. $3x^2 + 2x = 16$

37. $2 - \dfrac{5}{p} = \dfrac{3}{p^2}$

38. $\dfrac{4}{m^2} = 2 + \dfrac{7}{m}$

39. $ix^2 - 4x + i = 0$

40. $4p^2 - ip + 2 = 0$

Evaluate the discriminant for each of the following, and use it to predict the type of solutions for the equation.

41. $8y^2 = 2y - 6$

42. $6k^2 - 2k = 3$

43. $16r^2 + 3 = 26r$

44. $8p^2 + 10p = 7$

45. $25z^2 - 110z + 121 = 0$

46. $4y^2 - 8y + 17 = 0$

Solve each word problem.

47. Calvin wants to fence off a rectangular playground next to an apartment building. Since the building forms one boundary, he needs to fence only the other three sides. The area of the playground is to be 11,250 square meters. He has enough material to build 325 meters of fence. Find the length and width of the playground.

48. Steve and Paula sell pies. It takes Paula one hour longer than Steve to bake a day's supply of pies. Working together, it takes them 6/5 hour to bake the pies. How long would it take Steve working alone?

Solve each equation.

49. $4a^4 + 3a^2 - 1 = 0$

50. $2x^4 = x^2$

51. $(r + 1)^2 - 3(r + 1) = 4$

52. $4(y - 2)^2 - 9(y - 2) + 2 = 0$

53. $(2z + 3)^{2/3} + (2z + 3)^{1/3} = 6$

54. $\sqrt{x} - 7 = 10$

55. $\sqrt{2p + 1} = 8$

56. $5\sqrt{m} = \sqrt{3m + 2}$

57. $\sqrt{4y - 2} = \sqrt{3y + 1}$

58. $\sqrt{2x + 3} = x + 2$

59. $\sqrt{p + 2} = 2 + p$

60. $\sqrt{k} = \sqrt{k + 3} - 1$

61. $\sqrt{x^2 + 3x} - 2 = 0$

62. $\sqrt[3]{2r} = \sqrt[3]{3r + 2}$

63. $\sqrt[3]{6y + 2} = \sqrt[3]{4y}$

64. $(x - 2)^{2/3} = x^{1/3}$

65. $\sqrt{3 + x} = \sqrt{3x + 7} - 2$

66. $\sqrt{4 + 3y} = \sqrt{y + 5} + 1$

67. $\sqrt{8 - a} + 1 = \sqrt{10 - 6a}$

68. $2\sqrt{6 - r} = 2 + \sqrt{7 - 3r}$

Solve each of the following inequalities. Write solutions in interval notation.

69. $-9x < 4x + 7$

70. $11y \geq 2y - 8$

71. $-5z - 4 \geq 3(2z - 5)$

72. $-(4a + 5) < 3a - 2$

73. $3r - 4 + r > 2(r - 1)$

74. $7p - 2(p - 3) \leq 5(2 - p)$

75. $5 \leq 2x - 3 \leq 7$

76. $-8 < 3a - 5 < -1$

77. $-5 < \dfrac{2p - 1}{-3} \leq 2$

78. $3 < \dfrac{6z + 5}{-2} < 7$

79. $x^2 + 3x - 4 \leq 0$ **80.** $p^2 + 4p > 21$

81. $6m^2 - 11m < 10$ **82.** $k^2 - 3k - 5 \geq 0$

83. $z^3 - 16z \leq 0$

84. $2r^3 - 3r^2 - 5r < 0$

85. $\dfrac{3a - 2}{a} > 4$ **86.** $\dfrac{5p + 2}{p} < -1$

87. $\dfrac{3}{r - 1} \leq \dfrac{5}{r + 3}$ **88.** $\dfrac{3}{x + 2} > \dfrac{2}{x - 4}$

Work the following word problems.

89. Steve and Paula (from Exercise 48 above) have found that the profit from their pie shop is given by

$$P = -x^2 + 28x + 60,$$

where x is the number of units of pies sold daily. For what values of x is the profit positive?

90. A projectile is thrown upward. Its height in feet above the ground after t seconds is $320t - 16t^2$. (a) After how many seconds in the air will it hit the ground? (b) During what time interval is the projectile more than 576 feet above the ground?

Solve each equation.

91. $|a + 4| = 7$ **92.** $|3 - 2m| = 5$

93. $|2 - y| = 3$ **94.** $\left|\dfrac{r - 5}{3}\right| = 6$

95. $\left|\dfrac{7}{2 - 3a}\right| = 9$ **96.** $\left|\dfrac{8r - 1}{3r + 2}\right| = 7$

97. $|5r - 1| = |2r + 3|$

98. $|k + 7| = |k - 8|$

Solve each inequality. Write solutions in interval notation.

99. $|m| \leq 7$ **100.** $|z| > -1$

101. $|b| \leq -1$ **102.** $|5m - 8| \leq 2$

103. $|7k - 3| < 5$ **104.** $|2p - 1| > 2$

105. $|3r + 7| > 5$ **106.** $\left|\dfrac{1}{k + 3}\right| > 3$

107. $\left|\dfrac{3r}{r - 1}\right| \geq 3$ **108.** $\left|\dfrac{2p - 1}{p + 2}\right| \leq 1$

109. Let

$$f = 2x\left[\dfrac{1}{2}(x^2 + 1)^{-1/2}(2x)\right] + 2(x^2 + 1)^{1/2}.$$

Find all intervals where (a) $f > 0$; and (b) $f < 0$.

110. Let

$$g = \dfrac{(x^2 + 5)^{1/2} - x[(1/2)(x^2 + 5)^{-1/2}(2x)]}{x^2 + 5}$$

Find all intervals where (a) $g > 0$; and (b) $g < 0$.

Set up Exercises 111 and 112; do not *solve them.*

111. A book is to contain 36 square inches of printed material per page, with margins of 1 inch along the sides, and $1\frac{1}{2}$ inches along the top and bottom. Let x represent the width of the printed area, and write an expression for the area of the entire page.

112. A hunter is at a point on a riverbank. He wants to get to his cabin, located 3 miles north and 8 miles west (see the figure). He can travel 5 miles per hour on the river but only 2 miles per hour on this very rocky ground. If he travels x miles along the river and then walks in a straight line to the cabin, find an expression for the total time that he travels.

113. If $y = 2x + |2 - x|$, express x in terms of y.

114. Show that $(s + t + |s - t|)/2$ equals the larger of s and t.

115. Show that $(s + t - |s - t|)/2$ equals the smaller of s and t.

116. (a) Prove that

$$|A - B| \le |A - W| + |W - B|$$

for all real numbers A, B, and W. (b) Describe those situations in which the preceding "less-than-or-equal-to" statement is actually an equality.

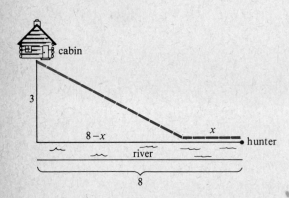

Cumulative Review Exercises, Chapters 1 and 2

Choose all words from the following list that describe the number given.

  (a) whole number   *(b) integer*
  (c) rational number   *(d) irrational number*
  (e) real number

1. $-11/3$ **2.** -5

3. $\sqrt{49}$ **4.** $\sqrt{3}$

Which of the following sets are closed with respect to the indicated operations?

5. Integers, subtraction

6. Integers, division

7. Odd integers, addition

8. Even integers, multiplication

Write the following numbers in numerical order, from smallest to largest.

9. $-|-7|, |-3|, -|2|$

10. $|-14|, -|-1|, |0|, -|-4|$

11. $|-8 + 2|, -|3|, -|-2|,$
    $-|-2| + (-3), -|-8| - |-6|$

12. $-2 - |-4|, -3 + |2|, -4, -5 + |-3|$

Write each of the following without absolute value bars.

13. $-|0|$

14. $|-5| - (-2)$

15. $|7 - \sqrt{5}|$

16. $|2 - \sqrt{11}|$

17. $|m - 3y|$, if $m/3 > y$

18. $|5 + y^2|$

Simplify each of the following. Write the results with only positive exponents. Assume all variables represent positive real numbers.

19. $(-5)^{-3}$

20. $2^{-4} + 3^{-2}$

21. $(r^{-3})(r^{-2})(r^5)$

22. $\dfrac{p^3 z^2}{p^{-1} z^{-3}}$

23. $[(3^{-2})^2]^{-1}$

24. $\dfrac{(2x^3)^{-2}(2^2 x^5)^{-1}}{(2x^4)^{-3}}$

25. $\dfrac{(5p^2 q)^{-1}(5p^3 q^{-2})^2}{5(pq)^{-3}(p^4 q^{-2})^{-1}}$

26. $\dfrac{.0000015 \times 40{,}000{,}000{,}000}{.00000000000003}$

Find each of the following.

27. $(-9m^2 + 11m - 2) + (4m^2 - 8m + 7)$

28. $(-7z^2 + 8z - 1) - (-4z^2 - 7z - 9)$

29. $(k - 7)(3k - 8)$

30. $(9w + 5)^2$

31. $(3k - 5)^3$

32. $(4k + 3)(2k^2 + 5k + 6)$

33. $(y + z + 2)(3y - 2z + 5)$

34. $(2r + 3s + 3)(3r - s + 2)$

35. $\dfrac{15x^4 + 30x^3 + 12x^2 - 9}{3x}$

36. $\dfrac{16a^6 + 24a^5 - 48a^4 + 12a}{8a^2}$

Factor as completely as possible.

37. $3m^3 + 9m + 15m^5$

38. $r^2 + 15r + 54$

39. $6q^2 - q - 12$

40. $10b^2 - 19b - 15$

41. $8a^3 + 125$

42. $64p^6 - 27q^9$

43. $rs + rt - ps - pt$

44. $2m + 6 - am - 3a$

45. $(z - 4)^2 - (z + 4)^2$

46. $6(r + s)^2 + 13(r + s) - 5$

Perform each of the following operations.

47. $\dfrac{2x - 2}{3} \cdot \dfrac{6x - 6}{(x - 1)^3}$

48. $\dfrac{3m - 15}{4m - 20} \cdot \dfrac{m^2 - 10m + 25}{12m - 60}$

49. $\dfrac{3z^2 + z - 2}{4z^2 - z - 5} \div \dfrac{3z^2 + 11z + 6}{4z^2 + 7z - 15}$

50. $\dfrac{1}{a + 1} - \dfrac{1}{a - 1}$

51. $\dfrac{m + 3}{m - 7} + \dfrac{m + 5}{2m - 14}$

52. $\dfrac{1}{x^2 + x - 12} - \dfrac{1}{x^2 - 7x + 12}$

53. $\dfrac{m^{-1} - n^{-1}}{(mn)^{-1}}$

54. $\dfrac{z^{-1} - y^{-1}}{z^{-1} + y^{-1}} \cdot \dfrac{z + y}{z - y}$

Simplify each of the following. Assume that all variables represent positive real numbers.

55. $\sqrt{1000}$

56. $\sqrt[3]{54}$

57. $-\sqrt[4]{32}$

58. $\sqrt{24 \cdot 3^2 \cdot 2^4}$

59. $\sqrt[3]{25 \cdot 3^4 \cdot 5^3}$

60. $\sqrt{50p^7 q^8}$

61. $\sqrt[4]{1875 h^5 y^6 q^9}$

62. $\dfrac{\sqrt[3]{a^3 b^7 c^7} \cdot \sqrt[3]{a^6 b^8 c^9}}{\sqrt[3]{a^7 b^3 c^5}}$

63. $\sqrt{8} + 5\sqrt{32} - 7\sqrt{128}$

64. $\dfrac{15}{\sqrt{3}} - \dfrac{2}{\sqrt{27}} + \dfrac{4}{\sqrt{12}}$

65. $\dfrac{1}{2 - \sqrt{7}}$

66. $\dfrac{\sqrt{p} + \sqrt{p+1}}{\sqrt{p} - \sqrt{p+1}}$

Rewrite each of the following, using only positive exponents. Assume that all variables represent positive real numbers, and that variables used as exponents represent rational numbers.

67. $32^{-6/5}$ **68.** $(625z^8)^{1/2}$

69. $(a - b)^{2/3} \cdot (a - b)^{-5/3}$ $(a > b)$

70. $(7k^{3/4}x^{1/8})(9k^{7/4}x^{3/8})$

71. $(3a^{-2/3}b^{5/3})(8a^2b^{-10/3})$

72. $z^{1+r} \cdot z^{3-2r}$

73. $\dfrac{r^{1/3} \cdot s^{5/3} \cdot t^{1/2}}{r^{-2/3} \cdot s^2 \cdot t^{-3/2}}$ **74.** $\dfrac{q^r \cdot q^{-5r}}{q^{-3r}}$

Perform the following operations.

75. $(-3 + 5i) - (-9 + 3i)$

76. $(-6 + 2i) + (-1 + 7i)$

77. $(1 + 3i)(2 - 5i)$ **78.** $(-2 + 5i)^2$

79. $i(2 + 3i)^2$ **80.** $(1 - 2i)^3$

81. $\dfrac{4 + 3i}{1 + i}$ **82.** $\dfrac{3 + 7i}{5 - 3i}$

83. $\dfrac{i}{3 + 2i}$

84. $\dfrac{5 - 2i}{12 + i} - \dfrac{2 - 9i}{4 + 3i}$

Simplify each of the following.

85. $\sqrt{-400}$ **86.** $-\sqrt{-39}$

87. $\sqrt{-3} \cdot \sqrt{-7}$ **88.** $\sqrt{-11} \cdot \sqrt{-6}$

89. Find i^{15} **90.** Find i^{245}.

Solve each equation.

91. $5(a + 3) + 4a - 5 = -(2a - 4)$

92. $\dfrac{x}{3} - 7 = 6 - \dfrac{3x}{4}$

93. $\dfrac{-7r}{2} + \dfrac{3r - 5}{4} = \dfrac{2r + 5}{4}$

94. $\dfrac{1}{z - 5} = 2 - \dfrac{3}{z - 5}$

95. $(3x - 4)^2 - 5 = 3(x + 5)(3x + 2)$

96. $\dfrac{x}{b - 1} = 5x + 3b$ (Solve for x.)

Solve each of the following for the indicated variable.

97. $v = v_0 + gt$ for t

98. $s = s_0 + gt^2 + k$ for t^2

99. $A = \dfrac{1}{2}(b + B)h$ for B

100. $S = 2\pi(r_1 + r_2)h$ for r_2

Solve each of the following.

101. A triangle has a perimeter of 54 cm. Two of the sides of the triangle are equal in length, with the third side 6 cm shorter than either of the two equal sides. Find the lengths of the three sides of the triangle.

102. After a lottery win, John has $90,000 to invest. He puts part of the money in a certificate of deposit at 10%, with $10,000 less than this amount put into a real estate scheme paying 12%. The total annual income from the investments is $9800. How much does he have invested at each rate?

103. Suppose $40,000 is invested at 7%. How much additional money would have to be invested at 11% to make the yield on the entire amount equal to 9.4%?

104. A student needs 25% acid for an experiment. How many ml of 50% acid should be mixed with 40 ml of 15% acid to get the necessary 25% acid?

105. How many pounds of coffee worth $6 per pound should be mixed with 20 pounds of coffee selling for $4.50 per pound to get a mixture that can be sold for $5 per pound?

106. Tom can run 6 miles per hour, while Roy runs 4 miles per hour. If they start running at the same time, how long would it take them to be 3/4 miles apart?

107. A boat can go 15 km upstream in the same time that it takes to go 27 km downstream. The speed of the current is 2 km per hour. Find the speed of the boat in still water.

108. An inlet pipe can fill a swimming pool in 1 day. An outlet can empty the pool in 36 hours. Suppose that both the inlet and the outlet are opened. How long would it take to fill the pool 5/8 full?

Solve each equation.

109. $(5r - 3)^2 = 7$ **110.** $(7q + 2)^2 = 40$

111. $8k^2 + 14k + 3 = 0$

112. $2s^2 + 2s = 3$

113. $4z^2 - 4z - 5 = 0$

114. $12r^2 = 4r$

115. $x^2 - 2x + 2 = 0$

116. $9k^2 - 12k + 8 = 0$

117. $12y^2 - 4y + 3 = 0$

118. $25z^2 + 30z + 11 = 0$

119. $im^2 + 2m + i = 0$

120. $w^2 - 7iw + 4 = 0$

Solve for the indicated variable.

121. $Z = 10a^2yb$ for a

122. $9k = 12qw + 5w^2$ for w

Solve each of the following problems.

123. Find two consecutive odd integers whose product is -1.

124. The area of a field is 9600 m². One side is 40 m longer than the other side. Find the length and width of the field.

125. Person A can do a job in 5 hours. Working with person B, the job takes 3 hours. How long would it take B working alone to do the job?

126. One leg of a right triangle is 3 cm longer than three times the length of the shorter leg. The hypotenuse is 1 cm longer than the longer leg. Find the lengths of the sides of the triangle.

Solve each equation.

127. $2z^4 - 7z^2 + 3 = 0$

128. $-(r + 1)^2 - 3(r + 1) + 3 = 0$

129. $(m - 1)^{2/3} + 3(m - 1)^{1/3} = 10$

130. $6z^{-2} - 7z^{-1} + 2 = 0$

131. $\sqrt{3s} - s = -6$

132. $\sqrt{r + 3} = \sqrt{2r - 1} - 1$

133. $(z^2 - 18z)^{1/4} = 0$

134. $p^{2/3} = 9p^{1/3}$

135. $(m^2 - 1)^2 - 5m^2 = 1$

136. $(y^2 + 2)^2 + y^2 = 10$

Solve each inequality. Write all solutions in interval notation.

137. $12m - 17 \geq 8m + 7$

138. $-h \leq 6h + 30$

139. $-15 < -2y + 3 < -1$

140. $z^2 + 6z + 16 < 8$

141. $y^2 + 6y \geq 0$

142. $p^2 < -1$

143. $2t^3 - 2t^2 - 12t \leq 0$

144. $\dfrac{a - 6}{a + 2} < -1$

145. $\dfrac{3}{y + 6} \geq \dfrac{1}{y - 2}$ **146.** $|z| > 8$

147. $|b| < 3$ **148.** $|x - 1/2| < 2$

149. $|2x + 5| > 3$ **150.** $|3 - 5k| > 2$

151. $\left|\dfrac{3x + 1}{x - 1}\right| \leq 2$ **152.** $\left|\dfrac{m + 4}{m - 3}\right| > 2$

Solve each of the following equations.

153. $|a - 2| = 1$ **154.** $\left|\dfrac{6y + 1}{y - 1}\right| = 3$

155. $|3z - 1| = |2z + 5|$

156. $|m + 2| = |m + 5|$

3 Functions and Graphs

The graph on the left in Figure 3.1 is from *Road and Track* magazine. The graph shows the speed in miles per hour at time t in seconds for a BMW 528e as it accelerates from rest. For example, by reading the graph we find that 15 seconds after starting, the car is going 80 miles per hour with a speed of 100 miles per hour after 18 seconds. The graph on the right in Figure 3.1 shows the variation in blood pressure for a typical person. (Systolic and diastolic pressures are the upper and lower limits in the periodic changes in pressure that produce the pulse. The length of time between peaks is called the period of the pulse.)

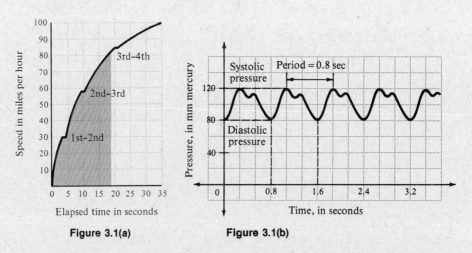

Figure 3.1(a)   **Figure 3.1(b)**

Both the graphs of Figure 3.1 are examples of **functions:** a rule or procedure giving just one value of one variable from a given value of the other variable. In each of these examples, a given value of time can be used to find just one value of the other variable, speed or blood pressure, respectively. Before we get to a complete discussion of functions, later in this chapter, we need to look at the topic of graphing, since functions are often studied by looking at their graphs.

3.1 A Two-Dimensional Coordinate System

In Chapter 1 we saw how each real number corresponds to a point on a number line. We set up this correspondence by establishing a coordinate system for the line. This idea can be extended to the two dimensions of a plane. The customary way to do this is by drawing two perpendicular lines, one horizontal and one vertical. These lines intersect at a point O called the **origin.** The horizontal line is often called the **x-axis,** and the vertical line is often called the **y-axis.**

Starting at the origin, we can make the x-axis into a number line by placing positive numbers to the right and negative numbers to the left. The y-axis can be made into a number line with positive numbers going up and negative numbers going down.

The x-axis and y-axis set up a **rectangular coordinate system,** or **Cartesian coordinate system** (named for one of its co-inventors René Descartes; the other co-inventor was Pierre de Fermat.) The plane into which the coordinate system is introduced is the **coordinate plane,** or **xy-plane.** The x-axis and y-axis divide the plane into four regions, or **quadrants,** labeled as shown in Figure 3.2. Points on the x-axis and y-axis themselves belong to no quadrant.

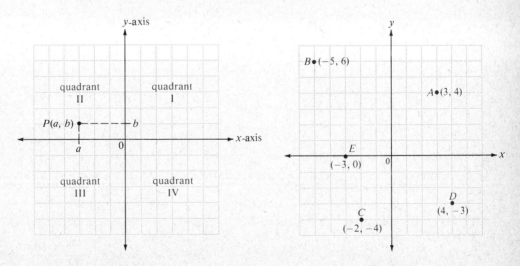

Figure 3.2

Figure 3.3

How can a given point P in the plane be labeled? As shown in Figure 3.2, start at P and draw a vertical line cutting the x-axis at a. Draw a horizontal line cutting the y-axis at b. Then point P has **coordinates** (a, b), where (a, b) is an *ordered pair* of numbers. An **ordered pair** of numbers consists of two numbers, written in parentheses, in which the sequence of the numbers is important. For example, $(4, 2)$ and $(2, 4)$ are not the same ordered pair since the sequence of the numbers is different. **Equality** for ordered pairs is defined as follows.

Equality for Ordered Pairs

$(a, b) = (c, d)$ if and only if $a = c$ and $b = d$

Note the similarity between this definition and the one for complex numbers given earlier.

A symbol such as $(3, 4)$ has now been used for two different purposes—as an interval on the number line and as an ordered pair of numbers. In virtually every case, however, it is easy to tell which use is intended from the context of the discussion.

To locate the point on the xy-plane corresponding to the ordered pair $(3, 4)$, draw a vertical line through 3 on the x-axis and a horizontal line through 4 on the y-axis. These two lines would cross at point A of Figure 3.3. This point A corresponds to the ordered pair $(3, 4)$. Also in Figure 3.3, B corresponds to the ordered pair $(-5, 6)$, C to $(-2, -4)$, D to $(4, -3)$, and E to $(-3, 0)$. Point O corresponds to the ordered pair $(0, 0)$.

Now we consider finding the distance between two points in a plane. To obtain a formula for this distance, let us start with two points on a horizontal line (see Figure 3.4(a)). Use the symbol $P(x_1, y_1)$ to represent point P, having coordinates (x_1, y_1). The distance between points $P(x_1, y_1)$ and $Q(x_2, y_1)$ can be found by subtracting the x-coordinates. (We use absolute value to make sure that the distance is not negative—recall our work with distance in Chapter 1.) The distance between points P and Q is

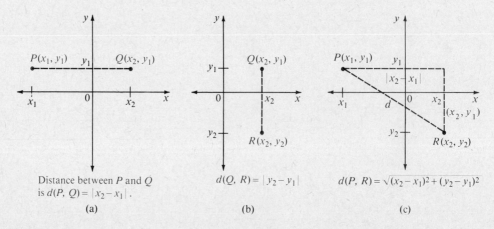

Distance between P and Q is $d(P, Q) = |x_2 - x_1|$.

(a)

$d(Q, R) = |y_2 - y_1|$

(b)

$d(P, R) = \sqrt{(x_2 - x_1)^2 + (y_2 - y_1)^2}$

(c)

Figure 3.4

thus $|x_1 - x_2|$. If we use $d(P, Q)$ to represent the distance between P and Q, then

$$d(P, Q) = |x_1 - x_2|.$$

Figure 3.4(b) shows points $Q(x_2, y_1)$ and $R(x_2, y_2)$ which are on a vertical line. To find the distance between Q and R, subtract the y-coordinates, finding

$$d(Q, R) = |y_1 - y_2|.$$

Finally, Figure 3.4(c) shows two points, $P(x_1, y_1)$ and $R(x_2, y_2)$, which are *not* on a horizontal or vertical line. To find $d(P, R)$, construct the right triangle shown in the figure. One side of this triangle is horizontal and has length $|x_1 - x_2|$. The other side is vertical and has length $|y_1 - y_2|$. By the Pythagorean theorem,

$$[d(P, R)]^2 = |x_1 - x_2|^2 + |y_1 - y_2|^2.$$

Since $d(P, R)$ must be nonnegative and since $|x_1 - x_2|^2 = (x_1 - x_2)^2$ and $|y_1 - y_2|^2 = (y_1 - y_2)^2$, we have the following result, called the **distance formula.**

Distance Formula

Suppose $P(x_1, y_1)$ and $R(x_2, y_2)$ are two points in a coordinate plane. Then the distance between P and R, written $d(P, R)$, is

$$d(P, R) = \sqrt{(x_1 - x_2)^2 + (y_1 - y_2)^2}.$$

Though the proof of the distance formula assumes that P and R are not on a horizontal or vertical line, the result is true for *any* choice of two points.

EXAMPLE 1 ☐ Find the distance between the following pairs of points.

(a) $P(-8, 4)$ and $Q(3, -2)$

Using the distance formula we get

$$d(P, Q) = \sqrt{(-8 - 3)^2 + [4 - (-2)]^2}$$
$$= \sqrt{(-11)^2 + 6^2}$$
$$= \sqrt{121 + 36} = \sqrt{157}.$$

Using a calculator we have $\sqrt{157} \approx 12.530$.

(b) $M(5, -11)$ and $N(-2, 13)$

Here $d(M, N) = \sqrt{[5 - (-2)]^2 + (-11 - 13)^2}$
$$= \sqrt{7^2 + (-24)^2}$$
$$= \sqrt{49 + 576} = \sqrt{625} = 25. \quad ☐$$

EXAMPLE 2 ☐ Are the points $M(-2, 5)$, $N(12, 3)$ and $Q(10, -11)$ the vertices of a right triangle?

To decide if this triangle is a right triangle, we can use the converse of the Pythagorean theorem: if the sides a, b, and c of a triangle satisfy $a^2 + b^2 = c^2$, then the triangle is a right triangle. A triangle with the three given points as vertices is shown in Figure 3.5. This triangle will be a right triangle if we can show the square of the length of the longest side equals the sum of the squares of the lengths of the other two sides. Using the distance formula, we obtain

$$d(M, N) = \sqrt{(-2 - 12)^2 + (5 - 3)^2} = \sqrt{196 + 4} = \sqrt{200}$$
$$d(M, Q) = \sqrt{(-2 - 10)^2 + [5 - (-11)]^2} = \sqrt{144 + 256} = \sqrt{400} = 20$$
$$d(N, Q) = \sqrt{(12 - 10)^2 + [3 - (-11)]^2} = \sqrt{4 + 196} = \sqrt{200}.$$

By these results,

$$[d(M, Q)]^2 = [d(M, N)]^2 + [d(N, Q)]^2,$$

proving the triangle is a right triangle with hypotenuse connecting M and Q. □

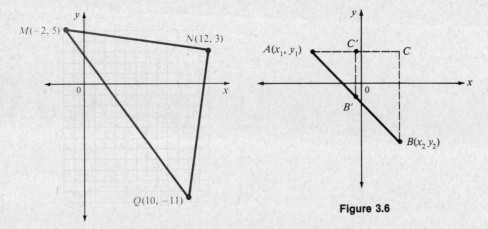

Figure 3.5

Figure 3.6

The distance formula is used to find the distance between any two points in a plane, while the **midpoint formula** is used to find the coordinates of the midpoint of a line segment.

To develop this formula, let $A(x_1, y_1)$ and $B(x_2, y_2)$ be two different points in a plane (see Figure 3.6). Assume that A and B are not on a horizontal or vertical line. Let C be the intersection of the horizontal line through A and the vertical line through B. Let B' (read "B-prime") be the midpoint of segment AB. Draw a line through B' and parallel to segment BC. Let C' be the point where this line cuts segment AC. If the coordinates of B' are (x', y'), then C' has coordinates (x', y_1). Since B' is the midpoint of AB, point C' must be the midpoint of segment AC (why?), and

$$d(C, C') = d(C', A),$$

or $|x_2 - x'| = |x' - x_1|.$

Because $d(C, C')$ and $d(C', A)$ must be positive the only solutions for this equation are found if

$$x_2 - x' = x' - x_1,$$

or $x_2 + x_1 = 2x'.$

Finally $x' = \dfrac{x_1 + x_2}{2}.$

Thus, the x-coordinate of the midpoint is the average of the x-coordinates of the endpoints of the segment. In a similar manner, the y-coordinate of the midpoint is $(y_1 + y_2)/2$, proving the following result.

Midpoint Formula

The midpoint of the line segment with endpoints (x_1, y_1) and (x_2, y_2) is

$$\left(\frac{x_1 + x_2}{2}, \frac{y_1 + y_2}{2} \right).$$

The above stated in words is: the coordinates of the midpoint of a segment are found by finding the *average* of the x-coordinates and the *average* of the y-coordinates of the endpoints of the segment.

EXAMPLE 3   Find the midpoint M of the segment with endpoints $(8, -4)$ and $(-9, 6)$. Use the midpoint formula to find that the coordinates of M are

$$\left(\frac{8 + (-9)}{2}, \frac{-4 + 6}{2} \right) = \left(-\frac{1}{2}, 1 \right). \quad \square$$

EXAMPLE 4   A line segment has an endpoint at $(2, -8)$ and a midpoint at $(-1, -3)$. Find the other endpoint of the segment.

The formula for the x-coordinate of the midpoint is $(x_1 + x_2)/2$. Here the x-coordinate of the midpoint is -1. We can let $x_1 = 2$, getting

$$-1 = \frac{2 + x_2}{2}$$

$$-2 = 2 + x_2$$

$$-4 = x_2.$$

In the same way, $y_2 = 2$ and the endpoint is $(-4, 2)$. $\quad \square$

The two formulas that we have derived in this section can be used to prove various properties from geometry.

EXAMPLE ☐ Prove that the diagonals of a parallelogram bisect each other.
5

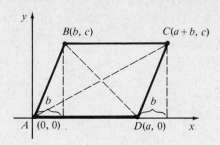

Figure 3.7

Figure 3.7 shows parallelogram $ABCD$ having diagonals AC and BD. The figure has been placed on a coordinate system with A at the origin and side AD along the x-axis. We assign coordinates (b, c) to B and $(a, 0)$ to D. Since DC is parallel to AB and is the same length as AB, we can use results from congruent triangles in geometry and write the coordinates of C as $(a + b, c)$. To show that the diagonals bisect each other, we show that they have the same midpoint. By the midpoint formula,

$$\text{midpoint of } AC = \left(\frac{a + b + 0}{2}, \frac{c + 0}{2}\right) = \left(\frac{a + b}{2}, \frac{c}{2}\right)$$

$$\text{midpoint of } BD = \left(\frac{a + b}{2}, \frac{c + 0}{2}\right) = \left(\frac{a + b}{2}, \frac{c}{2}\right).$$

Thus, AC and BD have the same midpoint and must bisect each other.   ☐

EXERCISES
3.1

Plot the following points in the xy-plane. Identify the quadrant for each.

1. $A(6, -5)$

2. $B(8, 3)$

3. $C(-4, 7)$

4. $D(-9, -8)$

5. $E(0, -5)$

6. $F(-8, 0)$

Graph the set of all points satisfying the following conditions for ordered pairs (x, y).

7. $x = 0$

8. $x > 0$

9. $y \leq 0$

10. $y = 0$

11. $xy < 0$

12. $\dfrac{x}{y} > 0$

13. $|x| = 4, y \geq 2$

14. $|y| = 3, x \geq 4$

15. $|y| < 2, x > 1$

16. $|x| < 3, y < -2$

17. $2 \leq |x| \leq 3, y \geq 2$

18. $1 \leq |y| \leq 4, x < 3$

Find the distance $d(P, Q)$ and the midpoint of segment PQ.

19. $P(5, 7), Q(13, -1)$

20. $P(-2, 5), Q(4, -3)$

21. $P(-8, -2), Q(-3, -5)$

22. $P(-6, -10), Q(6, 5)$

23. $P(\sqrt{2}, -\sqrt{5})$, $Q(3\sqrt{2}, 4\sqrt{5})$

24. $P(5\sqrt{7}, -\sqrt{3},)$, $Q(-\sqrt{7}, 8\sqrt{3})$

Give the distance between the following points rounded to the nearest thousandth. ⌐

25. (5, 7), (2, 14)   **26.** (−4, 6), (8, −5)

27. (3, −7), (−5, 19)

28. (−9, −2), (−1, −15)

Find the other endpoint of the segments having endpoints and midpoints as given.

29. endpoint (−3, 6), midpoint (5, 8)

30. endpoint (2, −8), midpoint (3, −5)

31. endpoint (6, −1), midpoint (−2, 5)

32. endpoint (−5, 3), midpoint (−7, 6)

Decide whether or not the following points are the vertices of a right triangle.

33. (−2, 5), (1, 5), (1, 9)

34. (−9, −2), (−1, −2), (−9, 11)

35. (−4, 0), (1, 3), (−6, −2)

36. (−8, 2), (5, −7), (3, −9)

37. $(\sqrt{3}, 2\sqrt{3} + 3)$, $(\sqrt{3} + 4, -\sqrt{3} + 3)$, $(2\sqrt{3}, 2\sqrt{3} + 4)$

38. $(4 - \sqrt{3}, -2\sqrt{3})$, $(2 - \sqrt{3}, -\sqrt{3})$, $(3 - \sqrt{3}, -2\sqrt{3})$

Use the distance formula to decide whether or not the following points lie on a straight line

39. (0, 7), (3, −5), (−2, 15)

40. (1, −4), (2, 1), (−1, −14)

41. (0, −9), (3, 7), (−2, −19)

42. (1, 3), (5, −12), (−1, 11)

Find all values of x or y such that the distance between the given points is as indicated.

43. (x, 7) and (2, 3) is 5

44. (5, y) and (8, −1) is 5

45. (3, y) and (−2, 9) is 12

46. (x, 11) and (5, −4) is 17

47. (x, x) and (2x, 0) is 4

48. (y, y) and (0, 4y) is 6

49. Show that the points (−2, 2), (13, 10), (21, −5), and (6, −13) are the vertices of a square.

50. Are the points $A(1, 1)$, $B(5, 2)$, $C(3, 4)$, $D(-1, 3)$ the vertices of a parallelogram? Of a rhombus (all sides equal in length)?

51. Use the distance formula and write an equation for all points that are 5 units from (0, 0). Sketch a graph showing these points.

52. Write an equation for all points 3 units from (−5, 6). Sketch a graph showing these points.

53. Find all points (x, y) with $x = y$ that are 4 units from (1, 3).

54. Find all points satisfying $x + y = 0$ that are 8 units from (−2, 3).

55. Write an equation for the points on the perpendicular bisector of the line segment with endpoints at (0, 0) and (−8, −10).

56. Let point A be (−3, 0) and point B be (3, 0). Write an expression for all points (x, y) such that the sum of the distances from A to (x, y) and from (x, y) to B is 8. Simplify the result so that no radicals are involved.

57. Let a be a positive number. Show that the distance between the points (ax_1, ay_1) and (ax_2, ay_2) is a times the distance between (x_1, y_1) and (x_2, y_2).

Use the midpoint formula and distance formula, as necessary, to prove each of the following.

58. The midpoint of the hypotenuse of a right triangle is equally distant from all three vertices.

59. The diagonals of a rectangle are equal in length.

60. The line segment connecting the midpoints of two adjacent sides of any quadrilateral is the same length as the line segment connecting the midpoints of the other two sides.

61. The diagonals of an isosceles trapezoid are equal.

62. If the diagonals of a parallelogram are equal in length, then the parallelogram is a rectangle.

3.2   Graphs

For any set of ordered pairs of real numbers, there is a corresponding set of points in a coordinate plane. If (x, y) is one of the ordered pairs in the set, then the corresponding point (x, y) can be located on the coordinate plane. This set of points is called the **graph** of the set of ordered pairs. For now, we shall find graphs by first identifying a reasonable number of ordered pairs. We locate the points corresponding to these ordered pairs on a coordinate plane and then try to decide on the shape of the entire graph. Later, we develop more useful methods of identifying particular graphs.

EXAMPLE 1   ☐   Draw the graph of $S = \{(x, y) | y = -4x + 3\}$.

As we have just said, the only way we now have of sketching this graph is by identifying various ordered pairs of S that can be obtained from the equation $y = -4x + 3$.

To find such ordered pairs, select a number of values for x (or y) and then find the corresponding values for the other variable. For example, if $x = -3$, then $y = -4(-3) + 3 = 15$, producing the ordered pair $(-3, 15)$. Typical ordered pairs such as this are given in the following table.

x	-3	-2	-1	0	1	2	3
y	15	11	7	3	-1	-5	-9
ordered pair	$(-3, 15)$	$(-2, 11)$	$(-1, 7)$	$(0, 3)$	$(1, -1)$	$(2, -5)$	$(3, -9)$

The ordered pairs from this table lead to the points that have been plotted in Figure 3.8(a). By studying these points, we would probably decide that the entire graph is a straight line, as drawn in Figure 3.8(b).   ☐

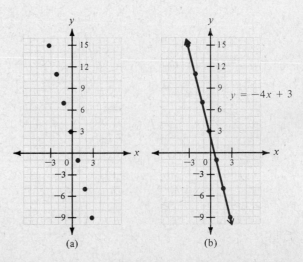

(a)   (b)

Figure 3.8

The set of ordered pairs in Example 1, $S = \{(x, y) | y = -4x + 3\}$ involves the equation $y = -4x + 3$. For each value of x that we might choose, we can use this equation to find a corresponding value of y. By using all possible such values of x, the equation $y = -4x + 3$ leads to a set of ordered pairs (a, b) such that $b = -4a + 3$. The ordered pairs (a, b) are called **solutions** of the equation $y = -4x + 3$. The set of all these solutions has a graph, called the **graph of the equation.** Thus, the graph of the equation $y = -4x + 3$ is the same as the graph of the set $\{(x, y) | y = -4x + 3\}$.

EXAMPLE 2 ☐ Graph the equation $y = x^2 + 2$.

Just as above, choose several values of x and find the corresponding values of y.

x	-4	-3	-2	-1	0	1	2	3	4
y	18	11	6	3	2	3	6	11	18

The ordered pairs obtained from this table were located in Figure 3.9(a), and a smooth curve was then drawn through the points as in Figure 3.9(b). This graph, called a **parabola,** will be studied in more detail later. The lowest point on this parabola, the point $(0, 2)$, is called the **vertex** of the parabola. In $y = x^2 + 2$, we may choose any value at all for x. However, for any real value of x, we have $x^2 \geq 0$, so that $x^2 + 2 \geq 2$. Hence, $y \geq 2$. ☐

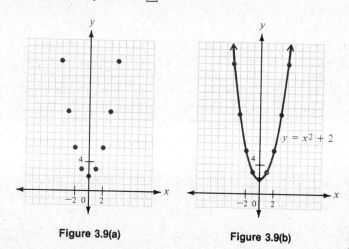

Figure 3.9(a)   **Figure 3.9(b)**

There is a danger in the method used here—we might chose a few values for x, find the corresponding values of y, begin to sketch a graph through these few points, but then make a completely wrong guess as to the shape of the graph. For example, if we had chosen only -1, 0, and 1 as values of x in Example 2 above, we would have only the three points $(-1, 3)$, $(0, 2)$, and $(1, 3)$. These three points would not at all be enough to give us sufficient information to decide on the proper graph for $y = x^2 + 2$. However, in this section we work only with elementary graphs, and when we get to more complicated graphs later, we shall develop more accurate methods of working with them.

EXAMPLE   ☐   Graph $x = y^2$.
3   Since y is squared, it is probably easier to choose values of y. If we choose the value 2 for y, we get $x = 2^2 = 4$. If we choose -2 for y, we get $x = (-2)^2 = 4$. The following table shows the results from choosing various values of y.

y	0	1	-1	2	-2	3	-3
x	0	1	1	4	4	9	9

The ordered pairs obtained from this table were used to get the points plotted in Figure 3.10. (Don't forget that x always goes first in the ordered pair.) A smooth curve was then drawn through the points. This curve is a parabola opening to the right with vertex (0, 0). Here, y can take on any value. Since $x = y^2$, we have $x \geq 0$.   ☐

EXAMPLE   ☐   Graph $y = |x|$.
4   Start with a table.

x	-4	-3	-2	-1	0	1	2	3	4
y	4	3	2	1	0	1	2	3	4

Use this table to get the points of Figure 3.11. The graph here is made up of portions of two straight lines. Here x may represent any real number while $y \geq 0$.   ☐

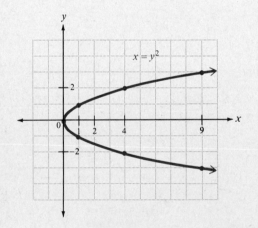

Figure 3.10

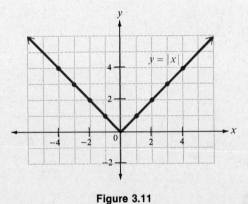

Figure 3.11

EXAMPLE   ☐   Graph $xy = -4$.
5   Make a table of values. Since this graph is more complicated than the ones above, it is a good idea to use more points. Also, neither x nor y can be equal to 0 so it is a good idea to make several choices for x that are close to 0.

x	-8	-4	-2	-1	$-\dfrac{1}{2}$	$-\dfrac{1}{4}$	$-\dfrac{1}{16}$	$-\dfrac{1}{64}$
y	$\dfrac{1}{2}$	1	2	4	8	16	64	256

x	$\dfrac{1}{64}$	$\dfrac{1}{16}$	$\dfrac{1}{4}$	$\dfrac{1}{2}$	1	2	4	8
y	-256	-64	-16	-8	-4	-2	-1	$-\dfrac{1}{2}$

As x approaches 0 from the left, y gets larger and larger. As x approaches 0 from the right, y gets more and more negative. If $x = 0$, there is no value of y; therefore the graph cannot cross the y-axis. Also, if $y = 0$, there is no value of x, so the graph cannot cross the x-axis either. The variables x and y can thus take on any values except 0; here $x \neq 0$ and $y \neq 0$. The table was used to get the points shown in Figure 3.12; a smooth curve was then drawn through them. ☐

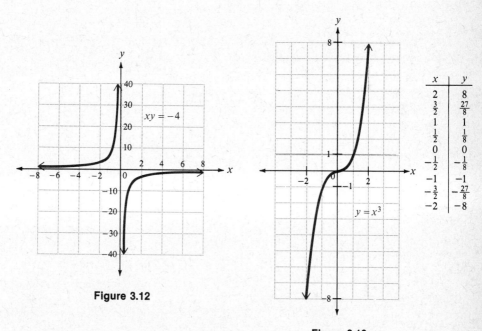

Figure 3.12

Figure 3.13

EXAMPLE 6 ☐ Graph $y = x^3$.

A table of values and the graph is shown in Figure 3.13. The variables x and y can take on any values at all. ☐

Circles

In the rest of this section we shall obtain a general equation for a circle. A **circle** is the set of points in a plane a fixed distance from a fixed point. The fixed distance is called the **radius,** and the fixed point is the **center.**

EXAMPLE 7 ☐ Find an equation for the circle having radius 6 and center at $(-3, 4)$.

This circle is shown in Figure 3.14. Its equation can be found by using the distance formula. Let (x, y) be any point on the circle. The distance from (x, y) to $(-3, 4)$ is given by

$$\sqrt{[x - (-3)]^2 + (y - 4)^2} = \sqrt{(x + 3)^2 + (y - 4)^2}.$$

Also, this same distance is given by the radius, 6. Therefore,

$$\sqrt{(x + 3)^2 + (y - 4)^2} = 6$$

or $(x + 3)^2 + (y - 4)^2 = 36.$

By studying this equation, the possible values of x are seen to be $-9 \le x \le 3$ while the possible values of y are $-2 \le y \le 10$.   ☐

Generalizing from the work of Example 7, the circle with center (h, k) and radius r has equation

**Circle–
Center-Radius
Form**

$$(x - h)^2 + (y - k)^2 = r^2.$$

This result is called the **center-radius form** of the equation of a circle. As a special case, a circle of radius r with center at the origin has equation

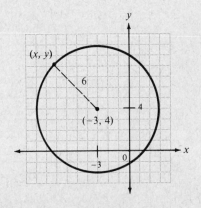

Figure 3.14

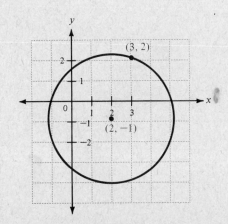

Figure 3.15

Circle–
Center at
Origin

$$x^2 + y^2 = r^2.$$

EXAMPLE 8   Find the equation of a circle that goes through the point (3, 2) and has its center at (2, -1).

To use the center-radius form of the equation of a circle, we need both the center and the radius of the circle. Here we know the center but not the radius. However, the radius, the distance from the center to any point on the circle, can be found by the distance formula. The distance between (2, -1) and (3, 2) gives the radius, r:

$$r = \sqrt{(2 - 3)^2 + (-1 - 2)^2} = \sqrt{1 + 9} = \sqrt{10}.$$

The equation of the circle with center at (2, -1) and going through (3, 2), shown in Figure 3.15, is thus

$$(x - 2)^2 + [y - (-1)]^2 = (\sqrt{10})^2,$$

or $(x - 2)^2 + (y + 1)^2 = 10.$ ☐

If we start with the center-radius form of the equation of a circle, $(x - h)^2 + (y - k)^2 = r^2$, and square $x - h$ and $y - k$, we get a result of the form

$$x^2 + y^2 + cx + dy + e = 0, \tag{*}$$

where c, d, and e are real numbers. Also, if we start with an equation of this form, we can use the process of **completing the square** to get an equation of the form

$$(x - h)^2 + (y - k)^2 = m$$

for some number m. If $m > 0$, then $r^2 = m$, and the graph is that of a circle. If $m = 0$, the graph is a single point, while there is no graph if $m < 0$.

EXAMPLE 9   Find the center and radius for $x^2 - 6x + y^2 + 10y + 25 = 0$.

Since this equation has the form of equation (*) above, it either represents a circle, a single point, or no points at all. To decide which, complete the square on both x and y as follows.

$$(x^2 - 6x + 9) + (y^2 + 10y + 25) = -25 + 9 + 25$$
$$(x - 3)^2 + (y + 5)^2 = 9$$

Since $9 > 0$, the equation represents a circle with center at (3, -5) and radius 3.   ☐

EXAMPLE 10   Graph $x^2 + 10x + y^2 - 4y + 33 = 0$.

Completing the square gives

$$(x^2 + 10x + 25) + (y^2 - 4y + 4) = -33 + 25 + 4$$
$$(x + 5)^2 + (y - 2)^2 = -4.$$

Since $-4 < 0$, there are no ordered pairs (x, y), with x and y real numbers, satisfying the equation. Thus, the graph has no points on it.   ☐

EXERCISES
3.2

Graph each of the following.

1. $y = 8x - 3$
2. $y = 2x + 7$
3. $y = 3x$
4. $y = -2x$
5. $3y + 4x = 12$
6. $5y - 3x = 15$
7. $y = 3x^2$
8. $y = 5x^2$
9. $y = -x^2$
10. $y = -2x^2$
11. $y = x^2 - 8$
12. $y = x^2 + 6$
13. $y = 4 - x^2$
14. $y = -2 - x^2$
15. $xy = -9$
16. $xy = 25$
17. $4x = y^2$
18. $9x = y^2$
19. $16y^2 = -x$
20. $4y^2 = -x$
21. $y^2 = x + 2$
22. $y^2 = -5 + x$
23. $y = x^3 - 3$
24. $y = x^3 + 4$
25. $y = 1 - x^3$
26. $y = -5 - x^3$
27. $2y = x^4$
28. $y = -x^4$
29. $y = \sqrt{x}$
30. $y = \sqrt{-x}$
31. $y = \sqrt{x + 5}$
32. $y = \sqrt{3 + x}$
33. $y = |x| + 4$
34. $y = |x| - 2$
35. $y = 8 - |x|$
36. $y = -3 - |x|$
37. $x^2 + y^2 = 36$
38. $x^2 + y^2 = 100$
39. $(x + 1)^2 + (y - 2)^2 = 25$
40. $(x - 3)^2 + (y + 7)^2 = 49$
41. $(x + 4)^2 + y^2 = 36$
42. $x^2 + (y - 2)^2 = 9$

Find equations for each of the following circles.

43. center $(1, 4)$, radius 3
44. center $(-2, 5)$, radius 4
45. center $(-8, 6)$, radius 5
46. center $(3, -2)$, radius 2
47. center $(-1, 2)$, passing through $(2, 6)$
48. center $(2, -7)$, passing through $(-2, -4)$
49. center $(-3, -2)$, tangent to the x-axis
50. center $(5, -1)$, tangent to the y-axis

Find the center and the radius of each of the following that are circles.

51. $x^2 + 6x + y^2 + 8y = -9$
52. $x^2 - 4x + y^2 + 12y + 4 = 0$
53. $x^2 - 12x + y^2 + 10y + 25 = 0$
54. $x^2 + 8x + y^2 - 6y = -16$
55. $x^2 + 8x + y^2 - 14y + 65 = 0$
56. $x^2 - 2x + y^2 + 1 = 0$
57. $x^2 + y^2 = 2y + 48$
58. $x^2 + 4x + y^2 = 21$
59. $x^2 - 2.84x + y^2 + 1.4y + 1.8664 = 0$
60. $x^2 + 7.4x + y^2 - 3.8y + 16.09 = 0$
61. Find the equation of the circle having the points $(3, -5)$ and $(-7, 2)$ as endpoints of a diameter.
62. Suppose a circle is tangent to both axes, has its center in the third quadrant, and has a radius of $\sqrt{2}$. Find its equation.
63. One circle has center at $(3, 4)$ and radius 5. A second circle has center at $(-1, -3)$ and radius 4. Do the circles cross?
64. Does the circle with radius 6 and center at $(0, 5)$ cross the circle with center at $(-5, -4)$ with radius 4?

3.3 Symmetry and Translations

In the graph of $y = x^2 + 2$, shown in Figure 3.9 of the previous section, you may have noticed that the graph could be cut in half by the y-axis with each half the mirror image of the other half. Such a graph is said to be symmetric with respect to the y-axis. As we shall see in this section, the concept of symmetry is helpful in drawing graphs.

Figure 3.16(a) shows a graph which is **symmetric with respect to the y-axis.** For a graph to be symmetric with respect to the y-axis, the point $(-x, y)$ must be on the graph whenever (x, y) is on the graph.

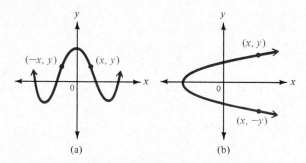

(a)   (b)

Figure 3.16

If the graph in Figure 3.16(b) were folded in half along the x-axis, the portion at the top would exactly match the portion at the bottom. Such a graph is **symmetric with respect to the x-axis.** As the graph suggests, symmetry with respect to the x-axis means that the point $(x, -y)$ must be on the graph whenever the point (x, y) is on the graph.

The following test tells when a graph is symmetric to the x-axis or y-axis.

Symmetry with Respect to an Axis

The graph of an equation is symmetric with respect to the y-axis if the replacement of x with $-x$ results in an equivalent equation.

The graph of an equation is symmetric with respect to the x-axis if the replacement of y with $-y$ results in an equivalent equation.

EXAMPLE 1   ☐ Test for symmetry with respect to the x-axis or y-axis.

(a) $y = x^2 + 4$

Replace x with $-x$:

$$y = x^2 + 4 \text{ becomes } y = (-x)^2 + 4 = x^2 + 4.$$

The result is the same as the original equation so that the graph, shown in Figure 3.17 on the next page, is symmetric with respect to the y-axis. Check that the graph is *not* symmetric with respect to the x-axis.

(b) $x = y^2 - 3$

Replace y with $-y$ to get $x = (-y)^2 - 3 = y^2 - 3$, the same as the original equation. The graph is symmetric to the x-axis as shown in Figure 3.18 on the next page. Is the graph symmetric to the y-axis?

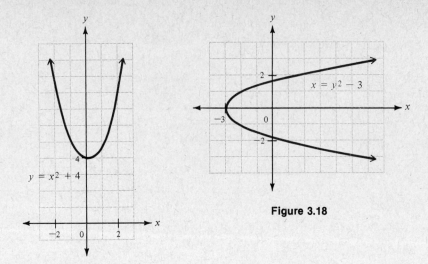

Figure 3.18

Figure 3.17

(c) $x^2 + y^2 = 16$

First replace x with $-x$ to see that the graph is symmetric with respect to the y-axis. Then, using $x^2 + y^2 = 16$, replace y with $-y$, to see that the graph, a circle of radius 4 centered at the origin, is also symmetric with respect to the x-axis.

(d) $2x + y = 4$

Replace x with $-x$ and y with $-y$; in neither case does an equivalent equation result. This graph is symmetric neither with respect to the x-axis nor the y-axis. ▢

Another kind of symmetry is found when it is possible to rotate a graph 180° about the origin and have it coincide with the original graph. Such symmetry is called **symmetry with respect to the origin.** Figures 3.19(a) and (b) show graphs that are symmetric to the origin; Figure 3.19(c) shows a graph that is not.

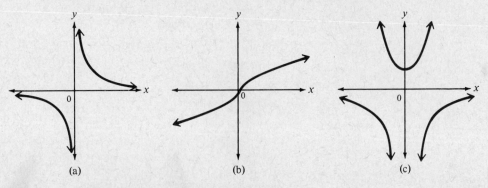

(a)   (b)   (c)

Figure 3.19

A graph is symmetric with respect to the origin if the point $(-x, -y)$ is on the graph whenever (x, y) is. This is restated in the following test for symmetry.

Symmetry with Respect to the Origin

The graph of an equation is symmetric with respect to the origin if the replacement of both x with $-x$ and y with $-y$ results in an equivalent equation.

EXAMPLE 2   ☐   Are the following graphs symmetric with respect to the origin?

(a) $x^2 + y^2 = 16$

Replace x with $-x$ and y with $-y$ to get

$$(-x)^2 + (-y)^2 = 16 \quad \text{or} \quad x^2 + y^2 = 16,$$

an equivalent equation. This graph, shown in Figure 3.20, is symmetric with respect to the origin.

(b) $y = x^3$

Replace x with $-x$ and y with $-y$ to get

$$-y = (-x)^3, \quad \text{or} \quad -y = -x^3, \quad \text{or} \quad y = x^3,$$

an equivalent equation. The graph, symmetric with respect to the origin, is shown in Figure 3.21.   ☐

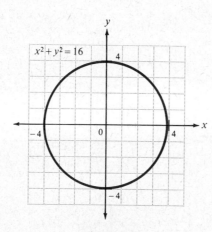

Figure 3.20

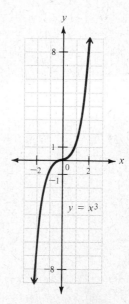

Figure 3.21

A graph symmetric with respect to both the x- and y-axes is automatically symmetric to the origin. (See Exercise 49.) However, a graph symmetric to the origin need not be symmetric to either axis. (See Figure 3.21 above.) Of the three types of symmetry—with respect to the x-axis, the y-axis, and the origin—a graph possessing any two must have the third type also. (See Exercise 50.)

One final type of symmetry is **symmetry with respect to the line $y = x$.** Figure 3.22(a) shows a graph that is symmetric to this line (a line through the origin making a 45° angle with the positive x-axis). Figure 3.22(b) shows a graph that is not symmetric to the line $y = x$. Use the following rule.

Symmetry with Respect to $y = x$

> The graph of an equation is symmetric with respect to the line $y = x$ if exchanging x and y results in an equivalent equation.

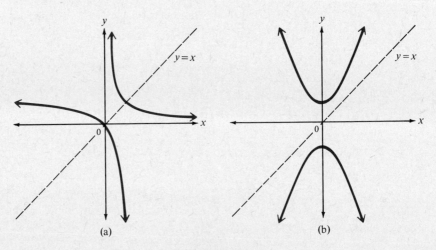

(a)   (b)

Figure 3.22

EXAMPLE 3   Are the following graphs symmetric with respect to the line $y = x$?

(a) $x^2 + y^2 = 100$

Exchange x and y to get

$$y^2 + x^2 = 100,$$

an equation equivalent to the original. This graph is symmetric to the line $y = x$ as shown in Figure 3.23.

(b) $y = x^2$

Exchange x and y to get $x = y^2$, which is not equivalent to the original equation. As shown in Figure 3.24, this graph is not symmetric to the line $y = x$.   ∎

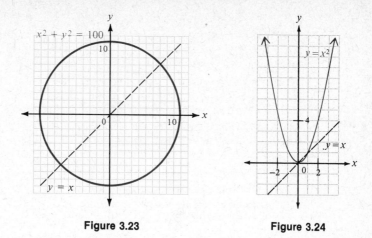

Figure 3.23   **Figure 3.24**

We can now summarize our tests for symmetry.

Tests for Symmetry

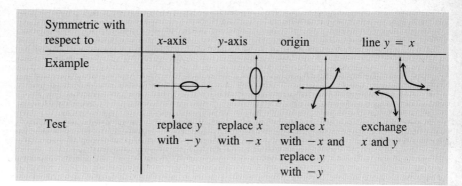

Symmetric with respect to	x-axis	y-axis	origin	line $y = x$
Example				
Test	replace y with $-y$	replace x with $-x$	replace x with $-x$ and replace y with $-y$	exchange x and y

Translations

We could find the graph of $y = x^2$ by choosing values of x and finding corresponding values for y. (See Figure 3.25, next page.) As we have seen, the graph of $y = x^2$ is called a parabola. Once we have this graph, how could we find the graph of $y = x^2 - 4$? For each point (x, y) on the graph of $y = x^2$, there will be a corresponding point $(x, y - 4)$ on the graph of $y = x^2 - 4$. Thus, each point on the graph of $y = x^2$ will be moved, or *translated*, 4 units downward to get the graph of $y = x^2 - 4$. (See Figure 3.25 again.) Such a vertical shift is called a **vertical translation.** As an example, Figure 3.26 shows a graph along with two different vertical translations of it.

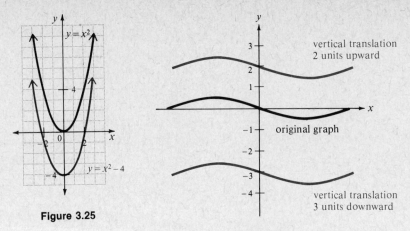

Figure 3.25

Figure 3.26

Figure 3.27 shows the graph of $y = x^2$ along with the graph of $y = (x - 3)^2$; this second graph is obtained from that of $y = x^2$ by a **horizontal translation.** Figure 3.28 shows a graph along with horizontal translations of the graph.

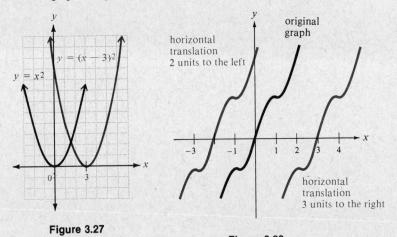

Figure 3.27

Figure 3.28

The following statements tell how to identify a translation.

Vertical Translation

Replacing y in an equation by $y - a$, where a is a constant, produces a vertical translation of the graph of $|a|$ units. The translation is upward if $a > 0$ and downward if $a < 0$.

Horizontal Translation

Replacing x in an equation by $x - b$, where b is a constant, produces a horizontal translation in the graph of $|b|$ units. The translation is to the right if $b > 0$ and to the left if $b < 0$.

EXAMPLE 4 ☐ Graph each of the following.

(a) $y = |x - 4|$.

Here x was replaced with $x - 4$ in the equation $y = |x|$. This results in a shift (or horizontal translation) 4 units to the right. See Figure 3.29.

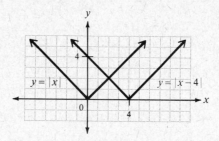

Figure 3.29

(b) $y = |x| - 5$

This equation can be written as the equivalent equation $y + 5 = |x|$. Here y was replaced with $y + 5$ in the equation $y = |x|$. Since $y + 5 = y - (-5)$, we have a shift of 5 units downward. See Figure 3.30.

(c) $y = |x + 3| + 2$

Rewrite $y = |x + 3| + 2$ as $y - 2 = |x + 3|$, to see that the graph involves both a horizontal translation of 3 units to the left and a vertical translation of 2 units upward. See Figure 3.31. ☐

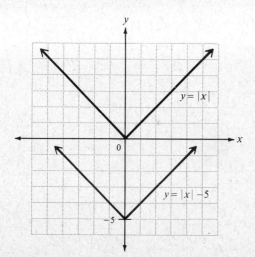

Figure 3.30

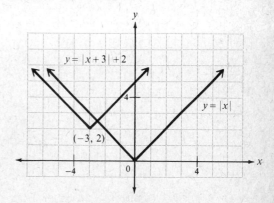

Figure 3.31

EXERCISES 3.3

Plot the following points, and then plot the points that are symmetric to the given point with respect to the (a) x-axis, (b) y-axis, (c) origin.

1. $(5, -3)$

2. $(-6, 1)$

3. $(-4, -2)$

4. $(-8, 3)$

5. $(-8, 0)$

6. $(0, -3)$

Use symmetry, point plotting, and translation to help you graph each of the following.

7. $x^2 + 2y^2 = 10$

8. $y^2 + x^2 = 5$

9. $3x^2 - y^2 = 8$

10. $5y^2 - x^2 = 6$

11. $y = 3|x|$

12. $y = -2|x|$

13. $|y| = 2x$

14. $-3|y| = 4x$

15. $|y| = |x + 2|$

16. $|x| = |y - 1|$

17. $xy = 4$

18. $\dfrac{x}{y} = 6$

19. $x^2 + y^2 = 9$

20. $y^2 = 25 - x^2$

21. $(x + 1)^2 + (y - 3)^2 = 9$

22. $(y + 2)^2 = 25 - (x - 3)^2$

A graph of $y = x^2$ is shown in the figure. Explain how each of the following graphs could be obtained from the graph shown. Sketch each graph.

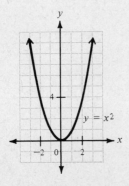

23. $y = x^2 + 2$

24. $y = x^2 - 7$

25. $y = -x^2$

26. $y = -x^2 + 8$

27. $y = -(x - 4)^2$

28. $y = -(x + 5)^2$

29. $y = (x - 2)^2 - 3$

30. $y = (x + 1)^2 + 5$

31. $y = -(x + 4)^2 + 2$

32. $y = -(x - 3)^2 - 1$

In Exercises 33 and 34, let F be some algebraic expression involving x as the only variable.

33. Suppose the equation $y = F$ is changed to $y = -F$. What is the relationship between the two graphs?

34. Suppose the equation $y = F$ is changed to $y = c \cdot F$, for some constant c. What is the effect on the graph of $y = F$? Discuss the effect depending on whether $c > 0$ or $c < 0$, and $|c| > 1$ or $|c| < 1$.

Decide which of the following are symmetric to the line $y = x$.

35. $xy = 2$

36. $xy = -5$

37. $\dfrac{x}{y} = 8$

38. $\dfrac{y}{x} = 2$

39. $x^2 + 2y^2 = 40$

40. $3x^2 + 2y^2 = 25$

41. $(xy)^2 + 4xy = 5$

42. $(8xy - 7)^2 = xy$

43. $3x(2 + y) = 1$

44. $-9y(8 - x) = 3$

Sketch examples of graphs which are

45. symmetric to the x-axis but not the y-axis

46. symmetric to the origin but to neither the x-axis nor the y-axis

47. symmetric to the line $y = x$ but not the origin

48. symmetric to the origin but not to the line $y = x$

Prove each statement below.

49. A graph symmetric with respect to both the *x*-axis and *y*-axis is also symmetric to the origin.

50. A graph possessing two of the three types of symmetry, with respect to the *x*-axis, *y*-axis, and origin, must possess the third type of symmetry also.

3.4   Functions

Suppose X is the set of all students studying this book every Monday evening at the local pizza parlor. Let Y be the set of integers between 0 and 125. To each student in set X we can associate a number from Y which represents the weight of the student to the nearest kilogram. Typical associations between students in set X and weights in set Y are shown in Figure 3.32.

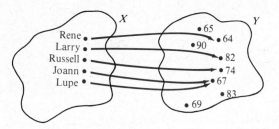

Figure 3.32

A correspondence such as the one shown in Figure 3.32 is called a **function** or a **mapping,** which is defined as follows.

Function

> A *function* from a set X to a set Y is a correspondence that assigns to each element of X exactly one element of Y.

The set X in this definition is called the **domain** of the function.

Three things should be noticed about the function shown in Figure 3.32 above.

1. There is a single weight associated with each student. That is, each element in X corresponds to exactly one element in Y.

2. The same weight may correspond to more than one student. In the example above, two students (Joann and Lupe) have a weight of 67 kilograms.

3. Not every element of Y need be used; for example, none of the students weighs 83 kilograms or 65 kilograms.

While the sets X and Y above were different, in many cases X and Y have many elements in common. In fact, it is not unusual at all for X and Y to be equal.

EXAMPLE 1   ☐   Decide whether or not the following diagrams represent functions from X to Y.

(a)

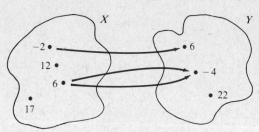

Figure 3.33

The diagram of Figure 3.33 does not represent a function from X to Y: there is no arrow leading from 17 in the domain, so that there is no element in Y corresponding to the element 17 in X.

(b)

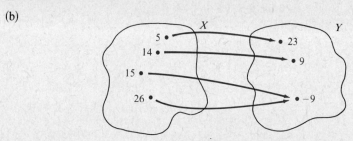

Figure 3.34

This diagram (Figure 3.34) does represent a function from X to Y since there is a single element in Y corresponding to each element in X.

(c)

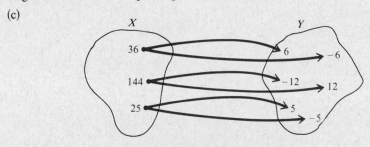

Figure 3.35

Here each element of set X corresponds to *two* elements in set Y, so that the diagram of Figure 3.35 does not represent a function from X to Y. ☐

It is common to use the letters f, g, and h to name functions. If f is a function and x is an element in the domain X, then $f(x)$ represents the element in Y that corresponds to x in X. For example in the diagram of Figure 3.34 above, if we use f to name this function, then

$$f(5) = 23, \qquad f(14) = 9, \qquad f(15) = -9, \qquad \text{and} \qquad f(26) = -9.$$

For a given element x in set X, the corresponding element $f(x)$ in set Y is called the **value** or **image** of f at x. The set of all possible values of $f(x)$ makes up the **range** of the function. Throughout this text, we assume that the domain and range of all functions are subsets of the set of real numbers, unless otherwise specified.

Functions are often expressed by simply giving a formula that shows how a given x in the domain is used to find the corresponding element $f(x)$ in the range. For example, if f is the function where a value of x is squared to find the corresponding value in the range, f could be expressed as

$$f(x) = x^2.$$

Throughout this book, if the domain for a function specified by an algebraic formula is not given, it will be assumed to be the largest possible set of real numbers for which the formula is meaningful unless otherwise specified. For example, suppose

$$f(x) = \frac{-4x}{2x - 3}.$$

Here any real number can be used for x except $x = 3/2$, which makes the denominator equal 0. Thus, the domain of this function will be assumed to be $\{x \mid x \neq 3/2\}$.

EXAMPLE 2   ☐   Find the domain and range for each of the following functions.

(a) $f(x) = x^2$

Any number may be squared, therefore the domain is the set of all real numbers. Since $x^2 \geq 0$ for every value of x, the range, written in interval notation, is $[0, +\infty)$.

(b) $f(x) = \sqrt{6 - x}$

Since $6 - x$ must be greater than or equal to 0, x can take on any value less than or equal to 6, so that the domain is $(-\infty, 6]$. The range is the interval $[0, +\infty)$. ☐

For most of the functions that we use, the domain can be found with the methods we have available. The range, however, must often be found by using graphing, complicated algebra, or calculus.

Based on the definition of a function, functions f and g are **equal** if and only if f and g have exactly the same domains and if $f(x) = g(x)$ for every value of x in the domain.

Suppose a function f is given by $f(x) = -3x + 2$. To emphasize that this statement is used to find values in the range of f, it is common to write

$$y = -3x + 2.$$

When a function is written in the form $y = f(x)$, x is called the **independent variable**, and y is the **dependent variable**.

EXAMPLE 3 ☐ Let $g(x) = 3\sqrt{x}$ and $h(x) = 1 + 4x$. Find each of the following.

(a) $g(16)$

   To find $g(16)$, replace x in $g(x) = 3\sqrt{x}$ with 16, getting

   $$g(16) = 3\sqrt{16} = 3 \cdot 4 = 12.$$

(b) $h(-3) = 1 + 4(-3) = -11$

(c) $g(-4)$ does not exist; -4 is not in the domain of g since $\sqrt{-4}$ is not a real number

(d) $h(\pi) = 1 + 4\pi$

(e) $g(m) = 3\sqrt{m}$, if m represents a nonnegative real number

(f) $g[h(3)]$

   First find $h(3)$, as follows.

   $$h(3) = 1 + 4 \cdot 3 = 1 + 12 = 13$$

Now, $g[h(3)] = g(13) = 3\sqrt{13}$. ☐

EXAMPLE 4 ☐ Let $f(x) = 2x^2 - 3x$, and find the quotient

$$\frac{f(x + h) - f(x)}{h},$$

which is important in calculus.

   To find $f(x + h)$, replace x with $x + h$, to get

$$\frac{f(x + h) - f(x)}{h} = \frac{2(x + h)^2 - 3(x + h) - (2x^2 - 3x)}{h}$$

$$= \frac{2(x^2 + 2xh + h^2) - 3x - 3h - 2x^2 + 3x}{h}$$

$$= \frac{2x^2 + 4xh + 2h^2 - 3x - 3h - 2x^2 + 3x}{h}$$

$$= \frac{4xh + 2h^2 - 3h}{h}$$

$$= 4x + 2h - 3. ☐$$

We have been finding the graphs of sets of ordered pairs throughout this chapter. A function can be thought of as a set of ordered pairs; in fact, a function f produces the set of ordered pairs $\{(x, f(x)) | x$ is in the domain of $f\}$. By the definition of function, for each x which appears in the first position of an ordered pair, there will be exactly one value of $f(x)$ in the second position.

In addition, if we have a set of ordered pairs in which any two ordered pairs that have the same x in the first entry also have equivalent values in the second entry, we must have a function. Based on this, we can give the following **alternate definition of a function.**

Alternate Definition of a Function

A *function* is a set of ordered pairs in which two ordered pairs cannot have the same first entries and different second entries.

The idea of a function as a set of ordered pairs lets us define the **graph of a function:** the graph of a function f is the set of all points in the plane of the form $(x, f(x))$, where x is in the domain of f. The graph of a function f is the same as the graph of the equation $y = f(x)$.

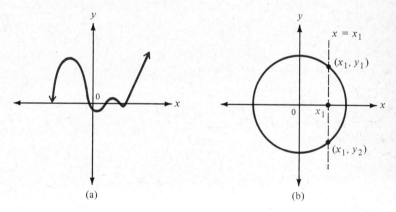

(a)   (b)

Figure 3.36

There is a quick way to tell if a given graph is the graph of a function or not. Figure 3.36 shows two graphs. In the graph of part (a), for any x that we might choose, there would be only one value of $f(x)$, or y. Thus, this graph is the graph of a function. On the other hand, the graph in part (b) is not the graph of a function. For example, if we choose $x = x_1$, the vertical line shows that there are two different values of y, namely y_1 and y_2.

We summarize this as the **vertical line test for a function.**

Vertical Line Test

If each vertical line cuts a graph in no more than one point, the graph is the graph of a function.

Intuitively, we consider a function *increasing* if its graph moves up and to the right as we move along the x-axis from left to right. The functions graphed in Figure 3.37(a) and (b) are increasing functions. On the other hand, a function is *decreasing* if its graph goes down and to the right as we move from left to right along the x-axis. The function graphed in Figure 3.37(c) is a decreasing function.

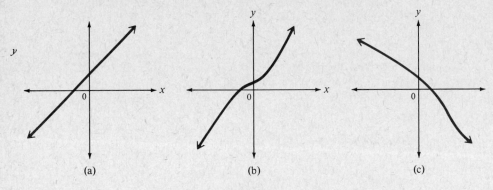

(a)   (b)   (c)

Figure 3.37

The function of Figure 3.38 is neither an increasing nor a decreasing function. However, this function is increasing if we look only at the interval $(-\infty, -2]$ or the interval $[4, +\infty)$; yet f is decreasing if we look only at the interval $[-2, 4]$.

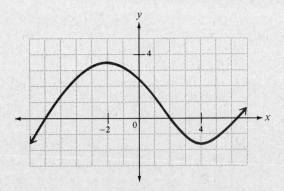

Figure 3.38

The idea of increasing and decreasing on an interval is defined in the following box, where I represents any interval of real numbers.

Increasing and Decreasing Functions

Let f be a function, with the interval I a subset of the domain of f. Let x_1 and x_2 be in I. Then

f is **increasing** on I if $f(x_1) < f(x_2)$ whenever $x_1 < x_2$,   and

f is **decreasing** on I if $f(x_1) > f(x_2)$ whenever $x_1 < x_2$.

EXAMPLE
5

Find where the following functions are increasing or decreasing.

(a) The function graphed in Figure 3.39(a) is increasing on the interval $[-2, 6]$. It is decreasing on $(-\infty, -2]$ and $[6, +\infty)$.

(b) The function of Figure 3.39(b) is never increasing or decreasing. This function is an example of a **constant function**, a function of the form $f(x) = k$, for k a real number.

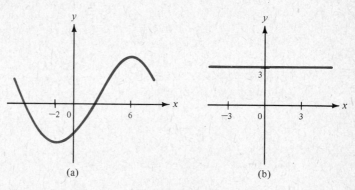

(a)   (b)

Figure 3.39

EXAMPLE
6

The symbol $[x]$ is used to represent the greatest integer less than or equal to x. For example, $[8.4] = 8$, $[-5] = -5$, $[\pi] = 3$, $[-6.9] = -7$, and so on. A graph of $y = [x]$ is shown in Figure 3.40, in which parentheses are used for endpoints that are not part of the graph and square brackets are used for endpoints that are part of the graph. As the vertical line test shows, this graph is the graph of a function. Also, there are no intervals where the function is increasing or decreasing.

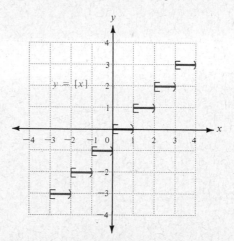

Figure 3.40

We can write our general rules for translations of a graph in terms of functions and $f(x)$ notation as follows.

**Translations
of the Graph
of a Function**

Let f be a function, and let c be a positive number.

To graph:	shift the graph of $y = f(x)$ by c units:
$y = f(x) + c$	upward
$y = f(x) - c$	downward
$y = f(x + c)$	left
$y = f(x - c)$	right

EXAMPLE 7   A graph of a function $y = f(x)$ is shown in Figure 3.41. Using this graph, find each of the following graphs.

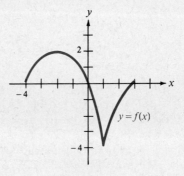

Figure 3.41

(a) $y = f(x) + 3$

This graph is the same as the graph of Figure 3.41, translated 3 units vertically upward. See Figure 3.42(a).

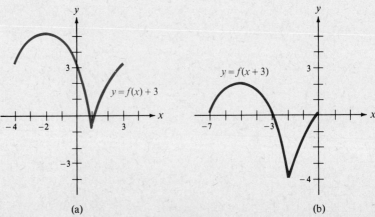

(a)   (b)

Figure 3.42

(b) $y = f(x + 3)$

To get the graph of $y = f(x + 3)$, translate the graph of $y = f(x)$ to the left 3 units. See Figure 3.42(b).

EXERCISES
3.4

For each of the following, find the indicated function values: (a) $f(-2)$, (b) $f(0)$, (c) $f(1)$, (d) $f(4)$.

1.

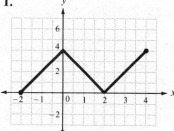

2.

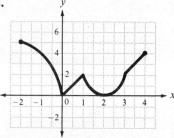

3.

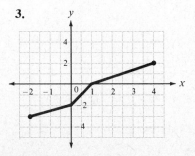

4.

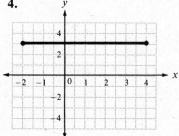

Let $f(x) = 3x - 1$ and $g(x) = x^2$. Find each of the following.

5. $f(0)$ **6.** $g(2)$

7. $f(a)$ **8.** $g(b)$

9. $f[f(1)]$ **10.** $g[g(-2)]$

11. $f(5a - 2)$ **12.** $f(3 + 2k)$

13. $g(5p - 2)$ **14.** $g(-6k + 1)$

15. $f[g(m)]$ **16.** $g[f(2r)]$

Let $f(x) = -4.6x^2 - 8.9x + 1.3$. Find each of the following.

17. $f(3)$ **18.** $f(-5)$

19. $f(-4.2)$ **20.** $f(-1.8)$

Give the domain of each of the following.

21. $f(x) = 2x$

22. $f(x) = x + 2$

23. $f(x) = x^4$

24. $f(x) = (x - 2)^2$

25. $f(x) = \sqrt{16 - x^2}$

26. $f(x) = |x - 1|$

27. $f(x) = (x - 3)^{1/2}$

28. $f(x) = (3x + 5)^{1/2}$

29. $f(x) = \dfrac{2}{x^2 - 4}$

30. $f(x) = \dfrac{-8}{x^2 - 36}$

31. $f(x) = \sqrt{\dfrac{3}{x^2 + 25}}$

32. $f(x) = \sqrt{\dfrac{6}{x^2 + 121}}$

33. $f(x) = -\sqrt{\dfrac{2}{x^2 + 9}}$

34. $f(x) = -\sqrt{\dfrac{5}{x^2 + 36}}$

35. $f(x) = \sqrt{x^2 - 4x - 5}$

36. $f(x) = \sqrt{x^2 + 7x + 10}$

37. $f(x) = -\sqrt{6x^2 + 7x - 5}$

38. $f(x) = \sqrt{15x^2 + x - 2}$

Give both the domain and the range of the following.

39.

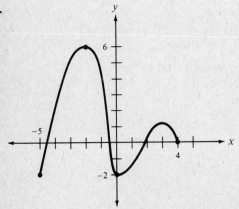

40.

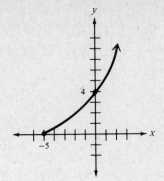

41.

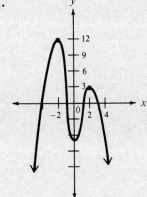

42.

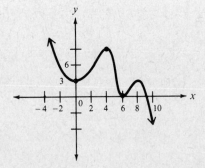

For each of the following, find: (a) $f(x + h)$; (b) $f(x + h) - f(x)$; (c) $\dfrac{f(x + h) - f(x)}{h}$.

43. $f(x) = x^2 - 4$ **44.** $f(x) = 8 - 3x^2$

45. $f(x) = 6x + 2$ **46.** $f(x) = 4x - 11$

47. $f(x) = 2x^3 + x^2$ **48.** $f(x) = -4x^3 - 8x$

Find where the following functions are increasing or decreasing. In Exercises 55–62, first graph the functions.

49.

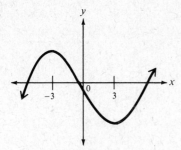

50.

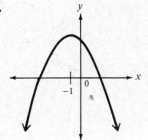

51.

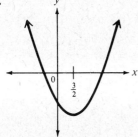

52.

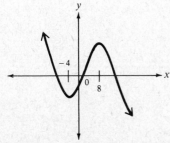

53.

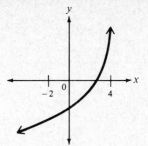

54.

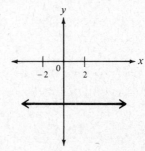

55. $y = -4x + 2$ **56.** $y = 5x - 1$

57. $y = x^2 + 4$ **58.** $y = -x^2 - 3$

59. $y = -|x + 2|$ **60.** $y = |3 - x|$

61. $y = x + |x|$ **62.** $y = |x| - x$

*A function f with the property that $f(-x) = f(x)$ for every value of x in its domain is called an **even function**. A function f with the property that $f(-x) = -f(x)$ for every x in its domain is called an **odd function**. Decide whether the following functions are even, odd, or neither.*

63. $f(x) = x^2$ **64.** $f(x) = x^3$

65. $f(x) = x^4 + x^2 + 5$

66. $f(x) = x^3 - x + 1$

67. $f(x) = 2x + 3$ **68.** $f(x) = |x|$

69. $f(x) = \dfrac{2}{x - 6}$ **70.** $f(x) = \dfrac{8}{x}$

Graph on the same coordinate axes, the graphs of f for the given values of c.

71. $f(x) = x^2 + c;\ c = -1,\ c = 2$

72. $f(x) = (x + c)^2;\ c = -1,\ c = 2$

73. $f(x) = |x - c|$; $c = -2$, $c = 1$

74. $f(x) = c - |x|$; $c = 2$, $c = -3$

Let the graph of a function $y = f(x)$ be as shown.

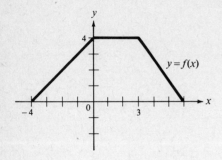

Sketch the graph of each of the following.

75. $y = f(x) + 4$ **76.** $y = 3 - f(x)$

77. $y = f(x - 1)$ **78.** $y = f(x + 2)$

79. $y = f(x + 3) - 2$ **80.** $y = f(x - 1) - 4$

81. Give the domain and range of the function $y = [x]$ of Example 6.

82. Suppose a chain-saw rental firm charges a fixed $4 sharpening fee plus $7 per day or fraction of a day. Let $S(x)$ represent the cost of renting a saw for x days. Find the value of (a) $S(1)$, (b) $S(1.25)$, (c) $S(3.5)$. (d) Graph $y = S(x)$. (e) Give the domain and range of S.

83. Assume that it costs 30¢ to mail a letter weighing one ounce or less, and then 27¢ for each additional ounce or fraction of an ounce. Let $L(x)$ be the cost of mailing a letter weighing x

ounces. Find (a) $L(.75)$, (b) $L(1.6)$, (c) $L(4)$. (d) Graph $y = L(x)$. (e) Give the domain and range of L.

84. Use the greatest integer function and write an expression for the number of ounces for which postage will be charged on a letter weighing x ounces (See Exercise 83).

85. In a recent year Washington, D.C. taxi rates were 90¢ for the first $1/9$ of a mile and 10¢ for each additional $1/9$ mile or fraction of a $1/9$. Let $C(x)$ be the cost for a taxi ride of x $1/9$'s of a mile. Find the following: (a) $C(1)$, (b) $C(2.3)$, (c) $C(8)$. (d) Graph $y = C(x)$. (e) Give the domain and range of C.

86. For a lift truck rental of no more than three days, the charge is $300. An additional charge of $75 is made for each day or portion of a day after three. Graph the ordered pairs (number of days, cost).

87. A car rental cost $37 for one day, which includes 50 free miles. Each additional 25 miles or portion costs $10. Graph the ordered pairs (miles, cost).

88. Prove that an even function has a graph which is symmetric to the y-axis.

89. Prove that an odd function has a graph which is symmetric to the origin.

Suppose a function is increasing for the interval $(-\infty, +\infty)$. Could the graph of the function be symmetric to the

90. x-axis?   **91.** y-axis?   **92.** origin?

3.5   Combining Functions

When a company accountant sits down to estimate the firm's overhead, the first step might be to find functions representing the cost of materials, labor charges, equipment maintenance, and so on. The sum of these various functions could then be used to find the total overhead for the company. We study methods of combining functions in this section.

Given two functions f and g, their **sum**, written $f + g$, is defined as

$$(f + g)(x) = f(x) + g(x),$$

for all x such that both $f(x)$ and $g(x)$ exist. Similar definitions can be given for the difference, $f - g$, product, $f \cdot g$, and quotient, f/g, of functions; however, the quotient,

$$\left(\frac{f}{g}\right)(x) = \frac{f(x)}{g(x)},$$

is defined for all x where both $f(x)$ and $g(x)$ exist, and in addition, $g(x) \neq 0$.

EXAMPLE 1 ☐ Let $f(x) = 8x - 9$ and $g(x) = \sqrt{2x - 1}$.

(a) $(f + g)(x) = f(x) + g(x) = 8x - 9 + \sqrt{2x - 1}$

(b) $(f - g)(x) = f(x) - g(x) = 8x - 9 - \sqrt{2x - 1}$

(c) $(f \cdot g)(x) = f(x) \cdot g(x) = (8x - 9)\sqrt{2x - 1}$

(d) $\left(\frac{f}{g}\right)(x) = \frac{f(x)}{g(x)} = \frac{8x - 9}{\sqrt{2x - 1}}$

The domain of f is the set of all real numbers, while the domain of $g(x) = \sqrt{2x - 1}$ includes just those real numbers that make $2x - 1 \geq 0$; the domain of g is the interval $[1/2, +\infty)$. The domain of $f + g$, $f - g$, and $f \cdot g$ is thus $[1/2, +\infty)$. With f/g, we cannot have a zero denominator, so that the value $1/2$ is excluded from the domain. The domain of f/g is $(1/2, +\infty)$. ☐

The following box summarizes the domains of $f + g$, $f - g$, $f \cdot g$, and f/g. (Recall: the intersection of two sets is the set of all elements belonging to *both* of the sets.)

> For functions f and g, the domains of $f + g$, $f - g$, and $f \cdot g$ are made up of all real numbers in the intersection of the domains of f and g, while the domain of f/g is made up of those real numbers in the intersection of the domains of f and g that do not make $g(x) = 0$.

Composition of Functions

The sketch in Figure 3.43 shows a function f which assigns to each element x of set X some element y of set Y. Suppose also that a function g takes each element of set Y and assigns a value z of set Z. Using both f and g, then, an element x in X is assigned to an element z in Z. The result of this process is a new function h, which takes an element x in X and assigns an element z in Z.

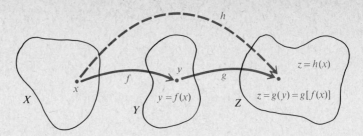

Figure 3.43

This function h is called the *composition* of functions g and f, written $g \circ f$, and defined as follows.

Composition of Functions

Let f and g be functions. The **composite function,** or **composition,** of g and f is

$$(g \circ f)(x) = g[f(x)],$$

for all x in the domain of f such that $f(x)$ is in the domain of g.

EXAMPLE 2

☐ Let $f(x) = 4x + 1$ and $g(x) = 2x^2 + 5x$. Find each of the following.

(a) $(g \circ f)(x)$

By definition, $(g \circ f)(x) = g[f(x)]$. Using the given functions, we have

$$
\begin{aligned}
(g \circ f)(x) &= g[f(x)] \\
&= g[4x + 1] \\
&= 2(4x + 1)^2 + 5(4x + 1) \\
&= 2(16x^2 + 8x + 1) + 20x + 5 \\
&= 32x^2 + 16x + 2 + 20x + 5 \\
&= 32x^2 + 36x + 7.
\end{aligned}
$$

(b) $(f \circ g)(x)$

If we use the definition above, with f and g interchanged, $(f \circ g)(x)$ becomes $f[g(x)]$, with

$$
\begin{aligned}
(f \circ g)(x) &= f[g(x)] \\
&= 4[g(x)] + 1 \\
&= 4(2x^2 + 5x) + 1 \\
&= 8x^2 + 20x + 1. \quad ☐
\end{aligned}
$$

As this example shows, it is not always true that $f \circ g = g \circ f$. In fact, these two composite functions are equal only for a special class of functions, discussed in the next section. In Example 2, the domain of both composite functions is the set of all real numbers.

EXAMPLE 3 ☐ Let $f(x) = \dfrac{1}{x}$ and $g(x) = \sqrt{3 - x}$. Find $f \circ g$ and $g \circ f$. Give the domain of each.

Let us first find $f \circ g$, getting

$$(f \circ g)(x) = f[g(x)]$$

$$= \frac{1}{g(x)}$$

$$= \frac{1}{\sqrt{3 - x}}.$$

In order for $\sqrt{3 - x}$ to be a nonzero real number, we must have $3 - x > 0$ or $x < 3$, so that the domain of $f \circ g$ is the interval $(-\infty, 3)$.

We use the same functions to find $g \circ f$, as follows.

$$(g \circ f)(x) = g[f(x)]$$

$$= g\left[\frac{1}{x}\right]$$

$$= \sqrt{3 - \frac{1}{x}}$$

$$= \sqrt{\frac{3x - 1}{x}}$$

The domain of $g \circ f$ is the set of all real numbers x such that

$$\frac{3x - 1}{x} \geq 0.$$

By the methods of Section 2.6, the domain of $g \circ f$ is the set $(-\infty, 0) \cup [1/3, +\infty)$.

☐

EXERCISES 3.5

For each of the following, find $f + g$, $f - g$, $f \cdot g$, and f/g. Give the domain of each.

1. $f(x) = 4x - 1$, $g(x) = 6x + 3$

2. $f(x) = 9 - 2x$, $g(x) = -5x + 2$

3. $f(x) = 3x^2 - 2x$, $g(x) = x^2 - 2x + 1$

4. $f(x) = 6x^2 - 11x$, $g(x) = x^2 - 4x - 5$

5. $f(x) = \sqrt{2x + 5}$, $g(x) = \sqrt{4x - 9}$

6. $f(x) = \sqrt{11x - 3}$, $g(x) = \sqrt{2x - 15}$

7. $f(x) = 4x^2 - 11x + 2$, $g(x) = x^2 + 5$

8. $f(x) = 15x^2 - 2x + 1$, $g(x) = 16 + x^2$

Let $f(x) = 4x^2 - 2x$ and let $g(x) = 8x + 1$. Find each of the following.

9. $(f + g)(3)$ **10.** $(f + g)(-5)$

11. $(f \cdot g)(4)$ **12.** $(f \cdot g)(-3)$

13. $\left(\dfrac{f}{g}\right)(-1)$ **14.** $\left(\dfrac{f}{g}\right)(4)$

15. $(f + g)(m)$ **16.** $(f - g)(2k)$

17. $(f \circ g)(2)$ **18.** $(f \circ g)(-5)$

19. $(g \circ f)(2)$ **20.** $(g \circ f)(-5)$

21. $(f \circ g)(k)$ **22.** $(g \circ f)(5z)$

Find $f \circ g$ and $g \circ f$ for each of the following.

23. $f(x) = 8x + 12, g(x) = 3x - 1$

24. $f(x) = -6x + 9, g(x) = 5x + 7$

25. $f(x) = 5x + 3, g(x) = -x^2 + 4x + 3$

26. $f(x) = 4x^2 + 2x + 8, g(x) = x + 5$

27. $f(x) = -x^3 + 2, g(x) = 4x$

28. $f(x) = 2x, g(x) = 6x^2 - x^3$

29. $f(x) = \dfrac{1}{x}, g(x) = x^2$

30. $f(x) = \dfrac{2}{x^4}, g(x) = 2 - x$

31. $f(x) = \sqrt{x + 2}, g(x) = 8x^2 - 6$

32. $f(x) = 9x^2 - 11x, g(x) = 2\sqrt{x + 2}$

33. $f(x) = \dfrac{1}{x - 5}, g(x) = \dfrac{2}{x}$

34. $f(x) = \dfrac{8}{x - 6}, g(x) = \dfrac{4}{3x}$

35. $f(x) = \sqrt{x + 1}, g(x) = \dfrac{-1}{x}$

36. $f(x) = \dfrac{8}{x}, g(x) = \sqrt{3 - x}$

For each of the following, show that $(f \circ g)(x) = x$ and $(g \circ f)(x) = x$.

37. $f(x) = 8x, g(x) = \dfrac{1}{8}x$

38. $f(x) = \dfrac{3}{4}x, g(x) = \dfrac{4}{3}x$

39. $f(x) = 8x - 11, g(x) = \dfrac{x + 11}{8}$

40. $f(x) = \dfrac{x - 3}{4}, g(x) = 4x + 3$

41. $f(x) = x^3 + 6, g(x) = \sqrt[3]{x - 6}$

42. $f(x) = \sqrt[5]{x - 9}, g(x) = x^5 + 9$

For each of the following, a function h is given. Find functions f and g such that $h(x) = (f \circ g)(x)$. Many such pairs of functions exist.

43. $h(x) = (6x - 2)^2$

44. $h(x) = (11x^2 + 12x)^2$

45. $h(x) = \sqrt{x^2 - 1}$ **46.** $h(x) = \dfrac{1}{x^2 + 2}$

47. $h(x) = \dfrac{8x + 2}{4x - 3}$

48. $h(x) = (x + 2)^3 - 3(x + 2)^2$

49. Suppose the population P of a certain species of fish depends on the number x (in hundreds) of a smaller kind of fish that serves as its food supply, so that

$$P(x) = 2x^2 + 1.$$

Suppose also that the number x (in hundreds) of the smaller species of fish depends on the amounts a (in appropriate units) of its food supply, a kind of plankton; that is, suppose

$$x = f(a) = 3a + 2.$$

Find $(P \circ f)(a)$, the relationship between the population P of the large fish and the amount a of plankton available.

50. Suppose the demand for a certain brand of vacuum cleaner is given by

$$D(p) = \dfrac{-p^2}{100} + 500,$$

where p is the price in dollars. If the price, in terms of the cost, c, is expressed as

$$p(c) = 2c - 10,$$

find the demand in terms of the cost.

51. An oil well off the Gulf Coast is leaking, with the leak spreading oil over the surface as a circle. At any time t, in minutes, after the beginning of the leak, the radius of the circular oil slick on the surface is $r(t) = 4t$ feet. Let $A(r) = \pi r^2$ represent the area of a circle of radius r. Find and interpret $(A \circ r)(t)$.

52. When a thermal inversion layer is over a city (such as happens often in Los Angeles), pollutants cannot rise vertically but are trapped below the layer and must disperse horizontally. Assume that a factory smokestack begins emitting a pollutant at 8 A.M. Assume that the pollutant disperses horizontally, forming a circle. If t represents the time, in hours, since the factory began emitting pollutants ($t = 0$ represents 8 A.M.), assume that the radius of the circle of pollution is $r(t) = 2t$ miles. Let $A(r) = \pi r^2$ represent the area of a circle of radius r. Find and interpret $(A \circ r)(t)$.

Let f and g be increasing functions on the same interval. Show that the following two functions are also increasing on that same interval.

53. $f + g$ **54.** $f \circ g$

55. What can you say about $f - g$?

56. Let $f(x) = x/(x - 1)$ for $x \neq 1$. Show that

$$f \circ f = x.$$

57. Let $f(x) = ax + b$ and $g(x) = cx + d$. Find conditions for c and d so that $f \circ g = g \circ f$.

Recall that we defined even and odd functions in the exercises for the previous section.

58. Let f be any function. Prove that the function $g(x) = \dfrac{1}{2}[f(x) + f(-x)]$ is even.

59. Let f be any function. Prove that the function $h(x) = \dfrac{1}{2}[f(x) - f(-x)]$ is odd.

60. Use the results of the previous two exercises to show that any function may be expressed as the sum of an odd function and an even function.

Prove that the

61. sum of two even functions is even.

62. product of two even functions is even.

63. sum of two odd functions is odd.

64. product of two odd functions is even.

65. product of an odd function and an even function is odd.

66. What can you say about the sum of an odd and an even function? Give examples.

3.6   Inverse Functions

For the function $y = 5x - 8$, two different values of x produce two different values of y. On the other hand, for the function $y = x^2$, two different values of x can lead to the *same* value of y; for example, both $x = 4$ and $x = -4$ give $y = 4^2 = (-4)^2 = 16$. A function, such as $y = 5x - 8$, where different elements from the domain lead to different elements from the range, is called a one-to-one function.

One-to-one Function

A function f is **one-to-one** if, for elements a and b from the domain of f,

$$a \neq b \text{ implies } f(a) \neq f(b).$$

EXAMPLE 1   ☐   Decide whether or not the following functions are one-to-one.

(a) $f(x) = -4x + 12$

     Suppose that $a \neq b$. Then $-4a \neq -4b$, and $-4a + 12 \neq -4b + 12$. Thus, the fact that $a \neq b$ implies that $f(a) \neq f(b)$, so that f is one-to-one.

(b) $f(x) = \sqrt{25 - x^2}$

     If $x = 3$, then $f(3) = \sqrt{25 - 3^2} = \sqrt{16} = 4$. For $x = -3$, we have $f(-3) = \sqrt{25 - (-3)^2} = \sqrt{16} = 4$. Both $x = 3$ and $x = -3$ lead to $y = 4$, so that f is not one-to-one.   ☐

     As shown in part (b) of this example, we can show that a function is *not* one-to-one by producing a pair of unequal numbers that lead to the same function value.

     In addition, there is a useful graphical test which tells whether or not a function is one-to-one. Figure 3.44(a) shows the graph of a function cut by a horizontal line. Each point where the horizontal line cuts the graph has the same y-value but a different x-value. Here, three different values of x lead to the same value of y; therefore, the function is not one-to-one. On the other hand, the graph of Figure 3.44(b) can be cut by any horizontal line in no more than one point, so that it is the graph of a one-to-one function.

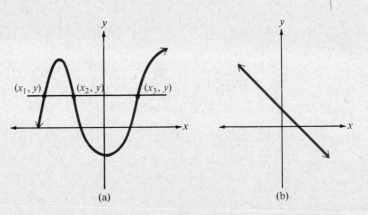

Figure 3.44

     In summary, we have the **horizontal line test** for one-to-one functions.

Horizontal Line Test   If every horizontal line cuts the graph of a function in no more than one point, then the function is one-to-one.

     In particular, functions that are increasing (or decreasing) are one-to-one.

Inverse functions

Suppose that f is a one-to-one function with domain X and range Y. Suppose too that we choose an x in X. The function f will then produce exactly one value $f(x)$ in Y. Since f is one-to-one, a different choice of x in X would produce a different value of $f(x)$ in Y.

Thus, there is a one-to-one pairing of values of x in X and values of $f(x)$ in Y. Because of the pairing, we could define a new function g that would start with an $f(x)$ in Y and produce the value x in X. That is,

if $f(x) = y$,   then $g(y) = x$.

See the sketch in Figure 3.45.

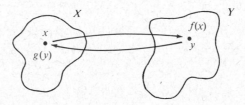

Figure 3.45

As suggested by Figure 3.45, the domain of f and the range of g are equal, while the range of f and the domain of g are also equal.

If $f(x) = y$, then $g(y) = x$; or

$$(g \circ f)(x) = g[f(x)] = x \qquad \text{for every } x \text{ in } X.$$

Also, $(f \circ g)(y) = f[g(y)] = y \qquad \text{for every } y \text{ in } Y.$

Since it is customary to use x for the domain element of a function, we replace y with x to change this last statement to

$$(f \circ g)(x) = f[g(x)] = x \qquad \text{for every } x \text{ in } Y.$$

Our two functions f and g have a special property: $(f \circ g)(x) = x$ and $(g \circ f)(x) = x$. In summary, we can make the following definition:

Inverse Functions

Let f be a one-to-one function with domain X and range Y. Let g be a function with domain Y and range X. Then g is the **inverse function** of f if

$$(f \circ g)(x) = x \qquad \text{for every } x \text{ in } Y,$$

and $(g \circ f)(x) = x \qquad \text{for every } x \text{ in } X.$

EXAMPLE 2   ☐   Let $f(x) = x^3 - 1$, and let $g(x) = \sqrt[3]{x + 1}$. Is g the inverse function of f? We use the definition to get

$$(f \circ g)(x) = f[g(x)] = (\sqrt[3]{x + 1})^3 - 1 = x + 1 - 1 = x$$

$$(g \circ f)(x) = g[f(x)] = \sqrt[3]{(x^3 - 1) + 1} = \sqrt[3]{x^3} = x.$$

Thus, g is the inverse function of f. Also, f is the inverse function of g.   ☐   .

A special notation is often used for inverse functions: If g is the inverse function of f, then g can be written as f^{-1} (read "f-inverse"). In Example 2,

$$f^{-1}(x) = \sqrt[3]{x + 1}.$$

Do not confuse the -1 in f^{-1} with a negative exponent. The symbol f^{-1} does not represent $1/f$; it represents the inverse function of function f.

Keep in mind that a function f can have an inverse function f^{-1} if and only if f is one-to-one. Since increasing and decreasing functions are one-to-one, they must have inverse functions.

Given a one-to-one function f and an x in the domain of f, we can find the corresponding value in the range of f by means of the equation $y = f(x)$. By the inverse function f^{-1}, we can take y and produce x: $x = f^{-1}(y)$. Therefore, to find the equation for f^{-1}, it is only necessary to solve $y = f(x)$ for x.

For example, let $f(x) = 7x - 2$. Then $y = 7x - 2$. This function is one-to-one, so that f^{-1} exists. Solve the equation for x, as follows:

$$y = 7x - 2$$
$$7x = y + 2$$
$$x = \frac{y + 2}{7}$$

or, since $x = f^{-1}(y)$,

$$f^{-1}(y) = \frac{y + 2}{7}.$$

For a given x in X (the domain of f), the equation $y = 7x - 2$ produces a value of $f(x)$ in Y (the range of f). The function $f^{-1}(y) = (y + 2)/7$ takes a value of y in Y and produces a value x in X. As we said, it is customary to use x for the domain element of a function, so we replace y with x in f^{-1} to get

$$f^{-1}(x) = \frac{x + 2}{7}.$$

Check that $(f \circ f^{-1})(x) = x$ and $(f^{-1} \circ f)(x) = x$, so that our function f^{-1} is indeed the inverse of f.

In summary, to find the equation of an inverse function, use the following steps.

Finding An Equation for f^{-1}

1. Let $y = f(x)$ be a one-to-one function.
2. Solve for x. Let $x = f^{-1}(y)$.
3. Exchange x and y to get $y = f^{-1}(x)$.
4. Check the domains and ranges: the domain of f and the range of f^- are equal, as are the domain of f^{-1} and the range of f.

EXAMPLE 3 ☐ For each of the following functions, find its inverse function where possible.

(a) $f(x) = \dfrac{4x + 6}{5}$

This function is one-to-one and thus has an inverse. Let $y = f(x)$, and solve for x, getting

$$y = \frac{4x + 6}{5}$$
$$5y = 4x + 6$$
$$5y - 6 = 4x$$
$$\frac{5y - 6}{4} = x.$$

Finally, exchange x and y, and let $y = f^{-1}(x)$, to get

$$\frac{5x - 6}{4} = y,$$

or $\quad f^{-1}(x) = \dfrac{5x - 6}{4}.$

(b) $f(x) = x^3 - 1$

If we choose two different values of x, we will get two different values of $x^3 - 1$, so that the function is one-to-one and has an inverse. Start by solving $y = x^3 - 1$ for x, as follows:

$$y = x^3 - 1$$
$$y + 1 = x^3$$
$$\sqrt[3]{y + 1} = x.$$

Exchange x and y, with

$$\sqrt[3]{x + 1} = y,$$

or $\quad f^{-1}(x) = \sqrt[3]{x + 1}.$

(c) $f(x) = x^2$

If we choose the two different x-values 4 and -4, we get the same value of y, namely 16. Thus the function is not one-to-one and has no inverse function.   □

Suppose f and f^{-1} are inverse functions of each other. By the definition of inverse function, if the ordered pair (a, b) belongs to the graph of f, then (b, a) will belong to the graph of f^{-1}. As shown in Figure 3.46, the points (a, b) and (b, a) are symmetric with respect to the 45° line $y = x$. Thus, the graph of f^{-1} can be obtained from the graph of f by reflecting the graph of f about the line $y = x$.

For example, Figure 3.47 shows the graph of $f(x) = x^3 - 1$ as a solid line and the graph of $f^{-1}(x) = \sqrt[3]{x + 1}$ as a dashed line. These graphs are symmetric with respect to the line $y = x$.

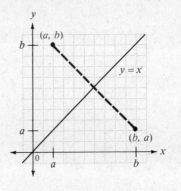

Figure 3.46   Figure 3.47

EXAMPLE 4   □   Let $f(x) = \sqrt{x + 5}$ with domain $[-5, +\infty)$. Find $f^{-1}(x)$.

Our function f is increasing and thus is one-to-one and has an inverse function. To find this inverse function, start with

$$y = \sqrt{x + 5}$$

and solve for x, to get

$$y = \sqrt{x + 5}$$
$$y^2 = x + 5$$
$$y^2 - 5 = x.$$

Exchanging x and y gives

$$x^2 - 5 = y.$$

We cannot just give $x^2 - 5$ as $f^{-1}(x)$. In the definition of f above, the domain was given as $[-5, +\infty)$. The range of f is $[0, +\infty)$. As we have said, the range of f equals the domain of f^{-1}, so that we must give f^{-1} as

$$f^{-1}(x) = x^2 - 5, \qquad \text{domain } [0, +\infty).$$

As a check, we have the range of f^{-1}, $[-5, +\infty)$, which equals the domain of f. Graphs of f and f^{-1} are shown in Figure 3.48; the line $y = x$ is included on the graph to show that the graphs of f and f^{-1} are mirror images with respect to this line. ☐

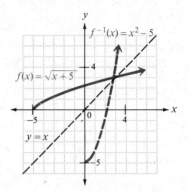

Figure 3.48

EXERCISES 3.6

Which of the following functions are one-to-one?

1.

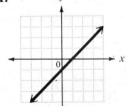

3.

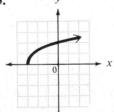

2.

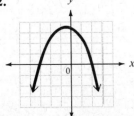

4. $y = 3x - 4$ **5.** $y = 4x - 5$

6. $x + y = 0$ **7.** $y = 6 - x$

8. $y = -x^2$ **9.** $y = (x - 2)^2$

10. $y = -(x + 3)^2 - 8$

11. $y = \sqrt{36 - x^2}$

12. $y = -\sqrt{100 - x^2}$ **13.** $y = |25 - x^2|$

14. $y = -|16 - x^2|$ **15.** $y = x^3 - 1$

16. $y = -\sqrt[3]{x + 5}$ **17.** $y = (\sqrt{x} + 1)^2$

18. $y = (3 - 2\sqrt{x})^2$ **19.** $y = \dfrac{1}{x + 2}$

20. $y = \dfrac{-4}{x - 8}$ **21.** $y = 9$

22. $y = -4$

Which of the following pairs of functions are inverses of each other?

23.

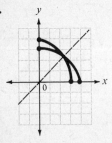

24.

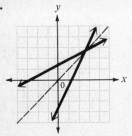

25.

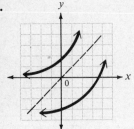

26. $f(x) = -8x$ and $g(x) = -\dfrac{1}{8}x$

27. $f(x) = 2x + 4$ and $g(x) = \dfrac{1}{2}x - 2$

28. $f(x) = 5x - 5$ and $g(x) = \dfrac{1}{5}x + 1$

29. $f(x) = 8x - 7$ and $g(x) = \dfrac{x + 8}{7}$

30. $f(x) = \dfrac{1}{x + 1}$ and $g(x) = \dfrac{x - 9}{12}$

31. $f(x) = \dfrac{1}{x + 1}$ and $g(x) = \dfrac{1 - x}{x}$

32. $f(x) = \dfrac{2}{x + 6}$ and $g(x) = \dfrac{6x + 2}{x}$

33. $f(x) = \dfrac{1}{x}$ and $g(x) = \dfrac{1}{x}$

34. $f(x) = 4x$ and $g(x) = -\dfrac{1}{4}x$

35. $f(x) = \sqrt{x + 8}$, domain $[-8, +\infty)$, and $g(x) = x^2 - 8$, domain $[0, +\infty)$.

36. $f(x) = x^2 + 3$, domain $[0, +\infty)$, and $g(x) = \sqrt{x - 3}$, domain $[3, +\infty)$.

37. $f(x) = |x - 1|$, domain $[-1, +\infty)$, and $g(x) = |x + 1|$, domain $[1, +\infty)$.

38. $f(x) = -|x + 5|$, domain $[-5, +\infty)$, and $g(x) = |x - 5|$, domain $[5, +\infty)$.

For each of the following functions $y = f(x)$ that is one-to-one, write an equation for the inverse function in the form of $y = f^{-1}(x)$ and then graph f and f^{-1}. Give the domain of both f and f^{-1}.

39. $y = 3x - 4$ **40.** $y = 4x - 5$

41. $y = \dfrac{1}{3}x$ **42.** $y = -\dfrac{2}{5}x$

43. $y = x^3 + 1$ **44.** $y = -x^3 - 2$

45. $x + 4y = 12$ **46.** $3x + y = 9$

47. $y = x^2$ **48.** $y = -x^2 + 2$

49. $y = \dfrac{1}{x}$ **50.** $xy = 4$

51. $y = \dfrac{8x + 3}{4x - 1}$ **52.** $y = \dfrac{-6x + 5}{3x - 1}$

53. $f(x) = 4 - x^2$, domain $(-\infty, 0]$ **54.** $f(x) = \sqrt{6 + x}$, domain $[-6, +\infty)$

Let $f(x) = -4x + 3$, while $g(x) = 2x^3 - 4$. Find each of the following.

55. $(f \circ g)^{-1}(x)$ **56.** $(f^{-1} \circ g^{-1})(x)$

57. $(g^{-1} \circ f^{-1})(x)$

58. Let f and g be functions having inverses f^{-1} and g^{-1} respectively. Let $f \circ g$ exist. Show that $(f \circ g)^{-1}$ is $(g^{-1} \circ f^{-1})$.

59. Show that a one-to-one function has exactly one inverse function.

60. Let f be an odd one-to-one function. What can you say about f^{-1}?

61. Let f be an even function. What can you say about f^{-1}?

62. Give an example of a function f such that $f = f^{-1}$.

3.7 Variation

In many applications of mathematics, it is necessary to write relationships between variables. For example, in chemistry the ideal gas law shows how temperature, pressure, and volume are related. In physics, various formulas in optics show how the focal length of a lens and the size of an image are related.

If the quotient of two variables is constant, then one variable *varies directly* or is *directly proportional* to the other. This idea can be stated in a different way as follows.

Directly Proportional

> y **varies directly** as x, or y **is directly proportional** to x, means that a nonzero real number k (called the **constant of variation**) exists such that
>
> $$y = kx.$$

EXAMPLE 1 Suppose the area of a certain rectangle varies directly as the length. If the area is 50 m² when the length is 10 m, find the area when the length is 25 m.

Since we know that the area varies directly as the length, we have

$$A = kL,$$

where A represents the area of the rectangle, L is the length, and k is a nonzero constant that we must determine. When $A = 50$ and $L = 10$, $A = kL$ becomes

$$50 = 10k \quad \text{or} \quad k = 5.$$

Thus, in this example, the relationship between area and length can be expressed as

$$A = 5L.$$

To find the area when the length is 25, replace L with 25 to get

$$A = 5L = 5(25) = 125.$$

The area of the rectangle is 125 m² when the length is 25 m. ◻

Sometimes y varies as a power of x.

Varies as nth Power

Let n be a positive real number. Then y **varies directly as the nth power** of x, or y is **directly proportional** to the nth power of x, if there exists a real number k such that

$$y = kx^n.$$

For example, the area of a square is given by the formula $A = x^2$, so that the area varies directly as the square of the length of a side. Here $k = 1$.

Sometimes y increases as x decreases. This is the case with *inverse variation*.

Inverse Variation

Let n be a positive real number. Then y **varies inversely as the nth power** of x means that there exists a real number k such that

$$y = \frac{k}{x^n}.$$

If $n = 1$, then $y = k/x$, and y **varies inversely** as x.

EXAMPLE 2   ☐   In a certain manufacturing process, the cost of producing a single item varies inversely as the square of the number of items produced. If 100 items are produced, each costs \$2. Find the cost per item if 400 items are produced.

We can let x represent the number of items produced and y the cost per item, and write

$$y = \frac{k}{x^2}$$

for some nonzero constant k. We know that $y = 2$ when $x = 100$. Hence,

$$2 = \frac{k}{100^2}$$

or $k = 20,000.$

Thus, the relationship between x and y is given by

$$y = \frac{20,000}{x^2}.$$

When 400 items are produced, the cost per item is given by

$$y = \frac{20,000}{400^2} = .125, \text{ or } 12.5\text{¢}. \quad ☐$$

One variable may depend on more than one other variable. Such variation is called *combined variation*. If a variable depends on the product of two or more other variables, we refer to that as *joint variation*.

Joint Variation

> *y* **varies jointly** as the *n*th power of *x* and the *m*th power of *z* if there exists a real number *k* such that
>
> $$y = kx^n z^m.$$

The next example shows combined variation.

EXAMPLE 3   ☐   Suppose *m* varies directly as the square of *p* and inversely as *q*. Suppose also that $m = 8$ when $p = 2$ and $q = 6$. Find *m* if $p = 6$ and $q = 10$.
Here *m* depends on two variables:

$$m = \frac{kp^2}{q}.$$

We know that $m = 8$ when $p = 2$ and $q = 6$. Thus,

$$8 = \frac{k \cdot 2^2}{6}$$

$$k = 12$$

and so

$$m = \frac{12p^2}{q}.$$

Letting, $p = 6$ and $q = 10$ we have

$$m = \frac{12 \cdot 6^2}{10}$$

$$m = \frac{216}{5}. \quad \square$$

Let us now summarize the steps involved in solving a problem in variation:

Solving Variation Problems

> 1. Write, in an algebraic form, the general relationship among the variables. Use the constant *k*.
> 2. Substitute given values of the variables and find the value of *k*.
> 3. Substitute this value of *k* into the formula of Step 1, thus obtaining a specific formula.
> 4. Solve for the required unknown.

EXERCISES
3.7

Express each of the following as an equation.

1. a varies directly as b.
2. m is proportional to n.
3. x is inversely proportional to y.
4. p varies inversely as y.
5. r varies jointly as s and t.
6. R is proportional to m and p.
7. w is proportional to x^2 and inversely proportional to y.
8. c varies directly as d and inversely as f^2 and g.

Solve each of the following.

9. If m varies directly as x and y, and $m = 10$ when $x = 4$ and $y = 7$; find m when $x = 11$ and $y = 8$.

10. Suppose m varies directly as z and p. If $m = 10$ when $z = 3$ and $p = 5$, find m when $z = 5$ and $p = 7$.

11. Suppose r varies directly as the square of m, and inversely as s. If $r = 12$ when $m = 6$ and $s = 4$, find r when $m = 4$ and $s = 10$.

12. Suppose p varies directly as the square of z, and inversely as r. If $p = \dfrac{32}{5}$ when $z = 4$ and $r = 10$, find p when $z = 2$ and $r = 16$.

13. Let a be proportional to m and n^2, and inversely proportional to y^3. If $a = 9$ when $m = 4$, $n = 9$, and $y = 3$, find a if $m = 6$, $n = 2$, and $y = 5$.

14. If y varies directly as x, and inversely as m^2 and r^2, and $y = \dfrac{5}{3}$ when $x = 1$, $m = 2$, and $r = 3$, find y if $x = 3$, $m = 1$, and $r = 8$.

15. Let r vary directly as p^2 and q^3, and inversely as z^2 and x. If $r = 334.6$ when $p = 1.9$, $q =$ 2.8, $z = .4$, and $x = 3.7$, find r when $p = 2.1$, $q = 4.8$, $z = .9$, and $x = 1.4$.

16. Let p vary directly as x and z^2, and inversely as m, n, and q. If $p = 9.25$ when $x = 3.2$, $z = 10.1$, $m = 2$, $n = 1.7$, and $q = 8.3$, find p when $x = 8.1$, $z = 4.3$, $m = 4$, $n = 2.1$, and $q = 6.2$.

17. The distance a body falls from rest varies directly as the square of the time it falls (disregarding air resistance). If an object falls 1024 feet in 8 seconds, how far will it fall in 12 seconds?

18. Hooke's law for an elastic spring states that the distance a spring stretches varies directly as the force applied. If a force of 15 pounds stretches a certain spring 8 inches, how much will a force of 30 pounds stretch the spring?

19. In electric current flow, it is found that the resistance (measured in units called ohms) offered by a fixed length of wire of a given material varies inversely as the square of the diameter of the wire. If a wire .01 inches in diameter has a resistance of .4 ohm, what is the resistance of a wire of the same length and material but .03 inches in diameter?

20. The illumination produced by a light source varies inversely as the square of the distance from the source. The illumination of a light source at 5 meters is 70 candela. What is the illumination 12 meters from the source?

21. The pressure exerted by a certain liquid at a given point is proportional to the depth of the point below the surface of the liquid. If the pressure 20 meters below the surface is 70 kilograms per square centimeter, what pressure is exerted 40 meters below the surface?

22. The distance that a person can see to the horizon from a point above the surface of the earth varies directly as the square root of the height. A

person on a hill 121 meters high can see for 15 kilometers to the horizon. How far is the horizon from a hill 900 meters high?

23. Simple interest varies jointly as principal and time. If $1000 left at interest for 2 years earned $110, find the amount of interest earned by $5000 for 5 years.

24. The volume of a right circular cylinder is jointly proportional to the square of the radius of the circular base and to the height. If the volume is 300 cubic centimeters when the height is 10.62 centimeters and the radius is 3 centimeters, find the volume for a cylinder with a radius of 4 centimeters and a height of 15.92 centimeters.

25. The Downtown Construction Company is designing a building whose roof rests on round concrete pillars. The company's engineers know that the maximum load a cylindrical column of circular cross section can hold varies directly as the fourth power of the diameter and inversely as the square of the height. If a 9 meter column 1 meter in diameter will support a load of 8 metric tons, how many metric tons will be supported by a column 12 meters high and 2/3 meter in diameter?

26. The company's engineers also know that the maximum load of a horizontal beam which is supported at both ends varies directly as the width and square of the height and inversely as the length between supports. If a beam 8 meters long, 12 centimeters wide, and 15 centimeters high can support a maximum of 400 kilograms, what will they find to be the maximum load of a beam of the same material 16 meters long, 24 centimeters wide, and 8 centimeters high?

27. The force needed to keep a car from skidding on a curve varies inversely as the radius of the curve and jointly as the weight of the car and the square of the speed. It takes 3000 pounds of force to keep a 2000 pound car from skidding on a curve of radius 500 feet at 30 mph. What force is needed to keep the same car from skidding on a curve of radius 800 feet at 60 mph?

28. The period of a pendulum varies directly as the square root of the length of the pendulum and inversely as the square root of the acceleration due to gravity. Find the period when the length is 121 centimeters and the acceleration due to gravity is 980 centimeters per second squared, if the period is 6π seconds when the length is 289 centimeters and the acceleration due to gravity is 980 centimeters per second squared.

29. The pressure on a point in a liquid is directly proportional to the distance from the surface to the point. In a certain liquid the pressure at a depth of 4 m is 60 kg per m^2. Find the pressure at a depth of 10 m.

30. The volume V of a gas varies directly as the temperature T and inversely as the pressure P. If V is 10 when T is 280 and P is 6, find V if T is 300 and P is 10.

31. A sociologist has decided to rank the cities in a nation by population, with population varying inversely as the ranking. If a city with a population of 1,000,000 ranks 8th, find the population of a city that ranks 2nd.

32. Under certain conditions, the length of time that it takes for fruit to ripen during the growing season varies inversely as the average maximum temperature during the season. If it takes 25 days for fruit to ripen with an average maximum temperature of 80°, find the number of days it would take at 75°.

33. The number of long distance phone calls between two cities in a certain time period varies directly as the populations p_1 and p_2 of the cities, and inversely as the distance between them. If 10,000 calls are made between two cities 500 miles apart, having populations of 50,000 and 125,000, find the number of calls between two cities 800 miles apart and having populations of 20,000 and 80,000.

34. The horsepower needed to run a boat through water varies as the cube of the speed. If 80 horsepower are needed to go 15 kilometers per hour in a certain boat, how many horsepower would be needed to go 30 kilometers per hour?

35. According to Poiseuille's Law, the resistance to flow of a blood vessel, R, is directly proportional to the length, l, and inversely proportional to the fourth power of the radius, r. If $R = 25$ when $l = 12$ and $r = 0.2$, find R as r increases to 0.3, while l is unchanged.

36. The Stefan-Boltzmann Law says that the radiation of heat from an object is directly proportional to the fourth power of the Kelvin temperature of the object. For a certain object, $R = 213.73$ at room temperature (293° Kelvin). Find R if the temperature increases to 335° Kelvin.

37. Suppose a nuclear bomb is detonated at a certain site. The effects of the bomb will be felt over a distance from the point of detonation that is directly proportional to the cube root of the yield of the bomb. Suppose a 100-kiloton bomb has certain effects to a radius of 3 km from the point of detonation. Find the distance that the effects would be felt for a 1500-kiloton bomb.

38. The maximum speed possible on a length of railroad track is directly proportional to the cube root of the amount of money spent on maintaining the track. Suppose that a maximum speed of 25 km per hour is possible on a stretch of track for which $450,000 was spent on maintenance. Find the maximum speed if the amount spent on maintenance is increased to $1,750,000.

39. Assume that a person's weight increases directly as the cube of their height. Find the weight of a 20 inch, 7 pound baby who grows up to be an adult 67 inches tall. (How reasonable does our assumption about weight and height seem?)

40. The cost of a pizza varies directly as the square of its radius. If a pizza with a radius of 15 cm costs $7, find the cost of a pizza having a radius of 22.5 cm. (You might want to do some research at a nearby pizza establishment and see if this assumption is reasonable.)

Chapter 3   Summary

Key Words

origin	radius	range
x-axis	center	independent variable
y-axis	center-radius form	dependent variable
rectangular coordinate system	completing the square	vertical line test for a function
cartesian coordinate system	symmetry with respect to the y-axis	increasing
coordinate plane	symmetry with respect to the x-axis	decreasing
xy-plane	symmetry with respect to the origin	constant function
quadrants	symmetry with respect to the line	composition of functions
coordinates	$y = x$	one-to-one function
ordered pair	translated	horizontal line test
distance formula	vertical translation	inverse functions
midpoint formula	horizontal translation	varies directly
graph	function	directly proportional
graph of the equation	mapping	constant of variation
parabola	domain	varies inversely
vertex	value	combined variation
circle	image	joint variation

Chapter 3   Review Exercises

Graph the set of points satisfying each condition below.

1. $x < 0$

2. $y \geq 0$

3. $xy > 0$

4. $x = 2$

Find the distance $d(P, Q)$ and the midpoint of segment PQ.

5. $P(3, -1)$ and $Q(-4, 5)$

6. $P(-8, 2)$ and $Q(3, -7)$

7. Find the other endpoint of a line segment having one end at $(-5, 7)$ and having midpoint at $(1, -3)$.

8. Are the points $(5, 7)$, $(3, 9)$, $(6, 8)$ the vertices of a right triangle?

9. Find all possible values of k so that $(-1, 2)$, $(-10, 5)$, and $(-4, k)$ are the vertices of a right triangle.

10. Find all possible values of x so that the distance between $(x, -9)$ and $(3, -5)$ is 6.

11. Find all points (x, y) with $x = 6$ so that (x, y) is 4 units from $(1, 3)$.

12. Find all points (x, y) with $x + y = 0$ so that (x, y) is 6 units from $(-2, 3)$.

Prove each of the following.

13. The medians to the two equal sides of an isosceles triangle are equal in length.

14. The line segment connecting midpoints of two sides of a triangle is half as long as the third side.

Graph each of the following. Give the domain of those that are functions.

15. $x + y = 4$

16. $3x - 5y = 20$

17. $y = \dfrac{1}{2}x^2$

18. $y = 3 - x^2$

19. $y = \dfrac{-8}{x}$

20. $y = -2x^3$

21. $y = \sqrt{x - 7}$

22. $(x - 3)^2 + y^2 = 16$

Find equations for each of the circles below.

23. Center $(-2, 3)$, radius 5

24. Center $(\sqrt{5}, -\sqrt{7})$, radius $\sqrt{3}$

25. Center $(-8, 1)$, passing through $(0, 16)$

26. Center $(3, -6)$, tangent to the x-axis

Find the center and radius of each of the following that are circles.

27. $x^2 - 4x + y^2 + 6y + 12 = 0$

28. $x^2 - 6x + y^2 - 10y + 30 = 0$

29. $x^2 + 7x + y^2 + 3y + 1 = 0$

30. $x^2 + 11x + y^2 - 5y + 46 = 0$

Decide whether the equations below have graphs that are symmetric to the x-axis, the y-axis, the origin, or the line $y = x$.

31. $3y^2 - 5x^2 = 15$

32. $x + y^2 = 8$

33. $y^3 = x + 1$

34. $x^2 = y^3$

35. $|y| = -x$

36. $|x + 2| = |y - 3|$

37. $|x| = |y|$

38. $xy = 8$

39. Graph $y = |x|$.

Using your graph from Exercise 39, explain how each of the following graphs could be obtained. Sketch each graph.

40. $y = |x| - 3$

41. $y = -|x|$

42. $y = -|x| - 2$

43. $y = -|x + 1| + 3$

44. $y = 2|x - 3| - 4$

Give the domain of each of the following.

45. $y = -4 + |x|$

46. $y = 3x^2 - 1$

47. $y = (x - 4)^2$

48. $y^2 = 8 - x$

49. $y = \dfrac{8 + x}{8 - x}$

50. $y = -\sqrt{\dfrac{5}{x^2 + 9}}$

51. $y = \sqrt{49 - x^2}$

52. $y = -\sqrt{x^2 - 4}$

The graph of a function f is shown in the figure below.

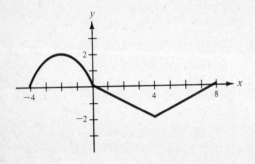

Sketch the graph of each of the following functions.

53. $y = f(x) + 3$

54. $y = f(x) - 4$

55. $y = f(x - 2)$

56. $y = f(x + 4)$

57. $y = f(x + 3) - 2$

58. $y = f(x - 1) + 4$

59. $y = |f(x)|$

For each of the following, find $\dfrac{f(x + h) - f(x)}{h}$.

60. $f(x) = -2x^2 + 4x - 3$

61. $f(x) = -x^3 + 2x^2$

62. $f(x) = 5$

63. A trailer hauling service charges $45, plus $2 per mile or part of a mile. Graph the ordered pairs (miles, cost).

64. Let f be a function which gives the cost to rent a floor polisher for x hours. The cost is a flat $3 for cleaning the polisher plus $4 per day or fraction of a day for using the polisher. (a) Graph f. (b) Give the domain and range of f.

Decide whether the following functions are odd, even, or neither.

65. $g(x) = \dfrac{x}{|x|}$ $(x \neq 0)$

66. $h(x) = 9x^6 + 5|x|$

67. $f(x) = x$

68. $f(x) = -x^4 + 9x^2 - 3x^6$

Let $f(x) = 3x^2 - 4$ *and* $g(x) = x^2 - 3x - 4$. *Find each of the following.*

69. $(f + g)(x)$

70. $(f \cdot g)(x)$

71. $(f - g)(4)$

72. $(f + g)(-4)$

73. $(f + g)(2k)$

74. $(f \cdot g)(1 + r)$

75. $(f/g)(3)$

76. $(f/g)(-1)$

77. Give the domain of $(f \cdot g)(x)$.

78. Give the domain of $(f/g)(x)$.

Let $f(x) = \sqrt{x - 2}$ *and* $g(x) = x^2$. *Find each of the following.*

79. $(f \circ g)(x)$

80. $(g \circ f)(x)$

81. $(f \circ g)(-6)$

82. $(f \circ g)(2)$

83. $(g \circ f)(3)$

84. $(g \circ f)(24)$

Find all intervals where the following functions are increasing or decreasing.

85. $y = -(x + 1)^2$

86. $y = -3 + |x|$

87. $y = \dfrac{|x|}{x}$

88. $y = |x| + x^2$

Which of the following functions are one-to-one?

89.

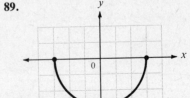

90.

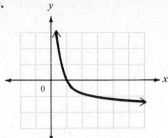

91.

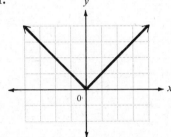

92. $y = \dfrac{8x - 9}{5}$ **93.** $y = -x^2 + 11$

94. $y = \sqrt{5 - x}$ **95.** $y = \sqrt{100 - x^2}$

96. $y = -\sqrt{1 - \dfrac{x^2}{100}}, \; x \geq 0$

For each of the following functions that is one-to-one, write an equation for the inverse function in the form $y = f^{-1}(x)$ and then graph f and f^{-1}.

97. $f(x) = 12x + 3$ **98.** $f(x) = \dfrac{2}{x - 9}$

99. $f(x) = x^3 - 3$ **100.** $f(x) = x^2 - 6$

101. $f(x) = \sqrt{25 - x^2}$, domain $[0, 5]$

102. $f(x) = -\sqrt{x - 3}$

Write each of the statements below as an equation.

103. m varies directly as the square of z

104. y varies inversely as r and directly as the cube of p

105. Y varies jointly as M and the square of N and inversely as the cube of X

106. A varies jointly as the third power of t and the fourth power of s, and inversely as p and the square of h.

Solve each problem below.

107. Suppose r varies directly as x and inversely as the square of y. If r is 10 when x is 5 and y is 3, find r when x is 12 and y is 4.

108. Suppose m varies jointly as n and the square of p, and inversely as q. If m is 20 when n is 5, p is 6, and q is 18, find m when n is 7, p is 11, and q is 2.

109. Suppose Z varies jointly as the square of J and the cube of M, and inversely as the fourth power of W. If Z is 125 when J is 3, M is 5 and W is 1, find Z if J is 2, M is 7, and W is 3.

110. The power a windmill obtains from the wind varies directly as the cube of the wind velocity. If a 10 kph wind produces 10,000 units of power, how much power is produced by a wind of 15 kph?

111. Hooke's Law for an elastic spring states that the distance a spring stretches varies directly as the force applied. If a force of 32 pounds stretches a certain spring 48 inches, how much will a force of 24 pounds stretch the spring?

112. A baseball diamond is a square, 90 feet on a side. Casey runs at a constant velocity of 30 ft/sec whether he hits a ground ball or a home run. Today, in his first time at bat, he hit a home run. Write an expression for the function that measures his line-of-sight distance from second base as a function of the time t, in sec, after he left home plate.

113. Alec, on vacation in Canada, found that he got a 12% premium on his U.S. money. When he returned, he discovered there was a 12% discount on converting his Canadian money back into U.S. currency. Describe each conversion function and then show that one is not the inverse of the other. In other words, show that after converting both ways, Alec lost money.

114. Let $f(x)$ be the fifth decimal place of the decimal expansion of x. For example, $f(1/64) = f(0.015625) = 2$, $f(98.7865433210) = 4$, $f(-78.90123456) = 3$, etc. Find the domain and range of f.

115. Prove that a graph that is symmetric with respect to any two perpendicular lines is also symmetric with respect to their point of intersection.

116. If the point $P(x, y)$ is on the line through $P_1(x_1, y_1)$ and $P_2(x_2, y_2)$ such that $d(P_1, P)/d(P_1, P_2) = k$, prove that the coordinates of P are given by $x = x_1 + k(x_2 - x_1)$ and $y = y_1 + k(y_2 - y_1)$, where

$0 < k < 1$ if P is between P_1 and P_2

$k > 1$ if P is not between P_1 and

         P_2 and is closer to P_2

$k < 0$ if P is not between P_1 and

         P_2 and is closer to P_1

117. Find formulas for $(f \circ g)(x)$ if

$$f(x) = \begin{cases} 0 & \text{if } x < 0 \\ 2x & \text{if } 0 \le x \le 1 \\ 0 & \text{if } x > 1 \end{cases} \quad \text{and}$$

$$g(x) = \begin{cases} 1 & \text{if } x < 0 \\ x/2 & \text{if } 0 \le x \le 1 \\ 1 & \text{if } x > 1. \end{cases}$$

118. Find formulas for $(g \circ f)(x)$ for the functions of Exercise 117.

119. Find formulas for $(f \circ g)(x)$ if

$$f(x) = \begin{cases} 0 & \text{if } x < 0 \\ x^2 & \text{if } 0 \le x \le 1 \\ 0 & \text{if } x > 1 \end{cases} \quad \text{and}$$

$$g(x) = \begin{cases} 1 & \text{if } x < 0 \\ 2x & \text{if } 0 \le x \le 1 \\ 1 & \text{if } x > 1. \end{cases}$$

120. From the origin, chords (lines cutting the circle in two points and perpendicular to a diameter) are drawn. Prove that the set of midpoints of these chords is a circle.

121. Given $f(x) = \begin{cases} x & \text{if } x < 1 \\ x^2 & \text{if } 1 \le x \le 9 \\ 27\sqrt{x} & \text{if } 9 < x, \end{cases}$

prove that f has an inverse function and find $f^{-1}(x)$.

122. Determine the value of the constant k so that the function defined by

$$f(x) = \frac{x + 5}{x + k}$$

will be its own inverse.

4

More Functions and Graphs

$\mathbf{I}$n this chapter we begin the study of *polynomial functions*. A **polynomial function of degree** n, where n is a nonnegative integer, is a function of the form

$$f(x) = a_n x^n + a_{n-1} x^{n-1} + \cdots + a_1 x + a_0,$$

where $a_n, a_{n-1}, \ldots, a_1$, and a_0 are real numbers and $a_n \neq 0$.

The first polynomial function that we study is degree 1; it is called a **linear function.** Then we discuss the polynomial function of degree 2, which is called a **quadratic function.** We then end the chapter with a discussion of some other useful graphs.

4.1 Linear Functions

$\mathbf{A}$s mentioned above, a linear function is a polynomial function of degree 1.

Linear Function

A function f is **linear** if

$$f(x) = ax + b,$$

for real numbers a and b with $a \neq 0$.

It is customary to write a linear function as $f(x) = ax + b$ rather than $f(x) = a_1 x + a_0$. We shall see that, as the name implies, every linear function has a graph that is a straight line.

Before we can show that this last statement is true, we need to discuss equations of straight lines. An important characteristic of a line is its *slope*, a numerical measure of the steepness of the line. To find this measure, let us consider the line through the two distinct points, (x_1, y_1) and (x_2, y_2), as shown in Figure 4.1(a). (We are assuming $x_1 \neq x_2$.) The difference

$$x_2 - x_1$$

is called the **change in x** and denoted Δx (read "delta *x*"), where Δ is the Greek letter *delta*. In the same way, the **change in y** can be written

$$\Delta y = y_2 - y_1.$$

Slope

> The **slope** of a nonvertical line (usually denoted *m*) is defined to be the quotient of the change in *y* and the change in *x*, or
>
> $$m = \frac{\Delta y}{\Delta x} = \frac{y_2 - y_1}{x_2 - x_1}.$$

We say that the slope of a line can be found only if the line is nonvertical. In this way we guarantee that $x_2 \neq x_1$ so that the denominator $x_2 - x_1 \neq 0$.

> The slope of a vertical line is not defined.

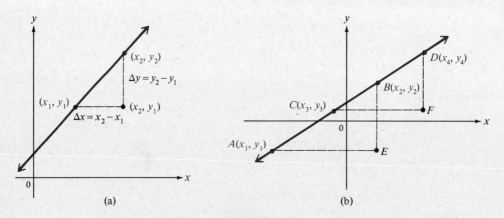

(a)   (b)

Figure 4.1

Figure 4.1(b) shows four points $A(x_1, y_1)$, $B(x_2, y_2)$, $C(x_3, y_3)$, and $D(x_4, y_4)$ all on the same straight line. Right triangles *ABE* and *CDF* have been completed. Since the lines *AE* and *CF* are parallel, as are the lines *EB* and *FD*, triangles ABE and *CDF* are similar. Similar triangles have corresponding sides proportional, so that

$$\frac{y_2 - y_1}{x_2 - x_1} = \frac{y_4 - y_3}{x_4 - x_3}.$$

The quotient on the left is the one we used in the definition of slope, and we have just proved that it equals the quotient on the right, for any choice of two distinct points on the line. Thus, the slope of a line is the same regardless of the choice of points made.

EXAMPLE 1   Find the slope of the line through each of the following pairs of points.

(a) $(-4, 8), (2, -3)$

We can choose $x_1 = -4$, $y_1 = 8$, $x_2 = 2$ and $y_2 = -3$. Then $\Delta y = -3 - 8 = -11$ and $\Delta x = 2 - (-4) = 6$. Thus the slope is $m = \Delta y/(\Delta x) = -11/6$.

(b) $(2, 7)$ and $(2, -4)$

A sketch shows that the line through $(2, 7)$ and $(2, -4)$ is vertical. As we said above, the slope of a vertical line is undefined. (An attempt to use the definition of slope here would produce a zero denominator.)

(c) $(5, -3)$ and $(-2, -3)$

By the definition of slope,

$$m = \frac{-3 - (-3)}{-2 - 5} = 0.$$

By drawing a graph, the line of Example 1(c) can be seen to be horizontal. The generalization we draw from this example is the following.

> The slope of a horizontal line is 0.

Figure 4.2 shows lines of various slopes. Note that a line with a positive slope goes up as we move from left to right but that a line with a negative slope goes down.

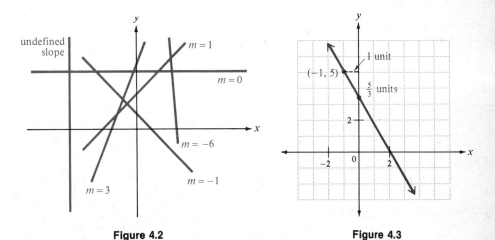

Figure 4.2   Figure 4.3

EXAMPLE 2   Graph the line going through $(-1, 5)$ and having slope $-5/3$.

We first locate the point $(-1, 5)$ as shown in Figure 4.3. Since the slope of this line is $-5/3$, a change of 1 unit horizontally produces a change of $-5/3$ units vertically. This gives a second point, which can then be used to complete the graph.

(Alternatively, a change of 3 units in a horizontal direction corresponds to a change of -5 units vertically.)   □

    Now we shall find formulas for the equation of a line. Let us first consider the formula for the equation of a vertical line. The vertical line through the point $(a, 0)$ will go through all the points of the form (a, y); this yields the following result.

Equation of a Vertical Line

> The equation of the vertical line through the point $(a, 0)$ is $x = a$.

    Let us now find the equation of the line with slope m that goes through the point (x_1, y_1). (By assuming that the line has a slope, we are assuming that it is nonvertical.) Let (x, y) represent any other point on the line. The point (x, y) can be on the line if and only if the slope of the line through (x_1, y_1) and (x, y) is m, or

$$\frac{y - y_1}{x - x_1} = m.$$

Multiplying both sides by $x - x_1$ gives

$$y - y_1 = m(x - x_1).$$

    This result, called the **point-slope form** of the equation of a line, lets us identify points on a given line: a point (x, y) lies on the line through (x_1, y_1) with slope m if and only if

$$y - y_1 = m(x - x_1).$$

In summary, we have the following.

Point-slope Form

> The line with slope m passing through the point (x_1, y_1) has an equation
> $$y - y_1 = m(x - x_1).$$
> This equation is called the **point-slope form** of the equation of a line.

EXAMPLE 3   □   Write an equation of the line through $(-4, 1)$ with slope -3.
    Here $x_1 = -4$, $y_1 = 1$, and $m = -3$. Using the point-slope form of the equation of a line, we get

$$y - 1 = -3[x - (-4)]$$
$$y - 1 = -3(x + 4)$$
$$y - 1 = -3x - 12$$
$$3x + y = -11. \quad □$$

EXAMPLE 4   □   Find an equation of the line through $(-3, 2)$ and $(2, -4)$.
    In this case we find the slope first. By the definition of slope, we have

$$m = \frac{-4 - 2}{2 - (-3)} = \frac{-6}{5}.$$

We can use either $(-3, 2)$ or $(2, -4)$ for (x_1, y_1). If we let $x_1 = -3$ and $y_1 = 2$ and use the point-slope form, we get

$$y - 2 = \frac{-6}{5}[x - (-3)]$$

$$5(y - 2) = -6(x + 3)$$

$$5y - 10 = -6x - 18$$

$$6x + 5y = -8.$$

Verify that we get the same equation if we use $(2, -4)$ instead of $(-3, 2)$ in the point-slope form.   ☐

Any value of x where a graph crosses the x-axis is called an **x-intercept** for the graph. Any value of y where the graph crosses the y-axis is called a **y-intercept** for the graph. The graph in Figure 4.4 has x-intercepts at x_1, x_2, and x_3 and a y-intercept at y_1. As suggested by the graph, x-intercepts can be found by letting $y = 0$ while letting $x = 0$ identifies y-intercepts.

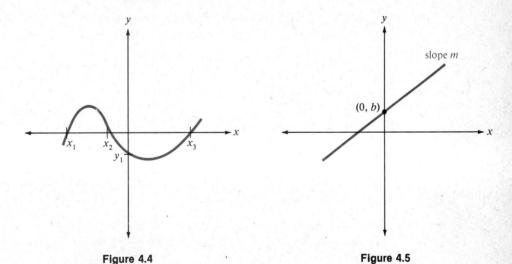

Figure 4.4   **Figure 4.5**

Suppose now a line has y-intercept b. This means that the line goes through the point $(0, b)$ as shown in Figure 4.5. If the line has slope m, then we can use the point-slope form to find an equation for this line:

$$y - y_1 = m(x - x_1)$$

$$y - b = m(x - 0)$$

$$y = mx + b.$$

This result, which shows both the slope and the y-intercept, is called the **slope-intercept form** of the equation of a line. By reversing these steps, we see that any equation of the form $y = mx + b$ has a graph that is a line with slope m and going through the point $(0, b)$. This is summarized in the box below.

Slope-Intercept Form

> The equation $y = mx + b$, called the **slope-intercept form** of the equation of a line, has a graph that is a line with slope m and y-intercept b.

By using this result, together with the fact that vertical lines have equations of the form $x = k$, we can see that every line has an equation of the form $ax + by + c = 0$, where a and b are not both 0. Conversely, by assuming $b \neq 0$ and solving $ax + by + c = 0$ for y, we get $y = (-a/b)x - c/b$. By the result above, this equation is a line with slope $-a/b$ and y-intercept $-c/b$. If $b = 0$, we solve for x to get $x = -c/a$, a vertical line. In any case, our equation $ax + by + c = 0$ has a straight line for its graph.

> If a and b are not both 0, then the equation $ax + by + c = 0$ has a line for its graph. Also, any line has an equation of the form $ax + by + c = 0$.

EXAMPLE 5

Graph $3x + 2y = 6$.

By the work above, this equation has a line for its graph. We can locate a line if we know two distinct points that the line goes through. The intercepts often provide the necessary points. To find the x-intercept, we can let $y = 0$ to get

$$3x + 2(0) = 6$$
$$3x = 6$$
$$x = 2.$$

Thus, the x-intercept is 2. Let $x = 0$ to find that the y-intercept is 3. These two intercepts were used to get the graph shown in Figure 4.6.

Alternatively, we could solve $3x + 2y = 6$ for y to get

$$3x + 2y = 6$$
$$2y = -3x + 6$$
$$y = -\frac{3}{2}x + 3.$$

By the slope intercept form, the graph of this equation is the line with y-intercept 3 and slope $-3/2$.

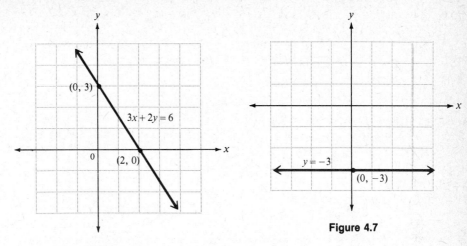

Figure 4.7

Figure 4.6

EXAMPLE 6   ☐   Graph $y = -3$.

To have an x-intercept there must be a value of x that makes $y = 0$. However, here $y = -3 \neq 0$. Hence the line has no x-intercept. This can happen only if the line is parallel to the x-axis (see Figure 4.7).   ☐

One application of slopes involves deciding whether two lines are parallel. Since two parallel lines are equally "steep," they should have the same slope. Also, two lines with the same "steepness" are parallel. We therefore have the following result.

Parallel Lines   Two nonvertical lines are parallel if and only if they have the same slope.

For a proof of this statement, see Exercises 56 and 57.

EXAMPLE 7   ☐   Find the equation of the line through the point $(3, 5)$ and parallel to the line $2x + 5y = 4$.

We know that the point $(3, 5)$ is on the line, and we need the slope to use the point-slope form of the equation of a line. We can find the slope by writing the equation of the given line in slope-intercept form, as follows.

$$2x + 5y = 4$$
$$y = -\frac{2}{5}x + \frac{4}{5}.$$

The slope is $-2/5$. Since the lines are parallel, $-2/5$ is also the slope of the line whose equation we are to find. Substituting $m = -2/5$, $x_1 = 3$, and $y_1 = 5$ into the point-slope form gives

$$y - y_1 = m(x - x_1)$$

$$y - 5 = -\frac{2}{5}(x - 3)$$

$$y - 5 = -\frac{2}{5}x + \frac{6}{5}$$

$$y = -\frac{2}{5}x + \frac{31}{5},$$

or, upon eliminating fractions,

$$2x + 5y = 31. \quad \square$$

Two lines with the same slope are parallel. As is stated below, two lines having slopes with a product of -1 are perpendicular.

Perpendicular Lines

Two lines, neither of which is vertical, are perpendicular if and only if their slopes have a product of -1.

Exercises 71–74 outline a proof of this fact.

EXAMPLE 8   □   Find the slope of the line L perpendicular to the line with equation $5x - y = 4$. To find the slope, solve $5x - y = 4$ for y, which gives

$$y = 5x - 4.$$

Hence, the slope is 5. Since the lines are perpendicular, if line L has slope m, then

$$5m = -1$$

$$m = -\frac{1}{5}. \quad \square$$

Some applications require a function that consists of parts of two or more lines. The next example shows how to graph such a function.

EXAMPLE 9   □   Graph the function

$$f(x) = \begin{cases} x + 1 & \text{if } x > 2 \\ -2x + 5 & \text{if } x \le 2. \end{cases}$$

For $x > 2$, we graph $f(x) = x + 1$. For $x \le 2$, we graph $f(x) = -2x + 5$, as shown in Figure 4.8. A single parenthesis is used at $(2, 3)$ to show that this point is not part of the graph, while the square bracket at $(2, 1)$ shows that the point is part of the graph.   □

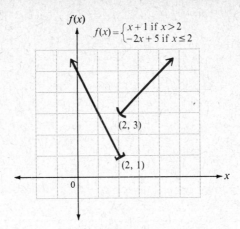

$$f(x) = \begin{cases} x+1 \text{ if } x > 2 \\ -2x+5 \text{ if } x \le 2 \end{cases}$$

(2, 3)

(2, 1)

Figure 4.8

EXERCISES
4.1

Find the slope of each of the following lines that has a slope.

1. through (4, 5) and (−1, 2)

2. through (5, −4) and (1, 3)

3. through (8, 4) and (8, −7)

4. through (1, 5) and (−2, 5)

5. $y = 2x$ **6.** $y = 3x - 2$

7. $5x - 9y = 11$ **8.** $4x + 7y = 1$

9. $x = -6$ **10.** the x-axis

11. the line parallel to $2y - 4x = 7$.

12. the line perpendicular to $6x = y - 3$.

13. through (−1.978, 4.806) and (3.759, 8.125)

14. through (11.72, 9.811) and (−12.67, −5.009)

Write an equation of each line below.

15. through (1, 3), $m = -2$

16. through (2, 4), $m = -1$

17. through (6, 1), $m = 0$

18. through (−8, 1), no slope

19. through (4, 2) and (1, 3)

20. through (8, −1) and (4, 3)

21. through (0, 3) and (4, 0)

22. through (−3, 0) and (0, −5)

23. x-intercept 3, y-intercept −2

24. x-intercept −2, y-intercept 4

25. vertical, through (−6, 5)

26. horizontal, through (8, 7)

27. through (−1.76, 4.25), with slope −5.081

28. through (5.469, 11.08), with slope 4.723

Graph the following lines.

29. through (−1, 3), $m = 3/2$

30. through (−2, 8), $m = -1$

31. through (3, −4), $m = -1/3$

32. through (−2, −3), $m = -3/4$

33. $3x + 5y = 15$

34. $2x - 3y = 12$

35. $4x - y = 8$

36. $x + 3y = 9$

37. $x + 2y = 0$

38. $3x - y = 0$

39. $x = -1$

40. $y + 2 = 0$

41. $y = -3$

42. $x = 5$

Write an equation for each of the following lines.

43. through $(-1, 4)$, parallel to $x + 3y = 5$

44. through $(2, -5)$, parallel to $y - 4 = 2x$

45. through $(3, -4)$, perpendicular to $x + y = 4$

46. through $(-2, 6)$, perpendicular to $2x - 3y = 5$

47. x-intercept -2, parallel to $y = 2x$

48. y-intercept 3, parallel to $x + y = 4$

49. the perpendicular bisector of the segment connecting $(4, -6)$ and $(3, 5)$

50. the line with x-intercept $-2/3$ and perpendicular to $2x - y = 4$

51. Do the points $(4, 3)$, $(2, 0)$, and $(-18, -12)$ lie on the same line? (Hint: find the equation of the line through two of the points.)

52. Do the points $(4, -5)$, $(3, -5/2)$, and $(-6, 18)$ lie on a line?

53. Find k so that the line through $(4, -1)$ and $(k, 2)$ is (a) parallel to $3y + 2x = 6$, (b) perpendicular to $2y - 5x = 1$.

54. Use slopes to show that the quadrilateral with vertices at $(1, 3)$, $(-5/2, 2)$, $(-7/2, 4)$, and $(2, 1)$ is a parallelogram.

55. Use slopes to show that the square with vertices at $(-2, 3)$, $(4, 3)$, $(4, -3)$, and $(-2, -3)$ has diagonals which are perpendicular.

Prove each of the following statements.

56. Two nonvertical parallel lines have the same slope.

57. Two lines with the same slope are parallel.

58. The linear function $f(x) = ax + b$ is increasing if $a > 0$ and decreasing if $a < 0$.

59. The line $y = x$ is the perpendicular bisector of the segment connecting (a, b) and (b, a), where $a \neq b$.

60. The line $ax + by = 0$, where $ab \neq 0$, goes through the origin. If $b \neq 0$, the slope of the line is $-a/b$.

Graph the following functions.

61. $y = \begin{cases} x - 1 & \text{if } x \leq 3 \\ 2 & \text{if } x > 3 \end{cases}$

62. $y = \begin{cases} 6 - x & \text{if } x \leq 3 \\ 3x - 6 & \text{if } x > 3 \end{cases}$

63. $y = \begin{cases} 4 - x & \text{if } x < 2 \\ 1 + 2x & \text{if } x \geq 2 \end{cases}$

64. $y = \begin{cases} -2 & \text{if } x \geq 1 \\ 2 & \text{if } x < 1 \end{cases}$

65. $y = \begin{cases} 2x + 1 & \text{if } x \geq 0 \\ x & \text{if } x < 0 \end{cases}$

66. $y = \begin{cases} 5x - 4 & \text{if } x \geq 1 \\ x & \text{if } x < 1 \end{cases}$

67. $y = \begin{cases} 2 + x & \text{if } x < -4 \\ -x & \text{if } -4 \leq x \leq 5 \\ 3x & \text{if } x > 5 \end{cases}$

68. $y = \begin{cases} -2x & \text{if } x < -3 \\ 3x + 1 & \text{if } -3 \leq x < 2 \\ -4x & \text{if } x \geq 2 \end{cases}$

69. When a diabetic takes long-acting insulin, the insulin reaches its peak effect on the blood sugar level in about three hours. This effect remains fairly constant for five hours, then declines, and is very low until the next injection. In a typical patient, the level of insulin might be given by the following function.

$$i(t) = \begin{cases} 40t + 100 & \text{if } 0 \leq t \leq 3 \\ 220 & \text{if } 3 < t \leq 8 \\ -80t + 860 & \text{if } 8 < t \leq 10 \\ 60 & \text{if } 10 < t \leq 24 \end{cases}$$

Here $i(t)$ is the blood-sugar level, in appropriate units, at time t measured in hours from the time of the injection. Suppose a patient takes insulin at 6 AM. Find the blood-sugar level at each of the following times: (a) 7 a.m. (b) 9 a.m. (c) 10 a.m. (d) noon (e) 2 p.m. (f) 5 p.m. (g) midnight. (h) Graph $y = i(t)$.

70. To rent a midsized car from Avis costs $27 per day or fraction of a day. If you pick up the car in Lansing, and drop it in West Lafayette, there is a fixed $25 dropoff charge. Let $C(x)$ represent the cost of renting the car for x days, taking it from Lansing to West Lafayette. Find each of the following: (a) $C(3/4)$ (b) $C(9/10)$ (c) $C(1)$ (d) $C\left(1\dfrac{5}{8}\right)$ (e) $C(2.4)$. (f) Graph $y = C(x)$.

(g) Is C a function? (h) Is C a linear function?

To prove that two perpendicular lines, neither of which is vertical, have slopes with a product of -1, go through the following steps. Let line L_1 have equation $y = m_1 x + b_1$, and let line L_2 have equa-

tion $y = m_2 x + b_2$. Assume that L_1 and L_2 are perpendicular and complete right triangle MPN as shown in the figure.

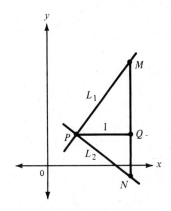

71. Show that MQ has a length m_1.

72. Show that QN has a length $-m_2$.

73. Show that triangles MPQ and PQN are similar.

74. Show that $m_1/1 = 1/-m_2$ and that $m_1 m_2 = -1$.

4.2 Quadratic Functions

Recall that we defined a polynomial function at the beginning of this chapter. A polynomial function of degree 1 is a linear function; a polynomial function of degree 2 is a **quadratic function.**

Quadratic Function

A **quadratic function** is a function of the form

$$f(x) = ax^2 + bx + c,$$

where a, b, and c are real numbers with $a \neq 0$.

The simplest quadratic function is $y = x^2$ with $a = 1$, $b = 0$, and $c = 0$. To find some points of this function, we choose values for x and find the corresponding values for y, as shown in the chart with Figure 4.9. These points are then graphed, and a smooth curve drawn through them. The reason that the curve is smooth depends on ideas from calculus. As we learned in Chapter 3, this curve is called a *parabola*. Every quadratic function has a graph that is a parabola.

Parabolas are examples of graphs having symmetry about a line (the y-axis in Figure 4.9). The line of symmetry for a parabola is called the **axis** of the parabola. The point where the axis intersects the parabola is the **vertex** of the parabola.

There are many real-world instances of parabolas. For example, cross sections of spotlight reflectors or radar dishes form parabolas.

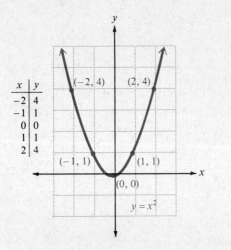

x	y
-2	4
-1	1
0	0
1	1
2	4

Figure 4.9

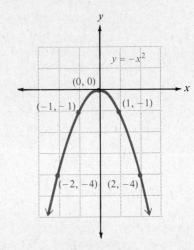

Figure 4.10

EXAMPLE 1

Graph $y = -x^2$.

For a given value of x, the corresponding value of y will be the negative of what it was for $y = x^2$. Thus, the graph of $y = -x^2$ is the same shape as that of $y = x^2$ except that it opens downward. See Figure 4.10. □

EXAMPLE 2

Graph $y = x^2 - 4$.

Each value of y will be 4 less than the corresponding value of y in $y = x^2$. Thus, $y = x^2 - 4$ has the same shape as $y = x^2$ but is shifted 4 units down. See Figure 4.11. The vertex of the parabola (on this parabola, the lowest point) is at $(0, -4)$.

The graph of $y = x^2 - 4$ has been *translated* 4 units down in relation to the graph of $y = x^2$. The axis of the parabola is the vertical line $x = 0$. □

EXAMPLE 3

Graph $y = (x - 4)^2$.

By choosing values of x and finding the corresponding values of y, this parabola can be seen to be translated 4 units to the right when compared to $y = x^2$. The vertex is at $(4, 0)$. The axis of this parabola is the vertical line $x = 4$. See Figure 4.12. □

EXAMPLE 4

Graph $y = -(x + 3)^2 + 1$.

This parabola is translated 3 units to the left and 1 unit up. It opens downward. The vertex, the point $(-3, 1)$, is the *highest* point on the graph. The axis is the line $x = -3$. See Figure 4.13. □

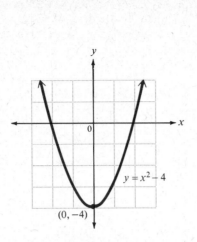

Figure 4.11

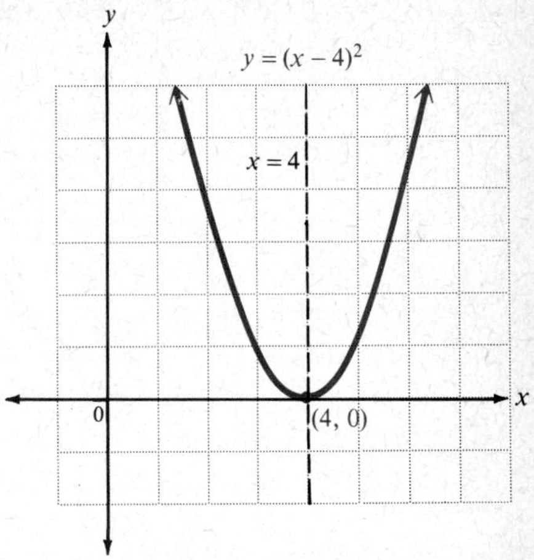

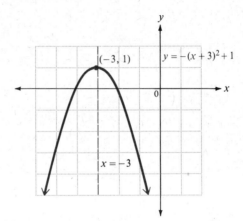

Figure 4.12

Figure 4.13

In graphing a parabola, it is helpful to find any x-intercepts that the graph might have. To get the x-intercepts of the parabola having equation $y = ax^2 + bx + c$ $(a \neq 0)$, let $y = 0$:

$$0 = ax^2 + bx + c.$$

As we saw in Chapter 2, the solutions of this equation are given by the quadratic formula:

$$x = \frac{-b \pm \sqrt{b^2 - 4ac}}{2a}.$$

The solutions are real numbers only if the discriminant $b^2 - 4ac$ is nonnegative; the graph will have no x-intercepts if $b^2 - 4ac$ is negative. This information is summarized below.

Intercepts of a Parabola

The graph of $y = ax^2 + bx + c\,(a \neq 0)$ has

(a) two different x-intercepts if $b^2 - 4ac > 0$,

(b) exactly one x-intercept if $b^2 - 4ac = 0$ (the graph just touches the x-axis as for $y = x^2$),

(c) no x-intercepts if $b^2 - 4ac < 0$.

The axis of the parabolas of this section is a vertical line halfway between any x-intercepts; we can find the equation of the axis by finding the midpoint of the line segment connecting the two x-intercepts. This midpoint is found by averaging the values of the x-intercepts, that is, taking half of

$$\left(\frac{-b + \sqrt{b^2 - 4ac}}{2a} \right) + \left(\frac{-b - \sqrt{b^2 - 4ac}}{2a} \right) = \frac{-2b}{2a} = -\frac{b}{a}.$$

Half of $-b/a$ is $-b/(2a)$: therefore, the axis of $y = ax^2 + bx + c$, $(a \neq 0)$, is the vertical line

$$x = -\frac{b}{2a}.$$

Since the axis of the parabola goes through the vertex, the x-coordinate of the vertex is $-b/(2a)$. To find the y-coordinate of the vertex, replace x with $-b/(2a)$ in the function $f(x) = ax^2 + bx + c$ to get the vertex

$$\left(-\frac{b}{2a}, f\left(-\frac{b}{2a} \right) \right).$$

Evaluating $f(x)$ for $x = -b/(2a)$ gives the results shown in the box.

Vertex and Axis of a Parabola

The parabola $f(x) = ax^2 + bx + c$ has as its axis the vertical line

$$x = -\frac{b}{2a}$$

and as its vertex the point

$$\left(-\frac{b}{2a}, f\left(-\frac{b}{2a} \right) \right) = \left(-\frac{b}{2a}, \frac{-b^2 + 4ac}{4a} \right).$$

The vertex leads to a maximum if $a < 0$ and a minimum if $a > 0$.

EXAMPLE □   Graph $f(x) = 2x^2 + 4x + 5$.
5   Here $a = 2$, $b = 4$, and $c = 5$, with $b^2 - 4ac = -24 < 0$, so that the graph has no x-intercepts. Using the results given in the box just above, we find the axis of the parabola to be the vertical line

$$x = -\frac{b}{2a} = -\frac{4}{2(2)} = -1.$$

The vertex is the point $(-1, f(-1))$, or

$$\left(-1, \frac{-4^2 + 4(2)(5)}{4(2)}\right) = \left(-1, \frac{-16 + 40}{8}\right) = (-1, 3).$$

Use the vertex and the axis, along with additional ordered pairs as necessary, to get the graph of Figure 4.14. Notice that the coefficient of 2 for x^2 makes the graph go up "faster" than the graph of $y = x^2$. □

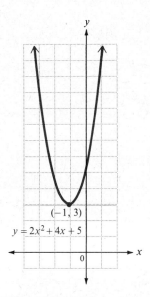

$(-1, 3)$

$y = 2x^2 + 4x + 5$

Figure 4.14

EXAMPLE □   Graph $y = -3x^2 - 2x + 1$.
6   We could use the results above. As an alternative, however, we can complete the square, just as we did in Chapter 2. First factor out -3 to get

$$y = -3\left(x^2 + \frac{2}{3}x \qquad\right) + 1.$$

Half the coefficient of x is $1/3$, and $(1/3)^2 = 1/9$. Add and subtract $1/9$ inside the parentheses as follows:

$$y = -3\left(x^2 + \frac{2}{3}x + \frac{1}{9} - \frac{1}{9}\right) + 1.$$

Using the distributive property and simplifying gives

$$y = -3\left(x^2 + \frac{2}{3}x + \frac{1}{9}\right) - 3\left(-\frac{1}{9}\right) + 1$$

$$y = -3\left(x^2 + \frac{2}{3}x + \frac{1}{9}\right) + \frac{4}{3}.$$

Factor to get $y = -3\left(x + \frac{1}{3}\right)^2 + \frac{4}{3}.$

This result shows that the axis is the vertical line

$$x + \frac{1}{3} = 0 \qquad \text{or} \qquad x = -\frac{1}{3}$$

and that the vertex is $(-1/3, 4/3)$. Use these results and additional ordered pairs as needed to get the graph of Figure 4.15. ☐

The examples of this section suggest the following result.

> When the equation of a parabola is written in the form
>
> $$y - k = a(x - h)^2,$$
>
> the vertex is at (h, k). The axis is the vertical line $x = h$.
> If $a > 0$, the parabola opens upward; if $a < 0$, it opens downward.

If $a > 0$, then k is the minimum; if $a < 0$, k is the maximum. If $|a| > 1$, the parabola goes up "faster" than $y = x^2$, but it goes up "slower" than $y = x^2$ if $0 < |a| < 1$.

The fact that the vertex of a parabola of the form $y = ax^2 + bx + c$ is the highest or lowest point on the graph can be used in applications to find a maximum or a minimum value.

EXAMPLE 7 ☐ Ms. Whitney owns and operates Aunt Emma's Pie Shop. She has hired a consultant to analyze her business operations. The consultant tells her that her profit $P(x)$ is given by

$$P(x) = 120x - x^2,$$

where x is the number of units of pies that she makes. How many units of pies should be made in order to maximize the profit? What is the maximum possible profit?

The profit function can be rewritten as $P(x) = -x^2 + 120x + 0$, a quadratic function with $a = -1$, $b = 120$, and $c = 0$. Use the methods of this section to find that the vertex of the parabola is (60, 3600). Since $a < 0$ the vertex is the highest point on the graph and produces a *maximum* rather than a minimum. Figure 4.16 shows that portion of the profit function in Quadrant I. (Why is Quadrant I the only one of interest here?) The maximum profit of $3600 is made when 60 units of pies are made. In this case, profit increases as more and more pies are made up to 60 units and then decreases as more and more pies are made past this point. ☐

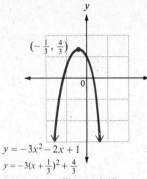

$y = -3x^2 - 2x + 1$

$y = -3(x + \frac{1}{3})^2 + \frac{4}{3}$

Figure 4.15

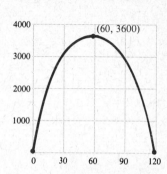

Figure 4.16

EXERCISES
4.2

1. Graph the following functions on the same co-ordinate system.

   (a) $f(x) = 2x^2$ (c) $f(x) = \frac{1}{2}x^2$

   (b) $f(x) = 3x^2$ (d) $f(x) = \frac{1}{3}x^2$

   (e) How does the coefficient affect the shape of the graph?

2. Graph the following functions on the same co-ordinate system.

   (a) $f(x) = x^2 + 2$ (c) $f(x) = x^2 + 1$

   (b) $f(x) = x^2 - 1$ (d) $f(x) = x^2 - 2$

   (e) How do these graphs differ from the graph of $y = x^2$?

3. Graph the following functions on the same co-ordinate system.

   (a) $f(x) = (x - 2)^2$ (c) $f(x) = (x + 3)^2$

   (b) $f(x) = (x + 1)^2$ (d) $f(x) = (x - 4)^2$

   (e) How do these graphs differ from the graph of $f(x) = x^2$?

4. Use the quadratic formula to find the values of x when $f(x) = 0$. Use the two values you get to locate the x value of the vertex. Find the axis of each parabola.

   (a) $f(x) = x^2 + 8x + 13$

   (b) $f(x) = x^2 - 12x + 30$

   (c) $f(x) = 3x^2 - 2x + 6$

   (d) $f(x) = 5x^2 + 6x - 3$

Graph each of the following parabolas. Give the vertex and axis of each.

5. $y = (x - 2)^2$

6. $y = (x + 4)^2$

7. $y = (x + 3)^2 - 4$

8. $y = (x - 5)^2 - 4$

9. $y = -2(x + 3)^2 + 2$

10. $y = -3(x - 2)^2 + 1$

11. $y = -\dfrac{1}{2}(x + 1)^2 - 3$

12. $y = \dfrac{2}{3}(x - 2)^2 - 1$

13. $y = x^2 - 2x + 3$

14. $y = x^2 + 6x + 5$

15. $y = -x^2 - 4x + 2$

16. $y = -x^2 + 6x - 6$

17. $y = 2x^2 - 4x + 5$

18. $y = -3x^2 + 24x - 46$

Find several points satisfying each of the following and then sketch the graph.

19. $y = .14x^2 + .56x - .3$

20. $y = .82x^2 + 3.24x - .4$

21. $y = -.09x^2 - 1.8x + .5$

22. $y = -.35x^2 + 2.8x - .3$

Give the intervals where the following quadratic functions are increasing.

23. $y = (x - 2)^2$ 24. $y = (x + 3)^2$

25. $y = -x^2 + 4$ 26. $y = -x^2 - 6$

27. $y = -2x^2 - 5x + 3$

28. $y = -3x^2 - 9x + 2$

29. Glenview Community College wants to construct a rectangular parking lot on land bordered on one side by a highway. It has 320 feet of fencing which it will use to fence off the other three sides. What should be the dimensions of the lot if the enclosed area is to be a maximum? (Hint: let x represent the width of the lot and let $320 - 2x$ represent the length. Graph the area parabola, $A = x(320 - 2x)$, and investigate the vertex.)

30. What would be the maximum area that could be enclosed by the college's 320 feet of fencing if it decided to close the entrance by enclosing all four sides of the lot? (See Exercise 29.)

31. George runs a sandwich shop. By studying data concerning his past costs, he has found that the cost of operating his shop is given by

$$C(x) = 2x^2 - 1200x + 180,100,$$

where $C(x)$ is the daily cost to make x sandwiches. Find the number of sandwiches George must sell to minimize the cost. What is the minimum cost?

32. The revenue of a charter bus company depends on the number of unsold seats. If the revenue $R(x)$, is given by

$$R(x) = 5000 + 50x - x^2,$$

where x is the number of unsold seats, find the maximum revenue and the number of unsold seats which produce maximum revenue.

33. The number of mosquitoes, $M(x)$, in millions, in a certain area of Kentucky depends on the June rainfall, x, in inches, approximately as follows.

$$M(x) = 10x - x^2$$

Find the rainfall that will produce the maximum number of mosquitoes.

34. If an object is thrown upward with an initial velocity of 32 feet per second, then its height after t seconds is given by

$$h = 32t - 16t^2.$$

Find the maximum height attained by the object. Find the number of seconds it takes the object to hit the ground.

35. Find two numbers whose sum is 20 and whose product is a maximum. (Hint: Let x and $20 - x$ be the two numbers, and write an equation for the product.)

36. A charter flight charges a fare of $200 per person plus $4 per person for each unsold seat on the plane. If the plane holds 100 passengers, and if x represents the number of unsold seats, find the following.

(a) An expression for the total revenue received

600

$2x(x - 600)$

for the flight. (Hint: Multiply the number of people flying, $100 - x$, by the price per ticket.)

(b) The graph for the expression of part (a).

(c) The number of unsold seats that will produce the maximum revenue.

(d) The maximum revenue.

37. The demand for a certain type of cosmetic is given by

$$p = 500 - x,$$

where p is the price when x units are demanded.

(a) Find the revenue, $R(x)$, that would be obtained at a price of x. (Hint: revenue = demand $\times$ price.)

(b) Graph the revenue function, $R(x)$.

(c) From the graph of the revenue function, estimate the price that will produce maximum revenue.

(d) What is the maximum revenue?

38. Between the months of June and October, the percent of maximum possible chlorophyll production in a leaf is approximated by $C(x)$, where

$$C(x) = 10x + 50.$$

Here x is time in months with $x = 1$ representing June. From October through December, $C(x)$ is approximated by

$$C(x) = -20(x - 5)^2 + 100,$$

with x as above. Find the percent of maximum possible chlorophyll production in each of the following months: (a) June (b) July (c) September (d) October (e) November (f) December.

39. Use your results from Exercise 38 to sketch a graph of $y = C(x)$, from June through December. In what month is chlorophyll production a maximum?

40. An arch is shaped like a parabola. It is 30 m wide at the base and 15 m high. How wide is the arch 10 m from the ground?

41. A culvert is shaped like a parabola, 18 cm across the top and 12 cm deep. How wide is the culvert 8 cm from the top?

42. (a) Graph the parabola $y = 2x^2 + 5x - 3$.

(b) Use the graph to find the solution of the quadratic inequality $2x^2 + 5x - 3 < 0$.

(c) Use the graph to find the solution of the quadratic inequality $2x^2 + 5x - 3 > 0$.

Exercises 43 and 44 refer to the quadratic function

$$y - k = a(x - h)^2.$$

43. Give the domain and range of this function if (a) $a > 0$ (b) $a < 0$.

44. For what intervals is the function increasing?

45. Find a value of c so that $y = x^2 - 10x + c$ has exactly one x-intercept.

46. Find b so that $y = x^2 + bx + 9$ has exactly one x-intercept.

47. Let x be in the interval $[0, 1]$. Show that the product $x(1 - x)$ is always less than or equal to $1/4$. For what values of x does the product equal $1/4$?

48. Let $f(x) = a(x - h)^2 + k$, and show that $f(h + x) = f(h - x)$. Why does this show that the parabola is symmetric to its axis?

49. A parabola can be defined as the set of all points in the plane equally distant from a given point and a given line not containing the point. See the figure.

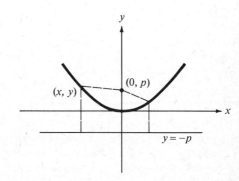

(a) Suppose (x, y) is any point on the parabola. Suppose the line mentioned in the definition is $y = -p$. Find the distance between (x, y) and the line $y = -p$. (The distance from a point to a line is the length of the perpendicular from the point to the line.)

(b) If $y = -p$ is the line mentioned in the definition, why is it reasonable to use $(0, p)$ as the given point? (See the figure.) Find the distance from (x, y) to $(0, p)$.

(c) Find an equation for the parabola of the figure.

50. Use the results of Exercise 49 to find the equation of a parabola with vertex at (h, k) and axis $x = h$.

4.3   Ellipses and Hyperbolas (Optional)

Ellipses

As the earth travels around the sun over a year's time, it traces out a curve called an *ellipse*.

<table>
<tr><td>Ellipse</td><td>An ellipse is the set of all points in a plane the sum of whose distances from two fixed points is constant. The two fixed points are called the foci of the ellipse.</td></tr>
</table>

For example, the ellipse of Figure 4.17 has foci at points F and F'. By the definition, the ellipse is made up of all points P such that the sum $d(P, F) + d(P, F')$ is constant. The ellipse of Figure 4.17 has its **center** at the origin.*

As the vertical line test shows, the graph of Figure 4.17 is not the graph of a function; here one value of x can lead to two values of y.

To obtain an equation for an ellipse centered at the origin, let the two foci have coordinates $(-c, 0)$ and $(c, 0)$, respectively. Let the sum of the distances from any point $P(x, y)$ on the ellipse to the two foci be $2a$. By the distance formula, segment PF has length

$$d(P, F) = \sqrt{(x - c)^2 + y^2},$$

while segment PF' has length

$$d(P, F') = \sqrt{[x - (-c)]^2 + y^2} = \sqrt{(x + c)^2 + y^2}.$$

*There are many applications of ellipses. In one interesting new application, patients with kidney stones are treated by being placed in a water bath in a tub with an elliptical cross section. Several hundred spark discharges are produced at one focus of the ellipse, with the kidney stone at the other focus. The discharges go through the water, causing the stone to break up into small pieces which can be readily excreted from the body.

We want the sum of the lengths $d(P, F)$ and $d(P, F')$ to be $2a$, or

$$\sqrt{(x - c)^2 + y^2} + \sqrt{(x + c)^2 + y^2} = 2a.$$

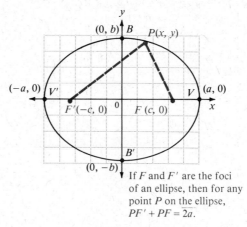

If F and F' are the foci of an ellipse, then for any point P on the ellipse, $PF' + PF = 2a$.

Figure 4.17

Using algebra (see Exercise 49) and letting $a^2 - b^2 = c^2$, we get

$$\frac{x^2}{a^2} + \frac{y^2}{b^2} = 1,$$

the **standard form** of the equation of an ellipse centered at the origin. Letting $y = 0$ in the standard form, we get

$$\frac{x^2}{a^2} + \frac{0^2}{b^2} = 1$$

$$\frac{x^2}{a^2} = 1$$

$$x^2 = a^2$$

$$x = \pm a$$

as the x-intercepts of the ellipse. The points $V'(-a, 0)$ and $V(a, 0)$ are the **vertices** of the ellipse; the segment VV' is called the **major axis**. In a similar manner, letting $x = 0$ shows that the y-intercepts are $\pm b$; the segment connecting $(0, b)$ and $(0, -b)$ is called the **minor axis**. We assumed throughout the work above that the foci were on the x-axis. If the foci were on the y-axis, an almost identical proof could be used to get the standard form

$$\frac{y^2}{a^2} + \frac{x^2}{b^2} = 1.$$

Do not be confused by the two standard forms—in one case a^2 is the denominator for x^2; in the other case a^2 is the denominator for y^2. However, in practice it is necessary only to find the intercepts of the graph—if the positive x-intercept is larger than the positive y-intercept, the major axis is horizontal, and otherwise it is vertical. When using the relationship $a^2 - c^2 = b^2$, or $a^2 - b^2 = c^2$, choose a^2 and b^2 so that $a^2 - b^2 > 0$. Let us now summarize our work with ellipses.

Equations for Ellipses

The ellipse with center at the origin and major axis along the x-axis has equation

$$\frac{x^2}{a^2} + \frac{y^2}{b^2} = 1 \ (a > b),$$

while if the ellipse has its major axis along the y-axis, the equation is

$$\frac{y^2}{a^2} + \frac{x^2}{b^2} = 1 \ (a > b).$$

Like a circle, an ellipse is symmetric about its center. An ellipse is also symmetric with respect to its major axis and its minor axis.

EXAMPLE 1   ☐   Graph $4x^2 + 9y^2 = 36$.

To obtain the standard form for the equation of an ellipse, multiply each side by $1/36$ to get

$$\frac{x^2}{9} + \frac{y^2}{4} = 1.$$

The x-intercepts of this ellipse are ± 3, and the y-intercepts ± 2. Additional ordered pairs satisfying the equation of the ellipse may be found if desired. The graph of the ellipse is shown in Figure 4.18.

Since $9 > 4$, we can find the foci by letting $c^2 = 9 - 4 = 5$ so that $c = \pm\sqrt{5}$. The major axis is along the x-axis so the foci are at $(-\sqrt{5}, 0)$ and $(\sqrt{5}, 0)$.   ☐

EXAMPLE 2   ☐   Find the equation of the ellipse having center at the origin, foci at $(0, 3)$ and $(0, -3)$, and major axis of length 8 units.

Since the major axis is 8 units long,

$$2a = 8$$

or $$a = 4.$$

To find b^2, use the relationship $a^2 - b^2 = c^2$. Here $a = 4$ and $c = 3$. Substituting for a and c gives

$$a^2 - b^2 = c^2$$
$$4^2 - b^2 = 3^2$$
$$16 - b^2 = 9$$
$$b^2 = 7.$$

Since the major axis is along the y-axis, the larger intercept (a) is used to find the denominator for y^2, giving the equation in standard form as $(y^2/16) + (x^2/7) = 1$. A graph of this ellipse is shown in Figure 4.19. ☐

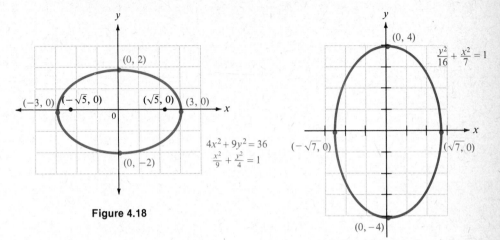

Figure 4.18

Figure 4.19

EXAMPLE 3   ☐ Graph $\dfrac{y}{4} = \sqrt{1 - \dfrac{x^2}{25}}$.

Square both sides to get

$$\frac{y^2}{16} = 1 - \frac{x^2}{25}$$

or $$\frac{x^2}{25} + \frac{y^2}{16} = 1,$$

the equation of an ellipse with x-intercepts ± 5 and y-intercepts ± 4. Since $\sqrt{1 - x^2/25} \geq 0$, we are restricted to values of y where $y/4 \geq 0$, giving the half-ellipse shown in Figure 4.20. While the graph of the ellipse $x^2/25 + y^2/16 = 1$ is not the graph of a function, the half-ellipse of Figure 4.20 *is* the graph of a function. The domain of this function is the interval $[-5, 5]$ and the range is $[0, 4]$. ☐

The results we derived above assumed an ellipse centered at the origin. The same idea, along with some complicated algebra, could be used to find the standard form of an ellipse centered at some point (h, k).

Ellipse Centered at (h, k)

An ellipse centered at (h, k) with horizontal major axis of length $2a$ has equation

$$\frac{(x - h)^2}{a^2} + \frac{(y - k)^2}{b^2} = 1.$$

There is a similar result for ellipses having major axis vertical.

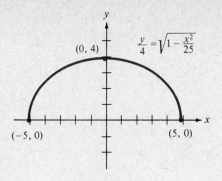

Figure 4.20

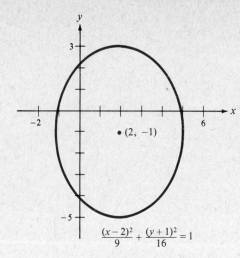

$$\frac{(x-2)^2}{9} + \frac{(y+1)^2}{16} = 1$$

Figure 4.21

**EXAMPLE
4**   ☐   Graph $\dfrac{(x - 2)^2}{9} + \dfrac{(y + 1)^2}{16} = 1$.

This equation represents an ellipse centered at $(2, -1)$. Since $16 > 9$, we have $a^2 = 16$ and $b^2 = 9$, with $a = 4$ and $b = 3$. Start at the center, $(2, -1)$, and locate two points each 3 units away from $(2, -1)$ on a horizontal line, one to the right of $(2, -1)$ and one to the left. Locate two other points on a vertical line through $(2, -1)$, one 4 units up and one 4 units down. Use additional points as needed to get the final graph shown in Figure 4.21. Here the major axis has endpoints $(2, 3)$ and $(2, -5)$.   ☐

Hyperbolas

An ellipse was defined as the set of all points in a plane having the *sum* of the distances from two fixed points a constant. Hyperbolas are defined in a similar way, except that the *difference* of the distances must be a constant.

Hyperbola

Let $F'(-c, 0)$ and $F(c, 0)$ be two points on the *x*-axis. A **hyperbola** is the set of all points $P(x, y)$ in a plane such that the difference of the distances $d(P, F')$ and $d(P, F)$ is a constant.

The midpoint of the segment $F'F$ is the **center** of the hyperbola. See Figure 4.22.

Suppose a hyperbola has center at the origin and foci at $F'(-c, 0)$ and $F(c, 0)$. If we choose $2a$ as the constant in the definition above, we have

$$d(P, F') - d(P, F) = 2a.$$

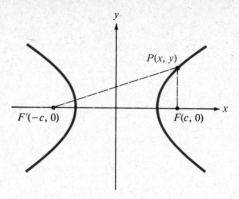

Figure 4.22

The distance formula and algebraic manipulation (see Exercise 48) produce the result

$$\frac{x^2}{a^2} - \frac{y^2}{b^2} = 1,$$

where $b^2 = c^2 - a^2$. By letting $y = 0$, the x-intercepts can be seen to be $\pm a$. If $x = 0$, the equation becomes

$$\frac{0^2}{a^2} - \frac{y^2}{b^2} = 1$$

$$-\frac{y^2}{b^2} = 1$$

$$y^2 = -b^2,$$

which has no real number solutions. Because of this, the hyperbola has no y-intercepts.

EXAMPLE 5 ▢ Graph $\dfrac{x^2}{16} - \dfrac{y^2}{9} = 1$.

This hyperbola has x-intercepts 4 and -4 and no y-intercepts. To sketch the graph, we can find other points that lie on the graph. For example, if $x = 6$, we get

$$\frac{6^2}{16} - \frac{y^2}{9} = 1$$

$$-\frac{y^2}{9} = 1 - \frac{6^2}{16}$$

$$\frac{y^2}{9} = \frac{20}{16}$$

$$y^2 = \frac{180}{16} = \frac{45}{4}$$

$$y = \frac{\pm 3\sqrt{5}}{2} \approx \pm 3.4.$$

The graph thus includes the points (6, 3.4) and (6, −3.4). Also, if $x = -6$, we would still have $y \approx \pm 3.4$ with the points (−6, 3.4) and (−6, −3.4) also on the graph. These points, along with other points on the graph, were used to help sketch the final graph shown in Figure 4.23.   ☐

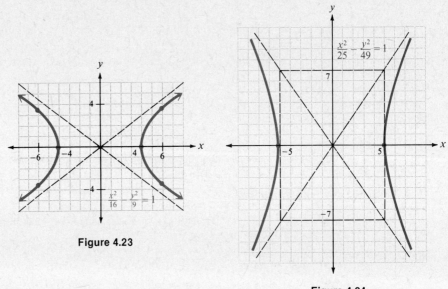

Figure 4.23

Figure 4.24

We summarize this information as follows.

Equations for Hyperbolas

A hyperbola with center at the origin and x-intercepts $\pm a$, has equation

$$\frac{x^2}{a^2} - \frac{y^2}{b^2} = 1.$$

If the hyperbola has y-intercepts $\pm a$, the equation is

$$\frac{y^2}{a^2} - \frac{x^2}{b^2} = 1.$$

If we start with $(x^2/a^2) - (y^2/b^2) = 1$ and solve for y, we get

$$\frac{x^2}{a^2} - 1 = \frac{y^2}{b^2}$$

$$\frac{x^2 - a^2}{a^2} = \frac{y^2}{b^2}$$

or $y = \pm \dfrac{b}{a}\sqrt{x^2 - a^2}.$ (1)

Suppose that x^2 is very large in comparison to a^2. In this case, the difference $x^2 - a^2$ would then be very close to just x^2. If this happens, then the points satisfying equation (1) would be very close to one of the lines

$$y = \pm \frac{b}{a} x.$$

Thus, as $|x|$ gets larger and larger, the points of the hyperbola $x^2/a^2 - y^2/b^2 = 1$ approach closer and closer to the lines $y = (\pm b/a) x$. These lines, called the **asymptotes** of the hyperbola, are very helpful when graphing the hyperbola.

EXAMPLE 6 ☐ Graph $\dfrac{x^2}{25} - \dfrac{y^2}{49} = 1$.

Here $a = 5$ and $b = 7$. Using these values, we see that $y = (\pm b/a) x$ becomes $y = (\pm 7/5) x$. Using these equations, if $x = 5$, then $y = \pm 7$, while $x = -5$ also leads to $y = \pm 7$. Thus, we have determined four points, $(5, 7)$, $(5, -7)$, $(-5, 7)$, and $(-5, -7)$. We use these four points to sketch a rectangle, as shown in Figure 4.24. The extended diagonals of this rectangle are the asymptotes of the hyperbola. The x-intercepts, ± 5, are at either side of the rectangle. The x-intercepts and asymptotes lead to the graph of Figure 4.24. We can plot additional points if greater precision is necessary.   ☐

The rectangle used to graph the hyperbola of Example 6 is called the **fundamental rectangle.**

All the hyperbolas graphed above have had foci on the x-axis. If the foci are on the y-axis, the equation of the hyperbola can be shown to be of the form

$$\frac{y^2}{a^2} - \frac{x^2}{b^2} = 1,$$

with asymptotes $y = \pm \dfrac{a}{b} x.$

If the foci of the hyperbola are on the x-axis, we found that the asymptotes have equations $y = \pm (b/a) x$, whereas foci on the y-axis lead to asymptotes $y = \pm (a/b) x$. There is an obvious chance for confusion here; to avoid mistakes write the equation of the hyperbola in either the form

$$\frac{x^2}{a^2} - \frac{y^2}{b^2} = 1 \quad \text{or} \quad \frac{y^2}{a^2} - \frac{x^2}{b^2} = 1,$$

and replace 1 with 0. Solving the resulting equation for y produces the proper equations for the asymptotes. (The reason why this process works is explained in more advanced courses.)

EXAMPLE 7 ☐ Graph $25y^2 - 4x^2 = 100$.

Multiply each side by $1/100$ to get

$$\frac{y^2}{4} - \frac{x^2}{25} = 1.$$

This hyperbola has foci on the y-axis, with y-intercepts ± 2. To find the asymptotes, replace 1 with 0, getting

$$\frac{y^2}{4} - \frac{x^2}{25} = 0,$$

or

$$\frac{y^2}{4} = \frac{x^2}{25}$$

$$y^2 = \frac{4x^2}{25},$$

from which

$$y = \pm\frac{2}{5}x.$$

Use the points $(5, 2)$, $(5, -2)$, $(-5, 2)$, and $(-5, -2)$ to get the fundamental rectangle shown in Figure 4.25. Use the diagonals of this rectangle to determine the asymptotes for the graph as shown in Figure 4.25. ▫

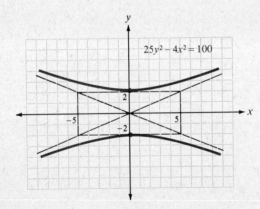

Figure 4.25

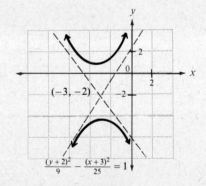

Figure 4.26

EXAMPLE 8 ▫ Graph $\dfrac{(y + 2)^2}{9} - \dfrac{(x + 3)^2}{4} = 1$.

This equation represents a hyperbola centered at $(-3, -2)$. See Figure 4.26. ▫

EXAMPLE 9 ▫ Graph $x = -\sqrt{1 + 4y^2}$.

Squaring both sides gives

$$x^2 = 1 + 4y^2,$$

or

$$x^2 - 4y^2 = 1.$$

Use the fact that $4 = 1/(1/4)$, and replace 1 with 0 to get (as equations of the asymptotes)

$$x^2 - \frac{y^2}{1/4} = 0$$

$$\frac{1}{4}x^2 = y^2$$

or, finally,

$$y = \pm\frac{1}{2}x.$$

Since the given equation $x = -\sqrt{1 + 4y^2}$ restricts x to nonpositive values, the graph is the left branch of a hyperbola, as shown in Figure 4.27. The graph is not the graph of a function. ☐

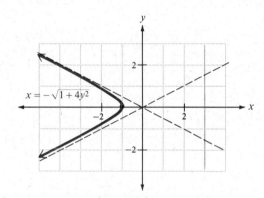

Figure 4.27

**EXERCISES
4.3**

Sketch the graph of each of the following. Give the endpoints of the major axis for ellipses and the equation of the asymptotes for hyperbolas. Give the center of each figure.

1. $\dfrac{x^2}{9} + \dfrac{y^2}{4} = 1$

2. $\dfrac{x^2}{16} + \dfrac{y^2}{36} = 1$

3. $\dfrac{x^2}{9} + y^2 = 1$

4. $\dfrac{y^2}{16} - \dfrac{x^2}{9} = 1$

5. $\dfrac{x^2}{6} + \dfrac{y^2}{9} = 1$

6. $\dfrac{x^2}{8} - \dfrac{y^2}{12} = 1$

7. $x^2 + 4y^2 = 16$

8. $25x^2 + 9y^2 = 225$

9. $x^2 = 9 + y^2$

10. $y^2 = 16 + x^2$

11. $2x^2 + y^2 = 8$

12. $9x^2 - 25y^2 = 225$

13. $25x^2 - 4y^2 = -100$

14. $4x^2 - y^2 = -16$

15. $\dfrac{x^2}{1/9} + \dfrac{y^2}{1/16} = 1$

16. $\dfrac{x^2}{4/25} + \dfrac{y^2}{9/49} = 1$

17. $\dfrac{64x^2}{9} + \dfrac{25y^2}{36} = 1$

18. $\dfrac{121x^2}{25} + \dfrac{16y^2}{9} = 1$

19. $\dfrac{(x - 1)^2}{9} + \dfrac{(y + 3)^2}{25} = 1$

20. $\dfrac{(x + 3)^2}{16} + \dfrac{(y - 2)^2}{36} = 1$

21. $\dfrac{(x - 3)^2}{16} - \dfrac{(y + 2)^2}{49} = 1$

22. $\dfrac{(y - 5)^2}{4} - \dfrac{(x + 1)^2}{9} = 1$

23. $\dfrac{(y + 1)^2}{25} - \dfrac{(x - 3)^2}{36} = 1$

24. $\dfrac{(x + 2)^2}{16} - \dfrac{(y + 2)^2}{25} = 1$

Sketch the graph of each of the following. Identify any functions.

25. $\dfrac{x}{4} = \sqrt{1 - \dfrac{y^2}{9}}$ **26.** $\dfrac{y}{2} = \sqrt{1 - \dfrac{x^2}{25}}$

27. $\dfrac{y}{3} = \sqrt{1 + \dfrac{x^2}{16}}$ **28.** $x = \sqrt{1 + \dfrac{y^2}{36}}$

29. $x = -\sqrt{1 - \dfrac{y^2}{64}}$ **30.** $y = \sqrt{1 - \dfrac{x^2}{100}}$

31. $y = -\sqrt{1 + \dfrac{x^2}{25}}$ **32.** $x = -\sqrt{1 + \dfrac{y^2}{9}}$

Find equations for each of the following ellipses.

33. x-intercepts ± 4; foci at $(-2, 0)$ and $(2, 0)$

34. y-intercepts ± 3; foci at $(0, \sqrt{3})$, $(0, -\sqrt{3})$

35. Endpoints of major axis at $(6, 0)$, $(-6, 0)$; $c = 4$

36. Vertices $(0, 5)$, $(0, -5)$; $b = 2$

37. Center $(3, -2)$, $a = 5$, $c = 3$, major axis vertical

38. Center $(2, 0)$, minor axis of length 6, major axis horizontal, and of length 9

Find equations for each of the following hyperbolas.

39. x-intercepts ± 3, foci at $(-4, 0)$, $(4, 0)$

40. y-intercepts ± 5, foci at $(0, 3\sqrt{3})$, $(0, -3\sqrt{3})$

41. Asymptotes $y = \pm(3/5)x$, y-intercepts $(0, 3)$, $(0, -3)$

42. Center at the origin, passing through $(5, 3)$ and $(-10, 2\sqrt{21})$, no y-intercepts

The orbit of Mars is an ellipse, with the sun at one focus. An approximate equation for the orbit is

$$\dfrac{x^2}{5013} + \dfrac{y^2}{4970} = 1,$$

where x and y are measured in millions of miles.

43. Find the length of the major axis.

44. Find the length of the minor axis.

45. Draftspeople often use the method shown on the sketch below to drawn an ellipse. Explain why the method works.

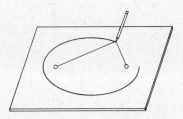

46. Ships and planes often use a location finding system called LORAN. With this system, a radio transmitter at M on the figure sends out a series of pulses. When each pulse is received at transmitter S, it then sends out a pulse. A ship at P receives pulses from both M and S. A receiver on the ship measures the difference in the arrival times of the pulses. The navigator then consults a special map, showing certain curves according to the differences in arrival times. In this way, the ship can be located as lying on a portion of which curve?

47. Microphones are placed at points $(-c, 0)$ and $(c, 0)$. An explosion occurs at point $P(x, y)$ having positive x-coordinate. (See the figure below.) The sound is detected at the closer microphone t seconds before being detected at the farther microphone. Assume that sound travels at a speed of 330 m per second, and show that P must be on the hyperbola

$$\frac{x^2}{330^2 t^2} - \frac{y^2}{4c^2 - 330^2 t^2} = \frac{1}{4}.$$

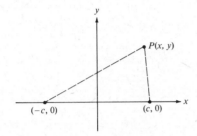

48. Suppose a hyperbola has center at the origin, foci at $F'(-c, 0)$ and $F(c, 0)$, and the value $d(P, F') - d(P, F) = 2a$. Let $b^2 = c^2 - a^2$, and show that an equation of the hyperbola is

$$\frac{x^2}{a^2} - \frac{y^2}{b^2} = 1.$$

49. Derive the standard form of the equation of an ellipse centered at the origin.

50. A rod of fixed length in the xy-coordinate plane is moved so that one end is always on the x-axis and the other end is always on the y-axis. Let P be any fixed point on the rod. Show that the path of P is an ellipse.

4.4 Conic Sections (Optional)

Before we give a general classification of the graphs of this chapter, we need to discuss the graph of an equation of the form

$$x = ay^2 + by + c,$$

where a, b, and c are real numbers and $a \neq 0$. This equation is the same as for the quadratic function, $y = ax^2 + bx + c$, with x and y interchanged. Thus, the graph of our new equation $x = ay^2 + by + c$ is the mirror image of that of $y = ax^2 + bx + c$. Hence, $x = ay^2 + by + c$ is a parabola with **horizontal** axis. That is,

Parabola with Axis Horizontal

$$x = a(y - k)^2 + h$$

is a parabola with vertex at (h, k) and axis the horizontal line $y = k$. The parabola opens to the right if $a > 0$ and to the left if $a < 0$.

EXAMPLE 1 Graph $x = 2y^2 + 6y + 5$.

To write this equation in the form $x = a(y - k)^2 + h$, we complete the square on y, as follows:

$$x = 2(y^2 + 3y \qquad) + 5$$

$$x = 2\left(y^2 + 3y + \frac{9}{4} - \frac{9}{4}\right) + 5$$

$$x = 2\left(y^2 + 3y + \frac{9}{4}\right) + 2\left(-\frac{9}{4}\right) + 5$$

$$x = 2\left(y + \frac{3}{2}\right)^2 + \frac{1}{2}.$$

As this result shows, the vertex of the parabola is the point $(1/2, -3/2)$. The axis is the horizontal line $y + (3/2) = 0$. By using the vertex and the axis and plotting a few more points, we get the graph of Figure 4.28.   ☐

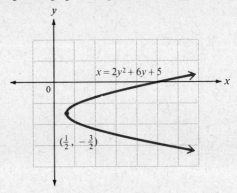

Figure 4.28

The graphs of parabolas, circles, hyperbolas, and ellipses are all called **conic sections** since each graph can be obtained by cutting a cone with a plane as shown in Figure 4.29.

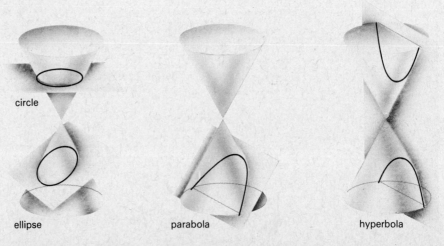

Figure 4.29

It turns out that all conic sections of the types we have studied have equations of the form

$$Ax^2 + Bx + Cy^2 + Dy + E = 0,$$

where either A or C must be nonzero. The special characteristics of each of the conic sections are summarized in the following box.

Equations of Conic Sections

Conic section	Characteristic	Example
Parabola	Either $A = 0$ or $C = 0$, but not both	$x^2 = y + 4$ $(y - 2)^2 = -(x + 3)$
Circle	$A = C \neq 0$	$x^2 + y^2 = 16$
Ellipse	$A \neq C, AC > 0$	$\dfrac{x^2}{16} + \dfrac{y^2}{25} = 1$
Hyperbola	$AC < 0$	$x^2 - y^2 = 1$

The graphs of the conic sections are summarized in Figure 4.30. Ellipses and hyperbolas having centers not at the origin can be shown in much the same way as we show circles and parabolas. (Figure 4.30 is continued on the next page.)

Equation   Graph

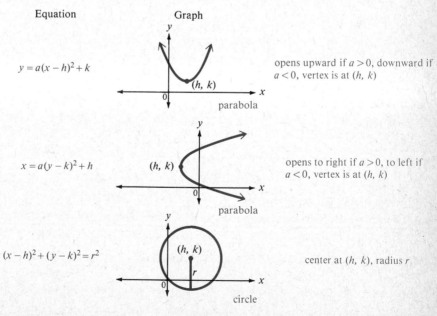

$y = a(x - h)^2 + k$

opens upward if $a > 0$, downward if $a < 0$, vertex is at (h, k)

parabola

$x = a(y - k)^2 + h$

opens to right if $a > 0$, to left if $a < 0$, vertex is at (h, k)

parabola

$(x - h)^2 + (y - k)^2 = r^2$

center at (h, k), radius r

circle

Figure 4.30

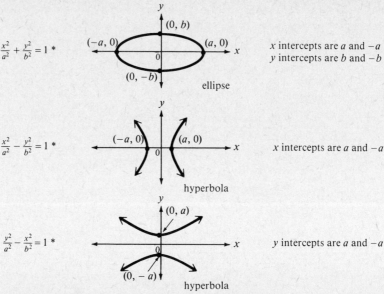

$\dfrac{x^2}{a^2} + \dfrac{y^2}{b^2} = 1$ *

x intercepts are a and $-a$
y intercepts are b and $-b$

ellipse

$\dfrac{x^2}{a^2} - \dfrac{y^2}{b^2} = 1$ *

x intercepts are a and $-a$

hyperbola

$\dfrac{y^2}{a^2} - \dfrac{x^2}{b^2} = 1$ *

y intercepts are a and $-a$

hyperbola

*Figures for the general case with center at (h, k) (like the circle) are similar.

Figure 4.30 (continued)

EXAMPLE 2 ☐ Decide on the type of conic section represented by each of the following equations, and sketch the graph.

(a) $x^2 = 25 + 5y^2$

Rewriting the equation as

$$x^2 - 5y^2 = 25$$

or $$\dfrac{x^2}{25} - \dfrac{y^2}{5} = 1$$

shows that the equation represents a hyperbola centered at the origin, with asymptotes

$$\dfrac{x^2}{25} - \dfrac{y^2}{5} = 0,$$

or $$y = \dfrac{\pm\sqrt{5}}{5}x.$$

The x-intercepts are ± 5; the graph is shown in Figure 4.31.

(b) $4x^2 - 16x + 9y^2 + 54y = -61$

Since the coefficients of the x^2 and y^2 terms are unequal and both positive, this equation might represent an ellipse. (It might also represent a single point or no points at all.) To find out, complete the square on x and y. Work as follows:

$$4(x^2 - 4x \qquad) + 9(y^2 + 6y \qquad) = -61$$
$$4(x^2 - 4x + 4 - 4) + 9(y^2 + 6y + 9 - 9) = -61$$
$$4(x^2 - 4x + 4) - 16 + 9(y^2 + 6y + 9) - 81 = -61$$
$$4(x - 2)^2 + 9(y + 3)^2 = 36$$
$$\frac{(x - 2)^2}{9} + \frac{(y + 3)^2}{4} = 1.$$

This equation represents an ellipse having center at $(2, -3)$ and graph as shown in Figure 4.32.

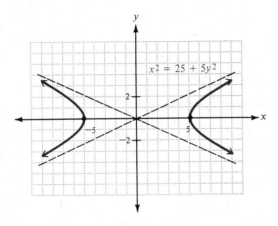

Figure 4.31   **Figure 4.32**

(c) $x^2 - 8x + y^2 + 10y = -41$

Complete the square on both x and y, as follows:

$$(x^2 - 8x + 16 - 16) + (y^2 + 10y + 25 - 25) = -41$$
$$(x - 4)^2 + (y + 5)^2 = 16 + 25 - 41$$
$$(x - 4)^2 + (y + 5)^2 = 0.$$

From this result we see that the equation is that of a circle of radius 0 or that of just a point—the point $(4, -5)$. Had we ended up with a negative number on the right (instead of 0), the equation would have represented no points at all, and there would be no graph.

(d) $x^2 - 6x + 8y - 7 = 0$

Since only one variable is squared (x, and not y), we have a parabola. Complete the square.

$$(x^2 - 6x + 9 - 9) + 8y - 7 = 0$$
$$(x - 3)^2 = 16 - 8y$$
$$(x - 3)^2 = -8(y - 2).$$

The parabola has vertex at (3, 2), and opens downward, as shown in Figure 4.33. ☐

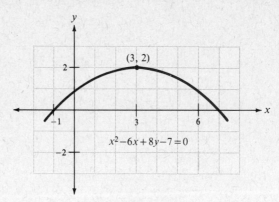

Figure 4.33

Eccentricity

We saw above how to classify a conic section as a parabola, circle, ellipse, or hyperbola by looking at its equation. An alternate way to classify conic sections is to find the *eccentricity* for the conic section. We shall first define eccentricity for an ellipse. We know that

$$\frac{x^2}{a^2} + \frac{y^2}{b^2} = 1$$

represents an ellipse centered at the origin. Thus $a^2 > b^2$ and $c = \sqrt{a^2 - b^2}$. The **eccentricity** of the ellipse, written e, is defined as

$$e = \frac{c}{a}.$$

By the way we defined c, we always have

$$0 < c < a,$$

so that

$$0 < \frac{c}{a} < 1,$$

and, for an ellipse,

$$0 < e < 1.$$

Since a is a constant, letting c approach 0 will force the ratio c/a to approach 0, which in turn forces b to approach a (so that $\sqrt{a^2 - b^2} = c$ can approach 0). Since b leads to the y-intercepts, this means that the x- and y-intercepts are almost the same, producing an ellipse very close in shape to a circle. In a similar manner, if e approaches 1, then b will approach 0, which gives a very flat ellipse. The path of the earth around the sun is an ellipse that is very nearly circular. In fact, for this ellipse, $e \approx .017$. On the other hand, the path of Halley's Comet is a very flat ellipse with $e \approx .98$.

EXAMPLE 3 Find the eccentricity of each ellipse.

(a) $\dfrac{x^2}{9} + \dfrac{y^2}{16} = 1$

Since $16 > 9$, let $a^2 = 16$, which gives $a = 4$. Also,

$c = \sqrt{a^2 - b^2}$

$c = \sqrt{16 - 9} = \sqrt{7}.$

Finally, $e = \dfrac{\sqrt{7}}{4}.$

(b) $5x^2 + 10y^2 = 50$

Multiply by $1/50$ to get

$\dfrac{x^2}{10} + \dfrac{y^2}{5} = 1.$

Here $a^2 = 10$, with $a = \sqrt{10}$. Now find c:

$c = \sqrt{10 - 5} = \sqrt{5}$

and $e = \dfrac{\sqrt{5}}{\sqrt{10}} = \dfrac{\sqrt{2}}{2}.$ ◻

The hyperbola

$$\dfrac{x^2}{a^2} - \dfrac{y^2}{b^2} = 1 \qquad \text{or} \qquad \dfrac{y^2}{a^2} - \dfrac{x^2}{b^2} = 1$$

where $c = \sqrt{a^2 + b^2}$, has eccentricity e defined again as

$$e = \dfrac{c}{a}.$$

By the definition of c, we must have

$c = \sqrt{a^2 + b^2} > a,$

so that $\dfrac{c}{a} > 1.$

Thus, for a hyperbola,

$e > 1.$

EXAMPLE 4 Find the eccentricity for the hyperbola

$$\dfrac{x^2}{9} - \dfrac{y^2}{4} = 1.$$

Here $a^2 = 9$; thus $a = 3$ and $c = \sqrt{9 + 4} = \sqrt{13}$. The eccentricity is

$$e = \dfrac{c}{a} = \dfrac{\sqrt{13}}{3}. \quad ◻$$

EXERCISES
4.4

Identify each equation of Exercises 1–20. Draw a graph of each that has a graph.

1. $x^2 = 25 + y^2$

2. $x^2 = 25 - y^2$

3. $9x^2 + 36y^2 = 36$

4. $x^2 = 4y - 8$

5. $\dfrac{(x + 3)^2}{16} + \dfrac{(y - 2)^2}{16} = 1$

6. $\dfrac{(x - 4)}{8} - \dfrac{(y + 1)^2}{2} = 0$

7. $y^2 - 4y = x + 4$

8. $11 - 3x = 2y^2 - 8y$

9. $(x + 7)^2 + (y - 5)^2 + 4 = 0$

10. $4(x - 3)^2 + 3(y + 4)^2 = 0$

11. $3x^2 + 6x + 3y^2 - 12y = 12$

12. $2x^2 - 8x + 5y^2 + 20y = 12$

13. $x^2 - 6x + y = 0$

14. $x - 4y^2 - 8y = 0$

15. $4x^2 - 8x - y^2 - 6y = 6$

16. $x^2 + 2x = y^2 - 4y - 2$

17. $4x^2 - 8x + 9y^2 + 54y = -84$

18. $3x^2 + 12x + 3y^2 = -11$

19. $6x^2 - 12x + 6y^2 - 18y + 25 = 0$

20. $4x^2 - 24x + 5y^2 + 10y + 41 = 0$

Find the eccentricity of each of the following ellipses or hyperbolas.

21. $12x^2 + 9y^2 = 36$

22. $8x^2 - y^2 = 16$

23. $x^2 - y^2 = 4$

24. $x^2 + 2y^2 = 8$

25. $4x^2 + 7y^2 = 28$

26. $9x^2 - y^2 = 1$

27. $x^2 - 9y^2 = 1$

28. $x^2 + 10y^2 = 10$

29. What is the eccentricity of a circle? (Think of a circle as an ellipse with $a = b$.)

30. The orbit of Mars around the sun is an ellipse with equation $\dfrac{x^2}{5013} + \dfrac{y^2}{4970} = 1$, where x and y are measured in millions of miles. Find the eccentricity of this ellipse.

31. Use the results of this section, along with the results of Exercise 29, to complete the following table. You will have to make a guess as to the eccentricity of a parabola. (For details on the eccentricity of a parabola, see any standard analytic geometry text.)

Conic section	Possible values of e
Circle	
Ellipse	$0 < e < 1$
Parabola	
Hyperbola	$e > 1$

32. Find the equation of a horizontal ellipse having $e = 1/2$ and $a = 4$.

4.5   Graphs of Polynomial Functions

As mentioned earlier, any function of the form

$$f(x) = a_n x^n + a_{n-1} x^{n-1} + \cdots + a_1 x + a_0,$$

is called a polynomial function of degree n, for real numbers $a_n \neq 0$, $a_{n-1}, \ldots,$ a_1, a_0, and any nonnegative integer n. Because the simplest polynomial functions to graph are those of the form $f(x) = x^n$, we discuss them first. To graph $f(x) = x^3$, find several ordered pairs that satisfy $y = x^3$, then plot them and connect the points with a smooth curve. The graph of $f(x) = x^3$ is shown as a solid curve in Figure 4.34.

Figure 4.34 also shows the graph of $f(x) = x^5$ as a broken line. Note that both graphs have symmetry about the origin. We can sketch graphs of $f(x) = x^4$ and $f(x) = x^6$ in a similar manner. Figure 4.35 shows $f(x) = x^4$ as a solid curve and $f(x) = x^6$ as a broken line. These graphs have symmetry about the y-axis as does the graph of $f(x) = ax^2$ for a nonzero real number a.

As we saw earlier with the graph of $f(x) = ax^2$, the value of a in $f(x) = ax^n$ affects the width of the graph. When $|a| > 1$, the graph is "thinner" than the graph of $f(x) = x^n$; when $0 < |a| < 1$, the graph is "fatter."

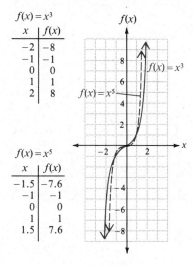

$f(x) = x^3$

x	$f(x)$
-2	-8
-1	-1
0	0
1	1
2	8

$f(x) = x^5$

x	$f(x)$
-1.5	-7.6
-1	-1
0	0
1	1
1.5	7.6

Figure 4.34

$f(x) = x^4$

x	y
-2	16
-1	1
0	0
1	1
2	16

x	y
-1.5	11.4
-1	1
0	0
1	1
1.5	11.4

Figure 4.35

EXAMPLE 1   Graph each of the following functions.

(a) $f(x) = \dfrac{1}{2}x^3$

The graph will be "fatter" than that of $f(x) = x^3$, but will have the same general shape. It goes through the points $(-2, -4)$, $(-1, -1/2)$, $(0, 0)$, $(1, 1/2)$, and $(2, 4)$. See Figure 4.36 on the next page.

(b) $f(x) = \dfrac{3}{2}x^4$

The following table gives some ordered pairs.

x	-2	-1	0	1	2
y	24	$3/2$	0	$3/2$	24

The graph is shown in Figure 4.37 on the next page. This graph is "thinner" than that of $f(x) = x^4$. ◻

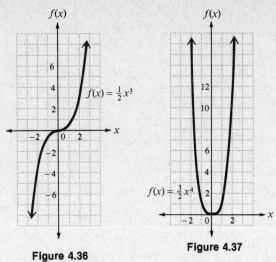

Figure 4.36

Figure 4.37

EXAMPLE
2   Graph each of the following.

(a) $f(x) = x^5 - 2$

The graph will be the same as that of $f(x) = x^5$, but translated down 2 units. See Figure 4.38.

(b) $f(x) = (x + 1)^6$

This function has a graph like that of $f(x) = x^6$, but translated 1 unit to the left as shown in Figure 4.39.

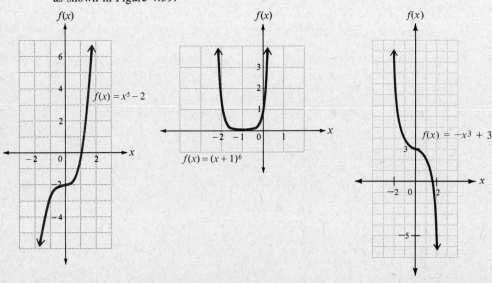

Figure 4.38   Figure 4.39   Figure 4.40

(c) $f(x) = -x^3 + 3$

The negative sign causes the graph to be reflected about the x-axis compared to the graph of $f(x) = x^3$. As shown in Figure 4.40 the graph is also translated up 3 units. ▨

Generalizing from the graphs of Example 2, the domain of a polynomial function is the set of all real numbers. The range of a polynomial function of odd degree is also the set of all real numbers. Some typical graphs of polynomial functions of odd degree are shown in Figure 4.41. These graphs suggest that for every polynomial function f of odd degree there is at least one real value of x that makes $f(x) = 0$. Such values of x are called the **real zeros** of f; these values are also the x-intercepts of the graph. In Chapter 10 we shall see how to find (or at least approximate) the zeros of a polynomial function.

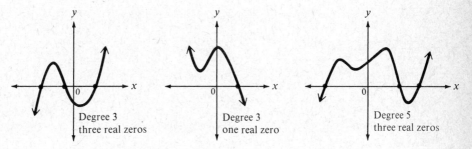

| Degree 3 | Degree 3 | Degree 5 |
| three real zeros | one real zero | three real zeros |

Figure 4.41

Polynomial functions of even degree have a range that takes the form $(-\infty, k]$ or else $[k, +\infty)$ for some real number k. Figure 4.42 shows two typical graphs of polynomial functions of even degree.

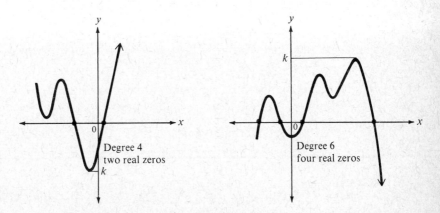

| Degree 4 | Degree 6 |
| two real zeros | four real zeros |

Figure 4.42

The graphs in Figures 4.41 and 4.42 show that polynomial functions usually have **turning points** where the function changes from increasing to decreasing or from decreasing to increasing. A polynomial function of degree n has at most $n - 1$ turning points. The graphs shown above illustrate this.

It is difficult to graph most polynomial functions without the use of calculus. A large number of points must be plotted to get a reasonably accurate graph. However, if a polynomial function can be factored, we can approximate its graph without plotting very many points; this method is shown in the next examples.

EXAMPLE 3 ☐ Graph $f(x) = (2x + 3)(x - 1)(x + 2)$.

Multiplying out the expression on the right would show that f is a third degree polynomial, also called a **cubic** polynomial. To sketch the graph of f, we first find its zeros by setting each of the three factors equal to 0 and solving the resulting equations.

$$2x + 3 = 0 \quad\text{or}\quad x - 1 = 0 \quad\text{or}\quad x + 2 = 0$$

$$x = -\frac{3}{2} \qquad\qquad x = 1 \qquad\qquad x = = -2$$

The three zeros, $-3/2$, 1, and -2, divide the x-axis into four regions:

$$x < -2, \quad -2 < x < -\frac{3}{2}, \quad -\frac{3}{2} < x < 1, \quad\text{and}\quad 1 < x.$$

The regions are shown in Figure 4.43.

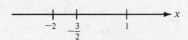

Figure 4.43

In any region, $f(x)$ is either always positive or always negative. We want to find the sign of $f(x)$ in each region. To do this, select an x-value in each region and substitute it into the function to determine if the function is positive or negative in that region. A typical selection of test points and the results of the tests are shown below.

Region	Test point	Sign of $f(x)$
$x < -2$	-3	negative
$-2 < x < -3/2$	$-7/4$	positive
$-3/2 < x < 1$	0	negative
$1 < x$	2	positive

When $f(x)$ is negative, the graph is below the x-axis, and when $f(x)$ is positive, the graph is above the x-axis. Thus, we know the graph looks something like the sketch in Figure 4.44. We could improve the sketch by plotting additional points in each region. ◻

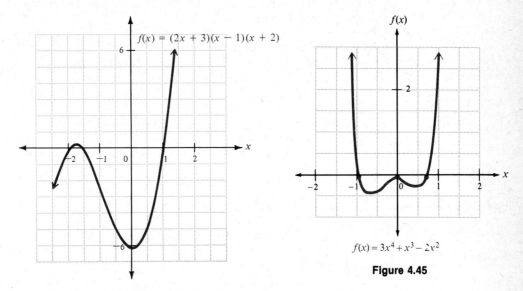

Figure 4.44

$f(x) = 3x^4 + x^3 - 2x^2$

Figure 4.45

EXAMPLE 4 ◻ Sketch the graph of $f(x) = 3x^4 + x^3 - 2x^2$.
The polynomial can be factored as follows:

$$3x^4 + x^3 - 2x^2 = x^2(3x^2 + x - 2)$$
$$= x^2(3x - 2)(x + 1).$$

The zeros are at $x = 0$, $x = 2/3$, and $x = -1$. They divide the x-axis into four regions:

$$x < -1, \quad -1 < x < 0, \quad 0 < x < 2/3, \quad \text{and} \quad 2/3 < x.$$

We determine the sign of $f(x)$ in each region by substituting an x-value from each region into the function to get the following information.

Region	Sign of $f(x)$	Location relative to axis
$x < -1$	positive	above
$-1 < x < 0$	negative	below
$0 < x < 2/3$	negative	below
$2/3 < x$	positive	above

With the values of x used for the test points and the corresponding values of y, we can sketch the graph as shown in Figure 4.45. ◻

EXERCISES
4.5

Each of the following polynomial functions is symmetric about a line or a point. For each function, (a) sketch the graph and (b) give the line or point of symmetry.

1. $f(x) = \frac{1}{4}x^6$ **2.** $f(x) = -\frac{2}{3}x^5$

3. $f(x) = \frac{-5}{4}x^5$ **4.** $f(x) = 2x^4$

5. $f(x) = \frac{1}{2}x^3 + 1$

6. $f(x) = -x^4 + 2$

7. $f(x) = -(x + 1)^3$

8. $f(x) = \frac{1}{3}(x + 3)^4$

9. $f(x) = (x - 1)^4 + 2$

10. $f(x) = (x + 2)^3 - 1$

Graph each of the following polynomial functions.

11. $f(x) = 2x(x - 3)(x + 2)$

12. $f(x) = x^2(x + 1)(x - 1)$

13. $f(x) = x^2(x - 2)(x + 3)^2$

14. $f(x) = x^2(x - 5)(x + 3)(x - 1)$

15. $f(x) = 3x^4 + 5x^3 - 2x^2$

16. $f(x) = 4x^3 + 2x^2 - 12x$

17. $f(x) = 2x^3(x^2 - 4)(x - 1)$

18. $f(x) = 5x^2(x^3 - 1)(x + 2)$

19. $f(x) = x^2(x - 3)^3(x + 1)$

20. $f(x) = x(x - 4)^2(x + 2)^2$

21. The polynomial function

$$A(x) = -0.015x^3 + 1.058x$$

gives the approximate alcohol concentration (in tenths of a percent) in an average person's bloodstream x hours after drinking about eight ounces of 100 proof whiskey. The function is approximately valid for x in the interval [0, 8].
  (a) Graph $A(x)$.
  (b) Using the graph you drew for part (a), estimate the time of maximum alcohol concentration.
  (c) In one state, a person is legally drunk if the blood alcohol concentration exceeds .15%. Use the graph of part (a) to estimate the period in which this average person is legally drunk.

Use the definition of odd and even functions given in the exercises for Section 3.4 to decide if the following polynomial functions are odd, even, or neither.

22. $f(x) = 2x^3$ **23.** $f(x) = -4x^5$

24. $f(x) = 0.2x^4$ **25.** $f(x) = -x^6$

26. $f(x) = -x^5$ **27.** $f(x) = (x - 1)^3$

28. $f(x) = 2x^3 + 3$

29. $f(x) = 4x^3 - x$

30. $f(x) = x^4 + 3x^2 + 5$

We can find approximate maximum or minimum values of polynomial functions for given intervals by first evaluating the function at the left endpoint of the given interval. Then add 0.1 to the value of x and reevaluate the polynomial. Keep doing this until the right endpoint of the interval is reached. Then identify the approximate maximum and minimum value for the polynomial on the interval.

31. $y = x^3 + 4x^2 - 8x - 8$, $[-3.8, -3]$

32. $y = x^3 + 4x^2 - 8x - 8$, $[0.3, 1]$

33. $y = 2x^3 - 5x^2 - x + 1$, $[-1, 0]$

34. $y = 2x^3 - 5x^2 - x + 1$, $[1.4, 2]$

35. $y = x^4 - 7x^3 + 13x^2 + 6x - 28$, $[-2, -1]$

36. $y = x^4 - 7x^3 + 13x^2 + 6x - 28$, $[2, 3]$

4.6   Rational Functions

A function of the form

$$f(x) = \frac{p(x)}{q(x)},$$

where $p(x)$ and $q(x)$ are polynomial functions, is called a **rational function.** Since any values of x such that $q(x) = 0$ are excluded from the domain, a rational function usually has a graph which has one or more breaks in it.

The simplest rational function with a variable denominator is

$$f(x) = \frac{1}{x}.$$

The domain of this function is the set of all nonzero real numbers. To graph the function, it is helpful to see what happens for values of x close to 0. In the following table, we show what happens to $f(x)$ as x gets closer and closer to 0 from either side.

x	-1	$-.1$	$-.01$	$-.001$	$.001$	$.01$	$.1$	1
$f(x)$	-1	-10	-100	-1000	1000	100	10	1

From the table, we see that $|f(x)|$ gets larger and larger as x gets closer and closer to 0. This is written in symbols as

$$|f(x)| \rightarrow \infty \text{ as } x \rightarrow 0.$$

(The symbol $x \rightarrow 0$ means that x approaches closer and closer to 0, without necessarily ever being equal to 0.) Since x cannot equal 0, the graph will never intersect the vertical line $x = 0$. This line is called a **vertical asymptote.**

Definition of a Vertical Asymptote

If $|f(x)| \rightarrow \infty$ as $x \rightarrow a$, then the line $x = a$ is a vertical asymptote.

On the other hand, as $|x|$ gets larger and larger, $f(x) = 1/x$ gets closer and closer to 0. (See the table.)

x	$-10,000$	-1000	-100	-10	10	100	1000	$10,000$
$f(x)$	$-.0001$	$-.001$	$-.01$	$-.1$	$.1$	$.01$	$.001$	$.0001$

Thus, as $|x|$ gets larger and larger without bound (written $|x| \rightarrow \infty$), the graph of $y = 1/x$ approaches closer and closer to the horizontal line $y = 0$. This line is called a **horizontal asymptote.**

If $f(x) \to a$ as $|x| \to \infty$, then the line $y = a$ is a horizontal asymptote.

From the equation, we see that $f(x) = 1/x$ is symmetric to the origin. If we find and plot points in the first quadrant and let the graph approach the asymptotes, we get the first quadrant part of the graph as shown in Figure 4.46. The other part of the graph (in the third quadrant) can be found by symmetry.

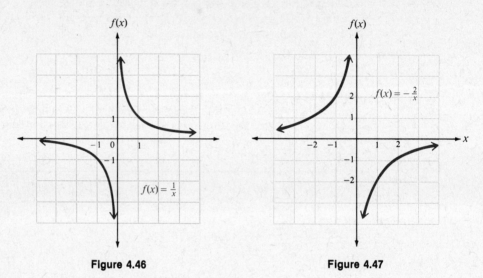

Figure 4.46   Figure 4.47

EXAMPLE 1 ☐ Graph $f(x) = \dfrac{-2}{x}$.

The function can be written as

$$f(x) = -2 \cdot \frac{1}{x}.$$

Compared to $f(x) = 1/x$, the graph will be reflected about the x-axis (because of the negative sign), and each point will be twice as far from the x-axis. See Figure 4.47. ☐

EXAMPLE 2 ☐ Graph $f(x) = \dfrac{2}{1 + x}$.

Here the domain is the set of all real numbers except -1. As shown in Figure 4.48, the graph is that of $f(x) = 1/x$, shifted 1 unit to the left, with each y-value doubled. ☐

EXAMPLE 3 ☐ Graph $y = \dfrac{1}{(x - 1)(x + 4)}$.

First, we note that this graph has no symmetry about either axis or the origin.

There are vertical asymptotes at $x = 1$ and $x = -4$, as shown in the following charts.

x	-4.1	-4.01	-4.001	-3.999	-3.99	-3.9
$f(x)$	1.96	19.96	199.96	-200.04	-20.04	-2.04

x	.9	.99	.999	1.001	1.01	1.1
$f(x)$	-2.04	-20.04	-200.04	199.96	19.96	1.96

We can find any vertical asymptotes algebraically by setting the denominator equal to 0 and solving for x. Here, we get

$$(x - 1)(x + 4) = 0$$
$$x = 1 \quad \text{or} \quad x = -4,$$

the equations of the two vertical asymptotes.

As $|x|$ gets larger and larger, $|(x - 1)(x + 4)|$ also gets larger and larger, making y closer and closer to 0. This means that the x-axis is a horizontal asymptote. The vertical asymptotes divide the x-axis into three regions. It is often convenient to consider each region separately. If we find the intercepts and plot a few points, we get the result shown in Figure 4.49. ▢

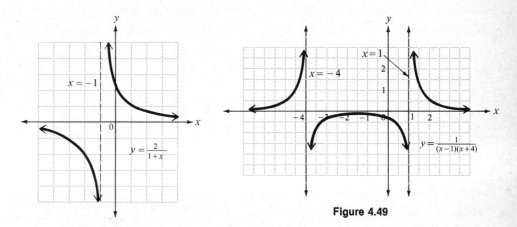

Figure 4.48

Figure 4.49

EXAMPLE 4 ▢ Graph $y = \dfrac{x}{(x - 1)(x + 3)}$.

There is no symmetry about either axis or the origin. There are two vertical asymptotes, $x = 1$ and $x = -3$. As $|x|$ gets larger and larger, both numerator and denominator get larger and larger. In this case, we cannot tell what happens to y. Therefore we must find selected ordered pairs of the function for values of x where $|x|$ is relatively large. (See the table next to the graph.) We plot the intercepts and selected ordered pairs and consider the vertical asymptotes. The graph is shown in Figure 4.50 on the next page. ▢

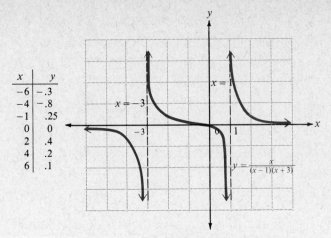

x	y
-6	$-.3$
-4	$-.8$
-1	$.25$
0	0
2	$.4$
4	$.2$
6	$.1$

Figure 4.50

The rational functions we have discussed so far all had numerators of lower degree than their denominators. In general, when this is the case, the horizontal asymptote is $y = 0$, the x-axis.

Rational functions in which the degree of the numerator and denominator are the same, as in Example 5 below, also have a horizontal asymptote. If the rational function is of the form

$$f(x) = \frac{a_n x^n + \cdots + a_0}{b_n x^n + \cdots + b_0}, \qquad b_n \neq 0$$

the horizontal asymptote is

$$y = \frac{a_n}{b_n}.$$

EXAMPLE 5   ☐   Graph $f(x) = \dfrac{2x + 1}{x - 3}$.

Since the degree of the numerator and denominator are the same, we can use the result just given to find that the graph has a horizontal asymptote at

$$y = \frac{2}{1} = 2.$$

The vertical asymptote is the line $x = 3$. By using the asymptotes, plotting the y-intercept, which is $-1/3$, and the x-intercept, which is $-1/2$, and a few other points as needed, we get the graph of Figure 4.51.   ☐

EXAMPLE 6   ☐   Graph $f(x) = \dfrac{x^2 + 1}{x - 2}$.

Since $x = 2$ makes the denominator zero, the line $x = 2$ is a vertical asymptote.

The numerator has higher degree than the denominator; therefore we divide to rewrite the function in another form. In this case, divide as follows:

$$
\begin{array}{r}
x + 2 \\
x - 2 \overline{)x^2 + 1} \\
\underline{x^2 - 2x } \\
2x + 1 \\
\underline{2x - 4} \\
5
\end{array}
$$

Thus, $\dfrac{x^2 + 1}{x - 2} = x + 2 + \dfrac{5}{x - 2}.$

For very large values of $|x|$, $5/(x - 2)$ is close to 0, and the graph approaches closer and closer to the line $y = x + 2$. This function has an **oblique asymptote** (neither vertical nor horizontal). The y-intercept is $-1/2$. There is no x-intercept. Using the asymptotes, the y-intercept and additional points as needed, leads to the graph of Figure 4.52. ▢

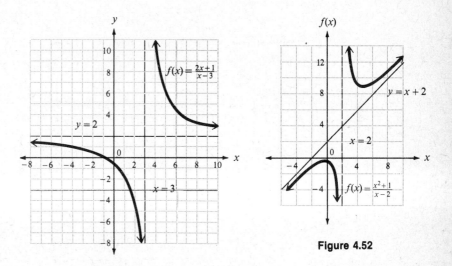

Figure 4.51

Figure 4.52

In general, if the degree of the numerator is greater than the degree of the denominator, the rational function may have an oblique asymptote. The equation of this asymptote is found by dividing the numerator by the denominator and dropping the remainder.

To summarize: use the following procedure when graphing a rational function.

Graphing
Rational
Functions

1. Find any vertical asymptotes by setting the denominator equal to 0 and solving for x. If a is a zero of the denominator, then $x = a$ is a vertical asymptote.
2. Determine any other asymptotes. There are three possibilities:
   (a) If the numerator has lower degree than the denominator, there is a horizontal asymptote, $y = 0$.
   (b) If the numerator and denominator have the same degree, and the function is of the form

   $$f(x) = \frac{a_n x^n + \cdots + a_0}{b_n x^n + \cdots + b_0} \qquad b_n \neq 0$$

   there is a horizontal asymptote,

   $$y = \frac{a_n}{b_n}.$$

   (c) If the numerator is of higher degree than the denominator, then there are no horizontal asymptotes but there may be an oblique asymptote. To find it, divide the numerator by the denominator and drop any remainder. The rest of the quotient gives the equation of the asymptote.
3. Find any intercepts.
4. Plot a few selected points—at least one in each region of the domain determined by the vertical asymptotes.
5. Complete the sketch.

EXERCISES 4.6

1. Sketch the following graphs on the same axes and compare them with the graph of $f(x) = \frac{1}{x}$.

   (a) $f(x) = \frac{1}{x^3}$ (b) $f(x) = \frac{1}{x^5}$

2. Sketch the following graphs on the same axes and compare them with each other.

   (a) $f(x) = \frac{1}{x^2}$ (b) $f(x) = \frac{1}{x^4}$

   (c) $f(x) = \frac{1}{x^6}$

3. Sketch the following graphs and compare them with the graph of $f(x) = \frac{1}{x}$.

   (a) $f(x) = \frac{1}{x + 2}$ (b) $f(x) = \frac{3}{x}$

   (c) $f(x) = \frac{3}{x + 2}$ (d) $f(x) = \frac{1}{x} + 2$

4. Sketch the following graphs and compare them with the graph of $f(x) = \frac{1}{x^2}$.

(a) $f(x) = \dfrac{1}{(x - 3)^2}$

(b) $f(x) = \dfrac{-2}{x^2}$

(c) $f(x) = \dfrac{-2}{(x - 3)^2}$

Graph each of the following.

5. $y = \dfrac{4}{5 + 3x}$

6. $y = \dfrac{1}{(x - 2)(x + 4)}$

7. $y = \dfrac{3}{(x + 4)^2}$

8. $y = \dfrac{3x}{(x + 1)(x - 2)}$

9. $y = \dfrac{2x + 1}{(x + 2)(x + 4)}$

10. $y = \dfrac{5x}{x^2 - 1}$

11. $y = \dfrac{-x}{x^2 - 4}$

12. $y = \dfrac{3x}{x - 1}$

13. $y = \dfrac{4x}{1 - 3x}$

14. $y = \dfrac{x + 1}{x - 4}$

15. $y = \dfrac{x - 3}{x + 5}$

16. $y = \dfrac{x - 5}{x + 3}$

17. $y = \dfrac{3x}{x^2 - 1}$

18. $y = \dfrac{x}{x^2 - 9}$

19. $y = \dfrac{x^2 - 5}{x + 2}$

20. $y = \dfrac{x^2 - 3x + 2}{x - 3}$

21. $y = \dfrac{x^2 + 1}{x + 3}$

22. $y = \dfrac{2x^2 + 3}{x - 4}$

23. $y = \dfrac{x^2 - x}{x + 2}$

24. $y = \dfrac{x^2 + 2x}{2x - 1}$

25. Suppose the average cost per unit, $C(x)$, to produce x units of margarine is given by

$$C(x) = \frac{500}{x + 30}.$$

(a) Find $C(10)$, $C(20)$, $C(50)$, $C(75)$, and $C(100)$.

(b) Would a more reasonable domain for C be $(0, +\infty)$ or $[0, +\infty)$? Why?

26. In a recent year, the cost per ton, y, to build an oil tanker of x thousand deadweight tons was approximated by

$$y = \frac{110,000}{x + 225}.$$

(a) Find y for $x = 25$, $x = 50$, $x = 100$, $x = 200$, $x = 300$, and $x = 400$.

(b) Graph the function.

27. Antique-car fans often enter their cars in a *concours d'elegance* in which a maximum of 100 points can be awarded to a particular car. Points are awarded for the general attractiveness of the car. The function

$$C(x) = \frac{10x}{49(101 - x)}$$

expresses the cost, in thousands of dollars, of restoring a car so that it will win x points. Graph the function.

28. In situations involving environmental pollution, a cost-benefit model expresses cost as a function of the percentage of pollutant removed from the environment. Suppose a cost-benefit model is expressed as

$$y = \frac{6.7x}{100 - x},$$

where y is the cost in thousands of dollars of removing x percent of a certain pollutant.

(a) Graph the function.

(b) Is it possible, according to this function, to remove all the pollutant?

CHAPTER SUMMARY
4

Key Words

polynomial functions	parallel lines	center
polynomial function of degree n	perpendicular lines	hyperbola
linear function	parabola	asymptotes
quadratic function	axis	fundamental rectangle
change in x	vertex	conic sections
change in y	ellipse	turning point
slope	foci	rational function
point-slope form	vertices of an ellipse	vertical asymptote
x-intercept	major axis	horizontal asymptote
y-intercept	minor axis	oblique asymptote
slope-intercept form		

CHAPTER REVIEW EXERCISES
4

Find the slope for each of the following lines that has a slope.

1. through $(8, 7)$ and $(1/2, -2)$

2. through $(2, -2)$ and $(3, -4)$

3. through $(5, 6)$ and $(5, -2)$

4. through $(0, -7)$ and $(3, -7)$

5. $9x - 4y = 2$

6. $11x + 2y = 3$

7. $x - 5y = 0$

8. $x - 2 = 0$

9. $y + 6 = 0$

10. $y = x$

Graph each of the following.

11. $3x + 7y = 14$

12. $2x - 5y = 5$

13. $3y = x$

14. $y = 3$

15. $x = -5$

16. $y = x$

For each of the following lines, write the equation in the form $ax + by = c$.

17. through $(-2, 4)$ and $(1, 3)$

18. through $(-2/3, -1)$ and $(0, 4)$

19. through $(3, -5)$ with slope -2

20. through $(-4, 4)$ with slope $3/2$

21. through $(1/5, 1/3)$ with slope $-1/2$

22. x-intercept -3, y-intercept 5

23. no x-intercept, y-intercept $3/4$

24. through $(2, -1)$, parallel to $3x - y = 1$

25. through $(0, 5)$, perpendicular to $8x + 5y = 3$

26. through $(2, -10)$, perpendicular to a line with no slope

27. through $(3, -5)$ parallel to $y = 4$

28. through $(-7, 4)$, perpendicular to $y = 8$

Graph each of the following lines.

29. through $(2, -4)$, $m = 3/4$

30. through $(0, 5)$, $m = -2/3$

31. through $(-4, 1)$, $m = 3$

32. through $(-3, -2)$, $m = -1$

Graph each of the following.

33. $y = \begin{cases} 3x + 1 & \text{if } x < 2 \\ -x + 4 & \text{if } x \geq 2 \end{cases}$

34. $y = \begin{cases} -4x + 2 & \text{if } x \leq 1 \\ 3x - 5 & \text{if } x > 1 \end{cases}$

35. $y = x^2 - 4$ 36. $y = 6 - x^2$

37. $y = 3(x + 1)^2 - 5$

38. $y = -\dfrac{1}{4}(x - 2)^2 + 3$

39. $y = x^2 - 4x + 2$

40. $y = -3x^2 - 12x - 1$

Give the vertex and axis of each of the following.

41. $y = -(x + 3)^2 - 9$

42. $y = (x - 7)^2 + 3$

43. $y = x^2 - 7x + 2$

44. $y = -x^2 - 4x + 1$

45. $y = -3x^2 - 6x + 1$

46. $y = 4x^2 - 4x + 3$

Use parabolas to work each of the following problems.

47. Find two numbers whose sum is 11 and whose product is a maximum.

48. Find two numbers having a sum of 40 such that the sum of the square of one and twice the square of the other is maximum.

49. Find the rectangular region of maximum area that can be enclosed with 180 meters of fencing.

50. Find the rectangular region of maximum area that can be enclosed with 180 meters of fencing if no fencing is needed along one side of the region.

Graph each of the following. Identify each graph.

51. $\dfrac{x^2}{25} + \dfrac{y^2}{4} = 1$

52. $\dfrac{x^2}{3} + \dfrac{y^2}{16} = 1$

53. $\dfrac{x^2}{4} - \dfrac{y^2}{9} = 1$

54. $\dfrac{y^2}{100} - \dfrac{x^2}{25} = 1$

55. $x^2 = 16 + y^2$

56. $4x^2 + 9y^2 = 36$

57. $\dfrac{25x^2}{9} + \dfrac{4y^2}{25} = 1$

58. $\dfrac{100x^2}{49} + \dfrac{9y^2}{16} = 1$

59. $\dfrac{(x - 2)^2}{9} + \dfrac{(y + 3)^2}{4} = 1$

60. $\dfrac{(x - 3)^2}{4} + (y + 1)^2 = 1$

61. $\dfrac{(y + 2)^2}{4} - \dfrac{(x + 3)^2}{9} = 1$

62. $\dfrac{(x + 1)^2}{16} - \dfrac{(y - 2)^2}{4} = 1$

63. $\dfrac{x}{3} = -\sqrt{1 - \dfrac{y^2}{16}}$

64. $x = -\sqrt{1 - \dfrac{y^2}{36}}$

65. $y = -\sqrt{1 - \dfrac{x^2}{25}}$

66. $y = -\sqrt{1 + x^2}$

Graph each of the following.

67. $f(x) = x^3 + 5$

68. $f(x) = 1 - x^4$

69. $f(x) = x^2(2x + 1)(x - 2)$

70. $f(x) = (4x - 3)(3x + 2)(x - 1)$

71. $f(x) = 2x^3 + 13x^2 + 15x$

72. $f(x) = x(x - 1)(x + 2)(x - 3)$

73. $f(x) = \dfrac{8}{x}$

74. $f(x) = \dfrac{2}{3x - 1}$

75. $f(x) = \dfrac{4x - 2}{3x + 1}$

76. $f(x) = \dfrac{6x}{(x - 1)(x + 2)}$

77. $f(x) = \dfrac{2x}{x^2 - 1}$

78. $f(x) = \dfrac{x^2 + 4}{x + 2}$

79. $f(x) = \dfrac{x^2 - 1}{x}$

80. $f(x) = \dfrac{x^2 + 6x + 5}{x - 3}$

81. Find an equation of the line with slope 2/3 that goes through the center of the circle $(x - 4)^2 + (y + 2)^2 = 9$.

82. Use slopes to show that the quadrilateral with vertices at $(-2, 2)$, $(4, 2)$, $(4, -1)$, and $(-2, -1)$ is a rectangle.

83. Find a value of k so that $5x - 3y = k$ goes through the point $(1, 4)$.

84. Find k so that $3x + ky = 2$ goes through $(3, -2)$.

85. Find all squares that have $(0, 7)$ and $(12, 12)$ as two of the vertices. (Hint: it is sufficient to identify a square by listing its vertices.)

86. Suppose $a > 0$ and $a + h > 0$. Show that the straight line through $(a, \sqrt{a})$ and $(a + h, \sqrt{a + h})$ has slope $1/(\sqrt{a + h} + \sqrt{a}.)$ (Hint: $(\sqrt{B} - \sqrt{A})(\sqrt{B} + \sqrt{A}) = B - A$ if $A > 0$ and $B > 0$.)

87. Suppose that $x^2 + y^2 + Ax + By + C = 0$ and $x^2 + y^2 + ax + by + c = 0$ are different circles that meet at two distinct points. Show that the line through those two points of intersection has the equation

$$(A - a)x + (B - b)y + (C - c) = 0.$$

88. Show that the line with equation

$$Ax + By = r^2$$

is tangent to (touches in only one point) the circle $x^2 + y^2 = r^2$ at (A, B).

Exercises 85–88: Stanley I. Grossman, *Calculus*, 2nd Edition (Academic Press, New York, 1981), pp. 22, 30, 31.

5 Exponential and Logarithmic Functions

In this chapter, we study two kinds of functions that are quite different from those we have studied before—different in that they do not involve just the basic operations of addition, subtraction, multiplication, division, and taking roots. The functions we have discussed up till now are examples of *algebraic* functions. The functions we study in this chapter are *transcendental* functions. Many applications of mathematics, particularly those pertaining to growth and decay of populations, involve exponential or logarithmic functions. We shall see that these two types of functions are inverses of each other.

5.1 Exponential Functions

We know that if $a > 0$, we can define the symbol a^m for any rational value of m. In this section, we want to extend the definition of a^m to include all *real,* and not just rational, values of the exponent m. For example, what is meant by $2^{\sqrt{3}}$? The exponent, $\sqrt{3}$, can be approximated more and more closely by the numbers 1.7, 1.73, 1.732, and so on. Thus, it seems reasonable that $2^{\sqrt{3}}$ should be approximated more and more closely by the numbers $2^{1.7}$, $2^{1.73}$, $2^{1.732}$, and so on. (Recall, for example, that $2^{1.7} = 2^{17/10}$, which means $\sqrt[10]{2^{17}}$.) In fact, this is exactly how $2^{\sqrt{3}}$ is defined (in a more advanced course). Figure 5.1, on the next page, shows the graphs of $f(x) = 2^x$ with three different domains. These graphs illustrate that the assumption we have made is reasonable.

We shall assume that the meaning given to real exponents is such that all our rules and theorems for exponents are valid for real number exponents as well as rational ones. In addition to the rules for exponents presented earlier, we will need some additional properties. For example, if $y = 2^x$, we note that each real value of x leads to exactly one value of 2^x. Thus,

$$2^2 = 4, \; 2^3 = 8, \; 2^{1/2} = \sqrt{2} \approx 1.4142, \text{ and so on.}$$

Furthermore, if $\qquad\qquad 3^x = 3^4$, then $x = 4$.

And, if $\qquad\qquad\qquad\quad p^2 = 3^2$, then $p = 3$.

Also note that $\qquad 4^2 < 4^3 \qquad$ but $\qquad \left(\dfrac{1}{2}\right)^2 > \left(\dfrac{1}{2}\right)^3,$

so that when $a > 1$, increasing the exponent on a leads to a *larger* number, but if $0 < a < 1$, increasing the exponent on a leads to a *smaller* number.

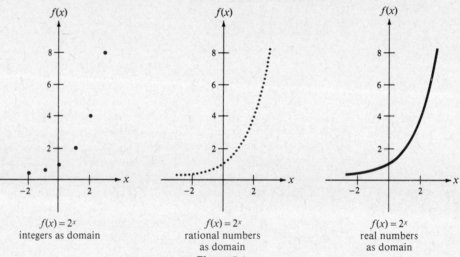

$f(x) = 2^x$
integers as domain

$f(x) = 2^x$
rational numbers
as domain

$f(x) = 2^x$
real numbers
as domain

Figure 5.1

These properties are generalized in the next theorem. We have not included a proof of these properties, since the proof requires more advanced mathematics than we have available.

Theorem

> For $a > 0$, $a \neq 1$, and any real number x
>
> (a) a^x is a unique real number.
> (b) $a^b = a^c$ if and only if $b = c$.
> (c) If $a > 1$ and $m < n$, then $a^m < a^n$.
> (d) If $0 < a < 1$ and $m < n$, then $a^m > a^n$.

Part (a) of the theorem requires $a > 0$ so that a^x is always defined. For example, $(-6)^x$ is not a real number if $x = 1/2$. This means that a^x will always be positive, since a is positive. For part (b) to hold, a must not equal 1 since $1^4 = 1^5$, even though $4 \neq 5$. Parts (c) and (d) are illustrated by the graphs of Figure 5.2.

As we have said, the expression a^x satisfies all the properties of exponents from

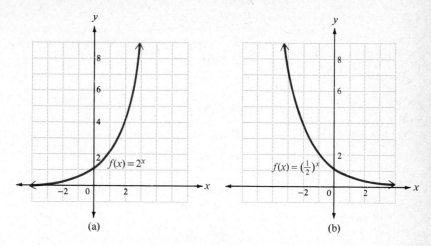

Figure 5.2

Chapter 1. We can now define a function, $f(x) = a^x$, whose domain is the set of all real numbers (and not just the rationals).

Exponential Function

The function

$$f(x) = a^x, \qquad a > 0 \text{ and } a \neq 1,$$

is the **exponential function** with base a.

(If $a = 1$, the function is the constant function $f(x) = 1$.)

EXAMPLE 1 If $f(x) = 2^x$, find each of the following.

(a) $f(-1)$

Replacing x with -1, we have

$$f(-1) = 2^{-1} = \frac{1}{2}.$$

(b) $f(3) = 2^3 = 8$

(c) $f(5/2) = 2^{5/2} = (2^5)^{1/2} = 32^{1/2} = \sqrt{32} = 4\sqrt{2}$ ∎

Figure 5.2(a) shows the graph of the exponential function $f(x) = 2^x$; the base of this exponential function is 2. This graph was found by obtaining a number of ordered pairs satisfying the function and then drawing a smooth curve through them. The domain of the function is the set of all real numbers, and the range is the set of all positive numbers. The function is increasing on its entire domain, and is thus a one-to-one function. The x-axis is a horizontal asymptote.* The graph of $f(x) = 2^x$ is

*Recall that an asymptote is a straight line that the graph approaches more and more closely.

typical of graphs of $f(x) = a^x$ where $a > 1$. For larger values of a, the graphs rise more steeply, but the general shape is similar to the graph in Figure 5.2(a). In Figure 5.2(b), the graph of $f(x) = (1/2)^x$ was obtained in a similar way. This graph is decreasing on its entire domain.

The graph of $f(x) = a^x$ is always increasing if $a > 1$, and always decreasing if $0 < a < 1$.

If we start with $f(x) = 2^x$, and replace x with $-x$, we get $f(-x) = 2^{-x} = (2^{-1})^x = (1/2)^x$. For this reason, the graphs of $f(x) = 2^x$ and $f(x) = (1/2)^x$ are symmetric with respect to the y-axis. This is also suggested by the graphs of Figures 5.2(a) and (b).

EXAMPLE 2   ☐   Graph $f(x) = 2^{-x^2}$.

Since $x^2 \geq 0$ for all x, we have $2^{x^2} \geq 2^0 = 1$ for all x, with $f(x) = 2^{-x^2} = 1/(2^{x^2}) \leq 1$ for all x. By plotting some typical points, such as $(-2, 1/16)$, $(-1, 1/2)$, $(0, 1)$, $(1, 1/2)$, and $(2, 1/16)$, and drawing a smooth curve through them, we get the graph of Figure 5.3. This graph is symmetric with respect to the y-axis and has the x-axis as a horizontal asymptote.   ☐

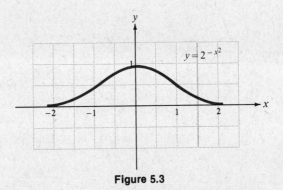

Figure 5.3

We can use the results of the last theorem to help solve equations, as shown by the next example.

EXAMPLE 3   ☐   Solve $\left(\dfrac{1}{3}\right)^x = 81$.

First, write $1/3$ as 3^{-1}, so that $(1/3)^x = (3^{-1})^x = 3^{-x}$. Since $81 = 3^4$, we have

$$\left(\frac{1}{3}\right)^x = 81$$

$$3^{-x} = 3^4$$

$$-x = 4$$

$$x = -4.$$

The solution set of the given equation is $\{-4\}$. (In Section 5.4 we discuss a method for solving equations of this type where this approach is not possible.) ☐

EXAMPLE 4 ☐ If $81 = b^{4/3}$, find b.

To solve for b, we need an exponent of 1 on b. Therefore, raise both sides of the equation to the 3/4 power.

$$81 = b^{4/3}$$
$$81^{3/4} = (b^{4/3})^{3/4}$$
$$(\sqrt[4]{81})^3 = b$$
$$3^3 = b$$
$$27 = b$$

Remember that the process of raising both sides of an equation to the same power may result in false "solutions." For this reason, it is necessary to check all proposed solutions. Replacing b with 27 gives

$$27^{4/3} = (\sqrt[3]{27})^4 = 3^4 = 81,$$

which checks; therefore the solution set is $\{27\}$. ☐

Perhaps the single most useful exponential function is the function $f(x) = e^x$, where e is an irrational number that occurs often in practical applications, as we shall see. To see how the number e comes up in a practical problem, let us begin with the formula for compound interest (interest paid on both principal and interest). If P dollars is deposited in an account paying a rate of interest i compounded (paid) m times per year, the account will contain

$$P\left(1 + \frac{i}{m}\right)^{nm}$$

dollars after n years.

For example, suppose \$1000 is deposited into an account paying 8% per year compounded quarterly, or four times a year. After 10 years the account will contain

$$P\left(1 + \frac{i}{m}\right)^{nm} = 1000\left(1 + \frac{.08}{4}\right)^{10(4)} = 1000(1 + .02)^{40} = 1000(1.02)^{40}$$

dollars. The number $(1.02)^{40}$ can be found in financial tables or by using a calculator with an x^y key. To five decimal places, $(1.02)^{40} = 2.20804$. The amount on deposit after 10 years is thus

$$1000(1.02)^{40} = 1000(2.20804) = 2208.04,$$

or \$2208.04.

Suppose now that a lucky investment you make produces annual interest of 100%, so that $i = 1.00$, or $i = 1$. Suppose also that you can deposit only \$1 at this rate, and for only one year. Then $P = 1$ and $n = 1$. Substituting into our formula for compound interest, we then have

$$P\left(1 + \frac{i}{m}\right)^{nm} = 1\left(1 + \frac{1}{m}\right)^{1(m)} = \left(1 + \frac{1}{m}\right)^m.$$

As interest is compounded more and more often, the value of this expression will increase. If $m = 1$ (interest is compounded annually), we have

$$\left(1 + \frac{1}{m}\right)^m = \left(1 + \frac{1}{1}\right)^1 = 2^1 = 2,$$

so that your $1 becomes $2 in one year.

Using a calculator with an x^y key, we can get the results shown in the following table. These results have been rounded to five decimal places.

m	$\left(1 + \dfrac{1}{m}\right)^m$
1	2
2	2.25
5	2.48832
10	2.59374
25	2.66584
50	2.69159
100	2.70481
500	2.71557
1000	2.71692
10,000	2.71815
1,000,000	2.71828

The table suggests that as m increases, the value of $(1 + 1/m)^m$ gets closer and closer to some fixed number. It turns out that this is indeed the case. This fixed number is called e. To nine decimal places,

e $e = 2.718281828.$

Table 3 in this book gives various powers of e. Also, some calculators will give values of e^x. In Figure 5.4, the functions $y = 2^x$, $y = e^x$, and $y = 3^x$ are graphed for comparison.

It can be shown that in many situations involving growth or decay of a population, the amount or number present at time t can be closely approximated by a function of the form

Growth or Decay $y = y_0 e^{kt},$

where y_0 is the amount or number present at time $t = 0$ and k is a constant. (In other words, once the numbers y_0 and k have been determined, population is a function of time.) The next example illustrates exponential growth.

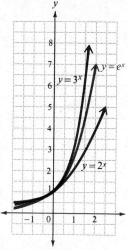

Figure 5.4

EXAMPLE 5

Suppose the population of a midwestern city is

$$P(t) = 10,000e^{0.04t},$$

where t represents time measured in years. The population at time $t = 0$ is

$$P(0) = 10,000e^{(0.04)0}$$
$$= 10,000e^0$$
$$= 10,000(1)$$
$$= 10,000.$$

Thus, the population of the city is 10,000 at time $t = 0$, written $P_0 = 10,000$. The population of the city at year $t = 5$ is

$$P(5) = 10,000e^{(0.04)5}$$
$$= 10,000e^{0.2}$$

The number $e^{0.2}$ can be found in Table 3 or by using a suitable calculator. By either of these methods, $e^{0.2} = 1.22140$ (to five decimal places), so that

$$P(5) = 10,000(1.22140) = 12,214.$$

Thus, in five years the population of the city will be about 12,200. ☐

In Example 5 we were given the exponential function $P(t) = 10,000e^{0.04t}$. In Example 6, on the other hand, we must actually produce such a function from given information.

EXAMPLE 6

There are 600 grams of a radioactive substance present initially. After 3 years, only 300 grams remain. How much of the substance will be present after 6 years?

Radioactive decay is described by a function of the form

$$y = y_0 e^{-kt},$$

for suitable values of y_0 and k. We know that $y = 600$ when $t = 0$, or

$$600 = y_0 e^{-k(0)}$$

$$600 = y_0,$$

giving the function

$$y = 600 e^{-kt}. \tag{*}$$

To find k, use the fact that $y = 300$ when $t = 3$, or

$$300 = 600 e^{-3k}$$

$$\frac{1}{2} = e^{-3k}.$$

At this time we cannot solve for k (see Section 5.4). However, we can solve for e^{-k}, by raising both sides of the equation to the 1/3 power.

$$\left(\frac{1}{2}\right)^{1/3} = (e^{-3k})^{1/3}$$

$$\left(\frac{1}{2}\right)^{1/3} = e^{-k}$$

Substituting into equation (*) gives

$$y = 600 e^{-kt}$$

$$= 600 (e^{-k})^{t}$$

$$= 600 \left(\frac{1}{2}\right)^{(1/3)t}$$

$$y = 600 \left(\frac{1}{2}\right)^{t/3}$$

as the equation expressing the decay of our radioactive substance. After 6 years,

$$y = 600 \left(\frac{1}{2}\right)^{6/3} = 600 \left(\frac{1}{2}\right)^{2} = 600 \left(\frac{1}{4}\right) = 150$$

grams of the substance remain. ▢

EXERCISES
5.1

1. Graph each of the following functions. Compare the graphs to that of $y = 2^x$.
   - (a) $y = 2^x + 1$
   - (b) $y = 2^x - 4$
   - (c) $y = 2^{x+1}$
   - (d) $y = 2^{x-4}$

2. Graph each of the following. Compare the graphs to that of $y = 3^{-x}$.
   - (a) $y = 3^{-x} - 2$
   - (b) $y = 3^{-x} + 4$
   - (c) $y = 3^{-x-2}$
   - (d) $y = 3^{-x+4}$

Graph each of the following functions.

3. $y = 10^{-x}$ **4.** $y = 10^x$

5. $y = e^{-x}$ **6.** $y = e^{2x}$

7. $y = 2^{|x|}$ **8.** $y = 2^{-|x|}$

9. $y = 2^x + 2^{-x}$

10. $y = \left(\dfrac{1}{2}\right)^x + \left(\dfrac{1}{2}\right)^{-x}$

11. $y = x \cdot 2^x$

12. $y = x^2 \cdot 2^x$

Solve each of the following equations.

13. $4^x = 2$ **14.** $125^r = 5$

15. $\left(\dfrac{1}{2}\right)^k = 4$ **16.** $\left(\dfrac{2}{3}\right)^x = \dfrac{9}{4}$

17. $2^{3-y} = 8$ **18.** $5^{2p+1} = 25$

19. $\dfrac{1}{27} = b^{-3}$ **20.** $\dfrac{1}{81} = k^{-4}$

21. $4 = r^{2/3}$ **22.** $z^{5/2} = 32$

23. $27^{4z} = 9^{z+1}$ **24.** $32^t = 16^{1-t}$

25. $125^{-x} = 25^{3x}$ **26.** $216^{3-a} = 36^a$

27. $\left(\dfrac{1}{8}\right)^{-2p} = 2^{p+3}$ **28.** $3^{-h} = \left(\dfrac{1}{27}\right)^{1-2h}$

29. For $a > 1$, how does the graph of $y = a^x$ change as a increases? What if $0 < a < 1$?

📧 *Use a calculator to help graph each of the following functions.*

30. $y = \dfrac{e^x - e^{-x}}{2}$ **31.** $y = \dfrac{e^x + e^{-x}}{2}$

32. $y = (1 - x)e^x$ **33.** $y = x \cdot e^x$

34. Suppose the population of a city is given by $P(t)$, where

$$P(t) = 1,000,000 e^{.02t},$$

where t represents time measured in years from some initial year. Find each of the following:

(a) P_0; (b) $P(2)$; (c) $P(4)$; (d) $P(10)$. (e) Graph $y = P(t)$.

35. Suppose the quantity in grams of a radioactive substance present at time t is

$Q(t) = 500 e^{-.05t}$.

Let t be time measured in days from some initial day. Find the quantity present at each of the following times:

(a) $t = 0$; (b) $t = 4$; (c) $t = 8$; (d) $t = 20$. (e) Graph $y = Q(t)$.

36. Experiments have shown that the sales of a product, under relatively stable market conditions, but in the absence of promotional activities such as advertising, tend to decline at a constant yearly rate. This rate of sales decline varies considerably from product to product, but seems to remain the same for any particular product. The sales decline can be expressed by a function of the form

$$S(t) = S_0 e^{-at},$$

where $S(t)$ is the rate of sales at time t measured in years, S_0 is the rate of sales at time $t = 0$, and a is the sales decay constant. (a) Suppose the sales decay constant for a particular product is $a = 0.10$. Let $S_0 = 50{,}000$ and find $S(1)$ and $S(3)$. (b) Find $S(2)$ and $S(10)$ if $S_0 = 80{,}000$ and $a = 0.05$.

37. It can be shown that P dollars compounded continuously (every instant) at an interest rate i per year would amount to

$$A = Pe^{ni}$$

dollars at the end of n years. How much would $20{,}000$ amount to at 8% compounded continuously for the following number of years? (a) 1 year; (b) 5 years.

If P dollars is deposited into an account paying a rate of interest i compounded m times a year, with the money left on deposit for n years, the account will end up with a total of   📧

$$A = P\left(1 + \frac{i}{m}\right)^{nm}$$

dollars on deposit. Find the total amount on deposit for each of the following.

38. 4292 at 6% compounded annually for 10 years

39. $965.43 at 9% compounded annually for 15 years

40. $10,765 at 11% compounded semiannually for 7 years

41. $1593.24 at 10½% compounded quarterly for 14 years

42. $68,922 at 10% compounded daily (365 days) for 4 years

43. $2964.58 at 11¼% compounded daily for 9 years (Ignore leap years.)

44. Suppose $10,000 is left at interest for 3 years at 12%. Find the final amount on deposit if the interest is compounded (a) annually; (b) quarterly; (c) daily (365 days).

45. *Escherichia coli* is a strain of bacteria that occurs naturally in many different organisms. Under certain conditions, the number of these bacteria present in a colony is

$$E(t) = E_0 \cdot 2^{t/30},$$

where $E(t)$ is the number of bacteria present t minutes after the beginning of an experiment, and E_0 is the number present when $t = 0$. Let $E_0 = 2,400,000$ and find the number of bacteria at the following times: (a) $t = 5$; (b) $t = 10$; (c) $t = 60$; (d) $t = 120$.

46. The higher a student's grade-point average, the fewer applications that the student need send to medical schools (other things being equal). Using information given in a guidebook for prospective medical students, we constructed the function $y = 540e^{-1.3x}$ for the number of applications a student should send out. Here y is the number of applications for a student whose grade-point average is x. The domain of x is the interval [2.0, 4.0]. Find the number of applications that should be sent out by students having a grade-point average of (a) 2.0; (b) 3.0; (c) 3.5; (d) 3.9.

Newton's Law of Cooling says that the rate at which a body cools is proportional to the difference in temperature between the body and the environment into

which it is introduced. The temperature $f(t)$ of the body at time t after being introduced into an environment having constant temperature T_0 is

$$f(t) = T_0 + Ce^{-kt},$$

where C and k are constants. Use this result in Exercises 47–50.

47. A piece of metal is heated to 300°C and then placed in a cooling liquid at 50°C. After 4 minutes, the metal has cooled to 175°C. Find its temperature after 12 minutes.

48. Boiling water, at 100°C, is placed in a freezer at 0°C. The temperature of the water is 50°C after 24 minutes. Find the temperature of the water after 96 minutes.

49. A volcano discharges lava at 800°C. The surrounding air has a temperature of 20°C. The lava cools to 410°C in five hours. Find its temperature after 15 hours.

50. Paisley refuses to drink coffee cooler than 95°F. She makes coffee with a temperature of 170°F in a room with a temperature of 70°F. The coffee cools to 120°F in 10 minutes. What is the longest time she can let the coffee sit before she drinks the coffee?

The pressure of the atmosphere, $p(h)$, in pounds per square inch, is given by

$$p(h) = p_0 e^{-kh},$$

where h is the height above sea level and p_0 and k are constants. The pressure at sea level is 15 pounds per square inch and the pressure is 9 pounds per square inch at a height of 12,000 feet.

51. Find the pressure at an altitude of 6000 feet.

52. What would be the pressure encountered by a spaceship at an altitude of 150,000 feet?

53. In our definition of exponential function, we ruled out negative values of a. However, in a textbook on mathematical economics, the author obtained a "graph" of $y = (-2)^x$ by plotting the following points.

x	-4	-3	-2	-1	0	1	2
y	$1/16$	$-1/8$	$1/4$	$-1/2$	1	-2	4

The graph, which occupies a half page in the book, oscillates very neatly from positive to negative values of y. Comment on this approach. (This example shows the dangers of relying solely on point plotting when drawing graphs.)

54. When defining an exponential function, why did we require $a > 0$?

At any points where the graphs of functions $f(x)$ and $g(x)$ cross, we find solutions of the equation $f(x) = g(x)$. Use this idea to estimate the number of solutions of the following equations.

55. $x = 2^x$

56. $2^{-x} = -x$

57. $3^{-x} = 1 - 2x$

58. $3x + 2 = 4^x$

Let $f(x) = a^x$ be an exponential function of base a.

59. Is f odd, even, or neither?

60. Prove that $f(m + n) = f(m) \cdot f(n)$ for any real numbers m and n.

Find examples of a function f satisfying the following conditions.

61. $f(2x) = [f(x)]^2$ **62.** $f(x + 1) = 2 \cdot f(x)$

In calculus, it is shown that

$$e^x = 1 + x + \frac{x^2}{2 \cdot 1} + \frac{x^3}{3 \cdot 2 \cdot 1} +$$

$$\frac{x^4}{4 \cdot 3 \cdot 2 \cdot 1} + \frac{x^5}{5 \cdot 4 \cdot 3 \cdot 2 \cdot 1} + \ldots .$$

63. Use the terms shown here and replace x with 1 to approximate $e^1 = e$ to three decimal places. Then check your results in Table 3 or with a calculator.

64. Use the terms shown here and replace x with $-.05$ to approximate $e^{-.05}$ to four decimal places. Check your results in Table 3 or with a calculator.

5.2 Logarithmic Functions

In the previous section we discussed exponential functions of the form $y = a^x$, for all $a > 0$, $a \neq 1$. Since these functions are one-to-one, they have inverse functions. We look at these inverse functions in this section.

The inverse of $y = a^x$ is written $y = \log_a x$; that is, for all real numbers y and all positive numbers a, where $a \neq 1$,

Definition of $\log_a x$

$$y = \log_a x \quad \text{if and only if} \quad x = a^y.$$

Log is an abbreviation for *logarithm*. Read $\log_a x$ as "logarithm of x to the base a."

EXAMPLE 1   In the chart below we show several pairs of equivalent statements. The same statement is written in both exponential and logarithmic forms.

Exponential form	Logarithmic form
$2^3 = 8$	$\log_2 8 = 3$
$(1/2)^{-4} = 16$	$\log_{1/2} 16 = -4$
$10^5 = 100,000$	$\log_{10} 100,000 = 5$
$3^{-4} = 1/81$	$\log_3(1/81) = -4$
$5^1 = 5$	$\log_5 5 = 1$
$(3/4)^0 = 1$	$\log_{3/4} 1 = 0$ □

Using the definition of logarithm, we define the logarithmic function of base a.

Logarithmic Function

If $a > 0$, $a \neq 1$, and $x > 0$, then

$$f(x) = \log_a x$$

is the **logarithmic function of base a.**

Exponential and logarithmic functions are inverses of each other. Since the domain of an exponential function is the set of all real numbers, the range of a logarithmic function will also be the set of all real numbers. In the same way, both the range of an exponential function and the domain of a logarithmic function are the set of all positive real numbers. Thus, we can find logarithms for positive numbers only.

The graph of $y = 2^x$ is shown in Figure 5.5(a). To get the graph of its inverse, reflect $y = 2$ about the 45° line $y = x$. The graph of the inverse, $y = \log_2 x$, is shown as a dashed curve. As the graph shows, $y = \log_2 x$ is increasing for all its domain, is one-to-one, and has the y-axis as a vertical asymptote.

The graph of $y = (1/2)^x$ is shown in Figure 5.5(b). Its inverse, $y = \log_{1/2} x$, shown as a dashed curve, is found by reflecting $y = (1/2)^x$ about the line $y = x$. The function $y = \log_{1/2} x$ is decreasing for all its domain, is one-to-one, and has the y-axis for a vertical asymptote.

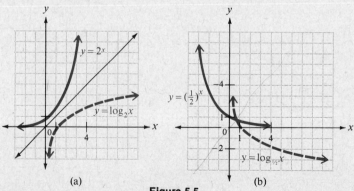

(a)   (b)

Figure 5.5

**EXAMPLE
2**   Graph $y = \log_2 (x - 1)$.

First notice that the domain here is $(1, +\infty)$, since we can only find logarithms of positive numbers. The graph of $y = \log_2 (x - 1)$ will be the graph of $y = \log_2 x$, translated one unit to the right. See Figure 5.6.

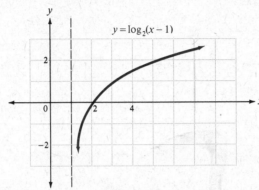

Figure 5.6

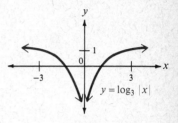

Figure 5.7

**EXAMPLE
3**   Graph $y = \log_3 |x|$.

There are two ways to graph this function: we can choose values of x and find the corresponding values of y, as shown in the table below, or we can first use the definition of logarithm to replace $y = \log_3 |x|$ with $|x| = 3^y$. In either case, we get the graph shown in Figure 5.7.

x	-3	-1	$-1/3$	$1/3$	1	3
y	1	0	-1	-1	0	1

**EXAMPLE
4**   Solve each of the following equations.

(a) $\log_x \dfrac{8}{27} = 3$

First, write the expression in exponential form.

$$x^3 = \frac{8}{27}$$

$$x^3 = \left(\frac{2}{3}\right)^3$$

$$x = \frac{2}{3}$$

The solution set is $\{2/3\}$.

(b) $\log_4 x = 5/2$

In exponential form, the given statement becomes

$$4^{5/2} = x$$
$$(4^{1/2})^5 = x$$
$$2^5 = x$$
$$32 = x.$$

The solution set is $\{32\}$.   ☐

Logarithms were originally important as an aid for numerical calculations, but the availability of inexpensive calculators has cut the need for this application of logarithms. Yet the principles behind the use of logarithms for calculation are important; these principles are based on the properties discussed in the next theorem.

Properties of Logarithms

> If x and y are any positive real numbers, r is any real number, and a is any positive real number, $a \neq 1$, then
>
> (a) $\log_a xy = \log_a x + \log_a y$ (b) $\log_a \dfrac{x}{y} = \log_a x - \log_a y$
>
> (c) $\log_a x^r = r \cdot \log_a x$ (d) $\log_a a = 1$
>
> (e) $\log_a 1 = 0$

To prove part (a) of the properties of logarithms, let $m = \log_a x$ and $n = \log_a y$. Then, by the definition of logarithm,

$$a^m = x \quad \text{and} \quad a^n = y.$$

Hence, $a^m \cdot a^n = xy.$

By a property of exponents,

$$a^{m+n} = xy.$$

Now use the definition of logarithm to write

$$\log_a xy = m + n.$$

Since $m = \log_a x$ and $n = \log_a y$,

$$\log_a xy = \log_a x + \log_a y.$$

To prove part (b) of the properties, use m and n as defined above. Then

$$\frac{a^m}{a^n} = \frac{x}{y}.$$

Since $\dfrac{a^m}{a^n} = a^{m-n},$

then $a^{m-n} = \dfrac{x}{y}.$

By the definition of logarithms,

$$\log_a \frac{x}{y} = m - n,$$

or $\log_a \frac{x}{y} = \log_a x - \log_a y.$

For part (c),

$$(a^m)^r = x^r \quad \text{or} \quad a^{mr} = x^r.$$

Again using the definition of logarithms,

$$\log_a x^r = mr,$$

or $\log_a x^r = r \cdot \log_a x.$

Finally, (d) and (e) follow directly from the definition of logarithm since $a^1 = a$ and $a^0 = 1$.

The properties of logarithms are useful for rewriting expressions with logarithms in different forms as shown in the next examples.

EXAMPLE 5 ☐ Assuming all variables represent positive real numbers, use the properties of logarithms to write each of the following in a different form.

(a) $\log_6 7 \cdot 9 = \log_6 7 + \log_6 9$

(b) $\log_9 \frac{15}{7} = \log_9 15 - \log_9 7$

(c) $\log_5 \sqrt{8} = \log_5 8^{1/2} = \frac{1}{2} \log_5 8$

(d) $\log_a \frac{mnq}{p^2} = \log_a m + \log_a n + \log_a q - 2 \log_a p$

(e) $\log_a \sqrt[3]{m^2} = \frac{2}{3} \log_a m.$

(f) $\log_b \sqrt[n]{\frac{x^3 y^5}{z^m}} = \frac{1}{n} \log_b \frac{x^3 y^5}{z^m}$

$$= \frac{1}{n} (\log_b x^3 + \log_b y^5 - \log_b z^m)$$

$$= \frac{3}{n} \log_b x + \frac{5}{n} \log_b y - \frac{m}{n} \log_b z. \quad ☐$$

EXAMPLE 6 ☐ Use the properties of logarithms to write each of the following as a single logarithm with a coefficient of 1. Assume all variables represent positive real numbers.

(a) $\log_3 (x + 2) + \log_3 x - \log_3 2 = \log_3 \frac{(x + 2)x}{2}$

(b) $2 \log_a m - 3 \log_a n = \log_a m^2 - \log_a n^3 = \log_a \dfrac{m^2}{n^3}$

(c) $\dfrac{1}{2} \log_b m + \dfrac{3}{2} \log_b 2n - \log_b m^2 n$

$$= \log_b m^{1/2} + \log_b (2n)^{3/2} - \log_b m^2 n$$

$$= \log_b \dfrac{m^{1/2}(2n)^{3/2}}{m^2 n}$$

$$= \log_b \dfrac{2^{3/2} n^{1/2}}{m^{3/2}} \quad \square$$

EXAMPLE 7   Assume $\log_{10} 2 = .3010$. Find the base 10 logarithms of 4 and 5. By the properties of logarithms,

$$\log_{10} 4 = \log_{10} 2^2 = 2 \log_{10} 2 = 2(.3010) = .6020$$

$$\log_{10} 5 = \log_{10} \dfrac{10}{2} = \log_{10} 10 - \log_{10} 2 = 1 - .3010 = .6990. \quad \square$$

We can use compositions of the exponential and logarithmic functions to get two more useful properties. If $f(x) = a^x$ and $g(x) = \log_a x$, then

$$f[g(x)] = a^{\log_a x}$$

and $g[f(x)] = \log_a a^x.$

In Section 3.6, we saw that for functions f and g that are inverses of each other, $f[g(x)] = g[f(x)] = x$. Since exponential and logarithmic functions of the same base are inverses of each other, we can apply this to get the following theorem.

Theorem $a^{\log_a x} = x$ and $\log_a a^x = x.$

EXAMPLE 8   Simplify each of the following.

(a) $\log_5 5^3 = 3$

(b) $7^{\log_7 10} = 10$

(c) $\log_r r^{k+1} = k + 1 \quad \square$

EXERCISES 5.2

Change each of the following to logarithmic form.

1. $3^4 = 81$ **2.** $2^5 = 32$ **5.** $10^{-4} = .0001$

3. $(1/2)^{-4} = 16$ **4.** $(2/3)^{-3} = 27/8$ **6.** $(1/100)^{-2} = 10,000$

Change each of the following to exponential form.

7. $\log_6 36 = 2$ **8.** $\log_5 625 = 4$

9. $\log_{1/2} 16 = -4$ **10.** $\log_4 \dfrac{1}{64} = -3$

Find the value of each of the following.

11. $\log_5 25$ **12.** $\log_3 81$

13. $\log_8 8$ **14.** $\log_7 1$

15. $\log_{10} 0.001$ **16.** $\log_6 \dfrac{1}{216}$

17. $\log_{25} 5$ **18.** $\log_{16} 2$

19. $\log_4 \dfrac{\sqrt[3]{4}}{2}$ **20.** $\log_9 \dfrac{\sqrt[4]{27}}{3}$

21. $\log_{1/3} \dfrac{9^{-4}}{3}$ **22.** $\log_{1/4} \dfrac{16^2}{2^{-3}}$

23. $\log_6 36^4$ **24.** $\log_5 125^2$

25. $2^{\log_2 9}$ **26.** $8^{\log_8 11}$

Solve each of the following equations.

27. $\log_x 25 = -2$ **28.** $\log_x \dfrac{1}{16} = -2$

29. $\log_9 27 = m$ **30.** $\log_8 4 = z$

31. $\log_y 8 = \dfrac{3}{4}$ **32.** $\log_r 7 = 1/2$

Write each of the following as a sum, difference, or product of logarithms. Simplify the result if possible. Assume all variables represent positive real numbers.

33. $\log_3 (2/5)$ **34.** $\log_4 (6/7)$

35. $\log_2 \dfrac{6x}{y}$ **36.** $\log_3 \dfrac{4p}{q}$

37. $\log_5 \dfrac{5\sqrt{7}}{3}$ **38.** $\log_2 \dfrac{2\sqrt{3}}{5}$

39. $\log_4 (2x + 5y)$ **40.** $\log_6 (7m - 3q)$

41. $\log_k \dfrac{pq^2}{m}$ **42.** $\log_z \dfrac{x^5 y^3}{3}$

43. $\log_m \sqrt{\dfrac{5r^3}{z^5}}$ **44.** $\log_p \sqrt[3]{\dfrac{m^5 n^4}{t^2}}$

Write each of the following expressions as a single logarithm. Assume that all variables represent positive real numbers.

45. $\log_a x + \log_a y - \log_a m$

46. $(\log_b k - \log_b m) - \log_b a$

47. $2 \log_m a - 3 \log_m b^2$

48. $\dfrac{1}{2} \log_y p^3 q^4 - \dfrac{2}{3} \log_y p^4 q^3$

49. $-\dfrac{3}{4} \log_x a^6 b^8 + \dfrac{2}{3} \log_x a^9 b^3$

50. $\log_a (pq^2) + 2 \log_a (p/q)$

51. $\log_b (x + 2) + \log_b 7x - \log_b 8$

52. $\log_h (2m - 3) + \log_h 2m - \log_h 3$

53. $2 \log_a (z - 1) + \log_a (3z + 2)$

54. $\log_b (2y + 5) - \dfrac{1}{2} \log_b (y + 3)$

55. $-\dfrac{2}{3} \log_5 5m^2 + \dfrac{1}{2} \log_5 25m^2$

56. $-\dfrac{3}{4} \log_3 16p^4 - \dfrac{2}{3} \log_3 8p^3$

57. Graph each of the following. Compare the graphs to that of $y = \log_2 x$.

(a) $y = (\log_2 x) + 3$ (b) $y = \log_2 (x + 3)$

(c) $y = |\log_2 (x + 3)|$

58. Graph each of the following. Compare the graphs to that of $y = \log_{1/2} x$.

(a) $y = (\log_{1/2} x) - 2$ (b) $y = \log_{1/2} (x - 2)$

(c) $y = |\log_{1/2} (x - 2)|$

Graph each of the following functions.

59. $y = \log_3 x$

60. $y = \log_{10} x$

61. $y = \log_{1/2} (1 - x)$

62. $y = \log_{1/3} (3 - x)$

63. $y = \log_2 x^2$

64. $y = \log_3 (x - 1)$

65. $y = x \cdot \log_{10} x$

66. $y = x^2 \cdot \log_{10} x$

Given $\log_{10} 2 = 0.3010$ *and* $\log_{10} 3 = 0.4771$, *find each of the following without using calculators or tables.*

67. $\log_{10} 6$

68. $\log_{10} 12$

69. $\log_{10} 9$

70. $\log_{10} 20$

71. $\log_{10} 30$

72. $\log_{10} 36$

73. The population of an animal species that is introduced into a certain area may grow rapidly at first but then grow more slowly as time goes on. A logarithmic function can provide an excellent description of such growth. Suppose that the population of foxes, $F(t)$, in an area t months after the foxes were introduced there is

$$F(t) = 500 \log_{10} (2t + 3).$$

Use a calculator with a log key to find the population of foxes at the following times: (a) when they are first released into the area (that is, when $t = 0$); (b) after 3 months; (c) after 15 months. (d) Graph $y = F(t)$.

The loudness of sounds is measured in a unit called a decibel. *To measure with this unit, we first assign an intensity of* I_0 *to a very faint sound, called the* threshold sound. *If a particular sound has intensity* I, *then the decibel rating of this louder sound is*

$$10 \cdot \log_{10} \frac{I}{I_0}.$$

74. Find the decibel ratings of sounds having the following intensities: (a) $100I_0$; (b) $1000I_0$; (c) $100,000I_0$; (d) $1,000,000I_0$

75. Find the decibel ratings of the following sounds, having intensities as given. (You will need a calculator with a log key.) Round answers to the nearest whole number.

(a) whisper, $115I_0$.

(b) busy street, $9,500,000I_0$

(c) heavy truck, 20 m away, $1,200,000,000I_0$

(d) rock music, $895,000,000,000I_0$

(e) jetliner at takeoff, $109,000,000,000,000I_0$

76. The intensity of an earthquake, measured on the *Richter Scale*, is given by

$$\log_{10} \frac{I}{I_0},$$

where I_0 is the intensity of an earthquake of a certain (small) size. Find the Richter Scale ratings of earthquakes having intensity (a) $1000I_0$; (b) $1,000,000I_0$; (c) $100,000,000I_0$.

77. The San Francisco earthquake of 1906 had a Richter Scale rating of 8.6. Use a calculator with an x^y key to express the intensity of this earthquake as a multiple of I_0 (see Exercise 76).

78. How much more powerful is an earthquake with a Richter Scale rating of 8.6 than one with a rating of 8.2?

79. Using a calculator*, evaluate $(3^{0.003})^{1001}$ and $3^{(0.003 \times 1001)}$ by computing the expression within parentheses first. Did you get the same results? If not, can you explain the difference? What does this tell you about the laws of exponents as applied to calculator arithmetic?

80. Using a calculator, evaluate $\log_{10} (2^{0.0001})$ and $0.0001 \times (\log_{10} 2)$ by computing the expression within parentheses first. Did you get the same results? If not, can you explain the difference? What does this tell you about the properties of logarithms as applied to calculator arithmetic?

*Exercises 79 and 80 taken with permission from Abe Mizrahi and Michael Sullivan, *Calculus and Analytic Geometry*, Wadsworth Publishing Company, Belmont, California, p. 395.

5.3   Natural Logarithms

Since our number system uses base 10, logarithms to base 10 are most convenient for numerical calculation, historically the main application of logarithms. Base 10 logarithms are called **common logarithms.** The common logarithm of the number x, or $\log_{10} x$, is often abbreviated as just $\log x$.

In most other practical applications of logarithms, however, the number $e \approx$ 2.718281828 is used as base. Logarithms to base e are called **natural logarithms,** since they occur in many natural-world applications, such as those involving growth and decay. The abbreviation $\ln x$ is used for the natural logarithm of x, so that $\log_e x = \ln x$.

Natural logarithms can be found with a calculator which has a ln key or with a table of natural logarithms. A table of natural logarithms is given in Table 4. From this table we find, for example, that

$$\ln 55 = 4.0073,$$

$$\ln 1.9 = 0.6419,$$

and $\ln 0.4 = -.9163.$

EXAMPLE 1   ▢   Use a calculator or Table 4 to find the following logarithms.

(a) ln 83

If you are using a calculator, simply press the keys for 83, then press the ln key, and read the result, 4.4188.

To use Table 4, note that Table 4 does not give ln 83. However, we can use the properties of logarithms to write

$$\ln 83 = \ln(8.3 \times 10)$$

$$= \ln 8.3 + \ln 10$$

$$\approx 2.1163 + 2.3026$$

$$= 4.4189.$$

This result is slightly different from the answer we got above when using a calculator. This difference is due to rounding error.

(b) ln 36

A calculator gives ln 36 = 3.5835. To use the table, first use properties of logarithms, since 36 is not listed directly in Table 4.

$$\ln 36 = \ln 6^2$$

$$= 2 \ln 6$$

$$\approx 2(1.7918)$$

$$= 3.5836.$$

Alternatively, we can find ln 36 as follows:

$$\ln 36 = \ln 9 \cdot 4 = \ln 9 + \ln 4 = 2.1972 + 1.3863 = 3.5835.   ▢$$

The next examples show applications of natural logarithms.

**EXAMPLE
2**
☐ Suppose the amount, y, in grams, of a certain radioactive substance at a time t is given by

$$y = y_0 e^{-.1t},$$

where y_0 is the amount of the substance present initially (when $t = 0$) and t is measured in days. Find the *half-life* of the substance—that is, the time it takes for half a given amount of the substance to decay.

We want to know the time t that must elapse for y to be reduced to a value equal to $y_0/2$. That is, we want to solve the equation

$$\frac{1}{2}y_0 = y_0 e^{-.1t}.$$

Multiplying both sides by $1/y_0$ yields $1/2 = e^{-0.1t}$. If we now take natural logarithms of both sides (see the next section for more detail), we have

$$\ln \frac{1}{2} = \ln e^{-.1t}$$

$$\ln \frac{1}{2} = -.1t \ln e.$$

Since $\ln e = 1$,

$$\ln \frac{1}{2} = -.1t,$$

$$t = \frac{-\ln 1/2}{.1}.$$

From Table 4 or a calculator, we have $\ln 1/2 = \ln 0.5 = -.6931$. Thus,

$$t \approx 6.9 \text{ days.} \quad ☐$$

**EXAMPLE
3**
☐ The population of a certain bird found in the delta of the Mississippi river is given by

$$p(t) = p_0 e^{.02t},$$

where t is time in years since the first count was made, and p_0 is the population when $t = 0$. How long would it take for the population of birds to double?

We want to find the value of t that makes $p(t) = 2p_0$. Substitute $2p_0$ for $p(t)$ and then solve for t.

$$2p_0 = p_0 e^{.02t}$$

$$2 = e^{.02t}$$

Taking natural logarithms of both sides gives

$$\ln 2 = \ln e^{.02t}$$

$$\ln 2 = .02t \ln e$$

$$\ln 2 = .02t \quad (\ln e = 1)$$

$$t = \frac{\ln 2}{.02} \approx 34.7.$$

It will take about 34.7 years for the population of birds to double. ∎

EXAMPLE 4 ∎ Carbon 14 is a radioactive isotope of carbon which has a half-life of about 5600 years. The atmosphere contains much carbon, mostly in the form of carbon dioxide, with small traces of carbon 14. Most of this atmospheric carbon is in the form of the nonradioactive isotope carbon 12. The ratio of carbon 14 to carbon 12 is virtually constant in the atmosphere. However, as a plant absorbs carbon dioxide from the air in the process of photosynthesis, the carbon 12 stays in the plant while the carbon 14 decays by conversion to nitrogen. Thus, the ratio of carbon 14 to carbon 12 is smaller in the plant than it is in the atmosphere. Even when the plant is eaten by an animal, this ratio will continue to decrease. Based on these facts, a method of dating objects called carbon 14 dating has been developed.

Let R be the (nearly constant) ratio of carbon 14 to carbon 12 found in the atmosphere, and let r be the ratio found in a fossil. It can be shown that the relationship between R and r is given by

$$\frac{R}{r} = e^{(t \ln 2)/5600},$$

where t is the age of the fossil in years.

(a) Verify the formula for $t = 0$.

If $t = 0$, we have

$$\frac{R}{r} = e^0 = 1.$$

Thus $R = r$, so that the ratio in the fossil is the same as the ratio in the atmosphere. This is true only when $t = 0$.

(b) Verify the formula for $t = 5600$.

Substitute 5600 for t. Then

$$\frac{R}{r} = e^{(5600 \ln 2)/5600}$$

$$\frac{R}{r} = e^{\ln 2} = 2 \qquad \text{(Recall that } a^{\log_a x} = x\text{)}$$

$$r = \frac{1}{2}R.$$

From this last result, we see that the ratio in the fossil is half the ratio in the atmosphere. Since the half-life of carbon 14 is 5600 years, we expect only half of it to remain at the end of that time. Thus, the formula gives the correct result for $t = 5600$. ∎

Logarithms to other bases

We can use a calculator or a table to find the values of either natural logarithms (base e) or common logarithms (base 10). However, sometimes it is convenient to use logarithms to other bases. The following theorem can be used to convert logarithms from one base to another.

Change of Base Theorem

If x is any positive number and if a and b are positive real numbers, $a \neq 1, b \neq 1$, then

$$\log_a x = \frac{\log_b x}{\log_b a}.$$

To prove this result, we can use the definition of logarithm to write $y = \log_a x$ as $x = a^y$ or $x = a^{\log_a x}$ (for positive x and positive a, $a \neq 1$). If we now take base b logarithms of both sides of this last equation, we have

$$\log_b x = \log_b a^{\log_a x}$$

or $\log_b x = (\log_a x)(\log_b a),$

from which we obtain

$$\log_a x = \frac{\log_b x}{\log_b a}.$$

EXAMPLE 5   ☐   Use natural logarithms to find each of the following. Round to the nearest hundredth.

(a) $\log_5 27$

Let $x = 27$, $a = 5$, and $b = e$. Substituting into the change of base theorem gives

$$\log_5 27 = \frac{\log_e 27}{\log_e 5}$$

$$= \frac{\ln 27}{\ln 5}.$$

Now use a calculator or Table 4.

$$\log_5 27 \approx \frac{3.2958}{1.6094}$$

$$\approx 2.05$$

To check, use a calculator with an x^y key, along with the definition of logarithm, to verify that $5^{2.05} \approx 27$.

(b) $\log_2 0.1$

$$\log_2 0.1 = \frac{\ln 0.1}{\ln 2} \approx \frac{-2.3026}{.6931} = -3.32 \quad ☐$$

EXAMPLE   ☐   One measure of the diversity of the species in an ecological community is given
6   by

$$H = -[P_1 \log_2 P_1 + P_2 \log_2 P_2 + \cdots + P_n \log_2 P_n],$$

where $P_1, P_2 \ldots, P_n$ are the proportions of a sample belonging to each of n species found in the sample. For example, in a community with two species, where there are 90 of one species and 10 of the other, $P_1 = 90/100 = 0.9$ and $P_2 = 10/100 = 0.1$. Thus,

$$H = -[0.9 \log_2 0.9 + 0.1 \log_2 0.1].$$

We found $\log_2 0.1$ in Example 5(b) above. Now find $\log_2 0.9$.

$$\log_2 0.9 = \frac{\ln 0.9}{\ln 2} \approx \frac{-.1054}{.6931} \approx -.152$$

Therefore,

$$H \approx -[(0.9)(-.152) + (0.1)(-3.32)] \approx .469. \quad ☐$$

**EXERCISES
5.3**

Find each of the following to four decimal places.

1. ln 4.2

2. ln 6.1

3. ln 17

4. ln 80

5. ln 350

6. ln 980

7. ln 49

8. ln 64

9. ln 4200

10. ln 72,000

Find each of the following to the nearest hundredth.

11. $\log_5 10$

12. $\log_9 12$

13. $\log_{15} 5$

14. $\log_6 8$

15. $\log_{1/2} 3$

16. $\log_{12} 62$

17. $\log_{100} 83$

18. $\log_{200} 375$

For Exercises 19 and 20, refer to Example 6.

19. Suppose a sample of a small community shows two species with 50 individuals each. Find the index of diversity H.

20. A virgin forest in northwestern Pennsylvania has four species of large trees with the follow-ing proportions of each: hemlock, .521; beech, .324; birch, .081; maple, .074. Find the index of diversity H.

21. The number of species in a sample is given by

$$S(n) = a \ln \left(1 + \frac{n}{a}\right),$$

Here n is the number of individuals in the sample and a is a constant that indicates the diversity of species in the community. If $a = .36$, find $S(n)$ for the following values of n: (a) 100, (b) 200, (c) 150, (d) 10.

22. In Exercise 21, find n if $S(n) = 9$ and $a = .36$.

23. Suppose the number of rabbits in a colony is

$$y = y_0 e^{.4t},$$

where t represents time in months and y_0 is the rabbit population when $t = 0$.

(a) If $y_0 = 100$, find the number of rabbits present at time $t = 4$.

(b) How long will it take for the number of rabbits to triple?

24. A Midwestern city finds its residents moving to the suburbs. Its population is declining according to the relationship

$$P = P_0 e^{-.04t},$$

where t is time measured in years and P_0 is the population at time $t = 0$. Assume $P_0 = 1,000,000$.

(a) Find the population at time $t = 1$.

(b) Estimate the time it will take for the population to be reduced to 750,000.

(c) How long will it take for the population to be cut in half?

Exercises 25–28 refer to Example 4 in the text.

25. Suppose an Egyptian mummy is discovered in which the ratio of carbon 14 to carbon 12 is only about half the ratio found in the atmosphere. About how long ago did the Egyptian die?

26. If the ratio of carbon 14 to carbon 12 in an object is 1/4 the atmospheric ratio, how old is the object? How old if the ratio is 1/8?

27. Verify the formula of Example 4 for $t = 11,200$.

28. Paint from the Lascaux caves of France contains 15% of the normal amount of carbon 14. Estimate the age of the caves.

29. If an object is fired vertically upward and is subject only to the force of gravity, g, and to air resistance, then the maximum height, H, attained by the object is

$$H = \frac{1}{K}\left(V_0 - \frac{g}{K}\ln\frac{g + V_0 K}{g}\right),$$

where V_0 is the initial velocity of the object and K is a constant. Find H if $K = 2.5$, $V_0 = 1000$

feet per second, and $g = 32$ feet per second per second.

30. The pull, P, of a tracked vehicle on dry sand under certain conditions is approximated by

$$P = W\left[.2 + .16 \ln \frac{G(bl)^{3/2}}{W}\right]$$

where G is an index of sand strength, W is the load on the vehicle, b is the width of the track, and l is the length of the track.* Find P if $W = 10$, $G = 5$, $b = 30.5$ cm, and $l = 61.0$ cm.

31. The following formula† can be used to estimate the population of the United States, where t is time in years measured from 1914 (times before 1914 are negative):

$$N = \frac{197,273,000}{1 + e^{-0.03134t}}$$

(a) Complete the following chart.

Year	Observed Population	Predicted Population
1790	3,929,000	3,929,000
1810	7,240,000	
1860	31,443,000	30,412,000
1900	75,995,000	
1970	204,000,000	
1980	224,000,000	

(b) Estimate the number of years after 1914 that it would take for the population to increase to 197,273,000.

32. Many environmental situations place effective limits on the growth of the number of an organism in an area. Many such limited growth situations are described by the *logistic function;*

*Gerald W. Turnage, *Prediction of Track Pull Performance in a Desert Sand*, unpublished MS thesis, The Florida State University, 1971.
†The formula is given in Alfred J. Lotka, *Elements of Mathematical Biology*, reprinted by Dover Press, 1957, p. 67.

$$G(t) = \frac{m \cdot G_0}{G_0 + (m - G_0)e^{-kmt}},$$

where G_0 is the initial number present, m is the maximum possible size of the population, and k is a positive constant. Assume $G_0 = 1000$, $m = 25{,}000$, and $k = .04$. (a) Find $G(5)$. (b) Find $G(10)$. (c) Find a value of t so that $G(t) = m/2$. ▣

To find the maximum permitted levels of certain pollutants in freshwater, the EPA has established the

functions of Exercises 33–34, where $M(h)$ is the maximum permitted level of pollutant for a water hardness of h milligrams per liter. Find $M(h)$ in each case. (These results give the maximum permitted average concentration in micrograms per liter for a 24-hour period.) ▣

33. Pollutant: copper; $M(h) = e^r$, where $r = 0.65 \cdot \ln h - 1.94$ and $h = 9.7$

34. Pollutant: Lead; $M(h) = e^r$, where $r = 1.51 \cdot \ln h - 3.37$ and $h = 8.4$

5.4 Exponential and Logarithmic Equations

In Section 5.1 we solved exponential equations such as $(1/3)^x = 81$ by writing both sides of the equation as a power of 3. However, we can't use that method to solve an equation such as $3^x = 5$ since we can't easily write 5 as a power of 3. To solve such equations, use the following result, which depends on the fact that a logarithmic function is one-to-one. Thus, if $x > 0$, $y > 0$, $b > 0$, and $b \neq 1$, then

$$x = y \quad \text{if and only if} \quad \log_b x = \log_b y.$$

EXAMPLE 1 ▢ Solve the equation $7^x = 12$.

If we take natural (base e) logarithms on both sides, we get

$$\ln 7^x = \ln 12$$
$$x \cdot \ln 7 = \ln 12$$
$$x = \frac{\ln 12}{\ln 7}.$$

If desired, we can get a decimal approximation for x. First use Table 4 or a calculator to get

$$x = \frac{\ln 12}{\ln 7} \approx \frac{2.4849}{1.9459}.$$

Using a calculator to divide 2.4849 by 1.9459 gives

$$x \approx 1.277.$$

A calculator with an x^y key can be used to check this answer. Evaluate $7^{1.277}$; the result should be approximately 12. This step verifies that, to the nearest thousandth, the solution set is $\{1.277\}$. ▢

EXAMPLE ☐ Solve $3^{2x-1} = 4^{x+2}$.
2   Taking natural logarithms on both sides gives

$$\ln 3^{2x-1} = \ln 4^{x+2}$$

$$(2x - 1) \ln 3 = (x + 2) \ln 4$$

$$2x \ln 3 - \ln 3 = x \ln 4 + 2 \ln 4$$

$$2x \ln 3 - x \ln 4 = 2 \ln 4 + \ln 3.$$

Factor out x on the left to get

$$x(2 \ln 3 - \ln 4) = 2 \ln 4 + \ln 3$$

or $$x = \frac{2 \ln 4 + \ln 3}{2 \ln 3 - \ln 4}.$$

Using the properties of logarithms, this can be expressed as

$$x = \frac{\ln 16 + \ln 3}{\ln 9 - \ln 4}$$

or finally, $$x = \frac{\ln 48}{\ln \dfrac{9}{4}}.$$

This quotient could be approximated by a decimal if desired:

$$x = \frac{\ln 48}{\ln 2.25} \approx \frac{3.8712}{.8109} \approx 4.774.$$

To the nearest thousandth, the solution set is $\{4.774\}$. To find $\ln 2.25$ with Table 4, use properties of logarithms to write $\ln 2.25$ as $\ln 1.5^2 = 2 \ln 1.5$.   ☐

Logarithmic Equations

The properties of logarithms given in Section 5.2 are useful in solving logarithmic equations, as shown in the next examples.

EXAMPLE ☐ Solve $\log_a (x + 4) - \log_a (x + 2) = \log_a x$.
3   Using a property of logarithms, we have

$$\log_a \frac{x + 4}{x + 2} = \log_a x$$

$$\frac{x + 4}{x + 2} = x$$

$$x + 4 = x(x + 2)$$

$$x + 4 = x^2 + 2x$$

$$x^2 + x - 4 = 0.$$

By the quadratic formula,

$$x = \frac{-1 \pm \sqrt{1 + 16}}{2}$$

so that $\quad x = \dfrac{-1 + \sqrt{17}}{2} \quad$ or $\quad x = \dfrac{-1 - \sqrt{17}}{2}.$

We cannot evaluate $\log_a x$ for $x = (-1 - \sqrt{17})/2$, since this number is negative and thus not in the domain of $\log_a x$. By substitution we can verify that $x = (-1 + \sqrt{17})/2$ is a solution, giving the solution set $\{(-1 + \sqrt{17})/2\}$.   □

EXAMPLE 4   □   Solve $\log (3x + 2) + \log (x - 1) = 1$.

Recall that $\log x$ is an abbreviation for $\log_{10} x$. Thus, using the fact that $1 = \log_{10} 10$ and the properties of logarithms, we have the following.

$$\log (3x + 2)(x - 1) = \log 10$$
$$(3x + 2)(x - 1) = 10$$
$$3x^2 - x - 2 = 10$$
$$3x^2 - x - 12 = 0.$$

Now use the quadratic formula to arrive at

$$x = \frac{1 \pm \sqrt{1 + 144}}{6}.$$

thus, $\quad x = \dfrac{1 + \sqrt{145}}{6} \quad$ or $\quad x = \dfrac{1 - \sqrt{145}}{6}.$

If $x = (1 - \sqrt{145})/6$, then $x - 1 < 0$; therefore, $\log (x - 1)$ does not exist. For this reason, $(1 - \sqrt{145})/6$ must be discarded as a solution. A calculator can help to show that $(1 + \sqrt{145})/6$ is a solution, with the solution set $\{(1 + \sqrt{145})/6\}$.   □

EXAMPLE 5   □   Suppose the function

$$P(t) = 10{,}000 e^{0.4t},$$

gives the population of a city at time t (in years). In how many years will the population double?

We want to find the value of t for which $P(t) = 20{,}000$. Thus,

$$20{,}000 = 10{,}000 e^{0.4t}$$
$$2 = e^{0.4t}.$$

To solve for t, take logarithms on both sides. Since the base is e, it is convenient to use natural logarithms.

$$\ln 2 = \ln e^{0.4t}$$
$$\ln 2 = 0.4t$$

In the last step, we used the fact that $\ln e^{0.4t} = 0.4t \cdot \ln e = 0.4t$. Finally,

$$t = \frac{\ln 2}{0.4}.$$

Using a calculator or Table 4 to find $\ln 2$ and then using a calculator to find the quotient to three-digit accuracy, we get $t \approx 1.73$. Thus, the population will double in about 1.73 years. ☐

EXAMPLE 6 ☐ Solve the equation $y = \dfrac{1 - e^x}{1 - e^{-x}}$ for x.

Begin by multiplying both sides by $1 - e^{-x}$.

$$y(1 - e^{-x}) = 1 - e^x$$

or $\qquad y - ye^{-x} = 1 - e^x$.

Get 0 alone on one side of the equation.

$$e^x + y - 1 - ye^{-x} = 0.$$

Multiply both sides by e^x.

$$e^{2x} + (y - 1)e^x - y = 0.$$

Rewrite this equation as

$$(e^x)^2 + (y - 1)e^x - y = 0,$$

a quadratic equation in e^x. We can solve this equation by using the quadratic formula with $a = 1$, $b = y - 1$, and $c = -y$.

$$e^x = \frac{-(y - 1) \pm \sqrt{(y - 1)^2 - 4(1)(-y)}}{2(1)}$$

$$= \frac{-(y - 1) \pm \sqrt{y^2 - 2y + 1 + 4y}}{2}$$

$$= \frac{-(y - 1) \pm \sqrt{y^2 + 2y + 1}}{2}$$

$$= \frac{-(y - 1) \pm \sqrt{(y + 1)^2}}{2}$$

$$e^x = \frac{-y + 1 \pm (y + 1)}{2}.$$

First, use the $+$ sign:

$$e^x = \frac{-y + 1 + y + 1}{2} = \frac{2}{2} = 1.$$

We cannot have $e^x = 1$ in the original equation since $e^x = 1$ forces $e^{-x} = 1$ also, giving a zero denominator.

Try the $-$ sign:

$$e^x = \frac{-y + 1 - y - 1}{2} = \frac{-2y}{2} = -y.$$

If $e^x = -y$, then we may take natural logarithms on both sides to get

$$\ln e^x = \ln(-y)$$
$$x \cdot \ln e = \ln(-y)$$
$$x \cdot 1 = \ln(-y)$$
$$x = \ln(-y).$$

This result will be satisfied by all values of y less than 0. ☐

Solving Exponential or Logarithmic Equations

In summary, to solve an exponential or logarithmic equation, first use the properties of algebra to change the given equation into one of the following forms, where a and b are real numbers.

1. $a^{f(x)} = b$
   To solve, take logarithms to base a on both sides.

2. $\log_a f(x) = b$
   Solve by changing to the exponential form $a^b = f(x)$.

3. $\log_a f(x) = \log_a g(x)$
   From the given equation, we obtain the equation $f(x) = g(x)$, which is then solved algebraically.

4. In a more complicated equation, such as the one in Example 6, it is necessary to first solve for $e^{f(x)}$ or $\log_a f(x)$ and then solve the resulting equation using one of the methods given above.

EXERCISES 5.4

Solve each of the following equations. If you have a calculator, give answers as decimals rounded to the nearest thousandth.

1. $3^x = 6$

2. $4^x = 12$

3. $7^x = 8$

4. $13^p = 55$

5. $e^{k-1} = 4$

6. $e^{2-y} = 12$

7. $2e^{5a+2} = 8$

8. $10e^{3z-7} = 5$

9. $\left(1 + \frac{r}{2}\right)^5 = 9$

10. $\left(1 + \frac{n}{4}\right)^3 = 12$

11. $\log(t - 1) = 1$

12. $\log q^2 = 1$

13. $\log(x - 3) = 1 - \log x$

14. $\log(z - 6) = 2 - \log(z + 15)$

15. $\ln(y + 2) = \ln(y - 7) + \ln 4$

16. $\ln p - \ln(p + 1) = \ln 5$

17. $\log_3(a - 3) = 1 + \log_3(a + 1)$

18. $\log w + \log(3w - 13) = 1$

19. $\log_2 \sqrt{2y^2 - 1} = 1/2$

20. $\log_2 (\log_2 x) = 1$

21. $\log z = \sqrt{\log z}$

22. $\log x^2 = (\log x)^2$

Find the solution of each of the following to three significant digits.

23. $\log_x 7 = 2$ **24.** $\log_x 10 = 3$

25. $1.8^{x+4} = 9.31$ **26.** $3.7^{5x-1} = 5.88$

27. The amount of a radioactive specimen present at time t (measured in seconds) is $A(t) = 5000(10)^{-.02t}$, where $A(t)$ is measured in grams. Find the half-life of the specimen, that is, the time it will take for half the specimen to remain.

A large cloud of radioactive debris from a nuclear explosion has floated over the Pacific Northwest, contaminating much of the hay supply. Consequently, farmers in the area are concerned that the cows who eat this hay will give contaminated milk. (The tolerance level for radioactive iodine in milk is 0.) The percent of the initial amount of radioactive iodine still present in the hay after t days is approximated by $P(t) = 100 \, e^{-.1t}$, where t is time measured in days.

28. Some scientists feel that the hay is safe after the percent of radioactive iodine has declined to 10% of the original amount. Find the number of days before the hay can be used.

29. Other scientists believe that the hay is not safe until the level of radioactive iodine has declined to only 1% of the level. Find the number of days this would take.

30. Use the formula $A = Pe^{ni}$ for continuous compounding of interest to find the number of years it will take for $1000, compounded continuously at 5%, to double.

In Section 5.3, the formula

$$\frac{R}{r} = e^{(t \ln 2)/5600}$$

was given to express the relationship between the ratio of carbon 14 to carbon 12 in the atmosphere

(R) compared to the ratio in a fossil (r), where t is the age of the fossil in years.

31. Solve the formula for t.

32. Suppose a specimen is found in which $r = \dfrac{2}{3} R$.

Estimate the age of the specimen.

Solve the following equations for x.

33. $y = \dfrac{1 + e^{-x}}{1 + e^x}$ **34.** $y = \dfrac{e^x}{1 - e^x}$

35. $y = \dfrac{e^x + e^{-x}}{2}$ **36.** $y = \dfrac{e^x - e^{-x}}{2}$

Solve each of the following equations for the indicated variables. Use logarithms to the appropriate bases.

37. $P = P_0 e^{kt/1000}$, for t

38. $I = \dfrac{E}{R}(1 - e^{-Rt/2})$, for t

39. $T = T_0 + (T_1 - T_0) \, 10^{-kt}$, for t

40. $A = \dfrac{Pi}{1 - (1 + i)^{-n}}$, for n

Recall (from the exercises for Section 5.2) the formula for the decibel rating of a sound:

$$d = 10 \log \frac{I}{I_0}.$$

41. Solve this formula for I.

42. A few years ago, there was a controversy about a proposed government limit on factory noise— one group wanted a maximum of 89 decibels, while another group wanted 86. This difference seemed very small to many people. Find the percent by which the 89 decibel intensity exceeds that for 86 decibels.

In Section 5.1 we gave the formula for compound interest:

$$A = P\left(1 + \frac{i}{m}\right)^{nm}.$$

Use natural logarithms and solve for

43. *i*

44. *m*

45. The turnover of legislators is a problem of interest to political scientists. One model of legislative turnover in the U.S. House of Representatives is given by

$$M = 434e^{-0.08t},$$

where M is the number of continuously serving members at time t.* This model is based on the 1965 membership of the House. Find the number of continuously serving members in each of the following years: (a) 1969, (b) 1973, (c) 1979.

46. Solve the formula of Exercise 45 for t.

47. The growth of bacteria in food products makes it necessary to time-date some products (such as milk) so that they will be sold and consumed before the bacteria count is too high. Suppose for a certain product that the number of bacteria present is given by

$$f(t) = 500e^{.1t},$$

under certain storage conditions, where t is time in days after packing of the product and the value of $f(t)$ is in millions. Find the number of bacteria present at each of the following times: (a) 2 days, (b) 1 week, (c) 2 weeks.

48. (a) Suppose the product of Exercise 47 cannot be safely eaten after the bacteria count reaches 3,000,000,000. How long will this take? (b) If $t = 0$ corresponds to January 1, what date should be placed on the product?

5.5 Common Logarithms (Optional)

As we said earlier, base 10 logarithms are called **common logarithms.** It is customary to abbreviate $\log_{10} x$ as simply $\log x$. This convention started when base 10 logarithms were used extensively for calculation. (But be careful: some advanced books use $\log x$ as an abbreviation for $\log_e x$.)

Examples of common logarithms include

$$\log 1000 = \log 10^3 = 3$$
$$\log 100 = \log 10^2 = 2$$
$$\log 10 = \log 10^1 = 1$$
$$\log 1 = \log 10^0 = 0$$
$$\log 0.1 = \log 10^{-1} = -1$$
$$\log 0.001 = \log 10^{-3} = -3.$$

Though it can be shown that there is no rational number x such that $10^x = 6$ (and thus no rational number x such that $x = \log 6$), we can use a table of common logarithms, or a calculator, to find a decimal approximation for $\log 6$. From Table 2 in the Appendix, a decimal approximation of $\log 6$ is

$$\log 6 \approx 0.7782,$$

*Thomas W. Casstevens, "Exponential Models of Legislative Turnover," *UMAP*, Unit 296.

or, equivalently, $10^{0.7782} \approx 6$. Since most logarithms are approximations anyway, it is common to replace $\approx$ with $=$ and write

$$\log 6 = 0.7782.$$

Table 2 gives the logarithms of numbers between 1 and 10. Since every positive number can be written in scientific notation as the product of numbers between 1 and 10 and a power of 10, we can find the logarithm of any positive number by using the table and the properties of logarithms.

EXAMPLE 1   ☐   Find log 6.24.

Locate the first two digits, 6.2, in the left column of the table. Then find the third digit, 4, across the top of the table. You should find

$$\log 6.24 = 0.7952. \quad ☐$$

EXAMPLE 2   ☐   Find log 6240.

Write 6240 using scientific notation, as

$$6240 = 6.24 \times 10^3.$$

Then use the properties of logarithms:

$$\begin{aligned}
\log 6240 &= \log (6.24 \times 10^3) \\
&= \log 6.24 + \log 10^3 \\
&= \log 6.24 + 3 \log 10 \\
&= \log 6.24 + 3.
\end{aligned}$$

We found log 6.24 = 0.7952 in Example 1. Thus,

$$\log 6240 = 0.7952 + 3 = 3.7952. \quad ☐$$

The decimal part of the logarithm, 0.7952 in Example 2, is called the **mantissa**, and the integer part, 3 here, is the **characteristic.** When using a table of logarithms, always make sure the mantissa is positive. The characteristic can be any integer, positive, negative, or zero.

EXAMPLE 3   ☐   Find log 0.00587.

Use scientific notation and the properties of logarithms to get

$$\begin{aligned}
\log 0.00587 &= \log (5.87 \times 10^{-3}) \\
&= \log 5.87 + \log 10^{-3} \\
&= 0.7686 + (-3) \\
&= 0.7686 - 3.
\end{aligned}$$

The logarithm is usually left in this form. A calculator would give the answer as -2.2314, the algebraic sum of 0.7686 and -3. Note that the mantissa in the calculator answer is a negative number. When using a table of logarithms, you should note that this is not the best form for the answer, since it is not clear as to which number is the mantissa and which is the characteristic.

It is possible to write the characteristic in other forms. For example, we could write log 0.00587 as

log 0.00587 = 7.7686 − 10.

The best choice depends on the anticipated use of the logarithm.   □

EXAMPLE 4   □   Find each of the following.

(a) log $(2.73)^4$

Use a property of logarithms to get

$$\log (2.73)^4 = 4 \log 2.73$$
$$= 4(.4362)$$
$$= 1.7448.$$

(b) log $\sqrt[3]{0.0762}$

We find that

$$\log \sqrt[3]{0.0762} = \log (0.0762)^{1/3}$$
$$= \frac{1}{3} \log 0.0762$$
$$= \frac{1}{3} (0.8820 - 2).$$

To preserve the characteristic as an integer, we must change the characteristic to a multiple of 3 before multiplying by 1/3. One way to do this is to add and subtract 1 (which adds up to 0) as follows.

$$0.8820 - 2 = 1 - 1 + 0.8820 - 2$$
$$= 1.8820 - 3$$

Now complete the work above.

$$\log \sqrt[3]{0.0762} = \frac{1}{3} (1.8820 - 3)$$
$$= 0.6273 - 1   □$$

Sometimes we know the logarithm of a number and wish to find the number. We call the number the **antilogarithm,** sometimes abbreviated **antilog.** For example, 0.756 is the antilogarithm of 0.8785 − 1, since

log 0.756 = 0.8785 − 1.

To find this result, look for 0.8785 in the body of the logarithm table. You should find 7.5 at the left and 6 at the top. Since the characteristic of 0.8785 − 1 is −1, the antilogarithm is

$$7.56 \times 10^{-1} = 0.756.$$

An antilogarithm can be thought of as an exponent. For example, since 0.756 is the (base 10) antilogarithm of $0.8785 - 1 = -0.1215$,

$$\log 0.756 = -0.1215$$

or $10^{-0.1215} = 0.756.$

EXAMPLE 5   ☐   Find each of the following antilogarithms.

(a) $\log x = 2.5340$

Find 0.5340 in the body of the table; 3.4 is at the left and 2 is at the top. Thus,

$$\log x = 2 + 0.5340$$
$$= \log 10^2 + \log 3.42 \qquad \text{(from Table 2)}$$
$$= \log (3.42 \times 10^2)$$
$$= \log 342.$$

Since $\log x = \log 342,$

$$x = 342,$$

and 342 is the antilogarithm of 2.5340.

(b) $\log x = 0.7536 - 3$

Table 2 shows that the antilogarithm of $0.7536 - 3$ is

$$5.67 \times 10^{-3} = 0.00567.$$

(c) $\log x = -4.0670$
$$= -4 + (-0.0670)$$
$$= -4 + (-1) + [1 + (-0.0670)]$$
$$= -5 + 0.9330$$
$$\log x = \log 0.0000857$$
$$x = 0.0000857 \qquad ☐$$

We can now use Table 2 and the properties of logarithms to help with a numerical calculation.

EXAMPLE 6   ☐   Find $(\sqrt[3]{42})(76.9)(0.00283)$.

Use the properties of logarithms and Table 2.

$$\log (\sqrt[3]{42})(76.9)(0.00283) = \frac{1}{3} \log 42 + \log 76.9 + \log 0.00283$$

$$= \frac{1}{3} (1.6232) + 1.8859 + (0.4518 - 3)$$

$$= 0.5411 + 1.8859 + (0.4518 - 3)$$
$$= 2.8788 - 3$$
$$= 0.8788 - 1.$$

From the logarithm table, 0.756 is the antilogarithm of 0.8788 − 1, and so

$$(\sqrt[3]{42})(76.9)(0.00283) \approx 0.756. \quad \square$$

EXAMPLE $\square$ The cost in dollars, $C(x)$, of manufacturing x picture frames, where x is
7   measured in thousands, is

$$C(x) = 5000 + 2000 \log (x + 1).$$

Find the cost of manufacturing 19,000 frames.
To find the cost of producing 19,000 frames, let $x = 19$. This gives

$$\begin{aligned}
C(19) &= 5000 + 2000 \log (19 + 1) \\
&= 5000 + 2000 \log 20 \\
&= 5000 + 2000 \,(1.3010) \\
&= 7602.
\end{aligned}$$

Thus, 19,000 frames cost a total of about $7600 to produce. $\square$

EXERCISES
5.5

*Find the characteristic of the logarithms of each of
the following.*

1. 2,400,000

2. 875,000

3. 0.00023

4. 0.098

5. 0.000042

6. 0.000000257

Find the common logarithms of the following numbers.

7. 875

8. 3750

9. 0.000893

10. 0.00376

11. 68,200

12. 103,000

13. 42,967

14. 1.51172

15. 0.00809432

16. 0.0000439761

Find the common antilogarithms of each of the following.

17. 1.5366

18. 2.9253

19. 0.8733

20. 0.2504 − 1

21. −3.4283

22. −4.2933

*Use common logarithms and Table 2 to find each of
the following. Round to the nearest hundredth. Use
the change of base formula from Section 5.4.*

23. ln 125

24. ln 63.1

25. ln 0.98

26. ln 1.53

27. ln 275

28. ln 39.8

29. $\log_2 10$

30. $\log_5 8$

31. $\log_4 12$

32. $\log_7 4$

Use common logarithms to solve the following equations. Round to the nearest thousandth.

33. $10^x = 2.5$

34. $10^{x-1} = 143$

35. $100^{2x+1} = 17$

36. $1000^{5x} = 2010$

37. $50^{x+2} = 8$

38. $200^{x+3} = 12$

39. $0.01^{3x} = 0.005$

40. $0.001^{2x} = 0.0004$

Use common logarithms to find approximations to three significant digits for each of the following.

41. $(8.16)^{1/3}$

42. $\sqrt{276}$

43. $\dfrac{(7.06)^3}{(31.7)(\sqrt{1.09})}$

44. $\dfrac{(2.51)^2}{(\sqrt{1.52})(3.94)}$

45. $(115)^{1/2} + (35.2)^{2/3}$

46. $(778)^{1/3} + (159)^{3/4}$

Use logarithms to solve each of the following applications.

47. The number of years, n, since two independently evolving languages split off from a common ancestral language is approximated by

$$n \approx -7600 \log r,$$

where r is the proportion of words from the ancestral language common to both languages. Find n if (a) $r = 0.9$; (b) $r = 0.3$; (c) How many years have elapsed since the split if half of the words of the ancestral language are common to both languages?

48. Midwest Creations finds that its total sales, $T(x)$, from the distribution of x catalogs, measured in thousands, is approximated by $T(x) = 5000 \log (x + 1)$. Find the total sales resulting from the distribution of (a) 0 catalogs; (b) 5000 catalogs; (c) 24,000 catalogs; (d) 49,000 catalogs.

In chemistry, the pH *of a solution is defined as*

$$pH = -\log [H_3O^+]$$

where $[H_3O^+]$ *is the hydronium ion concentration in moles per liter. Find the* pH *of each of the following, whose hydronium ion concentration is given.*

49. Grapefruit, 6.3×10^{-4}

50. Crackers, 3.9×10^{-9}

51. Limes, 1.6×10^{-2}

52. Sodium hydroxide (lye), 3.2×10^{-14}

Find $[H_3O^+]$ *for each of the following, whose* pH *is given.*

53. Soda pop, 2.7

54. Wine, 3.4

55. Beer, 4.8

56. Drinking water, 6.5

The area of a triangle having sides of length a, b, *and* c *is given by*

$$\sqrt{s(s - a)(s - b)(s - c)},$$

where $s = (a + b + c)/2$. *Use logarithms to find the areas of the following triangles.*

57. $a = 114$, $b = 196$, $c = 153$

58. $a = 0.0941$, $b = 0.0873$, $c = 0.0896$

59. A common problem in archaeology is to determine estimates of populations at a particular site. Several methods have been proposed to do this. One method relates the total surface area of a site to the number of occupants. If P represents the population of a site which covers an area of a square units, then

$$\log P = k \log a,$$

where k is an appropriate constant which varies for hilly, coastal, or desert environments, or for sites with single family dwellings or multiple family dwellings.* Find the population of sites with the following areas (use 0.8 for k): (a) 230 m², (b) 95 m², (c) 20,000 m².

60. In Exercise 59, find the population of a site with an area 100,000 m² for the following values of k: (a) 1.2, (b) 0.5, (c) 0.7.

*Cook, S. F., 1950, "The Quantitative Investigation of Indian Mounds." Berkeley: University of California Publications in American Archaeology and Ethnology, 40: 231–233.

Appendix   Interpolation

The table of logarithms included in this book contains decimal approximations of common logarithms to four significant digits. More accurate tables are available; however, more accuracy may be obtained from the table included in this book by the process of **linear interpolation.** As an example, let us use linear interpolation to approximate log 75.37.

First, we can establish that log 75.3 < log 75.37 < 75.4.

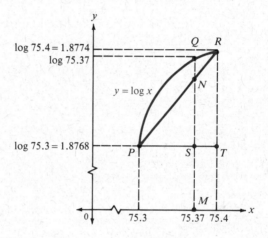

Figure 5.8

Figure 5.8 shows the portion of the curve $y = \log x$ between $x = 75.3$ and $x = 75.4$. We shall use the line segment PR to approximate the logarithm curve (this approximation is usually adequate for values of x that are close together.) From the figure, log 75.37 is given by the length of segment MQ, which we cannot find. We can, however, find MN, which we shall use as our approximation to log 75.37. By properties of similar triangles,

$$\frac{PS}{PT} = \frac{SN}{TR}.$$

In this case, $PS = 75.37 - 75.3 = .07$, $PT = 75.4 - 75.3 = .1$, and $RT = \log 75.4 - \log 75.3 = 1.8774 - 1.8768 = .0006$. Hence,

$$\frac{.07}{.1} = \frac{SN}{.0006}$$

or $SN = .7(.0006) \approx .0004$.

(This is 7/10 of the difference of the two logarithms.) Since log $75.37 = MN = MS + SN$, and since $MS = $ log $75.3 = 1.8768$,

log $75.37 = 1.8768 + .0004 = 1.8772$.

EXAMPLE 1 Find log 8726.

All the work of the example above can be condensed as follows.

$$10 \begin{Bmatrix} 6 \begin{Bmatrix} \log 8720 = 3.9405 \\ \log 8726 = \end{Bmatrix} x \\ \log 8730 = 3.9410 \end{Bmatrix} .0005$$

From this display,

$$\frac{6}{10} = \frac{x}{.0005}$$

$$x = \frac{6(.0005)}{10}$$

$$x = .0003.$$

Thus, log $8726 = 3.9405 + .0003 = 3.9408$.

EXAMPLE 2 Find log .0005958.

Work as above.

$$10 \begin{Bmatrix} 8 \begin{Bmatrix} \log .0005950 = -4 + .7745 \\ \log .0005958 = \end{Bmatrix} x \\ \log .0005960 = -4 + .7752 \end{Bmatrix} .0007$$

We get $\dfrac{8}{10} = \dfrac{x}{.0007}$,

or $x = .0006$,

and   log $.0005958 = (-4 + .7745) + .0006$

$$= -4 + .7751$$

$$= 6.7751 - 10.$$

We can also interpolate when finding an antilogarithm, as shown in the next example.

EXAMPLE 3 Find x such that log $x = .3275$.

From the logarithm table,

log $2.12 = .3263 < .3275 < .3284 = $ log 2.13

Set up the work as follows.

$$.01 \begin{Bmatrix} y \begin{Bmatrix} \log 2.12 = .3263 \\ \log x = .3275 \end{Bmatrix} .0012 \\ \log 2.13 = .3284 \end{Bmatrix} .0021$$

$$\frac{y}{.01} = \frac{.0012}{.0021}$$

$$y = \frac{(.01)(.0012)}{.0021}$$

$$y = .006$$

Thus, $x = 2.12 + .006 = 2.126$, and

$$\log 2.126 = .3275. \quad \square$$

APPENDIX EXERCISES

Interpolate to find logarithms of the following numbers.

1. 2345

2. 1.732

3. 48.26

4. 351.9

5. .06273

6. .003471

7. 27.05

8. 342.6

Use interpolation to find the antilogarithms of the following logarithms.

9. 1.7942

10. 3.9225

11. $7.6565 - 10$

12. $8.7296 - 10$

13. 5.6930

14. 12.6268

15. -3.7778

16. -4.1323

Use logarithms (and interpolation) to find approximations to four-digit accuracy for each of the following.

17. $\dfrac{(26.13)(5.427)}{101.6}$

18. $\dfrac{(32.68)(142.8)}{973.4}$

19. $\sqrt{\dfrac{6.532}{2.718}}$

20. $\left(\dfrac{27.46}{58.29}\right)^3$

21. $(.2374)^{.05}$

22. $(1.792)^{.23}$

23. $(49.83)^{1/2} + (2.917)^2$

24. $(38.42)^{1/3} + (86.13)^{1/4}$

25. The maximum load L that a cylindrical column can hold is given by

$$L = \frac{d^4}{h^2},$$

where d is the diameter of the cylindrical cross section and h is the height of the column. Find L if $d = 2.143$ feet and $h = 12.25$ feet.

26. Two electrons repel each other with a force $F = d^{-2}$, where d is the distance between the electrons. Find F if $d = .0005241$ cm.

CHAPTER 5   SUMMARY

Key Words

exponential function	common logarithm	characteristic
logarithm	natural logarithm	antilogarithm
logarithmic function	mantissa	interpolation

CHAPTER REVIEW EXERCISES
5

Graph each of the following.

1. $y = 2^x$

2. $y = 2^{-x}$

3. $y = (1/2)^{x+1}$

4. $y = 4^x + 4^{-x}$

Solve each of the following equations.

5. $8^p = 32$

6. $9^{2y-1} = 27^y$

7. $\dfrac{8}{27} = b^{-3}$

8. $\dfrac{1}{2} = \left(\dfrac{b}{4}\right)^{1/4}$

The amount of a certain radioactive material, in grams, present after t days, is given by

$$A(t) = 800\,e^{-0.04t}.$$

Find $A(t)$ if

9. $t = 0$.

10. $t = 5$.

Suppose P dollars is compounded continuously at a rate of interest i annually; the final amount on deposit would be $A = Pe^{ni}$ dollars after n years. How much would $1200 amount to at 10% compounded continuously for the following number of years?

11. 4 years

12. 10 years

13. Historically, the consumption of electricity has increased at a continuous rate of 6% per year. If it continued to increase at this rate, find the number of years before exactly twice as much electricity would be needed.

14. Suppose a conservation campaign together with higher rates caused demand for electricity to in-

crease at only 2% per year. (See Exercise 13.) Find the number of years before twice as much electricity would be needed as is needed today.

Write each of the following in logarithmic form.

15. $2^5 = 32$

16. $100^{1/2} = 10$

Write each of the following in exponential form.

17. $\log_{10} 0.001 = -3$

18. $\log_2 \sqrt{32} = 5/2$

Use properties of logarithms to write each of the following as a sum, difference, or product of logarithms.

19. $\log_3 \dfrac{mn}{p}$

20. $\log_2 \dfrac{\sqrt{5}}{3}$

21. $\log_5 x^2 y^4 \sqrt[5]{m^3 p}$

Find the common logarithm of each of the following numbers.

22. 8.47

23. 0.00421

Find the antilogarithm of each of the following numbers to three significant digits.

24. 3.4983

25. 9.7243

26. 0.6493 − 2

Use common logarithms to approximate each of the following.

27. $\dfrac{6^{2.1}}{\sqrt{52}}$

28. $2.43^{3.2}$

29. $\sqrt[5]{\dfrac{27.1}{4.33}}$

Solve each of the following equations.

30. $e^k = 12$

31. $10^{2r-3} = 17$

32. $\log_{64} y = 1/3$

33. $\log_2 (y + 3) = 5$

34. $\ln 6x - \ln (x + 1) = \ln 4$

35. $\log_{16} \sqrt{x + 1} = 4^{-1}$

Solve for x.

36. $y = \dfrac{5^x - 5^{-x}}{2}$

37. $y = \dfrac{1}{2(5^x - 5^{-x})}$

Find each of the following.

38. $\ln 89$

39. $\ln 0.000050$

40. $\ln 8$

41. $\log_{3.4} 15.8$

42. $\log_{1/2} 9.45$

43. $\log_3 769$

The height, in meters, of the members of a certain tribe is approximated by

$$h = 0.5 + \log t,$$

where t is the tribe member's age in years, and $1 \leq t \leq 20$. Find the height of a tribe member of age

44. 2 years

45. 5 years

46. 10 years

47. 20 years

The formula

$$A = P\left(1 + \frac{i}{m}\right)^{mn}$$

of Section 5.1 gives the amount of money in an account after n years if P dollars are deposited at a rate of interest i compounded m times per year. If A is known, then P is called the present value of A. (Think of present value as the value today of a sum of money to be received at some time in the future.) Find the present value of the following sums.

48. $4500 at 6% compounded annually for 9 years

49. $11,500 at 4% compounded annually for 12 years

50. $2000 at 4% compounded semiannually for 11 years

51. $2000 at 6% compounded quarterly for 8 years

52. How long would it take for $1000 at 5% compounded quarterly to double?

53. In Exercise 52, how long would it take at 10%?

54. If the inflation rate were 10%, use the formula for continuous compounding to find the number of years for a $1 item to cost $2.

55. In Exercise 54, find the number of years if the rate of inflation were 13%.

If R dollars is deposited at the end of each year in an account paying a rate of interest of i per year compounded annually, then after n years the account will contain a total of

$$R\left[\frac{(1 + i)^n - 1}{i}\right]$$

dollars. Find the final amount on deposit for each of the following. Use logarithms or a calculator.

56. $800, 12%, 10 years

57. $1500, 14%, 7 years

58. $375, 10%, 12 years

59. Manuel deposits $10,000 at the end of each year for 12 years in an account paying 12% compounded annually. He then puts this total amount on deposit in another account paying

10% compounded semiannually for another 9 years. Find the total amount on deposit after the entire 21-year period.

60. Scott Hardy deposits $12,000 at the end of each year for 8 years in an account paying 14% compounded annually. He then leaves the money alone with no further deposits for an additional 6 years. Find the total amount on deposit after the entire 14-year period.

61. Natural logarithms can be calculated on a hand calculator even if the calculator does not have an ln key. If $1/2 \leq x \leq 3/2$ and if $A = (x - 1)/(x + 1)$, then a good approximation to ln x is given by

ln $x \approx ((3A^2/5 + 1)A^2/3 + 1)2A$,

where $1/2 \leq x \leq 3/2$.

(a) Use this formula to calculate ln 0.8 and ln 1.2.

(b) Using facts about logarithms, use the formula to calculate (approximately)

ln $2 = $ ln $((3/2)(4/3))$.

(c) Using (b), calculate ln 3 and ln 8.

62. The exponential e^x can be estimated for x in $[-1/2, 1/2]$ by the formula

$$e^x \approx \left(\left(\left(\left(\frac{x}{5} + 1\right)\frac{x}{4} + 1\right)\frac{x}{3} + 1\right)\frac{x}{2} + 1\right)x + 1.$$

(a) Calculate an approximate value for $e^{0.13}$.

(b) Calculate an approximate value for $e^{-0.37}$.

(c) Calculate an approximate value for $e^{4.13}$.

(d) Calculate an approximate value for $e^{-2.63}$. (Hint: Use part (b).)

63. The quantity $n! = n(n - 1)(n - 2) \cdots 3 \cdot 2 \cdot 1$ grows very rapidly as n increases. According to *Stirling's Formula*, when n is large,

$$n! \approx \sqrt{2\pi n} \left(\frac{n}{e}\right)^n.$$

Use Stirling's Formula to estimate 100! and 200!. (Hint: Use common logarithms.)

64. Watch carefully. Suppose $0 < A < B$. Because the logarithm is an increasing function, we have

(a) log $A <$ log B; then

(b) $10A \cdot$ log $A < 10B \cdot$ log B;

(c) log $A^{10A} <$ log B^{10B};

(d) $A^{10A} < B^{10B}$;

On the other hand, we run into trouble with particular choices of A and B. For instance, choose $A = 1/10$ and $B = 1/2$. Clearly $0 < A < B$, but $A^{10A} = (1/10)^1 = 1/10$ is greater than $B^{10B} = (1/2)^5 = 1/32$. Where was the first false step made?

Exercises 61–64: Stanley I. Grossman, *Calculus*, 2nd Edition (Academic Press, New York, 1981), pp. 331–332.

6 Trigonometric Functions

We have discussed many different types of functions throughout this book, including linear, quadratic, polynomial, exponential, and logarithmic. In this chapter we introduce the trigonometric functions, which differ in a fundamental way from those previously studied: the trigonometric functions describe a *periodic* or *repetitive* relationship.

An example of a periodic relationship is shown by this electrocardiogram, a graph of the human heartbeat. The EKG shows electrical impulses from the heart. Each small square represents .04 seconds, and each large square represents .2 seconds. How often does this (abnormal) heart beat?

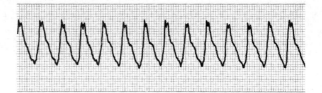

Trigonometric functions describe many natural phenomena, thus leading to their importance in the study of electronics, optics, heat, X-rays, acoustics, seismology, and many other areas. Trigonometric functions occur again and again in calculus and are key to the study of navigation and surveying.

6.1 The Sine and Cosine Functions

A circle with center at the origin and radius one unit is called a **unit circle.** Consider starting at the point (1, 0) on the unit circle and letting (x, y) be any other point on the circle. By measuring in a counterclockwise direction along the arc of the circle from (1, 0) to (x, y), we can find a unique positive real number s that gives the length of this arc. There is also a unique negative number that gives the length of the arc from (1, 0) to (x, y), measured in a clockwise direction. An arc with s > 0 is shown in Figure 6.1.

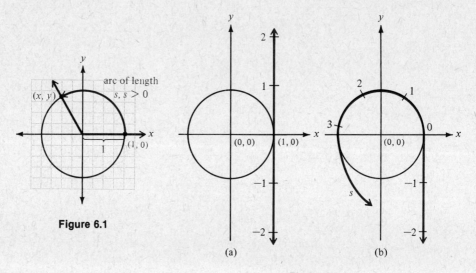

Figure 6.1

(a)   (b)

Figure 6.2

The circumference of a circle of radius r is given by the formula $C = 2\pi r$. Thus, the circumference of the unit circle is 2π, so that s may take values from 0 to 2π or from -2π to 0. We can extend the domain of s to include all real numbers by allowing arcs which wrap around the circle more than once. For example, $s = 3\pi$ would correspond to the same point on the circle as $s = \pi$ or $s = -\pi$. Figure 6.2 shows how s can take any real number as a value by wrapping a real number line around the unit circle.

EXAMPLE 1   ☐ Find the coordinates of the points on the unit circle corresponding to the following arc lengths.

(a) π

Since π is one-half the circumference of the unit circle, an arc of length π which starts at (1, 0) would end at the point $(-1, 0)$, as shown in Figure 6.3.

(b) $\pi/2$

An arc of length $\pi/2$ corresponds to the point (0, 1). See Figure 6.4.

(c) $\dfrac{3\pi}{2}$

As Figure 6.5 shows, the arc of length $\dfrac{3\pi}{2}$ corresponds to $(0, -1)$.   ☐

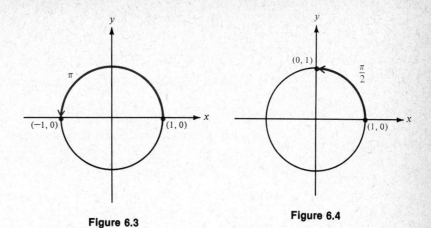

Figure 6.3

Figure 6.4

EXAMPLE 2   Find the coordinates of the point P on the unit circle which corresponds to $s = \pi/4$.

The point P on the unit circle which corresponds to $s = \pi/4$ is halfway between $(1, 0)$ and $(0, 1)$, as shown in Figure 6.6. This point also lies on the line $y = x$. Since the equation of the unit circle is $x^2 + y^2 = 1$, we have, replacing y with x,

$$x^2 + x^2 = 1,$$

or $$2x^2 = 1$$

$$x^2 = \frac{1}{2}.$$

We take the positive square root because P is in the first quadrant. This gives

$$x = \frac{1}{\sqrt{2}} = \frac{\sqrt{2}}{2}.$$

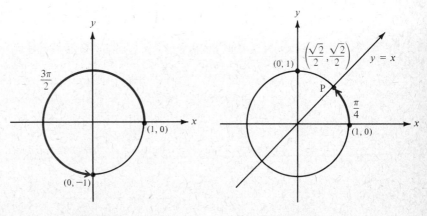

Figure 6.5

Figure 6.6

Also, since $y = x$, we have $\qquad y = \dfrac{\sqrt{2}}{2}.$

Thus, the point P on the unit circle which corresponds to $s = \pi/4$ is $(\sqrt{2}/2, \sqrt{2}/2)$. ☐

EXAMPLE
3 ☐   Find the coordinates of the point Q on the unit circle which corresponds to each of the following.

(a) $s = 3\pi/4$

Figure 6.7 shows the point Q and the point P from Example 2 above. Since $3\pi/4 = \pi - \pi/4$, by symmetry Q has the same y-coordinate as P, but an x-coordinate with opposite sign. Thus, the coordinates of Q are $(-\sqrt{2}/2, \sqrt{2}/2)$. ☐

(b) $s = -\pi/4$

As Figure 6.8 shows, this time the coordinates of Q are $(\sqrt{2}/2, -\sqrt{2}/2)$.

(c) $s = 5\pi/4$

Since $5\pi/4 = \pi + \pi/4$, use symmetry again to find that the coordinates are $(-\sqrt{2}/2, -\sqrt{2}/2)$. See Figure 6.9. ☐

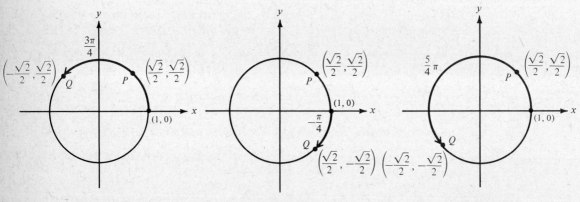

Figure 6.7   **Figure 6.8**   **Figure 6.9**

Using the correspondence between an arc of length s and a point (x, y) on the unit circle, we can define two functions. Each value of s leads to a unique value of x and a unique value of y. The function which associates each arc length s with the y-value of the corresponding point on the unit circle is called the **sine function,** abbreviated

sine function $\sin s = y.$

The function which associates each arc length s with the x-value of the corresponding point is called the **cosine function,** abbreviated

cosine function

$$\cos s = x.$$

By these definitions, the coordinates of the point associated with an arc of length s can be written as $(\cos s, \sin s)$.

Both the sine and cosine functions have the set of real numbers as domain. The values of $\cos s$ (or x) range from 1 at $s = 0$ to -1 at $s = \pi$. Cos s is 0 when s is $\pi/2$ or $3\pi/2$. Similarly, $\sin s$ ranges from -1 when s is $3\pi/2$ to 1 for $s = \pi/2$, with $\sin s = 0$ when s is 0 or π.

If 2π is added to any value of s, the corresponding values of $\sin s$ and $\cos s$ are unchanged. That is, for every value of s,

$$\sin s = \sin(s + 2\pi) \tag{1}$$
$$\text{and } \cos s = \cos(s + 2\pi). \tag{2}$$

A function of this type, which describes a cyclic relationship, is called a **periodic function.** Here, the number 2π is called the **period** of the function, because it is the smallest positive number which satisfies statements (1) and (2) above. It is also true, for example, that $\sin s = \sin (s + 4\pi)$ and $\sin s = \sin (s + 6\pi)$, but 2π is the period since 2π can be shown to be the *smallest* positive number that can be used.

EXAMPLE 4

Find each of the following function values.

(a) $\sin 0$

If $s = 0$, the corresponding point is $(1, 0)$. Since $\sin s = y$, $\sin 0 = 0$.

(b) $\cos \pi/2$

The point corresponding to $s = \pi/2$ is $(0, 1)$. We know $\cos s = x$, so $\cos \pi/2 = 0$.

(c) $\cos \pi/4$

In Example 2, we saw that $(\sqrt{2}/2, \sqrt{2}/2)$ corresponds to $\pi/4$. Thus, $\cos \pi/4 = \sqrt{2}/2$. ☐

While we shall not graph sine and cosine functions in detail until later in this chapter, it is beneficial to sketch the basic graphs now. To do this, we use the definitions of sine and cosine in terms of the unit circle. A unit circle with various arc lengths s is shown in Figure 6.10. The vertical dotted lines give the corresponding values of y, which is $\sin s$. By projecting horizontally, we get points on the graph of $y = \sin s$. Figure 6.10, on the next page, shows only *one period* of the graph; the complete graph would extend indefinitely to the right and to the left. A similar process can be used to get the graph of $y = \cos s$, shown in Figure 6.11 on the next page. Again, the portion shown here is only one period of the graph.

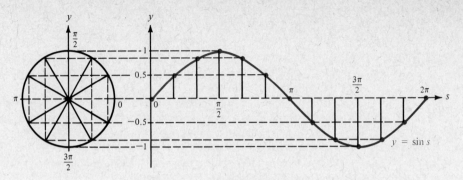

Figure 6.10

The following chart summarizes the intervals where sine and cosine are increasing and decreasing. The results in the chart can be found by inspecting the graphs of Figure 6.10 and Figure 6.11.

interval	$[0, \pi/2]$	$[\pi/2, \pi]$	$[\pi, 3\pi/2]$	$[3\pi/2, 2\pi]$
quadrant	I	II	III	IV
$\sin s$	increases from 0 to 1	decreases from 1 to 0	decreases from 0 to -1	increases from -1 to 0
$\cos s$	decreases from 1 to 0	decreases from 0 to -1	increases from -1 to 0	increases from 0 to 1

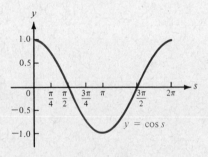

Figure 6.11

EXERCISES
6.1

For each of the following arc lengths, find the coordinates of the corresponding point on the unit circle.

1. $5\pi/4$

2. $7\pi/4$

3. π

4. -3π

5. 4π

6. 17π

7. -2π

8. $-\dfrac{\pi}{2}$

9. $\dfrac{5\pi}{2}$

10. $\dfrac{11\pi}{2}$

11. $\dfrac{-11\pi}{4}$

12. $\dfrac{-7\pi}{4}$

13. 2.25π

14. 1.5π

15. $.75\pi$

16. -2.25π

In each of the following exercises, the point which corresponds to arc length s is given. Use symmetry to find the coordinates of the point that corresponds to (a) $-s$; (b) $s + 2\pi$; (c) $s + \pi$; (d) $\pi - s$.

17. $\left(\dfrac{2}{3}, \dfrac{\sqrt{5}}{3}\right)$

18. $\left(\dfrac{5}{13}, \dfrac{12}{13}\right)$

19. $\left(\dfrac{4}{5}, \dfrac{3}{5}\right)$

20. $\left(\dfrac{3}{4}, \dfrac{\sqrt{7}}{4}\right)$

21. $\left(-\dfrac{1}{2}, \dfrac{\sqrt{3}}{2}\right)$

22. $\left(\dfrac{\sqrt{3}}{2}, -\dfrac{1}{2}\right)$

23. $\left(-\dfrac{2}{5}, \dfrac{-\sqrt{21}}{5}\right)$

24. $\left(\dfrac{-3}{5}, \dfrac{-4}{5}\right)$

For each of the following arc lengths, find sin s and cos s.

25. 0

26. $\dfrac{\pi}{2}$

27. $\dfrac{\pi}{4}$

28. $\dfrac{3\pi}{4}$

29. $\dfrac{3\pi}{2}$

30. 987π

31. -1423π

32. -2π

33. $-\dfrac{\pi}{4}$

34. $\dfrac{-\pi}{2}$

35. $\dfrac{-3\pi}{4}$

36. $\dfrac{-7\pi}{4}$

Identify the quadrant in which arcs having the following lengths would terminate.

37. $s = 10$

38. $s = 18$

39. $s = 36$

40. $s = 92$

41. $s = -896.1$

42. $s = -1046.001$

Show that the following points lie (approximately) on the unit circle.

43. $(.39852144, .91715902)$

44. $(-.81745602, -.57599102)$

Follow the method used in Examples 2 and 3 to work the following problems.

45. (a) Use a 30°–60°–90° triangle to show that the arc length $\pi/3$ corresponds to a point on the unit circle which is also on the line $y = \sqrt{3}\, x$.
  (b) Find the coordinates of the point on the unit circle which corresponds to $\pi/3$.
  (c) Use symmetry to find the coordinates of the points on the unit circle which correspond to $-\pi/3$, $2\pi/3$, and $4\pi/3$.

46. (a) Use a 30°–60°–90° triangle to show that the arc length $\pi/6$ corresponds to a point on the unit circle which is also on the line $\sqrt{3}\, y = x$.
  (b) Find the coordinates of the point on the unit circle which corresponds to $\pi/6$.
  (c) Use symmetry to find the coordinates of the points on the unit circle which correspond to $-\pi/6$, $5\pi/6$, and $7\pi/6$.

Use the results of Exercises 45 and 46 to find each of the following.

47. $\sin \pi/3$

48. $\cos \pi/3$

49. $\cos 2\pi/3$

50. $\sin 2\pi/3$

51. $\sin 4\pi/3$

52. $\cos -\pi/3$

53. $\sin 5\pi/6$

54. $\cos 5\pi/6$

6.2   Further Trigonometric Functions

In the previous section, we defined the sine and cosine functions as $\sin s = y$ and $\cos s = x$, where (x, y) is the point on the unit circle which corresponds to an arc of length s. We now define four additional functions derived from these two basic functions: the **tangent, cotangent, cosecant,** and **secant** functions, abbreviated as **tan, cot, csc,** and **sec,** respectively. (The definitions of sine and cosine are included for reference.)

Trigonometric or Circular Functions

If (x, y) is a point on the unit circle which corresponds to an arc of length s, then

$$\sin s = y \qquad\qquad \csc s = \frac{1}{y} \quad (y \neq 0)$$

$$\cos s = x \qquad\qquad \sec s = \frac{1}{x} \quad (x \neq 0)$$

$$\tan s = \frac{y}{x} \quad (x \neq 0) \qquad \cot s = \frac{x}{y} \quad (y \neq 0)$$

These six functions are called the **trigonometric functions** or the **circular functions.**

EXAMPLE 1   Find $\tan \pi/4$, $\cot \pi/4$, $\csc \pi/4$, and $\sec \pi/4$.

In the last section (Example 2), we saw that the point $(\sqrt{2}/2, \sqrt{2}/2)$ corresponds to an arc of length $s = \pi/4$. Thus $x = \sqrt{2}/2$ and $y = \sqrt{2}/2$. By their definitions,

$$\tan \frac{\pi}{4} = \frac{y}{x} = \frac{\sqrt{2}/2}{\sqrt{2}/2} = 1$$

$$\cot \frac{\pi}{4} = \frac{x}{y} = \frac{\sqrt{2}/2}{\sqrt{2}/2} = 1$$

$$\csc \frac{\pi}{4} = \frac{1}{y} = \frac{1}{\sqrt{2}/2} = \frac{2}{\sqrt{2}} = \sqrt{2}$$

$$\sec \frac{\pi}{4} = \frac{1}{x} = \frac{1}{\sqrt{2}/2} = \frac{2}{\sqrt{2}} = \sqrt{2}. \quad \square$$

Like the sine and cosine functions, the four new trigonometric functions are periodic. The cosecant and secant functions have the same period as sine and cosine, 2π. The tangent and cotangent functions, however, have a period of π. We shall prove this fact in the next chapter.

EXAMPLE 2   Find the values of the trigonometric functions for an arc of length $\pi/2$.

As shown in Figure 6.12, the point which corresponds to an arc of length $\pi/2$ is $(0, 1)$, so that $x = 0$ and $y = 1$. Use the definitions of the various functions.

$$\sin \frac{\pi}{2} = y = 1 \qquad\qquad \cot \frac{\pi}{2} = \frac{x}{y} = \frac{0}{1} = 0$$

$$\cos \frac{\pi}{2} = x = 0 \qquad\qquad \csc \frac{\pi}{2} = \frac{1}{y} = \frac{1}{1} = 1$$

$$\tan \frac{\pi}{2} = \frac{y}{x} = \frac{1}{0} \quad \text{(undefined)} \qquad \sec \frac{\pi}{2} = \frac{1}{x} = \frac{1}{0} \quad \text{(undefined)} \quad \square$$

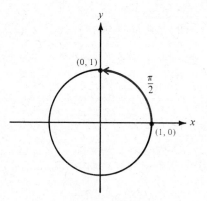

Figure 6.12

Several important properties of the trigonometric functions can be obtained from the definitions of the functions. First, recall that the *reciprocal* of a number a is $1/a$. (There is no reciprocal for 0.) The numbers $1/x$ and $x/1$ are reciprocals. From the definitions of the trigonometric functions, we see that $1/x$ is $\sec s$ and $x/1$ (or x) is $\cos s$. Thus, $\cos s$ and $\sec s$ are reciprocals. In the same way, sine and cosecant are reciprocals as are tangent and cotangent. In summary, we have the following relationships, called the **reciprocal identities.**

Reciprocal Identities

$$\sin s = \frac{1}{\csc s} \qquad \text{and} \qquad \csc s = \frac{1}{\sin s},$$

$$\cos s = \frac{1}{\sec s} \qquad \text{and} \qquad \sec s = \frac{1}{\cos s},$$

$$\tan s = \frac{1}{\cot s} \qquad \text{and} \qquad \cot s = \frac{1}{\tan s}.$$

These identities hold whenever the denominators are not zero.

EXAMPLE 3 $\square$ Suppose $\tan s = 3/4$. Find $\cot s$.
Since $\cot s = 1/\tan s$, we have

$$\cot s = \frac{1}{3/4} = \frac{4}{3}. \quad \square$$

We can use the definitions to determine the signs of the trigonometric functions in each of the four quadrants. For example, if s terminates in quadrant I, then both x and y are positive, and all the trigonometric functions have positive values. In quadrant II, x is negative, and y is positive. Thus, $\cos s = x$ is negative, $\sin s = y$ is positive, $\tan s = y/x$ is negative, and so on. The signs of the values of the trigonometric functions in the various quadrants are summarized in the following box.

Signs of Trigonometric Functions

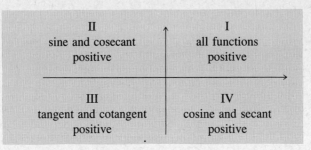

II sine and cosecant positive	I all functions positive
III tangent and cotangent positive	IV cosine and secant positive

EXAMPLE 4

☐ Suppose s terminates in quadrant II and $\sin s = 2/3$. Find the values of the other trigonometric functions.

Since $\sin s = y$, $y = 2/3$. To find x, use the fact that on the unit circle $x^2 + y^2 = 1$.

$$x^2 + \left(\frac{2}{3}\right)^2 = 1$$

$$x^2 + \frac{4}{9} = 1$$

$$x^2 = \frac{5}{9}$$

$$x = \frac{\pm\sqrt{5}}{3}$$

We know s terminates in quadrant II. Therefore, x must be negative, so $x = -\sqrt{5}/3$.

The remaining functions can now be found from their definitions.

$$\cos s = -\frac{\sqrt{5}}{3}$$

$$\tan s = \frac{2/3}{-\sqrt{5}/3} = \frac{2}{-\sqrt{5}} = \frac{-2\sqrt{5}}{5}$$

$$\cot s = \frac{-\sqrt{5}/3}{2/3} = \frac{-\sqrt{5}}{2}$$

$$\sec s = \frac{1}{-\sqrt{5}/3} = \frac{-3}{\sqrt{5}} = \frac{-3\sqrt{5}}{5}$$

$$\csc s = \frac{1}{2/3} = \frac{3}{2} \quad \square$$

We can also derive other relationships among the trigonometric functions by starting with the definitions of these functions. The equation of the unit circle is $x^2 + y^2 = 1$; since $\cos s = x$ and $\sin s = y$, we have

$$(\cos s)^2 + (\sin s)^2 = 1.$$

It is customary to write $(\sin s)^2$ as $\sin^2 s$, giving $\sin^2 s + \cos^2 s = 1$. Starting with $x^2 + y^2 = 1$, we can divide both sides by x^2, and then by y^2, to get two additional identities: $1 + \tan^2 s = \sec^2 s$, and $\cot^2 s + 1 = \csc^2 s$. Finally, we have two more identities, for $\tan s$ and $\cot s$, derived from the equalities $\cos s = x$ and $\sin s = y$. These last few identities (whose proofs are included as Exercises 43–46 below), make up the **fundamental identities.**

Fundamental Identities

$\sin^2 s + \cos^2 s = 1$	$\tan s = \dfrac{\sin s}{\cos s}$	$\cot s = \dfrac{\cos s}{\sin s}$
$1 + \tan^2 s = \sec^2 s$		
$\cot^2 s + 1 = \csc^2 s$	$(\cos s \neq 0)$	$(\sin s \neq 0)$

Using these relationships, we can find all values of the trigonometric functions for a particular value of s, given the value of one function and the quadrant in which the arc of length s terminates.

EXAMPLE 5   Suppose that $\tan t = 1/4$, and t terminates in quadrant III. Find the other five trigonometric function values for t.

Since $\cot t = 1/\tan t$,

$$\cot t = \frac{1}{1/4} = 4.$$

We know that $1 + \tan^2 t = \sec^2 t$. Since $\tan t = 1/4$,

$$1 + \left(\frac{1}{4}\right)^2 = \sec^2 t$$

$$1 + \frac{1}{16} = \sec^2 t$$

$$\frac{17}{16} = \sec^2 t$$

Before we take the square root of both sides to find $\sec t$, we need to know whether the result is positive or negative. We are told that the arc terminates in quadrant III, where secant is negative. Thus,

$$-\frac{\sqrt{17}}{4} = \sec t.$$

Since $\cos t = 1/\sec t$, we have $\cos t = -4\sqrt{17}/17$. Finally, we know that

$$\tan t = \frac{\sin t}{\cos t},$$

or $\sin t = (\tan t)(\cos t)$. Using our values,

$$\sin t = \left(\frac{1}{4}\right)\left(-\frac{4\sqrt{17}}{17}\right) = -\frac{\sqrt{17}}{17}.$$

Finally, $\csc t = 1/\sin t = -\sqrt{17}$.　☐

EXERCISES
6.2

For each of the following, find tan s, cot s, sec s, and csc s. (Do not use tables or calculator.)

1. $\sin s = 1/2$, $\cos s = \sqrt{3}/2$
2. $\sin s = 3/4$, $\cos s = \sqrt{7}/4$
3. $\sin s = 4/5$, $\cos s = -3/5$
4. $\sin s = -1/2$, $\cos s = -\sqrt{3}/2$
5. $\sin s = -\sqrt{3}/2$, $\cos s = 1/2$
6. $\sin s = 12/13$, $\cos s = 5/13$

For each of the following, find the values of the six trigonometric functions. For Exercises 13–18, use the results of Exercises 45 and 46 of Section 6.1. (Do not use tables or calculator.)

7. π	8. $3\pi/2$
9. $3\pi/4$	10. $5\pi/4$
11. $-\pi/4$	12. $-\pi/2$
13. $\pi/6$	14. $\pi/3$
15. $2\pi/3$	16. $5\pi/6$
17. $7\pi/6$	18. $4\pi/3$

Complete the following table of signs of the trigonometric functions. (Do not use tables or calculator.)

	Quadrant	sin	cos	tan	cot	sec	csc
19.	I	+	+			+	+
20.	II	+		−	−		
21.	III						
22.	IV			−			−

Decide what quadrant(s) s must terminate in to satisfy the following conditions for $0 \leq s < 2\pi$.

23. $\sin s > 0$, $\cos s < 0$

24. $\cos s > 0$, $\tan s > 0$

25. $\sec s < 0$, $\csc s < 0$

26. $\tan s > 0$, $\cot s > 0$

27. $\cos s < 0$ **28.** $\tan s > 0$

29. $\csc s < 0$ **30.** $\sin s > 0$

For each of the following find the values of the other trigonometric functions.

31. $\cos s = \dfrac{-4}{9}$, *s* terminates in quadrant II

32. $\csc s = 2$, *s* terminates in quadrant II

33. $\tan s = \dfrac{3}{2}$, $\csc s = \dfrac{\sqrt{13}}{3}$

34. $\sec s = -2$, $\cot s = \dfrac{\sqrt{3}}{3}$

35. $\sin t = \dfrac{\sqrt{5}}{5}$, $\cos t < 0$

36. $\tan t = -\dfrac{3}{5}$, $\sec t > 0$

37. $\sin s = a$, *s* is in quadrant I

38. $\tan t = m$, *t* is in quadrant III

Find the values of the other trigonometric functions.

39. $\cos s = -.428193$, *s* terminates in quadrant II

40. $\sec t = 28.4096$, *t* terminates in quadrant IV

41. $\csc t = -10.4349$, $\sec t > 0$

42. $\cot t = -.139725$, $\cos t > 0$

Use the definitions of the trigonometric functions to prove the following statements.

43. $\tan^2 s + 1 = \sec^2 s$

44. $1 + \cot^2 s = \csc^2 s$

45. $\tan s = \dfrac{\sin s}{\cos s}$, $\cos s \neq 0$

46. $\cot s = \dfrac{\cos s}{\sin s}$, $\sin s \neq 0$

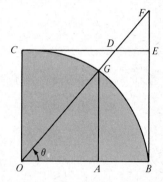

The figure shows a quarter circle of radius 1. Give each of the following lengths in terms of θ.

47. *OA**   **48.** *AG*

49. *OD*   **50.** *OF*

51. *BF*   **52.** *CD*

53. arc *BG*   **54.** arc *CG*

Prove that each of the following is true for any positive integer n.

55. $\cos n\pi = (-1)^n$

56. $\sin (2n + 1)\pi/2 = (-1)^n$

6.3 Angles and the Unit Circle

One basic idea in trigonometry is the angle, which we define in this section. Figure 6.13 on the next page shows a line through the two points *A* and *B*. This line is named **line *AB***.

*The distance from point O to point A.

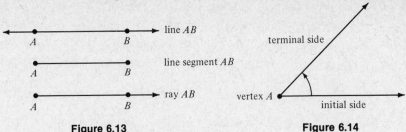

| Figure 6.13 | Figure 6.14 |

The portion of the line between A and B, including points A and B themselves, is called **line segment AB**. The portion of line AB that starts at A and continues through B, and on past B, is called **ray AB**. Point A is the endpoint of the ray. See Figure 6.13.

An **angle** is formed by rotating a ray around its endpoint. The initial position of the ray is called the **initial side** of the angle, while the location of the ray at the end of its rotation is called the **terminal side** of the angle. The endpoint of the ray is called the **vertex** of the angle. Figure 6.14 shows the initial and terminal sides of an angle with vertex A.

If the rotation of an angle is counterclockwise, we call the angle **positive.** If the rotation is clockwise, the angle is **negative.** Figure 6.15 shows both types.

An angle can be named by using the name of its vertex. For example, the angle on the right in Figure 6.15 can be called angle C. Also, an angle can be named by using three letters. For example, the angle on the right could also be named angle ACB. (Put the vertex in the middle.)

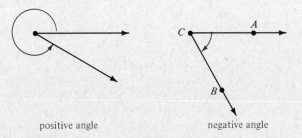

positive angle   negative angle

Figure 6.15

An angle is in **standard position** if its vertex is at the origin of a coordinate system and its initial side is along the positive x-axis. The two angles of Figure 6.16 are in standard position. An angle in standard position is said to lie in the quadrant where its terminal side lies.

EXAMPLE 1   ☐   Find the quadrants for the angles in Figure 6.16.

The angle on the left in Figure 6.16 is a quadrant I angle. The angle on the right is a quadrant II angle.   ☐

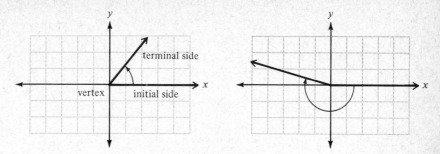

Figure 6.16

An angle in standard position whose terminal side coincides with the *x*-axis or *y*-axis is called a **quadrantal angle.** Two angles with the same initial side and the same terminal side, but different amounts of rotation, are called **coterminal angles.** Figure 6.17 shows examples of coterminal angles.

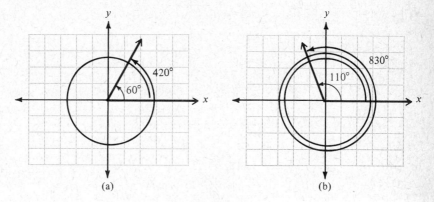

Figure 6.17

There are two systems commonly used to measure angles. In most work in applied trigonometry, angles are measured in degrees. Degree measure has remained unchanged since the Babylonians developed it over 4000 years ago. To use degree measure, assign 360 degrees to the rotation of a ray through a complete circle. As Figure 6.18 shows, the terminal side corresponds with its initial side when it makes a complete rotation. **One degree,** 1°, represents 1/360 of a rotation. One sixtieth of a degree is called a *minute,* and one sixtieth of a minute is a *second.* The measure 12° 42′ 38″ represents 12 degrees, 42 minutes, 38 seconds.

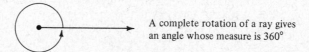

A complete rotation of a ray gives
an angle whose measure is 360°

Figure 6.18

An angle having a measure between 0° and 90° is called an *acute angle*. An angle whose measure is exactly 90° is a *right angle*. An angle measuring more than 90° but less than 180° is an *obtuse angle,* and an angle of exactly 180° is a *straight angle*.

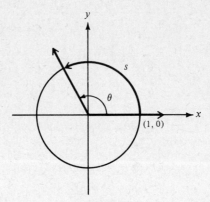

Figure 6.19

In advanced work in trigonometry, and in calculus, angles are measured in **radians.** Radian measure simplifies many formulas. Figure 6.19 shows an angle θ (θ is the Greek letter theta) in standard position. As shown in the figure, angle θ determines an arc of positive length *s* on the unit circle. Thus, it is reasonable to use the real number *s* as a measure of θ. We say that

$$\theta = s \text{ radians}$$

or simply θ = *s*.

Several angles and their radian measures are shown in Figure 6.20.

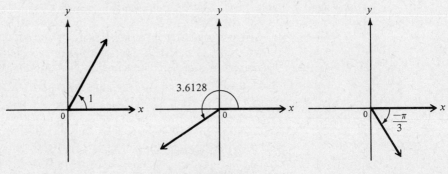

Figure 6.20

We defined the radian measure of an angle in terms of a unit circle. However, radian measures can be found for angles from a circle of any positive radius r. To see how, start with the **central angle** (an angle whose vertex is the center of a circle) θ in Figure 6.21. Let θ cut an arc of length s on the unit circle and an arc of length t on the circle of radius r, where $r > 0$. By the definition of radian measure above, we have $\theta = s$. From geometry we know that the arc lengths s and t have the same ratios as the radii of the respective circles, or

$$\frac{t}{s} = \frac{r}{1}.$$

Since $\theta = s$, we end up with

$$s = \frac{t}{r}, \quad \text{or} \quad \theta = \frac{t}{r}.$$

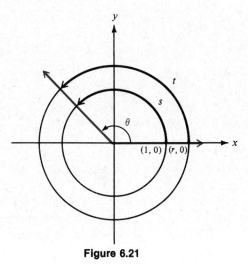

Figure 6.21

In summary,

Theorem
Radian
Measure

Suppose a circle has a radius $r > 0$. Let θ be a central angle of the circle. If θ cuts off an arc of length t on the circle, then the radian measure of θ is

$$\theta = \frac{t}{r}$$

The radian measure of an angle is the ratio of the arc length cut by the angle to the radius of the circle. In this ratio, the units of measure "divide out," leaving only a number. For this reason, radian measure is just a real number—there are no units associated with a radian measure.

EXAMPLE 2 ☐ Find the radian measure of a central angle which cuts off an arc of length 8 inches on a circle with a radius of 5 inches.

We know that $\theta = t/r$. Here, t is 8 inches and r is 5 inches. Thus,

$$\theta = \frac{8 \text{ inches}}{5 \text{ inches}} = 1.6 \text{ radians.} \quad ☐$$

An angle of measure 360° would correspond to an arc that went entirely around the unit circle. Thus, the radian measure of a 360° angle is 2π, or

$$360° = 2\pi \text{ radians.}$$

This result gives us a basis for comparing degree measure and radian measure. Dividing both sides by 2 gives

$$180° = \pi \text{ radians.}$$

Since π radians = 180°, we divide both sides by π to find that

1 radian

$$1 \text{ radian} = \frac{180°}{\pi},$$

or, approximately, 1 radian = 57° 17′ 45″.

On the other hand, dividing the equation 180° = π radians on both sides by 180 gives

1 degree

$$1° = \frac{\pi}{180} \text{ radians,}$$

or, approximately, 1° = .0174533 radians.

EXAMPLE 3 ☐ Convert each of the following degree measures to radians.

(a) 45°

Since $1° = \pi/180$ radians,

$$45° = 45\left(\frac{\pi}{180}\right) \text{ radians} = \frac{45\pi}{180} \text{ radians} = \frac{\pi}{4} \text{ radians.}$$

The word *radian* is often omitted, so that we could simply write 45° = $\pi/4$.

(b) 240°

$$240° = 240\left(\frac{\pi}{180}\right) = \frac{4\pi}{3}. \quad ☐$$

EXAMPLE 4 ☐   Convert each of the following radian measures to degrees.

(a) $\dfrac{9\pi}{4}$

We know that 1 radian $= 180°/\pi$. Thus,

$$\frac{9\pi}{4} \text{ radians} = \frac{9\pi}{4}\left(\frac{180°}{\pi}\right)$$

$$\frac{9\pi}{4} \text{ radians} = 405°.$$

(b) $\dfrac{11\pi}{3}$ radians $= \dfrac{11\pi}{3}\left(\dfrac{180°}{\pi}\right) = 660°.$ ☐

☐   Many calculators will convert back and forth from radian measure to degree measure. A difficulty that often comes up is that most calculators work with decimal degrees, rather than degrees, minutes, and seconds. The next example shows how to handle this.

EXAMPLE 5 ☐   (a) Convert 146° 18′ 34″ to radians.

Since $1' = 1/60°$ and $1'' = 1/60' = 1/3600°$,

$$146° \ 18' \ 34'' = 146° + \frac{18°}{60} + \frac{34}{3600}°$$

$$= 146° + .3° + .00944°$$

$$= 146.30944°.$$

Now, activate the calculator keys that convert from degrees to radians to get

146° 18′ 34″ $\approx$ 2.55358 radians.

(b) Convert .97682 radians to degrees.

Enter .97682 in your calculator and use the keys that convert from radians to degrees to get

.97682 radians $= 55.967663°.$

This result may be converted to degree-minute-second measure if desired.

$$55.967663° = 55° + (.967663)(60')$$

$$= 55° + 58.05978'$$

$$= 55° + 58' + (.05978)(60'')$$

$$= 55° + 58' + 4''$$

$$= 55° \ 58' \ 4'' \quad ☐$$

The relationship $\theta = t/r$, found above, gives us a way to find an arc length on a circle when the central angle and radius are known. This is useful in applications as the next example shows. (In this example, $\theta = t/r$ is rewritten as $t = r\theta$.)

EXAMPLE   □   Reno, Nevada, is approximately due north of Los Angeles. The latitude of Reno
6   is 40° N, while that of Los Angeles is 34° N. (The N means that the location is north
of the equator.) If the radius of the earth is 4.0×10^3 miles, find the north-south
distance between the two cities.

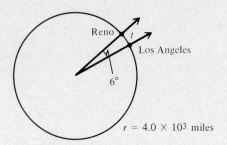

Figure 6.22

Latitude gives the measure of a central angle whose initial side goes through the
earth's equator and whose terminal side goes through the location in question. As
shown in Figure 6.22, the central angle between Reno and Los Angeles is 6°. The
distance between the two cities can thus be found by the formula $t = r\theta$, after 6° is
first converted to radians.

$$6° = 6\left(\frac{\pi}{180}\right) = \frac{\pi}{30} \text{ radians}$$

The distance between the two cities is thus

$$t = r\theta$$

$$t = (4.0 \times 10^3)\left(\frac{\pi}{30}\right) \text{ miles}$$

$$= \frac{400\pi}{3} \text{ miles}$$

$$t = 420 \text{ miles} \quad \text{(to two significant digits).} \quad □$$

**EXERCISES
6.3**

Find the angles of smallest positive measure coter-   *Convert each of the following degree measures to*
minal with the following angles.   *radians. Leave answers as multiples of π.*

1. −40°   **2.** −98°   **9.** 60°   **10.** 30°

3. −125°   **4.** −203°   **11.** 90°   **12.** 120°

5. 450°   **6.** 489°   **13.** 135°   **14.** 270°

7. 539°   **8.** 699°   **15.** 300°   **16.** 390°

17. 405°   **18.** 20°

19. 140°   **20.** 320°

Convert each of the following radian measures to degrees.

21. $\pi/3$ **22.** $8\pi/3$

23. $7\pi/4$ **24.** $2\pi/3$

25. $11\pi/6$ **26.** $15\pi/4$

27. $-\pi/6$ **28.** $-\pi/4$

29. 5π **30.** 7π

31. $7\pi/20$ **32.** $17\pi/20$

Convert as indicated. Write radian measure with four significant digits and degree measure to the nearest second.

33. 56° 25′ to radians

34. 289° 14′ 45″ to radians

35. .735114 to degrees

36. 12.69734 to degrees

Find the radian measure (as a multiple of π) of the smallest angle between the hands of a clock at the following times. Assume that the hour hand stays exactly on the given hour.

37. 12:15   **38.** 12:30

39. 2:40   **40.** 3:35

41. A tire is rotating 600 times per minute. How many degrees does a point on the edge of the tire revolve through in 1/2 second?

42. An airplane propeller rotates 1000 times per minute. Find the number of degrees that a point on the edge of the propeller would rotate through in 1 second.

Find the measure of the central angle in radians for each of the following.

43. $r = 8$ inches, $t = 12$ inches

44. $r = 18$ mm, $t = 6$ mm

45. $r = 16.4$ m, $t = 20.1$ m

46. $r = 5.80$ cm, $t = 12.3$ cm

47. $r = 1.93470$ cm, $t = 5.98421$ cm

48. $r = 294.893$ m, $t = 122.097$ m

Find the distance in miles between the following pairs of cities whose latitudes are given. Assume the cities are on a north-south line and that the radius of the earth is 4.0×10^3 miles. Give answers to two significant digits.

49. Grand Portage, Minnesota, 44° N, and New Orleans, Louisiana, 30° N

50. Charleston, South Carolina, 33° N, and Toronto, Ontario, 43° N

51. Panama City, Panama, 9° N, and Pittsburgh, Pennsylvania, 40° N

52. Farmersville, California, 36° N and Penticton, British Columbia 49° N

53. New York City, 41° N, and Lima, Peru, 12° S

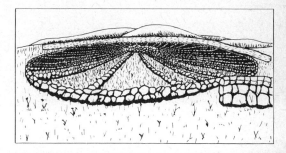

Medicine Wheel is a large Indian structure in northern Wyoming. The circular structure is perhaps 200 years old. There are 32 spokes in the wheel, all equally spaced.

54. Find the measure, in degrees and radians, of each central angle that consecutive pairs of spokes make.

55. If the radius of the wheel is 76 feet, find the circumference.

56. Find the length of each arc cut off by consecutive pairs of spokes.

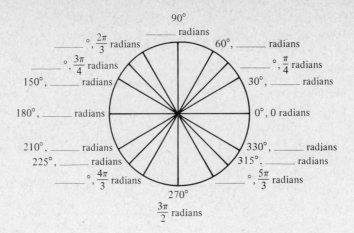

57. The figure above shows the same angles measured in both degrees and radians. Complete the missing measures.

Angular velocity is a measure of how fast the terminal side of an angle is rotating. The symbol ω (the Greek letter omega) is used for angular velocity. The distance traveled in t seconds by a point on the edge of a circle of radius r that is rotating through ω radians per second is given by rωt.

58. Find ω for the hour hand of a clock.

59. Find ω for the second hand of a clock.

60. Find ω for a line from the center to the edge of a phonograph record revolving 33 1/3 times per minute.

61. Find the distance traveled in 9 seconds by a point on the edge of a circle of radius 6 cm, which is rotating through $\pi/3$ radians per second.

62. A point on the edge of a circle travels $8\pi/9$ m in 12 seconds. The radius of the circle is 4/3 m. Find ω.

63. Find the number of seconds it would take for a point on the edge of a circle to move $12\pi/5$ m, if the radius of the circle is 1.5 m and the circle is rotating through $2\pi/5$ radians per second.

64. Eratosthenes (*ca.* 230 B.C.)* made a famous measurement of the earth. He observed at Syene [the modern Aswan], at noon and at the summer solstice, that a vertical stick had no shadow, while at Alexandria (on the same meridian as Syene) the sun's rays were inclined 1/50 of a complete circle to the vertical. [See the figure.] He then calculated the circumference of the earth from the known distance of 5000 stades between Alexandria and Syene. Obtain Eratosthenes' result of 250,000 stades for the circumference of the earth. There is reason to suppose that a stade is about equal to 516.7 feet. Assuming this, calculate from the above result the polar diameter of the earth in miles. (The actual polar diameter of the earth, to the nearest mile, is 7900 miles.)   ⌐

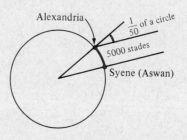

6.4   Trigonometric Functions of Angles

The trigonometric functions we have defined so far have had sets of real numbers for domains. We can extend this domain to include angles by using the radian measure of the angle. For example, from Figure 6.23, we have θ in degrees equal to s in radians, and thus $\sin \theta = \sin s$, $\cos \theta = \cos s$, and so on.

Thus, the trigonometric functions lead to the same function values whether the domain represents arc lengths or angle measures.

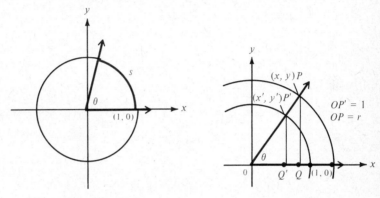

Figure 6.23   **Figure 6.24**

EXAMPLE 1   ☐   Find the following trigonometric function values.

(a) $\sin 90°$

We know that $90° = \pi/2$ radians. Then

$$\sin 90° = \sin \frac{\pi}{2} = 1.$$

(b) $\tan 45°$

Since $45° = \pi/4$ radians, and using results from Section 6.2,

$$\tan 45° = \tan \frac{\pi}{4} = 1. \quad \blacksquare$$

We defined the trigonometric functions in terms of the coordinates of point P on the unit circle. We now extend the definitions so that P need not be on the unit circle. Figure 6.24 shows an angle θ in standard position, a point P' on the unit circle, and a point P which lies r units from the origin. The distance r from O to P is called the *radius vector*. Triangles OPQ and $OP'Q'$ are both right triangles and OP has length r. From the figure,

$$\frac{y'}{1} = \frac{y}{r} \text{ and } \frac{x'}{1} = \frac{x}{r}.$$

Since $\sin \theta = y'$ and $\cos \theta = x'$, we have $\sin \theta = \dfrac{y}{r}$ and $\cos \theta = \dfrac{x}{r}$.

All the trigonometric functions can now be defined as follows.

Trigonometric Functions of an Angle

For angle θ in standard position (with θ not necessarily acute),

$$\sin \theta = \frac{y}{r} \qquad \cos \theta = \frac{x}{r} \qquad \tan \theta = \frac{y}{x}$$

$$\csc \theta = \frac{r}{y} \qquad \sec \theta = \frac{r}{x} \qquad \cot \theta = \frac{x}{y}.$$

As we can see by looking at Figure 6.24, the new definitions of the trigonometric functions allow us to think of the trigonometric functions of first quadrant angles as ratios of the sides of a right triangle. (A right triangle has a 90° angle.)

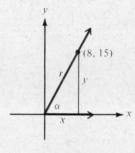

Figure 6.25

EXAMPLE 2 The terminal side of an angle α goes through the point (8, 15). Find the values of the trigonometric functions of α.

Figure 6.25 shows angle α and the triangle formed by dropping a perpendicular from the point (8, 15). We have $x = 8$ and $y = 15$. To find the radius vector r, use the Pythagorean theorem:

$$r^2 = x^2 + y^2 \qquad \text{or} \qquad r = \sqrt{x^2 + y^2}.$$

(Recall that $\sqrt{a}$ represents the *nonnegative* square root of a.)

$$r = \sqrt{8^2 + 15^2}$$
$$= \sqrt{64 + 225}$$
$$r = \sqrt{289}.$$

By using a calculator, or from Table 1 in the Appendix, we have $r = 17$. The values of the trigonometric functions of angle α are now found by the definitions given above.

$$\sin \alpha = \frac{y}{r} = \frac{15}{17} \qquad \cos \alpha = \frac{x}{r} = \frac{8}{17} \qquad \tan \alpha = \frac{y}{x} = \frac{15}{8}$$

$$\csc \alpha = \frac{r}{y} = \frac{17}{15} \qquad \sec \alpha = \frac{r}{x} = \frac{17}{8} \qquad \cot \alpha = \frac{x}{y} = \frac{8}{15} \qquad \square$$

Now that we have defined the trigonometric functions for angles, we can use some results from geometry to find the trigonometric functions for 30°, 45°, and 60°. (We could also use the methods of Exercises 45 and 46 of Section 6.1.) The trigonometric functions of 30° and 60° are found by considering a 30°–60° right triangle. Such a triangle can be obtained from an **equilateral triangle,** a triangle with all sides equal in length. Each angle of such a triangle has a measure of 60°.

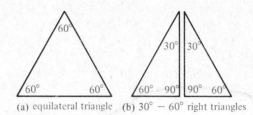

(a) equilateral triangle   (b) 30° − 60° right triangles

Figure 6.26

If we bisect one angle of an equilateral triangle, we obtain two right triangles, each of which has angles of 30°, 60°, and 90°, as shown in Figure 6.26(b). If the hypotenuse of one of these right triangles has a length of 2, then the shortest side will have a length of 1. (Why?) We use x to represent the length of the medium side, then use the Pythagorean theorem.

$$2^2 = 1^2 + x^2$$
$$4 = 1 + x^2$$
$$3 = x^2$$
$$\sqrt{3} = x.$$

The length of the medium side is thus $\sqrt{3}$. In summary, we have the following result.

30°–60° Right Triangles

In a 30°–60° right triangle, the hypotenuse is always twice as long as the shortest side (the side opposite the 30° angle), and the medium side is $\sqrt{3}$ times as long as the shortest side.

Now we can find the trigonometric functions for an angle of 30°. To do so, place a 30° angle in standard position, as shown in Figure 6.27 on the next page. Choose a point P on the terminal side of the angle so that $r = 2$. By the work above, P will have coordinates $(\sqrt{3}, 1)$. Thus, $x = \sqrt{3}$, $y = 1$, and $r = 2$. By the definitions of the trigonometric functions,

$$\sin 30° = \frac{1}{2} \qquad \tan 30° = \frac{\sqrt{3}}{3} \qquad \sec 30° = \frac{2\sqrt{3}}{3}$$

$$\cos 30° = \frac{\sqrt{3}}{2} \qquad \cot 30° = \sqrt{3} \qquad \csc 30° = 2.$$

If you have a calculator which finds trigonometric function values at the touch of a key, you may wonder why we spend so much time in finding values for special angles. We do this because a calculator gives only *approximate* values in most cases, while we need *exact* values. For example, a calculator might give the tangent of 30° as

$$\tan 30° \approx 0.5773502692$$

($\approx$ means "is approximately equal to"); however, we found the *exact* value:

$$\tan 30° = \frac{\sqrt{3}}{3}.$$

Since an exact value is frequently more useful than an approximation, you should be able to give exact values of all the trigonometric functions for the special angles.

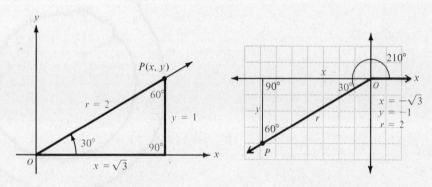

Figure 6.27   Figure 6.28

EXAMPLE 3   ☐ Find the values of the trigonometric functions for 210°.

Draw an angle of 210° in standard position, as shown in Figure 6.28. Choose point P on the terminal side of the angle so that $r = 2$. By our knowledge of 30°–60° right triangles, the coordinates of point P are $(-\sqrt{3}, -1)$. Thus, $x = -\sqrt{3}$, $y = -1$, and $r = 2$, with

$$\sin 210° = -\frac{1}{2} \qquad \tan 210° = \frac{\sqrt{3}}{3} \qquad \sec 210° = -\frac{2\sqrt{3}}{3}$$

$$\cos 210° = -\frac{\sqrt{3}}{2} \qquad \cot 210° = \sqrt{3} \qquad \csc 210° = -2. \qquad ■$$

In Section 6.1 we found the trigonometric function values for $\pi/4$, $3\pi/4$, $-\pi/4$, and so on, using a unit circle. Since $\pi/4$ corresponds to an angle of 45°, we can also find these function values using a 45°–45° right triangle, such as the one of Figure 6.29. This triangle has two sides of equal length (since it has two angles of equal measure) and thus is an **isosceles** triangle.

If we let the shorter sides each have length 1 and if r represents the length of the hypotenuse, then

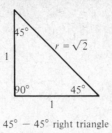

45° − 45° right triangle

Figure 6.29

$$1^2 + 1^2 = r^2$$
$$2 = r^2$$
$$\sqrt{2} = r.$$

Generalizing from this example, we have the following result.

45°–45° Right Triangles

In a 45°–45° right triangle, the hypotenuse is $\sqrt{2}$ times as long as either of the shorter sides.

To find the values of the trigonometric functions for 45°, place a 45° angle in standard position, as in Figure 6.30. Choose point P on the terminal side of the angle so that $r = \sqrt{2}$. Then the coordinates of P become $(1, 1)$. Hence, $x = 1$, $y = 1$, and $r = \sqrt{2}$, so that

$$\sin 45° = \frac{\sqrt{2}}{2} \qquad \tan 45° = 1 \qquad \sec 45° = \sqrt{2}$$

$$\cos 45° = \frac{\sqrt{2}}{2} \qquad \cot 45° = 1 \qquad \csc 45° = \sqrt{2}.$$

To find the values of sin 45° and cos 45°, we rationalized the denominators.

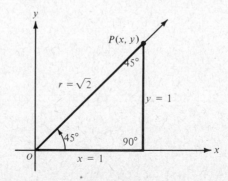

Figure 6.30

Figure 6.31 shows an acute angle θ in standard position. A right triangle has been drawn. By the work above, we know that sin θ = y/r. It is convenient to call y the length of the *side opposite* angle θ, with r the length of the *hypotenuse*. Also, x is the length of the *side adjacent* to θ. Using these terms, we have

Trigonometric Functions of an Acute Angle

If θ is an acute angle in standard position, then

$$\sin \theta = \frac{\text{side opposite}}{\text{hypotenuse}} \qquad \csc \theta = \frac{\text{hypotenuse}}{\text{side opposite}}$$

$$\cos \theta = \frac{\text{side adjacent}}{\text{hypotenuse}} \qquad \sec \theta = \frac{\text{hypotenuse}}{\text{side adjacent}}$$

$$\tan \theta = \frac{\text{side opposite}}{\text{side adjacent}} \qquad \cot \theta = \frac{\text{side adjacent}}{\text{side opposite}}.$$

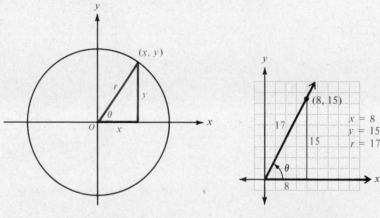

Figure 6.31   Figure 6.32

EXAMPLE 4   Find the values of sin θ, cos θ, and tan θ for angle θ of Figure 6.32.
As shown in the figure, the length of the hypotenuse is 17, the length of the side opposite θ is 15, and the length of the side adjacent to θ is 8. Thus,

$$\sin \theta = \frac{15}{17}, \qquad \cos \theta = \frac{8}{17}, \qquad \text{and} \qquad \tan \theta = \frac{15}{8}.$$

EXERCISES 6.4

Find the values of the six trigonometric functions for the following angles. Do not use tables or a calculator.

1. 120°

2. 135°

3. 150°

4. 225°

5. 240°

6. 300°

7. 330°

8. 390°

9. 420°

10. 495°

11. 510°   **12.** 570°

13. 180°   **14.** 270°

15. −90°   **16.** −180°

Complete the following table. Do not use tables or a calculator.

	θ	sin θ	cos θ	tan θ	cot θ	sec θ	csc θ
17.	30°	1/2	√3̄/2			2√3̄/3	2
18.	45°			1	1		
19.	60°		1/2	√3̄		2	
20.	120°	√3̄/2	−1/2	−√3̄			2√3̄/3
21.	135°	√2̄/2	−√2̄/2			−√2̄	√2̄
22.	150°		−√3̄/2	−√3̄/3			2
23.	210°	−1/2		√3̄/3	√3̄		−2
24.	240°	−√3̄/2	−1/2			−2	−2√3̄/3

Find the values of the six trigonometric functions of θ where the point given below is on the terminal side of angle θ in standard position.

25. (−3, 4)   **26.** (−4, −3)

27. (24, 7)   **28.** (−7, 24)

29. (−12, −5)   **30.** (6, 8)

31. (−9, −12)   **32.** (5, −12)

35.

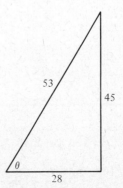

Find the values of the six trigonometric functions of θ.

33.

36.

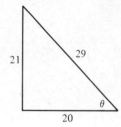

34.

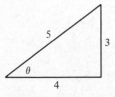

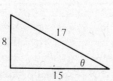

37.

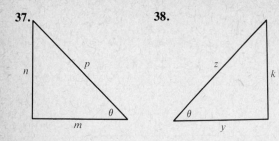

38.

45. $\sin^2 225° - \cos^2 270° + \tan 60°$

46. $\cot^2 90° - \sec^2 180° + \csc^2 135°$

47. $\cos^2 60° + \sec^2 150° - \csc^2 210°$

48. $\cot^2 135° + \tan^4 60° - \sin^4 180°$

49. $\sec 30° - \sin 60° + \cos 210°$

50. $\cot 30° + \tan 60° - \sin 240°$

Find sin A, cos A, and tan A for the following right triangles.

39.

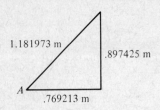

Answer true or false for each of the following.

51. $\sin 30° + \sin 60° = \sin(30° + 60°)$

52. $\sin(30° + 60°) = \sin 30° \cdot \cos 60° + \sin 60° \cdot \cos 30°$

53. $\cos 60° = 2 \cos^2 30° - 1$

54. $\cos 60° = 2 \cos 30°$

55. $\sin 120° = \sin 150° - \sin 30°$

56. $\sin 210° = \sin 180° + \sin 30°$

57. $\sin 120° = \sin 180° \cdot \cos 60° - \sin 60° \cdot \cos 180°$

58. $\cos 300° = \cos 240° \cdot \cos 60° - \sin 240° \cdot \sin 60°$

59. $\cos 150° = \cos 120° \cdot \cos 30° - \sin 120° \cdot \sin 30°$

60. $\sin 120° = 2 \sin 60° \cdot \cos 60°$

40.

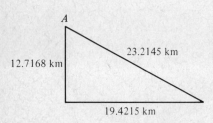

Evaluate each of the following. (Do not use tables or calculator.)

41. $\sin^2 120° + \cos^2 120°$

42. $\sin^2 225° + \cos^2 225°$

43. $2 \tan^2 120° + 3 \sin^2 150° - \cos^2 180°$

44. $\cot^2 135° - \sin 30° + 4 \tan 45°$

Use a calculator with sine and tangent keys and find each of the following. (Be sure to set the machine for degree measure.) Then explain why these answers are not really "correct" if the exact value has been requested.

61. $\sin 45°$ 62. $\tan 60°$

6.5   Values of Trigonometric Functions

Up to now, we have discussed values of the trigonometric functions only for certain special values of θ. The methods we used were primarily based on geometry. Certainly, there are many values of θ for which geometric methods are inadequate and for which more advanced methods, beyond the level of this course, are required.

Approximate values of the trigonometric functions can be found with a calculator or by using special tables, such as Table 5 in the Appendix. To use a calculator, first recall that

$$1 \text{ minute} = 1' = 60 \text{ seconds} = \frac{1}{60}^{\circ}$$

$$1 \text{ second} = 1'' = \frac{1}{60}' = \frac{1}{3600}^{\circ}.$$

For example, an angle of 12° 25′ 56″ is

$$12 + \frac{25}{60} + \frac{56}{3600} \text{ degrees.}$$

Most calculators work in **decimal degrees.** For example, 56.832° is

$$56\frac{832}{1000} \text{ of a degree.}$$

When using a calculator to find the values for an angle given in degrees, minutes, and seconds, it is often necessary to convert to decimal degrees.

EXAMPLE 1 ☐ Use a calculator with sine, cosine, and tangent keys to find each of the following. Round to five decimal places. Make sure the calculator is set for degree measure.

(a) sin 49° 12′

Convert 49° 12′ to decimal degrees.

$$49° \, 12' = 49\frac{12}{60}^{\circ} = 49.2°$$

Then push the sine button.

$$\sin 49.2° = 0.75700$$

(b) $\tan 132° \, 41' = \tan 132\frac{41}{60}^{\circ}$

$$= \tan \left(132 + \frac{41}{60}\right)^{\circ}$$

$$= \tan 132.68333°$$

$$= -1.08432$$

(c) $\sec 97° \, 58' \, 37'' = \sec \left(97 + \frac{58}{60} + \frac{37}{3600}\right)^{\circ} = \sec 97.97694°$

Calculators do not have secant keys. However,

$$\sec \theta = \frac{1}{\cos \theta}$$

for all angles θ when cos θ is not 0. Thus, sec 97.97694° is found by pushing the cosine key, and then taking the reciprocal. (Push the $1/x$ key.)

$$\sec 97.97694° = \frac{1}{\cos 97.97694°} = -7.20593$$

(d) cot 51.4283°

This angle is already in decimal degrees. Use the identity $\cot \theta = 1/\tan \theta$.

$$\cot 51.4283° = \frac{1}{\tan 51.4283°} = 0.79748 \quad \square$$

If your calculator does not have sine, cosine, and tangent keys, you can find the values of the trigonometric functions by using Table 5 in the back of the book. Table 5 gives the values of the trigonometric functions for values of θ where $0° \le \theta \le 90°$ or $0 \le \theta \le \pi/2$. Most of the values in Table 5 are four decimal place approximations. However, for convenience, we use the equality symbol with the understanding that the values are actually approximations. The table is designed to give the value of θ in the first two columns (in degrees, then radians) for $0° \le \theta \le 45°$ (or $0 \le \theta \le \pi/4$) and in the last two columns for $45° \le \theta \le 90°$ (or $\pi/4 \le \theta \le \pi/2$). When locating values for $45° \le \theta \le 90°$ (or $\pi/4 \le \theta \le \pi/2$), read *up* the table and refer to the names of the trigonometric functions at the bottom of the page. Function values of real numbers can be found by using the "radian" column of Table 5. A small portion of Table 5 is reproduced here.

θ (degrees)	θ (radians)	$\sin \theta$	$\cos \theta$	$\tan \theta$	$\cot \theta$	$\sec \theta$	$\csc \theta$	
36°00′	.6283	.5878	.8090	.7265	1.376	1.236	1.701	.9425	**54°00′**
10	.6312	.5901	.8073	.7310	1.368	1.239	1.695	.9396	50
20	.6341	.5925	.8056	.7355	1.360	1.241	1.688	.9367	40
30	.6370	.5948	.8039	.7400	1.351	1.244	1.681	.9338	30
40	.6400	.5972	.8021	.7445	1.343	1.247	1.675	.9308	20
50	.6429	.5995	.8004	.7490	1.335	1.249	1.668	.9279	10
37°00′	.6458	.6018	.7986	.7536	1.327	1.252	1.662	.9250	**53°00′**
		$\cos \theta$	$\sin \theta$	$\cot \theta$	$\tan \theta$	$\csc \theta$	$\sec \theta$	θ (radians)	θ (degrees)

EXAMPLE 2 $\square$ Use the portion of Table 5 above, when appropriate, to find the following.

(a) $\sin 36° \, 40'$

For angles between 0° and 45°, read down the *left* of the table and use the function names at the *top* of the table. Doing this here gives

$$\sin 36° \, 40' = .5972.$$

(b) $\csc 53° \, 40'$

Use the right "degree" column of the table for angles between 45° and 90°. Use the function names at the bottom. Notice that 53° 40′ is above the entry for 53°. We have

$$\csc 53° \, 40' = 1.241.$$

(c) $\tan 82° \, 00' = 7.115$

(d) $\sin .2676 = 0.2644$ (Use the "radian" column of the table.)

(e) $\cot 1.2043 = 0.3839$ $\square$

When the required value is not in the table, linear interpolation may be used, as shown in the next example. (The basic ideas of linear interpolation are explained in the appendix at the end of Chapter 5.)

EXAMPLE ☐   Find each of the following.

3   (a) tan 40° 52′

The value for tan 40° 52′ lies 2/10 or .2 of the way between the values for tan 40° 50′ and tan 41°00′.

$$10\left\{\begin{array}{l}2\left\{\begin{array}{l}\tan 40°\ 50′\ =\ .8642\\ \tan 40°\ 52′\ =\ ?\\ \tan 41°\ 00′\ =\ .8693\end{array}\right\}.0051\end{array}\right.$$

$$\begin{aligned}\tan 40°\ 52′ &= .2(.8693 - .8642) + .8642\\ &= .2(.0051) + .8642\\ &= .8652\end{aligned}$$

(b) cos 63° 34′

Work as follows.

$$10\left\{\begin{array}{l}4\left\{\begin{array}{l}\cos 63°\ 30′\ =\ .4462\\ \cos 63°\ 34′\ =\ ?\\ \cos 63°\ 40′\ =\ .4436\end{array}\right\}.0026\end{array}\right.$$

$$\begin{aligned}\cos 63°\ 34′ &= .4462 - (.4)(.4462 - .4436)\\ &= .4462 - .4(.0026)\\ &= .4452 \quad \blacksquare\end{aligned}$$

Here we had to *subtract* from .4462 rather than add because the cosine function *decreases* as θ increases for $0 < θ < π/2$.

EXAMPLE ☐   Find a value of θ satisfying each of the following.

4   (a) sin θ = .5807; θ in degrees

Use Table 5 and read columns having sine at either the top or the bottom. Here we find .5807 in a column having sine at the top. Thus, we use angles at the left to read the value of θ.

θ = 35° 30′

(b) tan θ = 2.699; θ in degrees

Here 2.699 is in a column having tangent at the bottom. Thus, we use angles at the right to read the value of θ.

θ = 69° 40′   ☐

Table 5 gives values only for angles between 0° and 90°, inclusive, or for real numbers between 0 and π/2, inclusive. For values outside this range, we need to use some of our earlier work with trigonometric functions. As we have seen, we can find the trigonometric function values for 90° < θ < 360° by referring to the appropriate value of θ in the interval 0° < θ < 90° and affixing the correct sign. For any value θ in the interval 90° < θ < 360°, the positive acute angle made by the terminal side of angle θ and the x-axis is called the **related** or **reference angle** for θ; this new angle is written θ′. See Figure 6.33 on the next page. For example, if θ = 135°, the related angle θ′ is 45°. If θ = 200°, then θ′ = 20°.

When working with radians, we could give a similar definition for **related numbers.** For example, if $s = -\pi/6$, then the related number s' is $s' = \pi/6$. Also, if $s = 5\pi/4$, then $s' = \pi/4$.

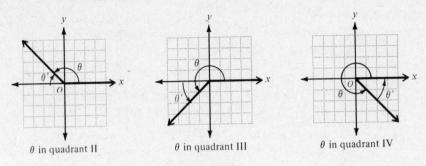

θ in quadrant II θ in quadrant III θ in quadrant IV

Figure 6.33

EXAMPLE 5   ☐   Find the related angle or related number for each of the following.

(a) 218°

As shown in Figure 6.34, the positive acute angle made by the terminal side of this angle and the x-axis is $218° - 180° = 38°$.

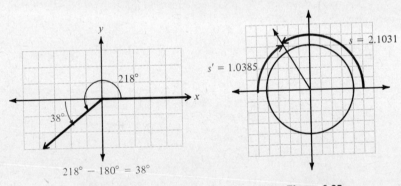

Figure 6.34   **Figure 6.35**

(b) 321° 10′

The positive acute angle made by the terminal side of this angle and the x-axis is $360° - 321° 10'$. Write 360° as 359° 60′ so that the related angle is

$$359° 60' - 321° 10' = 38° 50'$$

By the way, an angle of $-38° 50'$, which has the same terminal ray as 321° 10′, also has a related angle of 38° 50′.

(c) $s = 2.1031$

The related number is found by subtracting from π. (See Figure 6.35.) Using the four-decimal-place approximation 3.1416 for π gives

$$s' = 3.1416 - 2.1031 = 1.0385. \quad \square$$

Based on these examples, we can make up the following table to find the related angle θ' for any angle θ between $0°$ and $360°$, or the related number s' for any real number s between 0 and 2π.

Related Angles or Numbers

θ or s in quadrant	θ' is	or	s' is
I	θ		s
II	$180° - \theta$		$\pi - s$
III	$\theta - 180°$		$s - \pi$
IV	$360° - \theta$		$2\pi - s$

Figure 6.36 shows an angle θ and related angle θ', drawn so that θ' is in standard position. Point P, with coordinates (x_1, y_1), has been located on the terminal side of angle θ. Let r be the distance from O to P.

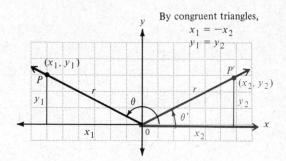

Figure 6.36

Choose point P' on the terminal side of angle θ', so that the distance from O to P' is also r. Let P' have coordinates (x_2, y_2). By congruent triangles, verify that

$$x_1 = -x_2 \quad \text{and} \quad y_1 = y_2.$$

Thus,

$$\sin \theta = \frac{y_1}{r} = \frac{y_2}{r} = \sin \theta'$$

$$\cos \theta = \frac{x_1}{r} = \frac{-x_2}{r} = -\cos \theta'$$

$$\tan \theta = \frac{y_1}{x_1} = \frac{y_2}{-x_2} = -\tan \theta',$$

and so on. Notice that the values of the trigonometric functions of the related angle θ' are the same as those of angle θ, except perhaps for signs. We have shown the truth of this statement for an angle θ in quadrant II; similar results can be proven for angles in the other quadrants.

Based on this work, the values of the trigonometric functions for any angle θ can be found by finding the function value for an angle between 0° and 90°. To do this, go through the following steps. (Similar steps can be used to find trigonometric function values for any real number *s*.)

Finding Trigonometric Function Values

1. If θ ≥ 360°, or if θ < 0°, add or subtract 360° as many times as needed to get an angle at least 0° but less than 360°.

2. Find the related angle θ′. (Use the table given above.)

3. Find the necessary values of the trigonometric functions for the related angle θ′.

4. Find the correct signs for the values found in Step 3. This gives you the trigonometric values for angle θ.

EXAMPLE 6   ☐   Find each of the following values.

(a) tan 315°

To begin, find the related angle for 315°. See Figure 6.37. Since 315° is in quadrant IV, we subtract 315° from 360°. The related angle is

$$360° - 315° = 45°$$

We know that tan 45° = 1 and that the tangents for quadrant IV angles are negative. Thus,

$$\tan 315° = -\tan 45° = -1.$$

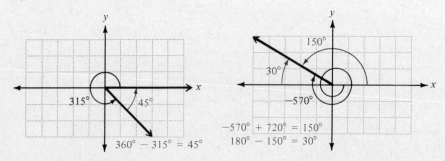

| Figure 6.37 | Figure 6.38 |

(b) cos(−570°)

To begin, find the smallest possible positive angle coterminal to −570°. See Figure 6.38. Add 2 · 360°, or 720°, to −570°.

$$-570° + 720° = 150°$$

Since 150° is in quadrant II, we find its related angle by subtracting 150° from 180°.

$$180° - 150° = 30°$$

We know that cos 30° = $\sqrt{3}/2$. The cosine for quadrant II angles is negative. Thus,

$$\cos(-570°) = -\cos 30° = -\sqrt{3}/2.$$

(c) cot 600°

$$600° - 360° = 240°$$

Verify that $\theta' = 240° - 180° = 60°$, and that

$$\cot 600° = \cot 60° = \sqrt{3}/3.$$

(d) sin 3.7845

Since $s = 3.7845$ is in Quadrant III, the reference number s' is found by subtracting π, or 3.1416, from s.

$$s' = 3.7845 - 3.1416 = 0.6429$$

The values of sine are negative in Quadrant III, so

$$\sin 3.7845 = -\sin 0.6429 = -0.5995. \quad \blacksquare$$

As we saw in the last section, the trigonometric functions can be obtained from the lengths of the sides of a right triangle. The next example shows another application of this fact.

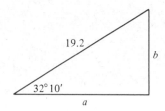

Figure 6.39

EXAMPLE 7 ☐ A surveyor gathered the data shown on the triangle of Figure 6.39. Find the lengths of sides a and b of the triangle.

Let us first find b. The sine of angle B (the angle opposite b) is given by

$$\sin B = \frac{\text{side opposite}}{\text{hypotenuse}}.$$

Substitute in the known values.

$$\sin 32° \ 10' = \frac{b}{19.2}.$$

Use Table 5 or a calculator.

$$.5324 = \frac{b}{19.2}$$

$$19.2(.5324) = b$$

$$10.2 = b$$

Use the cosine of angle B to find that $a = 16.3$. □

We shall study right triangle applications in more detail in Chapter 8.

EXERCISES
6.5

Find the related angle for each of the following. Use 3.1416 as an approximation for π.

1. 215° **2.** 550°

3. −143° **4.** −12°

5. −110° 10′ **6.** −429° 30′

7. 3.21 **8.** 2.4

9. 5.9690 **10.** −1.7861

11. −4.0230 **12.** −2.4580

Use Table 5 or a calculator to get a value for each of the following. Use 3.1416 as an approximation for π.

13. sin 39° 20′ **14.** cos 58° 40′

15. sin (−38° 40′) **16.** csc (−168° 30′)

17. cos (−124° 50′) **18.** sec 274° 30′

19. sec 1.9024 **20.** cot 3.1998

21. sin 7.5835 **22.** tan 6.4752

23. cos (−4.0230) **24.** cot (−3.8426)

Use interpolation to find each of the following values.

25. tan 29° 42′ **26.** sin 56° 38′

27. tan 49° 17′ **28.** sin 78° 32′

29. sin .1635 **30.** cos (−1.3870)

31. tan (−.3034) **32.** csc .2518

Use Table 5 or a calculator to find a value of θ in degrees for each of the following.

33. sin θ = 0.8480 **34.** cot θ = 1.257

35. sec θ = 2.759

Use Table 5 or a calculator to find a value of s in radians for each of the following.

36. cos s = 0.7826 **37.** sin s = 0.9936

38. tan s = 2.605

For small real number values of x, $\sin x \approx x$. (A graph that makes this plausible is given in the next section.) Use this fact to find approximate values of the following without using tables or a calculator.

39. sin .001 **40.** sin .009

41. sin .01 **42.** sin .05

As we know, $\tan x = \sin x/\cos x$. For small values of x, $\cos x \approx 1$ and $\sin x \approx x$. Thus, $\tan x \approx x$. Use this fact to find approximate values of the following without using tables or a calculator.

43. tan .01 **44.** tan .005

45. tan .10 **46.** tan .05

Approximate the value of the following without using tables or a calculator.

47. $\dfrac{\sin .01}{.01}$ **48.** $\dfrac{\sin .009}{.009}$

49. $\dfrac{\sin .05}{\tan .05}$

50. $\dfrac{\tan .10 + \sin .10}{\sin .10}$

In calculus courses sin x and cos x are expressed as infinite sums, where x is measured in radians. For all real numbers x and all natural numbers n, the following hold.

$$\cos x = 1 - \frac{x^2}{2!} + \frac{x^4}{4!} - \cdots + (-1)^{n-1}\frac{x^{2n-2}}{(2n-2)!} + \cdots$$

and

$$\sin x = x - \frac{x^3}{3!} + \frac{x^5}{5!} - \cdots + (-1)^{n-1}\frac{x^{2n-1}}{(2n-1)!} + \cdots$$

where $2! = 2 \cdot 1;$ $3! = 3 \cdot 2 \cdot 1;$ $4! = 4 \cdot 3 \cdot 2 \cdot 1;$
and $n! = n(n-1)(n-2) \ldots (2)(1).$

Using these formulas, sin x and cos x can be evaluated to any desired degree of accuracy. For example, let us find cos 0.5 using the first three terms of the definition and compare this with the result using the first four terms. Using the first three terms, we have

$$\cos 0.5 = 1 - \frac{(0.5)^2}{2!} + \frac{(0.5)^4}{4!}$$

$$= 1 - \frac{.25}{2} + \frac{.0625}{24}$$

$$= 1 - 0.125 + 0.00260$$

$$= .87760.$$

If we use four terms, the result is

$$\cos 0.5 = 1 - \frac{(0.5)^2}{2!} + \frac{(0.5)^4}{4!} - \frac{(0.5)^6}{6!}$$

$$= 1 - 0.125 + 0.00260 - 0.00002$$

$$= 0.87758.$$

The two results differ by only 0.00002, so that the first three terms give results which are accurate to the fourth decimal place. From the tables, using interpolation, cos 0.5 = .87754.

Use the first three terms of the definitions given above for sin x and cos x to find the following. Compare your results with the values in Table 5 in the Appendix.

51. sin 1

52. sin 0.1

53. sin 0.01

54. cos 0.1

55. cos 0.01

56.

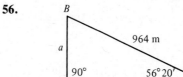

57.

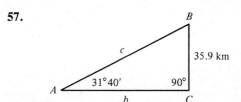

58.

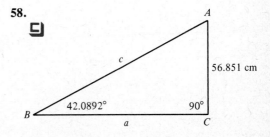

59.

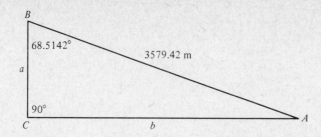

Find the lengths of the missing sides in each of the following right triangles.

60. A woman stands 59.7 m from the base of a tree. The angle her line of sight makes to the top of the tree is 39° 20′. If the woman's eyes are 1.74 m above the ground, find the height of the tree.

61. A ladder leans against a building, making an angle of 24° 10′ with the building. If the ladder is 4.88 m in length, find the distance between the bottom of the ladder and the building.

62. The length of the base of an isosceles triangle is 37 inches. Each base angle is 49°. Find the length of each of the two equal sides of the triangle. (Hint: break the triangle into two right triangles.)

6.6 Graphs of the Sine and Cosine Functions

In Section 6.1 we mentioned that sine and cosine were periodic functions with a period of 2π. The repeating nature of these functions is evident from their graphs, shown there and later in this section. By definition,

Periodic Function

> a **periodic function** is a function with the property that
>
> $$f(x) = f(x + p),$$
>
> for every real number x in the domain of f and for some positive real number p. The smallest possible value of p is called the **period.**

The sine function, $y = \sin x$, has all real numbers for its domain, and its range is $-1 \leq y \leq 1$.* Since the sine function has period 2π, we can sketch its graph by concentrating on the values of x between 0 and 2π. We can then repeat this portion of the graph for other values of x.

As we have seen, for x-values from 0 to $\pi/2$, $\sin x$ increases from 0 to 1, and for x-values from $\pi/2$ to π, $\sin x$ decreases from 1 back to 0.

We know that $\sin x$ is negative for $\pi < x < 2\pi$. Therefore, for x-values from π to $3\pi/2$, $\sin x$ decreases from 0 to -1, and for x-values from $3\pi/2$ to 2π, $\sin x$

*In this section, we use x as our variable (as in $\sin x$), instead of s or θ because such use allows us to draw graphs on the familiar xy-coordinate system.

increases from -1 back to 0. These facts are summarized in the table of values shown below. (Decimals have been rounded to the nearest tenth.)

x	0	$\pi/4$	$\pi/2$	$3\pi/4$	π	$5\pi/4$	$3\pi/2$	$7\pi/4$	2π
$\sin x$	0	.7	1	.7	0	$-.7$	-1	$-.7$	0

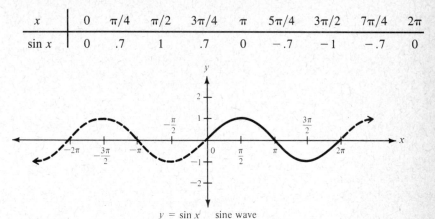

$y = \sin x$ sine wave

Figure 6.40

Plotting the points from the table of values and connecting them with a smooth line, we get the solid portion of the graph of Figure 6.40. Since $y = \sin x$ is periodic and has all real numbers as domain, the graph continues in both directions indefinitely, as indicated by the arrows. The same scale is used on both axes so as not to distort the shape of the graph. This graph is sometimes called a **sine wave** or **sinusoid.** You should memorize the shape of this key graph and be able to sketch it quickly. The main points of the graph are $(0, 0)$, $(\pi/2, 1)$, $(\pi, 0)$, $(3\pi/2, -1)$, and $(2\pi, 0)$. By plotting these five points, then connecting them with the characteristic sine wave, you can quickly sketch the graph.

The graph of $y = \cos x$ can be found by plotting points, just as we did for $y = \sin x$. The domain of cosine is the set of all real numbers, and the range of $y = \cos x$ is $-1 \le y \le 1$. Here the key points are $(0, 1)$, $(\pi/2, 0)$, $(\pi, -1)$, $(3\pi/2, 0)$, and $(2\pi, 1)$. The graph of $y = \cos x$ has the same shape as $y = \sin x$. In fact, it is the sine wave, shifted $\pi/2$ units to the left. See Figure 6.41.

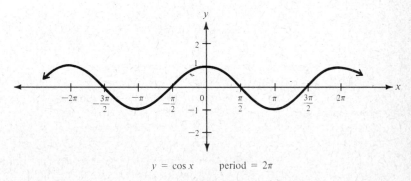

$y = \cos x$ period $= 2\pi$

Figure 6.41

**EXAMPLE
1**   Graph $y = 2 \sin x$.

For a given value of x, the value of y is twice as large as it would be for $y = \sin x$, as shown in the table of values. Thus, the only change in the graph is in the range, which becomes $-2 \le y \le 2$. (See Figure 6.42.)

x	0	$\pi/2$	π	$3\pi/2$	2π
$\sin x$	0	1	0	-1	0
$2 \sin x$	0	2	0	-2	0

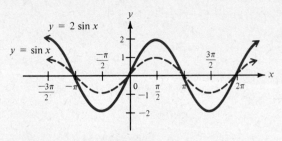

Figure 6.42

Generalizing from this example and assuming $a \ne 0$, we have the following result.

Amplitude

The graph of $y = a \sin x$ will have the same form as $y = \sin x$ except with range $-|a| \le y \le |a|$. The number $|a|$ is called the **amplitude.**

Also,

the graph of $y = a \cos x$ has the same shape as $y = \cos x$, but has a range of $-|a| \le y \le |a|$. Again, $|a|$ is called the **amplitude.** No matter the value of a, the period of $y = a \sin x$ and $y = a \cos x$ is still 2π.

**EXAMPLE
2**   Graph $y = \sin 2x$.

Start with a table of values.

x	0	$\pi/4$	$\pi/2$	$3\pi/4$	π	$5\pi/4$	$3\pi/2$	$7\pi/4$
$2x$	0	$\pi/2$	π	$3\pi/2$	2π	$5\pi/2$	3π	$7\pi/2$
$\sin 2x$	0	1	0	-1	0	1	0	-1

As the table shows, multiplying x by 2 shortens the period by half. The range is not changed. Figure 6.43 shows the graph of $y = \sin 2x$. ☐

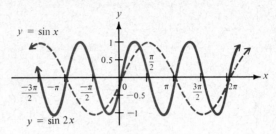

Figure 6.43

Generalizing, we have the following.

Period of Sine and Cosine	The graph of $y = \sin bx$ will look like that of $\sin x$, but with period $\left| 2\pi/b \right|$. Also, the graph of $y = \cos bx$ looks like that of $y = \cos x$, but with period $\left| 2\pi/b \right|$.

(See Exercises 51 and 52.)

EXAMPLE 3 ☐ Graph $y = \cos \dfrac{1}{2}x$.

The period here is $\left| 2\pi/(1/2) \right| = 4\pi$. The amplitude is 1. The graph is shown in Figure 6.44. ☐

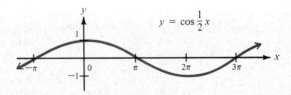

Figure 6.44

When graphing $y = a \sin bx$ or $y = a \cos bx$, we assume $b > 0$. If a function has $b < 0$, we can use the identities of the next chapter to change the function to one where $b > 0$. Let us now list some steps that can be used to graph $y = a \sin bx$ or $y = a \cos bx$, where $b > 0$.

Graphing Sine and Cosine

1. Find the period, $2\pi/b$. Start at 0 on the x-axis and lay off a distance $2\pi/b$.

2. Divide the interval from 0 to $2\pi/b$ into four equal parts.

3. Locate the points where the graph crosses the x-axis.

Function	Graph crosses x-axis at
$y = a \sin bx$	$0, \dfrac{\pi}{b}, \dfrac{2\pi}{b}$ (beginning, middle, and end of interval)
$y = a \cos bx$	$\dfrac{\pi}{2b}, \dfrac{3\pi}{2b}$ (one-fourth and three-fourths point of interval)

4. Locate the points where the graph reaches maximum and minimum values.

Function	Graph has a maximum when x is
$y = a \sin bx$	$\dfrac{\pi}{2b}$ (for $a > 0$) or $\dfrac{3\pi}{2b}$ (for $a < 0$)
$y = a \cos bx$	0 and $\dfrac{2\pi}{b}$ (for $a > 0$) or $\dfrac{\pi}{b}$ (for $a < 0$)

5. Use the tables or a calculator to find as many additional points as needed. Then sketch the graph.

6. Draw additional periods of the graph, to the right and to the left, as needed.

EXAMPLE 4   ☐   Graph $y = -2 \sin 3x$.

The period is $2\pi/3$. The amplitude is $|-2| = 2$. Sketch the graph using the steps shown in Figure 6.45.   ☐

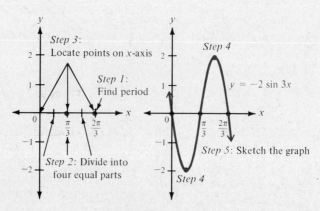

Figure 6.45

EXAMPLE
5

Graph $y = 3 \cos \frac{1}{2} x$.

The period is $2\pi/(1/2) = 4\pi$. Follow the steps shown in Figure 6.46. The amplitude is 3.

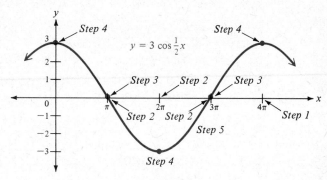

Figure 6.46

The next few examples show sine and cosine graphs that include horizontal translations.

EXAMPLE
6

Graph $y = \sin\left(x - \frac{\pi}{3}\right)$.

Based on our work with translations in Chapter 3, this graph is the same as the graph of $y = \sin x$, except that it is shifted $\pi/3$ units to the right. The graph of $y = \sin(x - \pi/3)$ is the solid line of Figure 6.47.

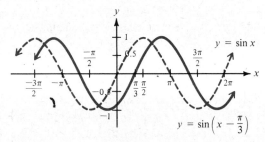

Figure 6.47

Phase Shift

The function $y = \sin(x - c)$ has the shape of the basic sine graph $y = \sin x$, but with a translation of $|c|$ units—to the left if $c < 0$ and to the right if $c > 0$. The number c is the **phase shift** of the graph.

Also, $y = \cos(x - c)$ has the shape of $y = \cos x$ but is translated $|c|$ units—to the left if $c < 0$ and to the right if $c > 0$. Again, c is the phase shift.

EXAMPLE
7

Graph $y = -2 \cos (3x + \pi)$.

The amplitude of this graph is $|-2| = 2$. The period is $|2\pi/3| = 2\pi/3$. To find the phase shift, factor as follows:

$$y = -2 \cos 3 \left(x + \frac{\pi}{3} \right)$$

This form of the equation indicates that its graph is the same as that of $y = -2 \cos 3x$, but shifted $\pi/3$ units to the left. See Figure 6.48.

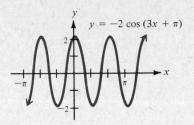

Figure 6.48

Let us now summarize what we have learned about sine and cosine graphs. (Assume $b > 0$.)

Sine and Cosine Graphs

Function $y = c + a \sin b(x - d)$
or
$y = c + a \cos b(x - d)$

Amplitude $|a|$

Period $\dfrac{2\pi}{b}$

Vertical translation   up c units if $c > 0$
down $|c|$ units if $c < 0$

Phase shift d units to the right if $d > 0$
(horizontal $|d|$ units to the left if $d < 0$
translation)   (To find d, set the argument equal to 0.)

EXAMPLE
8

Radio stations send out a carrier signal in the form of a sine wave having equation

$$y = A_0 \sin (2\pi\omega_0 t),$$

where A_0 is the amplitude of the carrier signal, ω_0 is the number of periods the signal oscillates through in one second (its *frequency*), and t is time. A carrier signal received by a radio would be a pure tone. To transmit music and voices, the station might change or *modulate* A_0 according to the function

$$A_0(t) = A_0 + mA_0 \sin (2\pi\omega t),$$

where ω is the frequency of a pure tone and m is a constant called the *degree of modulation*. The transmitted signal has equation

$$y = A_0 \sin (2\pi\omega_0 t) + A_0 m \sin (2\pi\omega t) \sin (2\pi\omega_0 t)$$

$$= A_0[1 + m \sin (2\pi\omega t)] \sin (2\pi\omega_0 t).$$

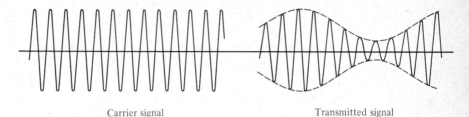

Carrier signal   Transmitted signal

Figure 6.49

A typical carrier signal and a typical graph of y are shown in Figure 6.49. This process of sending out a radio signal is called *amplitude modulation,* or AM, radio.

Frequency modulation, or FM, radio involves altering the frequency of the carrier signal rather than its amplitude. A typical graph is shown in Figure 6.50. ☐

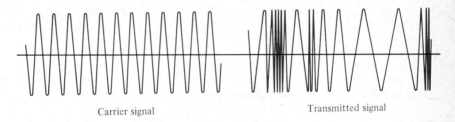

Carrier signal   Transmitted signal

Figure 6.50

EXAMPLE 9 ☐ One example of a phase shift occurs in electrical work. A simple alternating current circuit is shown in Figure 6.51. The relationship between voltage V and current I in the circuit is also shown in the figure.

As this graph shows, current and voltage are *out of phase* by 90°. In this example, current *leads* the voltage by 90°, or voltage *lags* by 90°. ☐

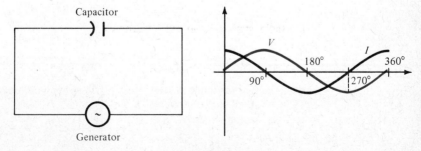

Figure 6.51

EXERCISES
6.6

Graph the following functions over the interval $-2\pi \le x \le 2\pi$. Identify the amplitude.

1. $y = 2 \cos x$ **2.** $y = 3 \sin x$

3. $y = \dfrac{2}{3} \sin x$ **4.** $y = \dfrac{3}{4} \cos x$

5. $y = -\cos x$ **6.** $y = -\sin x$

Graph each of the following functions over a two-period interval. Give the period and the amplitude.

7. $y = \sin \dfrac{1}{2} x$ **8.** $y = \sin \dfrac{2}{3} x$

9. $y = \cos \dfrac{1}{3} x$ **10.** $y = \cos \dfrac{3}{4} x$

11. $y = -\sin 4x$ **12.** $y = -\cos 6x$

13. $y = 2 \sin \dfrac{1}{4} x$ **14.** $y = 3 \sin 2x$

15. $y = -2 \cos 3x$ **16.** $y = -5 \cos 2x$

17. $y = \dfrac{1}{2} \sin 3x$ **18.** $y = \dfrac{2}{3} \cos \dfrac{1}{2} x$

For the following functions, find the amplitude, the period, any vertical translation, and any phase shift. Graph the functions over a one-period interval.

19. $y = \cos \left(x - \dfrac{\pi}{2} \right)$

20. $y = \sin \left(x + \dfrac{\pi}{4} \right)$

21. $y = \sin \left(x - \dfrac{\pi}{4} \right)$

22. $y = \cos \left(x + \dfrac{\pi}{3} \right)$

23. $y = 2 \cos \left(x - \dfrac{\pi}{3} \right)$

24. $y = 3 \sin \left(x + \dfrac{3\pi}{2} \right)$

25. $y = \dfrac{3}{2} \sin 2 \left(x - \dfrac{\pi}{4} \right)$

26. $y = -\dfrac{1}{2} \cos 4 \left(x + \dfrac{\pi}{2} \right)$

27. $y = -4 \sin (2x - \pi)$

28. $y = 3 \cos (4x + \pi)$

29. $y = \dfrac{1}{2} \cos \left(\dfrac{1}{2} x - \dfrac{\pi}{4} \right)$

30. $y = -\dfrac{1}{4} \sin \left(\dfrac{3}{4} x + \dfrac{\pi}{8} \right)$

31. $y = -3 + 2 \sin \left(x - \dfrac{\pi}{2} \right)$

32. $y = 4 - 3 \cos (x + \pi)$

33. $y = \dfrac{1}{2} + \sin 2 \left(x + \dfrac{\pi}{4} \right)$

34. $y = -\dfrac{5}{2} + \cos 3 \left(x - \dfrac{\pi}{6} \right)$

35. Is the function $y = \sin x$ even, odd, or neither?

36. Is the function $y = \cos x$ even, odd, or neither?

37. Graph $y = \sin \left(\theta - \dfrac{\pi}{2} \right)$. What trigonometric function has the same graph?

38. Graph $y = \cos \left(\theta - \dfrac{\pi}{2} \right)$. What trigonometric function has the same graph?

Pure sounds produce single sine waves on an oscilloscope. Find the amplitude and period of each sine wave in the following photographs. On the vertical scale, each square represents .5, and on the horizontal scale each square represents $\pi/6$.

39.

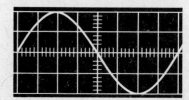

40.

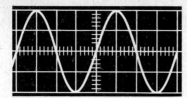

Scientists believe that the average annual temperature in a given location is periodic. The overall temperature at a given place during a given season fluctuates as time goes on, from colder to warmer, and back to colder. The graph below shows an idealized description of the temperature for the last few thousand years of a location at the latitude of Anchorage.

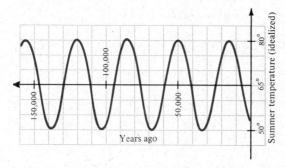

41. Find the period of this graph.

42. What is the trend of the temperature now?

Many of the activities of living organisms are periodic. For example, the graph below shows the time that flying squirrels begin their evening activity.

43. Find the amplitude of this graph.

44. Find the period.

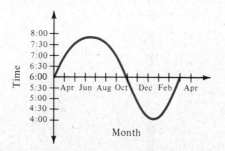

The graph shown here gives the variation in blood pressure for a typical person. Systolic and diastolic pressures are the upper and lower limits of the periodic changes in pressure which produce the pulse. The length of time between peaks is called the period of the pulse.

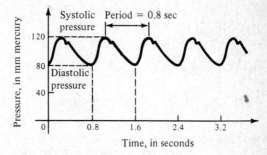

45. Find the amplitude of the graph.

46. Find the pulse rate for this person—the number of pulse beats in one minute.

47. The voltage E in an electrical circuit is given by

$$E = 5 \cos 120\pi t$$

where t is time measured in seconds.

(a) Find the amplitude and period.

(b) How many cycles are completed in one second? (The number of cycles completed in one second is the *frequency* of the function.)

(c) Find E when $t = 0$; 0.03; 0.06; 0.09; 0.12

(d) Graph E, for $0 \le t \le 1/30$.

48. For another electrical circuit, the voltage E is given by

$$E = 3.8 \cos 40\pi t$$

where t is time measured in seconds.

(a) Find the amplitude and the period.

(b) Find the frequency. See Exercise 47(b).

(c) Find E when $t = 0.02, 0.04, 0.08, 0.12, 0.14$.

(d) Graph one period of E.

49. Why does the graph of $y = \sin x$ suggest that sine is an odd function? Use the fact that sine is odd to show that $\sin(-s) = -\sin s$ for all s.

50. Why does the graph of $y = \cos x$ suggest that cosine is an even function? Use the fact that cosine is even to show that $\cos (-s) = \cos s$ for all s.

51. To find the period of $y = \sin bx$, where $b > 0$, first observe that as bx varies from 0 to 2π, we get one period of the graph of $y = \sin bx$. Show that x must therefore vary from 0 to $2\pi/b$, so that the period of $y = \sin bx$ is $2\pi/b$.

52. In Exercise 51, show that the period of $y = \sin bx$ is $|2\pi/b|$, no matter whether b is positive or negative.

53. Sketch the graph of $y = \sin x$ for real number values of x from 0 to .2 in increments of .02. On the same axes, draw $y = x$. Use this sketch to argue that for small values of x, $\sin x \approx x$.

54. The quotient $(\sin x)/x$ is very important in calculus. Find the value of this quotient, starting with $x = .1$. Then let x decrease by steps of .01 until $x = .01$ is reached. (Why can't we let x decrease all the way to 0?) What number does the quotient seem to be approaching as x gets closer and closer to 0?

Graph one period of each of the following.

55. $y = \sin^2 x$

56. $y = \sin^2 2x$

57. $y = \cos^2 x$

58. $y = 4 \cos^2 x$

6.7   Graphs of the Other Trigonometric Functions

In this section, we continue our discussion of the graphs of the trigonometric functions, beginning with $y = \tan x$. As we shall see in the next chapter, the period of $y = \tan x$ is π, so that we need to investigate the tangent function only within an interval of π units. A convenient interval for this purpose is $-\pi/2 < x < \pi/2$. Although the endpoints $-\pi/2$ and $\pi/2$ are not in the domain of tangent (why?), tan x exists for all other values in the interval.

There is no point on the graph of $y = \tan x$ at $x = -\pi/2$ or $x = \pi/2$. However, in the interval $0 < x < \pi/2$, tan x is positive. As x goes from 0 to $\pi/2$, we see from a calculator or from Table 5 that tan x gets larger and larger without bound. As x goes from $-\pi/2$ up to 0, the values of tan x approach 0. These results, along with similar results for $y = \cot x$ are summarized in the following chart.

As x increases from	tan x	cot x
0 to $\pi/2$	increases from 0, without bound	decreases to 0
$-\pi/2$ to 0	increases to 0	decreases from 0

As x approaches $\pi/2$ (through values of x less than $\pi/2$), tan x gets larger and larger; as x approaches $-\pi/2$ (through values of x greater than $-\pi/2$), tan x gets more and more negative, again without bound. For this reason, there is no point on the graph of $y = \tan x$ corresponding to $x = \pi/2$ or $x = -\pi/2$.

Based on these results, the graph will approach the vertical line $x = \pi/2$ but never touch it, so that the line $x = \pi/2$ is a **vertical asymptote.** The lines $x = \pi/2 + k\pi$, where k is any integer, are all vertical asymptotes. We indicate these asymptotes on the graph with a light dashed line. See Figure 6.52. In the interval $-\pi/2 < x < 0$, which corresponds to quadrant IV, tan x is negative, and as x goes from 0 to $-\pi/2$, tan x gets more and more negative. A table of values for tan x, where $-\pi/2 < x < \pi/2$, is given below.

x	$-\pi/3$	$-\pi/4$	$-\pi/6$	0	$\pi/6$	$\pi/4$	$\pi/3$
$\tan x$	-1.7	-1	$-.6$	0	.6	1	1.7

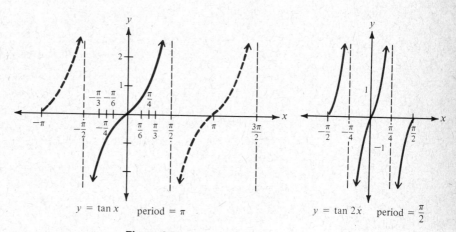

$y = \tan x$ period $= \pi$

Figure 6.52

$y = \tan 2x$ period $= \dfrac{\pi}{2}$

Figure 6.53

Plotting the points from the table and letting the graph approach the asymptotes at $x = \pi/2$ and $x = -\pi/2$, gives the portion of the graph shown with a solid line in Figure 6.52. We can sketch more of the graph by repeating the same curve, as shown in Figure 6.52. This graph, like the graphs for sine and cosine, should be memorized. Convenient main points are $(-\pi/4, -1)$, $(0, 0)$, $(\pi/4, 1)$. The vertical asymptotes are at $-\pi/2$ and $\pi/2$.

EXAMPLE 1   Graph $y = \tan 2x$.

Multiplying x by 2 changes the period to $\pi/2$. The effect on the graph is shown in Figure 6.53.   ◻

Period of Tangent

> If $b > 0$, the graph of $y = \tan bx$ has period π/b.

We can use the fact that $\cot x = 1/(\tan x)$ to find the graph of $y = \cot x$. The period of cotangent, like tangent, is π. The domain of $y = \cot x$ excludes $0 + k\pi$, where k is any integer (why?). Thus, the vertical lines $x = k\pi$ are asymptotes. Values of x that lead to asymptotes for $\tan x$ will make $\cot x = 0$, so that $\cot(-\pi/2) = 0$, $\cot \pi/2 = 0$, $\cot 3\pi/2 = 0$, and so on. Tan x increases as x goes from $-\pi/2$ to $\pi/2$. Since $\cot x = 1/(\tan x)$, the values of $\cot x$ decrease as x goes from $-\pi/2$ to 0 and from 0 to $\pi/2$. Using these facts and plotting points as necessary gives the graph of one period of $y = \cot x$ as shown with a solid curve in Figure 6.54. (An additional period is shown as a dashed curve.)

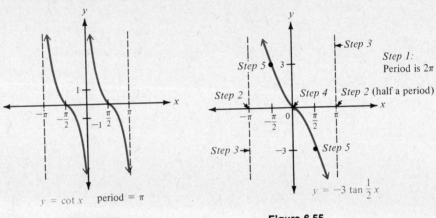

Figure 6.54   Figure 6.55

Let us summarize the steps in graphing one period of $y = a \tan bx$, where $b > 0$. (The steps for graphing $y = a \cot bx$ are similar.)

Graphing Tangent

1. Find the period, π/b.

2. Start at 0 on the x-axis and lay off two intervals, each with length half the period. One interval goes to the left and the other goes to the right of 0.

3. Draw the asymptotes as vertical dotted lines at the endpoints of the interval of Step 2.

4. Locate a point at $(0, 0)$.

5. Sketch the graph, finding additional points as needed.

6. Draw additional periods, both to the right and to the left, as needed.

EXAMPLE 2   Graph $y = -3 \tan \dfrac{1}{2} x$.

The period is $\pi/(1/2) = 2\pi$. Proceed as shown in Figure 6.55.   □

The graph of $y = \csc x$ is restricted to values of $x \neq k\pi$, where k is any integer (why?). Hence, the lines $x = k\pi$ are asymptotes. Since $\csc x = 1/\sin x$, the period is 2π, the same as for $\sin x$. When $\sin x = 1$, we have $\csc x = 1$, and when $0 < \sin x < 1$, then $\csc x > 1$. Thus, the graph takes the shape of the solid line shown in Figure 6.56. To show how the two curves are related, the graph of $y = \sin x$ is also shown, as a dashed curve.

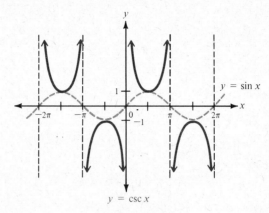

Figure 6.56

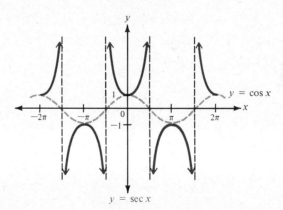

Figure 6.57

The graph of $y = \sec x$ in Figure 6.57 is related to the cosine graph in the same way that the graph of $y = \csc x$ is related to the sine graph.

EXAMPLE 3 ☐ Graph $y = \dfrac{3}{2} \csc\left(x - \dfrac{\pi}{2}\right)$.

Compared with $y = \csc x$, this graph is translated $\pi/2$ units to the right. Also, it has no values of y between $3/2$ and $-3/2$; note how this relates to the increased amplitude of $y = (3/2) \sin x$ as compared with $y = \sin x$. See Figure 6.58 on the next page. ☐

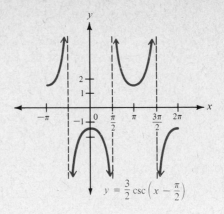

$$y = \frac{3}{2} \csc\left(x - \frac{\pi}{2}\right)$$

Figure 6.58

Additional Graphical Techniques (Optional)

To graph trigonometric functions which are the sum of two or more functions, we can use a method called **addition of ordinates.** An **ordinate** is the y-value of an ordered pair. For example, in the ordered pair $(\pi, -1)$, the number -1 is the ordinate. This graphing method is best described by examples.

EXAMPLE 4   Graph $y = x + \sin x$.

Begin by graphing the functions $y = x$ and $y = \sin x$ separately on the same coordinate axes. Figure 6.59 shows the two graphs. (We dashed $y = \sin x$ and $y = x$ to make them easier to see.) Then select some x-values, and for these values add the two corresponding ordinates to get the ordinate of the sum, $x + \sin x$. For example, when $x = 0$, both ordinates are 0, so that $P_1 = (0, 0)$ is a point on the graph of $y = x + \sin x$. When $x = \pi/2$, the ordinates are $\pi/2$ and 1, their sum is $\pi/2 + 1$, and $P_2 = (\pi/2, \pi/2 + 1)$ is on the graph. At $x = 3\pi/2$, the sum is $3\pi/2 + (-1)$, or $3\pi/2 - 1$. The point with ordinate $3\pi/2 - 1$ is indicated by P_3 on the graph. As many points as necessary can be located in this way. The graph is then completed by drawing a smooth curve through the points. The graph of $y = x + \sin x$ is shown in color in Figure 6.59.

As shown on the graph of Figure 6.59, we actually treat the ordinates as line segments. For example, the ordinate of P_2 is found by adding the lengths of the two line segments which represent the ordinates of $\sin x$ and x at $\pi/2$. The same is true for the ordinate of P_3 as well as for each of the other ordinates.

EXAMPLE 5   Graph $y = \cos x - \tan x$.

First graph $y = \cos x$ and $y = -\tan x$ on the same axes, as in Figure 6.60. At $x = 0$, the ordinates are 1 and 0, so the ordinate of $y = \cos x + (-\tan x)$ is $1 + 0$

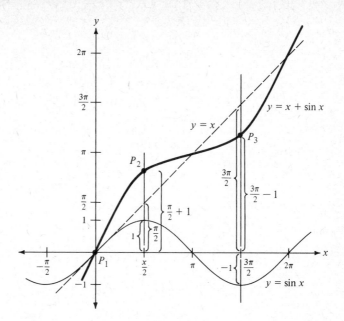

Figure 6.59

= 1, and the point $(0, 1)$ is on the graph. At any point where the graphs of $\cos x$ and $-\tan x$ cross, the ordinates are doubled. See P_1 and P_2 on the graph in Figure 6.60, for example.

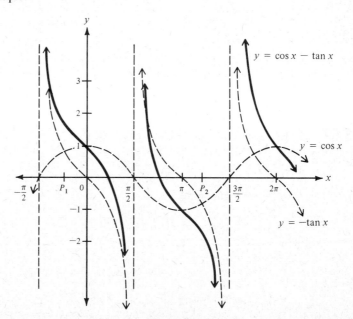

Figure 6.60

The domain of the function $y = \cos x - \tan x$ excludes $\pi/2 + k\pi$, where k is any integer, because these points are not in the domain of $\tan x$. Thus, the lines $x = \pi/2 + k\pi$ are asymptotes, so that, as x approaches $\pi/2$ and $3\pi/2$ from the left, y gets larger and larger. Also, when x approaches $\pi/2$ and $3\pi/2$ from the right, the value of y gets smaller and smaller. A portion of the graph is shown in color in Figure 6.60.   □

In Examples 4 and 5, we used *addition* of ordinates to find graphs; in Example 6 we use *multiplication* of ordinates.

EXAMPLE 6   □   Graph $y = \dfrac{1}{x} \sin x$ for x not zero.

We know that for all x, $-1 \le \sin x \le 1$. For $x > 0$, we also have $1/x > 0$, with

$$-1 \cdot \frac{1}{x} \le \frac{1}{x} \sin x \le \frac{1}{x} \cdot 1,$$

or

$$-\frac{1}{x} \le \frac{1}{x} \sin x \le \frac{1}{x}.$$

Thus, the graph of $y = (1/x)\sin x$ will always be bounded by the graphs of $y = 1/x$ and $y = -1/x$ (for positive x). The graph of $y = (1/x)\sin x$ will touch the graph of $y = 1/x$ or $y = -1/x$ whenever $|\sin x| = 1$, which happens for $x = \pi/2 + n\pi$, for n an integer. The x-intercepts occur when $\sin x = 0$ or when $x = 0 + n\pi$ for n an integer. Use a calculator to get additional points on the graph as needed. See

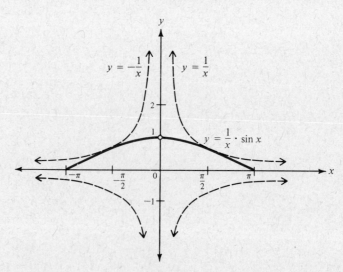

Figure 6.61

also Exercise 54 of the previous section, which suggests that $(1/x)\sin x$ approaches 1 as x approaches 0. The open circle on the final graph, shown in Figure 6.61, is used to show that the function is undefined when $x = 0$. ☐

EXERCISES
6.7

Graph each of the following over a two-period interval.

1. $y = \cot \dfrac{1}{2} x$

2. $y = 5 \tan x$

3. $y = \sec x$

4. $y = \cot 2x$

5. $y = -\tan x$

6. $y = 3 \csc x$

7. $y = \sec 2x$

8. $y = \cot \dfrac{1}{4} x$

9. $y = \csc x + 4$

10. $y = \sec \dfrac{1}{2} x - 1$

11. $y = \tan x + 1$

12. $y = 3 \sec x + 2$

Graph each of the following over a one-period interval.

13. $y = \tan\left(x - \dfrac{\pi}{4} \right)$

14. $y = \cot\left(x + \dfrac{3\pi}{4} \right)$

15. $y = \sec\left(x + \dfrac{\pi}{3} \right)$

16. $y = 3 \csc\left(x + \dfrac{3\pi}{2} \right)$

17. $y = 2 + \tan\left(\dfrac{1}{2}x + \dfrac{\pi}{3} \right)$

18. $y = \csc\left(\dfrac{1}{2}x - \dfrac{\pi}{4} \right) - 1$

Use the method of addition of ordinates to graph each of the following.

19. $y = x + \cos x$

20. $y = \sin x - 2x$

21. $y = 3x - \cos 2x$

22. $y = x + 2 - \sin x$

23. $y = \sin x + \sin 2x$

24. $y = \cos x - \cos \dfrac{1}{2} x$

25. $y = \sin x + \tan x$

26. $y = \sin x + \csc x$

27. $y = 2 \cos x - \sec x$

28. $y = 2 \sec x + \sin x$

29. $y = \cos x + \cot x$

30. $y = \sin x - 2 \cos x$

31. $y = -x + \sec x$

32. $y = x + \csc x$

Graph the following functions.

33. $y = x \sin x$

34. $y = x \cos x$

35. $y = 2^{-x} \sin x$

36. $y = x^2 \sin x$

37. The function $y = (6 \cos x)(\cos 8x)$ can come up in AM radio transmissions, as mentioned in Section 6.6. Graph this function on an interval from 0 to 2π. (Hint: first graph $y = 6 \cos x$ and $y = \cos 8x$ as dashed lines.)

38. Graph $y = \sin 8x + \sin 9x$. The period of this function is very long. By placing two engines side by side that are running at almost the same speed, we get an effect of "beats." See the caption on the next page.

The top two sine waves represent pure tones, such as those put out by a tuning fork or an electronic oscillator. When two such pure tones, having slightly different periods, are played side-by-side, the amplitudes add algebraically, instant by instant, producing a result such as shown in the bottom graph. The peaks here are called beats. *Beats result, for example, from engines on an airplane that are running at almost, but not quite, the same speeds. Blowers in different apartments can also cause such beats; these can be quite annoying.*

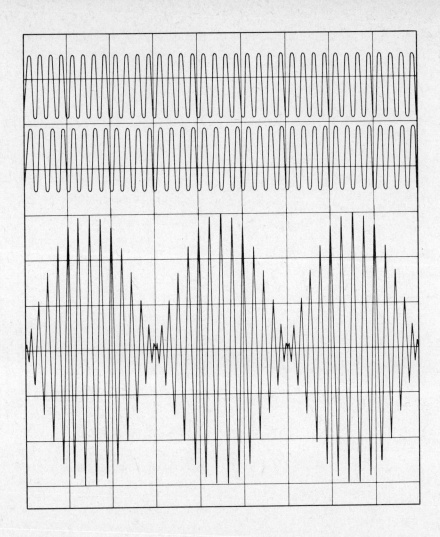

6.8   Inverse Trigonometric Functions

In Chapter 3 we defined the inverse of a one-to-one function. In this section we consider inverses of the trigonometric functions.

We begin with $y = \sin x$. Recall that a function must be one-to-one before it can have an inverse. The graph of sine in Figure 6.62 shows that $y = \sin x$ is not a one-to-one function since different values of x can lead to the *same* value of y. However, by suitably restricting the domain of the sine function, we can define a one-to-one function. It is customary to restrict the domain of $y = \sin x$ to $-\pi/2 \leq x \leq \pi/2$, which is the solid portion of the graph in Figure 6.62(a). Reflecting that portion of the sine graph about the 45° line $y = x$ gives the graph of the inverse function, shown in Figure 6.62(b).

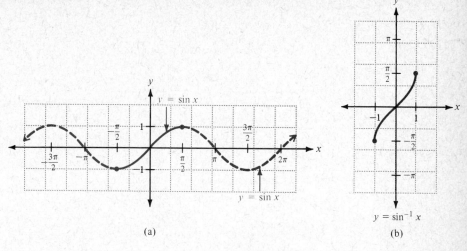

(a)   (b)

Figure 6.62

The inverse of the sine function is written $y = \sin^{-1} x$; that is,

Inverse Sine Function

$y = \sin^{-1} x$ if and only if $x = \sin y$,
where $1 \le x \le 1$
and $-\pi/2 \le y \le \pi/2$.

It is important to remember that for the trigonometric inverse functions, such as $\sin^{-1} x$, the $^{-1}$ indicates inverse and *not* reciprocal.

For each of the other trigonometric functions, we can define an inverse function by a suitable restriction on the domain, just as we did with sine. The **inverse trigonometric functions** and their ranges are:

Inverse Trigonometric Functions

Function	Range
$y = \sin^{-1}x$	$-\pi/2 \le y \le \pi/2$
$y = \cos^{-1} x$	$0 \le y \le \pi$
$y = \tan^{-1}x$	$-\pi/2 < y < \pi/2$
$y = \cot^{-1}x$	$0 < y < \pi$
$y = \sec^{-1}x$	$0 \le y < \pi/2$ or $\pi \le y < 3\pi/2$*
$y = \csc^{-1}x$	$0 < y \le \pi/2$ or $\pi < y \le 3\pi/2$*

*$\sec^{-1}$ and $\csc^{-1}$ are sometimes defined with a different range. The definition we have given is the most useful in calculus.

The graphs of $y = \cos^{-1} x$ and $y = \tan^{-1} x$ are shown in Figures 6.63 and 6.64.

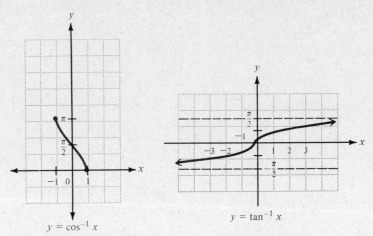

$y = \cos^{-1} x$

Figure 6.63

$y = \tan^{-1} x$

Figure 6.64

EXAMPLE ☐ Find $\sin^{-1} 1/2$.
1 Let $y = \sin^{-1} 1/2$. Then by the definition of the inverse sine function, $\sin y = 1/2$. Since $\sin \pi/6 = 1/2$ and $\pi/6$ is in the range of $y = \sin^{-1} x$, we have

$$\sin^{-1} 1/2 = \pi/6. \quad ☐$$

EXAMPLE ☐ Find $\cos^{-1}(-\sqrt{2}/2)$.
2 Since we have the restriction $0 \leq \cos^{-1} x \leq \pi$, the angle represented by $\cos^{-1} x$ can terminate only in quadrants I and II. Since $-\sqrt{2}/2$ is negative, we are restricted to quadrant II. Let $y = \cos^{-1}(-\sqrt{2}/2)$. Then $\cos y = -\sqrt{2}/2$. In quadrant II, $\cos 3\pi/4 = -\sqrt{2}/2$, so

$$\cos^{-1} -\frac{\sqrt{2}}{2} = \frac{3\pi}{4}. \quad ☐$$

EXAMPLE ☐ Find $\sin (\tan^{-1}(-3/2))$ without using tables or a calculator.
3 Let $y = \tan^{-1}(-3/2)$, so that $\tan y = -3/2$. Since $\tan^{-1} x$ is defined only in quadrants I and IV ($-\pi/2 < \tan^{-1} x < \pi/2$), and since we have $x = -3/2$, we must work in quadrant IV. Sketch y in quadrant IV and label a triangle as shown in Figure 6.65. The hypotenuse has a length of $\sqrt{13}$, so that $\sin y = -3/\sqrt{13}$, or

$$\sin \left(\tan^{-1} -\frac{3}{2} \right) = \frac{-3}{\sqrt{13}} = \frac{-3\sqrt{13}}{13}. \quad ☐$$

EXAMPLE ☐ Solve the equation $y = 2 \sin^{-1} (x + 4)$ for x.
4 First multiply by 1/2 to get

$$\frac{y}{2} = \sin^{-1} (x + 4).$$

Use the definition of the inverse sine function to get

$$\sin \frac{y}{2} = x + 4 \quad \text{or} \quad x = \left(\sin \frac{y}{2}\right) - 4. \quad \square$$

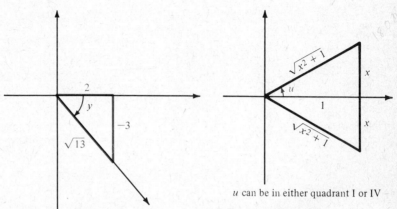

Figure 6.65

u can be in either quadrant I or IV

Figure 6.66

EXAMPLE □   Write sec (tan^{-1} x) as an algebraic expression.
5   Let tan^{-1} x = u. Then tan u = x. Since $-\pi/2 <$ tan^{-1} x $< \pi/2$, u is in quadrant I or quadrant IV. As shown in Figure 6.66, in either case

$$\sec (\tan^{-1} x) = \sec u = \sqrt{x^2 + 1}. \quad \square$$

> There is an alternate notation for inverse trigonometric functions used in some books: $y = $ sin^{-1} x is written $y = $ arcsin x, $y = $ cos^{-1} x is written $y = $ arccos x, and $y = $ tan^{-1} x is written $y = $ arctan x. In the exercise set below, we shall use these notations interchangeably.

EXERCISES
6.8

For the following, give the value of y in radians without using tables or a calculator.

1. $y = \sin^{-1} \dfrac{-\sqrt{3}}{2}$

2. $y = \cos^{-1} \dfrac{\sqrt{3}}{2}$

3. $y = \tan^{-1} 1$

4. $y = \tan^{-1} -1$

5. $y = \sin^{-1} -1$

6. $y = \cos^{-1} -1$

7. $y = \cos^{-1} 1/2$

8. $y = \sin^{-1} \dfrac{-\sqrt{2}}{2}$

9. $y = \arccos \dfrac{-\sqrt{2}}{2}$

10. $y = \arctan \dfrac{\sqrt{3}}{3}$

11. $y = \arctan -\sqrt{3}$

12. $y = \arccos -1/2$

For the following, give the value in degrees. (Round to the nearest ten minutes.)

13. $y = \sin^{-1}(-0.1334)$

14. $y = \cos^{-1}(-0.1334)$

15. $y = \cos^{-1}(-0.3987)$

16. $y = \sin^{-1} 0.7790$

17. $y = \cos^{-1} 0.9272$

18. $y = \tan^{-1} 1.767$

19. $y = \tan^{-1} 1.111$

20. $y = \sin^{-1} 0.8192$

21. $y = \arctan(-0.9217)$

22. $y = \arctan(-0.2867)$

Give the value of the following in radians to four significant digits.

23. $\sin^{-1} 0.7214$ 24. $\cos^{-1} 0.3004$

25. $\tan^{-1} -4.114$ 26. $\sin^{-1} -0.9946$

27. Enter 1.003 in your calculator and push the keys for inverse sine. The machine will tell you that something is wrong. What is wrong?

28. Enter 1.003 in your calculator and push the keys for inverse tangent. This time, unlike in Exercise 27, you get an answer easily. What is the difference?

Find each of the following to the nearest second.

29. $\sin^{-1}(-0.443981)$ 30. $\tan^{-1}(-0.394511)$

31. $\cos^{-1} 0.91441$ 32. $\tan^{-1} 14.76892$

Give the value of each of the following without using tables.

33. $\tan(\cos^{-1} 2/3)$ 34. $\sin(\cos^{-1} 1/4)$

35. $\cos(\tan^{-1} -2)$ 36. $\sec(\sin^{-1} -1/3)$

37. $\cot(\sin^{-1} -2/5)$ 38. $\cos(\tan^{-1} 8/5)$

39. $\sin(\arctan -3)$ 40. $\tan(\arccos -4/5)$

Use the inverse key of your calculator together with the trigonometric function keys to find each of the following in radians to six decimal places.

41. $\cos(\arctan .3)$ 42. $\sin(\arccos .75)$

43. $\tan(\arcsin .1225)$ 44. $\cot(\arccos .5823)$

Solve each of the following equations for x.

45. $y = 4 \sin^{-1} x$ 46. $3y = \cos^{-1} x$

47. $2y = \tan^{-1} 2x$ 48. $y = 3 \sin^{-1}(x/2)$

49. $y = \sin^{-1}(x + 2)$

50. $y = \tan^{-1}(2x - 1)$

Write each of the following as an algebraic expression.

51. $\sin(\cos^{-1} x)$ 52. $\tan(\cos^{-1} x)$

53. $\sec(\cos^{-1} x)$ 54. $\csc(\sin^{-1} x)$

55. $\cot(\sin^{-1} x)$ 56. $\cos(\sin^{-1} x)$

Graph each of the following and give the domain and range.

57. $y = 2 \cos^{-1} x$ 58. $y = \tan^{-1} 2x$

59. $y = \sin^{-1} x + 2$ 60. $y = \tan^{-1}(x + 1)$

61. We know that $\sin^{-1}(\sin x) = x$. Enter 1.74283 in your calculator (set for radians), and push the sine key. Then push the keys for $\sin^{-1}$. You get 1.398763 instead of 1.74283. What happened?

The following were used by the mathematicians who computed the value of π to 100,000 decimal places. Use a calculator to verify that each is (approximately) correct.

62. $\pi = 16 \tan^{-1} \dfrac{1}{5} - 4 \tan^{-1} \dfrac{1}{239}$

63. $\pi = 24 \tan^{-1} \dfrac{1}{8} + 8 \tan^{-1} \dfrac{1}{57} + 4 \tan^{-1} \dfrac{1}{239}$

64. $\pi = 48 \tan^{-1} \dfrac{1}{18} + 32 \tan^{-1} \dfrac{1}{57} - 20 \tan^{-1} \dfrac{1}{239}$

Answer true or false for each of the following.

65. $2 \cos^{-1} x = \cos^{-1} 2x$

66. $\cot^{-1} x = 1/\tan^{-1} x$

67. $\tan^{-1} x = \sin^{-1} x/\cos^{-1} x$

68. $\sin^{-1}(-x) = -\sin^{-1} x$

69. $y = \sin^{-1} x$ is an even function

70. $y = \cos^{-1} x$ is an even function

71. $y = x \cdot \sin^{-1} x$ is an odd function

72. $y = x^2 \cdot \cos^{-1} x$ is an even function

CHAPTER 6   SUMMARY

Key Words

unit circle	trigonometric functions	degree measure
sine	circular functions	radian measure
cosine	reciprocal identities	related angle
periodic function	angle	reference angle
period	vertex	amplitude
tangent	standard position	phase shift
cotangent	quadrantal angle	addition of ordinates
secant	coterminal angles	inverse trigonometric functions
cosecant		

CHAPTER 6   REVIEW EXERCISES

For each of the following arc lengths, find the coordinates of the corresponding point on the unit circle.

1. $-\pi/4$ **2.** $2\pi/3$

3. 3π

For each of the following values of s, find sin s, cos s, and tan s.

4. $5\pi/4$ **5.** $5\pi/3$

6. $-\pi/3$

The point $(-2/3, \sqrt{5}/3)$ is the endpoint of arc s on the unit circle. Find the coordinates of the endpoints of the following arcs.

7. $-s$ **8.** $\pi + s$

9. $\pi - s$

Give the quadrant in which θ terminates.

10. $\sin \theta > 0$, $\cos \theta < 0$

11. $\cos \theta > 0$, $\cot \theta < 0$

12. $\tan \theta > 0$, $\sin \theta < 0$

13. $\sec \theta > 0$, $\tan \theta < 0$

For each of the following, find the values of the other trigonometric functions.

14. $\sin s = 2/3$, s terminates in quadrant I

15. $\cos s = 4/5$, s terminates in quadrant IV

16. $\sec s = \sqrt{5}$, $\cot s = -1/2$

17. $\tan s = -4/3$, $\sec s = -5/3$

Convert radian measures to degrees, and degree measures to radians.

18. $3\pi/4$

19. $4\pi/5$

20. $31\pi/5$

21. $270°$

22. $480°$

Find the exact value of each of the following.

23. $\sin \pi/3$

24. $\cos 120°$

25. $\tan 240°$

26. $\cot 390°$

27. $\sec 900°$

28. $\cot (-1020°)$

Find the sine, cosine, and tangent of each of the following angles.

29.

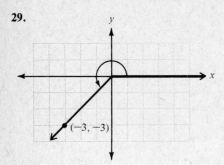

30.

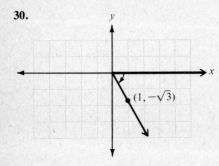

Find the values of each of the following. Use Table 5 or a calculator. Interpolate as necessary.

31. $\tan 235°$

32. $\sec(-87°)$

33. $\sin 247° \ 10'$

34. $\sec 28° \ 17'$

35. $\cos 58° \ 4'$

36. $\cos(-3.1998)$

Use Table 5 or a calculator to find θ in degrees for each of the following.

37. $\tan \theta = 5.226$

38. $\cos \theta = 0.4384$

39. $\sin \theta = 0.8871$

Use Table 5 or a calculator to find a value of s in radians for each of the following.

40. $\tan s = 0.7581$

41. $\sin s = 0.9596$

42. Find the lengths of the missing sides in this right triangle.

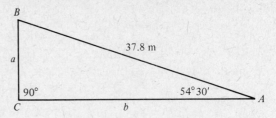

When a light ray travels from one medium, such as air, to another medium, such as water or glass, the speed of the light changes, and the direction that the ray is traveling changes. (This is why a fish under water is in a different position than it appears to us.) These changes are given by Snell's Law, from physics:

$$\frac{c_1}{c_2} = \frac{\sin \theta_1}{\sin \theta_2},$$

where c_1 is the speed in the first medium, c_2 is the speed in the second medium, and θ_1 and θ_2 are the angles shown in the figure.

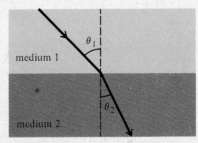

If this medium is less dense, light travels at a faster speed, c_1.

If this medium is more dense, light travels at a slower speed, c_2.

In the following exercises assume that $c_1 = 3 \times 10^8$ meters per second. Find the speed of light in the second medium.

43. $\theta_1 = 46°, \ \theta_2 = 31°$

44. $\theta_1 = 39°, \ \theta_2 = 28°$

Give the amplitude, period, and phase shift for each of the following.

45. $y = -2 \sin(x - \pi)$

46. $y = \dfrac{3}{4} \cos \left(x + \dfrac{\pi}{4} \right)$

47. $y = -\cos 4x$

48. $y = 6 \sin (11x - 2\pi)$

Graph the following functions over a one-period interval.

49. $y = 3 \sin x$

50. $y = -2 \cos x$

51. $y = \sin 2x$

52. $y = \tan 3x$

53. $y = -\tan x$

54. $y = \dfrac{1}{2} \sec x$

55. $y = 3 \cos 2x$

56. $y = \dfrac{1}{2} \cot 3x$

57. $y = 2 + \cot x$

58. $y = -1 + \csc x$

59. $y = \cos \left(x - \dfrac{\pi}{4} \right)$

60. $y = \tan \left(x - \dfrac{\pi}{2} \right)$

61. $y = \sec \left(2x + \dfrac{\pi}{3} \right)$

62. $y = \sin \left(3x + \dfrac{\pi}{2} \right)$

Graph each of the following using the method of addition of ordinates.

63. $y = -x + \tan x$

64. $y = \dfrac{1}{2}x + \cos x$

65. $y = \sin x + \cos x$

66. $y = \tan x + \cot x$

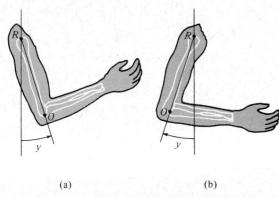

(a)   (b)

Schematic diagrams of a rhythmically moving arm. The upper arm RO rotates back and forth about the point R; the position of the arm is measured by the angle y between the actual position and the downward vertical position.

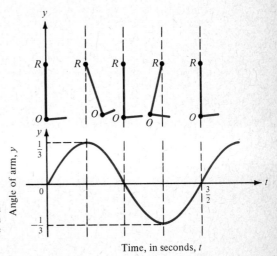

This graph shows the relationship between angle y and time t in seconds.

67. Find an equation of the form $y = a \sin kt$ for the graph immediately above.

68. How long does it take for a complete movement of the arm?

Figures for Exercises 67 and 68: From *Calculus for the Life Sciences* by Rodolfo De Sapio. W. H. Freeman and Company. Copyright © 1978.

Find each of the following. Give answers in radians.

69. $y = \sin^{-1} \sqrt{2}/2$ **70.** $y = \cos^{-1} -1/2$

71. $y = \tan^{-1} -\sqrt{3}$ **72.** $y = \arctan 1.780$

73. $y = \arcsin -0.6604$

74. $y = \cos^{-1} 0.8039$

Graph each of the following.

75. $y = 2 \cos^{-1} x$ **76.** $y = \sin^{-1} 3x$

77. $y = -1 + \tan^{-1} x$

78. Solve for x: $2y = 1 + \tan^{-1}(x + 1)$.

79. In the study of alternating current in electricity, instantaneous voltage in amps is given by

$$e = E_{max} \sin 2\pi ft,$$

where f is the number of cycles per second, E_{max} is the maximum voltage in amps, and t is time in seconds.

(a) Solve the equation for t.

(b) Find the smallest positive value of t in radians if $E_{max} = 12$, $e = 5$, and $f = 100$.

80. Many computer languages such as BASIC and FORTRAN have only the arctangent function available. To use the other inverse trigonometric functions, it is necessary to express them in terms of arctangent. This can be done as follows.

(a) Let $u = \arcsin x$. Solve the equation for x in terms of u.

(b) Use the result of part (a) to label the three sides of the triangle of the figure in terms of x.

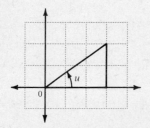

(c) Use the triangle from part (b) to write an equation for tan u in terms of x.

(d) Solve the equation from (c) for u.

(e) Use your equation from part (d) to calculate arcsin (1/2). Compare the answer with the actual value of arcsin (1/2).

Find each of the following without using tables.

81. $\tan (\sin^{-1} 1/3)$ **82.** $\cos^{-1} (\tan -\pi/4)$

The following exercises are taken with permission from a standard calculus book.*

83. Show that the area of a sector of a circle having central angle θ and radius r is $\frac{1}{2} r^2\theta$, if θ is measured in radians.

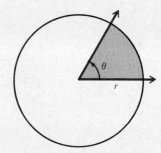

84. Let $A(r, 0)$ be the point where the positive x-axis cuts a circle of radius r, center at the origin O. Let $P(r \cos \theta, r \sin \theta)$ be a point on the circle in the first quadrant, with angle $AOP = \theta$ radians. Let AT be tangent to the circle at A and suppose it intersects the line OP at T. By considering the areas of triangle AOP, sector AOP, and triangle AOT, prove the following inequality.

$$\sin \theta < \theta < \tan \theta \qquad \text{if } 0 < \theta < \pi/2.$$

*George B. Thomas, Jr. and Ross L. Finney, *Calculus and Analytic Geometry*, © 1979. Addison—Wesley, Reading, Massachusetts, p. 101. Reprinted by Permission.

7 Trigonometric Identities and Equations

$\mathbf{A}$ *conditional equation,* such as $2x + 1 = 9$ or $m^2 - 2m = 3$ is true only for *certain* values in its domain; $2x + 1 = 9$ is true only for $x = 4$, and $m^2 - 2m = 3$ is true only for $m = 3$ and $m = -1$, for example. On the other hand, an **identity** is an equation which is true for *every* value in the domain of its variable. Examples of identities include

$$5(x + 3) = 5x + 15 \quad \text{and} \quad (a + b)^2 = a^2 + 2ab + b^2.$$

In this chapter we discuss identities which involve trigonometric functions. The variables in the trigonometric functions represent either angles or real numbers. The domain is assumed to be all values for which a given function is defined. We shall also look at conditional equations involving trigonometric functions.

7.1 Fundamental Identities

Let us begin by restating the fundamental identities introduced in Chapter 6. They hold for appropriate values of s.

Fundamental Identities

$$\sin s = \frac{1}{\csc s} \qquad \cos s = \frac{1}{\sec s} \qquad \tan s = \frac{1}{\cot s}$$

$$\cot s = \frac{1}{\tan s} \qquad \csc s = \frac{1}{\sin s} \qquad \sec s = \frac{1}{\cos s}$$

$$\tan s = \frac{\sin s}{\cos s} \qquad \cot s = \frac{\cos s}{\sin s}$$

$$\sin^2 s + \cos^2 s = 1 \qquad \tan^2 s + 1 = \sec^2 s$$

$$1 + \cot^2 s = \csc^2 s.$$

The unit circle in Figure 7.1 shows an arc of length s starting at $(1, 0)$, with endpoint (x, y). The same figure also shows an arc of length $-s$, again starting at $(1, 0)$. This arc has endpoint $(x, -y)$. Since $\sin s = y$, we have

$$\sin(-s) = -y = -\sin s,$$
$$\cos(-s) = x = \cos s$$

and $\quad \tan(-s) = \dfrac{-y}{x} = -\tan s.$

In summary,

Negative Angles

$$\sin(-s) = -\sin s \qquad \cos(-s) = \cos s$$
$$\tan(-s) = -\tan s.$$

These identities show that the sine and tangent functions are odd, while the cosine function is even.

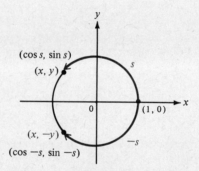

Figure 7.1

Trigonometric identities are useful in several ways. One use of trigonometric identities is in finding values of the other trigonometric functions from the value of a given trigonometric function. For example, if we are given a value of $\tan s$, we can find the value of $\cot s$ by using the identity $\cot s = 1/\tan s$. In fact, given $\tan s$ and the quadrant in which s lies, we can find the value of all the other trigonometric functions by using identities, as shown below.

EXAMPLE 1   ☐   If $\tan s = -5/3$ and s is in quadrant II, find the values of the other trigonometric functions.

We know $\cot s = 1/\tan s$, so that $\cot s = -3/5$. Also, $\tan^2 s + 1 = \sec^2 s$.

$$\left(-\frac{5}{3}\right)^2 + 1 = \sec^2 s$$

$$\frac{25}{9} + 1 = \sec^2 s$$

$$\frac{34}{9} = \sec^2 s$$

$$-\sqrt{\frac{34}{9}} = \sec s$$

$$-\frac{\sqrt{34}}{3} = \sec s$$

We choose the negative square root since the values of sec s are negative in quadrant II. Since cos s is the reciprocal of sec s,

$$\cos s = \frac{-3}{\sqrt{34}} = \frac{-3\sqrt{34}}{34},$$

after rationalizing the denominator. Now find sin s by using the identity $\sin^2 s + \cos^2 s = 1$.

$$\sin^2 s + \left(\frac{-3\sqrt{34}}{34}\right)^2 = 1$$

$$\sin^2 s = 1 - \frac{9}{34}$$

$$\sin^2 s = \frac{25}{34}$$

$$\sin s = \frac{5}{\sqrt{34}} \quad \text{or} \quad \sin s = \frac{5\sqrt{34}}{34}$$

Here we choose the positive square root because the values of sin s are positive in quadrant II. Finally, since csc s is the reciprocal of sin s, csc $s = \sqrt{34}/5$. ☐

EXAMPLE 2 ☐ Express cos x in terms of tan x.
Start with $\tan^2 x + 1 = \sec^2 x$. Then

$$\frac{1}{\tan^2 x + 1} = \frac{1}{\sec^2 x}$$

or $\dfrac{1}{\tan^2 x + 1} = \cos^2 x$.

Take the square root of both sides.

$$\pm\sqrt{\frac{1}{\tan^2 x + 1}} = \cos x$$

$$\cos x = \frac{\pm 1}{\sqrt{\tan^2 x + 1}}$$

Rationalize the denominator to get

$$\cos x = \frac{\pm\sqrt{\tan^2 x + 1}}{\tan^2 x + 1}.$$

Here, the $+$ or the $-$ sign is chosen, depending on the quadrant of x. ☐

Another use of identities is to simplify trigonometric expressions by substituting one half of an identity for the other half. For example, the expression $\sin^2 \theta + \cos^2 \theta$ can be simplified to just 1, as in the following example.

EXAMPLE 3 ☐ Use the fundamental identities to write $\tan \theta + \cot \theta$ in terms of $\sin \theta$ and $\cos \theta$. Then simplify.

From the fundamental identities,

$$\tan \theta + \cot \theta = \frac{\sin \theta}{\cos \theta} + \frac{\cos \theta}{\sin \theta}.$$

To simplify this expression, add the two fractions on the right side, using the common denominator $\cos \theta \sin \theta$.

$$\tan \theta + \cot \theta = \frac{\sin^2 \theta}{\cos \theta \sin \theta} + \frac{\cos^2 \theta}{\cos \theta \sin \theta}$$

$$= \frac{\sin^2 \theta + \cos^2 \theta}{\cos \theta \sin \theta}$$

Now substitute 1 for $\sin^2 \theta + \cos^2 \theta$, to get

$$\tan \theta + \cot \theta = \frac{1}{\cos \theta \sin \theta}.$$ ☐

EXERCISES 7.1

Find sin s for each of the following.

1. $\cos s = 3/4$, s in quadrant I

2. $\cot s = -1/3$, s in quadrant IV

3. $\cos s = \sqrt{5}/5$, $\tan s < 0$

4. $\tan s = -\sqrt{7}/2$, $\sec s > 0$

5. Find $\tan \theta$ if $\cos \theta = -2/5$, and $\sin \theta < 0$.

6. Find $\csc \alpha$ if $\tan \alpha = 6$, and $\cos \alpha > 0$.

Use the fundamental identities to find the remaining five trigonometric functions of θ.

7. $\sin \theta = \frac{2}{3}$, θ in quadrant II

8. $\cos \theta = \frac{1}{5}$, θ in quadrant I

9. $\tan \theta = -\frac{1}{4}$, θ in quadrant IV

10. $\tan \theta = \frac{2}{3}$, θ in quadrant III

11. $\sec \theta = -3$, θ in quadrant II

12. $\csc \theta = -\frac{5}{2}$, θ in quadrant III

13. $\cot \theta = \frac{4}{3}$, $\sin \theta > 0$

14. $\sin \theta = -\frac{4}{5}$, $\cos \theta < 0$

15. $\sec \theta = \frac{4}{3}$, $\sin \theta < 0$

16. $\cos \theta = -\frac{1}{4}$, $\sin \theta > 0$

For each trigonometric expression in Column I, choose the expression from Column II which completes a fundamental identity.

Column I   **Column II**

17. $\dfrac{\cos x}{\sin x}$ **(a)** $\sin^2 x + \cos^2 x$

18. $\tan x$ **(b)** $\cot x$

19. $\cos(-x)$ **(c)** $\sec^2 x$

20. $\tan^2 x + 1$ **(d)** $\dfrac{\sin x}{\cos x}$

21. 1 **(e)** $\cos x$

For each expression in Column I, choose the expression from Column II which completes an identity. You will have to rewrite one or both expressions, using a fundamental identity, to recognize the matches.

Column I   **Column II**

22. $-\tan x \cos x$ **(a)** $\dfrac{\sin^2 x}{\cos^2 x}$

23. $\sec^2 x - 1$ **(b)** $\dfrac{1}{\sec^2 x}$

24. $\dfrac{\sec x}{\csc x}$ **(c)** $\sin(-x)$

25. $1 + \sin^2 x$ **(d)** $\csc^2 x - \cot^2 x + \sin^2 x$

26. $\cos^2 x$ **(e)** $\tan x$

In each of the following, use the fundamental identities to get an equivalent expression involving only sines and cosines and then simplify it.

27. $\csc^2 \beta - \cot^2 \beta$

28. $\dfrac{\tan(-\theta)}{\sec \theta}$

29. $\tan(-\alpha) \cos(-\alpha)$

30. $\cot^2 x(1 + \tan^2 x)$

31. $\tan^2 \theta - \dfrac{\sec^2 \theta}{\csc^2 \theta}$

32. $\dfrac{\tan x \csc x}{\sec x}$

33. $\sec \theta + \tan \theta$

34. $\dfrac{\sec \alpha}{\tan \alpha + \cot \alpha}$

35. $\sec^2 t - \tan^2 t$

36. $\csc^2 \gamma + \sec^2 \gamma$

37. $\cot^2 \beta - \csc^2 \beta$

38. $1 + \cot^2 \alpha$

39. $\dfrac{1 + \tan^2 \theta}{\cot^2 \theta}$

40. $\dfrac{1 - \sin^2 t}{\csc^2 t}$

41. $\cot^2 \beta \sin^2 \beta + \tan^2 \beta \cos^2 \beta$

42. $\sec^2 x + \cos^2 x$

43. $\dfrac{\cot^2 \alpha + \csc^2 \alpha}{\cos^2 \alpha}$ **44.** $1 - \tan^4 \theta$

45. $1 - \cot^4 s$ **46.** $\tan^4 \gamma - \cot^4 \gamma$

Write all the trigonometric functions in terms of

47. $\sin x$ **48.** $\cos x$

49. Write $\tan x$ in terms of $\csc x$.

50. Write $\sec \alpha$ in terms of $\tan \alpha$.

51. Express $\cot s$ in terms of $\sec s$.

52. Express $\csc t$ in terms of $\tan t$.

53. Suppose $\cos \theta = x/(x + 1)$. Find $\sin \theta$.

54. Find $\tan \alpha$ if $\sec \alpha = (p + 4)/p$.

55. Prove that $|\tan x| \geq |\sin x|$ for all real numbers x in the domain of both functions.

56. Prove that $|\sec x| > |\tan x|$ for all real numbers x in the domain of both functions.

57. Let $\cos x = 1/5$. Find all possible values for

$$\dfrac{\sec x - \tan x}{\sin x}.$$

58. Let $\csc x = -3$. Find all possible values for

$$\dfrac{\sin x + \cos x}{\sec x}.$$

Prove that each of the following is not an identity.

59. $(\sin s + \cos s)^2 = 1$

60. $(\tan s + 1)^2 = \sec^2 s$

61. $2 \sin s = \sin 2s$

62. $\sin x = \sqrt{1 - \cos^2 x}$

63. $\sin^3 x + \cos^3 x = 1$

64. $\sin x + \sin y = \sin(x + y)$

Prove the following for first-quadrant values of s.

65. $\log \sin s = -\log \csc s$

66. $\log \tan s = \log \sin s - \log \cos s$

7.2   Verifying Trigonometric Identities

One of the skills required for more advanced work in mathematics (and especially in calculus) is the ability to use the trigonometric identities to write trigonometric expressions in alternate forms. To develop this skill, we use the fundamental identities to verify that a trigonometric equation is an identity (for those values of the variable for which it is defined). This process must be learned with practice. Here are some hints that may help you get started.

1. Memorize the fundamental identities given in the last section. Whenever you see either half of a fundamental identity, the other half should come to mind.

2. Try to rewrite the more complicated side of the equation so that it is identical to the simpler side.

3. It is often helpful to express all other trigonometric functions in the equation in terms of sine and cosine and then simplify the result.

4. You should usually perform any factoring or indicated algebraic operations. For example, the expression $\sin^2 x + 2 \sin x + 1$ can be factored as $(\sin x + 1)^2$. The sum or difference of two trigonometric expressions, such as

$$\frac{1}{\sin \theta} + \frac{1}{\cos \theta}$$

can be added or subtracted in the same way as any other rational expression. For example,

$$\frac{1}{\sin \theta} + \frac{1}{\cos \theta} = \frac{\cos \theta}{\sin \theta \cos \theta} + \frac{\sin \theta}{\sin \theta \cos \theta}$$

$$= \frac{\cos \theta + \sin \theta}{\sin \theta \cos \theta}.$$

5. Keep in mind the side you are not changing as you select substitutions. It represents your goal. For example, to verify the identity

$$\tan^2 x + 1 = \frac{1}{\cos^2 x},$$

try to think of an identity that relates $\tan x$ to $\cos x$. Here, since $\sec x = 1/\cos x$ and $\sec^2 x = \tan^2 x + 1$, the secant function is the best link between the two sides of the equation.

We use these hints in the following examples. (A word of warning: verifying identities is *not* the same as solving equations. Techniques used in solving equations, such as adding the same terms to both sides, are not valid when working with identities.)

EXAMPLE 1   Verify that

$$\cot s + 1 = \csc s(\cos s + \sin s)$$

is an identity.

We use the fundamental identities to rewrite one side of the equation so that it is identical to the other side. Since the right side is more complicated, we work with it. Here we use the method of changing all the trigonometric functions to sine or cosine.

$$\csc s(\cos s + \sin s) = \frac{1}{\sin s}(\cos s + \sin s)$$

$$= \frac{\cos s}{\sin s} + \frac{\sin s}{\sin s}$$

$$= \cot s + 1$$

The equation is an identity because the right side equals the left side, for any value of s. ☐

EXAMPLE 2   Verify that

$$\tan^2 \alpha \, (1 + \cot^2 \alpha) = \frac{1}{1 - \sin^2 \alpha}$$

is an identity.

Working with the left side gives

$$\tan^2 \alpha \, (1 + \cot^2 \alpha) = \tan^2 \alpha + \tan^2 \alpha \cot^2 \alpha$$

$$= \tan^2 \alpha + \tan^2 \alpha \cdot \frac{1}{\tan^2 \alpha}$$

$$= \tan^2 \alpha + 1$$

$$= \sec^2 \alpha$$

$$= \frac{1}{\cos^2 \alpha}$$

$$= \frac{1}{1 - \sin^2 \alpha}. ☐$$

EXAMPLE 3   Show that

$$\frac{\tan t - \cot t}{\sin t \cos t} = \sec^2 t - \csc^2 t.$$

Work with the left side.

$$\frac{\tan t - \cot t}{\sin t \cos t} = \frac{\tan t}{\sin t \cos t} - \frac{\cot t}{\sin t \cos t}$$

$$= \tan t \cdot \frac{1}{\sin t \cos t} - \cot t \cdot \frac{1}{\sin t \cos t}$$

$$= \frac{\sin t}{\cos t} \cdot \frac{1}{\sin t \cdot \cos t} - \frac{\cos t}{\sin t} \cdot \frac{1}{\sin t \cos t}$$

$$= \frac{1}{\cos^2 t} - \frac{1}{\sin^2 t}$$

$$= \sec^2 t - \csc^2 t \quad \square$$

EXERCISES
7.2

For each of the following, perform the indicated operations and simplify the result.

1. $\tan \theta + \dfrac{1}{\tan \theta}$

2. $\dfrac{\cos x}{\sin x} + \dfrac{\sin x}{\cos x}$

3. $\cot s(\tan s + \sin s)$

4. $\sec \beta(\cos \beta + \sin \beta)$

5. $\dfrac{1}{\csc^2 \theta} + \dfrac{1}{\sec^2 \theta}$

6. $\dfrac{1}{\sin \alpha - 1} - \dfrac{1}{\sin \alpha + 1}$

7. $\dfrac{\cos x}{\sec x} + \dfrac{\sin x}{\csc x}$

8. $\dfrac{\cos \gamma}{\sin \gamma} + \dfrac{\sin \gamma}{1 + \cos \gamma}$

9. $(1 + \sin t)^2 + \cos^2 t$

10. $(1 + \tan s)^2 - 2 \tan s$

Factor each of the following trigonometric expressions.

11. $\sin^2 \gamma - 1$

12. $\sec^2 \theta - 1$

13. $(\sin x + 1)^2 - (\sin x - 1)^2$

14. $(\tan x + \cot x)^2 - (\tan x - \cot x)^2$

15. $2 \sin^2 x + 3 \sin x + 1$

16. $4 \tan^2 \beta + \tan \beta - 3$

17. $4 \sec^2 x + 3 \sec x - 1$

18. $2 \csc^2 x + 7 \csc x - 30$

19. $\cos^4 x + 2 \cos^2 x + 1$

20. $\cot^4 x + 3 \cot^2 x + 2$

Use the fundamental identities to simplify each of the given expressions.

21. $\tan \theta \cos \theta$

22. $\cot \alpha \sin \alpha$

23. $\sec r \cos r$

24. $\cot t \tan t$

25. $\dfrac{\sin \beta \tan \beta}{\cos \beta}$

26. $\dfrac{\csc \theta \sec \theta}{\cot \theta}$

27. $\sec^2 x - 1$

28. $\csc^2 t - 1$

29. $\dfrac{\sin^2 x}{\cos^2 x} + \sin x \csc x$

30. $\dfrac{1}{\tan^2 \alpha} + \cot \alpha \tan \alpha$

Verify each of the following trigonometric identities.

31. $1 - \sec \alpha \cos \alpha = \tan \alpha \cot \alpha - 1$

32. $\csc^4 \theta = \cot^4 \theta + 2 \cot^2 \theta + 1$

33. $\dfrac{\sin^2 \theta}{\cos^2 \theta} = \sec^2 \theta - 1$

34. $\cot \beta \sin \beta = \cos \beta$

35. $\sin^2 \alpha + \tan^2 \alpha + \cos^2 \alpha = \sec^2 \alpha$

36. $\sin^2 s - 1 = -\cos^2 s$

37. $\dfrac{\sin^2 \gamma}{\cos \gamma} = \sec \gamma - \cos \gamma$

38. $(1 + \tan^2 x) \cos^2 x = 1$

39. $\cot s + \tan s = \sec s \csc s$

40. $\dfrac{\cos \alpha}{\sec \alpha} + \dfrac{\sin \alpha}{\csc \alpha} = \sec^2 \alpha - \tan^2 \alpha$

41. $\dfrac{\cos \alpha}{\sin \alpha \cot \alpha} = 1$

42. $\sin^4 \theta - \cos^4 \theta = 2 \sin^2 \theta - 1$

43. $\dfrac{1 + \sin x}{\cos x} = \dfrac{\cos x}{1 - \sin x}$ (Hint: rationalize the denominator on the right.)

44. $(1 - \cos^2 \alpha)(1 + \cos^2 \alpha) = 2 \sin^2 \alpha - \sin^4 \alpha$

45. $\dfrac{(\sec \theta - \tan \theta)^2 + 1}{\sec \theta \csc \theta - \tan \theta \csc \theta} = 2 \tan \theta$

46. $\dfrac{\cos \theta + 1}{\tan^2 \theta} = \dfrac{\cos \theta}{\sec \theta - 1}$

47. $\dfrac{1}{\sec \alpha - \tan \alpha} = \sec \alpha + \tan \alpha$

48. $\dfrac{1}{1 - \sin \theta} + \dfrac{1}{1 + \sin \theta} = 2 \sec^2 \theta$

49. $\dfrac{1 - \cos x}{1 + \cos x} = (\cot x - \csc x)^2$

50. $\dfrac{\tan s}{1 + \cos s} + \dfrac{\sin s}{1 - \cos s} = \cot s + \sec s \csc s$

51. $\dfrac{1}{\tan \alpha - \sec \alpha} + \dfrac{1}{\tan \alpha + \sec \alpha} = -2 \tan \alpha$

52. $\dfrac{\cot \alpha + 1}{\cot \alpha - 1} = \dfrac{1 + \tan \alpha}{1 - \tan \alpha}$

53. $\dfrac{\csc \theta + \cot \theta}{\tan \theta + \sin \theta} = \cot \theta \csc \theta$

54. $\sin^2 \alpha \sec^2 \alpha + \sin^2 \alpha \csc^2 \alpha = \sec^2 \alpha$

55. $\sec^4 x - \sec^2 x = \tan^4 x + \tan^2 x$

56. $\dfrac{1 - \sin \theta}{1 + \sin \theta} = \sec^2 \theta - 2 \sec \theta \tan \theta + \tan^2 \theta$

57. $\sin \theta + \cos \theta = \dfrac{\sin \theta}{1 - \dfrac{\cos \theta}{\sin \theta}} + \dfrac{\cos \theta}{1 - \dfrac{\sin \theta}{\cos \theta}}$

58. $\dfrac{\sin \theta}{1 - \cos \theta} - \dfrac{\sin \theta \cos \theta}{1 + \cos \theta} = \csc \theta + \csc \theta \cos^2 \theta$

59. $\dfrac{\sec^4 s - \tan^4 s}{\sec^2 s + \tan^2 s} = \sec^2 s - \tan^2 s$

60. $\dfrac{\cot^2 t - 1}{1 + \cot^2 t} = 1 - 2 \sin^2 t$

61. $\dfrac{\tan^2 t - 1}{\sec^2 t} = \dfrac{\tan t - \cot t}{\tan t + \cot t}$

62. $(1 + \sin x + \cos x)^2 = 2(1 + \sin x)(1 + \cos x)$

63. $(\sin s + \cos s)^2 \cdot \csc s = 2 \cos s + \dfrac{1}{\sin s}$

64. $\dfrac{\sin^3 t - \cos^3 t}{\sin t - \cos t} = 1 + \sin t \cos t$

65. $\dfrac{1 + \cos x}{1 - \cos x} - \dfrac{1 - \cos x}{1 + \cos x} = 4 \cot x \csc x$

66. $(\sec \alpha - \tan \alpha)^2 = \dfrac{1 - \sin \alpha}{1 + \sin \alpha}$

67. $(\sec \alpha + \csc \alpha)(\cos \alpha - \sin \alpha) = \cot \alpha - \tan \alpha$

68. $\dfrac{\sin^4 \alpha - \cos^4 \alpha}{\sin^2 \alpha - \cos^2 \alpha} = 1$

69. $\dfrac{\cot^2 x + \sec^2 x + 1}{\cot^2 x} = \sec^4 x$

70. $\dfrac{\cos x - (\sin x - 1)}{\cos x + (\sin x - 1)} = \dfrac{\sin x}{1 - \cos x}$

(Hint: multiply numerator and denominator on the left by $\cos x - (\sin x - 1)$.)

71. $\ln e^{|\sin x|} = |\sin x|$

72. $\ln |\tan x| = -\ln |\cot x|$

73. $\ln |\sec x - \tan x| = -\ln |\sec x + \tan x|$

74. $-\ln |\csc t - \cot t| = \ln |\csc t + \cot t|$

Given a complicated equation involving trigonometric functions, it is a good idea to decide whether it is likely to be an identity before trying to prove that it is. Substitute $s = 1$ and $s = 2$ into each of the *following (with the calculator set on **radian** mode). If you get the same results on both sides of the equation, it may be an identity. Then prove that it is.*

75. $\dfrac{2 + 5 \cos s}{\sin s} = 2 \csc s + 5 \cot s$

76. $1 + \cot^2 s = \dfrac{\sec^2 s}{\sec^2 s - 1}$

77. $\dfrac{\tan s - \cot s}{\tan s + \cot s} = 2 \sin^2 s$

78. $\dfrac{1}{1 + \sin s} + \dfrac{1}{1 - \sin s} = \sec^2 s$

79. $\dfrac{1 - \tan^2 s}{1 + \tan^2 s} = \cos^2 s - \sin s$

80. $\dfrac{\sin^3 s - \cos^3 s}{\sin s - \cos s} = \sin^2 s + 2 \sin s \cos s + \cos^2 s$

81. $\sin^2 s + \cos^2 s = \dfrac{1}{2}(1 - \cos 4s)$

82. $\cos 3s = 3 \cos s + 4 \cos^3 s$

Show that the following are not identities for real numbers s and t.

83. $\sin (\csc s) = 1$

84. $\sqrt{\cos^2 s} = \cos s$

85. $\csc t = \sqrt{1 + \cot^2 t}$

86. $\sin t = \sqrt{1 - \cos^2 t}$

87. $\sin s + \cos s = \sqrt{\sin^2 s + \cos^2 s}$

88. $\cos (-s) = -\cos s$

7.3 Sum and Difference Identities for Cosine

Sometimes we need to express a trigonometric function value for the sum or difference of two real numbers (or angles) in terms of the trigonometric function values for each real number (or angle). For example, let us express $\cos(s - t)$ in terms of trigonometric function values of s and t themselves.

To do this, let $A(1, 0)$, $B(x_1, y_1)$, $C(x_2, y_2)$, and $D(x_3, y_3)$ be points on the unit circle such that the length of arc AD, written AD is s, $AB = t$, and $AC = s - t$. (See Figure 7.2.) Here we assume that $0 < t < s < 2\pi$ although the results we get are valid for any values of s and t. Arc BD also has a length $s - t$, so that $AC = BD$. Since these arcs are equal in length, line segments AC and BD must also be equal in length. By the distance formula,

$$\sqrt{(x_2 - 1)^2 + (y_2 - 0)^2} = \sqrt{(x_3 - x_1)^2 + (y_3 - y_1)^2}.$$

Squaring both sides and simplifying gives

$$x_2^2 - 2x_2 + 1 + y_2^2 = x_3^2 - 2x_3x_1 + x_1^2 + y_3^2 - 2y_3y_1 + y_1^2. \tag{1}$$

Since points B, C, and D are on the unit circle,

$$x_1^2 + y_1^2 = 1, \qquad x_2^2 + y_2^2 = 1, \qquad \text{and } x_3^2 + y_3^2 = 1.$$

Substituting these results into equation (1) gives

$$2 - 2x_2 = 2 - 2x_3x_1 - 2y_3y_1$$

or $\qquad x_2 = x_3x_1 + y_3y_1.$

Since $x_2 = \cos (s - t)$, $x_3 = \cos s$, $x_1 = \cos t$, $y_3 = \sin s$, and $y_1 = \sin t$, we end up with

Cosine of Difference of Two Angles

$$\cos(s - t) = \cos s \cos t + \sin s \sin t.$$

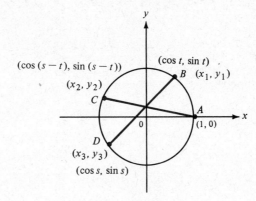

Figure 7.2

EXAMPLE 1   Find the value of cos 15°.

To find cos 15°, write 15° as the difference of two angles which have known function values. Since we know the trigonometric function values of both 45° and 30°, write 15° as 45° − 30°. Then use the identity for the cosine of the difference of two angles.

$$\cos 15° = \cos (45° - 30°)$$
$$= \cos 45° \cos 30° + \sin 45° \sin 30°$$
$$= \frac{\sqrt{2}}{2} \cdot \frac{\sqrt{3}}{2} + \frac{\sqrt{2}}{2} \cdot \frac{1}{2}$$
$$\cos 15° = \frac{\sqrt{6} + \sqrt{2}}{4}$$

We have found a formula for cos $(s - t)$. This formula can now be used to find cos $(s + t)$. To do so, use the result for cos $(s - t)$, and write $s + t$ as $s - (-t)$.

$$\cos (s + t) = \cos [s - (-t)]$$
$$= \cos s \cos (-t) + \sin s \sin (-t)$$

From earlier work, $\cos (-t) = \cos t$, and $\sin (-t) = -\sin t$. Making these substitutions gives

Cosine of Sum of Two Angles

$$\cos (s + t) = \cos s \cos t - \sin s \sin t.$$

EXAMPLE 2   Find $\cos \frac{5}{12} \pi$.

$$\cos \frac{5}{12}\pi = \cos\left(\frac{\pi}{6} + \frac{\pi}{4}\right)$$

$$= \cos\frac{\pi}{6}\cos\frac{\pi}{4} - \sin\frac{\pi}{6}\sin\frac{\pi}{4}$$

$$= \frac{\sqrt{3}}{2}\cdot\frac{\sqrt{2}}{2} - \frac{1}{2}\cdot\frac{\sqrt{2}}{2}$$

$$\cos\frac{5}{12}\pi = \frac{\sqrt{6} - \sqrt{2}}{4}\quad\square$$

EXAMPLE 3   ☐   Suppose $\sin x = 1/2$, $\cos y = -12/13$, and x and y are both in quadrant II. Find $\cos(x + y)$.

We know that $\cos(x + y) = \cos x \cos y - \sin x \sin y$. We are given the values of $\sin x$ and $\cos y$. To use the formula for $\cos(x + y)$, we need to find $\cos x$ and $\sin y$. To find $\cos x$, use the fact that $\sin^2 x + \cos^2 x = 1$, and then substitute $1/2$ for $\sin x$.

$$\sin^2 x + \cos^2 x = 1$$

$$\left(\frac{1}{2}\right)^2 + \cos^2 x = 1$$

$$\frac{1}{4} + \cos^2 x = 1$$

$$\cos^2 x = \frac{3}{4}$$

$$\cos x = \pm\frac{\sqrt{3}}{2}$$

Since x is in quadrant II, $\cos x$ is negative, so $\cos x = -\sqrt{3}/2$. Find $\sin y$ as follows.

$$\sin^2 y + \cos^2 y = 1$$

$$\sin^2 y + \left(-\frac{12}{13}\right)^2 = 1$$

$$\sin^2 y + \frac{144}{169} = 1$$

$$\sin^2 y = \frac{25}{169}$$

$$\sin y = \pm\frac{5}{13}$$

Since y is in quadrant II, $\sin y = 5/13$.

Now we find $\cos(x + y)$.

$$\cos(x + y) = \cos x \cos y - \sin x \sin y$$

$$= -\frac{\sqrt{3}}{2}\cdot\left(-\frac{12}{13}\right) - \frac{1}{2}\cdot\frac{5}{13}$$

$$= \frac{12\sqrt{3}}{26} - \frac{5}{26}$$

$$\cos(x + y) = \frac{12\sqrt{3} - 5}{26} \quad \square$$

The identities for the cosine of the sum and difference of two angles can be used to derive other identities. For example, if we substitute $\pi/2$ for s in the identity for $\cos(s - t)$, we have

$$\cos(\pi/2 - t) = \cos \pi/2 \cdot \cos t + \sin \pi/2 \cdot \sin t$$

$$= 0 \cdot \cos t + 1 \cdot \sin t$$

$$\cos(\pi/2 - t) = \sin t.$$

This result is true for any value of t since the identity for $\cos(s - t)$ is true for any value of s and t. The identity $\cos(\pi/2 - t) = \sin t$ is an example of a **cofunction identity.** The common cofunction identities are

Cofunction Identities

$$\cos(\pi/2 - t) = \sin t$$
$$\sin(\pi/2 - t) = \cos t$$
$$\tan(\pi/2 - t) = \cot t$$
$$\cot(\pi/2 - t) = \tan t.$$

The derivation of the cofunction identities is included in Exercises 55–58 below. Because of these identities, sine and cosine are called **cofunctions,** as are tangent and cotangent, and secant and cosecant.

EXAMPLE 4 $\square$ Find a number s which satisfies each of the following.

(a) $\cot s = \tan \pi/12$

Since tangent and cotangent are cofunctions, we have

$$\cot s = \tan(\pi/2 - s) = \tan \pi/12.$$

Thus, $\pi/2 - s = \pi/12$, and $s = 5\pi/12$.

(b) $\sin \theta = \cos(-30°)$

In a similar way,

$$\sin \theta = \cos(90° - \theta) = \cos(-30°).$$

$$90° - \theta = -30°$$

Thus, $\theta = 120°.$ $\square$

EXAMPLE 5 $\square$ Write $\cos(\pi - t)$ as a function of t.

Use the identity for $\cos(s - t)$. Replace s with π.

$$\cos(\pi - t) = \cos \pi \cdot \cos t + \sin \pi \cdot \sin t$$

$$= (-1) \cdot \cos t + (0) \cdot \sin t$$

$$= -\cos t \quad \square$$

EXERCISES
7.3

Write each of the following in terms of cofunctions.

1. $\tan 87°$ **2.** $\sin 15°$

3. $\cos \pi/12$ **4.** $\sin 2\pi/5$

5. $\csc (-14° \, 24')$ **6.** $\sin 142° \, 14'$

7. $\sin 5\pi/8$ **8.** $\cot 9\pi/10$

9. $\sec 146° \, 42'$ **10.** $\tan 174° \, 3'$

Identify each of the following as true or false.

11. $\cos 42° = \cos (30° + 12°)$

12. $\cos (-24°) = \cos 16° - \cos 40°$

13. $\cos 74° = \cos 60° \cos 14° + \sin 60° \sin 14°$

14. $\cos 140° = \cos 60° \cos 80° - \sin 60° \sin 80°$

15. $\cos (-10°) = \cos 90° \cos 80° +$
     $\sin 90° \sin 80°$

16. $\cos (10°) = \cos 90° \cos 80° + \sin 90° \sin 80°$

Use the cofunction identities to find the angle θ which makes each of the following true.

17. $\tan \theta = \cot (45° + 2\theta)$

18. $\sin \theta = \cos (2\theta - 10°)$

19. $\sec \theta = \csc (\theta/2 + 20°)$

20. $\cos \theta = \sin (3\theta + 10°)$

21. $\sin (3\theta - 15°) = \cos (\theta + 25°)$

22. $\cot (\theta - 10°) = \tan (2\theta + 20°)$

Use the sum and difference identities for cosine to find each of the following without using tables or calculator.

23. $\cos 285°$ **24.** $\cos (-15°)$

25. $\cos \left(-\dfrac{5\pi}{12}\right)$ **26.** $\cos \dfrac{7\pi}{12}$

27. $\cos 14° \cos 31° - \sin 14° \sin 31°$

28. $\cos 40° \cos 50° - \sin 40° \sin 50°$

29. $\cos 80° \cos 35° + \sin 80° \sin 35°$

30. $\cos (-10°) \cos 35° + \sin (-10°) \sin 35°$

Write as a function of x or θ.

31. $\cos (30° + \theta)$ **32.** $\cos (45° - \theta)$

33. $\cos (60° + \theta)$ **34.** $\cos (\theta - 30°)$

35. $\cos \left(\dfrac{3\pi}{2} - x\right)$ **36.** $\cos \left(x + \dfrac{\pi}{4}\right)$

For each of the following, find $\cos (s + t)$ and $\cos (s - t)$.

37. $\cos s = 3/5$ and $\sin t = 5/13$, s and t in quadrant I

38. $\cos s = -1/5$ and $\sin t = 3/5$, s and t in quadrant II

39. $\sin s = 2/3$ and $\sin t = -1/3$, s in quadrant II and t in IV

40. $\sin s = 3/5$ and $\sin t = -12/13$, s in quadrant I and t in III

41. $\cos s = -8/17$ and $\cos t = -3/5$, s and t in quadrant III

42. $\cos s = -15/17$ and $\sin t = 4/5$, s in quadrant II and t in I

43. $\sin s = -4/5$ and $\cos t = 12/13$, s in quadrant IV and t in I

44. $\sin s = -5/13$ and $\sin t = 3/5$, s in quadrant III and t in II

45. $\sin s = -8/17$ and $\cos t = -8/17$, s and t in quadrant III

46. $\sin s = 2/3$ and $\sin t = 2/5$, s and t in quadrant I

Verify each of the following identities.

47. $\cos (\pi/2 + x) = -\sin x$

48. $\sec (\pi - x) = -\sec x$

49. $\cos 2x = \cos^2 x - \sin^2 x$ (Hint: $\cos 2x = \cos (x + x)$)

50. $\cos (x + y) + \cos (x - y) = 2 \cos x \cos y$

51. $\dfrac{\cos (\alpha - \theta) - \cos (\alpha + \theta)}{\cos (\alpha - \theta) + \cos (\alpha + \theta)} = \tan \theta \tan \alpha$

52. $1 + \cos 2x - \cos^2 x = \cos^2 x$ (Hint: use the result in Exercise 49.)

53. $\cos (\pi + s - t) = -\sin s \sin t - \cos s \cos t$

54. $\cos (\pi/2 + s - t) = \sin (t - s)$

55. Use the identities for the cosine of the sum and difference of two angles to complete each of the following.

$\cos (0 - t) =$	$\cos (0 + t) =$
$\cos (\pi/2 - t) =$	$\cos (\pi/2 + t) =$
$\cos (\pi - t) =$	$\cos (\pi + t) =$
$\cos (3\pi/2 - t) =$	$\cos (3\pi/2 + t) =$

56. Use the identity $\cos (\pi/2 - t) = \sin t$; replace t with $\pi/2 - t$, and derive the identity $\cos t = \sin (\pi/2 - t)$.

57. Derive the identity $\tan t = \cot (\pi/2 - t)$.

58. Prove that $\csc t = \sec (\pi/2 - t)$.

Let $\sin s = -0.09463$ and $\cos t = 0.83499$, where s terminates in quadrant III and t in quadrant IV. Find each of the following.

59. $\cos (s - t)$ **60.** $\cos (s + t)$

61. $\cos 2s$ **62.** $\cos 2t$

63. Let $f(x) = \cos x$. Prove that $\dfrac{f(x + h) - f(x)}{h}$

$$= \cos x\left(\frac{\cos h - 1}{h}\right) - \sin x\left(\frac{\sin h}{h}\right).$$

Use the identities of this section to find each of the following.

64. $\cos (\tan^{-1} 5/12 - \cos^{-1} 4/5)$

65. $\cos (\sin^{-1} 8/17 + \tan^{-1} 3/4)$

66. $\cos (\sin^{-1} 1/3 - \cos^{-1} 2/5)$

7.4   Sum and Difference Identities for Sine and Tangent

Formulas for $\sin (s + t)$ and $\sin (s - t)$ can be developed from the results of the previous section. From a cofunction identity,

$$\sin (s + t) = \cos \left[\frac{\pi}{2} - (s + t)\right]$$

$$= \cos \left[\left(\frac{\pi}{2} - s\right) - t\right].$$

Using the identity for $\cos (s - t)$ from the previous section gives

$$\sin (s + t) = \cos \left(\frac{\pi}{2} - s\right) \cdot \cos t + \sin \left(\frac{\pi}{2} - s\right) \cdot \sin t$$

Sine of Sum of Two Angles

$$\sin (s + t) = \sin s \cos t + \cos s \sin t.$$

In the last step we substituted from the cofunction relationships.

If we now write $\sin (s - t)$ as $\sin [s + (-t)]$ and use the identity for $\sin (s + t)$ we have

Sine of Difference

$$\sin (s - t) = \sin s \cos t - \cos s \sin t.$$

Using the identities for sin $(s + t)$, cos $(s + t)$, sin $(s - t)$, and cos $(s - t)$, and the identity tan $x = \sin x/\cos x$, we can get the following identities.

Tangent of Sum and Difference of Two Angles

$$\tan (s + t) = \frac{\tan s + \tan t}{1 - \tan s \tan t}$$

$$\tan (s - t) = \frac{\tan s - \tan t}{1 + \tan s \tan t}$$

We show the proof for the first of these two identities. The proof of the other is very similar.

$$\tan (s + t) = \frac{\sin (s + t)}{\cos (s + t)}$$

$$= \frac{\sin s \cos t + \cos s \sin t}{\cos s \cos t - \sin s \sin t}$$

To express this result in terms of the tangent function, multiply both numerator and denominator by $1/(\cos s \cos t)$.

$$\tan (s + t) = \frac{\dfrac{\sin s \cos t + \cos s \sin t}{1}}{\dfrac{\cos s \cos t - \sin s \sin t}{1}} \cdot \frac{\dfrac{1}{\cos s \cos t}}{\dfrac{1}{\cos s \cos t}}$$

$$\tan (s + t) = \frac{\dfrac{\sin s \cos t}{\cos s \cos t} + \dfrac{\cos s \sin t}{\cos s \cos t}}{\dfrac{\cos s \cos t}{\cos s \cos t} - \dfrac{\sin s \sin t}{\cos s \cos t}}$$

Using the identity tan $x = \sin x/\cos x$ gives

$$\tan (s + t) = \frac{\tan s + \tan t}{1 - \tan s \tan t}.$$

Similar formulas can be found for the remaining trigonometric functions. However, they are seldom used, so we do not give them here.

EXAMPLE 1   Use identities to find the exact value of each of the following.

(a) sin 75°

$$\sin 75° = \sin (45° + 30°)$$

$$= \sin 45° \cos 30° + \cos 45° \sin 30°$$

$$= \frac{\sqrt{2}}{2} \cdot \frac{\sqrt{3}}{2} + \frac{\sqrt{2}}{2} \cdot \frac{1}{2}$$

$$= \frac{\sqrt{6} + \sqrt{2}}{4}$$

(b) $\tan \dfrac{7}{12}\pi = \tan \left(\dfrac{\pi}{3} + \dfrac{\pi}{4}\right)$

$$= \frac{\tan \dfrac{\pi}{3} + \tan \dfrac{\pi}{4}}{1 - \tan \dfrac{\pi}{3}\tan \dfrac{\pi}{4}}$$

$$= \frac{\sqrt{3} + 1}{1 - \sqrt{3}\cdot 1}$$

To simplify this result, rationalize the denominator by multiplying numerator and denominator by $1 + \sqrt{3}$.

$$= \frac{\sqrt{3} + 1}{1 - \sqrt{3}}\cdot\frac{1 + \sqrt{3}}{1 + \sqrt{3}}$$

$$= \frac{3 + 2\sqrt{3} + 1}{1 - 3}$$

$$\tan \frac{7}{12}\pi = -2 - \sqrt{3} \quad \blacksquare$$

EXAMPLE 2   ☐   If $\sin s = 4/5$ and $\cos t = -5/13$, where s is in quadrant II and t is in quadrant III, find each of the following.

(a) $\sin (s + t)$

Use the identity for the sine of the sum of two angles,

$$\sin (s + t) = \sin s \cos t + \cos s \sin t.$$

To use this identity we need to find $\cos s$ and $\sin t$. We can find $\cos s$ by using the identity $\sin^2 x + \cos^2 x = 1$.

$$\sin^2 s + \cos^2 s = 1$$

$$\frac{16}{25} + \cos^2 s = 1$$

$$\cos^2 s = \frac{9}{25}$$

$$\cos s = -\frac{3}{5} \qquad (s \text{ is in quadrant II})$$

In the same way, check that $\sin t = -12/13$. Now we have

$$\sin (s + t) = \frac{4}{5}\left(-\frac{5}{13}\right) + \left(-\frac{3}{5}\right)\left(-\frac{12}{13}\right)$$

$$= -\frac{20}{65} + \frac{36}{65}$$

$$\sin (s + t) = \frac{16}{65}.$$

(b) tan $(s + t)$

We know that

$$\tan (s + t) = \frac{\tan s + \tan t}{1 - \tan s \tan t}.$$

Here we have tan $s = -4/3$ and tan $t = 12/5$ (where did we get these values?) so that

$$\tan (s + t) = \frac{-\dfrac{4}{3} + \dfrac{12}{5}}{1 - \left(-\dfrac{4}{3}\right)\left(\dfrac{12}{5}\right)} = \frac{16}{63}. \quad \square$$

EXAMPLE 3 $\square$ Write each of the following as a function of θ.

(a) sin $(30° + \theta)$

Using the identity for sin $(s + t)$,

$$\sin (30° + \theta) = \sin 30° \cos \theta + \cos 30° \sin \theta$$

$$\sin (30° + \theta) = \frac{1}{2} \cos \theta + \frac{\sqrt{3}}{2} \sin \theta.$$

(b) $\tan (45° - \theta) = \dfrac{\tan 45° - \tan \theta}{1 + \tan 45° \tan \theta}$

$$\tan (45° - \theta) = \frac{1 - \tan \theta}{1 + \tan \theta} \quad \square$$

Reduction Identity

Expressions of the form $a \sin x + b \cos x$ occur so often in mathematics that it is useful to know how to rewrite these expressions in simpler form. In particular, these expressions must often be graphed, which can be done with the procedure developed here. Figure 7.3 shows a circle of radius r with the point (a, b) on the circle. The circle has equation $x^2 + y^2 = r^2$, so that

$$a^2 + b^2 = r^2, \quad \text{or} \quad r = \sqrt{a^2 + b^2}.$$

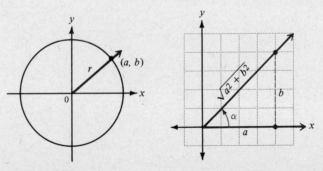

Figure 7.3   Figure 7.4

Let α be the angle shown in Figure 7.4. Then

$$\sin \alpha = \frac{b}{r} = \frac{b}{\sqrt{a^2 + b^2}} \quad \text{and} \quad \cos \alpha = \frac{a}{r} = \frac{a}{\sqrt{a^2 + b^2}}.$$

Now rewrite $a \sin x + b \cos x$:

$$a \sin x + b \cos x = \frac{a}{1} \cdot \frac{\sqrt{a^2 + b^2}}{\sqrt{a^2 + b^2}} \sin x + \frac{b}{1} \cdot \frac{\sqrt{a^2 + b^2}}{\sqrt{a^2 + b^2}} \cos x$$

$$= \sqrt{a^2 + b^2} \left(\frac{a}{\sqrt{a^2 + b^2}} \sin x + \frac{b}{\sqrt{a^2 + b^2}} \cos x \right)$$

$$= \sqrt{a^2 + b^2} \left(\sin x \frac{a}{\sqrt{a^2 + b^2}} + \cos x \frac{b}{\sqrt{a^2 + b^2}} \right).$$

Substitute $\sin \alpha$ and $\cos \alpha$ from above.

$$= \sqrt{a^2 + b^2} (\sin x \cos \alpha + \cos x \sin \alpha)$$

Using the identity for $\sin (s + t)$, we end up with

$$a \sin x + b \cos x = \sqrt{a^2 + b^2} \sin (x + \alpha).$$

In summary, we have the **reduction identity:**

Reduction Identity

$$a \sin x + b \cos x = \sqrt{a^2 + b^2} \sin (x + \alpha),$$

where $\quad \sin \alpha = \dfrac{b}{\sqrt{a^2 + b^2}} \quad$ and $\quad \cos \alpha = \dfrac{a}{\sqrt{a^2 + b^2}}.$

EXAMPLE 4

Simplify $\dfrac{1}{2} \sin \theta + \dfrac{\sqrt{3}}{2} \cos \theta$ using the reduction identity.

From the identity above, $a = \dfrac{1}{2}$ and $b = \dfrac{\sqrt{3}}{2}$, so that

$$a \sin \theta + b \cos \theta = \sqrt{a^2 + b^2} \sin (\theta + \alpha)$$

becomes

$$\frac{1}{2} \sin \theta + \frac{\sqrt{3}}{2} \cos \theta = 1 \cdot \sin (\theta + \alpha),$$

where angle α satisfies the conditions

$$\sin \alpha = \frac{b}{\sqrt{a^2 + b^2}} = \frac{\sqrt{3}}{2} \quad \text{and} \quad \cos \alpha = \frac{a}{\sqrt{a^2 + b^2}} = \frac{1}{2}.$$

The smallest possible positive value of α that satisfies both of these conditions is $\alpha = 60°$. Thus,

$$\frac{1}{2} \sin \theta + \frac{\sqrt{3}}{2} \cos \theta = \sin (\theta + 60°). \quad \square$$

The reduction identity of this section is useful when graphing functions which are sums of sine and cosine functions. It can be used instead of the method of addition of ordinates that we discussed in Chapter 6.

EXAMPLE 5 ☐ Graph $y = \sin x + \cos x$.

Reduce $\sin x + \cos x$ as follows. Since $a = b = 1$, we have $\sqrt{a^2 + b^2} = \sqrt{2}$, and

$$\sin x + \cos x = \sqrt{2} \sin (x + \alpha).$$

To find α, let

$$\sin \alpha = \frac{b}{\sqrt{a^2 + b^2}} = \frac{1}{\sqrt{2}} \quad \text{and} \quad \cos \alpha = \frac{a}{\sqrt{a^2 + b^2}} = \frac{1}{\sqrt{2}}.$$

The smallest positive angle satisfying these conditions is $\pi/4$, so that

$$y = \sin x + \cos x = \sqrt{2} \sin \left(x + \frac{\pi}{4} \right).$$

The graph of this function has an amplitude of $\sqrt{2}$, a period of 2π, and a phase shift of $\pi/4$ to the left, as shown in Figure 7.5.   ☐

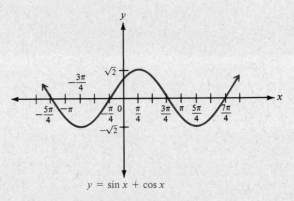

$$y = \sin x + \cos x$$

Figure 7.5

EXERCISES 7.4

Use the identities of this section to find the value of each of the following without using calculators or tables.

1. $\sin 15°$

2. $\sin 105°$

3. $\tan 15°$

4. $\tan (-105°)$

5. $\sin (-105°)$

6. $\tan \dfrac{5\pi}{12}$

7. $\sin \dfrac{5\pi}{12}$

8. $\sin 285°$

9. $\sin 76° \cos 31° - \cos 76° \sin 31°$

10. $\sin 40° \cos 50° + \cos 40° \sin 50°$

11. $\dfrac{\tan 80° + \tan 55°}{1 - \tan 80° \tan 55°}$

12. $\dfrac{\tan 80° - \tan (-55°)}{1 + \tan 80° \tan (-55°)}$

13. $\dfrac{\tan 100° + \tan 80°}{1 - \tan 100° \tan 80°}$

14. $\sin 100° \cos 10° - \cos 100° \sin 10°$

15. $\sin 80° \cos (-55°) - \cos 80° \sin (-55°)$

16. $\dfrac{\tan 40° + \tan 5°}{1 - \tan 40° \tan 5°}$

Write each of the following as a function of θ.

17. $\sin (45° + \theta)$ 18. $\sin (\theta - 30°)$
19. $\tan (\theta + 30°)$ 20. $\tan (60° - \theta)$
21. $\sin (180° - \theta)$ 22. $\sin (270° - \theta)$
23. $\tan (180° + \theta)$ 24. $\tan (0° - \theta)$
25. $\sin (180° + \theta)$ 26. $\tan (180° - \theta)$

For each of the following, find $\sin (s + t)$, $\sin (s - t)$, $\tan (s + t)$, *and* $\tan (s - t)$.

27. $\cos s = 3/5$ and $\sin t = 5/13$, s and t in quadrant I

28. $\cos s = -1/5$ and $\sin t = 3/5$, s and t in quadrant II

29. $\sin s = 2/3$ and $\sin t = -1/3$, s in quadrant II and t in quadrant IV

30. $\sin s = 3/5$ and $\sin t = -12/13$, s in quadrant I and t in quadrant III

31. $\cos s = -8/17$ and $\cos t = -3/5$, s and t in quadrant III

32. $\cos s = -15/17$ and $\sin t = 4/5$, s in quadrant II and t in quadrant I

33. $\sin s = -4/5$ and $\cos t = 12/13$, s in quadrant III and t in quadrant IV

34. $\sin s = -5/13$ and $\sin t = 3/5$, s in quadrant III and t in quadrant II

35. $\sin s = -8/17$ and $\cos t = -8/17$, s and t in quadrant III

36. $\sin s = 2/3$ and $\sin t = 2/5$, s and t in quadrant I

Verify that each of the following are identities.

37. $\sin \left(\dfrac{\pi}{2} + x \right) = \cos x$

38. $\sin \left(\dfrac{3\pi}{2} + x \right) = -\cos x$

39. $\tan \left(\dfrac{\pi}{2} + x \right) = -\cot x$

40. $\tan \left(\dfrac{\pi}{4} + x \right) = \dfrac{1 + \tan x}{1 - \tan x}$

41. $\sin 2x = 2 \sin x \cos x$
    (Hint: $\sin 2x = \sin (x + x)$)

42. $\sin (x + y) + \sin (x - y) = 2 \sin x \cos y$

43. $\sin (x + y) - \sin (x - y) = 2 \cos x \sin y$

44. $\tan (x - y) - \tan (y - x) =$
    $\dfrac{2 (\tan x - \tan y)}{1 + \tan x \tan y}$

45. $\sin (30° + \alpha) + \cos (60° + \alpha) = \cos \alpha$

46. $\sin (210° + x) - \cos (120° + x) = 0$

47. $\dfrac{\cos (\alpha - \beta)}{\cos \alpha \sin \beta} = \tan \alpha + \cot \beta$

48. $\dfrac{\sin (s + t)}{\cos s \cos t} = \tan s + \tan t$

A painting 1 meter high and 3 meters from the floor will cut off an angle θ *to an observer, where* $\theta = \tan^{-1} [x/(x^2 + 2)]$. *Assume the observer is x meters from the wall displaying the painting and that the eyes of the observer are 2 meters above the ground. (See the figure.) Find the value of* θ *for each of the following values of x. Round to the nearest degree.*

49. 1   50. 2

51. 3

52. Derive the formula for θ. Use the identity for $\tan (\theta + \alpha)$.

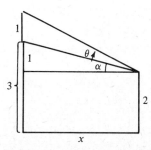

352   Trigonometric Identities and Equations

Use the reduction identity to simplify each of the following for angles between 0° and 360°. Use a calculator or Table 5 to find angles to the nearest degree. Choose the smallest possible positive value of α.

53. $-\sin x + \cos x$ **54.** $\sqrt{3}\sin x - \cos x$

55. $5\sin\theta - 12\cos\theta$ **56.** $12\sin A + 5\cos A$

57. $-15\sin x + 8\cos x$

58. $15\sin B - 8\cos B$

59. $-7\sin\theta - 24\cos\theta$

60. $24\cos t - 7\sin t$

61. $3\sin x + 4\cos x$

62. $-4\sin x + 3\cos x$

Graph each of the following by first changing to a function of the form $y = a\sin(x + \alpha)$.

63. $y = \sqrt{3}\sin x + \cos x$

64. $y = \sin x - \sqrt{3}\cos x$

65. $y = -\sin x + \cos x$

66. $y = -\sin x - \cos x$

67. Let $f(x) = \sin x$. Show that $\dfrac{f(x+h) - f(x)}{h} =$
$\sin x\left(\dfrac{\cos h - 1}{h}\right) + \cos x\left(\dfrac{\sin h}{h}\right).$

Let $\sin s = 0.599832$, *where s terminates in quadrant II. Let* $\sin t = -0.845992$, *where t terminates in quadrant III. Find each of the following.*

68. $\sin(s + t)$ **69.** $\sin(s - t)$

70. $\tan(s + t)$ **71.** $\sin 2s$

Use the identities of this section to find each of the following.

72. $\sin(\cos^{-1} 5/13 + \tan^{-1} 3/4)$

73. $\sin(\sin^{-1} 2/3 + \tan^{-1} 1/4)$

74. $\sin(\cos^{-1} 1/2 - \tan^{-1} -3)$

75. $\tan(\tan^{-1} 4/5 - \sin^{-1} 5/13)$

76. Use the results of Exercise 23 above to help show that the period of tangent is π.

7.5 Multiple-Angle Identities

Some special cases of the identities for the sum of two angles are used often enough to be expressed as separate identities. These are the identities that result from the addition identities when $s = t$, so that $s + t = 2s$. These identities are called **double-angle identities.** In the identity for $\cos(s + t)$, we let $t = s$ to derive an expression for $\cos 2s$.

$$\cos 2s = \cos(s + s)$$
$$= \cos s \cos s - \sin s \sin s$$

Cosine of Double Angle
$$\cos 2s = \cos^2 s - \sin^2 s$$

By substitution from either $\cos^2 s = 1 - \sin^2 s$ or $\sin^2 s = 1 - \cos^2 s$, we get

Cosine of Double Angle
$$\cos 2s = 1 - 2\sin^2 s$$
$$\cos 2s = 2\cos^2 s - 1.$$

We can do the same thing with $\sin(s + t)$.

$$\sin 2s = \sin(s + s)$$
$$= \sin s \cos s + \cos s \sin s$$

Sine of Double Angle

$$\sin 2s = 2 \sin s \cos s$$

We can also find $\tan 2s$ from the identity for $\tan(s + t)$.

$$\tan 2s = \tan(s + s)$$
$$= \frac{\tan s + \tan s}{1 - \tan s \tan s}$$

Tangent of Double Angle

$$\tan 2s = \frac{2 \tan s}{1 - \tan^2 s}$$

These identities, together with the addition identities, allow us to rewrite trigonometric functions of multiple values of s in terms of s.

EXAMPLE 1　Write $\sin 3s$ in terms of $\sin s$.

$$\sin 3s = \sin(2s + s)$$
$$= \sin 2s \cos s + \cos 2s \sin s$$
$$= (2 \sin s \cos s) \cos s + (\cos^2 s - \sin^2 s) \sin s$$
$$= 2 \sin s \cos^2 s + \cos^2 s \sin s - \sin^3 s$$
$$= 2 \sin s(1 - \sin^2 s) + (1 - \sin^2 s) \sin s - \sin^3 s$$
$$= 2 \sin s - 2 \sin^3 s + \sin s - \sin^3 s - \sin^3 s$$
$$\sin 3s = 3 \sin s - 4 \sin^3 s \quad \blacksquare$$

The double-angle identities for $2s$ can be used to find values of the trigonometric functions of s as shown in the following example.

EXAMPLE 2　Find the values of the six trigonometric functions of θ if $\cos 2\theta = 4/5$ and θ terminates in quadrant II.

Use one of the double-angle identities for cosine.

$$\cos 2\theta = 1 - 2 \sin^2 \theta$$
$$\frac{4}{5} = 1 - 2 \sin^2 \theta$$
$$-\frac{1}{5} = -2 \sin^2 \theta$$

$$\frac{1}{10} = \sin^2 \theta$$

$$\sin \theta = \frac{\sqrt{10}}{10}$$

We use the positive square root since θ terminates in quadrant II. Values of $\cos \theta$ and $\tan \theta$ can now be found using the fundamental identities.

$$\sin^2 \theta + \cos^2 \theta = 1$$

$$\frac{1}{10} + \cos^2 \theta = 1$$

$$\cos^2 \theta = \frac{9}{10}$$

$$\cos \theta = \frac{-3}{\sqrt{10}}$$

$$\cos \theta = \frac{-3\sqrt{10}}{10}$$

Verify that $\tan \theta = \sin \theta / \cos \theta = -1/3$. Find the other three functions using reciprocals.

$$\csc \theta = \frac{1}{\sin \theta} = \sqrt{10} \qquad \sec \theta = \frac{1}{\cos \theta} = \frac{-\sqrt{10}}{3}$$

$$\cot \theta = \frac{1}{\tan \theta} = -3 \quad \blacksquare$$

EXAMPLE 3 $\square$ Given $\cos \theta = 3/5$, where $3\pi/2 < \theta < 2\pi$, find $\cos 2\theta$, $\sin 2\theta$, and $\tan 2\theta$.

Since $\cos \theta = 3/5$, we can show that $\sin \theta = \pm 4/5$ by using the identity $\sin^2 \theta + \cos^2 \theta = 1$. Since θ terminates in quadrant IV, we choose $\sin \theta = -4/5$. Then, using the double-angle identities, we have

$$\sin 2\theta = 2 \sin \theta \cos \theta = 2 \left(-\frac{4}{5}\right)\left(\frac{3}{5}\right) = -\frac{24}{25}$$

$$\cos 2\theta = \cos^2 \theta - \sin^2 \theta = \frac{9}{25} - \frac{16}{25} = -\frac{7}{25}$$

$$\tan 2\theta = \frac{\sin 2\theta}{\cos 2\theta} = \frac{-24/25}{-7/25} = \frac{24}{7}.$$

We can also find $\tan 2\theta$ by noting that $\tan \theta = -4/3$, with

$$\tan 2\theta = \frac{2 \tan \theta}{1 - \tan^2 \theta} = \frac{2\left(-\frac{4}{3}\right)}{1 - \frac{16}{9}} = \frac{-\frac{8}{3}}{-\frac{7}{9}} = \frac{24}{7}. \quad \blacksquare$$

From the alternate forms of the double-angle identity for cosine, we can derive three additional identities. These **half-angle identities,** listed below, are used in the study of calculus.

Half-angle Identities

$$\cos \frac{s}{2} = \pm \sqrt{\frac{1 + \cos s}{2}} \qquad \sin \frac{s}{2} = \pm \sqrt{\frac{1 - \cos s}{2}}$$

$$\tan \frac{s}{2} = \pm \sqrt{\frac{1 - \cos s}{1 + \cos s}} \qquad \tan \frac{s}{2} = \frac{\sin s}{1 + \cos s}$$

$$\text{and} \qquad \tan \frac{s}{2} = \frac{1 - \cos s}{\sin s}$$

In these identities, the plus or minus sign is selected according to the quadrant in which $s/2$ terminates. For example, if s represents an angle of 324°, then $s/2 = 162°$, which lies in quadrant II. Thus, $\cos s/2$ and $\tan s/2$ would be negative, while $\sin s/2$ would be positive.

To derive the identity for $\sin s/2$, start with the identity

$$\cos 2x = 1 - 2 \sin^2 x.$$

Now solve for $\sin x$.

$$2 \sin^2 x = 1 - \cos 2x$$

$$\sin x = \pm \sqrt{\frac{1 - \cos 2x}{2}}$$

Let $2x = s$, so that $x = s/2$, and substitute into this last expression.

$$\sin \frac{s}{2} = \pm \sqrt{\frac{1 - \cos s}{2}}$$

The identity for $\cos s/2$ from the box above is derived in a very similar way, by starting with the double angle identity $\cos 2x = 2 \cos^2 x - 1$. The first identity from the box for $\tan s/2$ comes from the half-angle identities for sine and cosine.

$$\tan \frac{s}{2} = \frac{\pm \sqrt{\dfrac{1 - \cos s}{2}}}{\pm \sqrt{\dfrac{1 + \cos s}{2}}} = \pm \sqrt{\frac{1 - \cos s}{1 + \cos s}}$$

The other two identities for $\tan s/2$ that are given in the box above are proven in Exercises 46 and 47 below.

EXAMPLE 4 Find $\cos 112.5°$.

Since $112.5° = 225°/2$, we use the identity for $\cos s/2$. Also, $112.5°$ is in quadrant II, where cosine is negative, so that the minus sign must be used on the radical; we therefore have

$$\cos 112.5° = \cos \frac{225°}{2} = -\sqrt{\frac{1 + \cos 225°}{2}}$$

$$= -\sqrt{\frac{1 - \dfrac{\sqrt{2}}{2}}{2}}$$

$$= -\sqrt{\frac{2 - \sqrt{2}}{4}}$$

$$= -\frac{\sqrt{2 - \sqrt{2}}}{2} \quad \square$$

EXAMPLE 5 $\quad\square\quad$ Find tan 22.5°.

Use the identity $\tan \dfrac{s}{2} = \dfrac{\sin s}{1 + \cos s}$, with $s = 45°$.

$$\tan 22.5° = \tan \frac{45°}{2} = \frac{\sin 45°}{1 + \cos 45°}$$

$$= \frac{\dfrac{\sqrt{2}}{2}}{1 + \dfrac{\sqrt{2}}{2}}$$

$$= \frac{\sqrt{2}}{2 + \sqrt{2}}$$

$$= \sqrt{2} - 1$$

(Here we rationalized the denominator.) See also Exercises 81–82 below. $\quad\square$

EXAMPLE 6 $\quad\square\quad$ Given $\cos s = 2/3$, with $3\pi/2 < s < 2\pi$, find $\cos s/2$, $\sin s/2$ and $\tan s/2$.

Since $3\pi/2 < s < 2\pi$, we have $3\pi/4 < s/2 < \pi$. Thus, $s/2$ terminates in quadrant II and $\cos s/2$ and $\tan s/2$ are negative, while $\sin s/2$ is positive. Using the half-angle identities gives

$$\sin \frac{s}{2} = \sqrt{\frac{1 - \dfrac{2}{3}}{2}} = \sqrt{\frac{1}{6}} = \frac{\sqrt{6}}{6}$$

$$\cos \frac{s}{2} = -\sqrt{\frac{1 + \dfrac{2}{3}}{2}} = -\sqrt{\frac{5}{6}} = -\frac{\sqrt{30}}{6}$$

$$\tan \frac{s}{2} = \frac{\frac{\sqrt{6}}{6}}{-\frac{\sqrt{30}}{6}} = \frac{-\sqrt{5}}{5}. \quad \square$$

EXERCISES
7.5

Use the identities of this section to complete the following.

1. $2 \sin \dfrac{\pi}{5} \cos \dfrac{\pi}{5} = \sin$ _____

2. $\cos^2 10° - \sin^2 10° = \cos$ _____

3. $4 \cos^2 x - 2 = 2(\underline{\quad})$

4. $\dfrac{2 \tan \dfrac{\pi}{3}}{1 - \tan^2 \dfrac{\pi}{3}} = \tan$ _____

5. $\sin 320° = 2 \sin$ ____$\cos$ ____

6. $\tan 8k = \dfrac{2 \tan \underline{\quad}}{1 - \underline{\quad\quad}}$

7. $\cos 6x =$ ____$3x -$ ____$3x$

8. $\tan 10x = \dfrac{\underline{\quad}5x}{1 - \underline{\quad\quad}}$

9. $\sin$ ____ $= \sqrt{\dfrac{1 - \cos 18°}{2}}$

10. $\cos$ ____ $= \sqrt{\dfrac{\underline{\quad}\cos 44°}{2}}$

11. $\underline{\quad\quad} = -\sqrt{\dfrac{1 - \cos 340°}{1 + \cos 340°}}$

12. $\underline{\quad\quad} = \sqrt{\dfrac{1 + \cos 40°}{2}}$

Determine whether the positive or negative square root should be selected.

13. $\sin 195° = \pm \sqrt{\dfrac{1 - \cos 390°}{2}}$

14. $\cos 58° = \pm \sqrt{\dfrac{1 + \cos 116°}{2}}$

15. $\tan 225° = \pm \sqrt{\dfrac{1 - \cos 450°}{1 + \cos 450°}}$

16. $\sin (-10°) = \pm \sqrt{\dfrac{1 - \cos (-20°)}{2}}$

Find each of the following.

17. $\cos x$, if $\cos 2x = -5/12$ and $\pi/2 < x < \pi$

18. $\sin x$, if $\cos 2x = 2/3$ and $\pi < x < 3\pi/2$

19. $\cos \alpha/2$, if $\cos \alpha = -1/4$ and $\pi < \alpha < 3\pi/2$

20. $\sin \beta/2$, if $\cos \beta = 3/4$ and $3\pi/2 < \beta < 2\pi$

Use the identities of this section to find the sine, cosine, and tangent for each of the following.

21. $\theta = 22.5°$ **22.** $\theta = 15°$

23. $\theta = 195°$ **24.** $x = -\pi/8$

25. $x = 5\pi/2$ **26.** $x = 3\pi/2$

Use the identities of this section to find values of the six trigonometric functions for each of the following.

27. x, given $\cos 2x = -5/12$ and $\pi/2 < x < \pi$

28. t, given $\cos 2t = 2/3$ and $\pi/2 < t < \pi$

29. 2θ, given $\sin \theta = 2/5$ and $\cos \theta < 0$

30. 2β, given $\cos \beta = -12/13$ and $\sin \beta > 0$

31. $2x$, given $\tan x = 2$ and $\cos x > 0$

32. $2x$, given $\tan x = 5/3$ and $\sin x < 0$

33. $\alpha/2$, given $\cos \alpha = 1/3$ and $\sin \alpha < 0$

34. $\theta/2$, given $\cos \theta = -2/3$ and $\sin \theta > 0$

Verify that each of the following equations is an identity.

35. $(\sin \gamma + \cos \gamma)^2 = \sin 2\gamma + 1$

36. $\cos 2s = \cos^4 s - \sin^4 s$

37. $\sec 2x = \dfrac{\sec^2 x + \sec^4 x}{2 + \sec^2 x - \sec^4 x}$

38. $\cot s + \tan s = 2 \csc 2s$

39. $\sin 4\alpha = 4 \sin \alpha \cos \alpha \cos 2\alpha$

40. $\dfrac{1 + \cos 2x}{\sin 2x} = \cot x$

41. $\sec^2 \dfrac{x}{2} = \dfrac{2}{1 + \cos x}$

42. $\cot^2 \dfrac{x}{2} = \dfrac{(1 + \cos x)^2}{\sin^2 x}$

43. $\sin^2 \dfrac{x}{2} = \dfrac{\tan x - \sin x}{2 \tan x}$

44. $\dfrac{\sin 2x}{2 \sin x} = \cos^2 \dfrac{x}{2} - \sin^2 \dfrac{x}{2}$

45. $\dfrac{2}{1 + \cos x} - \tan^2 \dfrac{x}{2} = 1$

46. $\tan \dfrac{s}{2} = \dfrac{\sin s}{1 + \cos s}$

47. $\tan \dfrac{s}{2} = \dfrac{1 - \cos s}{\sin s}$

48. $\tan \dfrac{\gamma}{2} = \csc \gamma - \cot \gamma$

49. $\tan 8k - \tan 8k \cdot \tan^2 4k = 2 \tan 4k$

50. $\sin 2\gamma = \dfrac{2 \tan \gamma}{1 + \tan^2 \gamma}$

51. $-\tan 2\theta = \dfrac{2 \tan \theta}{\sec^2 \theta - 2}$

52. $\cos 2y = \dfrac{2 - \sec^2 y}{\sec^2 y}$

53. $\dfrac{2 \cos 2\alpha}{\sin 2\alpha} = \cot \alpha - \tan \alpha$

54. $\sin 2\alpha \cos 2\alpha = \sin 2\alpha - 4 \sin^3 \alpha \cos \alpha$

55. $\dfrac{\tan \dfrac{x}{2} + \cot \dfrac{x}{2}}{\cot \dfrac{x}{2} - \tan \dfrac{x}{2}} = \sec x$

56. $1 - \tan^2 \dfrac{\theta}{2} = \dfrac{2 \cos \theta}{1 + \cos \theta}$

57. $\cos x = \dfrac{1 - \tan^2 \dfrac{x}{2}}{1 + \tan^2 \dfrac{x}{2}}$

58. $\dfrac{\sin 2\alpha - 2 \sin \alpha}{2 \sin \alpha + \sin 2\alpha} = -\tan^2 \dfrac{\alpha}{2}$

Express each of the following as trigonometric functions of x. See Example 1.

59. $\tan^2 2x$ **60.** $\cos^2 2x$

61. $\cos 3x$ **62.** $\sin 4x$

63. $\cos 4x$ **64.** $\tan 4x$

An airplane flying faster than sound sends out sound waves that form a cone, as shown in the figure. The cone intersects the ground to form a hyperbola. As this hyperbola passes over a particular point on the ground, a sonic boom is heard at that point.

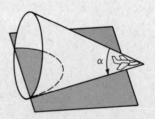

If α is the angle at the vertex of the cone, then

$$\sin \frac{\alpha}{2} = \frac{1}{m}$$

where m is the mach number of the plane. (We assume m > 1.) The mach number is the ratio of the speed of the plane and the speed of sound. Thus, a speed of mach 1.4 means that the plane is flying 1.4 times the speed of sound. Find α or m, as necessary, for each of the following.

65. $m = 3/2$ **66.** $m = 5/4$

67. $m = 2$ **68.** $m = 5/2$

69. $\alpha = 30°$ **70.** $\alpha = 60°$

Let sin s = −0.481143, with 7π/4 < s < 2π. Find each of the following. 🖂

71. $\sin 2s$ **72.** $\sin \frac{1}{2}s$

73. $\cos 2s$ **74.** $\cos \frac{1}{2}s$

75. $\tan 2s$ **76.** $\tan \frac{1}{2}s$

Let cos s = −0.592147, with π < s < 3π/2. Use the identities of this section to find each of the following. 🖂

77. $\sin \frac{1}{2}s$ **78.** $\cos \frac{1}{2}s$

79. $\tan \frac{1}{2}s$ **80.** $\csc \frac{1}{2}s$

In Example 5 we used the identity

$$\tan \frac{s}{2} = \frac{\sin s}{1 + \cos s}$$

to find that $\tan 22.5° = \sqrt{2} - 1$.

81. Find $\tan 22.5°$ with the identity $\tan s/2 = \pm \sqrt{(1 - \cos s)/(1 + \cos s)}$.

82. Show that both answers are the same.

Simplify each of the following.

83. $\sin^2 2x + \cos^2 2x$ **84.** $\sin^2 \frac{x}{2} + \cos^2 \frac{x}{2}$

85. $1 + \tan^2 4x$ **86.** $\cot^2 \frac{x}{3} + 1$

7.6 Sum and Product Identities

One group of identities in this section can be used to rewrite a product of two functions as a sum or difference. The other group can be used to rewrite a sum or difference of two functions as a product. Some of these identities can also be used to rewrite an expression involving both sine and cosine functions as one with only one of these functions. It the next section on conditional equations, the need for this kind of change will become clear. These identities are also useful in graphing and in calculus.

The identities that we discuss in this section all result from the sum and difference identities for sine and cosine. Adding the two addition identities for $\sin (s + t)$ and $\sin (s - t)$ gives

$$\begin{aligned}
\sin (s + t) &= \sin s \cos t + \cos s \sin t \\
\sin (s - t) &= \sin s \cos t - \cos s \sin t \\
\hline
\sin (s + t) + \sin (s - t) &= 2 \sin s \cos t,
\end{aligned}$$

or

$$\sin s \cos t = \frac{1}{2} [\sin (s + t) + \sin (s - t)].$$

If we subtract sin $(s - t)$ from sin $(s + t)$, the result is

$$\cos s \sin t = \frac{1}{2} [\sin (s + t) - \sin (s - t)].$$

If we use the identities for cos $(s + t)$ and cos $(s - t)$ in a similar manner we get

$$\cos s \cos t = \frac{1}{2} [\cos (s + t) + \cos (s - t)]$$

$$\sin s \sin t = \frac{1}{2} [\cos (s - t) - \cos (s + t)].$$

EXAMPLE 1   Rewrite $\cos 2\theta \sin \theta$ as the sum or difference of two functions.
Using the identity for cos s sin t gives

$$\cos 2\theta \sin \theta = \frac{1}{2} (\sin 3\theta - \sin \theta)$$

$$\cos 2\theta \sin \theta = \frac{1}{2} \sin 3\theta - \frac{1}{2} \sin \theta. \quad \square$$

EXAMPLE 2   Evaluate $\cos 15° \cos 45°$.
Use the identity for cos s cos t.

$$\cos 15° \cos 45° = \frac{1}{2} [\cos (15° + 45°) + \cos (15° - 45°)]$$

$$= \frac{1}{2} [\cos 60° + \cos (-30°)]$$

$$= \frac{1}{2} (\cos 60° + \cos 30°)$$

$$= \frac{1}{2} \left(\frac{1}{2} + \frac{\sqrt{3}}{2} \right)$$

$$\cos 15° \cos 45° = \frac{1 + \sqrt{3}}{4} \quad \square$$

Now we can use the identities above to obtain further identities that are used in calculus to rewrite a sum of trigonometric functions as a product. To begin, let $s + t = x$, and let $s - t = y$. Then we have

$$s = \frac{x + y}{2} \quad \text{and} \quad t = \frac{x - y}{2}.$$

(To get these results, add the two equations, and then subtract them.) With these results, the identity

$$\sin s \cos t = \frac{1}{2} [\sin (s + t) + \sin (s - t)]$$

becomes

$$\sin \left(\frac{x + y}{2}\right) \cos \left(\frac{x - y}{2}\right) = \frac{1}{2} (\sin x + \sin y),$$

or

$$\sin x + \sin y = 2 \sin \left(\frac{x + y}{2}\right) \cos \left(\frac{x - y}{2}\right).$$

Three other identities can be obtained in a similar way.

$$\sin x - \sin y = 2 \cos \left(\frac{x + y}{2}\right) \sin \left(\frac{x - y}{2}\right)$$

$$\cos x + \cos y = 2 \cos \left(\frac{x + y}{2}\right) \cos \left(\frac{x - y}{2}\right)$$

$$\cos x - \cos y = -2 \sin \left(\frac{x + y}{2}\right) \sin \left(\frac{x - y}{2}\right)$$

EXAMPLE 3   Write $\sin 2\gamma - \sin 4\gamma$ as a product of two functions. Use the identity for $\sin x - \sin y$.

$$\sin 2\gamma - \sin 4\gamma = 2 \cos \left(\frac{2\gamma + 4\gamma}{2}\right) \sin \left(\frac{2\gamma - 4\gamma}{2}\right)$$

$$= 2 \cos \frac{6\gamma}{2} \sin \frac{-2\gamma}{2}$$

$$= 2 \cos 3\gamma \sin (-\gamma)$$

$$\sin 2\gamma - \sin 4\gamma = -2 \cos 3\gamma \sin \gamma \quad \blacksquare$$

EXAMPLE 4   Verify that $\dfrac{\sin 3s + \sin s}{\cos s + \cos 3s} = \tan 2s$ is an identity.

Work as follows.

$$\frac{\sin 3s + \sin s}{\cos s + \cos 3s} = \frac{2 \sin \left(\dfrac{3s + s}{2}\right) \cos \left(\dfrac{3s - s}{2}\right)}{2 \cos \left(\dfrac{s + 3s}{2}\right) \cos \left(\dfrac{s - 3s}{2}\right)}$$

$$= \frac{\sin 2s \cos s}{\cos 2s \cos (-s)} = \frac{\sin 2s}{\cos 2s} = \tan 2s \quad \square$$

EXERCISES
7.6

Rewrite each of the following as a sum or difference of trigonometric functions.

1. $\cos 35° \sin 25°$

2. $2 \sin 2x \sin 4x$

3. $3 \cos 5x \cos 3x$

4. $2 \sin 74° \cos 114°$

5. $\sin (-\theta) \sin (-3\theta)$

6. $4 \cos (-32°) \sin 15°$

7. $-8 \cos 4y \cos 5y$

8. $2 \sin 3k \sin 14k$

Rewrite each of the following as a product of trigonometric functions.

9. $\sin 60° - \sin 30°$

10. $\sin 28° + \sin (-18°)$

11. $\cos 42° + \cos 148°$ **12.** $\cos 2x - \cos 8x$

13. $\sin 12\beta - \sin 3\beta$ **14.** $\cos 5x + \cos 10x$

15. $-3 \sin 2x + 3 \sin 5x$

16. $-\cos 8s + \cos 14s$

Verify that each of the following is an identity.

17. $\tan x = \dfrac{\sin 3x - \sin x}{\cos 3x + \cos x}$

18. $\dfrac{\sin 5t + \sin 3t}{\cos 3t - \cos 5t} = \cot t$

19. $\dfrac{\cot 2\theta}{\tan 3\theta} = \dfrac{\cos 5\theta + \cos \theta}{\cos \theta - \cos 5\theta}$

20. $\dfrac{\cos \alpha + \cos \beta}{\cos \alpha - \cos \beta} = -\cot \left(\dfrac{\alpha + \beta}{2}\right) \cot \left(\dfrac{\alpha - \beta}{2}\right)$

21. $\dfrac{1}{\tan 2s} = \dfrac{\sin 3s - \sin s}{\cos s - \cos 3s}$

22. $\dfrac{\sin^2 5\alpha - 2 \sin 5\alpha \sin 3\alpha + \sin^2 3\alpha}{\sin^2 5\alpha - \sin^2 3\alpha} = \dfrac{\tan \alpha}{\tan 4\alpha}$

23. $\sin 6\theta \cos 4\theta - \sin 3\theta \cos 7\theta = \sin 3\theta \cos \theta$

24. $\sin 8\beta \sin 4\beta + \cos 10\beta \cos 2\beta = \cos 6\beta \cos 2\beta$

25. $\sin^2 u - \sin^2 v = \sin (u + v) \sin (u - v)$

26. $\cos^2 u - \cos^2 v = -\sin (u + v) \sin (u - v)$

27. Show that the double-angle identity for sine can be considered a special case of the identity $\sin s \cos t = (1/2) [\sin (s + t) + \sin (s - t)]$.

28. Show that the double-angle identity $\cos 2s = 2 \cos^2 s - 1$ is a special case of the identity $\cos s \cos t = \dfrac{1}{2} [\cos (s + t) + \cos (s - t)]$.

7.7   Trigonometric Equations

So far in this chapter we have discussed trigonometric identities, statements which are true for every value in the domain of the variable. Here we discuss conditional equations which involve trigonometric functions. As we have said, a **conditional equation**

is an equation in which some replacements for the variable make the statement true, while others make it false. For example, $2x + 3 = 5$, $x^2 - 5x = 10$, and $2^x = 8$ are conditional equations. Conditional equations with trigonometric functions can usually be solved by using algebraic methods and trigonometric identities to simplify the equations. The next examples show methods for solving trigonometric equations.

EXAMPLE 1   ☐   Solve $3 \sin x = \sqrt{3} + \sin x$.

First, solve for $\sin x$. Collect all terms with $\sin x$ on one side of the equation.

$$3 \sin x = \sqrt{3} + \sin x$$
$$2 \sin x = \sqrt{3}$$
$$\sin x = \frac{\sqrt{3}}{2}$$

From earlier work, we know that $\sin 60° = \sqrt{3}/2$. However, $\sin 120° = \sqrt{3}/2$ also, as does $\sin 420°$, and so on. There are an infinite number of values of x which satisfy the equation $\sin x = \sqrt{3}/2$. We can express the infinite number of solutions by writing the solution set as

$$\{x \mid x = 60° + 360° \cdot n \text{ or } x = 120° + 360° \cdot n, \ n \text{ any integer}\}. \quad ☐$$

Usually, we need the solutions of a trigonometric equation only for some specified interval, as shown in the next examples.

EXAMPLE 2   ☐   Find all solutions of $\tan^2 x + \tan x - 2 = 0$ in the interval $[0, 2\pi)$.

If we let $y = \tan x$, the equation becomes $y^2 + y - 2 = 0$. Factoring gives $(y - 1)(y + 2) = 0$. Substituting $\tan x$ back for y, we have

$$(\tan x - 1)(\tan x + 2) = 0$$
$$\tan x - 1 = 0 \qquad \tan x + 2 = 0$$
$$\tan x = 1 \qquad\qquad \tan x = -2.$$

If $\tan x = 1$, then for $[0, 2\pi)$ we get $x = \pi/4$ or $x = 5\pi/4$. If $x = -2$, then using a calculator or Table 5, we find that $x = 2.0344$ or 5.1760 (approximately). The solution set for $[0, 2\pi)$ is $\{\pi/4, 5\pi/4, 2.0344, 5.1760\}$. ☐

When an equation involves more than one trigonometric function, it is often helpful to use a suitable identity to rewrite the equation in terms of just one trigonometric function as in the following example.

EXAMPLE 3   ☐   Find all solutions for $\sin x + \cos x = 0$ in the interval $[0, 360°)$.

Since $\sin x/\cos x = \tan x$ and $\cos x/\cos x = 1$, we can divide both sides of the equation by $\cos x$ (assuming $\cos x \neq 0$) to get

$$\sin x + \cos x = 0$$
$$\frac{\sin x}{\cos x} + \frac{\cos x}{\cos x} = \frac{0}{\cos x}$$
$$\tan x + 1 = 0$$
$$\tan x = -1.$$

For the last equation the solutions in the given interval are

$$x = 135° \quad \text{and} \quad x = 315°,$$

with solution set $\{135°, 315°\}$.

We assumed here that $\cos x \neq 0$. If $\cos x = 0$, our equation becomes $\sin x + 0 = 0$, or $\sin x = 0$. If $\cos x = 0$, it is not possible for $\sin x = 0$, so that no solutions were missed by assuming $\cos x \neq 0$. ☐

EXAMPLE 4 ☐ Find all solutions for $\sin x \tan x = \sin x$ in the interval $[0°, 360°)$.

Subtract $\sin x$ from both sides, then factor on the left.

$$\sin x \tan x = \sin x$$
$$\sin x \tan x - \sin x = 0$$
$$\sin x (\tan x - 1) = 0$$

Now set each factor equal to 0.

$$\sin x = 0 \qquad\qquad \tan x - 1 = 0$$
$$\tan x = 1$$
$$x = 0° \quad \text{or} \quad x = 180° \qquad x = 45° \quad \text{or} \quad x = 225°$$

The solution set is thus $\{0°, 180°, 45°, 225°\}$. ☐

There are four solutions to the equation of Example 4. If we had tried to solve the equation by dividing both sides by $\sin x$, we would have $\tan x = 1$, which would give $x = 45°$ or $x = 225°$. The other two solutions would not appear. The missing solutions are the ones which make the divisor, $\sin x$, equal 0. Thus, when dividing by a variable expression, it is necessary to check to see if the numbers which make that expression 0 are solutions.

Sometimes we can solve a trigonometric equation by first squaring both sides, then using a trigonometric identity. When we square both sides of an equation, we must remember to check for any numbers which satisfy the squared equation, but not the given equation.

EXAMPLE 5 ☐ Find all solutions for $\tan x + \sqrt{3} = \sec x$ in the interval $[0, 2\pi)$.

Square both sides, then express $\sec^2 x$ in terms of $\tan^2 x$.

$$\tan x + \sqrt{3} = \sec x$$
$$\tan^2 x + 2\sqrt{3} \tan x + 3 = \sec^2 x$$
$$\tan^2 x + 2\sqrt{3} \tan x + 3 = 1 + \tan^2 x$$
$$2\sqrt{3} \tan x = -2$$
$$\tan x = -\frac{1}{\sqrt{3}}$$

The possible solutions in the given interval are $5\pi/6$ and $11\pi/6$. Now check the possible solutions. Try $5\pi/6$ first.

$$\tan x + \sqrt{3} = \tan \frac{5\pi}{6} + \sqrt{3} = \frac{-\sqrt{3}}{3} + \sqrt{3} = \frac{2\sqrt{3}}{3}$$

$$\sec x = \sec \frac{5\pi}{6} = \frac{-2\sqrt{3}}{3}$$

Thus $5\pi/6$ is not a solution. Now try $11\pi/6$.

$$\tan \frac{11\pi}{6} + \sqrt{3} = \frac{-\sqrt{3}}{3} + \sqrt{3} = \frac{2\sqrt{3}}{3}$$

$$\sec \frac{11\pi}{6} = \frac{2\sqrt{3}}{3}$$

So $11\pi/6$ is a solution to the given equation, and the solution set, $\{11\pi/6\}$, contains only one element.   ☐

When a trigonometric equation which is quadratic in form cannot be factored, the quadratic theorem can be used to solve the equation.

▣ EXAMPLE 6   ☐   Find all solutions for $\cot^2 x + 3 \cot x = 1$ in the interval $[0°, 360°)$
Write the equation with 0 on one side.

$$\cot^2 x + 3 \cot x - 1 = 0$$

Since we cannot factor, use the quadratic formula with $a = 1$, $b = 3$, $c = -1$, and $\cot x$ as the variable.

$$\cot x = \frac{-3 \pm \sqrt{9 + 4}}{2} = \frac{-3 \pm \sqrt{13}}{2} = \frac{-3 \pm 3.606}{2}$$

$$\cot x = .303 \quad \text{or} \quad \cot x = -3.303$$

Use Table 5 or a calculator to find x to the nearest ten minutes, giving the solution set

$$\{73° \ 10', \ 253° \ 10', \ 163° \ 10', \ 343° \ 10'\}. \quad ☐$$

The methods for solving trigonometric equations illustrated in the examples can be summarized as follows.

Solving Trigonometric Equations

1. If only one trigonometric function is present, first solve the equation for that function.

2. If more than one trigonometric function is present, rearrange the equation so that one side equals 0. Then try to factor.

3. If Step 2 does not work, try using identities to change the form of the equation. It may be helpful to square both sides of the equation first.

4. If the equation is quadratic in form, but not easily factorable, use the quadratic formula.

Conditional trigonometric equations where a half angle or multiple angle is given, such as 2 sin $(x/2) = 1$, often require an additional step to solve. This extra step is shown in the following example.

EXAMPLE 7

Find all solutions for 2 sin $x/2 = 1$ in the interval $[0°, 360°)$.
We begin as before by solving for the trigonometric function.

$$2 \sin \frac{x}{2} = 1$$

$$\sin \frac{x}{2} = \frac{1}{2}$$

If we let $x/2 = \theta$, then

$$\sin \theta = \frac{1}{2}$$

$$\theta = 30° \quad \text{or} \quad \theta = 150°.$$

Since $x/2 = \theta$,

$$\frac{x}{2} = 30° \quad \text{or} \quad \frac{x}{2} = 150°,$$

from which $\qquad x = 60° \quad \text{or} \quad x = 300°,$

with solution set $\{60°, 300°\}$. ☐

EXAMPLE 8

Find all solutions for 4 sin x cos $x = \sqrt{3}$ in the interval $[0°, 360°)$.
The identity 2 sin x cos x = sin $2x$ is useful here.

$$4 \sin x \cos x = \sqrt{3}$$

$$2(2 \sin x \cos x) = \sqrt{3}$$

$$2 \sin 2x = \sqrt{3}$$

$$\sin 2x = \frac{\sqrt{3}}{2}$$

$$2x = 60°, 120°, 420°, \text{ or } 480°$$

$$x = 30°, 60°, 210°, \text{ or } 240°$$

Here the domain $0° \leq x < 360°$ implies $0° \leq 2x < 720°$, which allows four solutions: the solution set is $\{30°, 60°, 210°, 240°\}$. ☐

EXAMPLE 9

Find all solutions of cos $6x$ − cos $2x$ = −sin $4x$ in the interval $[0, 2\pi)$.
Use the identity

$$\cos x - \cos y = -2 \sin \left(\frac{x+y}{2}\right) \sin \left(\frac{x-y}{2}\right)$$

to rewrite the given equation as

$$-2 \sin\left(\frac{6x + 2x}{2}\right) \sin\left(\frac{6x - 2x}{2}\right) = -\sin 4x$$

$$-2 \sin 4x \sin 2x = -\sin 4x$$

or $2 \sin 4x \sin 2x - \sin 4x = 0.$

Factor $\sin 4x \, (2 \sin 2x - 1) = 0$

$\sin 4x = 0$ or $2 \sin 2x - 1 = 0$

$\sin 4x = 0$ or $\sin 2x = \dfrac{1}{2}.$

The solutions of $\sin 4x = 0$ are given by $4x = 0 + n \cdot \pi$. Thus,

$$x = 0 + n \cdot \frac{\pi}{4}.$$

For $n = 0, 1, 2, 3, 4, 5, 6, 7$ we get the solutions

$0, \pi/4, \pi/2, 3\pi/4, \pi, 5\pi/4, 3\pi/2,$ and $7\pi/4.$

From $\sin 2x = 1/2$, we get $2x = \pi/6 + n \cdot 2\pi$, and $2x = 5\pi/6 + n \cdot 2\pi$. The first of these produces the solutions $\pi/12$ and $13\pi/12$, while the second produces $5\pi/12$ and $17\pi/12$. In summary, the solution set of the original equation is

$$\{0, \pi/4, \pi/2, 3\pi/4, \pi, 5\pi/4, 3\pi/2, 7\pi/4, \pi/12, 13\pi/12, 5\pi/12, 17\pi/12\}. \qquad \square$$

The next example shows how to solve equations involving inverse trigonometric functions.

EXAMPLE 10 $\square$ Solve $\sin^{-1} x - \cos^{-1} x = \pi/6.$
Begin by adding $\cos^{-1} x$ to both sides of the equation to get

$$\sin^{-1} x = \cos^{-1} x + \frac{\pi}{6}.$$

By the definition of the inverse sine, this becomes

$$\sin\left(\cos^{-1} x + \frac{\pi}{6}\right) = x.$$

Let $u = \cos^{-1} x$. Then

$$\sin\left(u + \frac{\pi}{6}\right) = x.$$

Use the identity for $\sin (s + t)$, which gives

$$\sin u \cos \frac{\pi}{6} + \cos u \sin \frac{\pi}{6} = x. \qquad (1)$$

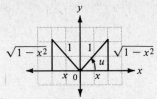

u can be in either quadrant I or II

Figure 7.6

Since $u = \cos^{-1} x$, we have $\cos u = x$. Because of the domain of $\cos^{-1} x$, u can be in either quadrant I or II. Sketch triangles in each of these quadrants and label them as shown in Figure 7.6. In either quadrant I or II, $\sin u$ is positive, with

$$\sin u = \sqrt{1 - x^2}.$$

Replace $\sin u$ with $\sqrt{1 - x^2}$, $\sin \pi/6$ with $1/2$, $\cos \pi/6$ with $\sqrt{3}/2$, and $\cos u$ with x, so that equation (1) becomes

$$\sqrt{1 - x^2} \cdot \frac{\sqrt{3}}{2} + x \cdot \frac{1}{2} = x$$

$$\sqrt{1 - x^2} \cdot \sqrt{3} + x = 2x$$

$$\sqrt{3} \cdot \sqrt{1 - x^2} = x.$$

Squaring both sides gives

$$3(1 - x^2) = x^2$$

$$3 - 3x^2 = x^2$$

$$3 = 4x^2$$

$$x = \pm \sqrt{\frac{3}{4}} = \pm \frac{\sqrt{3}}{2}.$$

Check that $\sqrt{3}/2$ is a solution of the original equation, while $-\sqrt{3}/2$ is not. The solution set is thus $\{\sqrt{3}/2\}$. ☐

**EXERCISES
7.7**

Find all solutions for the following equations in the interval $[0, 2\pi)$. Use 3.1416 as an approximation for π when you need values from Table 5.

1. $3 \tan x + 5 = 2$

2. $\tan x + 1 = 2$

3. $2 \sec x + 1 = \sec x + 3$

4. $\tan^2 x - 1 = 0$

5. $(\cot x - \sqrt{3})(2 \sin x + \sqrt{3}) = 0$

6. $(\tan x - 1)(\cos x - 1) = 0$

7. $(\sec x - 2)(\sqrt{3} \sec x - 2) = 0$

8. $(2 \sin x + 1)(\sqrt{2} \cos x + 1) = 0$

9. $\cos^2 x + 2 \cos x + 1 = 0$

10. $2 \cos^2 x - \sqrt{3} \cos x = 0$

11. $-2 \sin^2 x = 3 \sin x + 1$

12. $3 \sin^2 x - \sin x = 2$

13. $\cos^2 x - \sin^2 x = 0$

14. $\dfrac{2 \tan x}{3 - \tan^2 x} = 1$

15. $\sin 2x = 0$

16. $\cos 2x = 1$

17. $3 \tan 2x = \sqrt{3}$

18. $\cot 2x = \sqrt{3}$

19. $\sqrt{2} \cos 2x = -1$

20. $2 \sqrt{3} \sin 2x = -3$

21. $\sin \dfrac{x}{2} = \sqrt{2} - \sin \dfrac{x}{2}$

22. $\cos 2x - \cos x = 0$

Find all solutions for the following equations in the interval [0°, 360°). Find θ to the nearest ten minutes.

23. $\tan \theta + 6 \cot \theta = 5$

24. $\csc \theta = 2 \sin \theta + 1$

25. $\sec^2 \theta = 2 \tan \theta + 4$

26. $2 \tan^2 \theta \sin \theta - \tan^2 \theta = 0$

27. $5 \sec^2 \theta = 6 \sec \theta$

28. $\cos^2 \theta = \sin^2 \theta + 1$

29. $\csc^2 \theta - 2 \cot \theta = 0$

30. $3 \cot^3 \theta = \cot \theta$

31. $\sin^2 \theta \cos^2 \theta = 0$

32. $\sec^2 \theta \tan \theta = 2 \tan \theta$

33. $2 \sin 2\theta = \sqrt{3}$

34. $2 \cos 2\theta = \sqrt{2}$

35. $\cos \dfrac{\theta}{2} = 1$

36. $\sin \dfrac{\theta}{2} = 1$

37. $2 \sqrt{3} \sin \dfrac{\theta}{2} = 3$

38. $2 \sqrt{3} \cos \dfrac{\theta}{2} = -3$

39. $2 \sin \theta = 2 \cos 2\theta$

40. $\cos \theta - 1 = \cos 2\theta$

41. $\sin 2\theta = 2 \cos^2 \theta$

42. $\csc^2 \dfrac{\theta}{2} = 2 \sec \theta$

43. $\cos \theta = \sin^2 \dfrac{\theta}{2}$

44. $4 \cos 2\theta = 8 \sin \theta \cos \theta$

45. $2 \cos^2 2\theta = 1 - \cos 2\theta$

46. $\sin \theta = \cos \dfrac{\theta}{2}$

Give all solutions (to the nearest ten minutes) for each of the following. Write the solutions in the form used in Example 1.

47. $\sin x - \cos x = 1$

48. $\cos 2x = 1$

49. $\cos x \left(\sin x - \dfrac{1}{2} \right) = 0$

50. $(2 \sin x - \sqrt{3})(\cos x + 1) = 0$

51. $\tan^2 x + 2 \tan x = 3$

52. $\cot^2 x - 4 \cot x - 5 = 0$

53. $\sin 2x = \cos 2x$

54. $\tan \dfrac{1}{2} x = \cot \dfrac{1}{2} x$

To solve the following equations, you will need the quadratic formula. Find all solutions in the interval [0°, 360°). Give solutions to the nearest ten minutes.

55. $9 \sin^2 x - 6 \sin x = 1$

56. $4 \cos^2 x + 4 \cos x = 1$

57. $\tan^2 x + 4 \tan x + 2 = 0$

58. $3 \cot^2 x - 3 \cot x - 1 = 0$

59. $\sin^2 x - 2 \sin x + 3 = 0$

60. $2 \cos^2 x + 2 \cos x - 1 = 0$

61. $\cot x + 2 \csc x = 3$

62. $2 \sin x = 1 - 2 \cos x$

For the following equations, use the sum and product identities of Section 7.6. Give all solutions in the interval $[0, 2\pi)$.

63. $\sin x + \sin 3x = \cos x$

64. $\cos 4x - \cos 2x = \sin x$

65. $\sin 3x - \sin x = 0$

66. $\cos 2x + \cos x = 0$

67. $\sin 4x + \sin 2x = 2 \cos x$

68. $\cos 5x + \cos 3x = 2 \cos 4x$

In an electric circuit, let V represent the electromotive force in volts at t seconds. Assume $V = \cos 2\pi t$. Find the smallest positive value of t where $0 \le t \le 1/2$ for each of the following values of V.

69. $V = 0$

70. $V = .5$

A coil of wire rotating in a magnetic field induces a voltage given by

$$e = 20 \sin\left(\frac{\pi t}{4} - \frac{\pi}{2}\right),$$

where t is time in seconds. Find the smallest positive time to produce the following voltages.

71. 0

72. $10\sqrt{3}$

Solve each of the following equations.

73. $\sin^{-1} x = \tan^{-1} 3/4$

74. $\tan^{-1} x = \cos^{-1} 5/13$

75. $\text{Arccos } x = \text{Arcsin } 3/5$

76. $\text{Arctan } x = \text{Arcsin } -4/5$

77. $\sin^{-1} x - \tan^{-1} 1 = -\pi/4$

78. $\cos^{-1} x + \cos^{-1} 1 = \pi/2$

79. $\cos^{-1} x + 2 \sin^{-1} \sqrt{3}/2 = \pi$

80. $\sin^{-1} x + \tan^{-1} \sqrt{3} = 2\pi/3$

81. $\sin^{-1} 2x + \cos^{-1} x = \pi/6$

82. $\sin^{-1} 2x + \sin^{-1} x = \pi/2$

83. $\cos^{-1} x + \tan^{-1} x = \pi/2$

84. $\tan^{-1} x + \cos^{-1} x = \pi/4$

CHAPTER SUMMARY
7

Key Words

Identity	Cofunction identities	Half angle identities
Fundamental identities	Reduction identity	Sum and product identities
Sum and difference identities	Double angle identities	Conditional equation

---

CHAPTER REVIEW EXERCISES
7

1. Use the trigonometric identities to find the remaining five trigonometric functions of x, given that $\cos x = 3/5$ and x is in quadrant IV.

2. Given $\tan x = -5/4$, where x is in the interval $(\pi/2, \pi)$, use trigonometric identities to find the other trigonometric functions of x.

3. Given $\sin x = -1/4$, $\cos y = -4/5$, and both x and y are in quadrant III, find $\sin (x + y)$ and $\cos (x - y)$.

4. Given $\sin 2\theta = \sqrt{3}/2$ and 2θ terminates in quadrant II, use trigonometric identities to find $\tan \theta$.

5. Given $x = \pi/8$, use trigonometric identities to find $\sin x$, $\cos x$, and $\tan x$.

For each item in List I, give the letter of the item in List II which completes an identity.

List I	List II
6. $\sin 35°$	(a) $\sin(-35°)$
7. $\tan(-35°)$	(b) $\cos 55°$
8. $\cos 35°$	(c) $\sqrt{\dfrac{1 + \cos 150°}{2}}$
9. $\cos 75°$	
10. $\sin 75°$	
11. $\sin 300°$	(d) $2 \sin 150° \cos 150°$
12. $\cos 300°$	(e) $\cos 150° \cos 60° - \sin 150° \cos 60°$
	(f) $\cot(-35°)$
	(g) $\cos^2 150° - \sin^2 150°$
	(h) $\sin 15° \cos 60° + \cos 15° \sin 60°$
	(i) $\cos(-35°)$
	(j) $\cot 125°$

For each item in List I give the letter of the item in List II which completes an identity.

List I	List II
13. $\csc x$	(a) $\dfrac{1}{\sin x}$
14. $\tan x$	
15. $\cot x$	(b) $\dfrac{1}{\cos x}$
16. $\sin^2 x$	
17. $\tan^2 x + 1$	(c) $\dfrac{\sin x}{\cos x}$
18. $\tan^2 x$	
	(d) $\dfrac{1}{\cot^2 x}$
	(e) $\dfrac{1}{\cos^2 x}$
	(f) $\dfrac{\cos x}{\sin x}$
	(g) $\dfrac{1}{\sin^2 x}$
	(h) $1 - \cos^2 x$

Use identities to express each of the following in terms of $\sin \theta$ and $\cos \theta$ and simplify.

19. $\sec^2 \theta - \tan^2 \theta$

20. $\dfrac{\cot \theta}{\sec \theta}$

21. $\tan^2 \theta \,(1 + \cot^2 \theta)$

22. $\csc \theta + \cot \theta$

23. $\csc^2 \theta + \sec^2 \theta$

24. $\tan \theta - \sec \theta \csc \theta$

Show that each of the following is an identity.

25. $\dfrac{\sin 2x}{\sin x} = \dfrac{2}{\sec x}$

26. $2 \cos A - \sec A = \cos A - \dfrac{\tan A}{\csc A}$

27. $\dfrac{2 \tan B}{\sin 2B} = \sec^2 B$

28. $\tan \beta = \dfrac{1 - \cos 2\beta}{\sin 2\beta}$

29. $1 + \tan^2 \alpha = 2 \tan \alpha \csc 2\alpha$

30. $-\dfrac{\sin(A - B)}{\sin(A + B)} = \dfrac{\cot A - \cot B}{\cot A + \cot B}$

31. $\dfrac{\sin t}{1 - \cos t} = \cot \dfrac{t}{2}$

32. $2 \cos(A + B) \sin(A + B) = \sin 2A \cos 2B + \sin 2B \cos 2A$

33. $\dfrac{2 \cot x}{\tan 2x} = \csc^2 x - 2$

34. $\sin t = \dfrac{\cos t \sin 2t}{1 + \cos 2t}$

35. $\tan \theta \sin 2\theta = 2 - 2 \cos^2 \theta$

36. $\csc A \sin 2A - \sec A = \cos 2A \sec A$

37. $2 \tan x \csc 2x - \tan^2 x = 1$

38. $2 \cos^2 \theta - 1 = \dfrac{1 - \tan^2 \theta}{1 + \tan^2 \theta}$

39. $\sin^3 \theta = \sin \theta - \cos^2 \theta \sin \theta$

40. $\dfrac{\sin^2 x}{2 - 2 \cos x} = \cos^2 \dfrac{x}{2}$

41. $\cos^4 \theta = \dfrac{3}{8} + \dfrac{1}{2} \cos 2\theta + \dfrac{1}{8} \cos 4\theta$

42. $8 \sin^2 \dfrac{\gamma}{2} \cos^2 \dfrac{\gamma}{2} = 1 - \cos 2\gamma$

43. $\cos^2 \dfrac{x}{2} = \dfrac{1 + \sec x}{2 \sec x}$

44. $\tan \theta \cos^2 \theta = \dfrac{2 \tan \theta \cos^2 \theta - \tan \theta}{1 - \tan^2 \theta}$

45. $\tan 8k - \tan 8k \cdot \tan^2 4k = 2 \tan 4k$

46. $\sec^2 \alpha - 1 = \dfrac{\sec 2\alpha - 1}{\sec 2\alpha + 1}$

47. $\dfrac{\sin 3t + \sin 2t}{\sin 3t - \sin 2t} = \dfrac{\tan \dfrac{5t}{2}}{\tan \dfrac{t}{2}}$

48. $\sin 2\alpha = \dfrac{2(\sin \alpha - \sin^3 \alpha)}{\cos \alpha}$

49. $\tan 2\beta - \sec 2\beta = \dfrac{\tan \beta - 1}{\tan \beta + 1}$

50. $\dfrac{\sin^3 t - \cos^3 t}{\sin t - \cos t} = \dfrac{2 + \sin 2t}{2}$

51. $-\cot \dfrac{x}{2} = \dfrac{\sin 2x + \sin x}{\cos 2x - \cos x}$

52. $2 \cos^3 x - \cos x = \dfrac{\cos^2 x - \sin^2 x}{\sec x}$

Find all solutions for the following equations in the interval $[0, 2\pi)$.

53. $\sin^2 x = 1$

54. $2 \tan x - 1 = 0$

55. $2 \sin^2 x - 5 \sin x + 2 = 0$

56. $\tan x = \cot x$

57. $\sec^4 2x = 4$

58. $\tan^2 2x - 1 = 0$

59. $\sin \dfrac{x}{2} = \cos \dfrac{x}{2}$

60. $\sec \dfrac{x}{2} = \cos \dfrac{x}{2}$

61. $\cos 2x + \cos x = 0$

62. $\sin x \cos x = \dfrac{1}{4}$

Find all solutions for the following equations in the interval $[0°, 360°)$.

63. $2 \cos \theta = 1$

64. $(\tan \theta + 1)\left(\sec \theta - \dfrac{1}{2}\right) = 0$

65. $\sin^2 \theta + 3 \sin \theta + 2 = 0$

66. $\sin 2\theta = \cos 2\theta + 1$

67. $\dfrac{\sin \theta}{\cos \theta} = \tan^2 \theta$

68. $2 \sin 2\theta = 1$

Exact values of the trigonometric functions of 15° can be found by the following method, an alternative to the use of the half-angle formulas. Start with a right triangle ABC having a 60° angle at A and a 30° angle at B. Let the hypotenuse of this triangle have length 2. Extend side BC and draw a semicircle with diameter along BC extended, center at B, and radius AB. Draw segment AE. (See the figure.) Since any angle inscribed in a semicircle is a right angle, triangle AED is a right triangle.

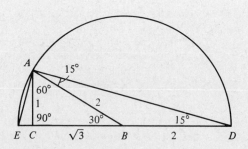

69. Show that triangle *ABD* is isosceles.

70. Show that angle *ABD* is 150°.

71. Show that angle *DAB* is 15°, as is angle *ADB*.

72. Show that *DC* has length $2 + \sqrt{3}$.

73. Since *AC* has length 1, the length of *AD* is given by

$(AD)^2 = 1^2 + (2 + \sqrt{3})^2$.

Reduce this to $\sqrt{8 + 4\sqrt{3}}$, and show that this result equals $\sqrt{6} + \sqrt{2}$.

74. Use angle ADB of triangle ADE and find $\cos 15°$.

75. Show that AE has length $\sqrt{6} - \sqrt{2}$.

76. Find $\sin 15°$.

77. Use triangle ACE and find $\tan 15°$.

78. Find $\cot 15°$.

The following exercises are taken with permission from a standard calculus book.

79. Let α and β be two given numbers.

(a) Prove that if $\sin(\alpha + \beta) = \sin(\alpha - \beta)$, then either α is an odd multiple of $\pi/2$ or β is a multiple of π, or both.

(b) Prove that if $\cos(\alpha + \beta) = \cos(\alpha - \beta)$, then either α is a multiple of π or β is a multiple of π, or both.

(c) Prove that if $\tan(\alpha + \beta) = \tan(\alpha - \beta)$, then β is a multiple of π.

80. Note that for $\alpha = 18°$, we have $\cos 3\alpha = \sin 2\alpha$. Use the formula

$$\cos 3\alpha = 3 \sin \alpha - 4 \sin^3 \alpha$$

and show that $\sin 18° = (\sqrt{5} - 1)/4$.

8 Applications; De Moivre's Theorem

Trigonometry is over 3000 years old. The ancient Egyptians, Babylonians, and Greeks developed trigonometry to find the lengths of the sides of triangles and the measures of their angles. Every triangle has three sides and three angles. In this chapter we see that if we know any three of the six measures of a triangle (if at least one measure is a side), then we can find the other three measures. This process is called **solving a triangle**.

8.1 Right-Triangle Applications

We briefly mentioned right-triangle applications of trigonometry in Chapter 6. We shall discuss more applications in this section. First, recall that if θ is an acute angle of a right triangle, then $\sin \theta$, $\cos \theta$, and $\tan \theta$ can be found from the lengths of the sides of the triangle:

$$\sin \theta = \frac{\text{side opposite}}{\text{hypotenuse}} \qquad \cos \theta = \frac{\text{side adjacent}}{\text{hypotenuse}} \qquad \tan \theta = \frac{\text{side opposite}}{\text{side adjacent}}.$$

The other three functions, $\cot \theta$, $\sec \theta$, and $\csc \theta$, can be defined in a similar way. (See Figure 8.1.)

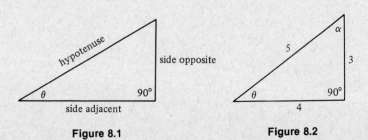

Figure 8.1   **Figure 8.2**

EXAMPLE 1   The right triangle of Figure 8.2 has sides of lengths 3, 4, and 5. Find each of the following.

(a) $\sin \theta$, $\cos \theta$, and $\tan \theta$.

Use the results given above.

$$\sin \theta = \frac{\text{side opposite}}{\text{hypotenuse}} = \frac{3}{5} \qquad \cos \theta = \frac{\text{side adjacent}}{\text{hypotenuse}} = \frac{4}{5}$$

$$\tan \theta = \frac{\text{side opposite}}{\text{side adjacent}} = \frac{3}{4}$$

(b) The degree measure of angle θ, to the nearest minute.

We know that $\sin \theta = 3/5 = .6000$. Using a calculator or interpolating in Table 5, we have

$$\theta = 36° \ 52'.$$

(c) $\sin \alpha$, $\cos \alpha$, and $\tan \alpha$.

$$\sin \alpha = \frac{4}{5}, \qquad \cos \alpha = \frac{3}{5} \qquad \text{and} \qquad \tan \alpha = \frac{4}{3}$$

(d) The degree measure of angle α.

We know that $\theta = 36° \ 52'$, and that $\alpha + \theta = 90°$. Thus,

$$\alpha = 90° - \theta$$
$$= 90° - 36° \ 52'$$
$$= 89° \ 60' - 36° \ 52'$$
$$\alpha = 53° \ 08'.$$

In using trigonometry to solve triangles or to find the measures of all sides and all angles, it is convenient to use a to represent the length of the side opposite angle A, b for the length of the side opposite angle B, and so on. The letter c is used for the hypotenuse in a right triangle.

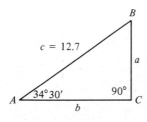

Figure 8.3

EXAMPLE 2   Solve right triangle ABC, with $A = 34° \ 30'$ and $c = 12.7$. See Figure 8.3.

To solve the triangle, we must find the measures of the remaining sides and angles. By the definitions given above, $\sin A = a/c$, where $A = 34° \ 30'$ and $c = 12.7$. Thus,

$$\sin A = \frac{a}{c}$$

$$\sin 34° \ 30' = \frac{a}{12.7}$$

or, upon multiplying both sides by 12.7,

$$a = 12.7 \sin 34° \ 30'$$
$$\approx 12.7(.5664)$$
$$a \approx 7.19.$$

We could use the Pythagorean theorem to find b. However, to do so, we would have to use the value of a just calculated. It is best to use the information given in the problem rather than a calculated result. If a mistake had been made in finding a, then b would also be incorrect. It is better to use cosine to find b.

$$\cos A = \frac{\text{side adjacent}}{\text{hypotenuse}} = \frac{b}{c}$$

$$\cos 34° \ 30' = \frac{b}{12.7}$$

$$b = 12.7 \cos 34° \ 30'$$

$$b \approx 12.7 \ (.8241)$$

$$b \approx 10.5$$

The Pythagorean theorem can be used as a check once b is found. All that remains to find is angle B. Since $A + B = 90°$, and $A = 34° \ 30'$, we have

$$A + B = 90°$$
$$B = 90° - A$$
$$B = 90° - 34° \ 30'$$
$$= 89° \ 60' - 34° \ 30'$$
$$B = 55° \ 30'.$$

Triangle ABC is now solved—we know the lengths of all sides and the measures of all angles.   ◻

In Example 2 above, we were told that $c = 12.7$. Since this measure for c is given to three significant digits, we must give both a and b to no more than three significant digits as the number of significant digits in an answer cannot exceed the least number of significant digits in any number used to find the answer. Also, an angle given to the nearest degree is assumed to have *two* significant digits; to the nearest ten minutes implies *three* significant digits; to the nearest minute implies *four* significant digits.

Figure 8.4 shows an angle of elevation and an angle of depression. For the **angle of elevation,** assume you are standing at point X and looking up at point Y. For the **angle of depression,** assume you are standing at point X and looking down at point Y.

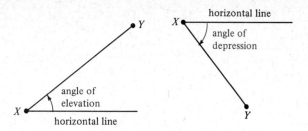

Figure 8.4

EXAMPLE 3 Wilma Spence knows that when she stands 123 feet from the base of a flagpole, the angle of elevation to the top is 26° 40′. If her eyes are 5.30 feet above the ground, find the height of the flagpole.

We know the length of the side adjacent to Spence (see Figure 8.5) and we want to find the length of the side opposite her. The ratio that involves these two values is the tangent. Thus,

$$\tan A = \frac{\text{side opposite}}{\text{side adjacent}}$$

$$\tan 26° 40′ = \frac{a}{123}$$

$$a = 123 \tan 26° 40′$$

$$a = 123(.5022)$$

$$a = 61.8 \text{ feet.}$$

Since Spence's eyes are 5.30 feet above the ground, the height of the flagpole is

$$61.8 + 5.30 = 67.1 \text{ feet.} \quad \blacksquare$$

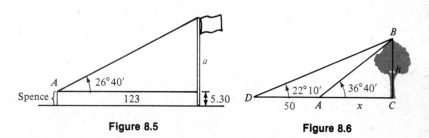

Figure 8.5 **Figure 8.6**

EXAMPLE 4 Francisco needs to know the height of a tree. From a given point on the ground he finds that the angle of elevation to the top of the tree is 36° 40′. He then moves back 50 ft. From the second point, the angle of elevation to the top of the tree is 22° 10′. See Figure 8.6. Find the height of the tree.

Here we have two unknowns, x, the distance from the center of the trunk of the tree to the point where the first observation was made, and h, the height of the tree. Since we know nothing about the length of the hypotenuse of either triangle ABC or triangle BCD, we use a ratio which doesn't involve the hypotenuse, that is, the tangent. We have in triangle ABC $\tan 36° \ 40' = \dfrac{h}{x}$ or $h = x \tan 36° \ 40'$

in triangle BCD $\tan 22° \ 10' = \dfrac{h}{50 + x}$ or $h = (50 + x) \tan 22° \ 10'.$

Since we have two expressions equaling h, these expressions must be equal.

$$x \tan 36° \ 40' = (50 + x) \tan 22° \ 10'$$

Now solve for x.

$$x \tan 36° \ 40' = 50 \tan 22° \ 10' + x \tan 22° \ 10'$$
$$x \tan 36° \ 40' - x \tan 22° \ 10' = 50 \tan 22° \ 10'$$
$$x(\tan 36° \ 40' - \tan 22° \ 10') = 50 \tan 22° \ 10'$$
$$x = \frac{50 \tan 22° \ 10'}{\tan 36° \ 40' - \tan 22° \ 10'}$$

We saw above that $h = x \tan 36° \ 40'$. Substituting for x,

$$h = \left(\frac{50 \tan 22° \ 10'}{\tan 36° \ 40' - \tan 22° \ 10'} \right)(\tan 36° \ 40').$$

From Table 5 or a calculator,

$$\tan 36° \ 40' = .7445$$
$$\tan 22° \ 10' = .4074.$$

Thus, $\tan 36° \ 40' - \tan 22° \ 10' = .7445 - .4074 = .3371,$

and

$$h = \left(\frac{50(.4074)}{.3371} \right)(.7445) = 45 \text{ ft.} \quad \square$$

Many applications of trigonometry involve **bearing,** an important idea in navigation and surveying. Bearing is used to give directions. There are two common systems used to express bearing. When a single angle is given, such as 164°, bearing is measured in a clockwise direction from due north. Several typical bearings are shown in Figure 8.7.

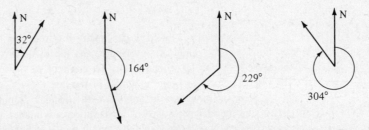

Figure 8.7

EXAMPLE 5   Radar stations A and B are on an east-west line, 3.7 kilometers apart. Station A detects a plane at C on a bearing of $61°$. Station B detects the same plane on a bearing of $331°$. Find the distance from A to C.

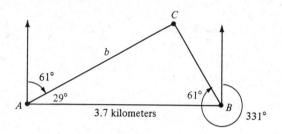

Figure 8.8

Draw a sketch showing the given information, as in Figure 8.8. Angle C is a right angle, since angles CAB and CBA are complementary. The necessary distance, b, can be found by using cosine.

$$\cos 29° = \frac{b}{3.7}$$

$$3.7 \cos 29° = b$$

$$3.7(.8746) = b$$

$$b = 3.2 \text{ kilometers}$$

The other common system for expressing bearing starts with a north-south line and uses an acute angle to show the direction, either east or west, from this line. Figure 8.9 shows several typical bearings using this system. The letter N or S always comes first, followed by an acute angle, and then E or W.

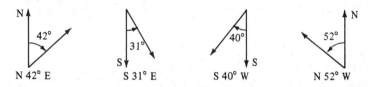

Figure 8.9

EXAMPLE 6   The bearing from A to C is S $52°$ E. The bearing from A to B is N $84°$ E. The bearing from B to C is S $38°$ W. A plane flying at 250 kilometers per hour takes 2.4 hours to go from A to B. Find the distance from A to C.

Figure 8.10 (next page) shows a sketch of the given information. Since the bearing from A to B is $84°$, angle ABE is $180° - 84° = 96°$. Thus, angle ABC is $46°$. Also, angle BAC is $180° - (84° + 52°) = 44°$. Thus, angle C is $90°$. We know that a plane flying at 250 kilometers per hour takes 2.4 hours to go from A to B so that the distance from A to B is $2.4(250) = 600$ kilometers. We need to find b, the distance from A to C. We use sine. (We also could have used cosine.)

$$\sin 46° = \frac{b}{c}$$

$$\sin 46° = \frac{b}{600}$$

$$600 \sin 46° = b$$

$$600(.7193) = b$$

$$b = 430 \text{ kilometers} \quad \square$$

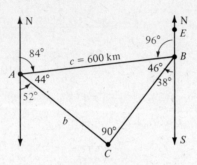

Figure 8.10

EXERCISES
8.1

Solve each of the following right triangles. Angle C is the right angle.

1. $A = 28°$, $c = 17.4$ m
2. $B = 46°$, $c = 29.7$ m
3. $B = 73°$, $b = 128$ inches
4. $A = 61°$, $b = 39.2$ cm
5. $a = 76.4$ yards, $b = 39.3$ yards
6. $a = 958$ m, $b = 489$ m
7. $a = 18.9$ cm, $c = 46.3$ cm
8. $b = 219$ m, $c = 647$ m
9. $B = 39° 10'$, $c = .623$ m
10. $B = 82° 50'$, $c = 4.82$ cm

Use a calculator to solve each of the following right triangles. Write answers with four significant digits.

11. $A = 59° 44'$, $c = 18.74$ feet

12. $B = 38° 9'$, $b = 469.8$ m
13. $A = 47.62°$, $a = 39.46$ cm
14. $B = 61.74°$, $a = 976.9$ mm
15. $a = 327.4$ cm, $b = 481.1$ cm
16. $a = 15,690$ mm, $b = 21,940$ mm

Solve each of the following.

17. A 39.4 meter fire-truck ladder is leaning against a wall. Find the distance the ladder goes up the wall if it makes an angle of $42° 30'$ with the ground.

18. A swimming pool is 50.0 feet long and 4.00 feet deep at one end. If it is 12.0 feet deep at the other end, find the total distance along the bottom.

19. A guy wire 87.4 meters long is attached to the top of a tower that is 69.4 meters high. Find the angle that the wire makes with the ground.

20. Find the length of a guy wire that makes an angle of 42° 10′ with the ground if the wire is attached to the top of a tower 79.6 meters high.

Work the following problems involving angles of elevation or depression.

21. Suppose the angle of elevation of the sun is 28° 10′. Find the length of the shadow cast by a man 6.0 feet tall.

22. The shadow of a vertical tower is 58.2 meters long when the angle of elevation of the sun is 36° 20′. Find the height of the tower.

23. Find the angle of elevation of the sun if a 53.9 foot flagpole casts a shadow 74.6 feet long.

24. The angle of depression from the top of a building to a point on the ground is 34° 50′. How far is the point on the ground from the top of the building if the building is 368 meters high?

25. Priscilla drives her Peterbilt up a straight road inclined at an angle of 4° 10′ with the horizontal. She starts at an elevation of 680 feet above sea level and drives 12,400 feet along the road. Find her final altitude.

26. The road into Death Valley is straight; it makes an angle of 4° 10′ with the horizontal. Starting at sea level, the road descends to − 121 feet. Find the distance it is necessary to travel along the road to reach bottom.

27. A tower stands on top of a hill. From a point on the ground 148 m from a point directly under the tower, the angle of elevation to the *bottom* of the tower is 18° 20′. From the same point, the angle of elevation to the *top* of the tower is 34° 10′. Find the height of the tower.

28. The angle of elevation from the top of an office building in New York City to the top of the World Trade Center is 68°, while the angle of depression from the top of the office building to the bottom of the Trade Center is 63°. The office building is 290 feet from the World Trade Center. Find the height of the World Trade Center.

29. Mt. Rogers, with an altitude of 5700 feet, is the highest point in Virginia. The angle of elevation from the top of Mt. Rogers to a plane flying overhead is 33°. The straight line distance from the mountaintop to the plane is 4600 feet. Find the altitude of the plane.

30. The highest point in Texas is Guadalupe Peak. The angle of depression from the top of this peak to a small miner's cabin at elevation 2000 feet is 26°. The cabin is 14,000 feet horizontally from a point directly under the top of the mountain. Find the altitude of the top of the mountain.

Find h in each of the following.

31.

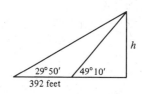

32.

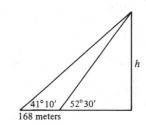

33. The angle of elevation from a point on the ground to the top of a pyramid is 35° 30′. The angle of elevation from a point 135 feet further back to the top of the pyramid is 21° 10′. Find the height of the pyramid.

34. A lighthouse keeper is watching a boat approach directly to the lighthouse. When she first begins watching the boat, the angle of depression of the boat is 15° 50′. Just as the boat turns away from the lighthouse, the angle of depression is 35° 40′. If the height of the lighthouse is 68.7 meters, find the distance traveled by the boat as it approaches the lighthouse.

35. A television antenna is on top of the center of a house. The angle of elevation from a point 28.0 meters from the center of the house to the top of the antenna is 27° 10′, and the angle of elevation to the bottom of the antenna is 18° 10′. Find the height of the antenna.

36. The angle of elevation from Lone Pine to the top of Mt. Whitney is 10° 50′. If I drive 7.00 kilometers along a straight level road toward Mt. Whitney, I find the angle of elevation to be 22° 40′. Find the height of the top of Mt. Whitney above the level of the road.

In each of the following exercises, generalize problems from above by finding formulas for h in terms of k, A, and B. Assume A < B in Exercise 37, and A > B in Exercise 38.

37.

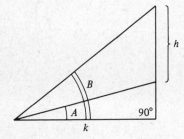

38.

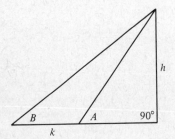

Solve each of the following problems involving bearing.

39. A ship leaves port and sails on a bearing of 28° 10′. Another ship leaves the same port at the same time and sails on a bearing of 118° 10′. If the first ship sails at 20.0 miles per hour and the second sails at 24.0 miles per hour, find the distance between the two ships after five hours.

40. Radio direction finders are set up at points *A* and *B*, which are 2.00 miles apart on an east-west line. From *A* it is found that the bearing of the signal from a radio transmitter is 36° 20′, while from *B* the bearing of the same signal is 306° 20′. Find the distance between the transmitter and *B*.

41. The bearing from Winston-Salem, North Carolina, to Danville, Virginia, is N 42° E. The bearing from Danville to Goldsboro, North Carolina; is S 48° E. A small plane traveling at 60 miles per hour takes 1 hour to go from Winston-Salem to Danville and 1.8 hours to go from Danville to Goldsboro. Find the distance from Winston-Salem to Goldsboro.

42. The bearing from Atlanta to Macon is S 27° E, while the bearing from Macon to Augusta is N 63° E. A plane traveling at 60 miles per hour needs 1¼ hours to go from Atlanta to Macon and 1¾ hours to go from Macon to Augusta. Find the distance from Atlanta to Augusta.

43. The airline distance from Philadelphia to Syracuse is 260 miles, on a bearing of 335°. The distance from Philadelphia to Cincinnati is 510 miles, on a bearing of 245°. Find the bearing from Cincinnati to Syracuse.

44. A ship travels 70 kilometers on a bearing of 27°, and then turns on a bearing of 117° for 180 kilometers. Find the distance of the end of the trip from the starting point.

45. A pendulum is *m* cm long. Suppose it is moved from its vertical position by an angle α (0° ≤ α ≤ 90°). By how much has the end of the pendulum been raised vertically?

46. A woman is standing on a ship at sea *h* meters above the water. If the radius of the earth is *r* kilometers, find the distance she can see to the horizon.

47. Atoms in metals can be arranged in patterns called **unit cells.** One such unit cell, called a **primitive cell,** is a cube with an atom at each corner. A right triangle can be formed from one edge of the cell, a face diagonal and a body diagonal as in the figure below. If each cell edge is 3.00×10^{-8} cm and the face diagonal is 4.24×10^{-8} cm, what is the angle between the cell edge and a body diagonal?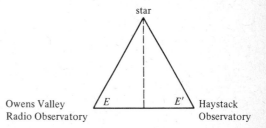

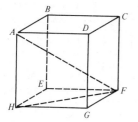

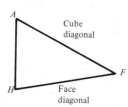

48. To determine the diameter of the sun, an astronomer might sight with a transit (a device used by surveyors for measuring angles) first to one edge of the sun and then to the other, finding that the included angle equals $1° 4'$. Assuming that the distance from the earth to the sun is 92,919,800 mi, calculate the diameter of the sun.

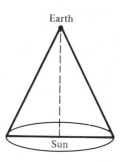

49. Very accurate measurements have shown that the distance between California's Owens Valley Radio Observatory and the Haystack Observatory in Massachusetts is 2441.2938 miles. Suppose the two observatories focus on a distant star and find that angles E and E' in the figure are both 89.99999°. Find the distance to the star from Haystack. (Assume the Earth is flat.)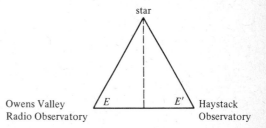

50. The figure shows a magnified view of the threads of a bolt. Find x if d is 2.894 mm.

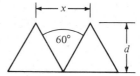

Use a right triangle to find each of the following.

51. $\sin(\cos^{-1} 1/4)$ **52.** $\tan(\sin^{-1} 4/7)$

53. $\cos(\sin^{-1} 2/3)$ **54.** $\sin(\tan^{-1} 1/5)$

8.2 Oblique Triangles and the Law of Sines

The methods of the previous section apply only to right triangles. In the next few sections we generalize these methods to include all triangles, not just right triangles. A triangle that is not a right triangle is called an **oblique triangle.** The measures of the three sides and the three angles of a triangle can be found if at least one side and any other two measures are known. There are four possible cases.

> **1.** One side and two angles are known.
>
> **2.** Two sides and one angle (not included between the two sides) are known. (This case may lead to more than one triangle.)
>
> **3.** Three sides are known.
>
> **4.** Two sides and the angle included between the two sides are known.

The first two cases require the **law of sines,** which is discussed in this section. The last two cases require the **law of cosines,** discussed in Section 8.3. To derive the law of sines, start with a general oblique triangle such as the one in Figure 8.11. (Here we assume that angle A is acute. For the cases in which $A = 90°$ or A is obtuse, see Exercise 53 below.)

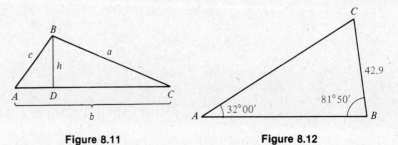

Figure 8.11   **Figure 8.12**

First, construct the perpendicular from B to side AC. Let h be the length of this perpendicular. Let c be the hypotenuse of right triangle ADB, and let a be the hypotenuse of right triangle BDC. By the definitions from the last section,

$$\text{in triangle } ADB \qquad \sin A = \frac{h}{c} \qquad \text{or} \qquad h = c \sin A,$$

$$\text{in triangle } BDC \qquad \sin C = \frac{h}{a} \qquad \text{or} \qquad h = a \sin C.$$

Since $h = c \sin A$, and $h = a \sin C$,

$$a \sin C = c \sin A$$

or, upon dividing both sides by $\sin A \sin C$,

$$\frac{a}{\sin A} = \frac{c}{\sin C}.$$

In a similar way,

$$\frac{a}{\sin A} = \frac{b}{\sin B} \qquad \text{and} \qquad \frac{b}{\sin B} = \frac{c}{\sin C}.$$

We have now proved the following theorem.

The Law of Sines

In any triangle ABC, with sides a, b, and c,

$$\frac{a}{\sin A} = \frac{b}{\sin B}, \quad \frac{a}{\sin A} = \frac{c}{\sin C}, \quad \text{and} \quad \frac{b}{\sin B} = \frac{c}{\sin C}.$$

The three formulas of the law of sines are sometimes written as

$$\frac{a}{\sin A} = \frac{b}{\sin B} = \frac{c}{\sin C}.$$

EXAMPLE 1

Solve triangle ABC if $A = 32°\ 00'$, $B = 81°\ 50'$, and $a = 42.9$ cm.

Start by drawing a triangle, roughly to scale, and labeling the given parts, as in Figure 8.12. Since we know A, B, and a, we use

$$\frac{a}{\sin A} = \frac{b}{\sin B}$$

to get

$$\frac{42.9}{\sin 32°\ 00'} = \frac{b}{\sin 81°\ 50'}.$$

After multiplying both sides of the equation by $\sin 81°\ 50'$, we get

$$b = \frac{42.9 \sin 81°\ 50'}{\sin 32°\ 00'}$$

$$b \approx \frac{42.9(.9899)}{.5299}$$

$$b \approx 80.1 \text{ centimeters.}$$

Use the fact that the sum of the angles of any triangle is $180°$ to find C.

$$A + B + C = 180°$$
$$C = 180° - A - B$$
$$= 180° - 32°\ 00' - 81°\ 50'$$
$$C = 66°\ 10'$$

Now we can use the law of sines again to find c. (Why shouldn't we use the Pythagorean theorem?) We have

$$\frac{a}{\sin A} = \frac{c}{\sin C}$$

$$\frac{42.9}{\sin 32°\ 00'} = \frac{c}{\sin 66°\ 10'}.$$

Now we find c.

$$c = \frac{42.9 \sin 66° \, 10'}{\sin 32° \, 00'} \approx \frac{42.9(.9147)}{.5299} \approx 74.1$$

The length of side c is approximately 74.1 cm.   ☐

EXAMPLE 2   ☐   Shawn Johnson wishes to measure the distance across the Big Muddy River. See Figure 8.13. She finds that $C = 112° \, 53'$, $A = 31° \, 06'$, and $b = 347.6$ ft. Find the required distance.

Before we can use the law of sines to find a, we must first find angle B.

$$B = 180° - A - C$$
$$= 180° - 31° \, 06' - 112° \, 53'$$
$$B = 36° \, 01'$$

Use the part of the law of sines involving A, B, and b.

$$\frac{a}{\sin A} = \frac{b}{\sin B}$$

Substitute the known values.

$$\frac{a}{\sin 31° \, 06'} = \frac{347.6}{\sin 36° \, 01'}$$

$$a = \frac{347.6 \sin 31° \, 06'}{\sin 36° \, 01'}$$

$$a = 305.3 \text{ ft}   ☐$$

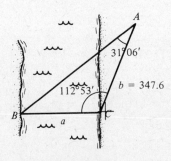

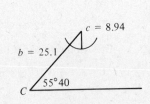

Figure 8.13   Figure 8.14

EXAMPLE 3   ☐   Solve triangle ABC if $C = 55° \, 40'$, $c = 8.94$ m, and $b = 25.1$ m.

Let us look first for angle B. We have

$$\frac{b}{\sin B} = \frac{c}{\sin C}$$

$$\frac{25.1}{\sin B} = \frac{8.94}{\sin 55° \, 40'}$$

$$\sin B = \frac{25.1 \sin 55° \, 40'}{8.94}$$

sin B = 2.3184.

Sin B is greater than 1. This is impossible, since $-1 \leq \sin B \leq 1$, for any angle B. Therefore, triangle ABC does not exist. See Figure 8.14. ☐

When any two angles and the length of a side of a triangle are given, the law of sines can be applied directly to solve the triangle. However, if only one angle and two sides are given, the triangle may not exist, as in Example 3, or there may be more than one triangle satisfying the given conditions. For example, suppose we know the measure of acute angle A of triangle ABC, the length of side a, and the length of side b. To show this information, draw angle A having a terminal side of length b. Now draw a side of length a opposite angle A. The chart in Figure 8.15 shows that there might be more than one possible outcome. This situation is called the **ambiguous case of the law of sines.**

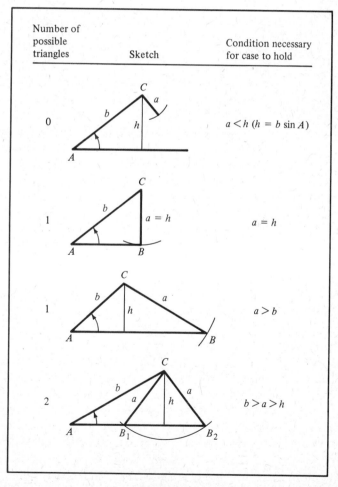

Number of possible triangles	Sketch	Condition necessary for case to hold
0		$a < h$ $(h = b \sin A)$
1		$a = h$
1		$a > b$
2		$b > a > h$

Figure 8.15

If angle A is obtuse, there are two possible outcomes as shown in the chart in Figure 8.16.

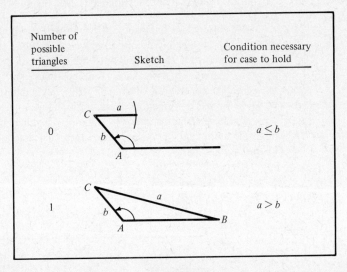

Number of possible triangles	Sketch	Condition necessary for case to hold
0		$a \leq b$
1		$a > b$

Figure 8.16

It is possible to derive formulas that show which of the various cases exist for a particular set of numerical data. However, this work is unnecessary if we use the law of sines. For example, if we use the law of sines and find that sin B is greater than 1, there is no triangle at all. (Why?) The case where we get two different triangles is illustrated in the next example.

EXAMPLE 4   Solve triangle ABC if $A = 55°\,20'$, $a = 22.8$, and $b = 24.9$.
To begin, let us use the law of sines to find angle B.

$$\frac{a}{\sin A} = \frac{b}{\sin B}$$

$$\frac{22.8}{\sin 55°\,20'} = \frac{24.9}{\sin B}$$

$$\sin B = \frac{24.9 \sin 55°\,20'}{22.8}$$

$$\sin B = .8982$$

Since sin $B < 1$, there is at least one triangle. Figure 8.17 shows the case if there are two triangles. We will assume there are two triangles and find the two values of B.
Since sin $B = .8982$, we find that one value of B is

$$B = 64°\,00'.$$

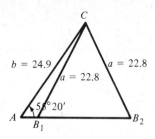

Figure 8.17

From the identity $\sin(180° - B) = \sin B$, another value of B is

$$B = 180° - 64° \, 00'$$
$$B = 116° \, 00'.$$

To keep track of these two different values of B, let

$$B_1 = 116° \, 00' \quad \text{and} \quad B_2 = 64° \, 00'.$$

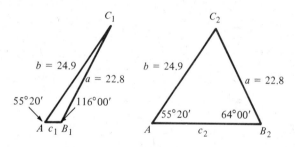

Figure 8.18

We must now separately solve triangles AB_1C_1 and AB_2C_2 shown in Figure 8.18. Since B_1 is the larger of the two values of B, we find C_1 next.

$$C_1 = 180° - A - B_1$$
$$C_1 = 8° \, 40'.$$

Had this answer been negative, there would have been only one triangle. This is why we used the larger angle first. Now, use the law of sines to find c_1.

$$\frac{a}{\sin A} = \frac{c_1}{\sin C_1}$$

$$\frac{22.8}{\sin 55° \, 20'} = \frac{c_1}{\sin 8° \, 40'}$$

$$c_1 = \frac{22.8 \sin 8° \, 40'}{\sin 55° \, 20'}$$

$$c_1 = 4.18$$

To solve triangle AB_2C_2, first find C_2.

$$C_2 = 180° - A - B_2$$

$$C_2 = 60° \ 40'$$

By the law of sines,

$$\frac{22.8}{\sin 55° \ 20'} = \frac{c_2}{\sin 60° \ 40'}$$

$$c_2 = \frac{22.8 \sin 60° \ 40'}{\sin 55° \ 20'}$$

$$c_2 = 24.2. \quad \blacksquare$$

The method we used to derive the law of sines can also be used to derive a useful formula for the area of a triangle. A familiar formula for the area of a triangle is $K = \frac{1}{2}bh$, where K represents the area, b the base, and h the height. This formula cannot always be used, since in practice h is often unknown. To find a more useful formula, refer to triangle ABC in Figure 8.19. (Again, we derive this formula only for an acute angle A. For the cases in which A is 90° or an obtuse angle, see Exercise 54 below.)

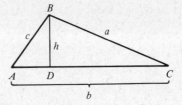

Figure 8.19

A perpendicular has been drawn from B to the base of the triangle. This perpendicular forms two smaller right triangles. Using triangle ABD, we get

$$\sin A = \frac{h}{c}, \quad \text{or} \quad h = c \sin A.$$

Substituting into the formula $K = \frac{1}{2}bh$, we get

$$K = \frac{1}{2} b(c \sin A)$$

$$K = \frac{1}{2} bc \sin A.$$

We can use any other pair of sides and the angle between them, as stated in the next theorem.

Area of a Triangle	The area of any triangle is given by half the product of the lengths of two sides and the sine of the angle between the two sides.

EXAMPLE 5   ☐   Find the area of triangle MNP if $m = 29.7$ meters, $n = 53.9$ meters, and $P = 28° 40'$.

By the theorem,

$$K = \frac{1}{2}(29.7)(53.9) \sin 28° 40',$$

where K represents the area. Using a calculator, we have

$$K \approx 384 \text{ square meters.} \quad \blacksquare$$

EXERCISES 8.2

Solve each of the following triangles that exist.

1. $A = 51°, B = 46°, c = 14$ m

2. $B = 57°, C = 38°, a = 32$ cm

3. $A = 46° 30', B = 52° 50', b = 87.3$ mm

4. $B = 124° 10', C = 18° 40', c = 94.6$ m

5. $A = 68° 10', b = 39.8$ feet, $a = 12.7$ feet

6. $C = 74° 00', b = 69.3$ m, $c = 15.4$ m

7. $B = 20° 50', C = 103° 10', AC = 132$ feet

8. $A = 61° 20', B = 65° 50', AC = 675$ feet

9. $A = 32° 17', B = 22° 42', a = 798.2$ m

10. $B = 47° 51', C = 122° 1', c = 31.82$ m

11. $C = 25.14°, c = 948.4$ cm, $b = 252.8$ cm

12. $A = 39.43°, b = 20.03$ m, $a = 15.72$ m

Find the missing angles in each of the following triangles.

13. $B = 29° 40', a = 39.6$ feet, $b = 28.4$ feet

14. $A = 52° 10', b = 981$ cm, $a = 796$ cm

15. $A = 74° 20', a = 659$ m, $b = 783$ m

16. $C = 82° 10', a = 10.9$ m, $c = 7.62$ m

17. $A = 142° 10', b = 5.43$ feet, $a = 7.29$ feet

18. $B = 113° 40', a = 189$ m, $b = 243$ m

19. $A = 58.42°, a = 56.78$ m, $c = 24.55$ m

20. $C = 102.33°, a = 1.005$ cm, $c = 2.031$ cm

Solve each of the following triangles.

21. $A = 42° 30', a = 15.6$ m, $b = 8.14$ m

22. $C = 52° 20', a = 32.5$ yards, $c = 59.8$ yards

23. $B = 72° 10', b = 78.3$ m, $c = 145$ m

24. $C = 29° 50', a = 8.61$ m, $c = 5.21$ m

25. $B = 32° 50', a = 7540$ cm, $b = 5180$ cm

26. $C = 22° 50', b = 159$ mm, $c = 132$ mm

27. $B = 39° 40', a = 29.8$ m, $b = 23.7$ m

28. $A = 51° 10', c = 798$ cm, $a = 720$ cm

29. $A = 29° 47', b = 292.7$ cm, $a = 289.6$ cm

30. $C = 58° 42', c = 39.87$ feet, $a = 42.41$ feet

Solve each of the following exercises. Recall that bearing was discussed in Section 8.1.

31. To find the distance AB across a river, a distance $BC = 354$ meters is laid off on one side of the river. In triangle ABC, it is found that $B = 112° 10'$ and $C = 15° 20'$. Find AB.

32. To determine the distance RS across a deep canyon, Joanna lays off a distance $TR = 582$ yards. She then finds that in triangle RST, $T = 32° 50'$ and $R = 102° 20'$. Find RS.

33. Radio direction finders are placed at points A and B, which are 3.46 miles apart on an east-west line, with A west of B. From A the bearing of a certain radio transmitter is $47° 40'$, while from B the bearing is $302° 30'$. Find the distance between the transmitter and A.

34. A ship is sailing due north. Captain Odjakjian notices that the bearing of a lighthouse 12.5 kilometers distant is $38° 50'$. Later on, the captain notices that the bearing of the lighthouse has become $135° 50'$. How far did the ship travel between the two observations of the lighthouse?

35. A hill slopes at an angle of $12° 20'$ with the horizontal. The angle of elevation from the base of the hill to the top of a 457-foot tower at the top of the hill is $35° 50'$. Find the distance from the base of the hill to the base of the tower.

36. Mark notices that the bearing of a tree on the opposite bank of a river is $115° 20'$. Lisa is on the same bank as Mark but 428 meters away. She notices that the bearing of the tree is $45° 20'$. The river is flowing north between parallel banks. What is the distance across the river?

37. Three gears are arranged as shown in the figure. Find angle θ.

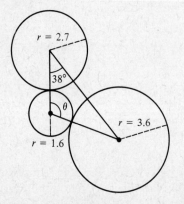

38. Three atoms with atomic radii of 2, 3, and 4.5 are arranged as in the figure. Find the distance between the centers of atoms A and C.

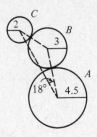

39. A surveyor reported the following data about a piece of property: "The property is triangular in shape, with dimensions as shown in the figure." Use the law of sines to see if such a piece of property could exist.

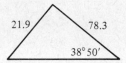

40. The surveyor tries again: "A second triangular piece of property has dimensions as shown." This time it turns out that the surveyor did not consider every possible case. Use the law of sines to show why.

Find the area of each of the following triangles.

41. $A = 46° 30'$, $b = 12.7$ feet, $c = 8.90$ feet

42. $C = 67° 40'$, $b = 46.7$ m, $a = 38.2$ m

43. $B = 124° 30'$, $a = 30.4$ cm, $c = 28.4$ cm

44. $C = 142° 50'$, $a = 21.9$ m, $b = 24.6$ m

45. $A = 56° 49'$, $b = 32.68$ m, $c = 52.82$ m

46. $A = 34° 57'$, $b = 47.83$ m, $c = 28.65$ m

Prove that each of the following statements are true

for any triangle ABC, with corresponding sides a, b, and c.

47. $\dfrac{a+b}{b} = \dfrac{\sin A + \sin B}{\sin B}$

48. $\dfrac{a-b}{a+b} = \dfrac{\sin A - \sin B}{\sin A + \sin B}$

49. $\dfrac{a+b}{c} = \dfrac{\cos \frac{1}{2}(A-B)}{\sin \frac{1}{2}C}$

50. $\dfrac{a-b}{c} = \dfrac{\sin \frac{1}{2}(A-B)}{\cos \frac{1}{2}C}$

51. In any triangle having sides a, b, and c, it must be true that $a + b > c$. Use this fact and the law of sines to show that $\sin A + \sin B > \sin (A + B)$ for any two angles A and B of a triangle.

52. Show that the area of a triangle having sides a, b, and c and corresponding angles A, B, and C is given by

$$\frac{a^2 \sin B \sin C}{2 \sin A}.$$

53. Prove the law of sines if angle A is a right angle or if angle A is an obtuse angle.

54. Derive the formula for the area of a triangle, $(1/2)bc \sin A$, if angle A is a right angle or an obtuse angle.

8.3 The Law of Cosines

If two sides and the angle between the two sides are given, the law of sines cannot be used to solve the triangle. (Try it.) Also, if all three of the sides of a triangle are given, the law of sines again cannot be used to find the unknown angles. For both these cases we need the law of cosines.

To obtain this law, let *ABC* be any oblique triangle. We choose a coordinate system so that the origin is at vertex *B* and side *BC* is along the positive *x*-axis. See Figure 8.20.

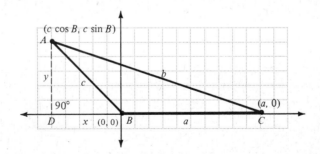

Figure 8.20

Let (x, y) be the coordinates of vertex A of the triangle. Verify that for angle B, whether obtuse or acute,

$$\sin B = \frac{y}{c} \quad \text{and} \quad \cos B = \frac{x}{c}.$$

(Here we assume x is negative if B is obtuse.) From these results,

$$y = c \sin B \quad \text{and} \quad x = c \cos B,$$

so that the coordinates of point A become

$$(c \cos B, \ c \sin B).$$

We know point C has coordinates $(a, 0)$ and AC has length b. By the distance formula

$$b = \sqrt{(c \cos B - a)^2 + (c \sin B)^2}.$$

Squaring both sides and simplifying gives

$$
\begin{aligned}
b^2 &= (c \cos B - a)^2 + (c \sin B)^2 \\
&= c^2 \cos^2 B - 2ac \cos B + a^2 + c^2 \sin^2 B \\
&= a^2 + c^2 (\cos^2 B + \sin^2 B) - 2ac \cos B \\
&= a^2 + c^2(1) - 2ac \cos B \\
b^2 &= a^2 + c^2 - 2ac \cos B.
\end{aligned}
$$

This result is one form of the **law of cosines.** (If $B = 90°$, this result reduces down to the Pythagorean theorem.) In the work above, we could just as easily have placed A or C at the origin. This would have given the same result but with the variables rearranged. These various forms of the law of cosines are summarized in the following theorem.

The Law of Cosines

In any triangle ABC, with sides a, b, and c,

$$a^2 = b^2 + c^2 - 2bc \cos A$$
$$b^2 = a^2 + c^2 - 2ac \cos B$$
$$c^2 = a^2 + b^2 - 2ab \cos C.$$

EXAMPLE 1 Solve triangle ABC if $A = 42°20'$, $b = 12.9$, and $c = 15.4$. See Figure 8.21.

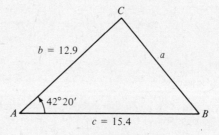

Figure 8.21

We can find a by using the law of cosines.

$$a^2 = b^2 + c^2 - 2bc \cos A$$
$$a^2 = (12.9)^2 + (15.4)^2 - 2(12.9)(15.4) \cos 42° \, 20'$$
$$= 166.41 + 237.16 - (397.32)(.7392)$$
$$= 403.57 - 293.70$$
$$a^2 = 109.87$$
$$a = 10.5$$

Now we know a, b, c, and A. We can use the law of sines to next find either angle B or angle C. If there is an obtuse angle in the triangle, it will be the larger of B and C. Since we can't tell from the sine of the angle whether it is acute or obtuse, it is a good idea to find the smaller angle (which will be acute) first. We know $B < C$ because $b < c$, so we use the law of sines to find B.

$$\frac{10.5}{\sin 42° \, 20'} = \frac{12.9}{\sin B}$$
$$\sin B = \frac{12.9 \sin 42° \, 20'}{10.5}$$
$$\sin B = .8274$$
$$B = 55° \, 50'$$

Finally, we can find C.

$$C = 180° - A - B$$
$$C = 81° \, 50' \quad \blacksquare$$

EXAMPLE 2   Solve triangle ABC if $C = 132° \, 40'$, $b = 259$, and $a = 423$.

Here we use the form of the law of cosines $c^2 = a^2 + b^2 - 2ab \cos C$. Inserting the given data gives

$$c^2 = a^2 + b^2 - 2ab \cos C$$
$$c^2 = (423)^2 + (259)^2 - 2(423)(259) \cos 132° \, 40'.$$

To find $\cos 132° \, 40'$, recall that $\cos C = -\cos(180° - C)$. Thus,

$$\cos 132° \, 40' = -\cos 47° \, 20' = -.6777.$$

Now we can finish finding c.

$$c^2 = (423)^2 + (259)^2 - 2(423)(259)(-.6777)$$
$$= 178{,}929 + 67{,}081 + 148{,}494$$
$$c^2 = 394{,}504$$

Finally,

$$c = 628.$$

The law of sines can be used to complete the solution. Check that $A = 29° 40'$ and $B = 17° 40'$. ☐

EXAMPLE 3 ☐ Solve triangle ABC if $a = 9.47$, $b = 15.9$, and $c = 21.1$.

Again we must use the law of cosines. When we use the law of cosines, we should find the largest angle first in case it is obtuse, so let us look for C. We have

$$c^2 = a^2 + b^2 - 2ab \cos C$$

or $\cos C = \dfrac{a^2 + b^2 - c^2}{2ab}$.

Substituting the given values leads to

$$\cos C = \frac{(9.47)^2 + (15.9)^2 - (21.1)^2}{2(9.47)(15.9)}$$

$$= \frac{-102.7191}{301.146}$$

$$\cos C = -.3411.$$

From this last result,

$$C = 110° 00'.$$

(We know that C is obtuse since $\cos C$ is negative.) Use the law of sines to find B. Verify that $B = 45° 00'$. Since $A = 180° - B - C$,

$$A = 25° 00'. \quad ☐$$

The law of cosines can be used to find a formula for the area of a triangle when only the lengths of the three sides of the triangle are known. This formula is given as the next theorem. (For a proof of this theorem, see Exercise 54 below.)

Heron's Area Formula

If a triangle has sides of lengths a, b, and c, and if

$$s = \frac{1}{2}(a + b + c),$$

then the area of the triangle is

$$K = \sqrt{s(s - a)(s - b)(s - c)}.$$

EXAMPLE 4 ☐ Find the area of the triangle having sides of lengths $a = 29.7$ ft, $b = 42.3$ ft, and $c = 38.4$ ft.

To use Heron's area formula, first find s.

$$s = \frac{1}{2}(a + b + c)$$

$$s = \frac{1}{2} (29.7 + 42.3 + 38.4)$$

$$s = 55.2$$

The area is then given by

$$K = \sqrt{s(s - a)(s - b)(s - c)}$$
$$= \sqrt{55.2(55.2 - 29.7)(55.2 - 42.3)(55.2 - 38.4)}$$
$$K = \sqrt{55.2(25.5)(12.9)(16.8)}$$
$$K = 552 \text{ ft}^2. \quad \square$$

As we have seen, there are four possible cases that can come up in solving an oblique triangle. These cases are summarized as follows. (The first two cases require the law of sines, while the second two require the law of cosines. In all four cases, we assume that the given information actually produces a triangle.)

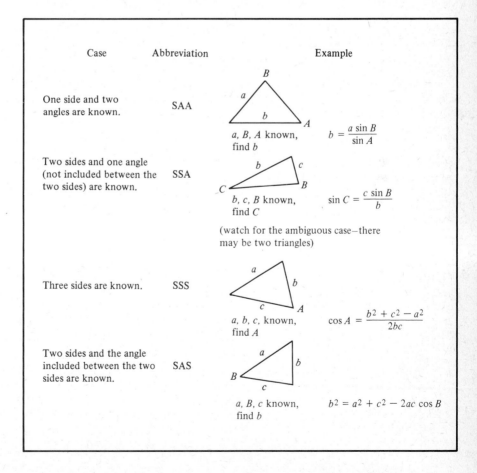

Case	Abbreviation	Example
One side and two angles are known.	SAA	a, B, A known, find b $\qquad b = \dfrac{a \sin B}{\sin A}$
Two sides and one angle (not included between the two sides) are known.	SSA	b, c, B known, find C $\qquad \sin C = \dfrac{c \sin B}{b}$ (watch for the ambiguous case—there may be two triangles)
Three sides are known.	SSS	a, b, c, known, find A $\qquad \cos A = \dfrac{b^2 + c^2 - a^2}{2bc}$
Two sides and the angle included between the two sides are known.	SAS	a, B, c known, find b $\qquad b^2 = a^2 + c^2 - 2ac \cos B$

EXERCISES
8.3

Solve each of the following triangles. Use a calculator or interpolate as necessary. ▣

1. $A = 39° 50'$, $b = 6.74$ in, $c = 5.92$ in
2. $B = 35° 10'$, $a = 5.78$ yd, $c = 4.87$ yd
3. $C = 45° 40'$, $b = 8.94$ m, $a = 7.23$ m
4. $A = 67° 20'$, $b = 37.9$ m, $c = 40.8$ m
5. $A = 80° 40'$, $b = 143$ cm, $c = 89.6$ cm
6. $C = 72° 40'$, $a = 327$ ft, $b = 251$ ft
7. $B = 74.80'$, $a = 8.919$ in, $c = 6.427$ in
8. $C = 59.70°$, $a = 3.725$ mi, $b = 4.698$ mi
9. $A = 112° 50'$, $b = 6.28$ m, $c = 12.2$ m
10. $B = 168° 10'$, $a = 15.1$ cm, $c = 19.2$ cm
11. $C = 24° 49'$, $a = 251.3$ m, $b = 318.7$ m
12. $B = 52° 28'$, $a = 7598$ in, $c = 6973$ in

Find all the angles in each of the following triangles. Round answers to the nearest ten minutes.

13. $a = 2$ ft, $b = 3$ ft, $c = 4$ ft
14. $a = 3$ m, $b = 4$ m, $c = 6$ m
15. $a = 9.3$ cm, $b = 5.7$ cm, $c = 8.2$ cm
16. $a = 28$ ft, $b = 47$ ft, $c = 58$ ft
17. $a = 42.9$ m, $b = 37.6$ m, $c = 62.7$ m
18. $a = 189$ yd, $b = 214$ yd, $c = 325$ yd
19. $AB = 1240$ ft, $AC = 876$ ft, $BC = 918$ ft
20. $AB = 298$ m, $AC = 421$ m, $BC = 324$ m

Use a calculator or interpolation to find all the angles in each of the following triangles to the nearest minute. ▣

21. $a = 18.92$ in, $b = 24.35$ in, $c = 22.16$ in
22. $a = 250.8$ ft, $b = 212.7$ ft, $c = 324.1$ ft
23. $a = 7.095$ m, $b = 5.613$ m, $c = 11.53$ m
24. $a = 15,250$ m, $b = 17,890$ m, $c = 27,840$ m

Solve each of the following problems.

25. Points A and B are on opposite sides of Lake Yankee. From a third point, C, the angle between the lines of sight to A and B is $46° 20'$. If AC is 350 m long and BC is 286 m long, find AB.

26. The sides of a parallelogram are 4.0 cm and 6.0 cm. One angle is 58° while another is 122°. Find the lengths of the diagonals of the figure.

27. Airports A and B are 450 km apart, on an east-west line. Tom flies in a northeast direction from A to airport C. From C he flies 359 km on a bearing of $128° 40'$ to B. How far is C from A?

28. Two ships leave a harbor together, traveling on courses that have an angle of $135° 40'$ between them. If they each travel 402 mi, how far apart are they?

29. Pearl took a plane from A to B, a distance of 350 mi. Then her plane continued from B to C, a distance of 400 mi. Finally, she returned to A, a distance of 300 mi. If A and B are on an east-west line, find the bearing from A to C. (Assume C is north of the line through A and B.)

30. A hill slopes at an angle of $12° 28'$ with the horizontal. From the base of the hill, the angle of inclination of a 459.0 ft tower at the top of the hill is $35° 59'$. How much rope would be required to reach from the top of the tower to the bottom of the hill?

31. Two factories blow their whistles at 5 o'clock exactly. A man hears the two blasts at 3 seconds and 6 seconds after 5, respectively. The angle between his lines of sight to the two factories is $42° 10'$. If sound travels 344 m per second, how far apart are the factories?

32. A parallelogram has sides of length 25.9 cm and 32.5 cm. The longer diagonal has a length of 57.8 cm. Find the angle opposite the diagonal.

33. A plane flying a straight course observes a mountain at a bearing 24° 10′ to the right of its course. At that time, the plane is 7.92 km from the mountain. After flying awhile, the bearing to the mountain becomes 32° 40′. How far is the plane from the mountain when the second bearing is found?

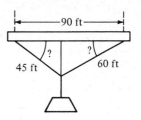

34. The aircraft carrier *Tallahassee* is traveling at sea on a steady course with a bearing of 30° at 32 miles per hour. Patrol planes on the carrier have enough fuel for 2.6 hours of flight when traveling at a speed of 520 miles per hour. One of the pilots takes off on a bearing of 338° and then turns and heads in a straight line, so as to be able to catch the carrier, landing on the deck at the exact instant that his fuel runs out. If the pilot left at 2 p.m., at what time did he turn to head for the carrier?

To help predict eruptions from the volcano Mauna Loa on the island of Hawaii, scientists keep track of the volcano's movement by using a "super triangle" with vertices on the three volcanos shown on the map below. (For example, in a recent year, Mauna Loa moved six inches north and northwest—a result of increasing internal pressure.) The data in the following exercises has been rounded.

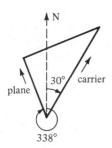

35. A crane with a counterweight is shown in the figure. Find the distance between points *A* and *B*.

36. A weight is supported by cables attached to both ends of a balance beam. See the figure above right. What angles are formed between the beams and the cables?

37. $AB = 22.47928$ mi, $AC = 28.14276$ mi, $A = 58.56989°$; find BC.

38. $AB = 22.47928$ mi, $BC = 25.24983$ mi, $A = 58.56989°$; find B.

Find the area of each of the following triangles.

39. $a = 15$ in, $b = 19$ in, $c = 24$ in

40. $a = 27$ m, $b = 40$ m, $c = 34$ m

41. $a = 154$ cm, $b = 179$ cm, $c = 183$ cm

42. $a = 25.4$ yd, $b = 38.2$ yd, $c = 19.8$ yd

43. $a = 76.3$ ft, $b = 109$ ft, $c = 98.8$ ft

44. $a = 15.89$ in, $b = 21.74$ in, $c = 10.92$ in

45. $a = 74.14$ ft, $b = 89.99$ ft, $c = 51.82$ ft

46. $a = 1.096$ km, $b = 1.142$ km, $c = 1.253$ km

47. Sam wants to paint a triangular region 75 by 68 by 85 m. A can of paint covers 75 m² of area. How many cans (to the next higher can) will he need?

48. How many cans would be needed if the region were 8.2 by 9.4 by 3.8 m?

Use the fact that $\cos A = (b^2 + c^2 - a^2)/(2bc)$ *to show that each of the following is true.*

49. $1 + \cos A = \dfrac{(b + c + a)(b + c - a)}{2bc}$

50. $1 - \cos A = \dfrac{(a - b + c)(a + b - c)}{2bc}$

51. $\cos \dfrac{A}{2} = \sqrt{\dfrac{s(s - a)}{bc}}$

$\left(\text{Recall: } \cos \dfrac{A}{2} = \sqrt{\dfrac{1 + \cos A}{2}} \right)$

52. $\sin \dfrac{A}{2} = \sqrt{\dfrac{(s - b)(s - c)}{bc}}$

$\left(\text{Recall: } \sin \dfrac{A}{2} = \sqrt{\dfrac{1 - \cos A}{2}} \right)$

53. The area of a triangle having sides b and c and angle A is given by $\dfrac{1}{2} bc \sin A$. Show that this result can be written as

$$\sqrt{\frac{1}{2} bc(1 + \cos A) \cdot \frac{1}{2} bc(1 - \cos A)}.$$

54. Use the results of Exercises 49–53 to prove Heron's Area Formula.

55. Let a and b be the equal sides of an isosceles triangle. Prove that $c^2 = 2a^2 (1 - \cos C)$.

56. Let point D on side AB of triangle ABC be such that CD bisects angle C. Show that $AD/DB = b/a$.

The following diagram is an engineering drawing used in the construction of Michigan's Mackinac Straits Bridge.*

57. Find the angles of the triangle formed by the Mackinac West Base, Green Island, and St. Ignace West Base.

58. Find the angles of the triangle formed by A_2, St. Ignace West Base, and St. Ignace East Base.

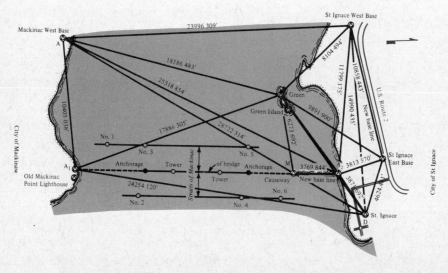

*Reproduced with permission from *Mackinac Bridge*, by G. Edwin Pidcoak, *Civil Engineering*, May 1956, p. 43.

8.4   Vectors and Applications

With what we have learned so far in this chapter, we can find the measure of all six of the parts of a triangle, given at least one side and any two other measures. In this section, we look at applications of this work to *vectors*.

Many quantities in mathematics involve magnitudes. For example, we have seen magnitudes such as 45 pounds or 60 miles per hour. These quantities are often called **scalars.** We sometimes need to work with quantities involving both magnitude and direction, called **vector quantities.** Typical vector quantities include velocities, accelerations, and forces, such as a force of 40 lb acting from the north.

Vector quantities are often represented with vectors. A **vector** is a directed line segment, that is, a line segment pointing in a particular direction. The length of the vector represents the magnitude of the vector quantity. The direction of the vector, indicated with an arrowhead, represents the direction of the quantity. For example, the vector in Figure 8.22 represents a force of 10 pounds applied at an angle of 30° from the horizontal.

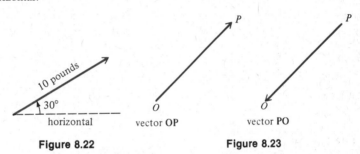

Figure 8.22   **Figure 8.23**

Vectors are often printed in boldface type. When writing vectors by hand, it is customary to use an arrow over the letter or letters. Thus **OP** and $\overrightarrow{OP}$ both represent vector OP. Vectors may be named by using either one lowercase or uppercase letter, or two uppercase letters. When two letters are used, the first indicates the *initial point* and the second indicates the *terminal point* of the vector. Knowing these points gives the direction of the vector. For example, vectors **OP** and **PO** of Figure 8.23 are not the same vectors. They have the same magnitudes, but opposite directions. The magnitude of vector **OP** is written $|\mathbf{OP}|$.

Two vectors are **equal** if and only if they have the same directions and the same magnitudes. In Figure 8.24, vectors **A** and **B** are equal, as are vectors **C** and **D**. Note that vectors **E** and **F** are not equal.

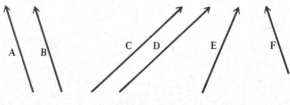

Figure 8.24

As Figure 8.24 shows, equal vectors need not coincide, but they must be parallel. Note that **A** $\neq$ **E** because they do not have the same direction, while **A** $\neq$ **F** because they have different magnitudes, as indicated by their different lengths.

To find the **sum** of two vectors **A** and **B**, written **A** + **B**, place the initial point of vector **B** at the terminal point of vector **A**, as shown in Figure 8.25. The vector with the same initial point as **A** and the same terminal pont as **B** is the sum **A** + **B**. The sum of two vectors is again a vector.

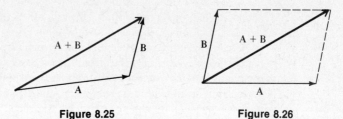

Figure 8.25   Figure 8.26

Another way to find the sum of two vectors is to use the **parallelogram rule.** Place vectors **A** and **B** so that their initial points coincide. Then complete a parallelogram which has **A** and **B** as two sides. The diagonal of the parallelogram with the same initial point as **A** and **B** is the same vector sum **A** + **B** that we found by the definition. See Figure 8.26.

Parallelograms can be used to show that vector **B** + **A** is the same as vector **A** + **B,** or that

A + **B** = **B** + **A.**

Thus, vector addition is **commutative.**

The vector sum **A** + **B** is called the **resultant** of vectors **A** and **B.** Each of the vectors **A** and **B** is called a **component** of vector **A** + **B.** In many practical applications, such as surveying, it is necessary to break a vector into its **vertical** and **horizontal components.** These components are two vectors, one vertical and one horizontal, whose resultant is the original vector. As shown in Figure 8.27, vector **OR** is the vertical component and vector **OS** is the horizontal component of **OP.**

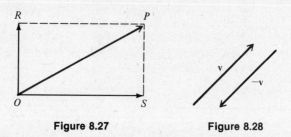

Figure 8.27   Figure 8.28

For every vector **v** there is a vector $-$**v**, which has the same magnitude as **v** but opposite direction. Vector $-$**v** is called the **opposite** of **v.** See Figure 8.28. The sum of **v** and $-$**v** has magnitude 0 and is called a **zero vector.** As with real numbers, to *subtract* vector **B** from vector **A**, find the vector sum **A** + ($-$**B**). See Figure 8.29.

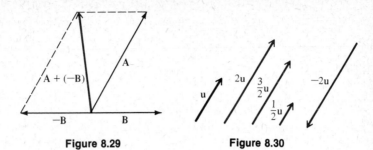

Figure 8.29   **Figure 8.30**

The **scalar product** of a real number (or scalar) k and a vector **u** is the vector $k \cdot \mathbf{u}$ which has magnitude $|k|$ times the magnitude of **u**. As shown in Figure 8.30, the vector $k \cdot \mathbf{u}$ has the same direction as **u** if $k > 0$, and the opposite direction if $k < 0$.

EXAMPLE 1   Vector **w** has magnitude 25.0 and is inclined at an angle of 40° from the horizontal. Find the magnitudes of the horizontal and vertical components of the vector.

In Figure 8.31, the vertical component is labeled **v** and the horizontal component is labeled **u**. Vectors **u**, **v**, and **w** form a right triangle. In this right triangle, $\sin 40° = |\mathbf{v}|/|\mathbf{w}| = |\mathbf{v}|/25.0$, from which

$$|\mathbf{v}| = 25.0 \sin 40° = 25.0(.6428) = 16.1.$$

In the same way, $\cos 40° = |\mathbf{u}|/25.0$, and

$$|\mathbf{u}| = 25.0 \cos 40° = 19.2. \quad \square$$

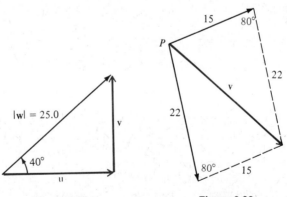

Figure 8.31   **Figure 8.32**

EXAMPLE 2   Two forces of 15 and 22 newtons (a newton is a unit of force used in physics) act on a point in the plane. If the angle between the forces is 100°, find the magnitude of the resultant force.

As shown in Figure 8.32, we can form a parallelogram with the forces as adjacent sides. The angles of the parallelogram adjacent to angle P each measure 80°, and the opposite sides of the parallelogram are equal in length. The resultant force divides the parallelogram into two triangles. Now we can use the law of cosines to get

$$|\mathbf{v}|^2 = 15^2 + 22^2 - 2(15)(22)\cos 80°$$
$$= 225 + 484 - 115$$
$$|\mathbf{v}|^2 = 594$$
$$|\mathbf{v}| = 24. \quad \square$$

Let vector **u** be placed in a plane so that the initial point of the vector is at the origin, $(0, 0)$, and the endpoint is at the point (a, b). A vector with initial point at the origin is called a **position vector** or (sometimes) a **radius vector.** A position vector having endpoint at the point (a, b) is called the **vector** (a, b). For simplicity, the vector (a, b) is written as $\langle a, b \rangle$. The numbers a and b are called the **x-component** and **y-component,** respectively. Figure 8.33 shows the vector $\mathbf{u} = \langle a, b \rangle$. Note that

$$|\mathbf{u}| = \sqrt{a^2 + b^2}.$$

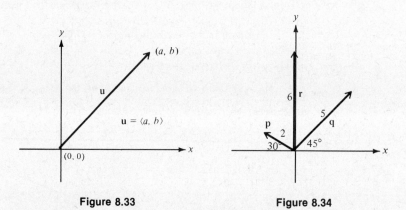

Figure 8.33   Figure 8.34

EXAMPLE 3   Write each of the position vectors of Figure 8.34 in the form $\langle a, b \rangle$.

Vector **p** has length 2 and makes an angle of 30° with the negative x-axis. Thus, $\mathbf{p} = \langle -\sqrt{3}, 1 \rangle$. Vector **q** has length 5 and makes an angle of 45° with the positive x-axis. Thus, $\mathbf{q} = \langle 5\sqrt{2}/2, 5\sqrt{2}/2 \rangle$. Finally, $\mathbf{r} = \langle 0, 6 \rangle$. $\quad \square$

Let vector **OM** in Figure 8.35 be given by $\langle a, b \rangle$, while **ON** is given by $\langle c, d \rangle$. Let **OP** be given by $\langle a + c, b + d \rangle$. Using facts from geometry, points O, N, M and P can be shown to form the vertices of a parallelogram. Since a diagonal of this parallelogram gives the resultant of **OM** and **ON**, we have **OP** = **OM** + **ON**, with the resultant of $\langle a, b \rangle$ and $\langle c, d \rangle$ given by $\langle a + c, b + d \rangle$. In the same way, for any real number k, we have $k \cdot \langle a, b \rangle = \langle ka, kb \rangle$. In summary,

$$\langle a, b \rangle + \langle c, d \rangle = \langle a + c, b + d \rangle$$
$$k \cdot \langle a, b \rangle = \langle ka, kb \rangle.$$

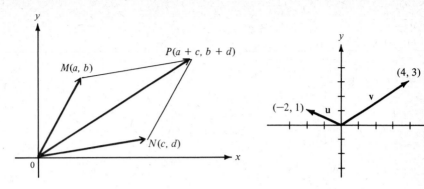

Figure 8.35   **Figure 8.36**

EXAMPLE 4   ☐   Let $\mathbf{u} = \langle -2, 1 \rangle$ and $\mathbf{v} = \langle 4, 3 \rangle$. Find each of the following. See Figure 8.36.

(a) $\mathbf{u} + \mathbf{v} = \langle -2, 1 \rangle + \langle 4, 3 \rangle = \langle -2 + 4, 1 + 3 \rangle = \langle 2, 4 \rangle$

(b) $-2\mathbf{u} = 2 \cdot \langle -2, 1 \rangle = \langle -2(-2), -2(1) \rangle = \langle 4, -2 \rangle$

(c) $4\mathbf{u} + 3\mathbf{v} = 4 \cdot \langle -2, 1 \rangle + 3 \cdot \langle 4, 3 \rangle = \langle -8, 4 \rangle + \langle 12, 9 \rangle$
$= \langle -8 + 12, 4 + 9 \rangle = \langle 4, 13 \rangle$ ☐

The angle between the positive x-axis and a vector, measured in a counterclockwise direction, is called the **direction angle** for the vector. In Figure 8.37, $\mathbf{u}$ has direction angle 60°, while $\mathbf{v}$ has direction angle 180°, and $\mathbf{w}$ has direction angle 280°. Generalizing and using earlier results, we have the following.

> If $\mathbf{u}$ has direction angle θ and magnitude r, then
> $$\mathbf{u} = \langle r \cos \theta, r \sin \theta \rangle.$$

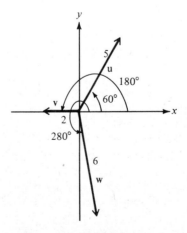

Figure 8.37

EXAMPLE 5

Write the vectors from Figure 8.37 in the form $\langle a, b \rangle$.
Vector **u** in Figure 8.37 has a magnitude of 5 and direction angle 60°. Thus,

$$\mathbf{u} = \langle 5 \cos 60°, 5 \sin 60° \rangle = \left\langle 5 \cdot \frac{1}{2}, 5 \cdot \frac{\sqrt{3}}{2} \right\rangle = \left\langle \frac{5}{2}, \frac{5\sqrt{3}}{2} \right\rangle.$$

Also,

$$\mathbf{v} = \langle 2 \cos 180°, 2 \sin 180° \rangle = \langle 2(-1), 2(0) \rangle = \langle -2, 0 \rangle.$$

Finally,

$$\mathbf{w} = \langle 6 \cos 280°, 6 \sin 280° \rangle$$
$$= \langle 6 \cos 80°, -6 \sin 80° \rangle.$$

(Here, we used reference angles.)

$$\approx \langle 6(.1736), -6(.9848) \rangle$$
$$\mathbf{w} \approx \langle 1.0416, -5.9088 \rangle. \quad \square$$

The next theorem summarizes the basic properties of vectors.

Theorem

Let a vector make an angle θ with the positive x-axis and have magnitude r. Then the x-component of the vector is

$$x = r \cos \theta,$$

and the y-component is

$$y = r \sin \theta.$$

Also, $x^2 + y^2 = r^2,$ and $\theta = \tan^{-1} \dfrac{y}{x} \; (x \neq 0).$

EXAMPLE 6

Figure 8.38 shows vector $\mathbf{u} = \langle 3, -2 \rangle$. The magnitude of **u** is given by $\sqrt{3^2 + (-2)^2} = \sqrt{13}$. To find the direction angle, first find angle θ'.

$$\theta' = \tan^{-1} \frac{-2}{3} \approx \tan^{-1} -0.6667,$$

so that $\theta' \approx -33° \; 40'.$

The direction angle is thus $\theta \approx 360° - 33° \; 40' = 326° \; 20'.$ $\square$

As shown in Figure 8.39, vector **u** can be thought of as the resultant of two vectors; one on the x-axis, having magnitude given by the absolute value of the x-component; and one on the y-axis, having magnitude given by the absolute value of the y-component.

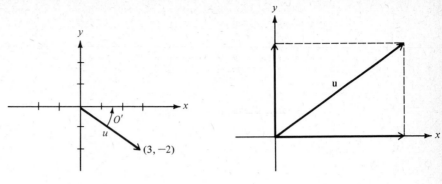

Figure 8.38   Figure 8.39

This same process can be carried out for any vector **u.** Let **u** $= \langle a, b \rangle$. Then

$$\mathbf{u} = \langle a, b \rangle = \langle a, 0 \rangle + \langle 0, b \rangle$$
$$= a \langle 1, 0 \rangle + b \langle 0, 1 \rangle.$$

The vector $\langle 1, 0 \rangle$ is called the **unit vector i,** while $\langle 0, 1 \rangle$ is the unit vector **j.** ("Unit vector" refers to the fact that the magnitude of both **i** and **j** is 1.) Thus, any vector **u** $= \langle a, b \rangle$ may be written as

$$\mathbf{u} = a\mathbf{i} + b\mathbf{j}.$$

Using unit vectors, vector **u** $= \langle 3, -2 \rangle$ can be written as

$$\mathbf{u} = 3\mathbf{i} - 2\mathbf{j}.$$

See Figure 8.40.

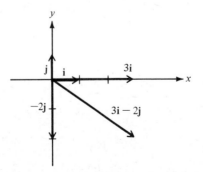

Figure 8.40

We have discussed methods for finding the resultant of two forces. If the resultant of two forces is **u,** then $-\mathbf{u}$ is called the equilibrant of the two forces. The **equilibrant** is the force necessary to counterbalance the joint action of the two forces.

EXAMPLE 7   Find the magnitude of the equilibrant of forces of 48 and 60 newtons acting on a point A, if the angle between the forces is 50°. Then find the angle between the equilibrant and the 48 newton force.

In Figure 8.41, the equilibrant is $-\mathbf{v}$. The magnitude of $\mathbf{v}$, and hence of $-\mathbf{v}$, is found by using triangle ABC and the law of cosines.

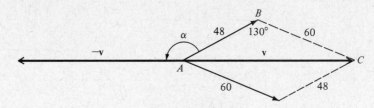

Figure 8.41

$$|\mathbf{v}|^2 = 48^2 + 60^2 - 2\,(48)(60)\cos 130°$$
$$= 2304 + 3600 - 5760\,(-.6428)$$
$$|\mathbf{v}|^2 = 9606.5,$$

or

$$|\mathbf{v}| = 98,$$

to two significant digits.

The angle we wish to find, labeled α in Figure 8.41, can be found by subtracting angle CAB from 180°. Using the law of sines to find angle CAB, we have

$$\frac{98}{\sin 130°} = \frac{60}{\sin CAB}$$
$$\sin CAB = .4690.$$

Using a calculator or table, we have

$$\text{angle } CAB = 28° \; 00'.$$

Thus, $\alpha = 180° - 28° \; 00' = 152° \; 00'$. ☐

EXAMPLE 8   Find the force required to pull a 50 lb weight up a ramp inclined at 20° to the horizontal.

In Figure 8.42, the vertical 50 lb force represents the force due to gravity. (We assume no friction in problems of this type.) The component $\mathbf{BC}$ represents the force with which the body pushes against the ramp, while the component $\mathbf{CA}$ represents the force which must be overcome in order to move the weight up the ramp. Since triangles ABC and DEB are similar, the angle between AB and BC is 20°. Using right triangle ABC,

$$\sin 20° = \frac{|\mathbf{AC}|}{50}$$
$$|\mathbf{AC}| = 50 \sin 20°$$
$$|\mathbf{AC}| = 17.1.$$

Thus, 17 lb of force, to the nearest pound, will be required to pull the weight up the ramp. ☐

Problems involving bearing can also be worked with vectors, as shown in the next example.

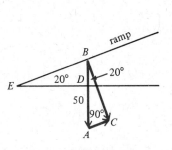

Figure 8.42   Figure 8.43

EXAMPLE 9   ☐   A ship leaves port on a bearing of 28° and travels 8.2 mi. The ship then turns due east and travels 4.3 mi. How far is the ship from port? What is its bearing from port?

In Figure 8.43, vectors **PA** and **AE** represent the ship's line of travel. We need to find the magnitude and bearing of the resultant **PE.** Triangle *PNA* is a right triangle, so angle *NAP* = 90° − 28° = 62°. Then angle *PAE* = 180° − 62° = 118°. Using the law of cosines, we can find $|$**PE**$|$, the magnitude of vector **PE.**

$$|\textbf{PE}|^2 = 8.2^2 + 4.3^2 - 2(8.2)(4.3) \cos 118°$$
$$= 67.24 + 18.49 - 70.52(-.4695)$$
$$|\textbf{PE}|^2 = 118.84$$

Therefore,

$$|\textbf{PE}| = 10.9.$$

To find the bearing of the ship from port, we need to find angle *APE*. Using the law of sines, we have

$$\frac{4.3}{\sin APE} = \frac{10.9}{\sin 118°}$$

$$\sin APE = \frac{4.3 \sin 118°}{10.9} \approx .3483.$$

Finally,   angle *APE* = 20° 20′.

The ship is 10.9 mi from port on a bearing of 28° + 20° 20′ = 48° 20′. ☐

In air navigation, the airspeed of a plane is its speed relative to the air, while the ground speed is its speed relative to the ground. Because of the wind, these two speeds are usually different. The ground speed of the plane is represented by the vector sum of the air speed and wind speed vectors. In Figure 8.44, on the next page, **OQ** represents the airspeed, **QP** the wind speed, and **OP** the ground speed.

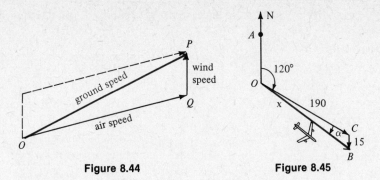

Figure 8.44   **Figure 8.45**

EXAMPLE 10   ☐   A plane with an airspeed of 190 mi per hour is headed on a bearing of 120°. A north wind is blowing (from north to south) at 15 mi per hour. Find the ground speed and the actual bearing of the plane.

In Figure 8.45 the ground speed is represented by $|\mathbf{x}|$. We must find the angle α to find the bearing, which will be $120° + \alpha$. From Figure 8.45, we have angle BCO equal to angle AOC, which equals 120°. We can find $|\mathbf{x}|$ by the law of cosines.

$$|\mathbf{x}|^2 = 190^2 + 15^2 - 2(190)(15) \cos 120°$$
$$= 36,100 + 225 - 5700(-.5)$$
$$|\mathbf{x}|^2 = 39,175$$

Therefore,

$$|\mathbf{x}| = 198.$$

Now find α by using the law of sines.

$$\frac{15}{\sin \alpha} = \frac{198}{\sin 120°}$$
$$\sin \alpha = .0656 = 3° 50'$$

The ground speed is about 198 mi per hour, on a bearing of 123° 50'.   ☐

EXERCISES 8.4

In Exercises 1–4 refer to the vectors below. Name all pairs of vectors

1. which appear to be equal.

2. which are opposites.

3. where the first is a scalar multiple of the second, with the scalar positive.

4. where the first is a scalar multiple of the second, with the scalar negative.

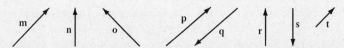

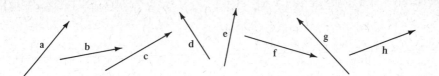

In Exercises 5–20 refer to the vectors pictured above. Draw a sketch to represent each of the following vectors.

5. −**b**   **6.** −**g**   **7.** 3**a**   **8.** 2**h**

9. **a** + **c**   **10.** **a** + **b**   **11.** **h** + **g**   **12.** **e** + **f**

13. **a** + **h**   **14.** **b** + **d**   **15.** **h** + **d**   **16.** **a** + **f**

17. **a** − **c**   **18.** **d** − **e**

19. **a** + (**b** + **c**)   **20.** (**a** + **b**) + **c**

Let **u** = ⟨−2, 5⟩, **v** = ⟨3, −2⟩, and **w** = ⟨−4, 6⟩. Find each of the following vectors.

21. **u** + **v**   **22.** **u** − **w**

23. 6**v** − 2**u**   **24.** −3**u** + 4**w**

In each of the following exercises, **v** has the given direction angle and magnitude. Find the x- and y-components of **v**.

25. θ = 45°, |**v**| = 20   **26.** θ = 75°, |**v**| = 100

27. θ = 60° 10′, |**v**| = 28.6

28. θ = 35° 50′, |**v**| = 47.8

29. θ = 128° 30′, |**v**| = 198

30. θ = 146° 10′, |**v**| = 238

31. θ = 251° 10′, |**v**| = 69.1

32. θ = 302° 40′, |**v**| = 7890

Find the magnitude and direction angle for each of the following vectors.

33. ⟨1, 1⟩   **34.** ⟨−4, 4 √3⟩

35. ⟨8 √2, −8 √2⟩   **36.** ⟨√3, −1⟩

37. ⟨15, −8⟩   **38.** ⟨−7, 24⟩

39. ⟨−6, 0⟩   **40.** ⟨0, −12⟩

Write each of the following vectors in the form a**i** + b**j**.

41. ⟨−5, 8⟩ **42.** ⟨6, −3⟩ **43.** ⟨2, 0⟩ **44.** ⟨0, −4⟩

45. direction angle 45°, magnitude 8

46. direction angle 210°, magnitude 3

47. direction angle 115°, magnitude .6

48. direction angle 208°, magnitude .9

Solve each of the following problems.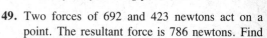

49. Two forces of 692 and 423 newtons act on a point. The resultant force is 786 newtons. Find the angle between the forces.

50. Two forces of 128 and 253 lb act on a point. The equilibrant force is 320 lb. Find the angle between the forces.

51. Find the force required to push a 100-lb box up a ramp inclined 10° with the horizontal.

52. Find the force required to keep a 3000-lb car parked on a hill which makes an angle of 15° with the horizontal.

53. A force of 25 lb is required to push an 80-lb lawn mower up a hill. What angle does the hill make with the horizontal?

54. A force of 500 lb is required to pull a boat up a ramp inclined at 18° with the horizontal. How much does the boat weigh?

55. Anna and Kerry are little dogs. Anna pulls on a rope attached to their doggie dish with a force of 3.89 lb. Kerry pulls on another rope with a force of 4.72 lb. The angle between the forces is 142.8°. Find the direction and magnitude of the equilibrant.

56. Two people are carrying a box. One person exerts a force of 150 lb at an angle of 62.4° with the horizontal. The other person exerts a force of 114 lb at an angle of 54.9°. Find the weight of the box.

57. A crate is supported by two ropes. One rope makes an angle of 46° 20′ with the horizontal and has a tension of 89.6 lb on it. The other rope is horizontal. Find the weight of the crate and the tension in the horizontal rope.

58. Three forces acting at a point are in equilibrium. The forces are 980 lb, 760 lb, and 1220 lb. Find the angles between the directions of the forces. (Hint: arrange the forces to form the sides of a triangle.)

59. A force of 176 lb makes an angle of 78° 50′ with a second force. The resultant of the two forces makes an angle of 41° 10′ with the first force. Find the magnitude of the second force and of the resultant.

60. A force of 28.7 lb makes an angle of 42° 10′ with a second force. The resultant of the two forces makes an angle of 32° 40′ with the first force. Find the magnitude of the second force and of the resultant.

61. A plane flies 650 mi per hour on a bearing of 175.3°. A 25 mi per hour wind, bearing 86.6°, blows against the plane. Find the resulting bearing of the plane.

62. A pilot wants to fly on a bearing of 74.9°. By flying due east, he finds that a 42 mi per hour wind, blowing from the south, puts him on course. Find the airspeed and the ground speed.

63. Starting at point A, a ship sails 18.5 km on a bearing of 189°, then turns and sails 47.8 km on a bearing of 317°. Find the distance of the ship from point A.

64. The distance between points A and B is 1.7 mi. In between A and B is a dark forest, containing a big woolly bear. To avoid the bear, John walks from A a distance of 1.1 mi on a bearing of 325°, and then turns and walks 1.4 mi to B. Find the bearing of B from A.

65. The airline route from San Francisco to Honolulu is on a bearing of 233°. A jet flying at 450 mi per hour on that bearing runs into a wind blowing at 39 mi per hour from a direction 114°. Find the resulting bearing and ground speed of the plane.

66. The bearing of the Evergreen Ranch from Galt is 57° 40′. Harriet is flying there to visit her sister. She flies at 168 mi per hour. A wind is blowing at 27.1 mi per hour from the south. Find the bearing she should fly and her ground speed.

67. What bearing and airspeed are required for a plane to fly 400 mi due north in 2.5 hours, if the wind is blowing from a direction 328° at 11 mi per hour?

68. Paula and Steve are pulling their daughter Jessie on a sled. Steve pulls with a force of 18 lb at an angle of 10°. Paula pulls with a force of 12 lb at an angle of 15°. Find the resultant. See the figure.

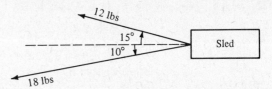

Let $\mathbf{u} = \langle a_1, b_1 \rangle$, $\mathbf{v} = \langle a_2, b_2 \rangle$, $\mathbf{w} = \langle a_3, b_3 \rangle$, *and* $\mathbf{0} = \langle 0, 0 \rangle$. *Let k and m be any real numbers. Prove each of the following statements.*

69. $\mathbf{u} + \mathbf{v} = \mathbf{v} + \mathbf{u}$

70. $\mathbf{u} + (\mathbf{v} + \mathbf{w}) = (\mathbf{u} + \mathbf{v}) + \mathbf{w}$

71. $-1(\mathbf{u}) = -\mathbf{u}$

72. $k(\mathbf{u} + \mathbf{v}) = k\mathbf{u} + k\mathbf{v}$

73. $\mathbf{u} + \mathbf{0} = \mathbf{0}$

74. $\mathbf{u} + (-\mathbf{u}) = \mathbf{0}$

Let $\mathbf{u} = a_1\mathbf{i} + b_1\mathbf{j}$ *and* $\mathbf{v} = a_2\mathbf{i} + b_2\mathbf{j}$. *The* dot product, *or* inner product, *of* $\mathbf{u}$ *and* $\mathbf{v}$, *written* $\mathbf{u} \cdot \mathbf{v}$, *is defined as*

$$\mathbf{u} \cdot \mathbf{v} = a_1 a_2 + b_1 b_2.$$

Find $\mathbf{u} \cdot \mathbf{v}$ *for each of the following pairs of vectors.*

75. $\mathbf{u} = 6\mathbf{i} - 2\mathbf{j}$, $\mathbf{v} = -3\mathbf{i} + 2\mathbf{j}$

76. $\mathbf{u} = 3\mathbf{i} + 2\mathbf{j}$, $\mathbf{v} = -3\mathbf{i} + 7\mathbf{j}$

77. $\mathbf{u} = \langle -6, 8 \rangle$ and $\mathbf{v} = \langle 3, -4 \rangle$

78. $\mathbf{u} = \langle 0, -2 \rangle$ and $\mathbf{v} = \langle -2, 6 \rangle$

79. Let α be the angle between the vectors $\mathbf{u}$ and $\mathbf{v}$, where $0° \leq \alpha \leq 180°$. Show that $\mathbf{u} \cdot \mathbf{v} = |\mathbf{u}| \cdot |\mathbf{v}| \cdot \cos \alpha$.

Use the result of Exercise 79 to find the angle between the following pairs of vectors.

80. $\mathbf{u} = \langle -2, 5 \rangle$ and $\mathbf{v} = \langle 3, -4 \rangle$

81. $\mathbf{u} = \langle 1, 8 \rangle$ and $\mathbf{v} = \langle 2, -5 \rangle$

82. $\mathbf{u} = \langle -6, -2 \rangle$ and $\mathbf{v} = \langle 3, -1 \rangle$

*Prove each of the following properties of the dot product. Assume that **u, v,** and **w** are vectors, and k is a nonzero real number.*

83. $\mathbf{u} \cdot \mathbf{v} = \mathbf{v} \cdot \mathbf{u}$

84. $\mathbf{u} \cdot \mathbf{u} = |\mathbf{u}|^2$

85. $\mathbf{u} \cdot (\mathbf{v} + \mathbf{w}) = \mathbf{u} \cdot \mathbf{v} + \mathbf{u} \cdot \mathbf{w}$

86. $(k \cdot \mathbf{u}) \cdot \mathbf{v} = k \cdot (\mathbf{u} \cdot \mathbf{v})$

87. If $\mathbf{u}$ and $\mathbf{v}$ are not $\mathbf{0}$, and if $\mathbf{u} \cdot \mathbf{v} = 0$, then $\mathbf{u}$ and $\mathbf{v}$ are perpendicular.

88. If $\mathbf{u}$ and $\mathbf{v}$ are perpendicular, then $\mathbf{u} \cdot \mathbf{v} = 0$.

8.5 Trigonometric Form of Complex Numbers

We shall now see how trigonometry and vectors can be used with the complex numbers that we first studied in Chapter 1. We have not yet graphed complex numbers such as $2 - 3i$; to do so we must modify our coordinate system. One way to do this is to call the horizontal axis the **real axis** and the vertical axis the **imaginary axis.** Then the complex number $2 - 3i$ can be graphed as shown in Figure 8.46.

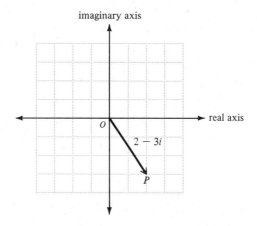

Figure 8.46

Each nonzero complex number graphed in this way determines a unique directed line segment, the segment from the origin to the point representing the complex number. Recall from the previous section that such directed line segments (like **OP** of Figure 8.46) are called vectors.

We know from Chapter 1 how to find the sum of the two complex numbers $4 + i$ and $1 + 3i$.

$$(4 + i) + (1 + 3i) = 5 + 4i$$

Graphically, the sum of two complex numbers is represented by the vector which is the resultant of the vectors corresponding to the two numbers. The vectors representing the complex numbers $4 + i$ and $1 + 3i$, and the resultant vector which represents their sum, $5 + 4i$, are shown in Figure 8.47.

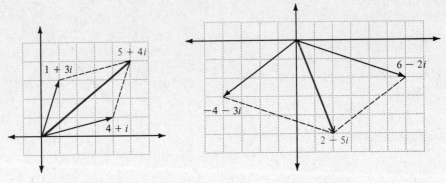

Figure 8.47   **Figure 8.48**

EXAMPLE 1   Find the resultant of $6 - 2i$ and $-4 - 3i$. Graph both complex numbers and their resultant.

The resultant is found by adding the two numbers.

$$(6 - 2i) + (-4 - 3i) = 2 - 5i$$

The graphs are shown in Figure 8.48.   ▢

Figure 8.49 shows a complex number $x + yi$ which determines a vector **OP**. Let r (with $r \geq 0$) represent the length of vector **OP**, and let θ be the smallest positive angle (measured in a counterclockwise direction) between the positive real axis and **OP**. Point P can be located uniquely if we know r and θ. Therefore, we can use r and θ as coordinates of point P. The ordered pair (r, θ) gives the **polar coordinates** of point P. The following relationships between r, θ, x, and y can be verified from Figure 8.49.

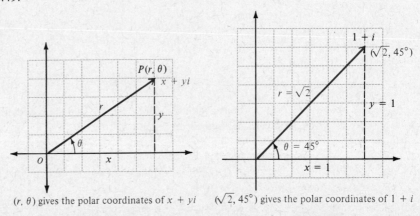

(r, θ) gives the polar coordinates of $x + yi$ $(\sqrt{2}, 45°)$ gives the polar coordinates of $1 + i$

Figure 8.49   **Figure 8.50**

$$x = r \cos \theta \qquad r = \sqrt{x^2 + y^2}$$
$$y = r \sin \theta \qquad \theta = \tan^{-1} y/x$$

**EXAMPLE
2**

☐ Find the polar coordinates of the complex number $1 + i$.

Here $x = 1$ and $y = 1$. Thus, $r = \sqrt{1^2 + 1^2} = \sqrt{2}$, and $\theta = \tan^{-1} 1/1 = \tan^{-1} 1$, so that $\theta = 45°$. (Angle θ cannot be 225° because the complex number lies in quadrant I.) Hence the polar coordinates of $1 + i$ are $(\sqrt{2}, 45°)$, as shown in Figure 8.50. ☐

From the box above, $x = r \cos \theta$ and $y = r \sin \theta$. Substituting these results into $x + yi$ gives

$$x + yi = r \cos \theta + (r \sin \theta) i$$

or

$$x + yi = r (\cos \theta + i \sin \theta).$$

The expression $r (\cos \theta + i \sin \theta)$ is called the **trigonometric form** or **polar form** of the complex number $x + yi$. The number r is called the **modulus** or **absolute value** of $x + yi$, while θ is called the **argument** of $x + yi$. Using the work from Example 2 above, we can express $1 + i$ in trigonometric form as

$$1 + i = \sqrt{2} (\cos 45° + i \sin 45°).$$

**EXAMPLE
3**

☐ Express $2 (\cos 300° + i \sin 300°)$ in standard form.

We know $\cos 300° = 1/2$, while $\sin 300° = -\sqrt{3}/2$. Hence

$$2 (\cos 300° + i \sin 300°) = 2 \left(\frac{1}{2} - i \frac{\sqrt{3}}{2} \right)$$
$$= 1 - i \sqrt{3}. \quad \square$$

**EXAMPLE
4**

☐ Write the following complex numbers in trigonometric form.

(a) $- \sqrt{3} + i$

Since $x = - \sqrt{3}$ and $y = 1$,

$$r = \sqrt{x^2 + y^2} = \sqrt{3 + 1} = 2,$$

$$\theta = \tan^{-1} \frac{y}{x} = \tan^{-1} \left(\frac{1}{- \sqrt{3}} \right) = \tan^{-1} \left(-\frac{\sqrt{3}}{3} \right).$$

As shown in Figure 8.51 next page, θ is in quadrant II, so that $\theta = 150°$. In trigonometric form,

$$x + yi = r(\cos \theta + i \sin \theta)$$
$$- \sqrt{3} + i = 2(\cos 150° + i \sin 150°).$$

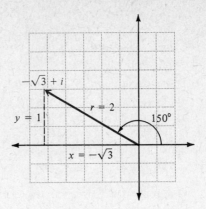

Figure 8.51

(b) $-2 - 2i$

First find r and θ.

$$r = \sqrt{x^2 + y^2} = \sqrt{4 + 4} = 2\sqrt{2}$$

$$\theta = \tan^{-1}\frac{y}{x} = \tan^{-1}\left(\frac{-2}{-2}\right) = \tan^{-1} 1$$

Since θ is in quadrant III, $\theta = 225°$. Thus, the trigonometric form of $-2 - 2i$ is

$$2\sqrt{2}(\cos 225° + i \sin 225°). \quad \blacksquare$$

We found the product of two complex numbers in Chapter 1. For example,

$$(1 + i\sqrt{3})(-2\sqrt{3} + 2i) = -2\sqrt{3} + 2i - 2i(3) + 2i^2\sqrt{3}$$
$$= -2\sqrt{3} + 2i - 6i - 2\sqrt{3}$$
$$(1 + i\sqrt{3})(-2\sqrt{3} + 2i) = -4\sqrt{3} - 4i.$$

We can also obtain this same product by first converting the complex numbers $1 + i\sqrt{3}$ and $-2\sqrt{3} + 2i$ to trigonometric form. Using the method explained above,

$$1 + i\sqrt{3} = 2(\cos 60° + i \sin 60°)$$

and $-2\sqrt{3} + 2i = 4(\cos 150° + i \sin 150°)$.

If we now multiply the trigonometric forms together and use the trigonometric identities for the cosine and the sine of the sum of two angles, we have

$$[2(\cos 60° + i \sin 60°)][4(\cos 150° + i \sin 150°)]$$
$$= 2 \cdot 4(\cos 60° \cdot \cos 150° + i \sin 60° \cdot \cos 150°$$
$$+ i \cos 60° \cdot \sin 150° + i^2 \sin 60° \cdot \sin 150°)$$
$$= 8[(\cos 60° \cdot \cos 150° - \sin 60° \cdot \sin 150°)$$
$$+ i(\sin 60° \cdot \cos 150° + \cos 60° \cdot \sin 150°)]$$
$$= 8[\cos(60° + 150°) + i \sin(60° + 150°)]$$
$$= 8(\cos 210° + i \sin 210°).$$

The modulus of the product, 8, is the *product* of the moduli of the factors, $2 \cdot 4$, and the argument of the product, 210°, is the *sum* of the arguments of the factors, 60° + 150°. Generalizing from this example, we have the following theorem.

Theorem

> If $r_1(\cos \theta_1 + i \sin \theta_1)$ and $r_2(\cos \theta_2 + i \sin \theta_2)$ are any two complex numbers written in trigonometric form, then
>
> $$[r_1(\cos \theta_1 + i \sin \theta_1)][r_2(\cos \theta_2 + i \sin \theta_2)] =$$
> $$r_1 r_2[\cos (\theta_1 + \theta_2) + i \sin (\theta_1 + \theta_2)].$$

The proof of this theorem is given as Exercise 93 below.

EXAMPLE 5 ▢ Find the product of $3(\cos 45° + i \sin 45°)$ and $2(\cos 135° + i \sin 135°)$. Using the theorem,

$$[3(\cos 45° + i \sin 45°)][2(\cos 135° + i \sin 135°)]$$
$$= 3 \cdot 2[\cos (45° + 135°) + i \sin (45° + 135°)]$$
$$= 6(\cos 180° + i \sin 180°),$$

which can be expressed as $6(-1 + i \cdot 0) = 6(-1) = -6$. Thus, the two complex numbers of this example are complex factors of -6. ▢

Now let's consider the quotient of two complex numbers. In standard form, the quotient of the complex numbers $1 + i\sqrt{3}$ and $-2\sqrt{3} + 2i$ is

$$\frac{1 + i\sqrt{3}}{-2\sqrt{3} + 2i} = \frac{(1 + i\sqrt{3})(-2\sqrt{3} - 2i)}{(-2\sqrt{3} + 2i)(-2\sqrt{3} - 2i)}$$

$$= \frac{-2\sqrt{3} - 2i - 6i - 2i^2\sqrt{3}}{12 - 4i^2}$$

$$= \frac{-8i}{16}$$

$$\frac{1 + i\sqrt{3}}{-2\sqrt{3} + 2i} = -\frac{1}{2} i.$$

If we write $1 + i\sqrt{3}$, $-2\sqrt{3} + 2i$, and $-\frac{1}{2} i$ in trigonometric form, we have

$$1 + i\sqrt{3} = 2(\cos 60° + i \sin 60°)$$
$$-2\sqrt{3} + 2i = 4(\cos 150° + i \sin 150°)$$

$$-\frac{1}{2} i = \frac{1}{2}[(\cos (-90°) + i \sin (-90°))].$$

The modulus of the quotient, 1/2, is the quotient of the two moduli, 2 and 4. The argument of the quotient, $-90°$, is the difference of the two arguments, $60° - 150° = -90°$. It is certainly easier to express the quotient of two complex numbers in trigonometric form than in standard form. Generalizing from this example leads to the following theorem. (For the proof of this result, see Exercise 94 below.)

Theorem

If $r_1(\cos \theta_1 + i \sin \theta_1)$ and $r_2(\cos \theta_2 + i \sin \theta_2)$ are complex numbers, where $r_2(\cos \theta_2 + i \sin \theta_2) \neq 0$, then

$$\frac{r_1(\cos \theta_1 + i \sin \theta_1)}{r_2(\cos \theta_2 + i \sin \theta_2)} = \frac{r_1}{r_2} [\cos (\theta_1 - \theta_2) + i \sin (\theta_1 - \theta_2)].$$

EXAMPLE 6

☐   Find the quotient of $10[\cos(-60°) + i \sin(-60°)]$ and $5(\cos 150° + i \sin 150°)$. Write the result in standard form.

By the theorem,

$$\frac{10[\cos(-60°) + i \sin(-60°)]}{5(\cos 150° + i \sin 150°)}$$

$$= \frac{10}{5}[\cos(-60° - 150°) + i \sin (-60° - 150°)]$$

$$= \frac{10}{5}[\cos(-210°) + i \sin(-210°)].$$

Since angles of $-210°$ and $150°$ are coterminal, we may replace $-210°$ with $150°$ to get

$$\frac{10[\cos(-60°) + i \sin(-60°)]}{5(\cos 150° + i \sin 150°)} = 2(\cos 150° + i \sin 150°).$$

Because $\cos 150° = -\sqrt{3}/2$ and $\sin 150° = 1/2$,

$$2(\cos 150° + i \sin 150°) = 2\left(\frac{-\sqrt{3}}{2} + i \cdot \frac{1}{2}\right)$$

$$= -\sqrt{3} + i.$$

The quotient in standard form is $-\sqrt{3} + i$.   ☐

EXAMPLE 7

☐   Note the use of polar coordinates in the following example, taken with permission from *Calculus and Analytic Geometry*, fifth edition, by George Thomas and Ross Finney (Addison-Wesley, 1979).

Karl von Frisch has advanced the following theory about how bees communicate information about newly discovered sources of food. A scout returning to the hive from a flower bed gives away samples of the food and then, if the bed is more than about a hundred yards away, performs a dance to show where the flowers are. The bee runs straight ahead for a centimeter or so, waggling from side to side, and circles

George B. Thomas, Jr./Ross L. Finney, *Calculus and Analytic Geometry*, 5/E, © 1979, Addison-Wesley, Reading, Massachusetts. (Pp. 462, 463.) Reprinted with permission.

back to the starting place. The bee then repeats the straight run, circling back in the opposite direction (See Figure 8.52) The dance continues this way in regular alternation. Exceptionally excited bees have been observed to dance for more than three and a half hours.

   If the dance is performed inside, it is performed on the vertical wall of a honeycomb, with gravity substituting for the sun's position. A vertical straight run means that the food is in the direction of the sun. A run 30° to the right of vertical means that the food is 30° to the right of the sun, and so on. Distance (more accurately, the amount of energy required to reach the food) is communicated by the duration of the straight-run portions of the dance. Straight runs lasting three seconds each are typical for distances of about a half-mile from the hive. Straight runs that last five seconds each mean about two miles. ◻

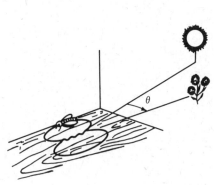 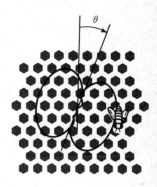

The waggle dance of a scout bee.

Figure 8.52

EXERCISES
8.5

Graph each of the following complex numbers.

1. $-2 + 3i$

2. $-4 + 5i$

3. $8 - 5i$

4. $6 - 5i$

5. $2 - 2i\sqrt{3}$

6. $4\sqrt{2} + 4i\sqrt{2}$

7. $-4i$

8. $3i$

9. -8

10. 2

Find the resultant of each of the following pairs of complex numbers.

11. $2 - 3i, -1 + 4i$

12. $-4 - 5i, 2 + i$

13. $-5 + 6i, 3 - 4i$

14. $8 - 5i, -6 + 3i$

15. $-2, 4i$

16. $5, -4i$

17. $2 + 6i, -2i$

18. $4 - 2i, 5$

19. $7 + 6i, 3i$

20. $-5 - 8i, -1$

Find the polar coordinates of each of the following complex numbers.

21. $2 - 2i$

22. $-1 - i$

23. $3i$

24. $-2i$

25. 5

26. -3

27. $\sqrt{3} + i$ **28.** $-\sqrt{2} - i\sqrt{2}$

29. $1 + i\sqrt{3}$ **30.** $-2 + 2i\sqrt{3}$

Rewrite in standard form the complex numbers whose polar coordinates are given.

31. $(2, 45°)$ **32.** $(3, 60°)$

33. $(1, 135°)$ **34.** $(2, 300°)$

35. $(3, 90°)$ **36.** $(2, 180°)$

37. $(4, 240°)$ **38.** $(5, 270°)$

39. $(1, 330°)$ **40.** $(4, 300°)$

Rewrite the following complex numbers in standard form.

41. $2(\cos 45° + i \sin 45°)$

42. $4(\cos 60° + i \sin 60°)$

43. $10(\cos 90° + i \sin 90°)$

44. $8(\cos 270° + i \sin 270°)$

45. $4(\cos 240° + i \sin 240°)$

46. $2(\cos 330° + i \sin 330°)$

47. $(\cos 30° + i \sin 30°)$

48. $3(\cos 150° + i \sin 150°)$

49. $5(\cos 300° + i \sin 300°)$

50. $6(\cos 135° + i \sin 135°)$

Rewrite each of the following complex numbers in trigonometric form.

51. $3 - 3i$ **52.** $-2 + 2i\sqrt{3}$

53. $-3 - 3i\sqrt{3}$ **54.** $1 + i\sqrt{3}$

55. $\sqrt{3} - i$ **56.** $4\sqrt{3} + 4i$

57. $-5 - 5i$

58. $-\sqrt{2} + i\sqrt{2}$

Using a calculator or Table 5, complete the following chart to the nearest ten minutes.

	Standard Form	Polar Coordinates	Trigonometric Form
59.	$2 + 3i$	_____	_____
60.	_____	$(1, 35°)$	_____
61.	_____	_____	$3(\cos 250° 10' + i \sin 250° 10')$
62.	$-4 + i$	_____	_____
63.	_____	$(2, 160°)$	_____
64.	_____	_____	$2(\cos 310° 20' + i \sin 310° 20')$

Find each of the following products. Write the result in standard form.

65. $[2(\cos 30° + i \sin 30°)][3(\cos 60° + i \sin 60°)]$

66. $[3(\cos 60° + i \sin 60°)][4(\cos 150° + i \sin 150°)]$

67. $[2(\cos 135° + i \sin 135°)][2(\cos 225° + i \sin 225°)]$

68. $[6(\cos 240° + i \sin 240°)][8(\cos 300° + i \sin 300°)]$

69. $[4(\cos 60° + i \sin 60°)][6(\cos 330° + i \sin 330°)]$

70. $[8(\cos 210° + i \sin 210°)][2(\cos 330° + i \sin 330°)]$

71. $[5(\cos 90° + i \sin 90°)][3(\cos 45° + i \sin 45°)]$

72. $[6(\cos 120° + i \sin 120°)][5(\cos 330° + i \sin 330°)]$

Find the following quotients. Write the results in standard form.

73. $\dfrac{3(\cos 60° + i \sin 60°)}{(\cos 30° + i \sin 30°)}$

74. $\dfrac{9(\cos 135° + i \sin 135°)}{3(\cos 45° + i \sin 45°)}$

75. $\dfrac{16(\cos 300° + i \sin 300°)}{8(\cos 60° + i \sin 60°)}$

76. $\dfrac{24(\cos 150° + i \sin 150°)}{2(\cos 30° + i \sin 30°)}$

77. $\dfrac{8}{\sqrt{3} + i}$ **78.** $\dfrac{2i}{-1 - i\sqrt{3}}$

79. $\dfrac{-i}{1 + i}$ **80.** $\dfrac{1}{2 - 2i}$

81. $\dfrac{2\sqrt{6} - 2i\sqrt{2}}{\sqrt{2} - i\sqrt{6}}$

82. $\dfrac{4 + 4i}{2 - 2i}$

Use your calculator to work each of the following problems. Leave your answers in standard form. ▣

83. $[3.7(\cos 27° 15' + i \sin 27° 15')][4.1(\cos 53° 42' + i \sin 53° 42')]$

84. $[2.81(\cos 54° 12' + i \sin 54° 12')][5.8(\cos 82° 53' + i \sin 82° 53')]$

85. $\dfrac{45.3(\cos 127° 25' + i \sin 127° 25')}{12.8(\cos 43° 32' + i \sin 43° 32')}$

86. $\dfrac{2.94(\cos 1.5032 + i \sin 1.5032)}{10.5(\cos 4.6528 + i \sin 4.6528)}$

Find and graph all complex numbers c satisfying the following conditions.

87. $|c| = 1$ **88.** $|c| > 1$

89. The real part of c is 1.

90. The imaginary part of c is 1.

91. The real and imaginary parts of c are equal.

92. The real part of c equals c itself.

93. Prove the first theorem in the section.

94. Prove the second theorem.

8.6   De Moivre's Theorem and nth Roots

We can use the results of the previous section to find the square of a complex number. For example, the square of $r(\cos \theta + i \sin \theta)$ is found by multiplying the number by itself.

$$[r(\cos \theta + i \sin \theta)]^2 = [r(\cos \theta + i \sin \theta)][r(\cos \theta + i \sin \theta)]$$
$$= r \cdot r[\cos (\theta + \theta) + i \sin (\theta + \theta)]$$
$$[r(\cos \theta + i \sin \theta)]^2 = r^2(\cos 2\theta + i \sin 2\theta)$$

In the same way, we can show

$$[r(\cos \theta + i \sin \theta)]^3 = r^3(\cos 3\theta + i \sin 3\theta).$$

These results suggest the plausibility of the following theorem.

De Moivre's Theorem

> If $r(\cos \theta + i \sin \theta)$ is a complex number expressed in trigonometric form, and if n is any positive integer, then
>
> $$[r(\cos \theta + i \sin \theta)]^n = r^n(\cos n\theta + i \sin n\theta).$$

De Moivre's Theorem can be proved by the method of mathematical induction, discussed in Chapter 11.

EXAMPLE 1   Find $(1 + i\sqrt{3})^8$.

We can use De Moivre's theorem if we first convert $1 + i\sqrt{3}$ into trigonometric form.

$$1 + i\sqrt{3} = 2(\cos 60° + i \sin 60°)$$

Now we can apply De Moivre's theorem.

$$(1 + i\sqrt{3})^8 = [2(\cos 60° + i \sin 60°)]^8$$
$$= 2^8[\cos (8 \cdot 60°) + i \sin (8 \cdot 60°)]$$
$$= 256(\cos 480° + i \sin 480°)$$
$$= 256(\cos 120° + i \sin 120°)$$
$$= 256\left(-\frac{1}{2} + i \frac{\sqrt{3}}{2}\right)$$
$$(1 + i\sqrt{3})^8 = -128 + 128i\sqrt{3} \quad \blacksquare$$

De Moivre's Theorem is also used to find the nth roots of complex numbers. By definition, for a positive integer n,

nth Root

> the complex number $a + bi$ is an **nth root** of the complex number $x + yi$ if
>
> $$(a + bi)^n = x + yi.$$

We can use De Moivre's theorem to find nth roots of complex numbers. For example, to find the cube roots of the complex number $8(\cos 135° + i \sin 135°)$, we look for a complex number, say $r(\cos \alpha + i \sin \alpha)$, that will satisfy

$$[r(\cos \alpha + i \sin \alpha)]^3 = 8(\cos 135° + i \sin 135°).$$

By De Moivre's theorem, this equation becomes

$$r^3(\cos 3\alpha + i \sin 3\alpha) = 8(\cos 135° + i \sin 135°).$$

One way to satisfy this equation is to set $r^3 = 8$ and

$$\cos 3\alpha + i \sin 3\alpha = \cos 135° + i \sin 135°.$$

The first condition implies that $r = 2$, and the second implies

$$\cos 3\alpha = \cos 135° \quad \text{and} \quad \sin 3\alpha = \sin 135°$$

or

$$\cos 3\alpha = \frac{-\sqrt{2}}{2} \quad \text{and} \quad \sin 3\alpha = \frac{\sqrt{2}}{2}.$$

This last pair of equations can be satisfied if and only if

$$3\alpha = 135° + 360° \cdot k, \quad k \text{ any integer,}$$

or

$$\alpha = \frac{135° + 360° \cdot k}{3}, \quad k \text{ any integer.}$$

If $k = 0$, $\alpha = \dfrac{135° + 0}{3} = 45°.$

For $k = 1$, $\alpha = \dfrac{135° + 360°}{3} = \dfrac{495°}{3} = 165°.$

When $k = 2$, $\alpha = \dfrac{135° + 720°}{3} = \dfrac{855°}{3} = 285°.$

In the same way, we get $\alpha = 405°$ when $k = 3$. But $\sin 405° = \sin 45°$ and $\cos 405° = \cos 45°$. Hence, we get all of the cube roots (3 of them) by letting $k = 0$, 1, and 2. When $k = 0$ we get the root

$$2(\cos 45° + i \sin 45°).$$

When $k = 1$ we have $2(\cos 165° + i \sin 165°)$, and when $k = 2$ we have $2(\cos 285° + i \sin 285°)$. These are the three cube roots of $8(\cos 135° + i \sin 135°)$.

We generalize this result in the following theorem. The proof is a generalization of the discussion above.

nth Root Theorem

If n is any positive integer and r is a positive real number, then the complex number $r(\cos \theta + i \sin \theta)$ has exactly n distinct nth roots. The roots are

$$r^{1/n}(\cos \alpha + i \sin \alpha),$$

$$\alpha = \frac{\theta + 360° \cdot k}{n}, \qquad k = 0, 1, 2, \ldots, n-1.$$

EXAMPLE 2   Find all 4th roots of $-8 + 8i\sqrt{3}$.
First write $-8 + 8i\sqrt{3}$ in trigonometric form as

$$-8 + 8i\sqrt{3} = 16(\cos 120° + i \sin 120°).$$

Here $r = 16$ and $\theta = 120°$. The 4th roots of this number have modulus $16^{1/4} = 2$ and arguments given as follows.

If $k = 0$, $\dfrac{120° + 360° \cdot 0}{4} = 30°$,

if $k = 1$, $\dfrac{120° + 360° \cdot 1}{4} = 120°$,

if $k = 2$, $\dfrac{120° + 360° \cdot 2}{4} = 210°$,

if $k = 3$, $\dfrac{120° + 360° \cdot 3}{4} = 300°$.

Using these angles, the 4th roots are

$2(\cos 30° + i \sin 30°)$
$2(\cos 120° + i \sin 120°)$
$2(\cos 210° + i \sin 210°)$
$2(\cos 300° + i \sin 300°)$.

We can also write these four roots in standard form as $\sqrt{3} + i$, $-1 + i\sqrt{3}$, $-\sqrt{3} - i$, and $1 - i\sqrt{3}$. The graphs of these roots are all on a circle which has center at the origin and radius 2, as shown in Figure 8.53.

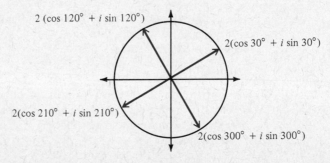

Figure 8.53

EXAMPLE □   Find all 5th roots of 1.
3   We can write 1 in trigonometric form as

$$1 = 1 + 0i = 1(\cos 0° + i \sin 0°).$$

The modulus of the 5th roots is $1^{1/5} = 1$, while the arguments are given by

$$\frac{0° + 360° \cdot k}{5}, \qquad k = 0, 1, 2, 3, \text{ or } 4.$$

Using these arguments the 5th roots become

$$1(\cos 0° + i \sin 0°)$$
$$1(\cos 72° + i \sin 72°)$$
$$1(\cos 144° + i \sin 144°)$$
$$1(\cos 216° + i \sin 216°)$$
$$1(\cos 288° + i \sin 288°).$$

The first of these roots equals 1, but the others cannot easily be expressed in standard form. The five 5th roots all lie on a unit circle and are equally spaced around it, as shown in Figure 8.54.   □

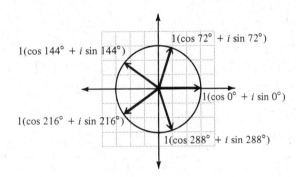

Figure 8.54

EXERCISES
8.6

Find the following powers. Write the result in standard form.

1. $[2(\cos 60° + i \sin 60°)]^3$

2. $[3(\cos 120° + i \sin 120°)]^4$

3. $(\cos 45° + i \sin 45°)^8$

4. $[2(\cos 120° + i \sin 120°)]^3$

5. $(\sqrt{3} + i)^5$

6. $(2\sqrt{2} - 2i\sqrt{2})^6$

7. $(2 - 2i\sqrt{3})^4$

8. $\left(\dfrac{\sqrt{2}}{2} - i\dfrac{\sqrt{2}}{2}\right)^8$

9. $(-2 - 2i)^5$

10. $(-1 + i)^7$

11. $(-.4283 + .5172i)^4$

12. $(1.87615 - 1.42213i)^3$

13. $\left[1.86 \left(\cos \dfrac{5\pi}{9} + i \sin \dfrac{5\pi}{9} \right) \right]^{15}$

14. $\left[24.3 \left(\cos \dfrac{7\pi}{12} + i \sin \dfrac{7\pi}{12} \right) \right]^3$

Find and graph all cube roots of the following complex numbers.

15. 1 **16.** i

17. $-8i$ **18.** $27i$

19. -64 **20.** 27

21. $1 + i\sqrt{3}$ **22.** $2 - 2i\sqrt{3}$

23. $-2\sqrt{3} + 2i$ **24.** $\sqrt{3} - i$

Find the following roots.

25. cube roots of $1.832 + 4.761i$

26. square roots of $11.489i$

27. fourth roots of 259.86

28. fourth roots of -38.476

Find and graph all the following roots of 1.

29. 2nd   **30.** 4th

31. 6th   **32.** 8th

Find and graph all the following roots of i.

33. 2nd   **34.** 4th

Find all solutions of each of the following equations.

35. $x^3 - 1 = 0$ **36.** $x^3 + 1 = 0$
(Hint: write this
equation as   **37.** $x^3 + i = 0$
$x^3 = 1$)   **38.** $x^4 + i = 0$

39. $x^3 - 8 = 0$ **40.** $x^3 + 27 = 0$

41. $x^4 + 1 = 0$ **42.** $x^4 + 16 = 0$

43. $x^4 - i = 0$ **44.** $x^5 - i = 0$

45. $x^3 - (4 + 4i\sqrt{3}) = 0$

46. $x^4 - (8 + 8i\sqrt{3}) = 0$

Use your calculator to find all solutions of each of the following equations.

47. $x^3 + 4 - 5i = 0$

48. $x^5 + 2 + 3i = 0$

49. $x^2 + (3.72 + 8.24i) = 0$

50. $x^4 - (5.13 - 4.27i) = 0$

Let $z = a + bi$. Solve the following equations for z.

51. $z^2 = 1 + i$

52. $z^2 = -\sqrt{2} + i\sqrt{2}$

53. $z^2 = 3 - 3i$ **54.** $z^2 = -\sqrt{3} - 1$

55. Let α be an nth root of 1 (that is, $\alpha^n = 1$). Show that $\alpha^{n-1} + \alpha^{n-2} + \cdots + \alpha + 1 = 0$. (Hint: divide $\alpha^n - 1$ by $\alpha - 1$.)

56. Prove the theorem on nth roots.

8.7 Polar Equations

In Section 8.5 we saw how to graph a point given its polar coordinates, (r, θ). An equation, such as $r = 3 \sin \theta$, with r and θ as variables, is called a **polar equation.** (Equations in x and y are called **rectangular** or **Cartesian equations.**) The simplest equation for many useful curves turns out to be a polar equation.

Graphing a polar equation is much the same as graphing a Cartesian equation— we obtain representative ordered pairs (r, θ) and then sketch the graph. For example, to graph $r = 1 + \cos \theta$, we first find and graph some ordered pairs, and then connect the points in order—from $(2, 0°)$ to $(1.9, 30°)$ to $(1.7, 45°)$, and so on. The final graph is shown in Figure 8.55. This curve is called a **cardioid** because of its heart shape.

θ	0°	30°	45°	60°	90°	120°	135°	150°	180°	270°	315°
$\cos\theta$	1	.9	.7	.5	0	−.5	−.7	−.9	−1	0	.7
$r = 1 + \cos\theta$	2	1.9	1.7	1.5	1	.5	.3	.1	0	1	1.7

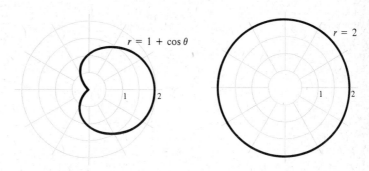

Figure 8.55   **Figure 8.56**

Once the pattern of values of r becomes clear, it is not necessary to find more ordered pairs. That is why we stopped with the ordered pair (1.7, 315°). From the pattern in the table, we see that the pair (1.9, 330°) would also satisfy the equation.

EXAMPLE 1   Graph $r = 2$.

Since r always has the value 2 for any value of θ, this equation represents a circle of radius 2 as shown in Figure 8.56. ▪

EXAMPLE 2   Graph $r^2 = \cos 2\theta$.

First, we complete a table of ordered pairs as shown and then sketch the graph as in Figure 8.57 on the next page. The point $(-1, 0°)$, with r negative, is plotted as $(1, 180°)$. Also, $(-2, 30°)$ is plotted as $(2, 210°)$, $(-5, 45°)$ as $(5, 225°)$, and so on. This curve is called a **lemniscate.**

θ	0°	30°	45°	135°	150°	180°
2θ	0	60	90	270	300	360
$\cos 2\theta$	1	.5	0	0	.5	1
$r = \pm\sqrt{\cos 2\theta}$	±1	±.7	0	0	±.7	±1

Values of θ for $45° < \theta < 135°$ are not included in the table because the corresponding values of $\cos 2\theta$ are negative (quadrants II and III) and so do not have real square roots. Values of θ larger than 180° give 2θ larger than 360°, so we would repeat the points we already have. ▪

Figure 8.57

EXAMPLE 3   ☐   Graph $r = \dfrac{4}{1 + \sin\theta}$.

We can again complete a table of ordered pairs, which leads to the graph shown in Figure 8.58.

θ	0°	30°	45°	60°	90°	120°	135°	150°	180°	210°	225°
$\sin\theta$	0	.5	.7	.9	1.0	.9	.7	.5	0	$-.5$	$-.7$
$r = \dfrac{4}{1 + \sin\theta}$	4	2.7	2.3	2.1	2.0	2.1	2.3	2.7	4.0	8.0	13.3

With the points given in the table, the pattern of the graph should be clear. If it is not, you should continue to find additional points.   ☐

We can convert the equation of Example 3,

$$r = \frac{4}{1 + \sin\theta},$$

to rectangular coordinates (x, y), using the results of Section 8.5. Work as follows.

$$r = \frac{4}{1 + \sin\theta}$$
$$r + r\sin\theta = 4$$
$$\sqrt{x^2 + y^2} + y = 4$$
$$\sqrt{x^2 + y^2} = 4 - y$$
$$x^2 + y^2 = (4 - y)^2$$
$$x^2 + y^2 = 16 - 8y + y^2$$
$$x^2 = -8y + 16$$
$$x^2 = -8\,(y - 2)$$

The final equation represents a parabola and can be graphed using rectangular coordinates.

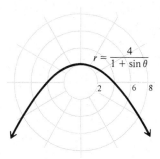

Figure 8.58

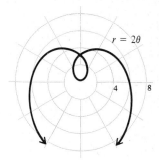

Figure 8.59

EXAMPLE
4
Graph $r = 2\theta$ (θ measured in radians).
Some ordered pairs are shown below. Since $r = 2\theta$, rather than a trigonometric function of θ, it is also necessary to consider negative values of θ. The radian measures have been rounded for simplicity.

θ(degrees)	-180	-90	-45	0	30	60	90	180	270	360
θ(radians)	-3.1	-1.6	$-.8$	0	.5	1	1.6	3.1	4.7	6.3
$r = 2\theta$	-6.2	-3.2	-1.6	0	1	2	3.2	6.2	9.4	12.6

Figure 8.59 shows this graph, called a **spiral of Archimedes.**

We have now seen how to convert an equation from polar coordinates to rectangular coordinates. The next example shows how to convert from rectangular coordinates to polar coordinates.

EXAMPLE
5
Convert $x^2 - y^2 = 4$ to polar coordinates.
Replace x with $r \cos \theta$ and y with $r \sin \theta$.

$$x^2 - y^2 = 4$$
$$(r \cos \theta)^2 - (r \sin \theta)^2 = 4$$
$$r^2 \cos^2 \theta - r^2 \sin^2 \theta = 4$$
$$r^2 (\cos^2 \theta - \sin^2 \theta) = 4$$

Since $\cos^2 \theta - \sin^2 \theta = \cos 2\theta$, our last equation becomes

$$r^2 \cos 2\theta = 4,$$

or $\qquad r^2 = 4 \sec 2\theta.$

EXERCISES
8.7

For each of the following equations, find an equiv-alent equation in rectangular coordinates and graph.

1. $r = 2 \sin \theta$ **2.** $r = 2 \cos \theta$

3. $r = \dfrac{2}{1 - \cos \theta}$ **4.** $r = \dfrac{3}{1 - \sin \theta}$

5. $r = -2 \cos \theta - 2 \sin \theta$

6. $r = \dfrac{3}{4 \cos \theta - \sin \theta}$

7. $r = 2 \sec \theta$ **8.** $r = -5 \csc \theta$

9. $r(\cos \theta + \sin \theta) = 2$

10. $r(2 \cos \theta + \sin \theta) = 2$

11. $r \sin \theta + 2 = 0$ **12.** $r \sec \theta = 5$

Graph each of the following for $0° \le \theta \le 180°$, unless other domains are specified.

13. $r = 2 + 2 \cos \theta$ **14.** $r = 2(4 + 3 \cos \theta)$

15. $r = 3 + \cos \theta$ (limaçon)

16. $r = 2 - \cos \theta$ (limaçon)

17. $r = \sin 2\theta$ (four-leaved rose)
(Hint: use $0° \le \theta < 360°$ every 15°.)

18. $r = 3 \cos 5\theta$ (five-leaved rose)
$0° \le \theta < 360°$

19. $r^2 = 4 \cos 2\theta$ (lemniscate)

20. $r^2 = 4 \sin 2\theta$ (lemniscate)
$0° \le \theta < 360°$

21. $r = 4(1 - \cos \theta)$ (cardioid)

22. $r = 3(2 - \cos \theta)$ (cardioid)

23. $r = 2 \sin \theta \tan \theta$ (cissoid)

24. $r = \dfrac{\cos 2\theta}{\cos \theta}$

25. $r = \dfrac{3}{2 + \sin \theta}$ **26.** $r = \sin \theta \cos^2 \theta$

Graph each of the following for $-\pi \le \theta \le \pi$, measuring θ in radians.

27. $r = 5\theta$ (spiral of Archimedes)

28. $r = \theta$ (spiral of Archimedes)

29. $r\theta = \pi$ (hyperbolic spiral)

30. $r^2 = \theta$ (parabolic spiral)

31. $\ln r = \theta$ (logarithmic spiral)

32. $\log r = \theta$ (logarithmic spiral)

Convert each of the following equations to polar co-ordinates.

33. $y = 2$ **34.** $x + 1 = 0$

35. $x^2 + y^2 = 49$ **36.** $y^2 = 25x$

37. $x^2 = 4y$ **38.** $x^2 + y = 1$

39. $2x + y = 4$ **40.** $3y - x = 5$

41. $x^2 + 9y^2 = 36$ **42.** $16x^2 + y^2 = 16$

43. Discuss the symmetry of $r = f(\theta)$ when r is replaced with $-r$; when θ is replaced with $-\theta$; and when θ is replaced with $\pi - \theta$.

44. Show that the distance between (r_1, θ_1) and (r_2, θ_2) is $\sqrt{r_1^2 + r_2^2 - 2r_1 r_2 \cos (\theta_1 - \theta_2)}$.

CHAPTER SUMMARY
8

Key Words

solving a triangle	magnitude	unit vector
angle of elevation	parallelogram rule	equilibrant
angle of depression	resultant	polar coordinates

oblique triangle
law of sines
ambiguous case of the law of sines
law of cosines
Heron's area formula
vector

opposite
zero vector
scalar product
position vector
x- and y-components

trigonometric form
modulus
argument
deMoivre's Theorem
nth root

CHAPTER REVIEW EXERCISES
8

Find the indicated parts in each of the following right triangles. Assume that the right angle is at C.

1. $A = 47° 20'$, $b = 39.6$ cm, find B and c

2. $A = 15° 20'$, $c = 301$ m, find B

3. $b = 68.6$ m, $c = 122.8$ m, find A and B

4. $A = 42°10'$, $a = 689$ cm, find b and c

5. $B = 88°20'$, $b = 402$ feet, find a and c

6. When the angle of elevation of the sun is $15°50'$, the shadow of a tower is 84.2 feet long. Find the height of the tower.

7. From the top of a cliff, the angle of depression to a river below is $32°10'$. The river is 850 feet from a point directly below the top of the cliff. How high is the cliff?

8. From a point at the base of a mountain, the angle of elevation to the top is $21°10'$. From a point 2000 feet back, the angle of elevation is $18°00'$. Find the height of the mountain.

Find the indicated parts of each of the following triangles.

9. $A = 100°10'$, $B = 25°00'$, $a = 165$ m, find b

10. $A = 82°50'$, $C = 62°10'$, $b = 12.8$ cm, find a

11. $B = 39°50'$, $b = 268$ m, $a = 430$ m, find A

12. $C = 79°20'$, $c = 97.4$ mm, $a = 75.3$ mm, find A

13. $A = 25°10'$, $a = 6.92$ feet, $b = 4.82$ feet, find B

14. $C = 74°10'$, $c = 96.3$ m, $B = 39°30'$, find b

Solve each of the following problems.

15. The angle of elevation from the top of a cliff to the top of a second cliff 290 ft away is 68°, while the angle of depression from the top of the first cliff to the bottom of the second cliff is 63°. Find the height of the second cliff.

16. A tree leans at an angle of 8° from the vertical. From a point 7 m from the bottom of the tree, the angle of elevation to the top of the tree is 68°. How tall is the tree?

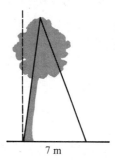

7 m

17. A hill makes an angle of $14°20'$ with the horizontal. From the base of the hill, the angle of elevation to the top of a tree on top of the hill is $27°10'$. The distance along the hill from the base to the tree is 212 feet. Find the height of the tree.

18. A ship is sailing east. At one point, the bearing of a submerged rock is $45°20'$. After sailing 15.2 miles, the bearing of the rock has become $308°40'$. Find the distance that the ship is then from the rock.

Find the indicated parts in each of the following triangles.

19. $A = 129°40'$, $a = 127$ feet, $b = 69.8$ feet, find B

20. $C = 51°20'$, $c = 68.3$ m, $b = 58.2$ m, find B

21. $a = 86.1$ inches, $b = 253$ inches, $c = 241$ inches, find A

22. $a = 14.8$ m, $b = 19.7$ m, $c = 31.8$ m, find B

23. $A = 46°10'$, $b = 18.4$ m, $c = 19.2$ m, find a

Find the area of each of the following triangles.

24. $b = 841$ m, $c = 716$ m, $A = 149°30'$

25. $a = 94.6$ yards, $b = 123$ yards, $c = 109$ yards

26. $a = 27.6$ cm, $b = 19.8$ cm, $C = 42°30'$

27. Raoul plans to paint a triangular wall in his A-frame cabin. Two sides measure 7 m each and the third side measures 6 m. How much paint will he need if a can of paint covers 7.5 m²?

In Exercises 28–30, use the vectors shown here. Find each of the following.

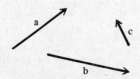

28. **a + b**   29. **a − b**   30. **a + 3c**

Find the horizontal and vertical components of each of the following vectors.

31. $\alpha = 45°$, magnitude 50

32. $\alpha = 75°$, magnitude 69.2

33. $\alpha = 154°20'$, magnitude 964

Suppose vector **v** *has the given direction angle and magnitude. Find the x- and y-components of* **v**.

34. $\theta = 45°$, $|\mathbf{v}| = 2\sqrt{2}$

35. $\theta = 210°$, $|\mathbf{v}| = 8$

36. $\theta = 302°$, $|\mathbf{v}| = 25$

Find the magnitude and direction angles for each of the following vectors.

37. $\langle -6, 2 \rangle$ 38. $\langle -6\sqrt{2}, 6\sqrt{2} \rangle$

39. $\langle 0, -2 \rangle$ 40. $\langle \pi, 0 \rangle$

Write each of the following vectors in the form $a\mathbf{i} + b\mathbf{j}$.

41. $\langle 2, -1 \rangle$ 42. $\langle -6, 3 \rangle$

43. direction angle 30°, magnitude 20

44. direction angle 162°, magnitude 5

Let **u** *and* **v** *be as given. Find* **u · v**.

45. $\mathbf{u} = \langle 2, 6 \rangle$, $\mathbf{v} = \langle -3, 2 \rangle$

46. $\mathbf{u} = \langle 4, -5 \rangle$, $\mathbf{v} = \langle -2, 1 \rangle$

47. $\mathbf{u} = 5\mathbf{i}$, $\mathbf{v} = 2\mathbf{i} + 3\mathbf{j}$

48. $\mathbf{u} = -\mathbf{i} + 8\mathbf{j}$, $\mathbf{v} = 4\mathbf{i} + 3\mathbf{j}$

Find the angle between each of the following pairs of vectors.

49. $\mathbf{u} = 6\mathbf{i} + 2\mathbf{j}$, $\mathbf{v} = 3\mathbf{i} - 2\mathbf{j}$

50. $\mathbf{u} = 4\mathbf{i} - 3\mathbf{j}$, $\mathbf{v} = 2\mathbf{i} + \mathbf{j}$

51. $\mathbf{u} = \langle 2\sqrt{3}, 2 \rangle$, $\mathbf{v} = \langle 5, 5\sqrt{3} \rangle$

52. $\mathbf{u} = \langle 2\sqrt{2}, 2\sqrt{2} \rangle$, $\mathbf{v} = \langle -1, 1 \rangle$

Solve each of the following problems.

53. A force of 150 pounds acts at a right angle to a force of 225 pounds. Find the magnitude of the equilibrant and the angle it makes with the 150 pound force.

54. Forces of 475 and 586 grams act on an object. The angle between the forces is 78°20'. Find the magnitude of the resultant.

55. A box of chickens is supported above the ground to keep the foxes out. The box hangs from two ropes. One makes an angle of 52°40' with the horizontal. The tension in this rope is 89.6 lbs. The second rope makes an angle of 82°30' with the first rope, and has a tension of 61.7 lbs. The box weighs 10 lbs. Find the

weight of the chickens. (Hint: add the y-components of each tension vector.)

56. A force of 186 pounds just keeps a 2800 pound Toyota from rolling down a hill. What angle does the hill make with the horizontal?

57. A plane has a still-air speed of 520 miles per hour. The pilot wishes to fly on a bearing of 310°. A wind of 37 miles per hour is coming from a bearing of 212°. What direction should the pilot fly, and what will be her actual speed?

58. A boat travels 15 kilometers per hour in still water. The boat is traveling across a large river, on a bearing of 130°. The current in the river, coming from the west, is at a speed of 7 kilometers per hour. Find the resulting speed of the boat and its resulting direction of travel.

Find the resultant of each of the following pairs of complex numbers.

59. $7 + 3i$ and $-2 + i$

60. $2 - 4i$ and $-1 - 2i$

Graph each of the following complex numbers.

61. $5i$ **62.** $-4 + 2i$

63. $3 - 3i\sqrt{3}$ **64.** $-5 + i\sqrt{3}$

Complete the following chart.

	Standard Form	Polar Coordinates	Trigonometric Form
65.	$-2 + 2i$	_____	_____
66.	_____	$(3, 90°)$	_____
67.	_____	_____	$2(\cos 315° + i \sin 315°)$
68.	$-4 + 4i\sqrt{3}$	_____	_____

Perform the indicated operations. Write answers in standard form.

69. $5(\cos 90° + i \sin 90°) \cdot 6(\cos 180° + i \sin 180°)$

70. $3(\cos 135° + i \sin 135°) \cdot 2(\cos 105° + i \sin 105°)$

71. $\dfrac{2(\cos 60° + i \sin 60°)}{8(\cos 300° + i \sin 300°)}$

72. $\dfrac{4(\cos 270° + i \sin 270°)}{2(\cos 90° + i \sin 90°)}$

73. $(\sqrt{3} + i)^3$

74. $(2 - 2i)^5$

75. $(\cos 100° + i \sin 100°)^6$

76. $(\cos 20° + i \sin 20°)^3$

77. Find the fifth roots of $-2 + 2i$.

78. Find the cube roots of $1 - i$.

79. Find the sixth roots of 1.

80. Find the fourth roots of $\sqrt{3} + i$.

Find all solutions for the following equations.

81. $x^3 + 125 = 0$ **82.** $x^4 + 16 = 0$

Graph the following.

83. $r = 4 \cos \theta$

84. $r = \dfrac{3}{1 + \cos \theta}$

85. $r = \sin \theta + \cos \theta$

86. $r = \dfrac{4}{2 \sin \theta - \cos \theta}$

87. $r = 1 + \cos \theta$ **88.** $r = 1 - \cos \theta$

89. $r = 2 \sin 4\theta$ **90.** $r = 3 \cos 5\theta$

91. $r^2 = \cos 2\theta$ **92.** $r = \sin^2 \theta \cos \theta$

93. $r = 2\theta$ **94.** $r = \pi\theta$

95. $r = 2$

96. $r = \pi$

The following identities involve all six parts of a triangle, ABC, and are thus useful for checking answers. Prove each of these results.

97. *Newton's Formula*
$$\frac{a + b}{c} = \frac{\cos \frac{1}{2}(A - B)}{\sin \frac{1}{2}C}$$

98. *Mollweide's Formula*
$$\frac{a - b}{c} = \frac{\sin \frac{1}{2}(A - B)}{\cos \frac{1}{2}C}$$

The Law of Sines can be used to prove the identity for sin (A + B). Let ABC be any triangle, and go through the following steps.

99. Show that $c = a \cos B + b \cos A$.

100. Multiply the terms in (1) by the corresponding terms in the Law of Sines:

$$\frac{\sin C}{c} = \frac{\sin A}{a} = \frac{\sin B}{b}$$

101. Since $A + B + C = \pi$, we have $C = \pi - (A + B)$. Use the fact that $\sin (\pi - s) = \sin s$ to show that $\sin C = \sin (A + B)$.

102. Finally, obtain the identity for sin *(A + B)*. *(This proof is only valid for values of A and B that might be angles of a triangle. Adjustments would have to be made to generalize this result for other angles. For more details, see the* Mathematics Magazine, *vol. 35 (1962), p. 229.)*

103. Let $\mathbf{v}_1 = \langle a_1, b_1 \rangle$ and $\mathbf{v}_2 = \langle a_2, b_2 \rangle$ be two nonzero vectors.

(a) Show that if $\mathbf{v}_1 = k\mathbf{v}_2$ for some scalar k, then $a_1 b_2 = a_2 b_1$.

(b) Show that if $a_1 b_2 = a_2 b_1$, then there exists a scalar k such that $\mathbf{v}_1 = k\mathbf{v}_2$.

104. Let $\mathbf{P}$ and $\mathbf{Q}$ be nonzero vectors. Prove that

(a) If $\mathbf{P}$ and $\mathbf{Q}$ are perpendicular, then $|\mathbf{P} + \mathbf{Q}|^2 = |\mathbf{P}|^2 + |\mathbf{Q}|^2$.

(b) If $|\mathbf{P} + \mathbf{Q}|^2 = |\mathbf{P}|^2 + |\mathbf{Q}|^2$, then $\mathbf{P}$ and $\mathbf{Q}$ are perpendicular.

105. Let A, B, C be the interior angles of a triangle in which no angle is a right angle. By calculating $\tan (A + B + C)$ and observing that $A + B + C = 180°$, show that the equation

$$\tan A + \tan B + \tan C = \tan A \tan B \tan C$$

must be satisfied.

Let $R(z)$ and $I(z)$ denote respectively the real and imaginary parts of a complex number z. Show that

106. $z + \bar{z} = 2R(z)$

107. $z - \bar{z} = 2iI(z)$

108. $|R(z)| \le |z|$

109. $|z_1 + z_2|^2 = |z_1|^2 + |z_2|^2 + 2R(z_1 \bar{z}_2)$

110. $|z_1 + z_2| \le |z_1| + |z_2|$

111. Find all pairs (x_0, y_0) that are both polar and rectangular coordinates for the same point.

112. A regular pentagon is inscribed in a circle with center at the origin and radius 3. One vertex is on the positive x-axis. Write the polar coordinates of all the vertices.

9 Systems of Equations and Inequalities

Many applications of mathematics require the simultaneous solution of a large number of equations or inequalities having many variables. Such a set of equations is called a **system of equations.** We now extend the definition of a linear equation given earlier to the following. Any equation with only first degree variables is a **linear equation.** If all the equations in a system are linear, the system is a **system of linear equations,** or a **linear system.** In this chapter, we discuss methods of solving these systems of equations or inequalities. The matrix techniques that we discuss have gained particular importance with the increasing availability of computers.

9.1 The Elimination Method

One way to solve a system of linear equations is to use properties of algebra to change the system until a simpler equivalent system is obtained. An **equivalent system** is one having the same solution set as the given system. There are three **transformations** that may be applied to a system to get an equivalent system.

Transformation of a Linear System

1. Any two equations of the system may be exchanged.

2. Both sides of any equation of the system may be multiplied by any nonzero real number.

3. Any equation of the system may be replaced by the sum of that equation and a multiple of another equation in the system.

We prove the third transformation, but only for a system of two linear equations in two unknowns,

$$a_1x + b_1y = c_1 \tag{1}$$

$$a_2x + b_2y = c_2; \tag{2}$$

the proof we give generalizes readily to equations with more variables.

Rewrite equation (1) as $a_1x + b_1y - c_1 = 0$, and, for simplicity, let $m = a_1x + b_1y - c_1$, so that (1) becomes $m = 0$. In the same way, write (2) as $n = 0$. Any solution of the original system will also be a solution of both $m = 0$ and $n = 0$. Form the new equation $km + n = 0$. Since both m and n become 0 when the numbers from the solution are substituted for x and y, the numbers from the solution must also make $km + n$ equal 0. Thus, a solution for the original system is also a solution for the transformed system

$$m = 0 \tag{3}$$

$$km + n = 0. \tag{4}$$

On the other hand, suppose we start with the system of equations (3) and (4). If some (x, y) makes $m = 0$, then $k(0) + n = 0$ so that $n = 0$. Thus, the original system is also satisfied by the solution (x, y).

EXAMPLE 1   ☐   Solve the system

$$3x - 4y = 1 \tag{5}$$

$$2x + 3y = 12. \tag{6}$$

We need to get a system of equations equivalent to the given system, but simpler. We do this by multiplying both sides of one equation by a real number and adding the result to the other equation so that we produce a new equation with only one variable. Start by multiplying both sides of equation (6) by -3, giving the system

$$3x - 4y = 1 \tag{5}$$

$$-6x - 9y = -36. \tag{7}$$

This system is equivalent to the original system. Now multiply both sides of equation (5) by 2, and add the result to equation (7) to get

$$3x - 4y = 1 \tag{5}$$

$$-17y = -34. \tag{8}$$

If we multiply both sides of equation (8) by $-1/17$, we have the equivalent system

$$3x - 4y = 1 \tag{5}$$

$$y = 2. \tag{9}$$

Now substitute 2 for y in equation (5) to get

$$3x - 4(2) = 1$$
$$3x - 8 = 1$$
$$3x = 9$$
$$x = 3.$$

The solution of the original system can be written as the ordered pair (3, 2), which gives the solution set {(3, 2)}. The graphs of both equations of the system are sketched in Figure 9.1. As the graph illustrates, (3, 2) satisfies both equations of the system. ☐

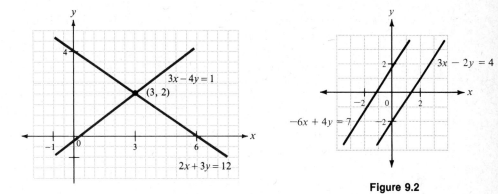

Figure 9.2

Figure 9.1

Since the solution shown in Example 1 results from the elimination of one variable from an equation of the system, this method of solution is called the **elimination method** for solving a system.

EXAMPLE 2 ☐ Solve the system

$$3x - 2y = 4 \qquad (10)$$
$$-6x + 4y = 7. \qquad (11)$$

Multiply both sides of equation (10) by 2, and add to equation (11), giving the equivalent system

$$3x - 2y = 4$$
$$0 = 15.$$

Since $0 = 15$ is never true, the system has no solution. As shown in Figure 9.2, this means that the graphs of the equation of the system never intersect (the lines are parallel). The solution set for this system is Ø. ☐

EXAMPLE 3 ☐ Solve the system

$$8x - 2y = -4 \qquad (12)$$
$$-4x + y = 2. \qquad (13)$$

Multiply both sides of equation (12) by 1/2, and add the result to equation (13), to get the equivalent system

$$8x - 2y = -4$$
$$0 = 0.$$

The second equation, $0 = 0$, is always true, which means that the equations of the original system are equivalent. Thus, any ordered pair (x, y) that satisfies either equation will satisfy the system. From equation (13), we get

$$-4x + y = 2,$$

or $$y = 2 + 4x$$

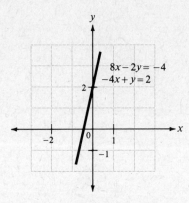

Figure 9.3

The solution of the given system is thus the (infinite) set of ordered pairs of the form $(x, 2 + 4x)$. The solution set can be written as $\{(x, y)|y = 2 + 4x\}$ or $\{(x, 2 + 4x)\}$. Typical ordered pairs belonging to the solution set are $(0, 2 + 4 \cdot 0) = (0, 2)$, $(-4, 2 + 4(-4)) = (-4, -14)$, $(3, 14)$, $(7, 30)$, and so on. As shown in Figure 9.3, both equations of the original system lead to the same graph, a straight line. ☐

Transformations can also be used to solve a system of three linear equations in three unknowns. It is a good idea to proceed systematically; use transformations to change a system having variables x, y, and z as follows.

1. If necessary, change the coefficient of x in the first equation to 1.

2. In the second equation, eliminate the x term by using the equation from step 1 and, if necessary, change the coefficient of y to 1.

3. In the third equation, eliminate the x and y terms by using the equations from steps 1 and 2 and, if necessary, change the coefficient of z to 1.

Of course, it is always possible to first rearrange the order of the given three equations if that will eliminate the need for some of the steps. The next example shows how we can use the transformations to accomplish these objectives.

EXAMPLE 4   ☐   Solve the system

$$x + 3y + 2z = 1 \qquad (14)$$
$$2x + y - z = 2 \qquad (15)$$
$$x + y + z = 2. \qquad (16)$$

Since the coefficient of x in equation (14) is already 1, we go on to the next step. To eliminate x in equation (15), multiply both sides of equation (14) by -2 and add the results to equation (15). This gives the system

$$x + 3y + 2z = 1 \tag{14}$$
$$-5y - 5z = 0 \tag{17}$$
$$x + y + z = 2. \tag{16}$$

Now multiply both sides of equation (17) by $-1/5$ to get

$$x + 3y + 2z = 1 \tag{14}$$
$$y + z = 0 \tag{18}$$
$$x + y + z = 2. \tag{16}$$

We can eliminate x in equation (16) by multiplying both sides of equation (14) by -1 and adding the results to equation (16). We have

$$x + 3y + 2z = 1 \tag{14}$$
$$y + z = 0 \tag{18}$$
$$-2y - z = 1. \tag{19}$$

To eliminate y in equation (19), multiply both sides of equation (18) by 2 and add the result to equation (19). The new system is

$$x + 3y + 2z = 1 \tag{14}$$
$$y + z = 0 \tag{18}$$
$$z = 1. \tag{20}$$

This last system is in what we call **triangular form.** From the last equation, we see that $z = 1$. Substitute $z = 1$ into equation (18) to get $y = -1$ and then substitute these values into equation (14) to find that $x = 2$. The solution of the system is the ordered triple $(2, -1, 1)$, which gives the solution set $\{(2, -1, 1)\}$. The process of substituting the known values of the variables back into the equations to find the remaining unknown variables is called **back-substitution.** ▢

A system of equations will sometimes have more variables than equations. In this case, there usually is no single solution, and in fact there might be no solution at all. The next example shows how to handle these systems.

EXAMPLE 5 ▢ Solve the system

$$x + 2y + z = 4 \tag{21}$$
$$3x - y - 4z = -9. \tag{22}$$

Eliminate x from equation (22) by multiplying both sides of equation (21) by -3 and adding the result to equation (22).

$$x + 2y + z = 4 \tag{21}$$
$$-7y - 7z = -21 \tag{23}$$

Multiply both sides of equation (23) by $-1/7$, giving the simplified system

$$x + 2y + z = 4 \tag{21}$$
$$y + z = 3. \tag{24}$$

It is not possible to eliminate any more variables here—any attempt to do so would reintroduce x into the second equation. Instead, solve equation (24) for y. (We could have used z.)

$$y = 3 - z$$

Solve equation (21) for x after first substituting $3 - z$ for y.

$$x + 2(3 - z) + z = 4$$
$$x + 6 - 2z + z = 4$$
$$x - z = -2$$
$$x = -2 + z$$

For any value of z that we might choose, y is found from the fact that $y = 3 - z$, and x from the fact that $x = -2 + z$. The solution set can be written as the set of ordered triples $\{(-2 + z, 3 - z, z)\}$. Typical values can be found by choosing values of z. For example, if z is 4, then $y = 3 - z = 3 - 4 = -1$, and $x = -2 + z = -2 + 4 = 2$, giving $(2, -1, 4)$. Thus, the solution set is an infinite set of ordered triples. Compare this with Example 3.

Had we solved equation (24) for z instead of y, the solution would have had a different form, but would have led to the same set of ordered triples. Our solution set, $\{(-2 + z, 3 - z, z)\}$, is said to have z **arbitrary.**   ☐

In the system of equations given in Example 6 below, each equation has a constant of 0. By inspection, the ordered triple $(0, 0, 0)$ is a solution of the system. However, in this case, there is an infinite number of other solutions, as shown in Example 5. The system

$$a_1x + b_1y + c_1z = 0$$
$$a_2x + b_2y + c_2z = 0$$
$$a_3x + b_3y + c_3z = 0$$

with all constants 0 is called a **homogeneous system** of three equations in three unknowns. The ordered triple $(0, 0, 0)$ is always a solution of a homogeneous system; this solution is the **trivial solution.**

EXAMPLE 6   ☐   Solve the system

$$x + 2y + z = 0 \tag{25}$$
$$4x - y + z = 0 \tag{26}$$
$$-x - 2y - z = 0. \tag{27}$$

Equation (27) is a multiple of equation (25), so the system can be reduced to

$$x + 2y + z = 0 \tag{25}$$
$$4x - y + z = 0. \tag{26}$$

This system of two equations in three variables can be solved by the method illustrated in Example 5. Verify that the solution set with z arbitrary is $\{(-z/3, -z/3, z)\}$. ☐

In Chapter 3 we found that the equation of the circle with center at (h, k) and radius r is

$$(x - h)^2 + (y - k)^2 = r^2.$$

By expanding both $(x - h)^2$ and $(y - k)^2$, this equation takes the form

$$x^2 + y^2 + Cx + Dy + E = 0,$$

which is called the **general form** of the equation of a circle. The general form is used in the next example.

EXAMPLE 7   ☐   Find an equation for all the circles going through $(2, 1)$ and $(-4, -1)$.

A circle is completely determined if we know three distinct points it goes through. Here we know only two points; there are an infinite number of circles through these two points. The set of all the circles going through these two points is called a **family of circles.** To find the equation of this family of circles, start with the general form of the equation of a circle,

$$x^2 + y^2 + Cx + Dy + E = 0.$$

Substitute the values from the two given points, $(2, 1)$, and $(-4, -1)$ into this equation, producing the following system of two equations in three unknowns.

$$4 + 1 + 2C + D + E = 0$$
$$16 + 1 - 4C - D + E = 0$$

These equations may be rewritten as

$$E + D + 2C = -5 \tag{28}$$
$$E - D - 4C = -17. \tag{29}$$

Multiplying both sides of equation (28) by -1 and adding the results to equation (29), we have

$$E + D + 2C = -5 \tag{28}$$
$$-2D - 6C = -12. \tag{30}$$

Multiply both sides of equation (30) by $-1/2$ to get

$$E + D + 2C = -5 \tag{28}$$
$$D + 3C = 6. \tag{31}$$

Solve equation (31) for D: $D = -3C + 6$. Substitute this value for D into equation (28) and solve for E.

$$E + (-3C + 6) + 2C = -5$$
$$E - C = -11$$
$$E = C - 11$$

The solution is $(C, -3C + 6, C - 11)$. The equation of the family of circles is thus

$$x^2 + y^2 + Cx + (-3C + 6)y + (C - 11) = 0.$$

To find the equation for a particular circle that goes through the two given points, choose a value for C. For example, if we choose $C = 3$, then $D = -3(3) + 6 = -3$ and $E = 3 - 11 = -8$, which gives the particular equation

$$x^2 + y^2 + 3x - 3y - 8 = 0.$$

Verify that this circle goes through the two given points by showing that both ordered pairs, $(2, 1)$ and $(-4, -1)$ satisfy the equation. ☐

In the last two examples, the systems we discussed had one more variable than equation. If there are two more variables than equations, there will be two arbitrary variables, and so on.

EXERCISES
9.1

Solve each system.

1. $3x + 2y = -6$
$\quad 5x - 2y = -10$

2. $3x - y = -4$
$\quad x + 3y = 12$

3. $2x - 3y = -7$
$\quad 5x + 4y = 17$

4. $4x + 3y = -1$
$\quad 2x + 5y = 3$

5. $5x + 7y = 6$
$\quad 10x - 3y = 46$

6. $12x - 5y = 9$
$\quad 3x - 8y = -18$

7. $6x + 7y = -2$
$\quad 7x - 6y = 26$

8. $2x + 9y = 3$
$\quad 5x + 7y = -8$

9. $\dfrac{x}{2} + \dfrac{y}{3} = 8$
$\quad \dfrac{2x}{3} + \dfrac{3y}{2} = 17$

10. $\dfrac{x}{5} + 3y = 31$
$\quad 2x - \dfrac{y}{5} = 8$

11. $\dfrac{3x}{2} - \dfrac{y}{3} = 5$
$\quad \dfrac{5x}{2} + \dfrac{2y}{3} = 12$

12. $\dfrac{4x}{5} + \dfrac{y}{4} = -2$
$\quad \dfrac{x}{5} + \dfrac{y}{8} = 0$

13. $0.05x - 0.02y = -0.18$
$\quad 0.04x + 0.06y = -0.22$

14. $0.08x + 0.03y = -0.24$
$\quad 0.04x - 0.01y = -0.32$

15. $0.6x + 0.3y = 0.087$
$\quad 0.5x - 0.4y = 0.378$

16. $0.7x - 0.5y = -0.884$
$\quad 0.1x + 0.3y = 0.13$

17. $\quad 1.9x - 4.8y = 11.2$
$\quad -4.37x + 11.04y = -7.6$

18. $\quad 3.7x + 1.82y = -9.7$
$\quad 2.96x + 1.456y = 4.8$

19. $\dfrac{2}{x} + \dfrac{1}{y} = \dfrac{3}{2}$
$\quad \dfrac{3}{x} - \dfrac{1}{y} = 1$

$\left(\text{Hint: let } \dfrac{1}{x} = t, \dfrac{1}{y} = u.\right)$

20. $\dfrac{1}{x} + \dfrac{3}{y} = \dfrac{16}{5}$
$\quad \dfrac{5}{x} + \dfrac{4}{y} = 5$

21. $\dfrac{2}{x} + \dfrac{1}{y} = 11$

$\dfrac{3}{x} - \dfrac{5}{y} = 10$

22. $\dfrac{2}{x} + \dfrac{3}{y} = 18$

$\dfrac{4}{x} - \dfrac{5}{y} = -8$

23. $\dfrac{4}{x+2} - \dfrac{3}{y-1} = 1$

$\dfrac{1}{x+2} + \dfrac{2}{y-1} = 1$

24. $\dfrac{1}{x-3} + \dfrac{5}{y+2} = -6$

$\dfrac{-3}{x-3} + \dfrac{2}{y+2} = 1$

25. $x + y + z = 2$

$2x + y - z = 5$

$x - y + z = -2$

26. $2x + y + z = 9$

$-x - y + z = 1$

$3x - y + z = 9$

27. $x + 3y + 4z = 14$

$2x - 3y + 2z = 10$

$3x - y + z = 9$

28. $4x - y + 3z = -2$

$3x + 5y - z = 15$

$-2x + y + 4z = 14$

29. $x + 2y + 3z = 8$

$3x - y + 2z = 5$

$-2x - 4y - 6z = 5$

30. $3x - 2y - 8z = 1$

$9x - 6y - 24z = -2$

$x - y + z = 1$

31. $3y + 2z = 6$

$x - y = 0$

$4x + z = 8$

32. $x + 3z = 0$

$2y - z = 9$

$4x + y = -7$

33. $\dfrac{2}{x} + \dfrac{3}{y} - \dfrac{2}{z} = -1$

$\dfrac{8}{x} - \dfrac{12}{y} + \dfrac{5}{z} = 5$

$\dfrac{6}{x} + \dfrac{3}{y} - \dfrac{1}{z} = 1$

34. $-\dfrac{5}{x} + \dfrac{4}{y} + \dfrac{3}{z} = 2$

$\dfrac{10}{x} + \dfrac{3}{y} - \dfrac{6}{z} = 7$

$\dfrac{5}{x} + \dfrac{2}{y} - \dfrac{9}{z} = 6$

35. $5.3x - 4.7y + 5.9z = 1.14$

$-2.5x + 3.2y - 1.4z = 7.22$

$2.25x - 2.88y + 1.26z = 4.88$

36. $7.77x - 8.61y + 4.2z = 15.96$

$11.6x - 4.9y + 0.8z = 12.4$

$3.7x - 4.1y + 2z = 7.6$

Solve each of the following systems in terms of an arbitrary variable.

37. $x - 2y + 3z = 6$

$2x - y = 5$

38. $3x + 4y - z = 8$

$x + 2z = 4$

39. $5x - 4y + z = 0$

$x + y = 0$

$-10x + 8y - 2z = 0$

40. $2x + y - 3z = 0$

$4x + 2y - 6z = 0$

$x - y + z = 0$

41. $4x + y + z = 6$

$y = 2x$

42. $3x - 5y - 4z = 6$

$3x = z$

Write a system of linear equations for each of the following, and then use the system to solve the problem.

43. At the Sharp Ranch, 6 goats and 5 sheep sell for $305 while 2 goats and 9 sheep cost $285. Find the cost of a goat and the cost of a sheep.

44. Linda Ramirez is a building contractor. If she hires 7 day laborers and 2 concrete finishers, her payroll for the day is $346, while 1 day laborer and 5 concrete finishers cost $238. Find the daily-wage charge of each type of worker.

45. During summer vacation Hector and Ann earned a total of $1088. Hector worked 8 days less than Ann and earned $2 per day less. Find the number of days he worked and the daily wage he made if the total number of days worked by both was 72.

46. The perimeter of a rectangle is 42 centimeters. The larger side has a length of 7 centimeters more than the shorter side. Find the length of the longer side.

47. Thirty liters of a 50% alcohol solution are to be made by mixing 70% solution and 20% solution. How many liters of each solution should be used?

48. A merchant wishes to make one hundred pounds of a coffee blend that can be sold for $4 per pound. This blend is to be made by mixing coffee worth $6 a pound with coffee worth $3 a pound. How many pounds of each will be needed?

49. Ms. Kelley inherits $25,000. She deposits part at 8% annual interest and part at 10%. Her total annual income from these investments is $2300. How much is invested at each rate?

50. Chuck Sullivan earned $100,000 in a lottery. He invested part of the money at 10% and part at 12%. His total annual income from the two investments is $11,000. How much did he have invested at each rate?

51. Mr. Caminiti has some money invested at 8% and three times as much invested at 12%. His total annual income from the two investments is $2200. How much is invested at each rate?

52. A cash drawer contains only fives and twenties. There are eight more fives than twenties. The total value of the money is $215. How many of each type of bill are there?

53. How many gallons of milk (4.5% butterfat) must be mixed with 250 gallons of skim milk (0% butterfat) to get low-fat milk (2% butterfat)?

54. By boat Tom can go 72 miles upstream to a fishing hole in 4 hours. Returning, he needs only 3 hours. What is the speed of the current in the stream?

55. Two cars start together and travel in opposite directions. At the end of four hours, the cars are 656 kilometers apart. If one car travels 20 kilometers per hour faster than the other, find the speed of each car.

56. Two trains leave towns 192 kilometers apart, traveling toward one another. One train travels 40 kilometers per hour faster than the other. They pass one another two hours later. What is the speed of each train?

57. The perimeter of a triangle is 33 centimeters. The longest side is 3 centimeters longer than the medium side. The medium side is twice the shortest side. Find the length of each side of the triangle.

58. Marcia Odjakjian invests in three ways the $30,000 she won in a lottery. With part of the money, she buys a mutual fund, paying 9% per year. The second part, $2000 more than the first, is used to buy utility bonds paying 10% per year. The rest is invested in a tax-free 5% bond. The first year her investments bring a return of $2500. How much is invested at each rate?

59. Find three numbers whose sum is 50, if the first is 2 more than the second, and the third is two-thirds of the second.

60. The sum of three numbers is 15. The second number is the negative of one-sixth of the first number, while the third number is 3 more than the second. Find the three numbers.

61. A wine merchant wishes to get 300 liters of wine that sells for $2 per liter. He wishes to mix wines selling for $3, $1, and $1.50, respectively. He must use twice as much of the $1.50 wine as the $1 wine. How many liters of each should he use?

62. A glue company needs to make some glue that it can sell for $120 per barrel. It wants to use 150 barrels of glue worth $100 per barrel, along with some glue worth $150 per barrel, and glue worth $190 per barrel. It must use the same amounts of $150 and $190 glue. How much of the $150 and $190 glue will be needed? How many barrels of $120 glue will be produced?

63. Find values of C, D, and E (with D arbitrary) so that the points $(2, -6)$, and $(-2, -2)$ lie on a circle with equation $x^2 + y^2 + Cx + Dy + E = 0$.

64. Find values of C, D, and E (with D arbitrary) so that the points $(0, -2)$, and $(3, 0)$ lie on a circle with equation $x^2 + y^2 + Cx + Dy + E = 0$.

65. Find a, b, and c (with a arbitrary) so that the graph of the equation $y = ax^2 + bx + c$ goes through the points $(0, 3)$, and $(-1, 6)$.

66. Find a, b, and c (with a arbitrary) so that the graph of the equation $y = ax^2 + bx + c$ goes through the points $(1, 1)$, and $(0, 3)$.

9.2   Nonlinear Systems

The **substitution method** of solving systems of equations is a variation of the elimination method discussed in Section 9.1. We shall first discuss the solution of linear systems by the substitution method and then extend this discussion to include the solution of nonlinear systems (those containing at least one nonlinear equation.) To solve solutions of linear equations by substitution, use the following steps.

Solving Linear Systems by Substitution

1. Solve any of the equations for one variable.

2. Substitute this result into the other equations of the system. Simplify. This produces a system with one less equation and one less variable.

3. Repeat steps 1 and 2 as necessary until only one equation, containing only one variable, is left. Solve this equation.

4. Use the solution from step 3 to find the values of all the variables of the system by back-substitution.

5. Check the solution in the original system.

The next few examples show how this method works.

EXAMPLE 1 ☐ Solve the system

$$2x - 5y = 15 \tag{1}$$

$$x = 4y. \tag{2}$$

From equation (2), we know that $x = 4y$. In equation (1), substitute $4y$ for x. This gives

$$2(4y) - 5y = 15$$
$$8y - 5y = 15$$
$$3y = 15$$
$$y = 5.$$

Since, from equation (2), $x = 4y$, we have $x = 4(5) = 20$. The solution set of the given system is thus $\{(20, 5)\}$.   ☐

EXAMPLE 2   ☐   Use substitution to solve the system

$$2x + y + z = 4 \tag{3}$$
$$x - 3y + 2z = -9 \tag{4}$$
$$3x - y - z = -9. \tag{5}$$

We shall use the steps given at the beginning of this section. In step 1, we solve any equation for any variable. Using one approach, let us solve equation (4) for x, getting

$$x = 3y - 2z - 9. \tag{6}$$

In step 2, this result is substituted for x into both equation (3) and equation (5). Substituting into equation (3) and simplifying gives

$$2(3y - 2z - 9) + y + z = 4$$
$$6y - 4z - 18 + y + z = 4$$
$$7y - 3z = 22. \tag{7}$$

Now substitute $3y - 2z - 9$ for x in equation (5) to get

$$3(3y - 2z - 9) - y - z = -9$$
$$8y - 7z = 18. \tag{8}$$

We now have a system of two equations with two unknowns,

$$7y - 3z = 22 \tag{7}$$
$$8y - 7z = 18. \tag{8}$$

To solve this system, repeat step 1. Let us solve equation (7) for y. We get

$$y = \frac{3z + 22}{7}. \tag{9}$$

In step 2, replacing y with $(3z + 22)/7$ in equation (8) gives

$$8\left(\frac{3z + 22}{7}\right) - 7z = 18.$$

Multiplying both sides by 7, we have

$$24z + 176 - 49z = 126$$
$$-25z = -50$$
$$z = 2.$$

From equation (9), check that $y = 4$, and from equation (6), check that $x = -1$. The solution set of the given system is $\{(-1, 4, 2)\}$. ☐

A system of equations in which at least one equation is *not* linear is called a **nonlinear system.** The substitution method works well for solving many such systems, as the next example shows.

EXAMPLE 3 ☐ Solve the system

$$x^2 - y = 4 \tag{10}$$
$$x + y = -2. \tag{11}$$

When one of the equations in a nonlinear system is linear, it is usually best to begin by solving the linear equation for any variable. Therefore, we begin by solving equation (11) for x, giving

$$x = -y - 2.$$

Now substitute this result for x in equation (10).

$$(-y - 2)^2 - y = 4$$
$$y^2 + 4y + 4 - y = 4$$
$$y^2 + 3y = 0$$
$$y(y + 3) = 0$$
$$y = 0 \quad \text{or} \quad y + 3 = 0$$
$$y = 0 \quad \text{or} \quad y = -3$$

If $y = 0$, then from equation (11), we have $x = -2$. Also, if $y = -3$, then $x = 1$. The solution set of the given system is thus $\{(-2, 0), (1, -3)\}$. A graph of the system is shown in Figure 9.4. ☐

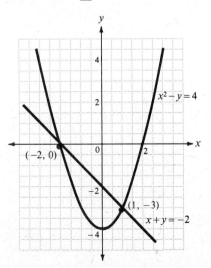

Figure 9.4

Some nonlinear systems are best solved by elimination, as shown in the next example.

EXAMPLE 4 ☐ Solve the system

$$x^2 + y^2 = 4 \tag{12}$$
$$2x^2 - y^2 = 8. \tag{13}$$

Replace equation (13) with the sum of the two equations to get

$$x^2 + y^2 = 4$$
$$3x^2 \quad\quad = 12.$$

Solve the last equation for x:

$$x^2 = 4$$
$$x = 2 \quad \text{or} \quad x = -2.$$

Find y by substituting back into equation (12). If $x = 2$, then $y = 0$, and if $x = -2$, then $y = 0$. The solution set of the given system is $\{(2, 0), (-2, 0)\}$, as shown in Figure 9.5.

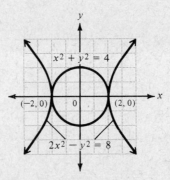

Figure 9.5

The elimination method works here since the system can be thought of as a system of linear equations where the variables are x^2 and y^2. In other words, the system is **linear in** x^2 and y^2. To see this, substitute u for x^2 and v for y^2. The resulting system is linear in u and v. ☐

Some systems require a combination of the elimination method and substitution, as shown in the next example.

EXAMPLE 5 ☐ Solve the system

$$x^2 + 3xy + y^2 = 22 \tag{14}$$
$$x^2 - xy + y^2 = 6. \tag{15}$$

Begin by multiplying both sides of equation (15) by -1, and adding the results to equation (14), giving

$$4xy = 16. \tag{16}$$

Now solve equation (16) for either x or y and substitute the result into one of the given equations. Let us solve for y.

$$y = \frac{4}{x} \quad (x \neq 0). \tag{17}$$

(Note that if $x = 0$ there is no value of y which satisfies the system.) Substituting back for y in equation (15) and simplifying gives

$$x^2 - x\left(\frac{4}{x}\right) + \left(\frac{4}{x}\right)^2 = 6$$

$$x^2 - 4 + \frac{16}{x^2} = 6$$

$$x^4 - 4x^2 + 16 = 6x^2$$

$$x^4 - 10x^2 + 16 = 0$$

$$(x^2 - 2)(x^2 - 8) = 0$$

$$x^2 = 2 \quad \text{or} \quad x^2 = 8$$

$$x = \sqrt{2} \quad \text{or} \quad x = -\sqrt{2} \quad \text{or} \quad x = 2\sqrt{2} \quad \text{or} \quad x = -2\sqrt{2}.$$

Substitute these values of x into equation (17) to find the corresponding values for y.

If $x = \sqrt{2}$, $y = \dfrac{4}{\sqrt{2}} = 2\sqrt{2}$.

If $x = -\sqrt{2}$, $y = -\dfrac{4}{\sqrt{2}} = -2\sqrt{2}$.

If $x = 2\sqrt{2}$, $y = \dfrac{4}{2\sqrt{2}} = \sqrt{2}$.

If $x = -2\sqrt{2}$, $y = -\dfrac{4}{2\sqrt{2}} = -\sqrt{2}$.

The solution set of the system is $\{(\sqrt{2}, 2\sqrt{2}), (-\sqrt{2}, -2\sqrt{2}), (2\sqrt{2}, \sqrt{2}), (-2\sqrt{2}, -\sqrt{2})\}$. ∎

EXERCISES
9.2

Solve each of the following systems by the substitution method.

1. $x - 5y = 8$

$ x = 4y$

2. $6x - y = 5$

$ y = 11x$

3. $4x - 5y = -11$

$ 2x + y = 5$

4. $-3x + 2y = 11$
    $2x + 7y = 11$

5. $x - 2y + \ z = 3$
    $3x + \ y - 2z = -1$
    $x - 4y + \ z = 3$

6. $-2x + 5y + 3z = -6$
    $4x - \ y - 2z = -6$
    $3x + 4y - 2z = -14$

7. $y = x^2$
    $x + y = 2$

8. $y = -x^2 + 2$
    $x - y = 0$

9. $y = x^2 - 2x + 1$
    $x - 3y = -1$

10. $y = x^2 + 6x + 9$
    $x + 2y = -2$

11. $3x^2 + 2y^2 = 5$
    $x - \ y = -2$

12. $x^2 + \ y^2 = 5$
    $-3x + 4y = 2$

Solve the following systems by any method.

13. $x^2 + y^2 = 8$
    $x^2 - y^2 = 0$

14. $x^2 + y^2 = 10$
    $2x^2 - y^2 = 17$

15. $5x^2 - \ y^2 = 0$
    $3x^2 + 4y^2 = 0$

16. $x^2 + \ y^2 = 4$
    $2x^2 - 3y^2 = -12$

17. $2x^2 + 3y^2 = 5$
    $3x^2 - 4y^2 = -1$

18. $3x^2 + 5y^2 = 17$
    $2x^2 - 3y^2 = 5$

19. $2x^2 + 2y^2 = 20$
    $3x^2 + 3y^2 = 30$

20. $x^2 + \ y^2 = 4$
    $5x^2 + 5y^2 = 28$

21. $xy = 6$
    $x + y = 5$

22. $xy = -4$
    $2x + y = -7$

23. $xy = -15$
    $4x + 3y = 3$

24. $xy = 8$
    $3x + 2y = -16$

25. $2xy + 1 = 0$
    $x + 16y = 2$

26. $-5xy + \ 2 \ = 0$
    $x - 15y = 5$

27. $x^2 + 4y^2 = 25$
    $xy = 6$

28. $5x^2 - 2y^2 = 6$
    $xy = 2$

29. $x^2 + 2xy - y^2 = 14$
    $x^2 - \qquad y^2 = -16$

30. $3x^2 + xy + 3y^2 = 7$
    $x^2 + \qquad y^2 = 2$

31. $x^2 - xy + y^2 = 5$
    $2x^2 + xy - y^2 = 10$

32. $3x^2 + 2xy - y^2 = 9$
    $x^2 - \ xy + y^2 = 9$

33. $x^2 + 2xy - y^2 + y = 1$
    $3x + y = 6$

34. $x^2 - 4x + y + xy = -10$
    $2x - y = 10$

35. $y = 3^x$
    $y = 9^{2x}$

36. $y = 5^{3x}$
    $y = 125^{4x}$

37. $y = \log (x - 2)$
    $y = -1 + \log (8x + 4)$

38. $y = -1 + \log (18x + 10)$
    $y = \log (x + 5)$

39. Find two numbers whose sum is 12 and whose product is 36.

40. Find two numbers whose sum is 17 and whose squares differ by 17.

41. Find two numbers whose ratio is 5:3 and whose product is 135.

42. Find two numbers whose squares have a sum of 100 and a difference of 28.

43. The longest side of a right triangle is 13 meters in length. One of the other sides is 7 meters longer than the shortest side. Find the length of each of the two shorter sides of the triangle.

44. Does the straight line $3x - 2y = 9$ intersect the circle $x^2 + y^2 = 25$? (To find out, try to solve the system made up of these two equations.)

45. Do the parabola $y = x^2 + 2$ and the ellipse $2x^2 + y^2 - 4x - 4y = 0$ have any points in common?

46. For what value of b will the line $x + 2y = b$ touch the circle $x^2 + y^2 = 9$ in only one point?

47. For what values of a do the circle $x^2 + y^2 = 25$ and the ellipse $x^2/a^2 + y^2/25 = 1$ have exactly two points in common?

48. For what values of k do the parabolas $y = x^2 - 4$ and $y = -x^2 + k$ have exactly two points in common?

9.3 Matrix Solution of Linear Systems

Systems of equations and, in particular, systems of linear equations occur in many practical situations. For this reason, several different methods for efficiently solving linear systems have been developed. Many of these methods depend on the computer. Computer solutions of linear systems almost always depend on the idea of a matrix. We shall discuss matrices in more detail later in this chapter. In this section we develop one method for solving linear systems using matrices.

A **matrix** is a rectangular array of numbers. Each number in a matrix is an **element** of the matrix. Matrices are classified by their dimension or order, that is, by the number of rows and columns that they contain. The matrix

$$\begin{bmatrix} 2 & 7 & -5 \\ 3 & -6 & 0 \end{bmatrix}$$

has two rows (horizontal) and three columns (vertical). This matrix is said to be of dimension 2×3 or of order 2×3; that is, a matrix with m rows and n columns is of **dimension** $m \times n$ or **order** $m \times n$. The number of rows is always given first.

Suppose we need to solve a system of n linear equations in n variables,

$$a_{11}x_1 + a_{12}x_2 + a_{13}x_3 + \cdots + a_{1n}x_n = b_1$$
$$a_{21}x_1 + a_{22}x_2 + a_{23}x_3 + \cdots + a_{2n}x_n = b_2$$
$$\vdots \qquad \vdots \qquad \vdots \qquad \qquad \vdots \qquad \vdots$$
$$a_{n1}x_1 + a_{n2}x_2 + a_{n3}x_3 + \cdots + a_{nn}x_n = b_n.$$

All the necessary information about the system is given by just the coefficients and the constants. We can write this information as an $n \times (n + 1)$ matrix, which we call the **augmented matrix** of the system.

Augmented Matrix

$$\begin{bmatrix} a_{11} & a_{12} & a_{13} & \cdots & a_{1n} & b_1 \\ a_{21} & a_{22} & a_{23} & \cdots & a_{2n} & b_2 \\ \cdot & \cdot & \cdot & & \cdot & \cdot \\ \cdot & \cdot & \cdot & & \cdot & \cdot \\ \cdot & \cdot & \cdot & & \cdot & \cdot \\ a_{n1} & a_{n2} & a_{n3} & \cdots & a_{nn} & b_n \end{bmatrix}$$

The dashed line, which is optional, is used only to separate the coefficients from the constants. Here, a_{11} represents the element in row 1, column 1, a_{12} represents the element in row 1, column 2, and a_{ij} represents the element in row i, column j.

We operate with the rows of this augmented matrix just as one would with the equations of a system of linear equations. Since the augmented matrix is nothing more than a short form of the system, we can perform any transformation of the matrix which results in an equivalent system of equations. Operations that produce such transformations are given below.

Matrix Row Transformations

For any augmented matrix of a system of linear equations, the following row transformations will result in the matrix of an equivalent system.

1. Any two rows may be interchanged.

2. The elements of any row may be multiplied by a nonzero real number.

3. Any row may be changed by adding to its elements a multiple of the elements of another row.

This is just a restatement in matrix form of the transformation of systems discussed in Section 9.1. From now on, when referring to the third transformation, we shall abbreviate "a multiple of the elements of a row" to "a multiple of a row."

EXAMPLE 1   Use the row transformations to solve the linear system

$$3x - 4y = 1$$
$$5x + 2y = 19.$$

Begin by forming the augmented matrix

$$\begin{bmatrix} 3 & -4 & 1 \\ 5 & 2 & 19 \end{bmatrix}.$$

Now we will use row transformations to transform this augmented matrix into a matrix in triangular form, as shown at the top of the next page.

Matrix in
Triangular
Form

$$\begin{bmatrix} 1 & m & n \\ 0 & 1 & p \end{bmatrix}$$

The necessary row transformations are performed as follows.

It is best to work in columns, beginning in each column with the element which is to become 1. This is the same order we used in Section 9.1 to arrange a system of equations in triangular form. In our augmented matrix, we have 3 in the first row, first column position. To get 1 in this position, use the second transformation and multiply each entry in the first row by 1/3.

$$\begin{bmatrix} 1 & -4/3 & 1/3 \\ 5 & 2 & 19 \end{bmatrix}$$

To get 0 in the second row, first column, add -5 times the first row to the second row.

$$\begin{bmatrix} 1 & -4/3 & 1/3 \\ 0 & 26/3 & 52/3 \end{bmatrix}$$

To get 1 in the second row, second column, multiply each element of the second row by 3/26, which gives

$$\begin{bmatrix} 1 & -4/3 & 1/3 \\ 0 & 1 & 2 \end{bmatrix}.$$

From this matrix, we get the system

$$x - \frac{4}{3}y = \frac{1}{3} \tag{1}$$

$$y = 2. \tag{2}$$

Substitute 2 for y in equation (1) to get $x = 3$. The solution set of the system is thus $\{(3, 2)\}$. ◻

EXAMPLE
2

◻ Use matrix methods to solve the system

$$x - y + 5z = -6$$
$$3x + 3y - z = 10$$
$$x + 3y + 2z = 5.$$

Begin by writing the augmented matrix of the linear system.

$$\begin{bmatrix} 1 & -1 & 5 & -6 \\ 3 & 3 & -1 & 10 \\ 1 & 3 & 2 & 5 \end{bmatrix}$$

We have 1 in row 1, column 1. The next thing we must do is get 0's in the rest of column 1. First, add to row 2 the results of multiplying row 1 by -3.

$$\begin{bmatrix} 1 & -1 & 5 & -6 \\ 0 & 6 & -16 & 28 \\ 1 & 3 & 2 & 5 \end{bmatrix}$$

Now add to row 3 the results of multiplying row 1 by -1.

$$\begin{bmatrix} 1 & -1 & 5 & \vdots & -6 \\ 0 & 6 & -16 & \vdots & 28 \\ 0 & 4 & -3 & \vdots & 11 \end{bmatrix}$$

To get 1 in row 2, column 2, multiply row 2 by 1/6.

$$\begin{bmatrix} 1 & -1 & 5 & \vdots & -6 \\ 0 & 1 & -8/3 & \vdots & 14/3 \\ 0 & 4 & -3 & \vdots & 11 \end{bmatrix}$$

Next, we need to get 0 in row 3, column 2. To do this, add to row 3 the results of multiplying row 2 by -4.

$$\begin{bmatrix} 1 & -1 & 5 & \vdots & -6 \\ 0 & 1 & -8/3 & \vdots & 14/3 \\ 0 & 0 & 23/3 & \vdots & -23/3 \end{bmatrix}$$

Finally, multiply the last row by 3/23 to get 1 in row 3, column 3.

$$\begin{bmatrix} 1 & -1 & 5 & \vdots & -6 \\ 0 & 1 & -8/3 & \vdots & 14/3 \\ 0 & 0 & 1 & \vdots & -1 \end{bmatrix}$$

The **main diagonal** of this matrix is the diagonal (to the left of the vertical bar) where all elements are 1. Since each element below the main diagonal is 0, the matrix is written in **echelon form.**

The final matrix above produces the system of equations

$$x - y + 5z = -6 \tag{3}$$

$$y - \frac{8}{3}z = \frac{14}{3} \tag{4}$$

$$z = -1. \tag{5}$$

By back-substitution into equations (3) and (4), we find the solution set of the system, $\{(1, 2, -1)\}$. ◻

EXAMPLE 3 ◻ Solve the system

$$x + y = 2$$
$$2x + 2y = 5.$$

Start with the augmented matrix.

$$\begin{bmatrix} 1 & 1 & \vdots & 2 \\ 2 & 2 & \vdots & 5 \end{bmatrix}$$

Next, add to row 2 the results of multiplying row 1 by -2.

$$\begin{bmatrix} 1 & 1 & \vdots & 2 \\ 0 & 0 & \vdots & 1 \end{bmatrix}$$

This matrix gives the system of equations

$$x + y = 2$$
$$0 = 1,$$

a system with no solution. Whenever a row of the augmented matrix is of the form $0\ 0\ 0\ \ldots\ .\,|a,$ (where $a \neq 0$), there will be no solution since this row corresponds to the equation $0 = a$. The solution set of the given system is $\emptyset$. ◻

**EXAMPLE
4**   ◻   Solve the system

$$x + y + 3z = 5$$
$$3x - 4y + z = 6.$$

The augmented matrix is

$$\begin{bmatrix} 1 & 1 & 3 & | & 5 \\ 3 & -4 & 1 & | & 6 \end{bmatrix}.$$

Get 0 in row 2, column 1.

$$\begin{bmatrix} 1 & 1 & 3 & | & 5 \\ 0 & -7 & -8 & | & -9 \end{bmatrix}$$

Now get 1 in row 2, column 2.

$$\begin{bmatrix} 1 & 1 & 3 & | & 5 \\ 0 & 1 & 8/7 & | & 9/7 \end{bmatrix}$$

We can't go any further since there are just two rows, so we write the system of equations

$$x + y + 3z = 5 \tag{6}$$
$$y + \frac{8}{7}z = \frac{9}{7}. \tag{7}$$

From equation (7), we can solve for y to get

$$y = -\frac{8}{7}z + \frac{9}{7}.$$

Substitute into equation (6) to get

$$x + \left(-\frac{8}{7}z + \frac{9}{7}\right) + 3z = 5$$
$$x = -\frac{13}{7}z + \frac{26}{7}.$$

The arbitrary variable is z; the solution set can be written

$$\left\{ \left(\frac{-13z + 26}{7}, \frac{-8z + 9}{7}, z \right) \right\}.$$ ◻

We can now summarize the cases that might come up when using matrix methods to solve a system of linear equations. When the matrix is written in echelon form:

1. If the number of rows having nonzero elements to the left of the vertical line is equal to the number of unknowns in the system, then the system has a single solution.

2. If the number of rows having nonzero elements to the left of the vertical line is less than the number of rows in the augmented matrix with nonzero elements, then the system has no solution.

For example, the system that produced the matrix

$$\begin{bmatrix} 1 & -2 & 4 & 2 \\ 0 & 1 & 3 & -1 \\ 0 & 0 & 0 & 4 \end{bmatrix}$$

has no solution. Here 2 rows to the left of the vertical line have nonzero elements, but 3 rows in the entire matrix have nonzero elements. See Example 3.

3. If the number of rows of the matrix containing nonzero elements is less than the number of unknowns, then there is an infinite number of solutions for the system. This infinite number of solutions should be given in terms of an arbitrary variable. See Example 4.

The method of using matrices to solve a system of linear equations that was developed in this section is called the **Gaussian reduction method** after the mathematician K. F. Gauss (1777–1855).

EXERCISES
9.3

Write the augmented matrix for each of the following systems. Do not solve.

1. $2x + 3y = 11$
$x + 2y = 8$

2. $3x + 5y = -13$
$2x + 3y = -9$

3. $x + 5y = 6$
$y = 1$

4. $2x + 7y = 1$
$5x = -15$

5. $2x + y + z = 3$
$3x - 4y + 2z = -7$
$x + y + z = 2$

6. $4x - 2y + 3z = 4$
$3x + 5y + z = 7$
$5x - y + 4z = 7$

7. $x + y = 2$
$2y + z = -4$
$z = 2$

8. $x = 6$
$y + 2z = 2$
$x - 3z = 6$

Write the system of equations associated with each of the following augmented matrices. Do not solve.

9. $\begin{bmatrix} 2 & 1 & \vdots & 1 \\ 3 & -2 & \vdots & -9 \end{bmatrix}$ **10.** $\begin{bmatrix} 1 & -5 & \vdots & -18 \\ 6 & 2 & \vdots & 20 \end{bmatrix}$

11. $\begin{bmatrix} 1 & 0 & 0 & \vdots & 2 \\ 0 & 1 & 0 & \vdots & 3 \\ 0 & 0 & 1 & \vdots & -2 \end{bmatrix}$ **12.** $\begin{bmatrix} 1 & 0 & 1 & \vdots & 4 \\ 0 & 1 & 0 & \vdots & 2 \\ 0 & 0 & 1 & \vdots & 3 \end{bmatrix}$

13. $\begin{bmatrix} 3 & 2 & 1 & \vdots & 1 \\ 0 & 2 & 4 & \vdots & 22 \\ -1 & -2 & 3 & \vdots & 15 \end{bmatrix}$

14. $\begin{bmatrix} 2 & 1 & 3 & \vdots & 12 \\ 4 & -3 & 0 & \vdots & 10 \\ 5 & 0 & -4 & \vdots & -11 \end{bmatrix}$

Use the Gaussian reduction method to solve each of the following systems of equations.

15. $x + y = 5$
$x - y = -1$

16. $x + 2y = 5$
$2x + y = -2$

17. $x + y = -3$
$2x - 5y = -6$

18. $3x - 2y = 4$
$3x + y = -2$

19. $2x - 3y = 10$
$2x + 2y = 5$

20. $6x + y = 5$
$5x + y = 3$

21. $2x - 5y = 10$
$3x + y = 15$

22. $4x - y = 3$
$-2x + 3y = 1$

23. $x + y \quad = -1$
$\quad y + z = 4$
$x \quad + z = 1$

24. $x \quad - z = -3$
$\quad y + z = 9$
$x \quad + z = 7$

25. $x + y - z = 6$
$2x - y + z = -9$
$x - 2y + 3z = 1$

26. $x + 3y - 6z = 7$
$2x - y + z = 1$
$x + 2y + 2z = -1$

27. $-x + y \quad = -1$
$\quad y - z = 6$
$x \quad + z = -1$

28. $x + y \quad = 1$
$2x \quad - z = 0$
$\quad y + 2z = -2$

29. $2x - y + 3z = 0$
$x + 2y - z = 5$
$2y + z = 1$

30. $4x + 2y - 3z = 6$
$x - 4y + z = -4$
$-x \quad + 2z = 2$

31. $x - 2y + z = 5$
$2x + y - z = 2$
$-2x + 4y - 2z = 2$

32. $3x + 5y - z = 0$
$4x - y + 2z = 1$
$-6x - 10y + 2z = 0$

33. $x + y + z = 6$
$2x - y - z = 3$

34. $5x - 3y + z = 1$
$2x + y - z = 4$

35. $x - 8y + z = 4$
$3x - y + 2z = -1$

36. $-3x + y - z = 8$
$2x + y + 4z = 0$

37. $x + 3y - z = 0$
$2y - z = 4$

38. $2x - y + 4z = 1$
$y + z = 3$

39. $3x + 2y \quad - w = 0$
$2x \quad + z + 2w = 5$
$x + 2y - z \quad = -2$
$2x - y + z + w = 2$

40. $x + 3y - 2z - w = 9$
$4x + y + z + 2w = 2$
$-3x - y + z - w = -5$
$x - y - 3z - 2w = 2$

41. $x - y + 2z + w = 4$
$y + z \quad = 3$
$z - w = 2$

42. $3x + y \qquad\qquad = 1$
$\qquad y - 4z + w = 0$
$\qquad\qquad z - 3w = -1$

43. $3.5x + 2.9y = 12.91$
$\qquad 1.7x - 3.8y = -6.23$

44. $-4.3x + 1.1y = 10.1$
$\qquad 4.8x - 2.7y = -17.31$

45. $9.1x + 2.3y - \qquad z = -3.06$
$\qquad 4.7x + \qquad y - 3.8z = -21.15$
$\qquad\qquad 5.1y - 4.7z = -31.77$

46. $7.2x + 4.8y + \quad 3.1z = -14.96$
$\qquad 3x - \qquad y + 11.5z = -15.6$
$\qquad 1.1x + 2.4y \qquad\qquad = -1.24$

Solve each of the following word problems.

47. A working couple earned a total of $4352. The wife earned $64 per day; the husband earned $8 per day less. Find the number of days each worked if the total number of days worked by both was 72.

48. Midtown Manufacturing Company makes two products, plastic plates and plastic cups. Both require time on two machines; plates: 1 hour on machine *A* and 2 hours on machine *B*, cups: 3 hours on machine *A* and 1 hour on machine *B*. Both machines operate 15 hours a day. How many of each product can be produced in a day under these conditions?

49. A company produces two models of bicycles, model 201 and model 301. Model 201 requires 2 hours of assembly time, and model 301 requires 3 hours of assembly time. The parts for model 201 cost $25 per bike; those for model 301 cost $30 per bike. If the company has a total of 34 hours of assembly time and $365 available per day for these two models, how many of each can be made in a day?

50. Juanita invests $10,000 in three ways. With one part, she buys mutual funds which offer a return of 8% per year. The second part, which amounts to twice the first, is used to buy government bonds at 9% per year. She puts the rest

in the bank at 5% annual interest. The first year her investments bring a return of $830. How much did she invest each way?

51. To get the necessary funds for a planned expansion, a small company took out three loans totaling $25,000. The company was able to borrow some of the money at 8%. They borrowed $2000 more than one-half the amount of the 8% loan at 10%, and the rest at 9%. The total annual interest was $2220. How much did they borrow at each rate?

At rush hours, substantial traffic congestion is encountered at the traffic intersections shown in the figure. (All streets are one-way.)

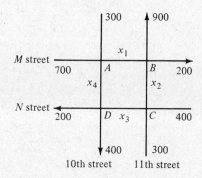

The city wishes to improve the signals at these corners so as to speed the flow of traffic. The traffic engineers first gather data. As the figure shows, 700 cars per hour come down M Street to intersection A; 300 cars per hour come to intersection A on 10th Street. A total of x_1 of these cars leave A on M Street, while x_4 cars leave A on 10th Street. The number of cars entering A must equal the number leaving, so that

$$x_1 + x_4 = 700 + 300$$

or $\quad x_1 + x_4 = 1000.$

For intersection B, x_1 cars enter B on M Street, and x_2 cars enter B on 11th Street. The figure shows that 900 cars leave B on 11th while 200 leave on M. We have

$$x_1 + x_2 = 900 + 200$$

$$x_1 + x_2 = 1100.$$

At intersection C, 400 cars enter on N Street, 300 on 11th Street, while x_2 leave on 11th Street and x_3 leave on N Street. This gives

$$x_2 + x_3 = 400 + 300$$

$$x_2 + x_3 = 700.$$

Finally, intersection D has x_3 cars entering at N and x_4 entering on 10th. There are 400 cars leaving D on 10th and 200 leaving on N.

52. Set up an equation for intersection D.

53. Use the four equations to set up an augmented matrix, and then use the Gaussian method to solve it.

54. Since you got a row of all zeros, the system of equations does not have a unique solution. Write three equations, corresponding to the three nonzero rows of the matrix. Solve each of the equations for x_4.

55. One of your equations should have been $x_4 = 1000 - x_1$. What is the largest possible value of x_1 so that x_4 is not negative?

56. Your second equation should have been $x_4 = x_2 - 100$. Find the smallest possible value of x_2 so that x_4 is not negative.

57. Find the largest possible values of x_3 and x_4 so that neither variable is negative.

58. Use the results of Exercises 52–57 to give a solution for the problem in which all the equations are satisfied and all variables are nonnegative. Is the solution unique?

Solve each of the following problems.

59. Three small boxes and 8 large boxes cost $13.30. If 6 small and 16 large boxes are purchased, the cost is $27.20. Find the cost of a small box and the cost of a large box.

60. The sum of three numbers is 20. One number is one more than another. Find the three numbers.

61. If three numbers are added, the result is -10. The second number, added to twice the third, gives 5. Find the three numbers.

62. If half of one number is added to another, the sum is twice a third number. The third number is half the second number. Find the three numbers.

9.4 Determinants

A matrix with the same number of rows as columns is called a **square matrix.** Given any square matrix A, there is a unique real number associated with A—this real number is the **determinant** of A. We define determinants in this section, look at their properties in the next, and then see how determinants can be used to solve a system of linear equations.

The determinant of a matrix A is written $|A|$. We shall first define $|A|$ for a 2×2 matrix:

Determinant of a 2 × 2 Matrix

If $A = \begin{bmatrix} a_{11} & a_{12} \\ a_{21} & a_{22} \end{bmatrix}$, then $|A| = \begin{vmatrix} a_{11} & a_{12} \\ a_{21} & a_{22} \end{vmatrix} = a_{11}a_{22} - a_{21}a_{12}.$

EXAMPLE 1 ☐ Let $A = \begin{bmatrix} -3 & 4 \\ 6 & 8 \end{bmatrix}$. Find $|A|$.

Use the definition above:

$$|A| = \begin{vmatrix} -3 & 4 \\ 6 & 8 \end{vmatrix} = -3(8) - 6(4) = -48. \quad \square$$

The determinant of the 3×3 matrix A, where

$$A = \begin{bmatrix} a_{11} & a_{12} & a_{13} \\ a_{21} & a_{22} & a_{23} \\ a_{31} & a_{32} & a_{33} \end{bmatrix},$$

is defined as

$$|A| = \begin{vmatrix} a_{11} & a_{12} & a_{13} \\ a_{21} & a_{22} & a_{23} \\ a_{31} & a_{32} & a_{33} \end{vmatrix} = (a_{11}a_{22}a_{33} + a_{12}a_{23}a_{31} + a_{13}a_{21}a_{32}) \\ - (a_{31}a_{22}a_{13} + a_{32}a_{23}a_{11} + a_{33}a_{21}a_{12}).$$

We can rearrange terms on the right side of the equation and factor to get

$$\begin{vmatrix} a_{11} & a_{12} & a_{13} \\ a_{21} & a_{22} & a_{23} \\ a_{31} & a_{32} & a_{33} \end{vmatrix} = a_{11}(a_{22}a_{33} - a_{32}a_{23}) - a_{21}(a_{12}a_{33} - a_{32}a_{13}) \\ + a_{31}(a_{12}a_{23} - a_{22}a_{13}).$$

Each of the quantities in parentheses above represents a determinant of a 2×2 matrix which is the part of the 3×3 matrix left when the row and column of the multiplier are eliminated. We can show this as follows.

$a_{11}(a_{22}a_{33} - a_{32}a_{23})$ $\begin{bmatrix} a_{11} & a_{12} & a_{13} \\ a_{21} & a_{22} & a_{23} \\ a_{31} & a_{32} & a_{33} \end{bmatrix}$

$a_{21}(a_{12}a_{33} - a_{32}a_{13})$ $\begin{bmatrix} a_{11} & a_{12} & a_{13} \\ a_{21} & a_{22} & a_{23} \\ a_{31} & a_{32} & a_{33} \end{bmatrix}$

$a_{31}(a_{12}a_{23} - a_{22}a_{13})$ $\begin{bmatrix} a_{11} & a_{12} & a_{13} \\ a_{21} & a_{22} & a_{23} \\ a_{31} & a_{32} & a_{33} \end{bmatrix}$

These determinants of 2×2 matrices are called **minors** of an element in the 3×3 matrix. The symbol M_{ij} represents the determinant of the matrix that results when row i and column j are eliminated. The following list gives some of the minors from the matrix above.

Element	Minor		Element	Minor
a_{11}	$M_{11} = \begin{vmatrix} a_{22} & a_{23} \\ a_{32} & a_{33} \end{vmatrix}$		a_{22}	$M_{22} = \begin{vmatrix} a_{11} & a_{13} \\ a_{31} & a_{33} \end{vmatrix}$
a_{21}	$M_{21} = \begin{vmatrix} a_{12} & a_{13} \\ a_{32} & a_{33} \end{vmatrix}$		a_{23}	$M_{23} = \begin{vmatrix} a_{11} & a_{12} \\ a_{31} & a_{32} \end{vmatrix}$

$$a_{31} \qquad M_{31} = \begin{vmatrix} a_{12} & a_{13} \\ a_{22} & a_{23} \end{vmatrix} \qquad a_{33} \qquad M_{33} = \begin{vmatrix} a_{11} & a_{12} \\ a_{21} & a_{22} \end{vmatrix}$$

Similarly, in a 4×4 matrix, the minors are determinants of 3×3 matrices, and an $n \times n$ matrix has minors that are determinants of $(n - 1) \times (n - 1)$ matrices.

To find the determinant of a 3×3 or larger matrix, first choose any row or column. Then the minor of each element in that row or column must be multiplied by $+1$ or -1, depending on whether the sum of the row numbers and column numbers is even or odd. The product of a minor and the number $+1$ or -1 is called a **cofactor.**

**Definition:
Cofactor**

> Let M_{ij} be the minor for element a_{ij} in an $n \times n$ matrix. The *cofactor* of a_{ij}, written A_{ij}, is
>
> $$A_{ij} = (-1)^{i+j} \cdot M_{ij}.$$

Finally, the determinant of a 3×3 or larger matrix is found by the following rule.

**Finding the
Determinant of
a Matrix**

> Multiply each element in any row or column of the matrix by its cofactor. The sum of these products gives the value of the determinant.

The process of forming this sum of products is called **expansion by a given row or column.** (See Exercise 35–36 below.)

**EXAMPLE
2**

Evaluate $\begin{vmatrix} 2 & -3 & -2 \\ -1 & -4 & -3 \\ -1 & 0 & 2 \end{vmatrix}$. Expand by the second column.

To find this determinant, first get the minors of each element in the second column.

$$M_{12} = \begin{vmatrix} -1 & -3 \\ -1 & 2 \end{vmatrix} = -1(2) - (-1)(-3) = -5$$

$$M_{22} = \begin{vmatrix} 2 & -2 \\ -1 & 2 \end{vmatrix} = 2(2) - (-1)(-2) = 2$$

$$M_{32} = \begin{vmatrix} 2 & -2 \\ -1 & -3 \end{vmatrix} = 2(-3) - (-1)(-2) = -8$$

Now find the cofactor of each of these minors.

$$A_{12} = (-1)^{1+2} \cdot M_{12} = (-1)^3 \cdot (-5) = (-1)(-5) = 5$$
$$A_{22} = (-1)^{2+2} \cdot M_{22} = (-1)^4 \cdot (2) = 1 \cdot 2 = 2$$
$$A_{32} = (-1)^{3+2} \cdot M_{32} = (-1)^5 \cdot (-8) = (-1)(-8) = 8$$

The determinant is found by multiplying each cofactor by its corresponding element in the matrix and finding the sum of these products.

$$\begin{vmatrix} 2 & -3 & -2 \\ -1 & -4 & -3 \\ -1 & 0 & 2 \end{vmatrix} = a_{12} \cdot A_{12} + a_{22} \cdot A_{22} + a_{32} \cdot A_{32}$$

$$= -3(5) + (-4)(2) + (0)(8)$$

$$= -15 + (-8) + 0$$

$$= -23 \quad \square$$

We would get exactly the same answer by using any row or column of the determinant. One reason that column 2 was used here is that it contains a 0 element. Thus, it was not really necessary to calculate M_{32} and A_{32}, as we did above. One learns quickly that 0's are friends when working with determinants.

Instead of calculating $(-1)^{i+j}$ for a given element, the following sign checkerboards can also be used.

Array of Signs

for 3 × 3 matrices			for 4 × 4 matrices			
+	−	+	+	−	+	−
−	+	−	−	+	−	+
+	−	+	+	−	+	−
			−	+	−	+

The signs alternate for each row and column, beginning with + in the first row, first column position. Thus, these arrays of signs can be reproduced as needed. If we expand a 3 × 3 matrix about row 3, for example, the first minor would have a + sign associated with it, the second minor a − sign, and the third minor a + sign. These arrays of signs can be extended in the same way for determinants of 5 × 5, 6 × 6, and larger matrices.

EXAMPLE 3 □ Evaluate $\begin{vmatrix} -1 & -2 & 3 & 2 \\ 0 & 1 & 4 & -2 \\ 3 & -1 & 4 & 0 \\ 2 & 1 & 0 & 3 \end{vmatrix}$.

Expand about the fourth row, and do the arithmetic that we are leaving out.

$$-2\begin{vmatrix} -2 & 3 & 2 \\ 1 & 4 & -2 \\ -1 & 4 & 0 \end{vmatrix} + 1\begin{vmatrix} -1 & 3 & 2 \\ 0 & 4 & -2 \\ 3 & 4 & 0 \end{vmatrix} - 0\begin{vmatrix} -1 & -2 & 2 \\ 0 & 1 & -2 \\ 3 & -1 & 0 \end{vmatrix} + 3\begin{vmatrix} -1 & -2 & 3 \\ 0 & 1 & 4 \\ 3 & -1 & 4 \end{vmatrix}$$

$$= -2(6) + 1(-50) - 0 + 3(-41)$$

$$= -185. \quad \square$$

EXERCISES
9.4

Find the value of each of the following determinants. All variables represent real numbers.

1. $\begin{vmatrix} 5 & 8 \\ 2 & -4 \end{vmatrix}$ **2.** $\begin{vmatrix} -3 & 0 \\ 0 & 9 \end{vmatrix}$

3. $\begin{vmatrix} -1 & -2 \\ 5 & 3 \end{vmatrix}$ **4.** $\begin{vmatrix} 6 & -4 \\ 0 & -1 \end{vmatrix}$

5. $\begin{vmatrix} 9 & 3 \\ -3 & -1 \end{vmatrix}$ **6.** $\begin{vmatrix} 0 & 2 \\ 1 & 5 \end{vmatrix}$

7. $\begin{vmatrix} 3 & 4 \\ 5 & -2 \end{vmatrix}$ **8.** $\begin{vmatrix} -9 & 7 \\ 2 & 6 \end{vmatrix}$

9. $\begin{vmatrix} 0 & 4 \\ 4 & 0 \end{vmatrix}$ **10.** $\begin{vmatrix} 1 & 0 \\ 0 & 2 \end{vmatrix}$

11. $\begin{vmatrix} y & 2 \\ 8 & y \end{vmatrix}$ **12.** $\begin{vmatrix} 3 & 8 \\ m & n \end{vmatrix}$

13. $\begin{vmatrix} x & y \\ y & x \end{vmatrix}$ **14.** $\begin{vmatrix} 2m & 8n \\ 8n & 2m \end{vmatrix}$

Find the cofactor of each element in the second row for the following determinants.

15. $\begin{vmatrix} -2 & 0 & 1 \\ 3 & 2 & -1 \\ 1 & 0 & 2 \end{vmatrix}$ **16.** $\begin{vmatrix} 0 & -1 & 2 \\ 1 & 0 & 2 \\ 0 & -3 & 1 \end{vmatrix}$

17. $\begin{vmatrix} 1 & 2 & -1 \\ 2 & 3 & -2 \\ -1 & 4 & 1 \end{vmatrix}$ **18.** $\begin{vmatrix} 2 & -1 & 4 \\ 3 & 0 & 1 \\ -2 & 1 & 4 \end{vmatrix}$

Find the value of each of the following determinants. All variables represent real numbers.

19. $\begin{vmatrix} 1 & 0 & 0 \\ 0 & -1 & 0 \\ 1 & 0 & 1 \end{vmatrix}$ **20.** $\begin{vmatrix} -2 & 0 & 1 \\ 0 & 1 & 0 \\ 0 & 0 & -1 \end{vmatrix}$

21. $\begin{vmatrix} -2 & 0 & 0 \\ 4 & 0 & 1 \\ 3 & 4 & 2 \end{vmatrix}$ **22.** $\begin{vmatrix} 7 & -1 & 1 \\ 1 & -7 & 2 \\ -2 & 1 & 1 \end{vmatrix}$

23. $\begin{vmatrix} 1 & -2 & 3 \\ 0 & 0 & 0 \\ 1 & 10 & -12 \end{vmatrix}$ **24.** $\begin{vmatrix} 2 & 3 & 0 \\ 1 & 9 & 0 \\ -1 & -2 & 0 \end{vmatrix}$

25. $\begin{vmatrix} 3 & 3 & -1 \\ 2 & 6 & 0 \\ -6 & -6 & 2 \end{vmatrix}$ **26.** $\begin{vmatrix} 5 & -3 & 2 \\ -5 & 3 & -2 \\ 1 & 0 & 1 \end{vmatrix}$

27. $\begin{vmatrix} 3 & 2 & 0 \\ 0 & 1 & x \\ 2 & 0 & 0 \end{vmatrix}$ **28.** $\begin{vmatrix} 0 & 3 & y \\ 0 & 4 & 2 \\ 1 & 0 & 1 \end{vmatrix}$

29. $\begin{vmatrix} i & j & k \\ -1 & 2 & 4 \\ 3 & 0 & 5 \end{vmatrix}$ **30.** $\begin{vmatrix} i & j & k \\ 0 & -4 & 2 \\ -1 & 3 & 1 \end{vmatrix}$

31. $\begin{vmatrix} 4 & 0 & 0 & 2 \\ -1 & 0 & 3 & 0 \\ 2 & 4 & 0 & 1 \\ 0 & 0 & 1 & 2 \end{vmatrix}$ **32.** $\begin{vmatrix} -2 & 0 & 4 & 2 \\ 3 & 6 & 0 & 4 \\ 0 & 0 & 0 & 3 \\ 9 & 0 & 2 & -1 \end{vmatrix}$

33. $\begin{vmatrix} 0.4 & -0.8 & 0.6 \\ 0.3 & 0.9 & 0.7 \\ 3.1 & 4.1 & -2.8 \end{vmatrix}$ **34.** $\begin{vmatrix} -0.3 & -0.1 & 0.9 \\ 2.5 & 4.9 & -3.2 \\ -0.1 & 0.4 & 0.8 \end{vmatrix}$

Let $A = \begin{bmatrix} a_{11} & a_{12} & a_{13} \\ a_{21} & a_{22} & a_{23} \\ a_{31} & a_{32} & a_{33} \end{bmatrix}$.

35. Find $|A|$ by expansion about row 3. Show that your result is really equal to $|A|$ given in equation (1) of this section.

36. Do the same thing for column 3.

37. Suppose that every element in one row of matrix B is 0. Prove that $|B| = 0$.

Let $M = \begin{bmatrix} a_{11} & a_{12} \\ a_{21} & a_{22} \end{bmatrix}$.

38. Form a matrix N by exchanging the rows of M. Prove that $|M| = -|N|$.

39. Form matrix R by multiplying every element of row 1 of M by the real number k. Prove that $|R| = k \cdot |M|$.

40. Let matrix S be formed by adding to row 1 the result of multiplying each element of row 2 of M by the real number k. Show that $|M| = |S|$.

41. Let $Z = \begin{bmatrix} a_{11} & a_{12} & a_{13} \\ a_{11} & a_{12} & a_{13} \\ a_{31} & a_{32} & a_{33} \end{bmatrix}$, in which two rows

are identical. Show that $|Z| = 0$.

42. Suppose matrix Y has no zero elements. Prove that $|Y|$ can be 0.

Determinants can be used to find the area of a triangle, given the coordinates of its vertices. Given a triangle PQR with vertices (x_1, y_1), (x_2, y_2), and (x_3, y_3), as shown in the following figure, we can introduce line segments PM, RN, and QS perpendicular to the x-axis, forming trapezoids PMNR, NSQR, and PMSQ. Recall that the area of a trapezoid is given by the product of half the sum of the parallel bases and the altitude. For example, the area of trapezoid PMSQ is $(1/2)(y_1 + y_2)(x_2 - x_1)$. The area of triangle PQR can be found by subtracting the area of PMSQ from the sum of the areas of PMNR and RNSQ. Thus, for the area A of triangle PQR,

$$A = \frac{1}{2}(x_3 y_1 - x_1 y_3 + x_2 y_3 - x_3 y_2 + x_1 y_2 - x_2 y_1).$$

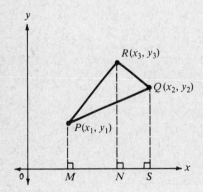

By evaluating the determinant

$$\begin{vmatrix} x_1 & y_1 & 1 \\ x_2 & y_2 & 1 \\ x_3 & y_3 & 1 \end{vmatrix}$$

it can be shown that the area of the triangle is given by A, where

$$A = \frac{1}{2} \begin{vmatrix} x_1 & y_1 & 1 \\ x_2 & y_2 & 1 \\ x_3 & y_3 & 1 \end{vmatrix}.$$

We assume that the points (x_1, y_1), (x_2, y_2), (x_3, y_3) are taken in counterclockwise order; if this is not done then A may have the wrong sign. Alternatively, we could define A as the absolute value of 1/2 the determinant shown above.

Use the formula given to find the area of the following triangles.

43. $P(0, 1)$, $Q(2, 0)$, $R(1, 3)$

44. $P(2, 5)$, $Q(-1, 3)$, $R(4, 0)$

45. $P(2, -2)$, $Q(0, 0)$, $R(-3, -4)$

46. $P(4, 7)$, $Q(5, -2)$, $R(1, 1)$

47. $P(3, 8)$, $Q(-1, 4)$, $R(0, 1)$

48. $P(-3, -1)$, $Q(4, 2)$, $R(3, -3)$

49. Prove that the straight line through the distinct points (x_1, y_1) and (x_2, y_2) has equation

$$\begin{vmatrix} x & y & 1 \\ x_1 & y_1 & 1 \\ x_2 & y_2 & 1 \end{vmatrix} = 0.$$

50. Use the result of Exercise 49 to show that three distinct points (x_1, y_1) and (x_2, y_2) and (x_3, y_3) lie on a straight line if

$$\begin{vmatrix} x_1 & y_1 & 1 \\ x_2 & y_2 & 1 \\ x_3 & y_3 & 1 \end{vmatrix} = 0.$$

51. Show that the lines $a_1 x + b_1 y = c_1$ and $a_2 x + b_2 y = c_2$, when $c_1 \neq c_2$, are parallel if

$$\begin{vmatrix} a_1 & b_1 \\ a_2 & b_2 \end{vmatrix} = 0.$$

52. Prove that

$$\begin{vmatrix} 1 & 1 & 1 \\ a & b & c \\ a^2 & b^2 & c^2 \end{vmatrix} = (a - b)(b - c)(c - a).$$

9.5 Properties of Determinants

Some determinants can only be evaluated after much tedious calculation. The more calculation involved, the more chance for error. To evaluate a determinant of a 3 × 3 matrix, it is necessary to evaluate three different determinants of a 2 × 2 matrix and then combine them correctly. For a determinant of a 4 × 4 matrix, twelve determinants of 2 × 2 matrices must be found. In fact, it turns out that to find the determinant of an $n \times n$ matrix, we must find $n!/2$ different determinants of 2 × 2 matrices, where $n!$ represents the product $n(n-1)(n-2)\cdots 3 \cdot 2 \cdot 1$.

In this section we give several theorems which make it easier to calculate certain determinants. These theorems help us reduce a determinant to one which is easier to evaluate. The theorems are true for square matrices of any order, but we prove them only for determinants of 3 × 3 matrices.

Theorem   If every element in a row (or column) of matrix A is 0, then $|A| = 0$.

EXAMPLE 1

$$\begin{vmatrix} -3 & 7 & 0 \\ 4 & 9 & 0 \\ -6 & 8 & 0 \end{vmatrix} = 0 \quad \square$$

To prove the theorem, expand the given determinant by the row (or column) of zeros. Each term of this expansion will have a zero factor, making the final determinant 0. For example,

$$\begin{vmatrix} a_{11} & a_{12} & a_{13} \\ 0 & 0 & 0 \\ a_{31} & a_{32} & a_{33} \end{vmatrix} = -0 \cdot \begin{vmatrix} a_{12} & a_{13} \\ a_{32} & a_{33} \end{vmatrix} + 0 \cdot \begin{vmatrix} a_{11} & a_{13} \\ a_{31} & a_{33} \end{vmatrix} - 0 \cdot \begin{vmatrix} a_{11} & a_{12} \\ a_{31} & a_{32} \end{vmatrix} = 0.$$

Theorem   If the rows of matrix A are the corresponding columns of matrix B, then $|B| = |A|$.

EXAMPLE 2

(a) Let $A = \begin{bmatrix} 2 & 1 \\ 3 & 4 \end{bmatrix}$. If we interchange the rows and columns of A, we get matrix B:

$$B = \begin{bmatrix} 2 & 3 \\ 1 & 4 \end{bmatrix}.$$

Check that $|A| = 5$ and $|B| = 5$, so that $|A| = |B|$.

(b)
$$\begin{vmatrix} 2 & 1 & 6 \\ 3 & 0 & 5 \\ -4 & 6 & 9 \end{vmatrix} = \begin{vmatrix} 2 & 3 & -4 \\ 1 & 0 & 6 \\ 6 & 5 & 9 \end{vmatrix} \quad \square$$

To prove the theorem, let

$$A = \begin{bmatrix} a_{11} & a_{12} & a_{13} \\ a_{21} & a_{22} & a_{23} \\ a_{31} & a_{32} & a_{33} \end{bmatrix} \quad \text{and} \quad B = \begin{bmatrix} a_{11} & a_{21} & a_{31} \\ a_{12} & a_{22} & a_{32} \\ a_{13} & a_{23} & a_{33} \end{bmatrix},$$

where B was obtained by interchanging the corresponding rows and columns of A. Find $|A|$ by expansion about row 1. Then find $|B|$ by expansion about column 1. You should find that $|A| = |B|$.

Theorem

If any two rows (or columns) of matrix A are interchanged to form matrix B, then $|B| = -|A|$.

EXAMPLE 3

☐ (a) Let $A = \begin{bmatrix} 2 & 5 \\ 3 & 4 \end{bmatrix}$. Exchange the two columns of A to get the matrix $B = \begin{bmatrix} 5 & 2 \\ 4 & 3 \end{bmatrix}$. Check that $|A| = -7$ and $|B| = 7$, so that $|B| = -|A|$.

(b) $\begin{vmatrix} 2 & 1 & 6 \\ 3 & 0 & 5 \\ -4 & 6 & 9 \end{vmatrix} = - \begin{vmatrix} -4 & 6 & 9 \\ 3 & 0 & 5 \\ 2 & 1 & 6 \end{vmatrix}$ ☐

This theorem is proved by steps very similar to those used to prove the previous theorem. (See Exercise 29.)

Theorem

Suppose matrix B is formed by multiplying every element of a row (or column) of matrix A by the real number k. Then $|B| = k \cdot |A|$.

EXAMPLE 4

☐ Let $A = \begin{bmatrix} 2 & -3 \\ 4 & 1 \end{bmatrix}$. Form the new matrix B by multiplying each element of the second row of A by -5:

$$B = \begin{bmatrix} 2 & -3 \\ 4(-5) & 1(-5) \end{bmatrix} = \begin{bmatrix} 2 & -3 \\ -20 & -5 \end{bmatrix}.$$

Check that $|B| = -5 \cdot |A|$. ☐

The proof of this theorem is left for Exercise 30 below.

Theorem

If two rows (or columns) of a matrix A are identical, then $|A| = 0$.

EXAMPLE 5 ☐ Since two rows are identical, $\begin{vmatrix} -4 & 2 & 3 \\ 0 & 1 & 6 \\ -4 & 2 & 3 \end{vmatrix} = 0.$ ☐

If two rows or columns of a matrix A are interchanged to form matrix B, then $|A| = -|B|$. If two rows of matrix A are identical and we interchange them, we still have matrix A. But we then also have $|A| = -|A|$. The only way this can happen is if $|A| = 0$.

The last theorem of this section is perhaps the most useful of all.

Theorem

> If matrix B is obtained from matrix A by adding to the elements of a row (or column) of A the results of multiplying the elements of any other row (or column) of A by the same number, then $|B| = |A|$.

This theorem is proved in much the same way as the others in this section. (See Exercises 31–32 below.)

EXAMPLE 6 ☐ Let $A = \begin{bmatrix} -3 & 5 \\ 1 & 2 \end{bmatrix}$. Add to the elements of the first row the results of multiplying the corresponding elements of the second row by 3, getting a new matrix, B:

$$B = \begin{bmatrix} -3 + 1(3) & 5 + 2(3) \\ 1 & 2 \end{bmatrix} = \begin{bmatrix} 0 & 11 \\ 1 & 2 \end{bmatrix}.$$

Check that $|A| = |B|$. ☐

EXAMPLE 7 ☐ Let $A = \begin{bmatrix} -2 & 4 & 1 \\ 2 & 1 & 5 \\ 3 & 0 & 2 \end{bmatrix}$. Obtain a new matrix by adding to the third column the results of multiplying each element of the first column by 3.

$$B = \begin{bmatrix} -2 & 4 & 1 + 3(-2) \\ 2 & 1 & 5 + 3(2) \\ 3 & 0 & 2 + 3(3) \end{bmatrix} = \begin{bmatrix} -2 & 4 & -5 \\ 2 & 1 & 11 \\ 3 & 0 & 11 \end{bmatrix}.$$

Check that $|A| = 37$ and $|B| = 37$, so that $|A| = |B|$. ☐

The following examples show how the properties of determinants can be used to simplify the calculation of determinants.

EXAMPLE 8 ☐ Without expanding, show that the value of the following determinant is 0.

$$\begin{vmatrix} 2 & 5 & -1 \\ 1 & -15 & 3 \\ -2 & 10 & -2 \end{vmatrix}$$

Examining the columns of the array, we note that each element in the second column is -5 times the corresponding element in the third column. Thus, if we use the last theorem above and add to the elements of the second column the results of multiplying the elements of the third column by 5, we have the equivalent determinant

$$\begin{vmatrix} 2 & 0 & -1 \\ 1 & 0 & 3 \\ -2 & 0 & -2 \end{vmatrix}.$$

The value of this determinant is 0. ■

EXAMPLE 9   ■ Find $|A|$ if

$$|A| = \begin{vmatrix} 4 & 2 & 1 & 0 \\ -2 & 4 & -1 & 7 \\ -5 & 2 & 3 & 1 \\ 6 & 4 & -3 & 2 \end{vmatrix}.$$

Our goal is to change the first row (we could have selected any row or column) to a row in which every element but one is 0. To begin, add to the elements of the first column the results of multiplying the elements of the second column by -2.

$$\begin{vmatrix} 0 & 2 & 1 & 0 \\ -10 & 4 & -1 & 7 \\ -9 & 2 & 3 & 1 \\ -2 & 4 & -3 & 2 \end{vmatrix}$$

Then replace the second column by the results we get when we add to the second column the results of multiplying the elements of column 3 by -2.

$$\begin{vmatrix} 0 & 0 & 1 & 0 \\ -10 & 6 & -1 & 7 \\ -9 & -4 & 3 & 1 \\ -2 & 10 & -3 & 2 \end{vmatrix}$$

The first row has only one nonzero number, and so we expand about the first row.

$$|A| = +1 \begin{vmatrix} -10 & 6 & 7 \\ -9 & -4 & 1 \\ -2 & 10 & 2 \end{vmatrix}$$

Now let us try to change the third column to a column with two zeros.

Add to the elements of row 1 the results of multiplying row 2 by -7.
$$\begin{vmatrix} 53 & 34 & 0 \\ -9 & -4 & 1 \\ -2 & 10 & 2 \end{vmatrix}$$

Add to row 3 the results of multiplying row 2 by -2.
$$\begin{vmatrix} 53 & 34 & 0 \\ -9 & -4 & 1 \\ 16 & 18 & 0 \end{vmatrix}$$

Now expand about the third column to find the value of $|A|$.

$$|A| = -1 \begin{vmatrix} 53 & 34 \\ 16 & 18 \end{vmatrix} = -1(954 - 544) = -410 \quad \blacksquare$$

Note that we worked with rows to get a column with only one nonzero number and with columns to get a row with one nonzero number.

EXERCISES
9.5

Tell why each of the following determinants has a value of 0. All variables represent real numbers.

1. $\begin{vmatrix} 2 & 3 \\ 2 & 3 \end{vmatrix}$　　　　**2.** $\begin{vmatrix} -5 & -5 \\ 6 & 6 \end{vmatrix}$

3. $\begin{vmatrix} 2 & 0 \\ 3 & 0 \end{vmatrix}$　　　　**4.** $\begin{vmatrix} -8 & 0 \\ -6 & 0 \end{vmatrix}$

5. $\begin{vmatrix} -1 & 2 & 4 \\ 4 & -8 & -16 \\ 3 & 0 & 5 \end{vmatrix}$　　**6.** $\begin{vmatrix} 1 & 0 & 0 \\ 1 & 0 & 1 \\ 3 & 0 & 0 \end{vmatrix}$

7. $\begin{vmatrix} m & 2 & 2m \\ 3n & 1 & 6n \\ 5p & 6 & 10p \end{vmatrix}$　**8.** $\begin{vmatrix} 7z & 8x & 2y \\ z & x & y \\ 7z & 7x & 7y \end{vmatrix}$

Use the appropriate theorems from this section to tell why each of the following is true. Do not evaluate the determinants. All variables represent real numbers.

9. $\begin{vmatrix} 2 & 1 & 6 \\ 3 & 0 & 2 \\ 4 & 1 & 8 \end{vmatrix} = \begin{vmatrix} 2 & 3 & 4 \\ 1 & 0 & 1 \\ 6 & 2 & 8 \end{vmatrix}$

10. $\begin{vmatrix} 4 & -2 \\ 3 & 8 \end{vmatrix} = \begin{vmatrix} 4 & 3 \\ -2 & 8 \end{vmatrix}$

11. $\begin{vmatrix} 2 & 6 \\ 3 & 5 \end{vmatrix} = - \begin{vmatrix} 3 & 5 \\ 2 & 6 \end{vmatrix}$

12. $\begin{vmatrix} -1 & 8 & 9 \\ 0 & 2 & 1 \\ 3 & 2 & 0 \end{vmatrix} = - \begin{vmatrix} 8 & -1 & 9 \\ 2 & 0 & 1 \\ 2 & 3 & 0 \end{vmatrix}$

13. $3 \begin{vmatrix} 6 & 0 & 2 \\ 4 & 1 & 3 \\ 2 & 8 & 6 \end{vmatrix} = \begin{vmatrix} 6 & 0 & 2 \\ 4 & 3 & 3 \\ 2 & 24 & 6 \end{vmatrix}$

14. $-\dfrac{1}{2} \begin{vmatrix} 5 & -8 & 2 \\ 3 & -6 & 9 \\ 2 & 4 & 4 \end{vmatrix} = \begin{vmatrix} 5 & 4 & 2 \\ 3 & 3 & 9 \\ 2 & -2 & 4 \end{vmatrix}$

15. $\begin{vmatrix} 3 & -4 \\ 2 & 5 \end{vmatrix} = \begin{vmatrix} 3 & -4 \\ 5 & 1 \end{vmatrix}$

16. $\begin{vmatrix} -1 & 6 \\ 3 & -5 \end{vmatrix} = \begin{vmatrix} -1 & 6 \\ 2 & 1 \end{vmatrix}$

17. $2 \begin{vmatrix} 4 & 2 & -1 \\ m & 2n & 3p \\ 5 & 1 & 0 \end{vmatrix} = \begin{vmatrix} 4 & 2 & -1 \\ 2m & 4n & 6p \\ 5 & 1 & 0 \end{vmatrix}$

18. $\begin{vmatrix} -4 & 2 & 1 \\ 3 & 0 & 5 \\ -1 & 4 & -2 \end{vmatrix} = \begin{vmatrix} -4 & 2 & 1 + (-4)k \\ 3 & 0 & 5 + 3k \\ -1 & 4 & -2 + (-1)k \end{vmatrix}$

Use the method of Examples 8 and 9 to find the value of each of the following determinants.

19. $\begin{vmatrix} 2 & 4 \\ 3 & 6 \end{vmatrix}$　　　　**20.** $\begin{vmatrix} -5 & 10 \\ 6 & -12 \end{vmatrix}$

21. $\begin{vmatrix} 4 & 8 & 0 \\ -1 & -2 & 1 \\ 2 & 4 & 3 \end{vmatrix}$　**22.** $\begin{vmatrix} 6 & 8 & -12 \\ -1 & 0 & 2 \\ 4 & 0 & -8 \end{vmatrix}$

23. $\begin{vmatrix} 3 & 1 & 2 \\ 2 & 0 & 1 \\ 1 & 0 & -2 \end{vmatrix}$　**24.** $\begin{vmatrix} -2 & 2 & 3 \\ 0 & 2 & 1 \\ -1 & 4 & 0 \end{vmatrix}$

25. $\begin{vmatrix} -4 & 2 & 3 \\ 2 & 0 & 1 \\ 0 & 4 & 2 \end{vmatrix}$　**26.** $\begin{vmatrix} 6 & 3 & 2 \\ 1 & 0 & 2 \\ -1 & 4 & 1 \end{vmatrix}$

27. $\begin{vmatrix} 1 & 0 & 2 & 2 \\ 2 & 4 & 1 & -1 \\ 1 & -3 & 1 & 0 \\ 1 & 1 & 0 & 1 \end{vmatrix}$　**28.** $\begin{vmatrix} 2 & -1 & 1 & 0 \\ 1 & 1 & 0 & 1 \\ 0 & -1 & 1 & 1 \\ 1 & 2 & 1 & 2 \end{vmatrix}$

$$Let\ A = \begin{bmatrix} a_{11} & a_{12} & a_{13} \\ a_{21} & a_{22} & a_{23} \\ a_{31} & a_{32} & a_{33} \end{bmatrix}.$$

29. Obtain matrix B by exchanging columns 1 and 3 of matrix A. Show that $|A| = -|B|$.

30. Obtain matrix B by multiplying each element of row 3 of A by the real number k. Show that $|B| = k \cdot |A|$.

31. Obtain matrix B by adding to column 1 of matrix A the result of multiplying each element of column 2 of A by the real number k. Show that $|A| = |B|$.

32. Obtain matrix B by adding to row 1 of matrix A the result of multiplying each element of row 3 of A by the real number k. Show that $|A| = |B|$.

33. Let A and B be any 2×2 matrices. Show that $|AB| = |A| \cdot |B|$, where $|AB|$ is the determinant of matrix AB.

34. Let A be an $n \times n$ matrix. Suppose matrix B is found by multiplying every element of A by the real number k. Express $|B|$ in terms of $|A|$.

35. Show that

$$\begin{vmatrix} a_{11} & a_{12} & a_{13} & a_{14} \\ 0 & a_{22} & a_{23} & a_{24} \\ 0 & 0 & a_{33} & a_{34} \\ 0 & 0 & 0 & a_{44} \end{vmatrix} = a_{11} a_{22} a_{33} a_{44}.$$

36. Show that

$$\begin{vmatrix} a_{11} & a_{12} & a_{13} & a_{14} \\ a_{21} & a_{22} & a_{23} & a_{24} \\ 0 & 0 & a_{33} & a_{34} \\ 0 & 0 & a_{43} & a_{44} \end{vmatrix} = \begin{vmatrix} a_{11} & a_{12} \\ a_{21} & a_{22} \end{vmatrix} \cdot \begin{vmatrix} a_{33} & a_{34} \\ a_{43} & a_{44} \end{vmatrix}.$$

9.6 Properties of Matrices

In Section 9.3 we saw how matrix notation is useful in solving a system of linear equations. Now, in this section, we look at the algebraic properties of matrices.

It is customary to use capital letters to name matrices. Subscript notation is used to name the elements of a matrix, as follows.

$$A = \begin{bmatrix} a_{11} & a_{12} & a_{13} & \cdots & a_{1n} \\ a_{21} & a_{22} & a_{23} & \cdots & a_{2n} \\ a_{31} & a_{32} & a_{33} & \cdots & a_{3n} \\ \cdot & \cdot & \cdot & & \cdot \\ \cdot & \cdot & \cdot & & \cdot \\ \cdot & \cdot & \cdot & & \cdot \\ a_{m1} & a_{m2} & a_{m3} & & a_{mn} \end{bmatrix}$$

As we have seen, the first row, first column element is denoted a_{11}; the second row, third column element is denoted a_{23}; and in general, the ith row, jth column element is denoted a_{ij}. As a shorthand notation, matrix A above is sometimes written as just $A = (a_{ij})$.

Certain matrices have special names. Earlier we said that an $n \times n$ matrix is a square matrix. A matrix with just one row is a **row matrix,** and a matrix with just one column is a **column matrix.**

Two matrices are **equal** if they are of the same order and if each corresponding element, position by position, is equal. Using this definition, the matrices

$$\begin{bmatrix} 2 & 1 \\ 3 & -5 \end{bmatrix} \quad \text{and} \quad \begin{bmatrix} 1 & 2 \\ -5 & 3 \end{bmatrix}$$

are *not* equal (even though they contain the same elements and are of the same order), since the corresponding elements differ.

EXAMPLE 1

From the definition of equality given above, the only way that the statement

$$\begin{bmatrix} 2 & 1 \\ p & q \end{bmatrix} = \begin{bmatrix} x & y \\ -1 & 0 \end{bmatrix}$$

can be true is if $2 = x$, $1 = y$, $p = -1$, and $q = 0$. ◻

EXAMPLE 2

The statement

$$\begin{bmatrix} x \\ y \end{bmatrix} = \begin{bmatrix} 1 \\ 4 \\ 0 \end{bmatrix}$$

can never be true, since the two matrices are of different order. (One is 2×1 and the other is 3×1.) ◻

To **add** two matrices of the same order, add corresponding elements. That is, adding 2×2 matrices is defined as follows.

$$\begin{bmatrix} a_{11} & a_{12} \\ a_{21} & a_{22} \end{bmatrix} + \begin{bmatrix} b_{11} & b_{12} \\ b_{21} & b_{22} \end{bmatrix} = \begin{bmatrix} a_{11} + b_{11} & a_{12} + b_{12} \\ a_{21} + b_{21} & a_{22} + b_{22} \end{bmatrix}$$

The definition of addition can be extended to matrices of any order as follows.

Definition: Addition of Matrices

Let $A = (a_{ij})$ and $B = (b_{ij})$ be two matrices of the same order. Then $A + B = (a_{ij}) + (b_{ij}) = (a_{ij} + b_{ij})$. Only matrices of the same order can be added.

EXAMPLE 3

Find each of the following sums.

(a) $\begin{bmatrix} 5 & -6 \\ 8 & 9 \end{bmatrix} + \begin{bmatrix} -4 & 6 \\ 8 & -3 \end{bmatrix} = \begin{bmatrix} 5 + (-4) & -6 + 6 \\ 8 + 8 & 9 + (-3) \end{bmatrix}$

$$= \begin{bmatrix} 1 & 0 \\ 16 & 6 \end{bmatrix}$$

(b) $\begin{bmatrix} 2 \\ 5 \\ 8 \end{bmatrix} + \begin{bmatrix} -6 \\ 3 \\ 12 \end{bmatrix} = \begin{bmatrix} -4 \\ 8 \\ 20 \end{bmatrix}$

(c) The matrices $A = \begin{bmatrix} 5 & 8 \\ 6 & 2 \end{bmatrix}$ and $B = \begin{bmatrix} 3 & 9 & 1 \\ 4 & 2 & 5 \end{bmatrix}$

are of different orders. Therefore, the sum $A + B$ does not exist. ◻

A matrix containing only zero elements is called a **zero matrix.** For example, [0 0 0] is the 1×3 zero matrix, while

$$\begin{bmatrix} 0 & 0 & 0 \\ 0 & 0 & 0 \end{bmatrix}$$

is the 2×3 zero matrix. We can have a zero matrix of any order. The matrix $A = (a_{ij})$ is a zero matrix if $a_{ij} = 0$ for every value of i and j.

In Chapter 1, we said that each real number has an additive inverse. That is, if a is a real number, we can find a real number $-a$ such that

$$a + (-a) = 0 \quad \text{and} \quad -a + a = 0.$$

What about matrices? If we are given the matrix

$$A = \begin{bmatrix} -5 & 2 & -1 \\ 3 & 4 & -6 \end{bmatrix},$$

can we find a matrix $-A$ such that

$$A + (-A) = 0 \quad \text{and} \quad -A + A = 0,$$

where 0 is the 2×3 zero matrix? To find matrix $-A$, we find the additive inverse of each element of A. (Remember, each element of A is a real number and, therefore, has an additive inverse.)

$$-A = \begin{bmatrix} 5 & -2 & 1 \\ -3 & -4 & 6 \end{bmatrix}$$

To check, first test that $A + (-A)$ equals 0.

$$A + (-A) = \begin{bmatrix} -5 & 2 & -1 \\ 3 & 4 & -6 \end{bmatrix} + \begin{bmatrix} 5 & -2 & 1 \\ -3 & -4 & 6 \end{bmatrix} = \begin{bmatrix} 0 & 0 & 0 \\ 0 & 0 & 0 \end{bmatrix}$$

Then test that $-A + A$ is also 0. Matrix $-A$ is called the **additive inverse,** or **negative,** of matrix A. Every matrix, no matter the order, has an additive inverse. In general terms, if $A = (a_{ij})$, then the additive inverse of A is $-A = (-a_{ij})$.

To subtract the real number b from the real number a, written $a - b$, we add a and the additive inverse of b. That is,

$$a - b = a + (-b).$$

The same definition works for **subtraction** of matrices.

Definition: Subtraction of Matrices	If $A = (a_{ij})$ and $B = (b_{ij})$ are two matrices of the same order, then $$A - B = A + (-B) = (a_{ij}) + (-b_{ij}) = (a_{ij} - b_{ij}).$$

It can be shown that matrix addition (for matrices of the same order) satisfies the commutative, associative, closure, identity, and inverse properties. (See Exercises 41–45.)

EXAMPLE 4 Find each of the following differences.

(a) $\begin{bmatrix} -5 & 6 \\ 2 & 4 \end{bmatrix} - \begin{bmatrix} -3 & 2 \\ 5 & -8 \end{bmatrix} = \begin{bmatrix} -5 & 6 \\ 2 & 4 \end{bmatrix} + \begin{bmatrix} 3 & -2 \\ -5 & 8 \end{bmatrix}$

(Here we found the additive inverse of each element in the second matrix.)

$$= \begin{bmatrix} -2 & 4 \\ -3 & 12 \end{bmatrix}$$

(b) $\begin{bmatrix} 8 & 6 & -4 \end{bmatrix} - \begin{bmatrix} 3 & 5 & -8 \end{bmatrix} = \begin{bmatrix} 5 & 1 & 4 \end{bmatrix}$

(c) The matrices

$$\begin{bmatrix} -2 & 5 \\ 0 & 1 \end{bmatrix} \quad \text{and} \quad \begin{bmatrix} 3 \\ 5 \end{bmatrix}$$

are of different orders and cannot be subtracted. ∎

In work with matrices, a real number is called a **scalar** to distinguish it from a matrix. The product of a scalar k and a matrix X is the matrix kX, each of whose elements is k times the corresponding element of X. That is, for matrix $A = (a_{ij})$,

Definition: Multiplication by a Scalar

$$k \cdot A = k \cdot (a_{ij}) = (k \cdot a_{ij}) \quad \text{and} \quad (a_{ij} \cdot k) = (a_{ij}) \cdot k = Ak.$$

EXAMPLE 5

(a) $5 \begin{bmatrix} 2 & -3 \\ 0 & 4 \end{bmatrix} = \begin{bmatrix} 10 & -15 \\ 0 & 20 \end{bmatrix}$

(b) $\dfrac{3}{4} \begin{bmatrix} 20 & 36 \\ 12 & -16 \end{bmatrix} = \begin{bmatrix} 15 & 27 \\ 9 & -12 \end{bmatrix}$ ∎

The proofs of the following properties of scalar multiplication are left for Exercises 49–52 below. (Assume that A and B are matrices of the same order and that c and d are real numbers.)

$$(c + d)A = cA + dA$$
$$c(A + B) = cA + cB$$
$$(cd)A = c(dA).$$

We have seen how to multiply a real number (scalar) and a matrix. Now we consider the product of two matrices. Let us first find the product of

$$A = \begin{bmatrix} -3 & 4 & 2 \\ 5 & 0 & 4 \end{bmatrix} \quad \text{and} \quad B = \begin{bmatrix} -6 & 4 \\ 2 & 3 \\ 3 & -2 \end{bmatrix}$$

First, locate *row* 1 of *A* and *column* 1 of *B,* shown shaded below.

$$A = \begin{bmatrix} -3 & 4 & 2 \\ 5 & 0 & 4 \end{bmatrix} \qquad B = \begin{bmatrix} -6 & 4 \\ 2 & 3 \\ 3 & -2 \end{bmatrix}$$

Multiply corresponding elements, and find the sum of the products.

$$(-3)(-6) + (4)(2) + (2)(3) = 32$$

This result is the element for row 1, column 1 of the product matrix.

Now use *row* 1 of *A* and *column* 2 of *B* (shown shaded below) to determine the element in row 1 and column 2 of the product matrix.

$$\begin{bmatrix} -3 & 4 & 2 \\ 5 & 0 & 4 \end{bmatrix} \qquad \begin{bmatrix} -6 & 4 \\ 2 & 3 \\ 3 & -2 \end{bmatrix}$$

Multiply corresponding elements and add the products:

$$(-3)(4) + (4)(3) + (2)(-2) = -4,$$

which is the row 1, column 2 element of the product matrix.

Next, use *row* 2 of *A* and *column* 1 of *B;* this will give the row 2, column 1 entry of the product matrix.

$$(5)(-6) + (0)(2) + (4)(3) = -18$$

Finally, use *row* 2 of *A* and *column* 2 of *B* to find the entry for row 2, column 2 of the product matrix.

$$(5)(4) + (0)(3) + (4)(-2) = 12$$

The product matrix can now be written.

$$\begin{bmatrix} -3 & 4 & 2 \\ 5 & 0 & 4 \end{bmatrix} \begin{bmatrix} -6 & 4 \\ 2 & 3 \\ 3 & -2 \end{bmatrix} = \begin{bmatrix} 32 & -4 \\ -18 & 12 \end{bmatrix}$$

As this example shows, the product of a 2×3 matrix and a 3×2 matrix is a 2×2 matrix.

Generalizing, the **product** *AB* of an $m \times n$ matrix *A* and an $n \times p$ matrix *B* is found as follows. Multiply each element of the first row of *A* by the corresponding element of the first column of *B*. The sum of these n products is the first row, first column element of *AB*.

Also, the sum of the products found by multiplying the elements of the first row of *A* times the corresponding elements of the second column of *B* gives the first row, second column element of *AB*, and so on.

To find the *i*th row, *j*th column element of *AB*, multiply each element in the *i*th row of *A* by the corresponding element in the *j*th column of *B* (note colored areas in matrices below). The sum of these products will give the element of row *i*, column *j* of *AB*.

$$A = \begin{bmatrix} a_{11} & a_{12} & a_{13} \cdots a_{1n} \\ a_{21} & a_{22} & a_{23} \cdots a_{2n} \\ & & \vdots \\ a_{i1} & a_{i2} & a_{i3} \cdots a_{in} \\ & & \vdots \\ a_{m1} & a_{m2} & a_{m3} \cdots a_{mn} \end{bmatrix} \qquad B = \begin{bmatrix} b_{11} & b_{12} \cdots & b_{1j} & \cdots b_{1p} \\ b_{21} & b_{22} \cdots & b_{2j} & \cdots b_{2p} \\ \vdots \\ b_{n1} & b_{n2} \cdots & b_{nj} & \cdots b_{np} \end{bmatrix}$$

To summarize:

Definition: **Matrix** **Multiplication**	Entry c_{ij} of the product matrix $C = AB$ is found as follows: $c_{ij} = a_{i1}b_{1j} + a_{i2}b_{2j} + \cdots + a_{in}b_{nj}$.

As this definition shows, matrix multiplication is defined only for two matrices A and B where the number of columns of A is the same as the number of rows of B. The final product will have as many rows as A and as many columns as B.

EXAMPLE 6   Suppose matrix A is 3×2, while matrix B is 2×4. Can the product AB be calculated? What is the order of the product? Can the product BA be calculated? What is the order of BA?

The following diagram helps us answer the questions about the product AB.

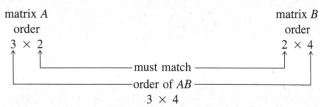

The product AB exists, since the two 2's are the same. The product has order 3×4.

Make a similar diagram for BA.

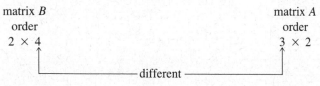

The product BA does not exist.   ▢

EXAMPLE 7   ☐   Find AB and BA, if possible, where

$$A = \begin{bmatrix} 1 & -3 \\ 7 & 2 \end{bmatrix} \quad \text{and} \quad B = \begin{bmatrix} 1 & 0 & -1 & 2 \\ 3 & 1 & 4 & -1 \end{bmatrix}.$$

Use the definition of matrix multiplication.

$$AB = \begin{bmatrix} 1 & -3 \\ 7 & 2 \end{bmatrix} \begin{bmatrix} 1 & 0 & -1 & 2 \\ 3 & 1 & 4 & -1 \end{bmatrix}$$

$$= \begin{bmatrix} 1(1) + (-3)(3) & 1(0) + (-3)1 & 1(-1) + (-3)4 & 1(2) + (-3)(-1) \\ 7(1) + 2(3) & 7(0) + 2(1) & 7(-1) + 2(4) & 7(2) + 2(-1) \end{bmatrix}$$

$$= \begin{bmatrix} -8 & -3 & -13 & 5 \\ 13 & 2 & 1 & 12 \end{bmatrix}$$

Since B is a 2×4 matrix, and A is a 2×2 matrix, we cannot find the product BA.   ☐

If $A = \begin{bmatrix} 1 & 3 \\ -2 & 5 \end{bmatrix}$ and $B = \begin{bmatrix} -2 & 7 \\ 0 & 2 \end{bmatrix}$, then we can use the definition of matrix multiplication to show that

$$AB = \begin{bmatrix} -2 & 13 \\ 4 & -4 \end{bmatrix} \quad \text{and} \quad BA = \begin{bmatrix} -16 & 29 \\ -4 & 10 \end{bmatrix}.$$

As this example shows, in general $AB \neq BA$. Thus, matrix multiplication is *not* commutative.

Matrix multiplication does, however, satisfy the associative and distributive properties: If A, B, and C are matrices such that all the following products and sums exist, then

$$(AB)C = A(BC)$$
$$A(B + C) = AB + AC$$
$$(B + C)A = BA + CA.$$

For proofs of these results for the special cases when A, B, and C are square matrices, see Exercises 46 and 47 below. The identity and inverse properties for matrix multiplication are discussed in the next section of this chapter.

EXERCISES 9.6

Find the values of the variables in each of the following.

1. $\begin{bmatrix} 2 & 1 \\ 4 & 8 \end{bmatrix} = \begin{bmatrix} x & 1 \\ y & z \end{bmatrix}$

2. $\begin{bmatrix} 2 & 5 & 6 \\ 1 & m & n \end{bmatrix} = \begin{bmatrix} z & y & w \\ 1 & 8 & -2 \end{bmatrix}$

3. $\begin{bmatrix} x + 6 & y + 2 \\ 8 & 3 \end{bmatrix} = \begin{bmatrix} -9 & 7 \\ 8 & k \end{bmatrix}$

4. $\begin{bmatrix} 9 & 7 \\ r & 0 \end{bmatrix} = \begin{bmatrix} m-3 & n+5 \\ 8 & 0 \end{bmatrix}$

5. $\begin{bmatrix} -7+z & 4r & 8s \\ 6p & 2 & 5 \end{bmatrix} + \begin{bmatrix} -9 & 8r & 3 \\ 2 & 5 & 4 \end{bmatrix} =$

$\begin{bmatrix} 2 & 36 & 27 \\ 20 & 7 & 12a \end{bmatrix}$

6. $\begin{bmatrix} a+2 & 3z+1 & 5m \\ 4k & 0 & 3 \end{bmatrix} + \begin{bmatrix} 3a & 2z & 5m \\ 2k & 5 & 6 \end{bmatrix} =$

$\begin{bmatrix} 10 & -14 & 80 \\ 10 & 5 & 9 \end{bmatrix}$

Perform each of the following operations, whenever possible.

7. $\begin{bmatrix} 6 & -9 & 2 \\ 4 & 1 & 3 \end{bmatrix} - \begin{bmatrix} -8 & 2 & 5 \\ 6 & -3 & 4 \end{bmatrix}$

8. $\begin{bmatrix} 9 & 4 \\ -8 & 2 \end{bmatrix} + \begin{bmatrix} -3 & 2 \\ -4 & 7 \end{bmatrix}$

9. $\begin{bmatrix} -6 & 8 \\ 0 & 0 \end{bmatrix} - \begin{bmatrix} 0 & 0 \\ -4 & -2 \end{bmatrix}$

10. $\begin{bmatrix} 1 & -4 \\ 2 & -3 \\ -8 & 4 \end{bmatrix} - \begin{bmatrix} -6 & 9 \\ -2 & 5 \\ -7 & -12 \end{bmatrix}$

11. $\begin{bmatrix} -8 & 4 & 0 \\ 2 & 5 & 0 \end{bmatrix} + \begin{bmatrix} 6 & 3 \\ 8 & 9 \end{bmatrix}$

12. $\begin{bmatrix} 2 \\ 3 \end{bmatrix} - \begin{bmatrix} 8 & 1 \\ 9 & 4 \end{bmatrix}$

13. $\begin{bmatrix} -4x+2y & -3x+y \\ 6x-3y & 2x-5y \end{bmatrix} +$

$\begin{bmatrix} -8x+6y & 2x \\ 3y-5x & 6x+4y \end{bmatrix}$

14. $\begin{bmatrix} 4k-8y \\ 6z-3x \\ 2k+5a \\ -4m+2n \end{bmatrix} - \begin{bmatrix} 5k+6y \\ 2z+5x \\ 4k+6a \\ 4m-2n \end{bmatrix}$

15. When John inventoried his screw collection, he found that he had 7 flat-head long screws, 9 flat-head medium, 8 flat-head short, 2 round-head long, no round-head medium, and 6 round-head short. Write this information first as a 3 × 2 matrix and then as a 2 × 3 matrix.

16. At the grocery store, Miguel bought 4 quarts of milk, 2 loaves of bread, 4 chickens, and an apple. Mary bought 2 quarts of milk, a loaf of bread, 5 chickens, and 4 apples. Write this information first as a 2 × 4 matrix and then as a 4 × 2 matrix.

Let $A = \begin{bmatrix} -2 & 4 \\ 0 & 3 \end{bmatrix}$ *and* $B = \begin{bmatrix} -6 & 2 \\ 4 & 0 \end{bmatrix}$. *Find each of the following.*

17. $2A$

18. $-3B$

19. $2A - B$

20. $-2A + 4B$

21. $-A + \dfrac{1}{2}B$

22. $\dfrac{3}{4}A - B$

Find each of the following matrix products, whenever possible.

23. $\begin{bmatrix} 1 & 2 \\ 3 & 4 \end{bmatrix} \begin{bmatrix} -1 \\ 7 \end{bmatrix}$

24. $\begin{bmatrix} 3 & -4 & 1 \\ 5 & 0 & 2 \end{bmatrix} \begin{bmatrix} -1 \\ 4 \\ 2 \end{bmatrix}$

25. $\begin{bmatrix} -1 & 2 & 0 \\ 0 & 3 & 2 \\ 0 & 1 & 4 \end{bmatrix} \begin{bmatrix} 2 & -1 & 2 \\ 0 & 2 & 1 \\ 3 & 0 & -1 \end{bmatrix}$

26. $\begin{bmatrix} -2 & -3 & -4 \\ 2 & -1 & 0 \\ 4 & -2 & 3 \end{bmatrix} \begin{bmatrix} 0 & 1 & 4 \\ 1 & 2 & -1 \\ 3 & 2 & -2 \end{bmatrix}$

27. $\begin{bmatrix} -2 & 1 & 4 \\ 0 & 1 & 2 \end{bmatrix} \begin{bmatrix} -2 & 1 & 0 \\ 0 & -2 & 0 \\ 4 & 1 & 2 \end{bmatrix}$

28. $\begin{bmatrix} -1 & 0 & 0 \\ 2 & 1 & 4 \end{bmatrix} \begin{bmatrix} 4 & -2 & 5 \\ 0 & 1 & 4 \\ 2 & -9 & 0 \end{bmatrix}$

29. $\begin{bmatrix} -3 & 0 & 2 & 1 \\ 4 & 0 & 2 & 6 \end{bmatrix} \begin{bmatrix} -4 & 2 \\ 0 & 1 \end{bmatrix}$

30. $\begin{bmatrix} -1 & 2 & 4 & 1 \\ 0 & 2 & -3 & 5 \end{bmatrix} \begin{bmatrix} 1 & 2 & 4 \\ -2 & 5 & 1 \end{bmatrix}$

31. $[-2 \quad 4 \quad 6] \begin{bmatrix} 3 \\ -2 \\ 1 \end{bmatrix}$ **32.** $[4 \quad 0 \quad 2] \begin{bmatrix} -5 \\ 1 \\ 6 \end{bmatrix}$

33. $\begin{bmatrix} 3 \\ -2 \\ 1 \end{bmatrix} [-2 \quad 4 \quad 6]$ **34.** $\begin{bmatrix} -5 \\ 1 \\ 6 \end{bmatrix} [4 \quad 0 \quad 2]$

35. The Bread Box, a small neighborhood bakery, sells four main items: sweet rolls, bread, cake, and pie. The amount of certain major ingredients required to make these items is given in matrix A.

	eggs	flour*	sugar*	short-ening*	milk*	
$A =$	1	4	$\frac{1}{4}$	$\frac{1}{4}$	1	rolls (dozen)
	0	3	0	$\frac{1}{4}$	0	bread (loaves)
	4	3	2	1	1	cake (1)
	0	1	0	$\frac{1}{3}$	0	pie (1)

The cost (in cents) for each ingredient when purchased in large lots and in small lots is given by matrix B.

	cost		
	large lot	small lot	
$B =$	5	5	eggs
	8	10	flour*
	10	12	sugar*
	12	15	shortening*
	5	6	milk*

(a) Use matrix multiplication to find a matrix representing the comparative costs per item under the two purchase options.

Suppose a day's orders consist of 20 dozen sweet rolls, 200 loaves of bread, 50 cakes, and 60 pies.

(b) Represent these orders as a 1 × 4 matrix and use matrix multiplication to write as a matrix the amount of each ingredient required to fill the day's orders.

*In cups.

(c) Use matrix multiplication to find a matrix representing the costs under the two purchase options to fill the day's orders.

Find each of the following products.

36. $\begin{bmatrix} .8 & .4 \\ -.7 & -.22 \end{bmatrix} \begin{bmatrix} -.72 & -.8 \\ .4 & -.2 \end{bmatrix}$

37. $\begin{bmatrix} -.6 & .93 \\ .8 & .47 \end{bmatrix} \begin{bmatrix} .9 & .4 \\ .6 & -.8 \end{bmatrix}$

38. $\begin{bmatrix} -.42 & .6 \\ .9 & .3 \end{bmatrix} \begin{bmatrix} .1 & .8 \\ -.4 & .11 \end{bmatrix}$

39. $\begin{bmatrix} .8 & .72 & .44 \\ -.41 & .83 & .29 \\ -.77 & .61 & -.42 \end{bmatrix} \begin{bmatrix} -3 & .8 & -.4 \\ -7 & -.1 & -.2 \\ -1 & .3 & .9 \end{bmatrix}$

40. $\begin{bmatrix} .71 & -.22 & .88 \\ .62 & -.79 & .33 \\ .43 & -.11 & .19 \end{bmatrix} \begin{bmatrix} .08 & -.73 & -.07 \\ .04 & -.64 & .90 \\ .91 & 1.33 & -.22 \end{bmatrix}$

For the following exercises, let

$$A = \begin{bmatrix} a_{11} & a_{12} \\ a_{21} & a_{22} \end{bmatrix}, B = \begin{bmatrix} b_{11} & b_{12} \\ b_{21} & b_{22} \end{bmatrix}, \text{ and}$$

$$C = \begin{bmatrix} c_{11} & c_{12} \\ c_{21} & c_{22} \end{bmatrix},$$

where all the elements are real numbers. Decide which of the following statements are true for these three matrices. If a statement is true, prove that it is true. If it is false, give a numerical example to show it false.

41. $A + B = B + A$ (commutative property)

42. $A + (B + C) = (A + B) + C$ (associative property)

43. $A + B$ is a 2 × 2 matrix (closure property)

44. There exists a matrix 0 such that $A + 0 = A$ and $0 + A = A$. (identity property)

45. There exists a matrix $-A$ such that $A + (-A) = 0$ and $-A + A = 0$. (inverse property)

46. $(AB)C = A(BC)$ (associative property)

47. $A(B + C) = AB + AC$ (distributive property)

48. AB is a 2 × 2 matrix (closure property)

49. $c(A + B) = cA + cB$ for any real number c

50. $(c + d)A = cA + dA$ for any real numbers c and d

51. $c(A)d = cd(A)$

52. $(cd)A = c(dA)$

53. $(A + B)(A - B) = A^2 - B^2$ (where $A^2 = AA$)

54. $(A + B)^2 = A^2 + 2AB + B^2$

9.7 Solution of Systems by Matrix Inverses

We have seen that the commutative, associative, closure, identity, and inverse properties hold for *addition* of matrices of the same order. Also, the associative and closure properties hold for *multiplication* of square matrices of the same order. There is no commutative property for multiplication, but the distributive property is valid for matrices of the proper order.

In this section we look at the two remaining properties, the identity and inverse properties for multiplication. Does the identity property hold? If it does hold, we must be able to find a matrix I such that

$$AI = A \qquad \text{and} \qquad IA = A$$

for any square matrix A. (Compare these products to the statement of the identity property for real numbers: $a \cdot 1 = a$ and $1 \cdot a = a$ for any real number a.) It turns out that for 2×2 matrices,

2 × 2 Identity

$$I_2 = \begin{bmatrix} 1 & 0 \\ 0 & 1 \end{bmatrix}.$$

In general, for any value of n there is an $n \times n$ identity matrix having 1's down the diagonal and 0's elsewhere. That is, the $n \times n$ **identity matrix** is given by I_n, where

n × n Identity

$$I_n = \begin{bmatrix} 1 & 0 & \cdots & 0 \\ 0 & 1 & \cdots & 0 \\ & & \ddots & \\ & & a_{ij} & \\ & & & \\ 0 & 0 & \cdots & 1 \end{bmatrix}$$

Here $a_{ij} = 1$ when $i = j$ (the diagonal elements) and $a_{ij} = 0$ otherwise.

For every nonzero real number a, there is a multiplicative inverse $1/a$ such that

$$a \cdot \frac{1}{a} = 1 \qquad \text{and} \qquad \frac{1}{a} \cdot a = 1.$$

(Recall: $1/a$ is also written a^{-1}.) In a similar way, if A is an $n \times n$ matrix, then its **multiplicative inverse,** written A^{-1}, must satisfy both

$$AA^{-1} = I_n \quad \text{and} \quad A^{-1}A = I_n.$$

For example, let A be the 2×2 matrix

$$A = \begin{bmatrix} a_{11} & a_{12} \\ a_{21} & a_{22} \end{bmatrix}.$$

To find a matrix A^{-1} such that $AA^{-1} = I_2$, let

$$A^{-1} = \begin{bmatrix} x & y \\ z & w \end{bmatrix}.$$

We want A^{-1} to satisfy

$$\begin{bmatrix} a_{11} & a_{12} \\ a_{21} & a_{22} \end{bmatrix}\begin{bmatrix} x & y \\ z & w \end{bmatrix} = \begin{bmatrix} 1 & 0 \\ 0 & 1 \end{bmatrix}.$$

By multiplying the two matrices on the left side of this equation and setting the elements of the product matrix equal to the corresponding elements of I_2, we get the following four equations.

$$a_{11}x + a_{12}z = 1$$

$$a_{11}y + a_{12}w = 0$$

$$a_{21}x + a_{22}z = 0$$

$$a_{21}y + a_{22}w = 1$$

The solution of this system of equations can be shown to be

$$x = \frac{a_{22}}{a_{11}a_{22} - a_{12}a_{21}}, \qquad y = \frac{-a_{12}}{a_{11}a_{22} - a_{12}a_{21}},$$

$$z = \frac{-a_{21}}{a_{11}a_{22} - a_{12}a_{21}}, \qquad w = \frac{a_{11}}{a_{11}a_{22} - a_{12}a_{21}}.$$

The number in the denominator, $a_{11}a_{22} - a_{12}a_{21}$, is the determinant of matrix A. Thus, if $|A| \neq 0$,

$$A^{-1} = \frac{1}{|A|} \begin{bmatrix} a_{22} & -a_{12} \\ -a_{21} & a_{11} \end{bmatrix}.$$

Verify by multiplication that A^{-1}, as defined here, really is the multiplicative inverse of matrix A.

EXAMPLE 1 ☐ Find A^{-1} if $A = \begin{bmatrix} 3 & -4 \\ -1 & -2 \end{bmatrix}$.

First find $|A|$.

$$|A| = 3(-2) - (-4)(-1) = -10.$$

Then, as shown above,

$$A^{-1} = \frac{1}{-10} \begin{bmatrix} -2 & 4 \\ 1 & 3 \end{bmatrix} = \begin{bmatrix} 1/5 & -2/5 \\ -1/10 & -3/10 \end{bmatrix}.$$

Verify this result by finding both AA^{-1} and $A^{-1}A$. ▢

A method of finding A^{-1} for $n \times n$ matrices $(n > 2)$ is given in the next theorem, which we state without proof. This method uses cofactors of the matrix A. Recall, the cofactor of a_{ij} is

$$A_{ij} = (-1)^{i+j} \cdot M_{ij},$$

where M_{ij} is the minor of a_{ij}.

Multiplicative Inverse Theorem

Given the $n \times n$ matrix

$$A = \begin{bmatrix} a_{11} & a_{12} & \cdots & a_{1n} \\ a_{21} & a_{22} & \cdots & a_{2n} \\ \cdot & \cdot & & \cdot \\ \cdot & \cdot & & \cdot \\ \cdot & \cdot & & \cdot \\ a_{n1} & a_{n2} & \cdots & a_{nn} \end{bmatrix}$$

with $|A| \neq 0$, then

$$A^{-1} = \frac{1}{|A|} \begin{bmatrix} A_{11} & A_{21} & \cdots & A_{n1} \\ A_{12} & A_{22} & \cdots & A_{n2} \\ \cdot & \cdot & & \cdot \\ \cdot & \cdot & & \cdot \\ \cdot & \cdot & & \cdot \\ A_{1n} & A_{2n} & \cdots & A_{nn} \end{bmatrix}.$$

In the theorem, note that the cofactors of elements in the *rows* of matrix A are used to obtain elements for the *columns* of the inverse matrix A^{-1}.

EXAMPLE 2 ▢ Find M^{-1}, given $M = \begin{bmatrix} 0 & 1 & 2 \\ 1 & 0 & 3 \\ -1 & -2 & 1 \end{bmatrix}$.

Use the methods of Section 9.4 to find that $|M| = -8$. Replace each entry in M with its cofactor to get

$$\begin{bmatrix} 6 & -4 & -2 \\ -5 & 2 & -1 \\ 3 & 2 & -1 \end{bmatrix}.$$

Now form the **transpose** of this matrix; that is, exchange the rows and the columns to get

$$\begin{bmatrix} 6 & -5 & 3 \\ -4 & 2 & 2 \\ -2 & -1 & -1 \end{bmatrix}.$$

Multiplying this transpose of the matrix of cofactors by $1/|M|$ gives M^{-1}.

$$M^{-1} = \frac{1}{-8}\begin{bmatrix} 6 & -5 & 3 \\ -4 & 2 & 2 \\ -2 & -1 & -1 \end{bmatrix} = \begin{bmatrix} -3/4 & 5/8 & -3/8 \\ 1/2 & -1/4 & -1/4 \\ 1/4 & 1/8 & 1/8 \end{bmatrix}$$

To check, multiply as follows.

$$M^{-1}M = \begin{bmatrix} -3/4 & 5/8 & -3/8 \\ 1/2 & -1/4 & -1/4 \\ 1/4 & 1/8 & 1/8 \end{bmatrix}\begin{bmatrix} 0 & 1 & 2 \\ 1 & 0 & 3 \\ -1 & -2 & 1 \end{bmatrix} = \begin{bmatrix} 1 & 0 & 0 \\ 0 & 1 & 0 \\ 0 & 0 & 1 \end{bmatrix}$$

Also find MM^{-1}.

$$MM^{-1} = \begin{bmatrix} 0 & 1 & 2 \\ 1 & 0 & 3 \\ -1 & -2 & 1 \end{bmatrix}\begin{bmatrix} -3/4 & 5/8 & -3/8 \\ 1/2 & -1/4 & -1/4 \\ 1/4 & 1/8 & 1/8 \end{bmatrix} = \begin{bmatrix} 1 & 0 & 0 \\ 0 & 1 & 0 \\ 0 & 0 & 1 \end{bmatrix}$$

Since both $M^{-1}M$ and MM^{-1} equal the identity matrix, we have verified that M^{-1} indeed is the multiplicative inverse of matrix M. ☐

Although the cofactor method can be used to find the inverse of any $n \times n$ matrix A for which $|A| \neq 0$, it can become quite a computational challenge for large values of n. Another way to get A^{-1} uses row operations. To use this method, first form the augmented matrix $[A|I_n]$, where A is any $n \times n$ matrix and I_n is the $n \times n$ multiplicative identity matrix. To find A^{-1}, perform row operations on $[A|I_n]$ until a matrix of the form $[I_n|B]$ is found. The matrix B is the desired matrix A^{-1}. If it is not possible to get a matrix of the form $[I_n|B]$, then $|A| = 0$ and the inverse of matrix A does not exist. (A proof of this row operation method of finding inverses is given in most standard linear algebra books.)

EXAMPLE 3 ☐ Find A^{-1} if $A = \begin{bmatrix} 1 & 0 & 1 \\ 2 & -2 & -1 \\ 3 & 0 & 0 \end{bmatrix}$.

Use row transformations as follows, going through as many steps as needed.

Step 1 Write the augmented matrix $[A|I_3]$:

$$\begin{bmatrix} 1 & 0 & 1 & 1 & 0 & 0 \\ 2 & -2 & -1 & 0 & 1 & 0 \\ 3 & 0 & 0 & 0 & 0 & 1 \end{bmatrix}.$$

Step 2 Since 1 is already in the upper left-hand corner as necessary, begin by selecting the row operation which will result in a 0 for the first element in the second row.

Add to each element in the second row the result of multiplying the first row by -2:

$$\begin{bmatrix} 1 & 0 & 1 & \vdots & 1 & 0 & 0 \\ 0 & -2 & -3 & \vdots & -2 & 1 & 0 \\ 3 & 0 & 0 & \vdots & 0 & 0 & 1 \end{bmatrix}.$$

Step 3 To get 0 for the first element in the third row, add to the third row the results of multiplying each element of the first row by -3. This gives the new matrix

$$\begin{bmatrix} 1 & 0 & 1 & \vdots & 1 & 0 & 0 \\ 0 & -2 & -3 & \vdots & -2 & 1 & 0 \\ 0 & 0 & -3 & \vdots & -3 & 0 & 1 \end{bmatrix}.$$

Step 4 To get 1 for the second element in the second row, multiply the second row by $-1/2$, obtaining the new matrix

$$\begin{bmatrix} 1 & 0 & 1 & \vdots & 1 & 0 & 0 \\ 0 & 1 & 3/2 & \vdots & 1 & -1/2 & 0 \\ 0 & 0 & -3 & \vdots & -3 & 0 & 1 \end{bmatrix}.$$

Step 5 To get 1 for the third element in the third row, multiply the third row by $-1/3$, with the result

$$\begin{bmatrix} 1 & 0 & 1 & \vdots & 1 & 0 & 0 \\ 0 & 1 & 3/2 & \vdots & 1 & -1/2 & 0 \\ 0 & 0 & 1 & \vdots & 1 & 0 & -1/3 \end{bmatrix}.$$

Step 6 To get 0 for the third element in the first row, add to the first row the results of multiplying each element in row three by -1:

$$\begin{bmatrix} 1 & 0 & 0 & \vdots & 0 & 0 & 1/3 \\ 0 & 1 & 3/2 & \vdots & 1 & -1/2 & 0 \\ 0 & 0 & 1 & \vdots & 1 & 0 & -1/3 \end{bmatrix}.$$

Step 7 To get 0 for the third element in the second row, add to the second row the results of multiplying each element of row three by $-3/2$:

$$\begin{bmatrix} 1 & 0 & 0 & \vdots & 0 & 0 & 1/3 \\ 0 & 1 & 0 & \vdots & -1/2 & -1/2 & 1/2 \\ 0 & 0 & 1 & \vdots & 1 & 0 & -1/3 \end{bmatrix}.$$

From the last transformation, we get the desired inverse:

$$A^{-1} = \begin{bmatrix} 0 & 0 & 1/3 \\ -1/2 & -1/2 & 1/2 \\ 1 & 0 & -1/3 \end{bmatrix}.$$

Confirm this by forming the products $A^{-1}A$, and AA^{-1}, each of which should equal I_3. ☐

As with the Gaussian reduction method, the most efficient order of steps is to make the changes column by column from left to right so that for each column the required 1 is the result of the first change. Next, perform the steps that obtain the zeros in that column. Then proceed to another column. The advantage of this method is that these same steps can be adapted for a computer program. A computer can produce the inverse of a large matrix in a few seconds.

EXAMPLE   ☐   Find A^{-1} given $A = \begin{bmatrix} 2 & -4 \\ 1 & -2 \end{bmatrix}$.
4

Using row operations to transform the first column of the augmented matrix

$$\begin{bmatrix} 2 & -4 & | & 1 & 0 \\ 1 & -2 & | & 0 & 1 \end{bmatrix},$$

results in the following matrices:

$$\begin{bmatrix} 1 & -2 & | & 1/2 & 0 \\ 1 & -2 & | & 0 & 1 \end{bmatrix},$$

$$\begin{bmatrix} 1 & -2 & | & 1/2 & 0 \\ 0 & 0 & | & -1/2 & 1 \end{bmatrix}.$$

At this point, we wish to change the matrix so that the second row, second column element will be 1. Since that element is now 0, there is no way to complete the desired transformation. Thus, matrix A^{-1} does not exist.   ☐

Matrix inverses can be used to solve square linear systems of equations. (A square system has the same number of equations as variables.) For example, if we are given the linear system

$$a_{11}x + a_{12}y + a_{13}z = d_1$$

$$a_{21}x + a_{22}y + a_{23}z = d_2$$

$$a_{31}x + a_{32}y + a_{33}z = d_3,$$

we can use the definition of matrix multiplication to rewrite the system as

$$\begin{bmatrix} a_{11} & a_{12} & a_{13} \\ a_{21} & a_{22} & a_{23} \\ a_{31} & a_{32} & a_{33} \end{bmatrix} \cdot \begin{bmatrix} x \\ y \\ z \end{bmatrix} = \begin{bmatrix} d_1 \\ d_2 \\ d_3 \end{bmatrix}. \tag{1}$$

(To see this, multiply the matrices on the left.)

Let $A = \begin{bmatrix} a_{11} & a_{12} & a_{13} \\ a_{21} & a_{22} & a_{23} \\ a_{31} & a_{32} & a_{33} \end{bmatrix}$, $X = \begin{bmatrix} x \\ y \\ z \end{bmatrix}$,   and $B = \begin{bmatrix} d_1 \\ d_2 \\ d_3 \end{bmatrix}$,

so that the system given in (1) becomes

$$AX = B.$$

If we assume that A^{-1} exists, we can multiply both sides of $AX = B$ on the left by A^{-1} to get

$$A^{-1}(AX) = A^{-1}B$$
$$(A^{-1}A)X = A^{-1}B$$
$$I_3X = A^{-1}B$$
$$X = A^{-1}B.$$

Matrix $A^{-1}B$ gives the solution of the system.

EXAMPLE 5   ☐   Use the method of matrix inverses to solve the system

$$2x - 3y = 4$$
$$x + 5y = 2.$$

To represent the system as a matrix equation, use one matrix for the coefficients, one for the variables, and one for the constants, as follows.

$$A = \begin{bmatrix} 2 & -3 \\ 1 & 5 \end{bmatrix}, \qquad X = \begin{bmatrix} x \\ y \end{bmatrix}, \qquad \text{and} \qquad B = \begin{bmatrix} 4 \\ 2 \end{bmatrix}.$$

The system can then be written in matrix form as the equation $AX = B$, since

$$AX = \begin{bmatrix} 2 & -3 \\ 1 & 5 \end{bmatrix}\begin{bmatrix} x \\ y \end{bmatrix} = \begin{bmatrix} 2x - 3y \\ x + 5y \end{bmatrix} = \begin{bmatrix} 4 \\ 2 \end{bmatrix} = B.$$

To solve the system, first find A^{-1}. Verify that

$$A^{-1} = \begin{bmatrix} 5/13 & 3/13 \\ -1/13 & 2/13 \end{bmatrix}.$$

Next, find the product $A^{-1}B$.

$$A^{-1}B = \begin{bmatrix} 5/13 & 3/13 \\ -1/13 & 2/13 \end{bmatrix}\begin{bmatrix} 4 \\ 2 \end{bmatrix} = \begin{bmatrix} 2 \\ 0 \end{bmatrix}.$$

Since $X = A^{-1}B$,

$$X = \begin{bmatrix} x \\ y \end{bmatrix} = \begin{bmatrix} 2 \\ 0 \end{bmatrix}.$$

Thus, the solution set of the system is $\{(2, 0)\}$.   ☐

This method of using matrix inverses to solve systems of equations is useful when the inverse is already known or when many systems must be solved where the coefficients are the same and only the constants differ.

EXERCISES 9.7

Decide whether the given matrices are inverses of each other.

1. $\begin{bmatrix} 1 & -2 & -3 \\ 2 & -2 & -5 \\ -1 & 1 & 4 \end{bmatrix}$ and $\begin{bmatrix} -1 & 5/3 & 4/3 \\ -1 & 1/3 & -1/3 \\ 0 & 1/3 & 2/3 \end{bmatrix}$ **2.** $\begin{bmatrix} 1 & 2 & -1 \\ 2 & -1 & 3 \\ 3 & -2 & 3 \end{bmatrix}$ and $\begin{bmatrix} 3/10 & -2/5 & 1/2 \\ 3/10 & 3/5 & -1/2 \\ -1/10 & 4/5 & -1/2 \end{bmatrix}$

3. $\begin{bmatrix} 1 & 2 & -1 \\ 0 & 1 & 3 \\ 2 & 1 & -2 \end{bmatrix}$ and $\begin{bmatrix} 1 & 1 & 2 \\ 1 & 1 & 1 \\ 2 & 3 & 4 \end{bmatrix}$

4. $\begin{bmatrix} 2 & -1 & 4 \\ 0 & 5 & 0 \\ 3 & 2 & -1 \end{bmatrix}$ and $\begin{bmatrix} 1 & 0 & 1 \\ 6 & 4 & 2 \\ 1 & 1 & 0 \end{bmatrix}$

Find the inverses of the following matrices if they exist.

5. $\begin{bmatrix} 1 & 5 \\ 2 & 0 \end{bmatrix}$

6. $\begin{bmatrix} 10 & 0 \\ 5 & 3 \end{bmatrix}$

7. $\begin{bmatrix} -6 & 4 \\ -3 & 2 \end{bmatrix}$

8. $\begin{bmatrix} -1 & 2 \\ 1 & -2 \end{bmatrix}$

9. $\begin{bmatrix} 1 & 2 & 3 \\ -3 & -2 & -1 \\ -1 & 0 & 1 \end{bmatrix}$

10. $\begin{bmatrix} 2 & 0 & 4 \\ 3 & 1 & 5 \\ -1 & 1 & -2 \end{bmatrix}$

11. $\begin{bmatrix} 2 & 1 & 3 \\ 7 & 4 & -2 \\ 0 & 1 & -1 \end{bmatrix}$

12. $\begin{bmatrix} -2 & -1 & -5 \\ 4 & 5 & 0 \\ 0 & 1 & 3 \end{bmatrix}$

13. $\begin{bmatrix} 1 & 1 & 0 & 2 \\ 2 & -1 & 1 & -1 \\ 3 & 3 & 2 & -2 \\ 1 & 2 & 1 & 0 \end{bmatrix}$

14. $\begin{bmatrix} 1 & -2 & 3 & 0 \\ 0 & 2 & -1 & 1 \\ -2 & 2 & 1 & 4 \\ 0 & 2 & -3 & 1 \end{bmatrix}$

15. $\begin{bmatrix} .6 & .2 \\ .5 & .1 \end{bmatrix}$

16. $\begin{bmatrix} .8 & -.3 \\ .5 & -.2 \end{bmatrix}$

17. $\begin{bmatrix} -.4 & 1 & .2 \\ 0 & .6 & .8 \\ .3 & 0 & -.2 \end{bmatrix}$

18. $\begin{bmatrix} .8 & .2 & .1 \\ -.2 & 0 & .3 \\ 0 & 0 & .5 \end{bmatrix}$

Solve the matrix equation $AX = B$ for X, given the following matrices.

19. $A = \begin{bmatrix} 1 & 0 & -1 \\ 2 & 1 & 1 \\ 3 & 2 & -1 \end{bmatrix}$ $B = \begin{bmatrix} 1 \\ 4 \\ 3 \end{bmatrix}$

20. $A = \begin{bmatrix} 2 & 3 & 1 \\ -3 & -4 & 0 \\ -1 & 2 & -2 \end{bmatrix}$ $B = \begin{bmatrix} 0 \\ 2 \\ -4 \end{bmatrix}$

Use the method of matrix inverses to solve each of the following systems. Some of the necessary inverses have been found in the text and exercises of this section.

21. $-x + y = 1$
    $2x - y = 1$

22. $3x - y = 1$
    $-5x + 2y = -1$

23. $2x - y = -8$
    $3x + y = -2$

24. $x + 3y = -12$
    $2x - y = 11$

25. $x \quad\quad + z = 3$
    $2x - 2y - z = -2$
    $3x \quad\quad = 3$

26. $4x + 3y + 3z = 11$
    $-x \quad\quad - z = -4$
    $-4x - 4y - 3z = -10$

27. $2x + y + 3z = 13$
    $7x + 4y - 2z = -4$
    $\quad\quad y - z = -3$

28. $2x \quad\quad + 4z = 14$
    $3x + y + 5z = 19$
    $-x + y - 2z = -7$

29. $x + y \quad\quad + 2w = 3$
    $2x - y + z - w = 3$
    $3x + 3y + 2z - 2w = 5$
    $x + 2y + z \quad\quad = 3$

30.
$$x - 2y + 3z \qquad = -3$$
$$2y - z + w = 3$$
$$-2x + 2y + z + 4w = 7$$
$$2y - 3z + w = 5$$

31. The Bread Box Bakery sells three types of cakes, each requiring the amounts of the basic ingredients shown in the following matrix.

types of cake

	I	II	III
flour (cups)	2	4	2
sugar (cups)	2	1	2
eggs	2	1	3

To fill its daily orders for the three kinds of cake, the bakery uses 72 cups of flour, 48 cups of sugar, and 60 eggs. How many daily orders for each type of cake does the bakery receive?

(Let the order matrix be a 3×1 matrix X and solve a matrix equation.)

32. Give an example of two matrices A and B where $(AB)^{-1} \neq A^{-1}B^{-1}$.

33. Suppose A and B are matrices where A^{-1}, B^{-1}, and AB all exist. Show that $(AB)^{-1} = B^{-1}A^{-1}$.

34. Let $A = \begin{bmatrix} a & 0 \\ 0 & b \end{bmatrix}$. Under what conditions on a and b, does A^{-1} exist?

35. Derive a formula for the inverse of $\begin{bmatrix} a & 0 \\ 0 & d \end{bmatrix}$, where $ad \neq 0$.

36. Let $A = \begin{bmatrix} a & 0 & 0 \\ 0 & b & 0 \\ 0 & 0 & c \end{bmatrix}$, where a, b and c are non-zero real numbers. Find A^{-1}.

9.8 Cramer's Rule

We have now seen how to solve a system of n linear equations with n variables using the following methods: elimination, substitution, row transformations of matrices, and matrix inverses. These systems can also be solved with determinants, as we shall see in this section. The derivation of this method is a little messy, but the actual method itself is not too complicated to state.

To see how determinants arise in solving a system, write the linear system

$$a_{11}x + a_{12}y = b_1$$
$$a_{21}x + a_{22}y = b_2,$$

where each equation has at least one nonzero coefficient. We shall solve this system using matrix methods, just as we did in Section 9.3. Begin by writing the augmented matrix

$$\begin{bmatrix} a_{11} & a_{12} & b_1 \\ a_{21} & a_{22} & b_2 \end{bmatrix}.$$

Multiply each element of row 1 by $1/a_{11}$. (Here we assume $a_{11} \neq 0$—see Exercise 42 below.) This gives the matrix of an equivalent system:

$$\begin{bmatrix} 1 & \dfrac{a_{12}}{a_{11}} & \dfrac{b_1}{a_{11}} \\ a_{21} & a_{22} & b_2 \end{bmatrix}.$$

Multiply each element of row 1 by $-a_{21}$, and add the result to the corresponding element of row 2.

$$\begin{bmatrix} 1 & \dfrac{a_{12}}{a_{11}} & \vdots & \dfrac{b_1}{a_{11}} \\[2em] 0 & a_{22} - \dfrac{a_{21}a_{12}}{a_{11}} & \vdots & b_2 - \dfrac{a_{21}b_1}{a_{11}} \end{bmatrix}$$

Multiply each element of row 2 by a_{11}.

$$\begin{bmatrix} 1 & \dfrac{a_{12}}{a_{11}} & \vdots & \dfrac{b_1}{a_{11}} \\[1.5em] 0 & a_{11}a_{22} - a_{21}a_{12} & \vdots & a_{11}b_2 - a_{21}b_1 \end{bmatrix}$$

This matrix leads to the systems of equations

$$x + \frac{a_{12}}{a_{11}}y = \frac{b_1}{a_{11}}$$

$$(a_{11}a_{22} - a_{21}a_{12})y = a_{11}b_2 - a_{21}b_1.$$

From the second equation of this system,

$$y = \frac{a_{11}b_2 - a_{21}b_1}{a_{11}a_{22} - a_{21}a_{12}}.$$

Both the numerator and denominator here may be written as determinants:

$$y = \frac{\begin{vmatrix} a_{11} & b_1 \\ a_{21} & b_2 \end{vmatrix}}{\begin{vmatrix} a_{11} & a_{12} \\ a_{21} & a_{22} \end{vmatrix}}. \tag{1}$$

By inserting this value of y into the first equation above, we find that x can also be written with determinants as

$$x = \frac{\begin{vmatrix} b_1 & a_{12} \\ b_2 & a_{22} \end{vmatrix}}{\begin{vmatrix} a_{11} & a_{12} \\ a_{21} & a_{22} \end{vmatrix}}. \tag{2}$$

The denominator of both x and y is just the determinant of the matrix of coefficients of the original system. This determinant is often denoted D, so that

$$D = \begin{vmatrix} a_{11} & a_{12} \\ a_{21} & a_{22} \end{vmatrix}.$$

In equation (1), the numerator is the determinant of a matrix obtained by replacing the coefficients of y in D with the respective constants: D_y is defined as

$$D_y = \begin{vmatrix} a_{11} & b_1 \\ a_{21} & b_2 \end{vmatrix}.$$

In the same way, from equation (2), D_x is defined as

$$D_x = \begin{vmatrix} b_1 & a_{12} \\ b_2 & a_{22} \end{vmatrix}.$$

With this notation, the solution of the given system is

$$x = \frac{D_x}{D} \quad \text{and} \quad y = \frac{D_y}{D}.$$

The system has a single solution as long as $D \neq 0$. We have now proved much of the next theorem, called **Cramer's Rule.**

Cramer's Rule

Let $a_{11}x + a_{12}y = b_1$
     $a_{21}x + a_{22}y = b_2$

be a system of equations, where each equation has at least one nonzero coefficient. The system has a unique solution if $D \neq 0$; this solution is

$$x = \frac{D_x}{D} \quad \text{and} \quad y = \frac{D_y}{D}.$$

If $D = 0$, the system is dependent if $D_x = 0$ and $D_y = 0$; otherwise it is inconsistent.

(For proof of this last statement, see Exercises 41–42 below.)

EXAMPLE 1

☐ Use Cramer's Rule to solve the system

$5x + 7y = -1$
$6x + 8y = 1.$

To use Cramer's Rule, we need to evaluate D, D_x, and D_y.

$$D = \begin{vmatrix} 5 & 7 \\ 6 & 8 \end{vmatrix} = 5(8) - 6(7) = -2.$$

$$D_x = \begin{vmatrix} -1 & 7 \\ 1 & 8 \end{vmatrix} = (-1)(8) - (1)(7) = -15.$$

$$D_y = \begin{vmatrix} 5 & -1 \\ 6 & 1 \end{vmatrix} = 5(1) - (6)(-1) = 11.$$

By Cramer's Rule, $x = -15/(-2) = 15/2$, and $y = 11/(-2) = -11/2$. The solution set is $\{(15/2, -11/2)\}$, as can be verified by substituting within the given system. $\square$

By much the same method we used above, Cramer's Rule can be generalized to a system of n linear equations with n variables.

General Form of Cramer's Rule

Let an $n \times n$ system have linear equations of the form

$$a_{11}x_1 + a_{12}x_2 + a_{13}x_3 + \cdots + a_{1n}x_n = b_1.$$

Define D as the determinant of the $n \times n$ matrix of all coefficients. Define D_{x_1} as the determinant obtained from D by replacing the entries in column 1 of D with the constants of the system. Define D_{x_i} as the determinant obtained from D by replacing the entries in column i with the constants of the system. If $D \neq 0$, the unique solution of the system is given by

$$x_1 = \frac{D_{x_1}}{D}, \; x_2 = \frac{D_{x_2}}{D}, \; x_3 = \frac{D_{x_3}}{D}, \; \ldots, \; x_n = \frac{D_{x_n}}{D}.$$

EXAMPLE 2

Use Cramer's Rule to solve the system

$$x + y - z + 2 = 0$$
$$2x - y + z + 5 = 0$$
$$x - 2y + 3z - 4 = 0.$$

To use Cramer's Rule, the system must be rewritten in the form

$$x + y - z = -2$$
$$2x - y + z = -5$$
$$x - 2y + 3z = 4.$$

The determinant of coefficients, D, is

$$D = \begin{vmatrix} 1 & 1 & -1 \\ 2 & -1 & 1 \\ 1 & -2 & 3 \end{vmatrix}.$$

To find D_x, replace the elements of the first column of D with the constants of the system. Find D_y and D_z in a similar way.

$$D_x = \begin{vmatrix} -2 & 1 & -1 \\ -5 & -1 & 1 \\ 4 & -2 & 3 \end{vmatrix}, \quad D_y = \begin{vmatrix} 1 & -2 & -1 \\ 2 & -5 & 1 \\ 1 & 4 & 3 \end{vmatrix}, \quad D_z = \begin{vmatrix} 1 & 1 & -2 \\ 2 & -1 & -5 \\ 1 & -2 & 4 \end{vmatrix}$$

Verify that

$$D = -3, D_x = 7, D_y = -22, \text{ and } D_z = -21.$$

By Cramer's Rule,

$$x = \frac{D_x}{D} = \frac{7}{-3} = \frac{-7}{3}, y = \frac{D_y}{D} = \frac{-22}{-3} = \frac{22}{3}, \text{ and } z = \frac{D_z}{D} = \frac{-21}{-3} = 7.$$

The solution set of the system is $\{(-7/3, 22/3, 7)\}$. ▪

EXAMPLE 3 ▪ Use Cramer's Rule to solve the system

$$2x - 3y + 4z = 10$$
$$6x - 9y + 12z = 24$$
$$x + 2y - 3z = 5.$$

Here $D = 0$. Also, $D_x = 6$. Since $D = 0$ and at least one of the other determinants is not zero, the system is inconsistent. (Had all the determinants been 0, the system would have been dependent.) ▪

We have now seen several different methods for solving systems of equations. In general, if a small system of linear equations must be solved by pencil and paper, substitution is the best method if the various equations can easily be solved in terms of each other. This happens rarely. The next choice, perhaps the best choice of all, is the elimination method. Some people like the Gaussian reduction method, which is really just a systematic way of doing the elimination method. The Gaussian reduction method is probably superior where four or more equations are involved. Cramer's Rule is seldom the method of choice simply because it involves more calculations than any other method.

**EXERCISES
9.8**

Use Cramer's Rule to solve each of the following systems of linear equations.

1. $x + y = 4$
   $2x - y = 2$

2. $3x + 2y = -4$
   $2x - y = -5$

3. $4x + 3y = -7$
   $2x + 3y = -11$

4. $3x + 2y = -4$
   $5x - y = 2$

5. $2x - 3y = -5$
   $x + 5y = 17$

6. $x + 9y = -15$
   $3x + 2y = 5$

7. $5x + 2y = 7$
   $6x + y = 8$

8. $7x + 3y = 5$
   $2x + 4y = 3$

9. $10x - 8y = 1$
   $-15x + 12y = 4$

10. $8x - 6y = 10$
    $20x - 15y = 6$

11. $4x - y + 3z = -3$
    $3x + y + z = 0$
    $2x - y + 4z = 0$

12. $5x + 2y + z = 15$
    $2x - y + z = 9$
    $4x + 3y + 2z = 13$

13. $2x - y + 4z = -2$
    $3x + 2y - z = -3$
    $x + 4y + 2z = 17$

14. $x + y + z = 4$
    $2x - y + 3z = 4$
    $4x + 2y - z = -15$

15. $4x - 3y + z = -1$
$5x + 7y + 2z = -2$
$3x - 5y - z = 1$

16. $2x - 3y + z = 8$
$-x - 5y + z = -4$
$3x - 5y + 2z = 12$

17. $x + 2y + 3z = 4$
$4x + 3y + 2z = 1$
$-x - 2y - 3z = 0$

18. $2x - y + 3z = 1$
$-2x + y - 3z = 2$
$5x - y + z = 2$

19. $x - 2y + 3z = 4$
$5x + 7y - z = 2$
$2x + 2y - 5z = 3$

20. $-3x - 2y - z = 4$
$4x + y + z = 5$
$3x - 2y + 2z = 1$

21. $2x + 3y = 13$
$2y - z = 5$
$x + 2z = 4$

22. $3x - z = -10$
$y + 4z = 8$
$x + 2z = -1$

23. $5x - y = -4$
$3x + 2z = 4$
$4y + 3z = 22$

24. $3x + 5y = -7$
$2x + 7z = 2$
$4y + 3z = -8$

25. $x + 2y = 10$
$3x + 4z = 7$
$-y - z = 1$

26. $5x - 2y = 3$
$4y + z = 8$
$x + 2z = 4$

27. $x + z = 0$
$y + 2z + w = 0$
$2x - w = 0$
$x + 2y + 3z = -2$

28. $x - y + z + w = 6$
$2y - w = -7$
$x - z = -1$
$y + w = 1$

29. $.4x - .6y = .4$
$.3x + .2y = -.22$

30. $-.5x + .4y = .43$
$.1x + .2y = .11$

31. $.4x - .2y + .3z = -.08$
$-.5x + .3y - .7z = .19$
$.8x - .7y + .1z = -.21$

32. $-.6x + .2y - .5z = -.16$
$.8x - .3y - .2z = -.48$
$.4x + .6y - .8z = -.8$

Use Cramer's Rule to solve each of the following.

33. Paying $96, Lucy Day bought 5 shirts and 2 pairs of pants. Later, paying $66, she bought one more shirt and 3 more pairs of pants. Find the cost of a shirt and the cost of a pair of pants.

34. Barbara Maring received $50,000 from the sale of a business. She invested part of the money at 16% and the rest at 18%, which resulted in her earning $8550 per year in interest. Find the amount invested at each rate.

35. A gold merchant has some 12-carat gold (12/24 pure gold), and some 22-carat gold (22/24 pure). How many grams of each should be mixed to get 25 grams of 15-carat gold?

36. A chemist has some 40% acid solution and some 60% solution. How many liters of each should be used to get 40 liters of a 45% solution?

37. How many pounds of tea worth $4.60 a pound should be mixed with 8 pounds of tea worth

$6.50 a pound to get a mixture worth $5.20 a pound?

38. A solution of a drug with a strength of 5% is to be mixed with some of a 15% solution to get 15 ml of an 8% solution. How many ml of each solution should be used?

39. The cashier at an amusement park has a total of $2480, made up of fives, tens, and twenties. The total number of bills is 290, and the value of the tens is $60 more than the value of the twenties. How many of each type of bill does the cashier have?

40. Ms. Levy invests $50,000 three ways—at 8%, $8\frac{1}{2}$%, and 11%. In total, she receives $4436.25

per year in interest. The interest from the 11% investment is $80 more than the interest on the 8% investment. Find the amount she has invested at each rate.

For the following two exercises, use the system of equations

$$a_1x + b_1y = c_1$$
$$a_2x + b_2y = c_2.$$

41. Assume $D_x = 0$ and $D_y = 0$. Show that if $c_1c_2 \neq 0$, then $D \neq 0$, and the equations are consistent.

42. Assume $D = 0$, $D_x = 0$, and $a_1a_2 \neq 0$. Show that $D_y = 0$.

9.9 Systems of Inequalities

Many mathematical descriptions of real world situations are best expressed as inequalities, rather than equalities. For example, a firm might be able to use a machine *no more* than 12 hours a day, while a production of *at least* 500 cases of a certain product might be required to meet a contract. Perhaps the simplest way to see the solution of an inequality in two variables is to draw its graph.

A line divides a plane into three sets of points—the points of the line itself and the points belonging to the two regions determined by the line. Each of these two regions is called a **half-plane**. In Figure 9.6 line *r* divides the plane into three different sets of points, line *r*, half-plane *P* and half-plane *Q*. The points of *r* belong to neither *P* nor *Q*. Line *r* is the boundary of each half-plane.

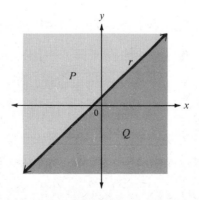

Figure 9.6

A **linear inequality in two variables** is an inequality of the form

$$Ax + By \leq C,$$

where A, B, and C are real numbers, with A and B not both equal to 0. (We could replace $\leq$ with $\geq$, $<$, or $>$.) The graph of a linear inequality turns out to be made up of a half-plane, perhaps with its boundary. For example, to graph the linear inequality $3x - 2y \leq 6$, first graph the boundary, $3x - 2y = 6$, as shown in Figure 9.7.

Since the points of the line $3x - 2y = 6$ satisfy $3x - 2y \leq 6$, this line is part of the solution.

To decide which half-plane—the one above the line $3x - 2y = 6$ or the one below the line—is part of the solution, solve the original inequality for y:

$$3x - 2y \leq 6$$
$$-2y \leq -3x + 6$$
$$y \geq \frac{3}{2}x - 3. \qquad \text{Change } \leq \text{ to } \geq.$$

If we choose a particular value of x, the inequality will be satisfied by all values of y which are *greater than* or equal to $(3/2)x - 3$. This means that the solution contains the half-plane *above* the line, as shown in Figure 9.8.

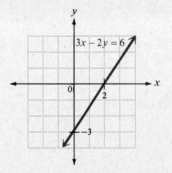

Figure 9.7

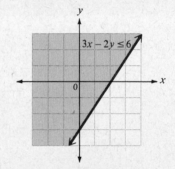

Figure 9.8

EXAMPLE 1   Graph $x + 4y > 4$.

The boundary here is the straight line $x + 4y = 4$. Since the points on this line do not satisfy $x + 4y > 4$, it is customary to make the line dashed, as in Figure 9.9. To decide on which half-plane makes up the solution, solve for y.

$$x + 4y > 4$$
$$4y > -x + 4$$
$$y > -\frac{1}{4}x + 1$$

Since y is *greater than* $-\frac{1}{4}x + 1$, the solution is made up of the half-plane above the boundary, as shown in Figure 9.9.   □

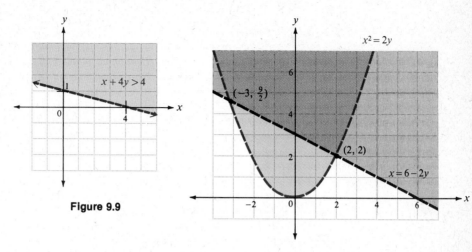

Figure 9.9

Figure 9.10

The methods we have used to graph linear inequalities can be used for other inequalities of the form $y \leq f(x)$ as summarized here:

Graphing Inequalities

For a function f,
the graph of $y < f(x)$ is made up of all the points which are *below* the graph of $y = f(x)$:
the graph of $y > f(x)$ is made up of all the points which are *above* the graph of $y = f(x)$.

The solution set of a **system of inequalities,** such as

$$x > 6 - 2y$$
$$x^2 < 2y,$$

is the intersection of the solution sets of its members. This intersection is found by graphing the solution sets of both inequalities on the same coordinate axes and identifying, by shading, the region common to all graphs.

EXAMPLE 2   □   Graph the solution set of the system above.

Figure 9.10 shows the graphs of both $x > 6 - 2y$ and $x^2 < 2y$. The methods of this chapter can be used to show that the graphs cross at $(2, 2)$ and $(-3, 9/2)$. The solution set of the system includes all points in the heavily shaded area. The points on the boundaries of $x > 6 - 2y$ and $x^2 < 2y$ do not belong to the graph of the solution. For this reason, the boundaries are dashed lines.   □

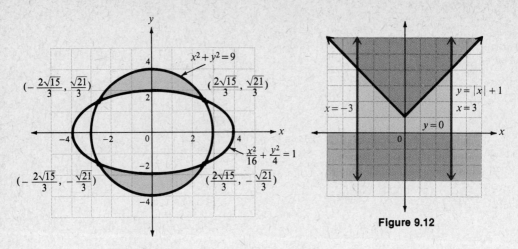

Figure 9.11

Figure 9.12

EXAMPLE 3   Graph the solution set of the system

$$x^2 + y^2 \leq 9$$

$$\frac{x^2}{16} + \frac{y^2}{4} \geq 1.$$

The graph is found by graphing both inequalities on the same axes and shading the common region as shown in Figure 9.11. The graphs meet in four points, $(2\sqrt{15}/3, \sqrt{21}/3)$, $(2\sqrt{15}/3, -\sqrt{21}/3)$, $(-2\sqrt{15}/3, \sqrt{21}/3)$, and $(-2\sqrt{15}/3, -\sqrt{21}/3)$. ☐

EXAMPLE 4   Graph the solution set of the system

$$|x| \leq 3$$

$$y \leq 0$$

$$y \geq |x| + 1.$$

The graph is shown in Figure 9.12. From the graph, we see that the solution sets of $y \leq 0$ and $y \geq |x| + 1$ have no points in common; therefore the solution set for the system is $\emptyset$. ☐

EXAMPLE 5   Midtown Manufacturing Company makes plastic plates and cups, both of which require time on two machines. Plates require one hour on machine A and two hours on machine B. Cups require three hours on machine A and one hour on machine B. Suppose that the two machines, A and B, are operated for at most 15 hours per day. How many plates and cups can be made under those conditions?

Let x represent the number of plates, and y the number of cups produced in one day. Each plate requires one hour on machine A, while each cup requires three hours

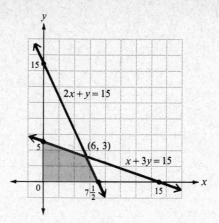

Figure 9.13

on this same machine. The total number of hours required on machine A is thus $x + 3y$. Since this machine operates for at most 15 hours per day,

$x + 3y \leq 15$.

In the same way, the total number of hours per day required on machine B is $2x + y$. It, too, operates at most 15 hours per day, so that

$2x + y \leq 15$.

Both x and y must be nonnegative integers (why?) so that

$x \geq 0$ and $y \geq 0$.

These four inequalities form the system

$x + 3y \leq 15$
$2x + y \leq 15$
$x \geq 0$
$y \geq 0$.

A graph of the solution of this system is shown in Figure 9.13. Any point in the shaded region satisfies all the conditions of the original problem.   ▢

**EXERCISES
9.9**

Graph the solution of each of the following systems of inequalities.

1. $x + y \leq 4$
$x - 2y \geq 6$

2. $2x + y > 2$
$x - 3y < 6$

3. $4x + 3y < 12$
$y + 4x > -4$

4. $3x + 5y \leq 15$
$x - 3y \geq 9$

5. $x + 2y \leq 4$
$y \geq x^2 - 1$

6. $4x - 3y \leq 12$
$y \leq x^2$

7. $y \le -x^2$
$y \ge x^2 - 6$

8. $x^2 + y^2 \le 9$
$x \le -y^2$

9. $x^2 - y^2 < 1$
$-1 < y < 1$

10. $x^2 + y^2 \le 36$
$-4 \le x \le 4$

11. $2x^2 - y^2 > 4$
$2y^2 - x^2 > 4$

12. $y \ge x^2 + 4x + 4$
$y < -x^2$

13. $\dfrac{x^2}{16} + \dfrac{y^2}{9} \le 1$

$\dfrac{x^2}{4} - \dfrac{y^2}{16} \ge 1$

14. $y \ge (x - 2)^2 + 3$
$16x^2 + y^2 \le 16$

15. $x + y \le 4$
$x - y \le 5$
$4x + y \le -4$

16. $3x - 2y \ge 6$
$x + y \le -5$
$y \le 4$

17. $-2 < x < 3$
$-1 \le y \le 5$
$2x + y < 6$

18. $-2 < x < 2$
$y > 1$
$x - y > 0$

19. $2y + x \ge -5$
$y \le 3 + x$
$x \le 0$
$y \le 0$

20. $2x + 3y \le 12$
$2x + 3y > -6$
$3x + y < 4$
$x \ge 0$
$y \ge 0$

21. $\dfrac{x^2}{4} + \dfrac{y^2}{9} > 1$

$x^2 - y^2 \ge 1$
$-4 \le x \le 4$

22. $2x - 3y < 6$
$4x^2 + 9y^2 < 36$
$x \ge -1$

23. $y \ge 3^x$
$y \ge 2$

24. $y \le \left(\dfrac{1}{2}\right)^x$

$y \ge 4$

25. $|x| \ge 2$
$|y| \ge 4$
$y < x^2$

26. $|x| + 2 \ge 4$
$|y| \le 1$
$\dfrac{x^2}{9} + \dfrac{y^2}{16} \le 1$

27. $y \le |x + 2|$
$\dfrac{x^2}{16} - \dfrac{y^2}{9} \le 1$

28. $y \le \log x$
$y \ge |x - 2|$

29. $y \ge |4 - x|$
$y \ge |x|$

30. $|x + 2| < y$
$|x| \le 3$

Write a system of inequalities for each of the following and then graph the solution of the system.

31. A pizza company makes two kinds of pizza, basic and plain. Basic contains cheese and beef, while plain contains onions and beef. The company sells at least three units a day of basic, and at least two units of plain. The beef costs $5 per unit for basic, and $4 per unit for plain. They can spend no more than $50 per day on beef. Dough for basic is $2 per unit, while dough for plain is $1 per unit. The company can spend no more than $16 per day on dough.

32. A farmer raises only pigs and geese. She wants to raise no more than 16 animals, including no more than 12 geese. She spends $5 to raise a pig and $2 to raise a goose. She has available $50 for this purpose.

33. George takes vitamin pills. Each day he must have at least 16 units of vitamin A, 5 units of vitamin B_1, and 20 units of vitamin C. He can choose between red pills which contain 8 units of A, 1 of B_1, and 2 of C, and green pills which contain 2 units of A, 1 of B_1 and 7 of C.

34. Sue is on a diet and wishes to restrict herself to no more than 1600 calories per day. Her diet consists of foods chosen from three food groups: I, meats and dairy products; II, fruits and vegetables; III, breads and other starches. These three groups contain calories per serving as shown below.

Group	I	II	III
Calories	170	50	140

Sue wishes to include four servings daily from group I and more servings from group II than from the other groups combined.

35. A manufacturer of refrigerators must ship at least 100 refrigerators to its two West Coast warehouses. Each warehouse holds a maximum of 100 refrigerators. Warehouse A holds 25 re-frigerators already, while warehouse B has 20 on hand. It costs $12 to ship a refrigerator to warehouse A and $10 to ship one to warehouse B. The total shipping cost is budgeted at a maximum of $1200.

36. The California Almond Growers have 2400 boxes of almonds to be shipped from their plant in Sacramento to Des Moines and San Antonio. The Des Moines market needs at least 1000 boxes, while the San Antonio market must have at least 800 boxes.

CHAPTER   SUMMARY
9

Key Words

system of equations	matrix	cofactor
system of linear equations	dimension	expansion
linear system	order	zero matrix
equivalent system	element	scalar
transformations	augmented matrix	row matrix
elimination method	row transformations	column matrix
triangular form	main diagonal	identity matrix
back-substitution	echelon form	multiplicative inverse
homogeneous system	Gaussian reduction method	transpose
trivial solution	arbitrary variable	Cramer's Rule
family of circles	square matrix	constraint
substitution method	determinant	region of feasible solutions
nonlinear system	minor	vertex

CHAPTER   REVIEW EXERCISES
9

Use the elimination or substitution method to solve each of the following linear systems. Identify any dependent or inconsistent systems.

1. $3x - 5y = -18$
$2x + 7y = 19$

2. $6x + 5y = 53$
$4x - 3y = 29$

3. $\dfrac{2}{3}x - \dfrac{3}{4}y = 13$
$\dfrac{1}{2}x + \dfrac{2}{3}y = -5$

4. $3x + 7y = 10$
$18x + 42y = 50$

5. $\dfrac{1}{x} + \dfrac{1}{y} = \dfrac{7}{10}$

$\dfrac{3}{x} - \dfrac{5}{y} = \dfrac{1}{2}$

6. $.9x - .2y = .8$

$.3x + .7y = 4.1$

7. $2x - 3y + z = -5$

$x + 4y + 2z = 13$

$5x + 5y + 3z = 14$

8. $x - 3y \quad\quad = 12$

$2y + 5z = 1$

$4x \quad\quad + z = 25$

9. A student bought some candy bars, paying 25¢ each for some and 50¢ for others. The student bought a total of 22 bars, paying a total of $8.50. How many of each kind of bar did he buy?

10. Ink worth $25 a bottle is to be mixed with ink worth $18 per bottle to get 12 bottles of ink worth $20 each. How much of each type of ink should be used?

11. Donna Sharp wins $50,000 in a lottery. She invests part of the money at 6%, twice as much at 7%, with $10,000 more than the amount invested at 6% invested at 9%. Total annual interest is $3800. How much is invested at each rate?

12. The sum of three numbers is 23. The second number is 3 more than the first. The sum of the first and twice the third is 4. Find the three numbers.

Find solutions for the following systems in terms of an arbitrary variable.

13. $3x - 4y + z = 2$

$2x + y - 4z = 1$

14. $ax + by + cz = 5$

$dx + y \quad\quad = 1$

Solve each of the following systems.

15. $x^2 - 4y^2 = 19$

$x^2 + y^2 = 29$

16. $xy = 4$

$x - 6y = 2$

17. $x^2 + 2xy + y^2 = 4$

$x = 3y - 2$

18. $y = 5^{x+5}$

$y = 25^{3x}$

19. Do the circle $x^2 + y^2 = 144$ and the line $x + 2y = 8$ have any points in common? If so, what are they?

20. Find a value of b so that the straight line $3x - y = b$ touches the circle $x^2 + y^2 = 25$ in only one point.

Find the values of all variables in the following.

21. $\begin{bmatrix} 5 & x+2 \\ -6y & z \end{bmatrix} = \begin{bmatrix} a & 3x-1 \\ 5y & 9 \end{bmatrix}$

22. $\begin{bmatrix} -6+k & 2 & a+3 \\ -2+m & 3p & 2r \end{bmatrix} +$

$\begin{bmatrix} 3-2k & 5 & 7 \\ 5 & 8p & 5r \end{bmatrix} = \begin{bmatrix} 5 & y & 6a \\ 2m & 11 & -35 \end{bmatrix}$

Perform each of the following operations whenever possible.

23. $\begin{bmatrix} 3 & -4 & 2 \\ 5 & -1 & 6 \end{bmatrix} + \begin{bmatrix} -3 & 2 & 5 \\ 1 & 0 & 4 \end{bmatrix}$

24. $\begin{bmatrix} 3 \\ 2 \\ 5 \end{bmatrix} - \begin{bmatrix} 8 \\ -4 \\ 6 \end{bmatrix} + \begin{bmatrix} 1 \\ 0 \\ 2 \end{bmatrix}$

25. $\begin{bmatrix} 2 & 5 & 8 \\ 1 & 9 & 2 \end{bmatrix} - \begin{bmatrix} 3 & 4 \\ 7 & 1 \end{bmatrix}$

26. $\begin{bmatrix} -3 & 4 \\ 2 & 8 \end{bmatrix} \begin{bmatrix} -1 & 0 \\ 2 & 5 \end{bmatrix}$

27. $\begin{bmatrix} 3 & 2 & -1 \\ 4 & 0 & 6 \end{bmatrix} \begin{bmatrix} -2 & 0 \\ 0 & 2 \\ 3 & 1 \end{bmatrix}$

28. $\begin{bmatrix} 1 & -2 & 4 & 2 \\ 0 & 1 & -1 & 8 \end{bmatrix} \begin{bmatrix} -1 \\ 2 \\ 0 \\ 1 \end{bmatrix}$

Use row operations to solve each of the following.

29. $2x + 3y = 10$

$-3x + y = 18$

30. $5x + 2y = -10$
$3x - 5y = -6$

31. $2x - y + 4z = -1$
$-3x + 5y - z = 5$
$2x + 3y + 2z = 3$

32. $5x - 8y + z = 1$
$3x - 2y + 4z = 3$
$10x - 16y + 2z = 3$

Find the inverse of each of the following matrices that have inverses.

33. $\begin{bmatrix} 2 & 1 \\ 5 & 3 \end{bmatrix}$

34. $\begin{bmatrix} -4 & 2 \\ 0 & 3 \end{bmatrix}$

35. $\begin{bmatrix} 2 & -1 & 0 \\ 1 & 0 & 1 \\ 1 & -2 & 0 \end{bmatrix}$

36. $\begin{bmatrix} 2 & 3 & 5 \\ -2 & -3 & -5 \\ 1 & 4 & 2 \end{bmatrix}$

Use the method of matrix inverses to solve each of the following.

37. $2x + y = 5$
$3x - 2y = 4$

38. $x + y + z = 1$
$2x - y = -2$
$3y + z = 2$

39. $x = -3$
$y + z = 6$
$2x - 3z = -9$

40. $3x - 2y + 4z = 1$
$4x + y - 5z = 2$
$-6x + 4y - 8z = -2$

Find each of the following determinants.

41. $\begin{vmatrix} -1 & 8 \\ 2 & 9 \end{vmatrix}$

42. $\begin{vmatrix} -2 & 4 \\ 0 & 3 \end{vmatrix}$

43. $\begin{vmatrix} -2 & 4 & 1 \\ 3 & 0 & 2 \\ -1 & 0 & 3 \end{vmatrix}$

44. $\begin{vmatrix} -1 & 2 & 3 \\ 4 & 0 & 3 \\ 5 & -1 & 2 \end{vmatrix}$

45. $\begin{vmatrix} -1 & 0 & 2 & -3 \\ 0 & 4 & 4 & -1 \\ -6 & 0 & 3 & -5 \\ 0 & -2 & 1 & 0 \end{vmatrix}$

Explain why each of the following statements is ture.

46. $\begin{vmatrix} 8 & 9 & 2 \\ 0 & 0 & 0 \\ 3 & 1 & 4 \end{vmatrix} = 0$

47. $\begin{vmatrix} 4 & 6 \\ 3 & 5 \end{vmatrix} = \begin{vmatrix} 4 & 3 \\ 6 & 5 \end{vmatrix}$

48. $\begin{vmatrix} 8 & 2 \\ 4 & 3 \end{vmatrix} = 2 \begin{vmatrix} 4 & 1 \\ 4 & 3 \end{vmatrix}$

49. $\begin{vmatrix} 4 & 6 & 2 \\ -3 & 8 & -5 \\ 4 & 6 & 2 \end{vmatrix} = 0$

50. $\begin{vmatrix} 5 & -1 & 2 \\ 3 & -2 & 0 \\ -4 & 1 & 2 \end{vmatrix} = \begin{vmatrix} 5 & -1 & 2 \\ 8 & -3 & 2 \\ -4 & 1 & 2 \end{vmatrix}$

51. $\begin{vmatrix} 8 & 2 & -5 \\ -3 & 1 & 4 \\ 2 & 0 & 5 \end{vmatrix} = - \begin{vmatrix} 8 & -5 & 2 \\ -3 & 4 & 1 \\ 2 & 5 & 0 \end{vmatrix}$

Solve each of the following systems by Cramer's Rule. Identify any dependent or inconsistent systems.

52. $3x + y = -1$
$5x + 4y = 10$

53. $3x + 7y = 2$
$5x - y = -22$

54. $3x + 2y + z = 2$
$4x - y + 3z = -16$
$x + 3y - z = 12$

55. $5x - 2y - z = 8$
$-5x + 2y + z = -8$
$x - 4y - 2z = 0$

Graph the solution of each of the following systems of inequalities.

56. $x + y \leq 6$
$2x - y \geq 3$

57. $x - 3y \geq 6$
$y^2 \leq 16 - x^2$

58. $9x^2 + 16y^2 \geq 144$
$x^2 - y^2 \leq 16$

59. A bakery makes both cakes and cookies. Each batch of cakes requires two hours in the oven and three hours in the decorating room. Each batch of cookies needs one and a half hours in the oven and two thirds of an hour in the decorating room. The oven is available no more than 16 hours a day, while the decorating room can be used no more than 12 hours per day. Set up a system of inequalities expressing this information, and then graph the system.

60. A candy company has 100 kilograms of chocolate-covered nuts and 125 kilograms of chocolate-covered raisins to be sold as two different mixtures. One mix will contain 1/2 nuts and 1/2 raisins, while the other mix will contain 1/3 nuts and 2/3 raisins. Set up a system of inequalities expressing this information, and then graph the system.

61. In Exercise 59, a batch of cookies produces a profit of $20; the profit on a batch of cakes is $30. Find the number of batches of each item which will maximize profit.

62. In Exercise 60, how much of each mixture should be made to maximize revenue if the first mix sells for $6.00 per kilogram and the second mix sells for $4.80 per kilogram?

63. Given the line $y = mx + b$ tangent to the circle $x^2 + y^2 = r^2$, find an equation involving m, b, and r.

64. Find an equation of the circle having as its diameter the common chord of the two circles $x^2 + y^2 + 2x - 2y - 14 = 0$ and $x^2 + y^2 - 4x + 4y - 2 = 0$.

Exercises 63, 64: Louis Leithold, *The Calculus with Analytic Geometry*, 4th Edition (Harper and Row, New York, 1981), pp. 40, 62.

10 Zeros of Polynomials

$\mathbf{R}$ecall that a **polynomial function of degree** n is defined as a function of the form

$$f(x) = a_n x^n + a_{n-1} x^{n-1} + \cdots + a_1 x + a_0,$$

where all the a's are complex numbers and $a_n \neq 0$. The number a_n is the **leading coefficient.** If all the coefficients are 0, the polynomial is 0 and is called the **zero polynomial.** The zero polynomial has no degree. However, a polynomial of the form $f(x) = a_0$, for a complex number $a_0 \neq 0$, has degree 0.

A complex number c is a **zero** of a polynomial $f(x)$ whenever $f(c) = 0$. The number c is also a **root** or **solution** of the equation $f(x) = 0$. We shall discuss methods of finding zeros of polynomials in this chapter. The chapter ends with a discussion of partial fractions, which are useful in calculus.

10.1 Polynomial Division

$\mathbf{T}$o find the quotient of two polynomials, we use a **division algorithm** which is very similar to that used for dividing whole numbers. (An algorithm is a step-by-step procedure for working a problem.)

EXAMPLE 1

Divide $4m^3 - 8m^2 + 4m + 6$ by $2m - 1$.
Work as follows.

$$
\begin{array}{r}
2m^2 - 3m + \frac{1}{2} \\
2m - 1 \overline{\smash{\big)}\ 4m^3 - 8m^2 + 4m + 6} \\
\underline{4m^3 - 2m^2} \\
-6m^2 + 4m \\
\underline{-6m^2 + 3m} \\
m + 6 \\
\underline{m - \frac{1}{2}} \\
\frac{13}{2}
\end{array}
$$

Hence, $\dfrac{4m^3 - 8m^2 + 4m + 6}{2m - 1} = 2m^2 - 3m + \dfrac{1}{2} + \dfrac{13/2}{2m - 1}$,

or $4m^3 - 8m^2 + 4m + 6 = (2m - 1)\left(2m^2 - 3m + \dfrac{1}{2}\right) + \dfrac{13}{2}$.

At the first step in dividing these polynomials, we subtracted $4m^3 - 2m^2$ from $4m^3 - 8m^2 + 4m + 6$. The result, $-6m^2 + 4m + 6$, should be written under the line. However, it is customary to save work and "bring down" only the $4m$, the only term needed for the next step. □

A polynomial such as $3x^3 - 2x^2 - 150$ has a missing term, the term where the power of x is 1. When dividing a polynomial with a missing term, it is helpful to allow space for that term as shown in the next example.

EXAMPLE 2 □ Divide $3x^3 - 2x^2 - 150$ by $x - 4$.

$$
\begin{array}{r}
3x^2 + 10x + 40 \\
x - 4\overline{\smash{\big)}\ 3x^3 - 2x^2 - 150} \\
\underline{3x^3 - 12x^2} \\
10x^2 \\
\underline{10x^2 - 40x} \\
40x - 150 \\
\underline{40x - 160} \\
10
\end{array}
$$

The result of this division can be written as

$$\frac{3x^3 - 2x^2 - 150}{x - 4} = 3x^2 + 10x + 40 + \frac{10}{x - 4}$$

or $3x^3 - 2x^2 - 150 = (x - 4)(3x^2 + 10x + 40) + 10$. □

The following theorem generalizes the division process we have illustrated above.

Division Algorithm

Let $f(x)$ and $g(x)$ be polynomials with $g(x)$ of lower degree than $f(x)$ and $g(x) \neq 0$. There exist unique polynomials $q(x)$ and $r(x)$ such that

$$f(x) = g(x) \cdot q(x) + r(x),$$

where $r(x) = 0$ or $r(x)$ is of degree less than the degree of $g(x)$.

We call the polynomial $f(x)$ in the division algorithm the **dividend** and $g(x)$ the **divisor**. The polynomial $q(x)$ is called the **quotient polynomial** or the **quotient**, while $r(x)$ is the **remainder polynomial** or the **remainder**.

The division algorithm applies to the polynomials of Examples 1 and 2. In Example 1, we had

$$\frac{4m^3 - 8m^2 + 4m + 6}{2m - 1} = 2m^2 - 3m + \frac{1}{2} + \frac{13/2}{2m - 1}$$

from which we see

$$4m^3 - 8m^2 + 4m + 6 = (2m - 1)\left(2m^2 - 3m + \frac{1}{2}\right) + \frac{13}{2}$$

$$\quad f(x) \qquad\qquad = \qquad\quad g(x) \cdot q(x) \qquad\qquad + \ r(x).$$

Identify $q(x)$ and $r(x)$ in Example 2.

Synthetic Division

We often need to divide a polynomial by a first-degree binomial of the form $x - k$, where the coefficient of x is 1. We can develop a shortcut for division problems of this type. To illustrate, we will rework Example 2. The steps of this example are repeated on the left below. On the right we can simplify the division process by omitting all variables and writing only coefficients. We use 0 to represent the coefficient of any missing terms. Since the coefficient of x in the divisor is always 1 in problems of this type, we can omit it too. These omissions simplify the problem as shown on the right below.

$$
\begin{array}{r}
3x^2 + 10x + 40 \\
x - 4\overline{\smash{\big)}\,3x^3 - 2x^2 - 150} \\
\underline{3x^3 - 12x^2} \\
10x^2 \\
\underline{10x^2 - 40x} \\
40x - 150 \\
\underline{40x - 160} \\
10
\end{array}
\qquad
\begin{array}{r}
3 \quad 10 \quad 40 \\
-4\overline{\smash{\big)}\,3 - 2 + 0 - 150} \\
\underline{3 - 12} \\
10 \\
\underline{10 - 40} \\
40 - 150 \\
\underline{40 - 160} \\
10
\end{array}
$$

The numbers in color are repetitions of the numbers directly above and can be omitted.

$$
\begin{array}{r}
3 \quad 10 \quad 40 \\
-4\overline{\smash{\big)}\,3 - 2 + 0 - 150} \\
\underline{- 12} \\
10 \\
\underline{- 40} \\
40 - 150 \\
\underline{- 160} \\
10
\end{array}
$$

We can now vertically condense the entire problem and omit the top row of numbers since it duplicates the bottom row.

$$
\begin{array}{r}
-4\overline{\smash{\big)}\,3 \quad -2 \quad 0 \quad -150} \\
\underline{-12 \quad -40 \quad -160} \\
3 \quad 10 \quad 40 \quad 10
\end{array}
$$

We obtained the bottom row by subtracting -12, -40, and -160 from the corresponding terms above. For reasons that will become clear in the discussion of the next theorem, we change the -4 at the left to 4, which also changes the sign of the numbers in the second row, and then add. Doing this, we have the result below.

$$\begin{array}{r|rrr} 4 & 3 & -2 & 0 & -150 \\ & & 12 & 40 & 160 \\ \hline & 3 & 10 & 40 & 10 \end{array}$$

This abbreviated process is called **synthetic division.**

In summary, to use synthetic division to divide a polynomial $f(x)$ by a binomial of the form $x - k$, begin by writing the coefficients of $f(x)$ in decreasing powers of the variable, using 0 as the coefficient of any missing powers. The number k is written to the left in the same row. In the example above, $x - k$ is $x - 4$ so k is 4. Next bring down the leading coefficient of $f(x)$, 3 in the example above, as the first number in the last row. Multiply the 3 by 4 to get the first number in the second row, 12. Add 12 to -2; this gives 10, the second number in the third row. Multiply 10 by 4 to get 40, the next number in the second row. Add 40 to 0 to get the third number in the third row, and so on. This process of multiplying each result in the third row by k and adding the product to the number in the next column is repeated until there is a number in the last row for each coefficient in the first row.

EXAMPLE 3   ☐   Use synthetic division to divide $5m^3 - 6m^2 - 28m - 2$ by $m + 2$.

We begin by writing

$$\begin{array}{r|rrr} -2 & 5 & -6 & -28 & -2. \end{array}$$

Note that we have changed the 2 to -2 since we find k by writing $m + 2$ as $m - (-2)$. To begin, we bring down the 5.

$$\begin{array}{r|rrr} -2 & 5 & -6 & -28 & -2 \\ \hline & 5 \end{array}$$

Now, multiply -2 by 5 to get -10, and add it to the -6 in the first row. The result is -16.

$$\begin{array}{r|rrr} -2 & 5 & -6 & -28 & -2 \\ & & -10 \\ \hline & 5 & -16 \end{array}$$

Next, we have $(-2)(-16) = 32$. We add this to the -28 in the first row.

$$\begin{array}{r|rrr} -2 & 5 & -6 & -28 & -2 \\ & & -10 & 32 \\ \hline & 5 & -16 & 4 \end{array}$$

Finally, since $(-2)(4) = -8$, we have

$$\begin{array}{r|rrr} -2 & 5 & -6 & -28 & -2 \\ & & -10 & 32 & -8 \\ \hline & 5 & -16 & 4 & -10. \end{array}$$

The coefficients of the quotient polynomial and the remainder are read directly from the bottom row. The degree of the quotient will always be one less than the degree of the polynomial to be divided. Therefore,

$$\frac{5m^3 - 6m^2 - 28m - 2}{m + 2} = 5m^2 - 16m + 4 + \frac{-10}{m + 2}.\quad\square$$

Evaluating $f(k)$

The division algorithm tells us that if we divide $f(x)$ by $x - k$, then

$$f(x) = (x - k) \cdot q(x) + r,$$

for some polynomial $q(x)$ and complex number r. This equality is true for all complex values of x so that it is true for $x = k$. Thus

$$f(k) = (k - k) \cdot q(k) + r$$
$$f(k) = r.$$

We have now proved the following theorem.

Remainder Theorem	If the polynomial $f(x)$ is divided by $x - k$, then the remainder is equal to $f(k)$.

For example, we found in Example 2 above that $f(x) = 3x^3 - 2x^2 - 150$ can be written as

$$f(x) = (x - 4)(3x^2 + 10x + 40) + 10.$$

To find $f(4)$, replace x with 4.

$$f(4) = (4 - 4)(3 \cdot 4^2 + 10 \cdot 4 + 40) + 10$$
$$= 0(3 \cdot 4^2 + 10 \cdot 4 + 40) + 10$$
$$f(4) = 10.$$

Thus, instead of replacing x with 4 in $f(x) = 3x^3 - 2x^2 - 150$, we need only divide $f(x)$ by $x - 4$ using synthetic division. Then $f(4)$ will be equal to the remainder, 10.

EXAMPLE 4 $\square$ Let $f(x) = -x^4 + 3x^2 - 4x - 5$. Find $f(-2)$.
Use the remainder theorem and synthetic division, as follows:

```
-2 | -1   0    3   -4   -5
   |       2   -4    2    4
   -----------------------
     -1   2   -1   -2   -1
```

Therefore, $f(-2) = -1$, the remainder when $f(x)$ is divided by $x - (-2)$ or $x + 2$. $\square$

By the remainder theorem, if $f(k) = 0$, then $x - k$ is a factor of $f(x)$, and conversely, if $x - k$ is a factor of $f(x)$, then $f(k)$ must equal 0. Thus, we have the following theorem.

Factor Theorem

> The polynomial $x - k$ is a factor of the polynomial $f(x)$ if and only if $f(k) = 0$. That is,
>
> $$f(k) = 0 \text{ if and only if } f(x) = (x - k) \cdot q(x),$$
> for some polynomial $q(x)$.

EXAMPLE 5

☐ Is $x - 3$ a factor of $f(x) = 2x^3 - 4x^2 + 5x - 3$?

By the factor theorem, $x - 3$ will be a factor of $f(x)$ only if $f(3) = 0$. Use synthetic division and the remainder theorem to decide.

$$
\begin{array}{r|rrrr}
3 & 2 & -4 & 5 & -3 \\
  &   & 6 & 6 & 33 \\
\hline
  & 2 & 2 & 11 & 30
\end{array}
$$

The remainder is 30, and not 0, so that $x - 3$ is not a factor of $f(x)$. ☐

EXAMPLE 6

☐ Is $x - i$ a factor of $f(x) = x^4 - x^3 - x^2 - x - 2$?

Use synthetic division to find out.

$$
\begin{array}{r|rrrrr}
i & 1 & -1 & -1 & -1 & -2 \\
  &   & i & -1 - i & 1 - 2i & 2 \\
\hline
  & 1 & -1 + i & -2 - i & -2i & 0
\end{array}
$$

The remainder is 0. Thus, $f(i) = 0$, and so $x - i$ is a factor of $f(x)$. Also, $x = i$ is a solution of the equation $x^4 - x^3 - x^2 - x - 2 = 0$. ☐

As we know, a complex number k is a *zero* of a polynomial $f(x)$ if $f(k) = 0$. By the results above, any time we have a zero of $f(x)$, we also have found a factor of $f(x)$. That is, if k is a zero of $f(x)$, then $x - k$ is a factor of $f(x)$, and conversely. Also, k is then a root or solution of the equation $f(x) = 0$.

EXAMPLE 7

☐ Find a polynomial having zeros 4, -3, and 2.

Let $f(x)$ be a polynomial having the three given zeros. By the factor theorem, $f(x)$ must have factors of $x - 4$, $x - (-3) = x + 3$, and $x - 2$. Thus, one polynomial with the desired zeros is

$$f(x) = (x - 4)(x + 3)(x - 2)$$
$$f(x) = x^3 - 3x^2 - 10x + 24.$$

This polynomial is not the only possible one—in fact, we could obtain an infinite number of such polynomials by multiplying $f(x)$ by any of an infinite number of non-zero polynomials. ☐

EXERCISES
10.1

Perform the following divisions. Find the quotient and the remainder for each. Use synthetic division as appropriate.

1. $\dfrac{2x^3 - 11x^2 + 19x - 10}{2x - 5}$

2. $\dfrac{3p^3 - 11p^2 + 5p + 3}{3p + 1}$

3. $\dfrac{x^4 + 2x^3 + 2x^2 - 2x - 3}{x^2 - 1}$

4. $\dfrac{2y^5 + y^3 - 2y^2 - 1}{2y^2 + 1}$

5. $\dfrac{4z^5 - 4z^2 - 5z + 3}{2z^2 + z + 1}$

6. $\dfrac{12z^4 - 25z^3 + 35z^2 - 26z + 10}{4z^2 - 3z + 5}$

7. $\dfrac{p^4 - 3p^3 - 5p^2 + 2p - 16}{p + 2}$

8. $\dfrac{3x^3 - 11x^2 - 20x + 3}{x - 5}$

9. $\dfrac{4m^3 - 3m - 2}{m + 1}$

10. $\dfrac{3q^3 - 4q + 2}{q - 1}$

11. $\dfrac{x^5 + 3x^4 + 2x^3 + 2x^2 + 3x + 1}{x + 2}$

12. $\dfrac{m^6 - 3m^4 + 2m^3 - 6m^2 - 5m + 3}{m + 2}$

Use synthetic division to decide whether or not the given number is a zero of the given polynomial.

13. $2; f(x) = x^2 + 2x - 8$

14. $-1; f(m) = m^2 + 4m - 5$

15. $2; f(g) = g^3 - 3g^2 + 4g - 4$

16. $-3; f(m) = m^3 + 2m^2 - m + 6$

17. $1 - i; f(y) = y^2 - 2y + 2$

18. $3 - 2i; f(r) = r^2 - 6r + 13$

19. $2 + i; f(k) = k^2 + 3k + 4$

20. $1 - 2i; f(z) = z^2 - 3z + 5$

21. $i; f(x) = x^3 + 2ix^2 + 2x + i$

22. $-i; f(p) = p^3 - ip^2 + 3p + 5i$

23. $1 + i; f(p) = p^3 + 3p^2 - p + 1$

24. $2 - i; f(r) = 2r^3 - r^2 + 3r - 5$

Use the Remainder Theorem to find $f(k)$.

25. $k = 5; f(x) = -x^2 + 2x + 7$

26. $k = -3; f(x) = 3x^2 + 8x + 5$

27. $k = 3; f(x) = x^2 - 4x + 5$

28. $k = -2; f(x) = x^2 + 5x + 6$

29. $k = 2 + i; f(x) = x^2 - 5x + 1$

30. $k = 3 - 2i; f(x) = x^2 - x + 3$

31. $k = 1 - i; f(x) = x^3 + x^2 - x + 1$

32. $k = 2 - 3i; f(x) = x^3 + 2x^2 + x - 5$

Use the Factor Theorem to find the value of c that makes the second polynomial a factor of the first.

33. $4x^2 + 2x - c; x - 3$

34. $-3x^2 + cx + 2; x + 2$

35. $3x^3 - 12x^2 - 11x + c; x - 5$

36. $4x^3 + 6x^2 - 5x + c; x + 2$

37. $2x^4 + 5x^3 - 2x^2 + cx + 3; x + 3$

38. $5x^4 + 16x^3 - 15x^2 + cx + 16; x + 4$

Find a polynomial of lowest degree having the given zeros.

39. $-1, 2, -5$

40. $-4, 6, 3$

41. $\sqrt{2}, -\sqrt{2}, 3$

42. $-\sqrt{5}, \sqrt{5}, -1$

43. $\dfrac{1 + \sqrt{11}}{2}, \dfrac{1 - \sqrt{11}}{2}, -2, 3$

44. $\dfrac{2 - \sqrt{7}}{3}, \dfrac{2 + \sqrt{7}}{3}, 4, 2$

*To evaluate a polynomial such as $f(x) = x^3 - 4x^2 + 2x - 5$ for a particular value of x, we can replace x with the particular value. To simplify such evaluation with calculators or computers, we can use the **nested form** of the polynomial. The nested form of our polynomial $f(x)$ is*

$$f(x) = x^3 - 4x^2 + 2x - 5$$
$$= x(x^2 - 4x + 2) - 5$$
$$f(x) = x(x(x - 4) + 2) - 5$$

Use this last form to evaluate $f(x)$ for

45. $x = 2$

46. $x = -3$

47. $x = -.08$

48. $x = .47$

Write the following polynomials in nested form.

49. $f(x) = x^3 + 2x^2 - 4x + 7$

50. $f(x) = x^3 - x^2 + 5x + 3$

51. $f(x) = -2x^3 + 3x^2 - x - 1$

52. $f(x) = -x^3 - 4x^2 + 2x + 3$

53. The function

$$f(x) = 1.012x^3 - 24.67x^2 + 216.9x$$

approximates the change in pressure of the oil in a certain reservoir, where x is time in years. Use the Remainder Theorem to find the following: (a) $f(3)$, (b) $f(5)$, (c) $f(10)$.

Show that each of the following is true.

54. $f(x) = 2x^4 + 4x^2 + 1$ can have no factor $x - k$ where k is a real number.

55. $f(x) = -x^4 - 5x^2 - 3$ can have no factor $x - k$ where k is a real number.

56. $x - c$ is a factor of $x^n - c^n$ for all positive integers n.

57. $x + c$ is a factor of $x^n + c^n$ for all odd positive integers n.

58. Let $f(x)$ be a polynomial having a zero of c. Let a be a number "close" to c. Write $f(x)$ as $(x - a) \cdot q(x) + r$. Find the following approximation for c:

$$c \approx a - \frac{f(a)}{q(a)}.$$

Use the result of Exercise 58 along with the given value of a to find an approximation to the nearest hundredth of a zero for the following polynomials.

59. $f(x) = x^3 + 7x^2 - 2x - 14; a = -1.4$

60. $f(x) = x^3 + 4x^2 - 5x - 5; a = 1.5$

61. $f(x) = 2x^3 + 3x^2 - 22x - 33; a = 3.2$

62. $f(x) = 2x^3 - 7x^2 + 2x + 6; a = -0.7$

10.2 Complex Zeros of Polynomial Functions

Although every linear or quadratic polynomial equation can be solved, it is usually quite difficult to solve a polynomial equation of degree greater than two. In this and the next two sections, we develop methods of finding, or approximating, the zeros of polynomial functions and, hence, the roots of polynomial equations.

We have seen that if a polynomial function $f(x)$ can be divided with remainder 0 by $x - k$, then $f(k) = 0$, and the converse is also true. The next theorem says that every polynomial function of degree at least 1 has a zero, meaning that every such polynomial can be factored. This theorem, called the **Fundamental Theorem of Algebra,** was first proved by the mathematician K. F. Gauss in his doctoral thesis of 1799 when he was 22 years old. Although many proofs of this result have been given over the years, none involve only the algebra of this book so no proof is included here.

Fundamental Theorem of Algebra

> Every polynomial function of degree at least one has at least one complex zero.

By the fundamental theorem, and the factor theorem, we can write any polynomial $f(x)$ as

$$f(x) = (x - k) \cdot q(x).$$

In the same way, we could then factor $q(x)$. If we assume $f(x)$ is of degree n and continue this process n times, we get

$$f(x) = a(x - k_1)(x - k_2) \cdots (x - k_n), \tag{1}$$

where a is the leading coefficient of $f(x)$.

Using this result, we can prove that a polynomial function of degree n has at most n distinct zeros. To prove this, we use an **indirect proof.** To prove a statement indirectly, we assume that the opposite of the statement is true. We then show that if we assume the opposite statement to be true, we are led to a contradiction. Here we wish to prove that a polynomial function of degree $n \geq 1$ has at most n distinct zeros; therefore, we begin by assuming the opposite—we assume that the polynomial has *more* than n distinct zeros.

Let us assume that a polynomial function $f(x)$ of degree n has $n + 1$ distinct zeros. We saw above that we can write $f(x)$ as

$$f(x) = a(x - k_1)(x - k_2) \cdots (x - k_n).$$

From this factored form of the polynomial function, we see that n of the zeros are the numbers $k_1, k_2 \cdots, k_n$. We assumed that $f(x)$ has $n + 1$ distinct zeros, so there must be another zero, say k, that is not in the list $k_1, k_2, \cdots, k_n$. Since k is a zero of $f(x)$,

$$f(k) = 0.$$

From equation (1), replacing x with k gives

$$f(k) = a(k - k_1)(k - k_2) \cdots (k - k_n)$$

or

$$0 = a(k - k_1)(k - k_2) \cdots (k - k_n). \tag{2}$$

Since $k \neq k_1$, $k \neq k_2$, $\cdots$, $k \neq k_n$, we have $k - k_1 \neq 0$, $k - k_2 \neq 0$, $\cdots$, $k - k_n \neq 0$, so that the right side of equation (2) is made up of a product of factors, all of which are nonzero. Since the product of nonzero factors cannot be 0, we have a contradiction. This contradiction shows that our original assumption is false. We have now proved the following result.

Theorem   A polynomial of degree n has at most n distinct zeros.

The theorem says that there exist *at most* n distinct zeros. For example, the polynomial function $f(x) = x^3 + 3x^2 + 3x + 1 = (x + 1)^3$ is of degree $n = 3$ but has only one zero, -1. Actually, the zero -1 occurs three times, since there are three factors of $x + 1$. We sometimes call this a **zero of multiplicity 3.**

EXAMPLE 1   ☐   Find a polynomial function $f(x)$ of degree 3 that satisfies the following conditions.

(a) Zeros of -1, 2, and 4; $f(1) = 3$

From these three zeros, we know that $x - (-1) = x + 1$, $x - 2$, and $x - 4$ must be factors of $f(x)$. Since $f(x)$ is to be of degree 3, these are the only possible factors by the theorem just above. Therefore, $f(x)$ has the form

$$f(x) = a(x + 1)(x - 2)(x - 4),$$

for some real number a. To find a, use the fact that $f(1) = 3$. We then have

$$f(1) = a(1 + 1)(1 - 2)(1 - 4) = 3$$
$$a(2)(-1)(-3) = 3$$
$$6a = 3$$
$$a = \frac{1}{2}$$

Thus $f(x) = \frac{1}{2}(x + 1)(x - 2)(x - 4)$

or $f(x) = \frac{1}{2}x^3 - \frac{5}{2}x^2 + x + 4.$

(b) -2 is a root of multiplicity 3; $f(-1) = 4$.

We know that $f(x)$ has the form

$$f(x) = a(x + 2)(x + 2)(x + 2)$$
$$= a(x + 2)^3.$$

Since $f(-1) = 4$, we have

$$f(-1) = a(-1 + 2)^3 = 4$$
or $$a(1)^3 = 4$$
$$a = 4.$$

Therefore, $f(x) = 4(x + 2)^3 = 4x^3 + 24x^2 + 48x + 32.$ ☐

Using the remainder theorem, we can show that both $2 + i$ and $2 - i$ are zeros of $f(x) = x^3 - x^2 - 7x + 15$. It is not just coincidence that both $2 + i$ and its conjugate $2 - i$ are zeros of this polynomial. We can prove that if $a + bi$ is a zero of a polynomial function with real coefficients, then so is its conjugate, $a - bi$.

To prove this, we need the following properties of complex conjugates. Let $z = a + bi$, and write $\overline{z}$ for the conjugate of z, so that $\overline{z} = a - bi$. For example, if $z = -5 + 2i$, then $\overline{z} = -5 - 2i$. We leave the proof of the following equalities for the exercises. For any complex numbers c and d,

Properties of Conjugates

$$\overline{c + d} = \overline{c} + \overline{d}$$
$$\overline{c \cdot d} = \overline{c} \cdot \overline{d}$$
$$\overline{c^n} = (\overline{c})^n.$$

Now consider the polynomial having real coefficients,

$$f(x) = a_n x^n + a_{n-1} x^{n-1} + \cdots + a_1 x + a_0.$$

If $z = a + bi$ is a zero of $f(x)$, then

$$f(z) = a_n z^n + a_{n-1} z^{n-1} + \cdots + a_1 z + a_0 = 0.$$

If we take the conjugate of both sides of this last equation, we have

$$\overline{a_n z^n + a_{n-1} z^{n-1} + \cdots + a_1 z + a_0} = \overline{0}.$$

This becomes $\quad \overline{a_n z^n} + \overline{a_{n-1} z^{n-1}} + \cdots + \overline{a_1 z} + \overline{a_0} = \overline{0}$

or $\quad \overline{a_n}\,\overline{z_n} + \overline{a_{n-1}}\,\overline{z^{n-1}} + \cdots + \overline{a_1}\,\overline{z} + \overline{a_0} = \overline{0}.$

Here we have used generalizations of the properties $\overline{c + d} = \overline{c} + \overline{d}$ and $\overline{cd} = \overline{c} \cdot \overline{d}$; now use the third property above and the fact that for any real number a, we have $\overline{a} = a$, to get

$$a_n (\overline{z})^n + a_{n-1} (\overline{z})^{n-1} + \cdots + a_1 (\overline{z}) + a_0 = 0.$$

Hence, $\overline{z}$ is also a zero of $f(x)$, and we have proved the following result.

Conjugate Zeros Theorem

If $f(x)$ is a polynomial having real coefficients and if $a + bi$ is a zero of $f(x)$, where a and b are real numbers, then $a - bi$ is also a zero of $f(x)$.

The requirement that the polynomial have real coefficients is very important. For example, $f(x) = x - (1 + i)$ has $1 + i$ as a zero, but the conjugate, $1 - i$, is not a zero.

EXAMPLE 2 ☐ Find a polynomial of lowest degree having real coefficients and zeros 3 and $2 + i$.

We know the number $2 - i$ must also be a zero. Thus, there must be at least three zeros, 3, $2 + i$, and $2 - i$. By the factor theorem there must then be three factors, $x - 3$, $x - (2 + i)$, and $x - (2 - i)$. A polynomial of lowest degree is

$$f(x) = (x - 3)[x - (2 + i)][x - (2 - i)]$$
$$= (x - 3)(x - 2 - i)(x - 2 + i)$$
$$f(x) = x^3 - 7x^2 + 17x - 15.$$

There are other polynomials, such as $3(x^3 - 7x^2 + 17x - 15)$ or $\sqrt{5}(x^3 - 7x^2 + 17x - 15)$, that satisfy the given conditions as to zeros. The information given in the statement of the example is not enough for us to find a single value of the leading coefficient.   ☐

The theorem on conjugate zeros is important in helping predict the number of real zeros of a polynomial with real coefficients. A polynomial with real coefficients of odd degree n must have at least one real zero (since we have just seen that zeros of the form $a + bi$, where $b \neq 0$, occur in pairs). On the other hand, a polynomial with real coefficients of even degree n need have no real zeros but may have up to n real zeros.

**EXAMPLE
3**   ☐   Find all zeros of $x^4 - 7x^3 + 18x^2 - 22x + 12$, given that $1 - i$ is a zero.
Since $1 - i$ is a zero and since the polynomial has all real coefficients, then the conjugate $1 + i$ is also a zero. Thus, we should first divide the original polynomial by $1 - i$ and then divide the quotient by $1 + i$, as follows:

$$
\begin{array}{r|rrrrr}
1 - i & 1 & -7 & 18 & -22 & 12 \\
 &   & 1-i & -7+5i & 16-6i & -12 \\
\hline
 & 1 & -6-i & 11+5i & -6-6i & 0
\end{array}
$$

$$
\begin{array}{r|rrrr}
1 + i & 1 & -6-i & 11+5i & -6-6i \\
 &   & 1+i & -5-5i & 6+6i \\
\hline
 & 1 & -5 & 6 & 0
\end{array}
$$

Now find the zeros of the quadratic polynomial $x^2 - 5x + 6$. By factoring, the zeros are 2 and 3, so that the four zeros of $x^4 - 7x^3 + 18x^2 - 22x + 12$ are $1 - i$, $1 + i$, 2, and 3.   ☐

**EXERCISES
10.2**

For each of the following, determine a polynomial of degree 3 with real coefficients that satisfies the given conditions.

1. zeros of -3, -1, and 4; and $f(2) = 5$

2. zeros of 1, -1, and 0; and $f(2) = -3$

3. -2 is a zero of multiplicity 2, -3 is a zero; and $f(1) = 4$

4. 5 is a zero of multiplicity 2, -1 is a zero; and $f(2) = 3$

5. zeros of 3 and i; and $f(2) = 50$.

6. zeros of -2 and $-i$; and $f(-3) = 30$

For each of the following, determine a polynomial of lowest degree with real coefficients having the given zeros.

7. $5 + i$ and $5 - i$

8. $3 - 2i$ and $3 + 2i$

9. $2, 1 - i,$ and $1 + i$

10. $-3, 2 - i,$ and $2 + i$

11. $1 + \sqrt{2}, 1 - \sqrt{2},$ and 1

12. $1 - \sqrt{3}, 1 + \sqrt{3},$ and -2

13. $2 + i, 2 - i, 3,$ and -1

14. $3 + 2i, 3 - 2i, -1,$ and 2

15. 2 and $3 + i$

16. -1 and $4 - 2i$

17. $1 - \sqrt{2}, 1 + \sqrt{2},$ and $1 - i$

18. $2 + \sqrt{3}, 2 - \sqrt{3},$ and $2 + 3i$

19. $2 - i, 3 + 2i$

20. $5 + i, 4 - i$

For each of the following polynomials, one zero is given. Determine all others.

21. $f(x) = x^3 - x^2 - 4x - 6;\ x = 3$

22. $f(x) = x^3 - 5x^2 + 17x - 13;\ x = 1$

23. $f(x) = x^3 + x^2 - 4x - 24;\ x = -2 + 2i$

24. $f(x) = x^3 + x^2 - 20x - 50;\ x = -3 + i$

25. $f(x) = 2x^3 - 2x^2 - x - 6;\ x = 2$

26. $f(x) = 2x^3 - 5x^2 + 6x - 2;\ x = 1 + i$

27. $f(x) = x^4 - 3x^3 + 6x^2 + 2x - 60;$
     $x = 1 + 3i$

28. $f(x) = x^4 - 6x^3 - x^2 + 86x + 170;$
     $x = 5 + 3i$

29. $f(x) = x^4 - 6x^3 + 15x^2 - 18x + 10;$
     $x = 2 + i$

30. $f(x) = x^4 + 8x^3 + 26x^2 + 72x + 153;$
     $x = -3i$

31. The displacement at time t of a particle moving along a line is given by

$$s(t) = t^3 - 2t^2 - 5t + 6,$$

where t is in seconds and s is measured in centimeters. The displacement is 0 after 1 second has elapsed. At what other times is the displacement 0?

32. Show that 1 is a zero of multiplicity 4 of $f(x) = x^6 - 8x^5 + 27x^4 - 48x^3 + 47x^2 - 24x + 5$, and find all other complex zeros. Then write $f(x)$ in factored form.

33. Show that 2 is a zero of multiplicity 3 of $f(x) = x^5 - 6x^4 + 13x^3 - 14x^2 + 12x - 8$, and find all other complex zeros. Then write $f(x)$ in factored form.

34. If c and d are complex numbers, show that (a) $\overline{c + d} = \overline{c} + \overline{d},$ (b) $\overline{cd} = \overline{c} \cdot \overline{d}$. (Hint: Let $c = a + bi$ and $d = m + ni$ and form the conjugates, the sums, and the products.)

35. Show that $\overline{a} = a$ for any real number a.

36. Explain why it is not possible for a polynomial of degree 3 with real coefficients to have zeros of $1, 2,$ and $1 + i$.

37. Show that the zeros of $f(x) = x^3 + ix^2 - (7 - i)x + (6 - 6i)$ are $1 - i, 2,$ and -3. Why doesn't the Conjugate Zeros Theorem apply?

38. Show that the equation

$$\frac{1}{x - 3} + \frac{2}{x + 3} - \frac{6}{x^2 - 9} = 0$$

has no solution. Why doesn't this contradict the fundamental theorem of algebra?

39. Show that if

$$f(x) = a_n x^n + a_{n-1}x^{n-1} + \cdots + a_1 x + a_0$$

has $n + 1$ distinct zeros, then each a_i must equal 0.

40. Show that $f(x) = 0$ and $g(x) = 0$ have exactly the same solutions (with the same multiplicity) if and only if $f(x) = c \cdot g(x)$ for some complex number c.

41. Let $f(x)$ and $q(x)$ be polynomials each of which has c as a zero. Show that $x - c$ is a factor of the remainder when $f(x)$ is divided by $q(x)$.

42. Let $f(x)$ be a polynomial of odd degree having only real coefficients. Show that a real number a exists such that $f(a) = 0$.

10.3   Rational Zeros of Polynomials

We know by the fundamental theorem of algebra that every polynomial function of degree at least one has a zero. However, the fundamental theorem merely tells us that such a zero exists. It gives no help at all in identifying zeros. However, there are theorems that we can use to find any rational zeros of polynomials with rational coefficients and decimal approximations of any irrational zeros.

The next theorem gives a useful test for determining whether or not a given rational number is a possible zero of a polynomial function $f(x)$ having integer coefficients.

Rational Zeros Theorem

Let $f(x) = a_n x^n + a_{n-1} x^{n-1} + \cdots + a_1 x + a_0$ be a polynomial function of degree n with integer coefficients. If p/q is a rational number written in lowest terms with the property that $f(p/q) = 0$, then p is a factor of a_0 and q is a factor of a_n.

To prove this theorem, use the fact that $f(p/q) = 0$ and replace x with p/q to get

$$a_n(p/q)^n + a_{n-1}(p/q)^{n-1} + \cdots + a_1(p/q) + a_0 = 0,$$

which can also be written as

$$a_n(p^n/q^n) + a_{n-1}(p^{n-1}/q^{n-1}) + \cdots + a_1(p/q) + a_0 = 0.$$

Multiply both sides of this last result by q^n and add $-a_0 q^n$ to both sides to get

$$a_n p^n + a_{n-1} p^{n-1} q + \cdots + a_1 p q^{n-1} = -a_0 q^n.$$

Factoring out p gives

$$p(a_n p^{n-1} + a_{n-1} p^{n-2} q + \cdots + a_1 q^{n-1}) = -a_0 q^n.$$

This result shows that $-a_0 q^n$ equals the product of the two factors, p and $(a_n p^{n-1} + \cdots + a_1 q^{n-1})$. For this reason, p must be a factor of $-a_0 q^n$. Since we have assumed that p/q is written in lowest terms, p and q have no common factor other than 1. Hence, p is not a factor of q^n. Thus we must conclude that p is a factor of a_0. In a similar way we can show that q is a factor of a_n.

EXAMPLE 1   ☐   Find all rational zeros of $f(x) = 2x^4 - 11x^3 + 14x^2 - 11x + 12$.

If p/q is to be a rational zero of $f(x)$, we know by the Rational Zeros Theorem that p must be a factor of $a_0 = 12$ and q must be a factor of $a_4 = 2$. Hence, p must be ± 1, ± 2, ± 3, ± 4, ± 6, or ± 12, while q must be ± 1 or ± 2. From this we see that any rational zero of $f(x)$ will come from the list

$$\pm 1, \ \pm 1/2, \ \pm 2, \ \pm 3, \ \pm 3/2, \ \pm 4, \ \pm 6, \ \text{or} \ \pm 12.$$

Though we are not sure that any of these numbers are zeros, if $f(x)$ has any rational zeros at all, they will be in the list above. We can check these proposed zeros by synthetic division. Doing so, we find that 4 is a zero.

$$
\begin{array}{r|rrrrr}
4 & 2 & -11 & 14 & -11 & 12 \\
  &   & 8 & -12 & 8 & -12 \\
\hline
  & 2 & -3 & 2 & -3 & 0
\end{array}
$$

As a fringe benefit of this calculation, we now only need to look for zeros of the simpler polynomial $q(x) = 2x^3 - 3x^2 + 2x - 3$. Any rational zero of $q(x)$ will have a numerator of ± 3 or ± 1 with a denominator of ± 1 or ± 2. Hence, any rational zeros of $q(x)$ will come from the list

$$\pm 3, \ \pm 3/2, \ \pm 1, \ \pm 1/2.$$

Again we use synthetic division and find that $3/2$ is a zero.

$$
\begin{array}{r|rrrr}
3/2 & 2 & -3 & 2 & -3 \\
    &   & 3 & 0 & 3 \\
\hline
    & 2 & 0 & 2 & 0
\end{array}
$$

The quotient is $2x^2 + 2$, which, by the quadratic formula, has i and $-i$ as zeros. They are, however, imaginary zeros. The *rational* zeros of the polynomial function

$$f(x) = 2x^4 - 11x^3 + 14x^2 - 11x + 12$$

are 4 and $3/2$. ▨

To find any rational zeros of a polynomial function with rational coefficients, first multiply the polynomial function by a number that will clear it of all fractional coefficients. Then use the Rational Zeros Theorem.

EXERCISES
10.3

Determine all rational zeros of the following polynomial functions.

1. $f(x) = x^3 - 2x^2 - 13x - 10$

2. $f(x) = x^3 + 5x^2 + 2x - 8$

3. $f(x) = x^3 + 6x^2 - x - 30$

4. $f(x) = x^3 - x^2 - 10x - 8$

5. $f(x) = x^3 + 9x^2 - 14x - 24$

6. $f(x) = x^3 + 3x^2 - 4x - 12$

7. $f(x) = x^4 + 9x^3 + 21x^2 - x - 30$

8. $f(x) = x^4 + 4x^3 - 7x^2 - 34x - 24$

9. $f(x) = 6x^3 + 17x^2 - 31x - 12$

10. $f(x) = 15x^3 + 61x^2 + 2x - 8$

11. $f(x) = 12x^3 + 20x^2 - x - 6$

12. $f(x) = 12x^3 + 40x^2 + 41x + 12$

13. $f(x) = 2x^3 + 7x^2 + 12x - 8$

14. $f(x) = 2x^3 + 20x^2 + 68x - 40$

15. $f(x) = x^4 + 4x^3 + 3x^2 - 10x + 50$

16. $f(x) = x^4 - 2x^3 + x^2 + 18$

17. $f(x) = x^4 + 2x^3 - 13x^2 - 38x - 24$

18. $f(x) = 6x^4 + x^3 - 7x^2 - x + 1$

19. $f(x) = 3x^4 + 4x^3 - x^2 + 4x - 4$

20. $f(x) = x^4 + 8x^3 + 16x^2 - 8x - 17$

21. $f(x) = x^5 + 3x^4 - 5x^3 - 11x^2 + 12$

22. $f(x) = 4x^5 + 4x^4 - 37x^3 - 37x^2 + 9x + 9$

23. $f(x) = x^3 - \dfrac{4}{3}x^2 - \dfrac{13}{3}x - 2$

24. $f(x) = x^3 + x^2 - \dfrac{16}{9}x + \dfrac{4}{9}$

25. $f(x) = x^4 + \dfrac{1}{4}x^3 + \dfrac{11}{4}x^2 + x - 5$

26. $f(x) = \dfrac{10}{7}x^4 - x^3 - 7x^2 + 5x - \dfrac{5}{7}$

27. $f(x) = \dfrac{1}{3}x^5 + x^4 - \dfrac{5}{3}x^3 - \dfrac{11}{3}x^2 + 4$

28. $f(x) = x^5 + x^4 - \dfrac{37}{4}x^2 + \dfrac{9}{4}x + \dfrac{9}{4}$

29. Show that $f(x) = x^2 - 2$ has no rational zeros, so that $\sqrt{2}$ must be irrational.

30. Show that $f(x) = x^2 - 5$ has no rational zeros, so that $\sqrt{5}$ must be irrational.

31. Show that $f(x) = x^4 + 5x^2 + 4$ has no rational zeros.

32. Show that $f(x) = x^5 - 3x^3 + 5$ has no rational zeros.

33. Show that any integer zeros of a polynomial function must be factors of the constant term a_0.

10.4 Approximate Zeros of Polynomials

We know that every polynomial function of degree 1 or more has a zero. However, we do not know whether or not the function has real zeros; and even if it does have real zeros, we often have no way to find them. In this section we discuss methods of approximating any real zeros a polynomial function may have. These methods work well using a computer or a calculator.

Much of our work in locating real zeros uses the following result, which is related to the fact that graphs of polynomial functions are unbroken curves, with no gaps or sudden jumps. The proof requires advanced methods so we do not give it here.

The Intermediate Value Theorem for Polynomials

If $f(x)$ is a polynomial function with only real coefficients and if $f(a) \neq f(b)$ for $a < b$, then in the interval $[a, b]$, $f(x)$ takes every value between $f(a)$ and $f(b)$.

The intermediate value theorem tells us that for every value c between a and b,

$f(c)$ will be between $f(a)$ and $f(b)$. See Figure 10.1. In particular, as Figure 10.1 shows, if $f(a)$ and $f(b)$ are opposite in sign, then there must be some number c, between a and b, such that $f(c) = 0$. This helps us to find the intervals where the zeros of the function are located.

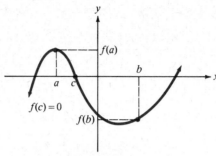

Figure 10.1

EXAMPLE 1 Does $f(x) = x^3 - 2x^2 - x + 1$ have any real zeros between 2 and 3? Use synthetic division to find $f(2)$ and $f(3)$.

$$
\begin{array}{r|rrrr}
2 & 1 & -2 & -1 & 1 \\
  &   & 2 & 0 & -2 \\
\hline
  & 1 & 0 & -1 & -1 \\
\end{array}
\qquad
\begin{array}{r|rrrr}
3 & 1 & -2 & -1 & 1 \\
  &   & 3 & 3 & 6 \\
\hline
  & 1 & 1 & 2 & 7 \\
\end{array}
$$

Since $f(2)$ is negative but $f(3)$ is positive, there is a real zero between 2 and 3.

The next theorem gives a good method for narrowing the search for real zeros. It depends on the fact that in any polynomial function, as $|x|$ gets larger and larger, so does $|y|$. This is illustrated by the typical polynomial functions sketched in Figure 10.2.

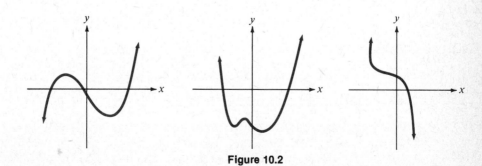

Figure 10.2

Theorem

If $f(x) = a_n x^n + a_{n-1}x^{n-1} + \cdots + a_1 x + a_0$ is a polynomial function of degree $n \geq 1$ with real coefficients and $a_n > 0$, and if $f(x)$ is divided synthetically by $x - c$, then

(a) if $c > 0$ and all numbers in the bottom row of the synthetic division are nonnegative, then c is greater than or equal to any zero of $f(x)$;

(b) if $c < 0$ and the numbers in the bottom row of the synthetic division alternate in sign (with 0 considered positive or negative, as needed), then c is less than or equal to any zero of $f(x)$.

EXAMPLE 2

☐ Approximate the real zeros of $f(x) = x^4 - 6x^3 + 8x^2 + 2x - 1$.

Let us begin to look for zeros by trying $c = -2$. To do this, we divide $f(x)$ by $x + 2$.

$$
\begin{array}{r|rrrrr}
-2 & 1 & -6 & 8 & 2 & -1 \\
   &   & -2 & 16 & -48 & 92 \\
\hline
   & 1 & -8 & 24 & -46 & 91
\end{array}
$$

The leading coefficient of $f(x)$ is positive, and the numbers in the bottom row alternate in sign. Since $-2 < 0$, we know that -2 is less than or equal to any zero of $f(x)$. If we divide $f(x)$ by $x + 1$, we get

$$
\begin{array}{r|rrrrr}
-1 & 1 & -6 & 8 & 2 & -1 \\
   &   & -1 & 7 & -15 & 13 \\
\hline
   & 1 & -7 & 15 & -13 & 12.
\end{array}
$$

By this result, -1 is also less than or equal to any real zero of $f(x)$. Note that $f(-1) = 12 > 0$, but $f(0) = -1 < 0$. Hence, there is at least one real number zero between -1 and 0.

Let us try $c = -0.5$. If we divide $f(x)$ by $x + 0.5$, we have

$$
\begin{array}{r|rrrrr}
-0.5 & 1 & -6 & 8 & 2 & -1 \\
     &   & -0.5 & 3.25 & -5.625 & 1.8125 \\
\hline
     & 1 & -6.5 & 11.25 & -3.625 & 0.8125
\end{array}
$$

Since $f(-0.5) > 0$ and $f(0) < 0$, there is a real number zero between -0.5 and 0.

Now let us try $c = -0.4$.

$$
\begin{array}{r|rrrrr}
-0.4 & 1 & -6 & 8 & 2 & -1 \\
     &   & -0.4 & 2.56 & -4.224 & 0.8896 \\
\hline
     & 1 & -6.4 & 10.56 & -2.224 & -0.1104
\end{array}
$$

Note that $f(-0.5) > 0$, but $f(-0.4) < 0$. Since $f(-0.4)$ is closer to zero than $f(-0.5)$ we are probably safe in saying that, to one decimal place of accuracy, -0.4 is a real number zero of $f(x)$. We could get further decimal places of accuracy by continuing this process. In the same way, we can show that $f(x)$ has three more real zeros. Use synthetic division to verify that, to one decimal place of accuracy, these three other zeros are 0.3, 2.4, and 3.7. ☐

Descartes' rule of signs, stated in the next theorem, gives a useful, practical test for finding the number of positive or negative real zeros of a given polynomial function. The terms of the polynomial are assumed to be in descending order.

Descartes' Rule of Signs

Let $f(x)$ be a polynomial function with real number coefficients.

(a) The number of positive real zeros of $f(x)$ is either equal to the number of variations in sign occurring in the coefficients of $f(x)$ or else is less than the number of variations by a positive even integer.

(b) The number of negative real zeros of $f(x)$ either equals the number of variations in sign of $f(-x)$ or else is less than the number of variations by a positive even integer.

When applying the theorem, any terms with 0 coefficients are deleted. A variation in sign occurs whenever two consecutive terms have opposite signs. For the purposes of this theorem, zeros of multiplicity k count as k zeros. For example,

$$f(x) = (x - 1)^4 = + x^4 - 4x^3 + 6x^2 - 4x + 1$$
$$ 1 \qquad 2 \qquad 3 \qquad 4$$

has 4 changes of sign. By Descartes' rule of signs, $f(x)$ has 4 or 2 or 0 positive real zeros. In this case, there are 4 positive real zeros each equal to 1.

The polynomial of Example 2,

$$f(x) = x^4 - 6x^3 + 8x^2 + 2x - 1,$$

has three variations in sign:

$$+ x^4 - 6x^3 + 8x^2 + 2x - 1.$$
$$ 1 \qquad 2 \qquad\qquad 3$$

Thus, by Descartes' rule of signs, $f(x)$ has either 3 or $3 - 2 = 1$ positive real zeros. We found in Example 2 that $f(x)$ has three positive real zeros. Since $f(x)$ is of degree 4 and has 3 positive real zeros, we know it must have 1 negative real zero, which corresponds to the result above. We could also verify this with part (b) of Descartes' rule of signs. Note that

$$f(-x) = (-x)^4 - 6(-x)^3 + 8(-x)^2 + 2(-x) - 1$$
$$= x^4 + 6x^3 + 8x^2 - 2x - 1$$

has only one variation in sign. Hence, $f(x)$ has only one negative real zero, which again corresponds to our result from above.

EXAMPLE 3 ☐ Find the number of positive and negative real zeros of

$$g(x) = x^5 + 5x^4 + 3x^2 + 2x + 1.$$

The polynomial $g(x)$ has no variations in sign and hence has no positive real zeros. Here

$$g(-x) = -x^5 + 5x^4 + 3x^2 - 2x + 1,$$

which has 3 variations in sign, so that $g(x)$ has either 3 or 1 negative real zeros. The other zeros are complex numbers. ▣

EXAMPLE 4

▣ Graph $f(x) = x^4 - 6x^3 + 8x^2 + 2x - 1$.

In Example 1, we found that the approximate zeros of this function are -0.4, 0.3, 2.4, and 3.7. These four zeros divide the x-axis into five regions. Using synthetic division to find a function value in each region gives the following information.

Region	Sign of $f(x)$	Location relative to x-axis
$x < -0.4$	+	above
$-0.4 < x < 0.3$	−	below
$0.3 < x < 2.4$	+	above
$2.4 < x < 3.7$	−	below
$x > 3.7$	+	above

Plot the four zeros together with additional ordered pairs as necessary and connect them with a smooth curve. The result is the graph of Figure 10.3. ▣

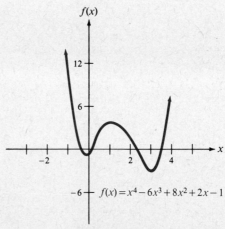

$$f(x) = x^4 - 6x^3 + 8x^2 + 2x - 1$$

Figure 10.3

Use the Intermediate Value Theorem for Polynomials to show that the following polynomial functions have a real zero between the numbers given

1. $f(x) = x^3 + 3x^2 - 2x - 6$; 1 and 2
2. $f(x) = x^3 + x^2 - 5x - 5$; 2 and 3

3. $f(x) = 2x^3 - 8x^2 + x + 16$; 2 and 2.5
4. $f(x) = 3x^3 + 7x^2 - 4$; 1/2 and 1
5. $f(x) = 2x^4 - 4x^2 + 3x - 6$; 1.5 and 2
6. $f(x) = x^4 - 4x^3 - x + 1$; .3 and 1
7. $f(x) = -3x^4 - x^3 + 2x^2 + 4$; 1 and 1.2
8. $f(x) = x^5 + 2x^4 + x^3 + 3$; −2 and −1.3

Show that the real zeros of each of the following polynomial functions satisfy the given conditions.

9. $f(x) = x^4 - x^3 + 3x^2 - 8x + 8$;
no real zeros greater than 2

10. $f(x) = 2x^5 - x^4 + 2x^3 - 2x^2 + 4x - 4$;
no real zero greater than 1

11. $f(x) = x^4 + x^3 - x^2 + 3$;
no real zero less than -2

12. $f(x) = x^5 + 2x^3 - 2x^2 + 5x + 5$;
no real zero less than -1

13. $f(x) = 3x^4 + 2x^3 - 4x^2 + x - 1$;
no real zero greater than 1

14. $f(x) = 3x^4 + 2x^3 - 4x^2 + x - 1$;
no real zero less than -2

15. $f(x) = x^5 - 3x^3 + x + 2$;
no real zero greater than 2

16. $f(x) = x^5 - 3x^3 + x + 2$;
no real zero less than -3

▣ *For each of the following polynomial functions (a) find the number of positive and negative real zeros, (b) approximate each zero as a decimal to the nearest tenth, and (c) graph the function.*

17. $f(x) = x^3 + 3x^2 - 2x - 6$

18. $f(x) = x^3 + x^2 - 5x - 5$

19. $f(x) = x^3 - 4x^2 - 5x + 14$

20. $f(x) = x^3 + 9x^2 + 34x + 13$

21. $f(x) = x^3 + 6x - 13$

22. $f(x) = x^3 - 3x^2 + 4x - 5$

23. $f(x) = x^4 - 8x^3 + 17x^2 - 2x - 14$

24. $f(x) = x^4 - 4x^3 - x^2 + 8x - 2$

25. A technique for measuring cardiac output depends on the concentration of a dye in the blood after a known amount is injected into a vein near the heart. In a normal heart, the concentration of the dye at time x (in seconds) is given by the function

$$g(x) = -0.006x^4 + 0.140x^3 - 0.053x^2 + 1.79x.$$

Graph $g(x)$.

▣ *The following polynomial functions have zeros in the given intervals. Approximate these zeros to the nearest hundredth.*

26. $f(x) = x^4 + x^3 - 6x^2 - 20x - 16$;
[3.2, 3.3] and [−1.4, −1.1]

27. $f(x) = x^4 - 2x^3 - 2x^2 - 18x + 5$;
[0.2, 0.4] and [3.7, 3.8]

28. $f(x) = x^4 - 4x^3 - 20x^2 + 32x + 12$;
[−4, −3], [−1, 0], [1, 2], and [6, 7]

29. $f(x) = x^4 - 4x^3 - 44x^2 + 160x - 80$;
[−7, −6], [0, 1], [2, 3], and [7, 8]

30. $f(x) = x^4 - 4x^3 - 4x^2 + 20x - 5$;
[−3, −2], [0, 1], [2, 3], [3, 4]

31. Suppose a polynomial $P(x)$ has only rational zeros. Show how to express $P(x)$ with only rational coefficients; with only irrational coefficients.

32. Let $P(x) = x^n + a_{n-1}x^{n-1} + a_{n-2}x^{n-2} + \cdots + a_1 x + a_0$. Let m be the largest of the absolute values of the a's. Show that all real zeros of $P(x)$ are no greater than $m + 1$.

10.5 Partial Fractions

In Section 1.6 we found sums of rational expressions by combining two or more rational expressions into one rational expression. Here we consider the reverse problem. Given one rational expression, we want to express it as the sum of two or more rational expressions. The sum of rational expressions is called the **partial fraction decomposition**; each term in the sum is a **partial fraction.**

To form a partial fraction decomposition of a rational expression, use the following steps.

Partial Fraction Decomposition of $\dfrac{f(x)}{g(x)}$

1. If $f(x)/g(x)$ is not a proper fraction (a fraction with the numerator of lower degree than the denominator), divide $f(x)$ by $g(x)$. For example,

$$\frac{x^4 - 3x^3 + x^2 + 5x}{x^2 + 3} = x^2 - 3x - 2 + \frac{14x + 6}{x^2 + 3}.$$

Then we would apply the remaining steps to the remainder, which is a proper fraction.

2. Factor $g(x)$ completely into factors of the form $(ax + b)^m$ or $(cx^2 + dx + e)^n$, where $cx^2 + dx + e$ is prime.

3. For each factor of the form $(ax + b)^m$ the decomposition must include the terms

$$\frac{A_1}{ax + b} + \frac{A_2}{(ax + b)^2} + \cdots + \frac{A_m}{(ax + b)^m}.$$

4. For each factor of the form $(cx^2 + dx + e)^n$, the decomposition must include the terms

$$\frac{B_1x + C_1}{cx^2 + dx + e} + \frac{B_2x + C_2}{(cx^2 + dx + e)^2} + \cdots + \frac{B_nx + C_n}{(cx^2 + dx + e)^n}.$$

5. Use algebraic techniques to solve for the constants in the numerators of the decomposition.

The following examples illustrate an algebraic technique used to find the constants in step 5.

Distinct Linear Factors

EXAMPLE 1 ▢ Find the partial fraction decomposition of

$$\frac{5x - 2}{x^3 - 4x}.$$

The denominator can be factored as $x^3 - 4x = x(x + 2)(x - 2)$. Since the factors are distinct linear factors, use step 3 to write the decomposition as

$$\frac{5x - 2}{x^3 - 4x} = \frac{A}{x} + \frac{B}{x + 2} + \frac{C}{x - 2}, \tag{1}$$

where A, B, and C are constants that we must find. Multiply both sides of equation (1) by $x(x + 2)(x - 2)$, getting

$$5x - 2 = A(x + 2)(x - 2) + Bx(x - 2) + Cx(x + 2). \tag{2}$$

Equation (1) is an identity, since both sides represent the same rational expression. Thus, equation (2) is also an identity. Equation (1) holds for all values of x except 0, -2, and 2. However, equation (2) holds for all values of x. In particular, if we substitute 0 for x in equation (2), we have

$$-2 = -4A, \quad \text{so that} \quad A = \frac{1}{2}.$$

Similarly, if $x = -2$, then

$$-12 = 8B, \quad \text{so that} \quad B = -\frac{3}{2}.$$

Finally, if $x = 2$, then

$$8 = 8C, \quad \text{so that} \quad C = 1.$$

Thus, the rational expression can be written as the sum of partial fractions.

$$\frac{5x - 2}{x^3 - 4x} = \frac{1}{2x} + \frac{-3}{2(x + 2)} + \frac{1}{x - 2} \quad \blacksquare$$

Repeated Linear Factors

EXAMPLE 2   Find the partial fraction decomposition of

$$\frac{2x}{(x - 1)^3}.$$

The denominator is already factored. Using Step 3 above, write the decomposition as

$$\frac{2x}{(x - 1)^3} = \frac{A}{x - 1} + \frac{B}{(x - 1)^2} + \frac{C}{(x - 1)^3}.$$

Clear the denominators by multiplying both sides of this equation by $(x - 1)^3$.

$$2x = A(x - 1)^2 + B(x - 1) + C$$

By substituting 1 for x, we find that $C = 2$, so that

$$2x = A(x - 1)^2 + B(x - 1) + 2. \tag{3}$$

We have substituted the only root and we still need to find values for A and B. However, we can substitute *any* number for x. For example, if we let $x = -1$, equation (3) becomes

$$-2 = 4A - 2B + 2$$
$$-4 = 4A - 2B$$
$$-2 = 2A - B; \tag{4}$$

if we substitute 0 for x, we have

$$0 = A - B + 2$$
$$2 = -A + B. \tag{5}$$

Now solve the system of equations (4) and (5) to get $A = 0$ and $B = 2$. The partial fraction decomposition is

$$\frac{2x}{(x-1)^3} = \frac{2}{(x-1)^2} + \frac{2}{(x-1)^3}.$$

We had to make three substitutions because there were three constants to evaluate, A, B, and C.

To check this result, combine terms on the right. ■

Distinct Linear and Quadratic Factors

EXAMPLE 3 ■ Find the partial fraction decompostion of

$$\frac{x^2 + 3x - 1}{(x+1)(x^2-2)}.$$

This denominator has one linear factor and one quadratic factor where neither is repeated. By step 3 above, the partial fraction decomposition is

$$\frac{x^2 + 3x - 1}{(x+1)(x^2-2)} = \frac{A}{x+1} + \frac{Bx+C}{x^2-2}.$$

Multiply both sides by $(x+1)(x^2-2)$ to get

$$x^2 + 3x - 1 = A(x^2 - 2) + (Bx + C)(x + 1). \tag{6}$$

First substitute -1 for x to get

$$(-1)^2 + 3(-1) - 1 = A[(-1)^2 - 2] + 0$$
$$-3 = -A$$
$$A = 3.$$

Replace A with 3 in equation (6) and choose any number to substitute for x. If $x = 0$,

$$0^2 + 3(0) - 1 = 3(0^2 - 2) + (B \cdot 0 + C)(0 + 1)$$
$$-1 = -6 + C$$
$$C = 5.$$

Now, letting $A = 3$ and $C = 5$, substitute again in equation (6) using another number for x. For $x = 1$, we get

$$3 = -3 + (B + 5)(2)$$
$$6 = 2B + 10$$
$$B = -2.$$

Using $A = 3$, $B = -2$, and $C = 5$, the partial fraction decomposition is

$$\frac{x^2 + 3x - 1}{(x + 1)(x^2 - 2)} = \frac{3}{x + 1} + \frac{-2x + 5}{x^2 - 2}.$$

Again, this work can be checked by combining terms on the right. ☐

For quadratic factors another method is often more convenient. For instance, in Example 3, after multiplying both sides of the equation by the common denominator, we had

$$x^2 + 3x - 1 = A(x^2 - 2) + (Bx + C)(x + 1).$$

If we multiply on the right and collect like terms, we get

$$x^2 + 3x - 1 = Ax^2 - 2A + Bx^2 + Bx + Cx + C$$
$$x^2 + 3x - 1 = (A + B)x^2 + (B + C)x + (C - 2A).$$

Now, equating the coefficients of like powers of x gives the three equations

$$1 = A + B$$
$$3 = B + C$$
$$-1 = C - 2A$$

By solving this system of equations for A, B, and C, we can write the partial fraction decomposition. The last example uses a combination of the two methods.

Repeated Quadratic Factors

EXAMPLE 4 ☐ Find the partial fraction decomposition of

$$\frac{2x}{(x^2 + 1)^2(x - 1)}.$$

Here we have a linear factor and a repeated quadratic factor. By step 3 from the box above,

$$\frac{2x}{(x^2 + 1)^2(x - 1)} = \frac{Ax + B}{x^2 + 1} + \frac{Cx + D}{(x^2 + 1)^2} + \frac{E}{x - 1}.$$

Multiplication of both sides by $(x^2 + 1)^2(x - 1)$ leads to

$$2x = (Ax + B)(x^2 + 1)(x - 1) + (Cx + D)(x - 1) + E(x^2 + 1)^2. \quad (7)$$

If $x = 1$, equation (7) reduces to

$$2 = 4E,$$

or $E = \dfrac{1}{2}.$

Substituting $1/2$ for E in equation (7) and combining terms on the right gives

$$2x = (A + 1/2)x^4 + (-A + B)x^3 + (A - B + C + 1)x^2 +$$
$$(-A + B + D - C)x + (-B - D + 1/2).$$

By setting corresponding coefficients equal, we first get

$$0 = A + \dfrac{1}{2} \quad \text{or} \quad A = -\dfrac{1}{2}.$$

From the corresponding coefficients of x^3,

$$0 = -A + B.$$

Since $A = -1/2$, we have

$$B = -\dfrac{1}{2}.$$

Using the coefficients of x^2,

$$0 = A - B + C + 1.$$

Since $A = -1/2$ and $B = -1/2$, we find

$$C = -1.$$

Finally, from the coefficients of x,

$$2 = -A + B + D - C.$$

Substituting for A, B, and C gives

$$D = 1.$$

Using $A = -1/2$, $B = -1/2$, $C = -1$, $D = 1$, and $E = 1/2$, the given fraction has the partial fraction decomposition

$$\frac{2x}{(x^2 + 1)^2(x - 1)} = \frac{-\frac{1}{2}x - \frac{1}{2}}{x^2 + 1} + \frac{-x + 1}{(x^2 + 1)^2} + \frac{\frac{1}{2}}{x - 1}$$

or $$\frac{2x}{(x^2 + 1)^2(x - 1)} = \frac{-(x + 1)}{2(x^2 + 1)} + \frac{-x + 1}{(x^2 + 1)^2} + \frac{1}{2(x - 1)}. \quad \blacksquare$$

The two methods discussed in this section are summarized here.

> *Method 1* For linear factors
>
> **1.** Multiply both sides of the rational expression by the common denominator.
>
> **2.** Substitute the root of each factor in the resulting equation. For repeated linear factors, substitute as many other numbers as necessary to find all the constants in the numerators.
>
> *Method 2* For quadratic factors
>
> **1.** Multiply both sides of the rational expression by the common denominator.
>
> **2.** Collect terms on the right side of the resulting equation.
>
> **3.** Equate the coefficients of like terms to get a system of equations.
>
> **4.** Solve the system to find the constants in the numerators.

EXERCISES 10.5

Find the partial fraction decomposition for the following.

1. $\dfrac{3x - 1}{x(x + 1)}$

2. $\dfrac{5}{3x(2x + 1)}$

3. $\dfrac{2}{x(x + 3)}$

4. $\dfrac{1}{3x(x - 1)}$

5. $\dfrac{x + 2}{(x + 1)(x - 1)}$

6. $\dfrac{4x + 2}{(x + 2)(2x - 1)}$

7. $\dfrac{-7}{x(3x - 1)(x + 1)}$

8. $\dfrac{-2}{x^2(x - 2)}$

9. $\dfrac{2}{x^2(x + 3)}$

10. $\dfrac{2x}{(x + 1)(x + 2)^2}$

11. $\dfrac{x - 1}{(x + 2)(x - 3)^2}$

12. $\dfrac{x + 2}{(x - 1)^2(x + 5)}$

13. $\dfrac{2x + 1}{(x + 2)^3}$

14. $\dfrac{4x^2 - x - 15}{x(x + 1)(x - 1)}$

15. $\dfrac{3}{x^2 + 4x + 3}$

16. $\dfrac{1}{x^2 + 5x + 4}$

17. $\dfrac{x}{2x^2 + 5x + 2}$

18. $\dfrac{2x}{3x^2 + 2x - 1}$

19. $\dfrac{x^3 + 2}{x^3 - 3x^2 + 2x}$

20. $\dfrac{x^3 + 4}{9x^3 - 4x}$

21. $\dfrac{5}{x(x^2 + 3)}$

22. $\dfrac{-3}{x^2(x^2 + 5)}$

23. $\dfrac{2x + 1}{(x + 1)(x^2 + 2)}$

24. $\dfrac{3x - 2}{(x + 4)(3x^2 + 1)}$

25. $\dfrac{3}{x(x + 1)(x^2 + 1)}$

26. $\dfrac{1}{x(2x + 1)(3x^2 + 4)}$

27. $\dfrac{x^4 + 1}{x(x^2 + 1)^2}$

28. $\dfrac{3x - 1}{x(2x^2 + 1)^2}$

29. $\dfrac{3x^4 + x^3 + 5x^2 - x + 4}{(x - 1)(x^2 + 1)^2}$

30. $\dfrac{-x^4 - 8x^2 + 3x - 10}{(x + 2)(x^2 + 4)^2}$

31. $\dfrac{11x - 24}{5(x^2 + x - 2)}$ 32. $\dfrac{25x - 25}{3(x^2 - 2x - 15)}$

33. $\dfrac{2x^2}{x^4 - 1}$ 34. $\dfrac{3x}{x^4 + x^2}$

35. $\dfrac{4x^2 + 13x - 9}{x^3 + 2x^2 - 3x}$ 36. $\dfrac{2x^2 - 12x + 4}{x^3 - 4x^2}$

Find constants a, b, and c so that the following

statements are true. (Hint: multiply to eliminate denominators and then equate like powers of x.)

37. $\dfrac{x - 1}{x^3 - x^2 - 4x} = \dfrac{a}{x} + \dfrac{bx + c}{x^2 - x - 4}$

38. $\dfrac{x + 2}{x^3 + 2x^2 + x} = \dfrac{a}{x} + \dfrac{bx + c}{x^2 + 2x + 1}$

39. $\dfrac{x + 3}{x^3 + 64} = \dfrac{a}{x + 4} + \dfrac{bx + c}{x^2 - 4x + 16}$

40. $\dfrac{2x + 1}{(x + 1)(x^2 - 3x + 5)} = \dfrac{a}{x + 1} + \dfrac{bx + c}{x^2 - 3x + 5}$

CHAPTER 10   SUMMARY

Key Words

division algorithm	synthetic division	Fundamental Theorem of Algebra
dividend	remainder theorem	indirect proof
divisor	factor theorem	multiplicity
quotient	zero	Descartes' rule of signs
remainder	nested form	partial fractions

CHAPTER 10   REVIEW EXERCISES

For each of the following divisions, find $q(x)$ and $r(x)$.

1. $\dfrac{4x^3 - 2x^2 + 3x + 5}{2x - 1}$

2. $\dfrac{3x^3 + 5x^2 - 5x + 2}{3x + 2}$

3. $\dfrac{x^4 - 3x^2 + 5x - 1}{x^2 + 2}$

4. $\dfrac{2x^4 - x^3 + x^2 + 5x}{2x^2 - x}$

Use synthetic division to find $q(x)$ and r for each of the following.

5. $\dfrac{2x^3 + 3x^2 - 4x + 1}{x - 1}$

6. $\dfrac{4x^3 + 3x^2 + 3x + 5}{x - 5}$

7. $\dfrac{2x^3 - 4x + 6}{x - 2}$

8. $\dfrac{3x^4 - x^2 + x - 1}{x + 1}$

Use synthetic division to find $f(2)$ for each of the following.

9. $f(x) = x^3 + 3x^2 - 5x + 1$

10. $f(x) = 2x^3 - 4x^2 + 3x - 10$

11. $f(x) = 5x^4 - 12x^2 + 2x - 8$

12. $f(x) = x^5 - 3x^2 + 2x - 4$

Find a polynomial of lowest degree having the following zeros.

13. $-1, 4, 7$

14. $8, 2, 3$

15. $-\sqrt{7}, \sqrt{7}, 2, -1$

16. $1 + \sqrt{5}, 1 - \sqrt{5}, -4, 1$

17. Is -1 a zero of $f(x) = 2x^4 + x^3 - 4x^2 + 3x + 1$?

18. Is -2 a zero of $f(x) = 2x^4 + x^3 - 4x^2 + 3x + 1$?

19. Is $x + 1$ a factor of $f(x) = x^3 + 2x^2 + 3x - 1$?

20. Is $x + 1$ a factor of $f(x) = 2x^3 - x^2 + x + 4$?

21. Find a polynomial of degree 3 with -2, 1, and 4 as zeros, and $f(2) = 16$.

22. Find a polynomial of degree 4 with 1, -1, and $3i$ as zeros, and $f(2) = 39$.

23. Find a lowest-degree polynomial with real coefficients having zeros 2, -2, and $-i$.

24. Find a lowest-degree polynomial with real coefficients having zeros of 2, -3, and $5i$.

25. Find the polynomial of lowest degree with real coefficients having -3 and $1 - i$ as zeros.

26. Find all zeros of $f(x) = x^4 - 3x^3 - 8x^2 + 22x - 24$, given that $1 - i$ is a zero.

27. Find all zeros of $f(x) = x^4 - 6x^3 + 14x^2 - 24x + 40$, given that $3 + i$ is a zero.

28. Find all zeros of $f(x) = x^4 + x^3 - x^2 + x - 2$, given that 1 is a zero.

Find all rational zeros of the following.

29. $f(x) = 2x^3 - 9x^2 - 6x + 5$

30. $f(x) = 3x^3 - 10x^2 - 27x + 10$

31. $f(x) = x^3 - \dfrac{17}{6}x^2 - \dfrac{13}{3}x - \dfrac{4}{3}$

32. $f(x) = 8x^4 - 14x^3 - 29x^2 - 4x + 3$

33. $f(x) = -x^4 + 2x^3 - 3x^2 + 11x - 7$

34. $f(x) = x^5 - 3x^4 - 5x^3 + 15x^2 + 4x - 12$

35. Use a polynomial to show that $\sqrt{11}$ is irrational.

36. Show that $f(x) = x^3 - 9x^2 + 2x - 5$ has no rational zeros.

Show that the following polynomials have real zeros satisfying the given conditions.

37. $f(x) = 3x^3 - 8x^2 + x + 2$, zero in $[-1, 0]$ and $[2, 3]$

38. $f(x) = 4x^3 - 37x^2 + 50x + 60$, zero in $[2, 3]$ and $[7, 8]$

39. $f(x) = x^3 + 2x^2 - 22x - 8$, zero in $[-1, 0]$ and $[-6, -5]$

40. $f(x) = 2x^4 - x^3 - 21x^2 + 51x - 36$ has no real zero greater than 4

41. $f(x) = 6x^4 + 13x^3 - 11x^2 - 3x + 5$ has no zero greater than 1 or less than -3

Approximate the real zeros of each of the following as a decimal to the nearest tenth. Then graph the function.

42. $f(x) = 2x^3 - 11x^2 - 2x + 2$

43. $f(x) = x^4 - 4x^3 - 5x^2 + 14x - 15$

44. (a) Find the number of positive and negative zeros of $f(x) = x^3 + 3x^2 - 4x - 2$.

   (b) Show that $f(x)$ has a zero between -4 and -3. Approximate this zero to the nearest tenth.

   (c) Graph $f(x)$.

Write each of the following as the sum of fractions.

45. $\dfrac{5x - 2}{x^2 - 4}$

46. $\dfrac{1}{x^3 + 3x^2}$

47. $\dfrac{8}{x^2(x^2 + 2)}$

48. $\dfrac{3x - 1}{(x + 1)^2(x^2 + 5)}$

11 Further Topics in Algebra

11.1 Mathematical Induction

Many results in mathematics are claimed to be true for every positive integer. Any of these results could be checked for $n = 1$, $n = 2$, $n = 3$, and so on, but since the set of positive integers is infinite it would be impossible to check every possible case. For example, let S_n represent the statement that the sum of the first n positive integers is $n(n + 1)/2$.

$$S_n: 1 + 2 + 3 + \cdots + n = \frac{n(n + 1)}{2}.$$

We can easily check the truth of this statement for the first few values of n:

If $n = 1$, then S_1 is $\qquad\qquad 1 = \dfrac{1(1 + 1)}{2}$ which is true.

If $n = 2$, then S_2 is $\qquad\quad 1 + 2 = \dfrac{2(2 + 1)}{2}$ which is true.

If $n = 3$, then S_3 is $\qquad 1 + 2 + 3 = \dfrac{3(3 + 1)}{2}$ which is true.

If $n = 4$, then S_4 is $\quad 1 + 2 + 3 + 4 = \dfrac{4(4 + 1)}{2}$ which is true.

We could continue in this way as long as we wanted, but we could still never prove that S_n is true for *every* positive integer value of n. To prove that such statements are true for every positive integer value of n, we often use the following principle.

Principle of Mathematical Induction

Let S_n be a statement concerning the positive integer n. Suppose that

(a) S_1 is true

(b) If S_n is true for the positive integer k, then S_n is also true for the integer $k + 1$. (That is, the truth of S_k implies the truth of S_{k+1}.)

Then S_n is true for every positive integer value of n.

A proof by mathematical induction can be explained as follows. By (a) above, the statement is true when $n = 1$. By (b) above, the fact that the statement is true for $n = 1$ implies that it is true for $n = 1 + 1 = 2$. Using (b) again, it is thus true for $2 + 1 = 3$, for $3 + 1 = 4$, for $4 + 1 = 5$, and so on. By continuing in this way, the statement must be true for *every* positive integer, no matter how large.

The situation is similar to that of a number of dominoes lined up as shown in Figure 11.1. If the first domino is pushed over, it pushes the next, which pushes the next, and so on until all are down.

Figure 11.1

Another example of the principle of mathematical induction might be an infinite ladder. Suppose the rungs are spaced so that, whenever you are on a rung, you know you can move to the next rung. Then *if* you can get to the first rung, you can go as high up the ladder as you wish.

Two separate steps are required for a proof by mathematical induction:

1. Prove that the statement S_n is true for $n = 1$.

2. Assume that S_n is true for the positive integer k. Show that this implies that S_n is also true for the positive integer $k + 1$.

In the next example we use mathematical induction to prove the statement S_n mentioned at the beginning of this section.

EXAMPLE 1

Let S_n represent the statement

$$1 + 2 + 3 + \cdots + n = \frac{n(n + 1)}{2}.$$

Prove that S_n is true for every positive integer n.

The proof by mathematical induction is as follows.

Step 1 Show that the statement is true when $n = 1$. If $n = 1$, S_1 becomes

$$1 = \frac{1(1 + 1)}{2}$$

which is true.

Step 2 Assume that S_n is true for the statement $n = k$. That is, assume that

$$1 + 2 + 3 + \cdots + k = \frac{k(k + 1)}{2}$$

is true. Show that this implies the truth of S_{k+1}, where S_{k+1} is the statement

$$1 + 2 + 3 + \cdots + k + (k + 1) = \frac{(k + 1)[(k + 1) + 1]}{2}.$$

We have assumed that

$$1 + 2 + 3 + \cdots + k = \frac{k(k + 1)}{2}.$$

By adding $k + 1$ to both sides of the known equation, we have

$$1 + 2 + 3 + \cdots + k + (k + 1) = \frac{k(k + 1)}{2} + (k + 1).$$

Then, factoring on the right gives

$$= (k + 1)\left(\frac{k}{2} + 1\right)$$

$$= (k + 1)\left(\frac{k + 2}{2}\right)$$

$$1 + 2 + 3 + \cdots + k + (k + 1) = \frac{(k + 1)[(k + 1) + 1]}{2}.$$

This final result is the statement we wished to establish for $n = k + 1$; therefore the truth of S_k implies the truth of S_{k+1}. The two steps required for a proof by mathematical induction have now been completed, so that our statement S_n is true for every positive integer value of n. ☐

EXAMPLE 2 ☐ Prove that if x is a real number between 0 and 1 for every positive integer n,

$$0 < x^n < 1.$$

Here S_1 is: if $0 < x < 1$, then $0 < x^1 < 1$, which is true. Assume now that S_k is true:

if $0 < x < 1$, then $0 < x^k < 1$.

We must now show that this implies the truth of S_{k+1}. Multiply all members of $0 < x^k < 1$ by x to get

$$x \cdot 0 < x \cdot x^k < x \cdot 1.$$

(Here we use the fact that $0 < x$.) Simplify to get

$$0 < x^{k+1} < x.$$

We know that $x < 1$. Thus,

$$x^{k+1} < x < 1,$$

and $0 < x^{k+1} < 1$.

By this work, the truth of S_k implies the truth of S_{k+1}, so that the given statement is true for every positive integer n. ▢

Some statements S_n are not true for the first few values of n, but are true for all values of n that are at least equal to some fixed integer j. The following slightly generalized form of the Principle of Mathematical Induction takes care of these cases.

Generalized Principle of Mathematical Induction

Let S_n be a statement concerning the positive integer n. Let j be a fixed positive integer. Suppose that

(a) S_j is true

(b) If S_n is true for the positive integer k, where $k \geq j$, then S_n is also true for the positive integer $k + 1$.

Then S_n is true for all positive integers n, where $n \geq j$.

EXAMPLE 3   ▢   Let S_n represent the statement

$$2^n > 2n + 1.$$

Show that S_n is true for all values of n such that $n \geq 3$.
   (Check that S_n is false for $n = 1$ and $n = 2$.) As before, our proof requires two steps.
Step 1   Show that S_n is true for $n = 3$. If $n = 3$, S_n is

$$2^3 > 2 \cdot 3 + 1$$

or $8 > 7$

which is true.
Step 2   Assume that S_n is true for some positive integer k, where $k \geq 3$. That is, assume the truth of

$$S_k: 2^k > 2k + 1.$$

We must now show that the truth of S_k implies the truth of S_{k+1}, or

S_{k+1}: $2^{k+1} > 2(k + 1) + 1$.

Our assumption is that $2^k > 2k + 1$. Multiply both sides by 2, obtaining

$2 \cdot 2^k > 2(2k + 1)$

or $2^{k+1} > 4k + 2$.

Rewrite $4k + 2$ as $2(k + 1) + 2k$, giving

$2^{k+1} > 2(k + 1) + 2k$.   (1)

Since k is a positive integer,

$2k > 1$.   (2)

Adding $2(k + 1)$ to both sides of inequality (2) gives

$2(k + 1) + 2k > 2(k + 1) + 1$.   (3)

From inequalities (1) and (3),

$2^{k+1} > 2(k + 1) + 2k > 2(k + 1) + 1$.

or $2^{k+1} > 2(k + 1) + 1$.

Thus, the truth of S_k implies the truth of S_{k+1}, and this, together with the fact that S_3 is true, shows that S_n is true for every positive integer value of n greater than or equal to 3. ☐

EXAMPLE 4 ☐ There is a flaw in the following "proof." See if you can spot it.

Prove: In every set of n dogs, if at least one is male, then they are all male.

S_1 is obviously true: in a set of one dog, if one is male then all are male. Assume that S_k is true. Let us show that S_{k+1} is true. Suppose

$\{D_1, D_2, \ldots, D_k, D_{k+1}\}$

is any set of $k + 1$ dogs, and suppose that one of them is male; say D_1 is male (if it happens that the male was D_2 or D_6, or whatever, just rearrange them so that D_1 is a male). Thus $\{D_1, D_2, \ldots, D_k\}$ is a set of k dogs and one of them is a male, so *all* are males, because S_k is assumed true. But this means that $\{D_2, D_3, \ldots, D_k, D_{k+1}\}$ is now a set of k dogs (we left D_1 out) and one is male (in fact, $D_2, \ldots, D_k$ are males) so they *all* are, including D_{k+1}. Thus, all $k + 1$ dogs are male. This completes our induction argument. (The error appears as a footnote* below.) ☐

*Answer to the question posed in Example 4: The deduction that S_k implies S_{k+1} is correct whenever $k \geq 2$. However, *it does not hold for $k = 1$*; that is, the argument fails to prove that if S_1 is true, then S_2 is also true.

EXERCISES
11.1

Write out in full and verify each of the statements $S_1, S_2, S_3, S_4,$ and S_5 for each of the following. Then use mathematical induction to prove that each of the given statements is true for every positive integer n.

1. $2 + 4 + 6 + \cdots + 2n = n(n + 1)$

2. $1 + 3 + 5 + \cdots + (2n - 1) = n^2$

Use the method of mathematical induction to prove that each of the following statements is true for every positive integer n.

3. $2 + 4 + 8 + \cdots + 2^n = 2^{n-1} - 2$

4. $1^2 + 2^2 + 3^2 + \cdots + n^2 =$
$$\frac{n(n + 1)(2n + 1)}{6}$$

5. $1^3 + 2^3 + 3^3 + \cdots + n^3 = \dfrac{n^2(n + 1)^2}{4}$

6. $3 + 3^2 + 3^3 + \cdots + 3^n = \dfrac{3(3^n - 1)}{2}$

7. $5 \cdot 6 + 5 \cdot 6^2 + 5 \cdot 6^3 + \cdots + 5 \cdot 6^n =$
$6(6^n - 1)$

8. $\dfrac{1}{1 \cdot 2} + \dfrac{1}{2 \cdot 3} + \dfrac{1}{3 \cdot 4} + \cdots + \dfrac{1}{n(n + 1)} =$
$$\frac{n}{n + 1}$$

9. $\dfrac{1}{1 \cdot 4} + \dfrac{1}{4 \cdot 7} + \dfrac{1}{7 \cdot 10} + \cdots +$
$$\frac{1}{(3n - 2)(3n + 1)} = \frac{n}{3n + 1}$$

10. $\dfrac{1}{2} + \dfrac{1}{2^2} + \dfrac{1}{2^3} + \cdots + \dfrac{1}{2^n} = 1 - \dfrac{1}{2^n}$

11. $1 \cdot 2 + 2 \cdot 3 + 3 \cdot 4 + \cdots + n(n + 1) =$
$$\frac{n(n + 1)(n + 2)}{3}$$

12. $1 \cdot 4 + 2 \cdot 9 + 3 \cdot 16 + \cdots + n(n + 1)^2 =$
$$\frac{n(n + 1)(n + 2)(3n + 5)}{12}$$

13. $x^{2n} + x^{2n-1}y + \cdots + xy^{2n-1} + y^{2n} =$
$$\frac{x^{2n+1} - y^{2n+1}}{x - y}$$

14. $x^{2n-1} + x^{2n-2}y + \cdots + xy^{2n-2} + y^{2n-1} =$
$$\frac{x^{2n} - y^{2n}}{x - y}$$

15. $(a^m)^n = a^{mn}$ (Assume a and m are constant.)

16. $(ab)^n = a^n b^n$ (Assume a and b are constant.)

17. $2^n > 2n$ if $n \geq 3$

18. $3^n > 2n + 1$, if $n \geq 2$

19. $3n^3 + 6n$ is divisible by 9

20. $n^2 + n$ is divisible by 2

21. $3^{2n} - 1$ is divisible by 8

22. If $a > 1$, then $a^n > 1$

23. If $a > 1$, then $a^n > a^{n-1}$

24. If $0 < a < 1$, then $a^n < a^{n-1}$

25. $2^n > n^2$, for $n > 4$

26. If $n \geq 4$, $n! > 2^n$, where $n! =$
$n(n - 1)(n - 2) \cdots (3)(2)(1)$

27. $4^n > n^4$, for $n \geq 5$

28. $(1 + x)^n \geq 1 + nx$ for every $n > 1$ and fixed $x \geq -1$

29. What is wrong with the following "proof" by mathematical induction?

Prove: Any natural number equals the next natural number.

To begin, we assume the statement true for some natural number k:

$k = k + 1$.

We must now show that the number is true for $n = k + 1$. If we add 1 to both sides, we have

$k + 1 = k + 1 + 1$
$k + 1 = k + 2$.

Hence, if the statement is true for $n = k$, it is also true for $n = k + 1$. Thus, the theorem is proved.

30. In the country of Pango, the government prints only three-glok banknotes and five-glok notes. Prove that any purchase of 8 gloks or more can be paid for with only three-glok notes and five-glok notes. (Hint: Consider two cases: k gloks being paid with only three-glok notes, and k gloks requiring at least one five-glok note.)

31. Suppose that n straight lines (with $n \geq 2$) are drawn in a plane, where no two lines are parallel and no three lines pass through the same point. Show that the number of points of intersection of the lines is $(n^2 - n)/2$.

32. The series of sketches below starts with an equilateral triangle having sides of length 1. In the following steps, equilateral triangles are constructed on each side of the preceding figure. The lengths of the sides of these new triangles is 1/3 the length of the sides of the preceding triangles. Develop a formula for the number of sides of the nth figure. Use mathematical induction to prove your answer.

33. Find the perimeter of the nth figure in Exercise 32.

34. Show that the area of the nth figure in Exercise 32 is

$$\sqrt{3}\left[\frac{2}{5} - \frac{3}{20}\left(\frac{4}{9}\right)^{n-1}\right].$$

35. A pile of n rings, each smaller than the one below it, is on a peg. Two other pegs are attached to a board with this peg. In the game called the *Tower of Hanoi* puzzle, all the rings must be moved to a different peg, with only one ring moved at a time, and with no ring ever placed on top of a smaller ring. Find the least number of moves that would be required. Prove your result with mathematical induction.

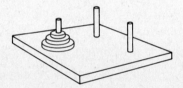

36. Prove De Moivre's theorem: for every positive integer n, $[r(\cos \theta + i \sin \theta)]^n = r^n(\cos n\theta + i \sin n\theta)$.

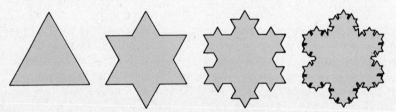

11.2 The Binomial Theorem

In this section, we look at a method of expanding the binomial expression $(x + y)^n$. We begin by listing expansions of $(x + y)^n$ for various positive integer values of n. For example,

$$(x + y)^1 = x + y$$
$$(x + y)^2 = x^2 + 2xy + y^2$$
$$(x + y)^3 = x^3 + 3x^2y + 3xy^2 + y^3$$
$$(x + y)^4 = x^4 + 4x^3y + 6x^2y^2 + 4xy^3 + y^4$$
$$(x + y)^5 = x^5 + 5x^4y + 10x^3y^2 + 10x^2y^3 + 5xy^4 + y^5,$$

and so on. By studying these results, we can find a pattern. Let us try to write down this pattern so that we can write a general expression for $(x + y)^n$.

First, notice that each expression begins with x raised to the same power as the binomial itself. That is, the expansion of $(x + y)^1$ has a first term of x^1, $(x + y)^2$ has a first term of x^2, $(x + y)^3$ has a first term of x^3, and so on. Also, the last term in each expansion is y to the same power as the binomial. We can see that the expression of $(x + y)^n$ should begin with the term x^n and end with the term y^n.

Also, the exponents on x decrease by one in each term after the first, while the exponents on y, beginning with y in the second term, increase by one in each succeeding term. Thus, the *variables* in the expansion of $(x + y)^n$ seem to have the following pattern.

$$x^n, \; x^{n-1}y, \; x^{n-2}y^2, \; x^{n-3}y^3, \; \ldots, \; xy^{n-1}, \; y^n$$

In this pattern, we note that the sum of the exponents on x and y in each term is n. For example, in the third term in the list above, the variable is $x^{n-2}y^2$, and the sum of the exponents is $n - 2 + 2 = n$.

Now examine the pattern for the *coefficients* in the terms of the expansions shown above. Writing the coefficients alone gives the following pattern.

Pascal's Triangle

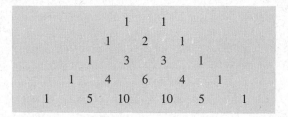

$$\begin{array}{ccccccccccc}
 & & & & 1 & & 1 & & & & \\
 & & & 1 & & 2 & & 1 & & & \\
 & & 1 & & 3 & & 3 & & 1 & & \\
 & 1 & & 4 & & 6 & & 4 & & 1 & \\
1 & & 5 & & 10 & & 10 & & 5 & & 1 \\
\end{array}$$

With the coefficients arranged in this way it can be seen that each number in the triangle is the sum of the two numbers directly above it (one to the right and one to the left.) For example, in the fourth row from the top, 1 is the sum of 1, the only number above it, 4 is the sum of 1 and 3, 6 is the sum of 3 and 3, and so on. This triangular array of numbers is called **Pascal's Triangle**, in honor of the famous seventeenth-century mathematician Blaise Pascal, one of the first to use it extensively.

To get the coefficients for $(x + y)^6$, include a sixth row on the array of numbers given above. By adding adjacent numbers the sixth row is found to be

$$1 \quad 6 \quad 15 \quad 20 \quad 15 \quad 6 \quad 1.$$

Using these coefficients the expansion of $(x + y)^6$ is

$$(x + y)^6 = x^6 + 6x^5y + 15x^4y^2 + 20x^3y^3 + 15x^2y^4 + 6xy^5 + y^6.$$

Although it is possible to use Pascal's Triangle to find the coefficients of $(x + y)^n$ for any positive integer value of n, this becomes impractical for large values of n due to the very large numbers involved. A more efficient way of finding these coefficients uses factorial notation. We define $n!$ (read "n-factorial"), by

n-factorial

$$n! = n(n - 1)(n - 2) \cdots (3)(2)(1)$$

for any positive integer n. For example, $5! = 5 \cdot 4 \cdot 3 \cdot 2 \cdot 1 = 120$, $7! = 7 \cdot 6 \cdot 5 \cdot 4 \cdot 3 \cdot 2 \cdot 1 = 5040$, $2! = 2 \cdot 1 = 2$, and so on. To simplify certain formulas later on, $0!$ is defined as

$$0! = 1.$$

Now let us look at the coefficients of the expression

$$(x + y)^5 = x^5 + 5x^4y + 10x^3y^2 + 10x^2y^3 + 5xy^4 + y^5.$$

The coefficient on the second term, $5x^4y$, is 5, and the exponents on the variables are 4 and 1. Note that

$$5 = \frac{5!}{4!1!}.$$

The coefficient on the third term is 10, with exponents of 3 and 2, and

$$10 = \frac{5!}{3!2!}.$$

The last term (the sixth term) can be written as $y^5 = 1x^0y^5$, with coefficient 1, and exponents of 0 and 5. Since $0! = 1$, check that

$$1 = \frac{5!}{0!5!}.$$

We can generalize to state that the coefficient for the term of the expansion of $(x + y)^n$ in which the variable part is x^ry^{n-r} (where $r \leq n$) will be

$$\frac{n!}{r!(n - r)!}.$$

This number, called a **binomial coefficient,** is often symbolized $\binom{n}{r}$, (read "n above r") so that

**Binomial
Coefficient**

$$\binom{n}{r} = \frac{n!}{r!(n - r)!}.$$

**EXAMPLE
1**   □   (a) $\binom{6}{2} = \frac{6!}{2!(6 - 2)!} = \frac{6!}{2!4!} = \frac{6 \cdot 5 \cdot 4 \cdot 3 \cdot 2 \cdot 1}{2 \cdot 1 \cdot 4 \cdot 3 \cdot 2 \cdot 1} = 15$

(b) $\binom{8}{0} = \frac{8!}{0!(8 - 0)!} = \frac{8!}{0!8!} = \frac{8!}{1 \cdot 8!} = 1$

(c) $\dbinom{10}{10} = \dfrac{10!}{10!(10 - 10)!} = \dfrac{10!}{10!0!} = 1$ $\square$

Using binomial coefficients, we can summarize our conjectures in the following **binomial theorem,** which gives the expansion of $(x + y)^n$ for any positive integer value of n.

Binomial Theorem

For any positive integer n and any complex numbers x and y,

$$(x + y)^n = x^n + \binom{n}{n - 1}x^{n-1}y + \binom{n}{n - 2}x^{n-2}y^2 +$$

$$\binom{n}{n - 3}x^{n-3}y^3 + \cdots + \binom{n}{n - r}x^{n-r}y^r + \cdots$$

$$+ \binom{n}{1}xy^{n-1} + y^n. \tag{4}$$

The binomial theorem can be proved by mathematical induction. Let S_n be statement (4) above. Begin by verifying S_n for $n = 1$.

S_1: $(x + y)^1 = x^1 + y^1,$

which is true.

Now assume that S_n is true for the positive integer k. Statement S_k becomes (using the definition of the binomial coefficient)

S_k: $(x + y)^k = x^k + \dfrac{k!}{1!(k - 1)!}x^{k-1}y + \dfrac{k!}{2!(k - 2)!}x^{k-2}y^2$

$$+ \cdots + \dfrac{k!}{(k - 1)!1!}xy^{k-1} + y^k. \tag{5}$$

Multiply both sides of equation (5) by $x + y$.

$(x + y)^k \cdot (x + y) = x(x + y)^k + y(x + y)^k$

$$= \left[x \cdot x^k + \dfrac{k!}{1!(k - 1)!}x^k y + \dfrac{k!}{2!(k - 2)!}x^{k-1}y^2 \right.$$

$$\left. + \cdots + \dfrac{k!}{(k - 1)!1!}x^2 y^{k-1} + xy^k \right]$$

$$+ \left[x^k \cdot y + \dfrac{k!}{1!(k - 1)!}x^{k-1}y^2 + \cdots \right.$$

$$\left. + \dfrac{k!}{(k - 1)!1!}xy^k + y \cdot y^k \right].$$

Rearrange terms to get

$$(x + y)^{k+1} = x^{k+1} + \left[\frac{k!}{1!(k-1)!} + 1\right]x^k y +$$

$$\left[\frac{k!}{2!(k-2)!} + \frac{k!}{1!(k-1)!}\right]x^{k-1}y^2 + \cdots +$$

$$\left[1 + \frac{k!}{(k-1)!1!}\right]xy^k + y^{k+1}. \tag{6}$$

The first expression in brackets in equation (6) simplifies to $\binom{k+1}{1}$. To see this, note that

$$\binom{k+1}{1} = \frac{(k+1)(k)(k-1)(k-2)\cdots 1}{1 \cdot (k)(k-1)(k-2)\cdots 1} = k + 1.$$

Also $\quad \dfrac{k!}{1!(k-1)!} + 1 = \dfrac{k!}{(k-1)!} + 1$

$$= \frac{k! + (k-1)!}{(k-1)!}$$

$$= \frac{k(k-1)! + (k-1)!}{(k-1)!}$$

$$= \frac{(k+1)(k-1)!}{(k-1)!}$$

$$= k + 1.$$

The second expression becomes $\binom{k+1}{2}$, the last $\binom{k+1}{k}$, and so on. The result of equation (6) is just equation (5) with every k replaced by $k + 1$. Thus, the truth of S_n when $n = k$ implies the truth of S_n for $n = k + 1$, proving the theorem by mathematical induction.

EXAMPLE 2   Write out the binomial expansion of $(x + y)^9$.

Using the binomial theorem,

$$(x + y)^9 = x^9 + \binom{9}{8}x^8 y + \binom{9}{7}x^7 y^2 + \binom{9}{6}x^6 y^3 + \binom{9}{5}x^5 y^4$$

$$+ \binom{9}{4}x^4 y^5 + \binom{9}{3}x^3 y^6 + \binom{9}{2}x^2 y^7 + \binom{9}{1}xy^8 + y^9.$$

Now evaluating each of the binomial coefficients, we have

$$= x^9 + \frac{9!}{1!8!}x^8 y + \frac{9!}{2!7!}x^7 y^2 + \frac{9!}{3!6!}x^6 y^3 + \frac{9!}{4!5!}x^5 y^4$$

$$+ \frac{9!}{5!4!}x^4 y^5 + \frac{9!}{6!3!}x^3 y^6 + \frac{9!}{7!2!}x^2 y^7 + \frac{9!}{8!1!}xy^8 + y^9$$

$$= x^9 + 9x^8y + 36x^7y^2 + 84x^6y^3 + 126x^5y^4 + 126x^4y^5$$
$$+ 84x^3y^6 + 36x^2y^7 + 9xy^8 + y^2. \quad \blacksquare$$

EXAMPLE 3   Expand $\left(a - \dfrac{b}{2} \right)^5$.

Again we use the binomial theorem. Let $x = a$, $y = -\dfrac{b}{2}$, and $n = 5$, to get

$$\left(a - \frac{b}{2} \right)^5 = a^5 + \binom{5}{4} a^4 \left(-\frac{b}{2} \right) + \binom{5}{3} a^3 \left(-\frac{b}{2} \right)^2 + \binom{5}{2} a^2 \left(-\frac{b}{2} \right)^3$$

$$+ \binom{5}{1} a \left(-\frac{b}{2} \right)^4 + \left(-\frac{b}{2} \right)^5$$

$$= a^5 + 5a^4 \left(-\frac{b}{2} \right) + 10a^3 \left(-\frac{b}{2} \right)^2 + 10a^2 \left(-\frac{b}{2} \right)^3$$

$$+ 5a \left(-\frac{b}{2} \right)^4 + \left(-\frac{b}{2} \right)^5$$

$$= a^5 - \frac{5}{2} a^4 b + \frac{5}{2} a^3 b^2 - \frac{5}{4} a^2 b^3 + \frac{5}{16} ab^4 - \frac{1}{32} b^5. \quad \blacksquare$$

EXAMPLE 4   Expand $\left(\dfrac{3}{m^2} - 2\sqrt{m} \right)^4$. (Assume $m > 0$.)

Using the binomial theorem,

$$\left(\frac{3}{m^2} - 2\sqrt{m} \right)^4 = \left(\frac{3}{m^2} \right)^4 + \binom{4}{1} \left(\frac{3}{m^2} \right)^3 (-2\sqrt{m})^1 + \binom{4}{2} \left(\frac{3}{m^2} \right)^2 (-2\sqrt{m})^2$$

$$+ \binom{4}{3} \left(\frac{3}{m^2} \right)^1 (-2\sqrt{m})^3 + (-2\sqrt{m})^4$$

$$= \frac{81}{m^8} + 4 \left(\frac{27}{m^6} \right) (-2m^{1/2}) + 6 \left(\frac{9}{m^4} \right) (4m)$$

$$+ 4 \left(\frac{3}{m^2} \right) (-8m^{3/2}) + 16m^2.$$

Here we used the fact that $\sqrt{m} = m^{1/2}$. Finally,

$$\left(\frac{3}{m^2} - 2\sqrt{m} \right)^4 = \frac{81}{m^8} - \frac{216}{m^{11/2}} + \frac{216}{m^3} - \frac{96}{m^{1/2}} + 16m^2. \quad \blacksquare$$

We can also write any single term of a binomial expansion. For example, to find the tenth term of $(x + y)^n$, where $n \geq 9$, first notice that in the tenth term we have y raised to the ninth power (since y has the power 1 in the second term, the power 2 in the third term, and so on). Since the exponents on x and y in any term must have a sum of n, the exponent on x in the tenth term is $n - 9$. Thus, the tenth term of the expansion is

$$\binom{n}{n-9}x^{n-9}y^9 = \frac{n!}{(n-9)!9!}x^{n-9}y^9.$$

This same idea can be used to obtain the result given in the following theorem.

rth Term of the Binomial Expansion	The rth term of the binomial expansion of $(x + y)^n$, where $n \geq r - 1$, is $$\binom{n}{n-(r-1)}x^{n-(r-1)}y^{r-1}.$$

EXAMPLE 5   Find the seventh term of $(a + 2b)^{10}$.

In the seventh term $2b$ has an exponent of 6, while a has an exponent of $10 - 6$, or 4. The seventh term is thus

$$\binom{10}{4}a^4(2b)^6 = 210a^4(64b^6)$$

$$= 13{,}440a^4b^6.$$

EXERCISES 11.2

Write out the binomial expansion for each of the following.

1. $(x + y)^6$

2. $(m + n)^4$

3. $(p - q)^5$

4. $(a - b)^7$

5. $(r^2 + s)^5$

6. $(m + n^2)^4$

7. $(p + 2q)^4$

8. $(3r - s)^6$

9. $\left(\dfrac{m}{2} - 1\right)^6$

10. $\left(3 + \dfrac{y}{3}\right)^5$

11. $\left(2p + \dfrac{q}{3}\right)^4$

12. $\left(\dfrac{r}{6} - \dfrac{m}{2}\right)^3$

13. $(m^{-2} + m^2)^4$

14. $\left(\sqrt{x} + \dfrac{4}{\sqrt{x}}\right)^5$

15. $(p^{2/3} - 5p^{4/3})^5$

16. $(3y^{3/2} + 5y^{1/2})^4$

Write only the first four terms in each of the following expansions.

17. $(x + 6)^{21}$

18. $(y - 8)^{17}$

19. $(3m^{-2} + 5m^{-4})^9$

20. $\left(\dfrac{8}{k^2} + 6k\right)^{17}$

For each of the following write the indicated term of the binomial expansion.

21. 5th term of $(m - 2p)^{12}$

22. 4th term of $(3x + y)^6$

23. 9th term of $(2m + n)^{10}$

24. 7th term of $(3r - 5s)^{12}$

25. 17th term of $(p^2 + q)^{20}$

26. 10th term of $(2x^2 + y)^{14}$

27. 8th term of $(x^3 + 2y)^{14}$

28. 13th term of $(a + 2b^3)^{12}$

29. Find the middle term of $(3x^7 + 2y^3)^8$.

30. Find the two middle terms of $(-2m^{-1} + 3n^{-2})^{11}$.

Use the binomial expansion to evaluate each of the following to three decimal places.

31. $(1.01)^{10}$ (Hint: $1.01 = 1 + .01$)

32. $(0.99)^{15}$ (Hint: $0.99 = 1 - 0.01$)

33. $(1.99)^8$

34. $(2.99)^3$

35. $(3.02)^6$

36. $(1.01)^9$

In later courses, it is shown that

$$(1 + x)^n = 1 + nx + \frac{n(n - 1)}{2!}x^2$$

$$+ \frac{n(n - 1)(n - 2)}{3!}x^3 + \cdots$$

for any real number n (not just positive integer values) and any real number x where $|x| < 1$. This result, a generalized binomial theorem, may be used to find approximate values of powers and roots. For example,

$$\sqrt[4]{630} = (625 + 5)^{1/4} =$$

$$\left[625\left(1 + \frac{5}{625}\right)\right]^{1/4} = 625^{1/4}\left(1 + \frac{5}{625}\right)^{1/4}.$$

37. Use the result above to approximate $(1 + 5/625)^{1/4}$ to the nearest thousandth. Then approximate $\sqrt[4]{630}$.

38. Approximate $\sqrt[3]{9.42}$, using this method.

39. Approximate $(1.02)^{-3}$.

40. Approximate $1/(1.04)^5$.

41. In the formula above, let $n = -1$ and find $(1 + x)^{-1}$.

42. Find the first four terms when you divide $1 + x$ into 1. Compare with the results of Exercise 41.

43. Work out the first four terms in the expansion of $(1 + 3)^{1/2}$ using $x = 3$ and $n = 1/2$ in the formula above. Do you get a result close to $(1 + 3)^{1/2} = 4^{1/2} = 2$? Why not?

44. Use the result above to show that for small values of x, $\sqrt{1 + x} \approx 1 + \frac{1}{2}x$.

Prove that each of the following properties of factorial notation are true for any positive integer n.

45. $n(n - 1)! = n!$

46. $\dfrac{n!}{n(n - 1)} = (n - 2)!$

Find all values of n for which

47. $(n + 2)! = 56 \cdot n!$ **48.** $(n - 4)! = n - 4$

11.3   Arithmetic Sequences

A sequence function is a function whose domain is a set of positive integers. The formal definition depends on whether the domain is finite or infinite. A **finite sequence function** is a function whose domain is a set of the form $\{1, 2, 3, \ldots, n\}$ where n is a positive integer. An **infinite sequence function** has the set of all positive integers as its domain.

The sequence function of positive even integers,

2, 4, 6, 8, 10, 12, 14, . . .

is infinite, while the sequence function of days in June,

1, 2, 3, 4, 5, 6, . . . , 29, 30,

is finite.

The elements in the range of a sequence function are called the **terms** of the sequence function. When the terms of a sequence function are listed in order, the result is called a **sequence.** The terms are denoted a_1, a_2, a_3, and so on.

It is common to specify the terms of a sequence by giving the **general** or **nth term** of the sequence, written a_n. In this way, any particular term may be found.

EXAMPLE 1   ☐   Find the first five terms for the sequence having general term

$$a_n = \frac{n + 1}{n + 2}.$$

Replace n, in turn, by 1, 2, 3, 4, and 5, to get

$$a_1 = \frac{2}{3}, \quad a_2 = \frac{3}{4}, \quad a_3 = \frac{4}{5}, \quad a_4 = \frac{5}{6}, \quad \text{and} \quad a_5 = \frac{6}{7}. \quad ☐$$

EXAMPLE 2   ☐   Find a_{100} for the sequences having the following general terms.

(a) $a_n = (-1)^n(2n)$

Replacing n with 100, we have

$$a_{100} = (-1)^{100}(2 \cdot 100) = 200.$$

(b) $a_n = \dfrac{(-1)^{n+1} \cdot 5n}{7n - 4}$

Here $a_{100} = \dfrac{(-1)^{101} \cdot 5(100)}{7(100) - 4} = \dfrac{-500}{696} = -\dfrac{125}{174}.$ ☐

A sequence in which successive terms differ by a constant number is called an **arithmetic sequence** (or **arithmetic progression**). The constant difference is called the **common difference, d.** For example, the sequence

5, 9, 13, 17, 21, . . .

is an arithmetic sequence since each pair of consecutive terms differ by 4. That is,

$$9 - 5 = 4$$
$$13 - 9 = 4$$
$$17 - 13 = 4$$
$$21 - 17 = 4$$

and so on. Here 4 is the common difference, written $d = 4$.

EXAMPLE 3 □ Find the common difference, d, for the arithmetic sequence

$$-9, \ -12, \ -15, \ -18, \ -21, \dots$$

Since we know that this sequence is arithmetic, we can find d by choosing any two adjacent terms and subtracting the first from the second. If we choose -15 and -18 here, we have $d = -18 - (-15) = -3$.

If we had chosen -9 and -12, we would have found $d = -12 - (-9) = -3$, the same result. □

EXAMPLE 4 □ Write the first five terms for each of the following arithmetic sequences.

(a) The first term is 7, and each succeeding term is found by adding -3 to the preceding term.

Here $a_1 = 7$

$a_2 = 7 + (-3) = 4$

$a_3 = 4 + (-3) = 1$

$a_4 = 1 + (-3) = -2$

$a_5 = -2 + (-3) = -5.$

(b) $a_1 = -12, \ d = 5$

We have $a_1 = -12$

$a_2 = -12 + d = -12 + 5 = -7$

$a_3 = -7 + d = -7 + 5 = -2$

$a_4 = -2 + d = -2 + 5 = 3$

$a_5 = 3 + d = 3 + 5 = 8.$ □

If a_1 is the first term of an arithmetic sequence and d is the common difference, then the terms of the sequence are given by

$$a_1 = a_1$$

$$a_2 = a_1 + d$$

$$a_3 = a_2 + d = a_1 + d + d = a_1 + 2d$$

$$a_4 = a_3 + d = a_1 + 2d + d = a_1 + 3d$$

$$a_5 = a_1 + 4d$$

$$a_6 = a_1 + 5d,$$

and, by this pattern,

$$a_n = a_1 + (n - 1)d.$$

This result could be proven by mathematical induction. (See Exercise 95.) In summary,

nth Term of an Arithmetic Sequence

In an arithmetic sequence with first term a_1 and common difference d, the nth term, a_n, is given by

$$a_n = a_1 + (n - 1)d.$$

EXAMPLE 5   Find a_{13} and a_n for the arithmetic sequence

$$-3, 1, 5, 9, \ldots .$$

Here $a_1 = -3$ and $d = 4$. (We found d by the method of Example 3.) Now substitute 13 for n in the formula above.

$$a_{13} = a_1 + (13 - 1)d$$
$$= -3 + (12)4$$
$$= -3 + 48$$
$$a_{13} = 45.$$

To find a_n, substitute values for a_1 and d, which gives

$$a_n = -3 + (n - 1) \cdot 4$$
$$a_n = -3 + 4n - 4$$
$$a_n = 4n - 7. \quad \square$$

EXAMPLE 6   Find a_{18} and a_n for the arithmetic sequence having $a_2 = 9$ and $a_3 = 15$.
Here $d = a_3 - a_2 = 15 - 9 = 6$. Since $a_2 = a_1 + d$, we have

$$9 = a_1 + 6$$
$$a_1 = 3.$$

Thus, $a_{18} = 3 + (18 - 1) \cdot 6$
$$a_{18} = 105$$

and $a_n = 3 + (n - 1) \cdot 6$
$$a_n = 6n - 3. \quad \square$$

Arithmetic sequences are useful in solving applications as illustrated by the next example.

EXAMPLE 7   A child building a tower with blocks uses 15 for the first row. Each row has 2 blocks less than the previous row. If there are 8 rows in the tower, how many blocks are used for the top row?
The number of blocks in each row forms an arithmetic sequence with $a_1 = 15$ and $d = -2$. We want to find a_n for $n = 8$. Using the formula,

$$a_n = a_1 + (n - 1)d$$
$$a_8 = 15 + (8 - 1)(-2) = 1. \quad \square$$

EXAMPLE 8   Suppose that an arithmetic sequence has $a_8 = -16$ and $a_{16} = -40$. Find a_1.
We know that $a_8 = a_1 + (8 - 1)d$. Therefore,

$$-16 = a_1 + 7d \quad \text{or} \quad a_1 = -16 - 7d.$$

Similarly, we find that

$$-40 = a_1 + 15d \quad \text{or} \quad a_1 = -40 - 15d.$$

From the two equations for a_1 we have

$$-16 - 7d = -40 - 15d$$
$$d = -3.$$

Since $-16 = a_1 + 7d$, we have $a_1 = 5$. ◻

Suppose now that a man borrows $3000 and agrees to pay $100 per month plus interest of 1% per month on the unpaid balance until the loan is paid off. How much interest will he pay?

The first month he pays $100 to reduce the loan plus interest of $(.01)3000 = 30$ dollars. The second month he pays another $100 toward the loan and interest of $(.01)2900 = 29$ dollars. Since the loan is reduced by $100 each month, his interest payments decrease by $(.01)100 = 1$ dollar each month. Thus, the interest payments form the arithmetic sequence

$$30, 29, 28, \ldots, 2, 1.$$

We need to find the sum of the terms of the sequence. There is a way to do this without directly adding these thirty numbers. We shall develop a formula to find this sum and then return to the problem.

Suppose a sequence has terms $a_1, a_2, a_3, a_4, \ldots$. Then S_n is defined as the sum of the first n terms of the sequence. That is,

$$S_n = a_1 + a_2 + a_3 + \cdots + a_n.$$

Given a_1 and d we can find the sum of the first n terms of an arithmetic sequence. To begin, write the sum of the first n terms as follows:

$$S_n = a_1 + [a_1 + d] + [a_1 + 2d] + \cdots + [a_1 + (n - 1)d].$$

Next, write this same sum in reversed order, as

$$S_n = [a_1 + (n - 1)d] + [a_1 + (n - 2)d] + \cdots + [a_1 + d] + a_1.$$

Now add the respective sides of these two equations, to get

$$S_n + S_n = (a_1 + [a_1 + (n - 1)d]) + ([a_1 + d] + [a_1 + (n - 2)d])$$
$$+ \cdots + ([a_1 + (n - 1)d] + a_1).$$

From this,

$$2S_n = [2a_1 + (n - 1)d] + [2a_1 + (n - 1)d]$$
$$+ \cdots + [2a_1 + (n - 1)d].$$

Since there are n of the $[2a_1 + (n - 1)d]$ terms on the right,

$$2S_n = n[2a_1 + (n - 1)d]$$

$$S_n = \frac{n}{2}[2a_1 + (n - 1)d].$$

Since $a_n = a_1 + (n - 1)d$, we also have $S_n = \dfrac{n}{2}[a_1 + a_1 + (n - 1)d]$, or

$$S_n = \frac{n}{2}(a_1 + a_n).$$

The following theorem summarizes our work with sums of terms in an arithmetic sequence.

Sum of the First n Terms of an Arithmetic Sequence

If an arithmetic sequence has first term a_1 and common difference d, then the sum of the first n terms, S_n, is given by

$$S_n = \frac{n}{2}(a_1 + a_n)$$

or $\quad S_n = \dfrac{n}{2}[2a_1 + (n - 1)d].$

The first formula of the theorem is used when we know the first and last term, otherwise the second formula is used.

Now, we can use this formula to find out how much interest the man above will pay on the $3000 loan. In the sequence of interest payments, $a_1 = 30$, $d = -1$, $n = 30$, $a_n = 1$. We can use either form of the formula. Choosing

$$S_n = \frac{n}{2}(a_1 + a_n),$$

we have $\quad S_n = \dfrac{30}{2}(30 + 1) = 15(31) = 465.$

Thus, the man will pay a total of $465 interest over 30 months.

EXAMPLE 9   (a) Find S_{12} for the arithmetic sequence $-9, -5, -1, 3, 7, \ldots$.
Here $a_1 = -9$, $d = 4$, and $n = 12$. We use the second formula above.

$$S_{12} = \frac{12}{2}[2(-9) + (12 - 1)(4)]$$

$$= 6[-18 + 44]$$

$$S_{12} = 156.$$

(b) Find the sum of the first 60 positive integers.
Here $n = 60$, $a_1 = 1$, and $a_{60} = 60$. This time it is more convenient to use the first of the two formulas. We have

$$S_{60} = \frac{60}{2}(1 + 60) = 30 \cdot 61 = 1830. \quad \blacksquare$$

EXAMPLE 10   The sum of the first 17 terms of an arithmetic sequence is 187. If $a_{17} = -13$, find a_1 and d.
Using the first formula for S_n developed above to find a_1, we have

$$187 = \frac{17}{2}(a_1 - 13)$$

$$374 = 17(a_1 - 13)$$

$$22 = a_1 - 13$$

$$a_1 = 35.$$

We know $a_{17} = a_1 + (17 - 1)d$. Hence.

$$-13 = 35 + 16d$$

$$-48 = 16d$$

$$d = -3. \quad \blacksquare$$

Sigma Notation

We sometimes use a special shorthand notation for the sum of n terms of a sequence. The symbol Σ, the Greek letter *sigma*, is used to indicate a sum. For example,

$$\sum_{i=1}^{n} a_i$$

represents the sum

$$\sum_{i=1}^{n} a_i = a_1 + a_2 + a_3 + \cdots + a_n.$$

The letter i is called the **index of summation.** Do not confuse this use of i with the use of i to represent an imaginary number.

EXAMPLE 11 $\quad\blacksquare\quad$ Find each of the following sums.

(a) $\displaystyle\sum_{i=1}^{10} (4i + 8)$

This sum can be written out as

$$\sum_{i=1}^{10} (4i + 8) = [4(1) + 8] + [4(2) + 8] + [4(3) + 8]$$

$$+ \cdots + [4(10) + 8]$$

$$= 12 + 16 + 20 + \cdots + 48.$$

Thus, it represents the sum of the first ten terms of the arithmetic sequence having $a_1 = 12$ and $a_n = a_{10} = 48$. Using the first formula for the sum of 10 terms of an arithmetic sequence, we have

$$\sum_{i=1}^{10} (4i + 8) = S_{10} = \frac{10}{2}(12 + 48)$$

$$= 5(60)$$

$$= 300.$$

$$\text{(b)} \sum_{i=1}^{15} (9 - i) = S_{15} = \frac{15}{2}[8 + (-6)]$$

$$= \frac{15}{2}[2]$$

$$= 15. \quad \square$$

EXERCISES
11.3

Write the first five terms in each of the following sequences.

1. $a_n = 6n + 4$ 2. $a_n = 3n - 2$

3. $a_n = 2^{-n+1}$

4. $a_n = (-1)^n(n + 2)$

5. $a_n = \dfrac{2}{n + 3}$ 6. $a_n = \dfrac{8n - 4}{2n + 1}$

7. $a_n = (-2)^n(n)$ 8. $a_n = (-\frac{1}{2})^n(n^{-1})$

9. $a_n = \dfrac{n^2 + 1}{n^2 + 2}$

10. $a_n = \dfrac{(-1)^{n-1}(n + 1)}{n + 2}$

11. a_n is the nth prime number

12. b_n is the number of positive integers that divide into n

Find the first ten terms for the sequences defined as follows.

13. $a_1 = 4$, $a_n = a_{n-1} + 5$, for $n > 1$

14. $a_1 = -3$, $a_n = a_{n-1} + 2$, for $n > 1$

15. $a_1 = 2$, $a_n = 2 \cdot a_{n-1}$, for $n > 1$

16. $a_1 = 3$, $a_n = -2 \cdot a_{n-1}$, for $n > 1$

17. $a_1 = 1$, $a_2 = 1$, $a_n = a_{n-1} + a_{n-2}$, for $n \geq 3$ (the Fibonacci sequence)

18. $a_1 = 1$, $a_n = n \cdot \cos(\pi \cdot a_{n-1})$, for $n > 1$

Write the terms of the arithmetic sequences satisfying each of the following conditions.

19. $a_1 = 4$, $d = 2$, $n = 5$

20. $a_1 = 6$, $d = 8$, $n = 4$

21. $a_2 = 9$, $d = -2$, $n = 4$

22. $a_3 = 7$, $d = -4$, $n = 4$

23. $a_1 = 4 - \sqrt{5}$, $a_2 = 4$, $n = 5$

24. $a_1 = -8$, $a_2 = -8 + \sqrt{7}$, $n = 5$

For each of the following arithmetic sequences, find d and a_n.

25. 12, 17, 22, 27, 32, 37, . . .

26. 8, 17, 26, 35, 44, 53, . . .

27. 18, 15, 12, 9, 6, . . .

28. 30, 24, 18, 12, . . .

29. $-6 + \sqrt{2}$, $-6 + 2\sqrt{2}$, $-6 + 3\sqrt{2}$, . . .

30. $4 - \sqrt{11}$, $5 - \sqrt{11}$, $6 - \sqrt{11}$, . . .

31. x, $x + m$, $x + 2m$, $x + 3m$, $x + 4m$, . . .

32. $k + p$, $k + 2p$, $k + 3p$, $k + 4p$, . . .

33. $2z + m$, $2z$, $2z - m$, $2z - 2m$, $2z - 3m$, . . .

34. $3r - 4z$, $3r - 3z$, $3r - 2z$, $3r - z$, $3r$, $3r + z$, . . .

Find a_8 and a_n for each of the following arithmetic sequences.

35. $a_1 = 5$, $d = 2$

36. $a_1 = -3$, $d = -4$

37. $a_3 = 2$, $d = 1$ 38. $a_4 = 5$, $d = -2$

39. $a_1 = 8$, $a_2 = 6$ 40. $a_1 = 6$, $a_2 = 3$

41. $a_1 = 12$, $a_3 = 6$ 42. $a_2 = 5$, $a_4 = 1$

43. $a_5 = 4.2$, $a_6 = 4.5$

44. $a_3 = -9.5$, $a_4 = -10.6$

45. $a_1 = x$, $a_2 = x + 3$

46. $a_2 = y + 1$, $d = -3$

47. $a_6 = 2m$, $a_7 = 3m$

48. $a_5 = 4p + 1$, $a_7 = 6p + 7$

Find the sum of the first ten terms for each of the following arithmetic sequences.

49. $a_1 = 8$, $d = 3$

50. $a_1 = 2$, $d = 6$

51. $a_3 = 5$, $a_4 = 8$

52. $a_2 = 9$, $a_4 = 13$

53. $5, 9, 13, \ldots$

54. $8, 6, 4, \ldots$

55. $a_1 = 9.428$, $d = -1.723$

56. $a_1 = -3.119$, $d = 2.422$

57. $a_4 = 2.556$, $a_5 = 3.004$

58. $a_7 = 11.192$, $a_9 = 4.812$

Evaluate each of the following sums.

59. $\displaystyle\sum_{i=1}^{3} (i + 4)$

60. $\displaystyle\sum_{i=1}^{5} (i - 8)$

61. $\displaystyle\sum_{i=1}^{10} (2i + 3)$

62. $\displaystyle\sum_{i=1}^{15} (5i - 9)$

63. $\displaystyle\sum_{i=1}^{12} (-5 - 8i)$

64. $\displaystyle\sum_{i=1}^{19} (-3 - 4i)$

65. $\displaystyle\sum_{i=1}^{1000} i$

66. $\displaystyle\sum_{i=1}^{2000} i$

67. $\displaystyle\sum_{i=6}^{15} (4i - 2)$

68. $\displaystyle\sum_{i=9}^{20} (8i + 3)$

69. $\displaystyle\sum_{i=7}^{12} (6 - 2i)$

70. $\displaystyle\sum_{i=4}^{17} (-3 - 5i)$

Find a_1 for each of the following arithmetic sequences.

71. $a_9 = 47$, $a_{15} = 77$

72. $a_{10} = 50$, $a_{20} = 110$

73. $a_{15} = 168$, $a_{16} = 180$

74. $a_{10} = -54$, $a_{17} = -89$

Find a_1 for each of the following arithmetic sequences.

75. $S_{20} = 1090$, $a_{20} = 102$

76. $S_{31} = 5580$, $a_{31} = 360$

77. Find the sum of all the integers from 51 to 71.

78. Find the sum of all the integers from -8 to 30.

79. Find the sum of the first n positive integers.

80. Find the sum of the first n odd positive integers.

81. If a clock strikes the proper number of bongs each hour on the hour, how many bongs will it bong in a month of 30 days?

82. A stack of telephone poles has 30 in the bottom row, 29 in the next, and so on, with one pole in the top row. How many poles are in the stack?

83. A sky diver falls 10 meters during the first second, 20 meters during the second, 30 meters during the third, and so on. How many meters will the diver fall during the tenth second? During the first ten seconds?

84. Deepwell Drilling Company charges a flat $500 set-up charge, plus $5 for the first foot of well drilled, $6 for the second, $7 for the third, and so on. Find the total charge for a 70-foot well.

85. An object falling under the force of gravity falls about 16 feet the first second, 48 feet during the second, 80 feet during the third, and so on. How far would the object fall during the eighth second? What is the total distance the object would fall in eight seconds?

86. The population of a city was 49,000 five years ago. Each year, the zoning commission permits an increase of 580 in the population of the city. What will the population be five years from now?

87. The sum of four terms in an arithmetic sequence is 66. The sum of the squares of the terms is 1214. Find the terms.

88. The sum of five terms of an arithmetic sequence is 5. If the product of the first and second is added to the product of the fourth and fifth, the result is 326. Find the terms.

89. A super slide of uniform slope is to be built on a level piece of land. There are to be twenty equally spaced supports, with the longest support 15 m in length, and the shortest 2 m in length. Find the total length of all the supports.

90. How much material would be needed for the rungs of a ladder of 31 rungs, if the rungs taper from 18 inches to 28 inches? Assume that the lengths of the rungs form the terms of an arithmetic sequence.

91. Find all arithmetic sequences a_1, a_2, a_3, . . . , such that $a_1{}^2$, $a_2{}^2$, $a_3{}^2$, . . . , is also an arithmetic sequence.

92. Suppose that a_1, a_2, a_3, . . . and b_1, b_2, b_3, . . . are each arithmetic sequences. Let $d_n = a_n + c \cdot b_n$, for any real number c and every positive integer n. Show that d_1, d_2, d_3, . . . is an arithmetic sequence.

Explain why each of the following sequences is arithmetic.

93. log 2, log 4, log 8, log 16, log 32, . . .
94. log 12, log 36, log 108, log 324, . . .

Prove each of the following results by mathematical induction.

95. $a_n = a_1 + (n - 1)d$

96. $S_n = \dfrac{n}{2}(a_1 + a_n)$

Decide which of the following statements are true for every real number x_i, real constant k, and natural number n.

97. $\displaystyle\sum_{i=1}^{n} k \cdot x_i = k^n \cdot \sum_{i=1}^{n} x_i$

98. $\displaystyle\sum_{i=1}^{n} k \cdot x_i = k \cdot \sum_{i=1}^{n} x_i$

99. $\displaystyle\sum_{i=1}^{n} k = nk$

100. $\displaystyle\sum_{i=1}^{n} k = k^n$

101. $\displaystyle\left[\sum_{i=1}^{n} x_i\right]^2 = \sum_{i=1}^{n} x_i{}^2$

102. $\displaystyle\sum_{i=1}^{n} (x_i - k) = \sum_{i=1}^{n} x_i - nk$

11.4   Geometric Sequences

A geometric sequence (or **geometric progression**) is a sequence in which the ratio of successive terms is constant. This constant ratio is called the **common ratio.** Thus,

2, 8, 32, 128, . . .

is a geometric sequence whose first term is 2 and whose common ratio is 4.

If the common ratio of a geometric sequence is r, then by the definition of a geometric sequence

$$\frac{a_{n+1}}{a_n} = r$$

for every positive integer n. Therefore, the common ratio can be found by choosing any term except the first and dividing it by the preceding term.

In the geometric sequence

2, 8, 32, 128, . . .

$r = 8/2 = 4$. Notice that

$$8 = 2 \cdot 4$$
$$32 = 8 \cdot 4 = (2 \cdot 4) \cdot 4 = 2 \cdot 4^2$$
$$128 = 32 \cdot 4 = (2 \cdot 4^2) \cdot 4 = 2 \cdot 4^3.$$

To generalize this, assume that a geometric sequence has first term a_1 and common ratio r. The second term can be written as $a_2 = a_1 r$, the third as $a_3 = a_2 r = (a_1 r)r = a_1 r^2$, and so on. Thus, following this pattern, the nth term is $a_n = a_1 r^{n-1}$. Again, this result is proven by mathematical induction. (See Exercise 77.)

nth Term of a Geometric Sequence

In the geometric sequence with first term a_1 and common ratio r, the nth term is

$$a_n = a_1 r^{n-1}.$$

EXAMPLE 1 ☐ Find a_5 and a_n for each of the following geometric sequences.

(a) 4, 12, 36, 108, . . .

In this sequence, $a_1 = 4$. To find r, choose any term except the first and divide it by the preceding term. Here

$$r = 36/12 = 3.$$

Thus, $a_n = a_1 r^{n-1}$

$$a_5 = 4 \cdot (3)^{5-1}$$
$$= 4 \cdot 3^4$$
$$a_5 = 324.$$

Also, $a_n = 4 \cdot 3^{n-1}.$

(b) 64, 32, 16, 8, . . .

Here $r = 8/16 = 1/2$, and $a_1 = 64$. Thus,

$$a_5 = 64(1/2)^{5-1}$$
$$= 64(1/16)$$
$$a_5 = 4.$$

Also, $a_n = 64(1/2)^{n-1}.$ ☐

EXAMPLE 2 ☐ Suppose that the third term of a geometric sequence is 20 and the sixth term is 160. Find the first term a_1 and the common ratio r.

Use the formula for the nth term of a geometric sequence.

for $n = 3$, $a_3 = a_1 r^2 = 20$;

for $n = 6$, $a_6 = a_1 r^5 = 160.$

Since $a_1 r^2 = 20$, then $a_1 = 20/r^2$. Substituting this in the second equation gives

$$a_1 r^5 = 160$$

$$\frac{20}{r^2} r^5 = 160$$

$$20 r^3 = 160$$

$$r^3 = 8$$

$$r = 2.$$

Since $a_1 r^2 = 20$ and $r = 2$, we end up with $a_1 = 5$.   ☐

EXAMPLE 3   ☐   An insect population is growing in such a way that each generation is 1.5 times as large as the last generation. Suppose there were 100 insects in the first generation. How many would there be in the fourth generation?

The population can be written as a geometric sequence in which a_1 is the first generation population, a_2 is the second generation population, and so on. Then the fourth generation population will be a_4, with $a_1 = 100$ and $r = 1.5$. Using the formula for a_n, where $n = 4$, we have

$$a_4 = a_1 r^3 = 100(1.5)^3 = 100(3.375) = 337.5.$$

In the fourth generation, the population will number about 338 insects.   ☐

In applications of geometric sequences, it is often necessary to know the sum of the first n terms for the sequence. For example, we might want to know the total number of insects in four generations of the population discussed in Example 3. To find a formula for S_n, the sum of the first n terms of a geometric sequence, we start with

$$S_n = a_1 + a_2 + a_3 + \cdots + a_n,$$

which can also be written as

$$S_n = a_1 + a_1 r + a_1 r^2 + \cdots + a_1 r^{n-1}. \tag{1}$$

If $r = 1$, we have $S_n = n a_1$, which is a correct formula for this case. If $r \neq 1$, we can multiply both sides of (1) by r, obtaining

$$r S_n = a_1 r + a_1 r^2 + a_1 r^3 + \cdots + a_1 r^n. \tag{2}$$

If we subtract (2) from (1), we have

$$S_n - r S_n = a_1 - a_1 r^n$$

or $S_n(1 - r) = a_1(1 - r^n),$

which finally gives $S_n = \dfrac{a_1(1 - r^n)}{1 - r} \qquad (r \neq 1).$

The following theorem summarizes our work with sums of the terms of a geometric sequence.

Sum of n Terms of a Geometric Sequence

If a geometric sequence has first term a_1 and common ratio r, then the sum of the first n terms, S_n, is

$$S_n = \frac{a_1(1 - r^n)}{1 - r} \qquad (r \neq 1).$$

We can use this formula to find the total insect population in Example 3 over the four-generation period. We have $n = 4$, $a_1 = 100$, and $r = 1.5$, and

$$S_n = \frac{100(1 - 1.5^4)}{1 - 1.5}$$

$$= \frac{100(1 - 5.0625)}{-.5}$$

$$= 812.5.$$

Thus, the total population for the four generations will amount to about 813 insects.

EXAMPLE 4 ☐ Find S_5 for the geometric sequence

3, 6, 12, 24, 48.

Here $a_1 = 3$ and $r = 2$. Thus,

$$S_5 = \frac{3(1 - 2^5)}{1 - 2}$$

$$= \frac{3(1 - 32)}{-1}$$

$$= \frac{3(-31)}{-1}$$

$$S_5 = 93. \quad ☐$$

EXAMPLE 5 ☐ Find $\displaystyle\sum_{i=1}^{6} 2 \cdot 3^i$.

This sum is the sum of the first six terms of a geometric sequence having $a_1 = 2 \cdot 3^1 = 6$ and $r = 3$. Thus,

$$\sum_{i=1}^{6} 2 \cdot 3^i = S_6 = \frac{6(1 - 3^6)}{1 - 3}$$

$$= \frac{6(1 - 729)}{-2}$$

$$= \frac{6(-728)}{-2}$$

$$= 2184. \quad ☐$$

Now let us look at the infinite geometric sequence

$$2, 1, \frac{1}{2}, \frac{1}{4}, \frac{1}{8}, \frac{1}{16}, \frac{1}{32}, \ldots$$

The first term of this sequence is $a_1 = 2$, and the common ratio is $r = 1/2$. Using the formula above or direct calculation, we can find S_n for any value of n. Each of these results is called a **partial sum.** Letting $n = 1, 2, 3$, and so on, we get the following sequence of partial sums.

$$S_1 = 2, S_2 = 3, S_3 = \frac{7}{2}, S_4 = \frac{15}{4}, S_5 = \frac{31}{8}, S_6 = \frac{63}{16}, \ldots$$

These partial sums seem to be getting closer and closer to the number 4. In fact, by selecting a value of n large enough, we can make S_n as close as we might wish to 4. This is expressed as

$$\lim_{n \to \infty} S_n = 4.$$

(Read: "the limit of S_n as n increases without bound is 4.") For no value of n is $S_n = 4$. However, if n is large enough, then S_n is as close to 4 as we might wish.*

Since $$\lim_{n \to \infty} S_n = 4,$$

the number 4 is called the **sum** of the infinite geometric sequence

$$2, 1, \frac{1}{2}, \frac{1}{4}, \ldots$$

and $2 + 1 + \frac{1}{2} + \frac{1}{4} + \frac{1}{8} + \cdots = 4.$

EXAMPLE 6   ☐   Find $1 + \frac{1}{3} + \frac{1}{9} + \frac{1}{27} + \ldots$.

Here we use the formula for the sum of the first n terms of a geometric sequence to get the sequence of partial sums

$$S_1 = 1, S_2 = \frac{4}{3}, S_3 = \frac{13}{9}, S_4 = \frac{40}{27},$$

and so on. In general, if $a_1 = 1$ and $r = 1/3$, we have

$$S_n = \frac{1 \left[1 - \left(\frac{1}{3} \right)^n \right]}{1 - \frac{1}{3}}.$$

---

*These phrases "large enough" and "as close as we wish" are not nearly precise enough for mathematicians; much of a standard calculus course is devoted to making them more precise.

The following chart shows the value of $(1/3)^n$ for larger and larger values of n.

n	1	10	100	200
$\left(\dfrac{1}{3}\right)^n$	$\dfrac{1}{3}$	.0000169	1.9×10^{-48}	3.76×10^{-96}

As the chart suggests, $(1/3)^n$ gets closer and closer to 0 as n gets larger and larger. This is written

$$\lim_{n \to \infty} \left(\frac{1}{3}\right)^n = 0.$$

Thus, it seems reasonable that

$$\lim_{n \to \infty} S_n = \frac{1(1 - 0)}{1 - \frac{1}{3}} = \frac{1}{\frac{2}{3}} = \frac{3}{2}.$$

Hence, $\quad 1 + \dfrac{1}{3} + \dfrac{1}{9} + \dfrac{1}{27} + \cdots = \dfrac{3}{2}.$ $\quad\blacksquare$

If a geometric sequence has a first term a_1 and a common ratio r, then

$$S_n = \frac{a_1(1 - r^n)}{1 - r}$$

for every positive integer n. If $-1 < r < 1$, then $\lim_{n \to \infty} r^n = 0$, and

$$\lim_{n \to \infty} S_n = \frac{a_1(1 - 0)}{1 - r}$$

$$\lim_{n \to \infty} S_n = \frac{a_1}{1 - r}.$$

This quotient, $a_1/(1 - r)$ is called the **sum of the infinite geometric sequence.** The limit, $\lim_{n \to \infty} S_n$, is often expressed as S_∞ or $\sum_{i=1}^{\infty} a_i$.

In summary, we have the following result.

Sum of an Infinite Geometric Sequence

The sum of the infinite geometric sequence with first term a_1 and common ratio r, where $-1 < r < 1$, is given by

$$S_\infty = \sum_{i=1}^{\infty} a_i = \lim_{n \to \infty} S_n = \frac{a_1}{1 - r}.$$

EXAMPLE 7 $\quad\square\quad$ (a) Find the sum $-\dfrac{3}{4} + \dfrac{3}{8} - \dfrac{3}{16} + \dfrac{3}{32} - \dfrac{3}{64} + \cdots.$

Here $a_1 = -3/4$. To find r, divide any term by the preceding term. For example,

$$r = \frac{-\dfrac{3}{16}}{\dfrac{3}{8}} = -\frac{1}{2}.$$

Since $-1 < r < 1$, the formula of this section applies, and

$$S_\infty = \frac{a_1}{1 - r} = \frac{-\dfrac{3}{4}}{1 - \left(-\dfrac{1}{2}\right)} = -\frac{1}{2}.$$

(b) $\displaystyle\sum_{i=1}^{\infty} \left(\frac{3}{5}\right)^i = \frac{\dfrac{3}{5}}{1 - \dfrac{3}{5}} = \frac{3}{2}$ ▢

EXERCISES 11.4

Write the terms of the geometric sequences that satisfy each of the following conditions.

1. $a_1 = 2, r = 3, n = 4$
2. $a_1 = 4, r = 2, n = 5$
3. $a_1 = 1/2, r = 4, n = 4$
4. $a_1 = 2/3, r = 6, n = 3$
5. $a_3 = 6, a_4 = 12, n = 5$
6. $a_2 = 9, a_3 = 3, n = 4$

Find a_5 and a_n for each of the following geometric sequences.

7. $a_1 = 4, r = 3$
8. $a_1 = 8, r = 4$
9. $a_1 = -3, r = -5$
10. $a_1 = -4, r = -2$
11. $a_2 = 3, r = 2$
12. $a_3 = 6, r = 3$
13. $a_4 = 64, r = -4$
14. $a_4 = 81, r = -3$

For each of the following sequences that are geometric, find r and a_n.

15. 6, 12, 24, 48, . . .
16. 4, 16, 64, 256, . . .
17. 3/4, 3/2, 3, 6, 12, . . .
18. $-7, -5, -3, -1, 1, 3, \ldots$
19. $a_3 = 9, r = 2$
20. $a_5 = 6, r = 1/2$
21. $a_3 = -2, r = 3$
22. $a_2 = 5, r = 1/2$

Find the sum of the first five terms for each of the following geometric sequences.

23. 3, 6, 12, 24, . . .
24. 5, 20, 80, 320, . . .
25. 12, -6, 3, $-3/2$, . . .
26. 18, -3, 1/2, $-1/12$, . . .
27. $a_1 = 8.423, r = 2.859$
28. $a_1 = -3.772, r = -1.553$

Find each of the following sums.

29. $\displaystyle\sum_{i=1}^{4} 2^i$

30. $\displaystyle\sum_{i=1}^{6} 3^i$

31. $\displaystyle\sum_{i=1}^{8} 64(1/2)^i$

32. $\displaystyle\sum_{i=1}^{6} 81(2/3)^i$

33. $\displaystyle\sum_{i=3}^{6} 2^i$ **34.** $\displaystyle\sum_{i=4}^{7} 3^i$

Find r for each of the following infinite geometric sequences. Identify any whose sums would exist.

35. 9, 18, 36, 72, 144, . . .

36. 3, 9, 27, 81, . . .

37. 10, 100, 1000, 10,000, . . .

38. $-8, -16, -32, -64, \ldots$

39. 12, 6, 3, 3/2, . . .

40. $1, -0.9, 0.81, -0.729, \ldots$

Find each of the following sums which exist by using the formula of this section where it applies.

41. $16 + 4 + 1 + \cdots$

42. $81 + 27 + 9 + 3 + 1 + \cdots$

43. $100 + 10 + 1 + \cdots$

44. $128 + 64 + 32 + \cdots$

45. $\dfrac{3}{4} + \dfrac{3}{8} + \dfrac{3}{16} + \cdots$

46. $\dfrac{4}{5} + \dfrac{2}{5} + \dfrac{1}{5} + \cdots$

47. $\dfrac{1}{3} - \dfrac{2}{9} + \dfrac{4}{27} - \dfrac{8}{81} + \cdots$

48. $1 + \dfrac{1}{1.01} + \dfrac{1}{(1.01)^2} + \cdots$

49. $\displaystyle\sum_{i=1}^{\infty} (1/4)^i$ **50.** $\displaystyle\sum_{i=1}^{\infty} (-1/4)^i$

51. $\displaystyle\sum_{i=1}^{\infty} (.3)^i$ **52.** $\displaystyle\sum_{i=1}^{\infty} 10^{-i}$

53. Mitzi drops a ball from a height of 10 meters and notices that on each bounce the ball returns to about 3/4 of its previous height. About how far will the ball travel before it comes to rest? (Hint: consider the sum of two sequences.)

54. A sugar factory receives an order for 1000 units of sugar. The production manager thus orders production of 1000 units of sugar. He forgets, however, that the production of sugar requires some sugar (to prime the machines, for example), and so he ends up with only 900 units of sugar. He then orders an additional 100 units, and receives only 90 units. A further order for 10 units produces 9 units. Finally seeing he is wrong, the manager decides to try mathematics. He views the production process as an infinite geometric progression with $a_1 = 1000$ and $r = .1$. Using this, find the number of units of sugar that he should have ordered originally.

55. After a person pedaling a bicycle removes his or her feet from the pedals, the wheel rotates 400 times the first minute. As it continues to slow down, it rotates in each minute only 3/4 as many times as in the previous minute. How many times will the wheel rotate before coming to a complete stop?

56. A pendulum bob swings through an arc 40 cm long on its first swing. Each swing thereafter, it swings only 80% as far as on the previous swing. How far will it swing altogether before coming to a complete stop?

57. Suppose you could save $1 on January 1, $2 on January 2, $4 on January 3, and so on. What amount would you save on January 31? What would be the total amount of your savings during January? (Hint: $2^{31} = 2,147,483,648$.)

58. Richland Oil has a well which produced $4,000,000 of income its first year. Each year thereafter, the well produced half as much income as the previous year. What is the total amount of income produced by the well in six years?

59. The final step in processing a black and white photographic print is to immerse the print in a chemical called "fixer." The print is then washed in running water. Under certain conditions, 98% of the fixer in a print will be removed with 15 minutes of washing. How much of the original fixer would be left after one hour of washing?

60. A sequence of equilatral triangles is constructed. The first triangle has sides 2 m in length. To get the second triangle, midpoints of the sides of the original triangle are connected. What is the length of the side of the eighth such triangle? See the figure below.

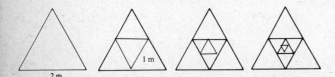

61. A sequence of equilateral triangles is constructed. The first triangle has sides 2 m in length. To get the second triangle, midpoints of the sides of the original triangle are connected. (See Exercise 60.) If this process could be continued indefinitely, what would be the total perimeter of all the triangles?

62. What would be the total area of all the triangles of Exercise 61, disregarding the overlapping?

63. A scientist has a vat containing 100 liters of a pure chemical. Twenty liters are drained and replaced with water. After complete mixing, twenty liters of the mixture is drained and replaced with water. What will be the strength of the mixture after nine such drainings?

64. The half-life of a radioactive substance is the time it takes for half the substance to decay. Suppose the half-life of a substance is 3 years, and that 10^{15} molecules of the substance are present initially. How many molecules will be present after 15 years?

65. Each year a machine loses 20% of the value it had at the beginning of the year. Find the value of the machine at the end of 6 years if it cost $100,000 new.

66. A bicycle wheel rotates 400 times in one minute. If the rider removes his or her feet from the pedals, the wheel will start to slow down. Each minute, it will rotate only 3/4 as many times as in the preceeding minute. How many times will the wheel rotate in the fifth minute after the rider's feet were removed from the pedals? (Compare your answer to that of Exercise 55.)

67. A piece of paper is .008 inch thick. Suppose the paper is folded in half, so that its thickness doubles, for 12 times in a row. How thick would the final stack of paper be?

68. Fruit-and-vegetable dealer Marcia Odjakjian paid 10¢ per pound for 10,000 pounds of onions. Each week the price she charges increases by .1¢ per pound, while the onions lose 5% of their weight. If she sells the onions after six weeks, did she make or lose money? How much?

69. Find three numbers x, y, and z that are consecutive terms of both an arithmetic sequence and a geometric sequence.

70. Let $a_1, a_2, a_3, \ldots$ and $b_1, b_2, b_3, \ldots$ be geometric sequences. Let $d_n = c \cdot a_n \cdot b_n$ for any real number c and every positive integer n. Show that $d_1, d_2, d_3, \ldots$ is a geometric sequence.

Find a_1 and r for each of the following geometric sequences.

71. $a_2 = 6$, $a_6 = 486$

72. $a_3 = -12$, $a_6 = 96$

73. $a_2 = 64$, $a_8 = 1$ 　　74. $a_2 = 100$, $a_5 = .1$

Explain why the following sequences are geometric.

75. log 6, log 36, log 1296, log 1,679,616, . . .

76. log 2, log 4, log 16, log 256, . . .

Prove each of the following results by mathematical induction.

77. $a_n = a_1 \cdot r^{n-1}$ 　　78. $S_n = \dfrac{a_1(1 - r^n)}{1 - r}$

11.5 Counting Problems

If there are 3 roads from Albany to Baker and 2 roads from Baker to Creswich, in how many ways can we travel from Albany to Creswich by way of Baker? For each of the 3 roads from Albany to Baker, there are 2 different roads from Baker to Creswich. Hence, there are $3 \cdot 2 = 6$ different ways to make the trip, as shown in the **tree diagram** of Figure 11.2.

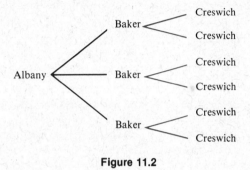

Figure 11.2

This example illustrates the following fundamental principle of counting.

Successive Events Property

If one event can occur in m ways and a second event can occur in n ways, then both events can occur in mn ways, provided the outcome of the first event does not influence the outcome of the second.

The successive events property can be extended to any finite number of events, provided the outcome of no one event influences the outcome of another. Such events are called **independent events**.

EXAMPLE 1   ☐   Maxwell's Restaurant offers 3 salads, 5 main dishes, and 2 desserts. Use the successive events property to find the number of different meals that can be selected.

We have three independent events: selecting a salad, selecting a main dish, and selecting a dessert. The first event can occur in 3 ways, the second event can occur in 5 ways, and the third event can occur in 2 ways; thus there are

$$3 \cdot 5 \cdot 2 = 30 \text{ different meals.} \quad \blacksquare$$

EXAMPLE 2   ☐   David Krause has 5 different history books which he wishes to arrange in a row on his desk. How many different arrangements are possible?

We have five events: selecting a book for the first space, selecting a book for the second space, and so on. Here the outcome of the first event *does* influence the outcome of the other events (since one book has already been chosen). However, we can say that for the second space Krause has only 4 choices, for the third space 3 choices, and so on, so that the outcome of any one event no longer influences the outcome of another. Now we can use the successive events property to find that there are

$$5 \cdot 4 \cdot 3 \cdot 2 \cdot 1 = 120 \text{ different arrangements.} \quad \square$$

Earlier in this chapter, we wrote products of this type with factorial notation. The product $5 \cdot 4 \cdot 3 \cdot 2 \cdot 1$ was written $5!$ ("5-factorial"). If David Krause of Example 2 had 7 books, instead of 5, he could arrange them in

$$7! = 7 \cdot 6 \cdot 5 \cdot 4 \cdot 3 \cdot 2 \cdot 1 = 5040$$

different ways. Recall also that we defined $0! = 1$.

EXAMPLE 3 $\square$ Suppose David wishes to place only 3 of the 5 history books on his desk. How many arrangements of 3 books are possible?

He again has 5 ways to fill the first space, 4 ways to fill the second space, and 3 ways to fill the third. Since he wants to use only 3 books, there are only 3 spaces to be filled (3 events) instead of 5. Thus there are

$$5 \cdot 4 \cdot 3 = 60 \text{ arrangements.} \quad \square$$

The number 60 in the example above is called the number of permutations of 5 things taken 3 at a time, written $P(5, 3) = 60$. In Example 2 we found the number of ways of arranging 5 elements from a set of 5 elements, written $P(5, 5) = 120$.

By the successive events property we have the following result.

Permutations of n Elements n at a Time

If $P(n, n)$ denotes the number of permutations of n elements taken n at a time, then

$$P(n, n) = n!.$$

A **permutation** of n elements taken r at a time is the number of ways we can *arrange* r elements from a set of n elements. Generalizing from the examples above, we have

$$P(n, r) = n(n - 1)(n - 2) \cdots (n - r + 1)$$
$$= \frac{n(n - 1)(n - 2) \cdots (n - r + 1)(n - r) \cdots (2)(1)}{(n - r)(n - r - 1) \cdots (2)(1)}$$
$$= \frac{n!}{(n - r)!}.$$

Summarizing, we have the following result.

**Permutations
of n Elements
r at a Time**

If $P(n, r)$ denotes the number of permutations of n elements taken r at a time, then

$$P(n, r) = \frac{n!}{(n - r)!}.$$

**EXAMPLE
4**

Suppose 8 people enter a swimming meet. In how many ways could the gold, silver, and bronze prizes be awarded?

Using the successive events property, there are 3 choices to be made, so that $P(8, 3) = 8 \cdot 7 \cdot 6 = 336$. However, we can also use the formula given above for $P(n, r)$ to get the same result.

$$P(8, 3) = \frac{8!}{5!} = \frac{8 \cdot 7 \cdot 6 \cdot 5 \cdot 4 \cdot 3 \cdot 2 \cdot 1}{5 \cdot 4 \cdot 3 \cdot 2 \cdot 1}$$

$$= 8 \cdot 7 \cdot 6$$

$$= 336 \quad \blacksquare$$

**EXAMPLE
5**

In how many ways can 6 students be seated in a row of 6 desks?
Here we have

$$P(6, 6) = 6! = 6 \cdot 5 \cdot 4 \cdot 3 \cdot 2 \cdot 1 = 720. \quad \blacksquare$$

In the work above, we discussed a method for finding the number of ways we can arrange r elements from a set of n elements. Sometimes we are not interested in the arrangement (or order) of the elements. For example, suppose three people, Ms. Opelka, Mr. Adams, and Ms. Jacobs, apply for 2 identical jobs. Ignoring all other factors, in how many ways can the personnel officer select 2 people from the 3 applicants? Here the arrangement (or order) of the people is unimportant. Selecting Ms. Opelka and Mr. Adams is the same as selecting Mr. Adams and Ms. Opelka. Therefore, there are only 3 ways to select 2 of the three applicants:

Ms. Opelka, Mr. Adams

Ms. Opelka, Ms. Jacobs

Mr. Adams, Ms. Jacobs.

These three choices are called the combinations of 3 elements taken 2 at a time. In general, a **combination** of n elements taken r at a time is the number of ways in which we *choose* r elements from n elements.

Each combination of r elements forms $r!$ permutations. Therefore, we can find the number of combinations of n elements taken r at a time by dividing the number of permutations, $P(n, r)$, by $r!$ to get

$$\frac{P(n, r)}{r!}$$

combinations. This expression can be rewritten as

$$\frac{P(n,\ r)}{r!} = \frac{\dfrac{n!}{(n-r)!}}{r!} = \frac{n!}{(n-r)!\,r!}.$$

The symbol $\binom{n}{r}$ is used to represent the number of combinations of n things taken r at a time. Using this symbol and the results above, we have the following theorem.

Combinations of n Elements r at a Time

If $\binom{n}{r}$ represents the number of combinations of n things taken r at a time, then

$$\binom{n}{r} = \frac{n!}{(n-r)!\,r!}.$$

This same result was obtained in our work with the binomial theorem in Section 2 of this chapter. There we found $\binom{n}{r}$ as the coefficients in the expansion of a binomial.

In the discussion above, we found that $\binom{3}{2} = 3$. We can get the same answer using the formula. We have

$$\binom{3}{2} = \frac{3!}{(3-2)!\,2!} = \frac{3\cdot2\cdot1}{1\cdot2\cdot1} = 3.$$

EXAMPLE 6 ☐ How many different committees of 3 people can be chosen from a group of 8 people?

We are not interested in the order in which the committee is chosen, so we use combinations to get

$$\binom{8}{3} = \frac{8!}{5!3!} = \frac{8\cdot7\cdot6\cdot5\cdot4\cdot3\cdot2\cdot1}{5\cdot4\cdot3\cdot2\cdot1\cdot3\cdot2\cdot1} = 56.\quad ☐$$

EXAMPLE 7 ☐ From a group of 30 employees, 3 are to be selected to work on a special project.

(a) In how many different ways can the employees be selected?

Here we wish to know the number of 3-element combinations from a set of 30 elements. (We want combinations, not permutations, because we don't care about order within the group of 3.) Using the formula gives

$$\binom{30}{3} = \frac{30!}{27!3!} = 4060.$$

There are 4060 ways to select the project group.

(b) In how many different ways can the group of 3 be selected if it has already been decided that a certain man must work on the project?

Since one man has already been selected for the project, the problem is reduced to selecting 2 more from the remaining 29 employees. Thus, we have

$$\binom{29}{2} = \frac{29!}{27!2!} = 406.$$

In this case, the project group can be selected in 406 different ways. ☐

EXERCISES
11.5

Evaluate each of the following.

1. $P(7, 7)$ **2.** $P(5, 3)$

3. $P(6, 5)$ **4.** $P(4, 2)$

5. $P(10, 2)$ **6.** $P(8, 2)$

7. $P(8, 3)$ **8.** $P(11, 4)$

9. $P(7, 1)$ **10.** $P(18, 0)$

11. $P(9, 0)$ **12.** $P(14, 1)$

13. $\binom{6}{5}$ **14.** $\binom{4}{2}$

15. $\binom{8}{5}$ **16.** $\binom{10}{2}$

17. $\binom{15}{4}$ **18.** $\binom{9}{3}$

19. $\binom{10}{7}$ **20.** $\binom{10}{3}$

21. $\binom{14}{1}$ **22.** $\binom{20}{2}$

23. $\binom{18}{0}$ **24.** $\binom{13}{0}$

25. In how many ways can 6 people be seated in a row of 6 seats?

26. In how many ways can 7 out of 10 people be assigned to 7 seats?

27. In how many ways can 5 bank tellers be assigned to 5 different windows? In how many ways can 10 tellers be assigned to the 5 windows?

28. A couple has narrowed down their choice of names for a new baby to 3 first names and 5 middle names. How many different first- and middle-name arrangements are possible?

29. How many different homes are available if a builder offers a choice of 5 basic plans, 3 roof styles, and 2 types of siding?

30. An automaker produces 7 models, each available in 6 colors, with 4 upholstery fabrics and 5 interior colors. How many varieties of the auto are available?

31. A concert is to consist of 5 works: two modern, two romantic, and one classical. In how many ways can the program be arranged?

32. If the program in Exercise 31 must be shortened to 3 works chosen from the 5, how many arrangements are possible?

33. In Exercise 31, how many different programs are possible if the two modern works are to be played first, then the two romantic, and then the classical?

34. How many 4-letter radio-station call letters can be made if the first letter must be K or W and no letter may be repeated? How many if repeats are allowed?

35. How many of the 4-letter call letters in Exercise 34 with no repeats end in K?

36. A business school gives courses in typing, shorthand, transcription, business English, technical writing, and accounting. How many ways can a student arrange his program if he takes 3 courses?

37. A club has 30 members. If a committee of 4 is to be selected at random, how many different committees are possible?

38. How many different samples of 3 apples can be drawn from a crate of 25 apples?

39. A group of 3 students is to be selected randomly from a group of 12 to participate in an experimental class. In how many ways can this be done? In how many ways can the group which will not participate be selected?

40. Hal's Hamburger Heaven sells hamburgers with cheese, relish, lettuce, tomato, mustard or ketchup. How many different hamburgers can be made using any 3 of the extras?

41. How many different 2-card hands can be dealt from a deck of 52 cards?

42. How many different 13-card bridge hands can be dealt from a deck of 52 cards?

43. Five cards are marked with the numbers 1, 2, 3, 4, and 5, shuffled, and 2 cards are then drawn. How many different 2-card combinations are possible?

44. If a bag contains 15 marbles, how many samples of 2 marbles can be drawn from it? How many samples of 4 marbles?

45. In Exercise 44, if the bag contains 3 yellow, 4 white, and 8 blue marbles, how many samples of 2 can be drawn in which both marbles are blue?

46. In Exercise 38, if it is known that there are 5 rotten apples in the crate:
   (a) How many samples of 3 could be drawn in which all 3 are rotten?
   (b) How many samples of 3 could be drawn in which there are 1 rotten apple and 2 good apples?

47. Glendale Heights City Council is composed of 5 liberals and 4 conservatives. Three members are to be selected randomly as delegates to a convention.
   (a) How many delegations are possible?
   (b) How many delegations could have all liberals?
   (c) How many delegations could have 2 liberals and 1 conservative?
   (d) If 1 member of the council serves as mayor, how many delegations are possible which include the mayor?

48. Seven factory workers decide to send a delegation of 2 to their supervisor to discuss their grievances.
   (a) How many different delegations are possible?
   (b) If it is decided that a certain employee must be in the delegation, how many different delegations are possible?
   (c) If there are 2 women and 5 men in the group, how many delegations would include a woman?

A poker hand is made up of 5 cards drawn at random from a deck of 52 cards. Any 5 cards in one suit are called a flush. The 5 highest cards, that is, the A, K, Q, J, and 10 of any one suit are called a royal flush. Use combinations to set up each of the following. Do not evaluate.

49. Find the total number of all possible poker hands.

50. How many royal flushes in hearts are possible?

51. How many royal flushes in any of the four suits are possible?

52. How many flushes in hearts are possible?

53. How many flushes in any of the four suits are possible?

54. How many combinations of 3 aces and 2 eights are possible?

Solve the following problems by using either combinations or permutations.

55. In how many ways can the letters of the word TOUGH be arranged?

56. If Matthew has 8 courses to choose from, how many ways can he arrange his schedule if he must pick 4 of them?

57. How many samples of 3 pineapples can be drawn from a crate of 12?

58. Velma specializes in making different vegetable soups with carrots, celery, beans, peas, mushrooms, and potatoes. How many different soups can she make using any 4 ingredients?

Solve each of the following equations for n.

59. $P(n, 3) = 8 \cdot P(n - 1, 2)$

60. $30 \cdot P(n, 2) = P(n + 2, 4)$

61. $\dbinom{n + 2}{4} = 15 \cdot \dbinom{n}{4}$

62. $\dbinom{n}{n - 2} = 66$

Prove each of the following statements for positive integers n and r, with r ≤ n.

63. $P(n, n - 1) = P(n, n)$

64. $P(n, 1) = n$

65. $P(n, 0) = 1$

66. $\dbinom{n}{n} = 1$

67. $\dbinom{n}{0} = 1$

68. $\dbinom{n}{n-1} = n$

69. $\dbinom{n}{n-r} = \dbinom{n}{r}$

70. $\dbinom{2n}{r} \le \dbinom{2n}{n}$ for all r, $0 \le r \le 2n$

71. $\dbinom{n}{r} = \dbinom{n-2}{r-2} + 2 \cdot \dbinom{n-2}{r-1} + \dbinom{n-2}{r}$ for $2 \le r \le n-2$.

72. If $n > 1$, then $r \cdot \dbinom{n}{r} = n \cdot \dbinom{n-1}{r-1}$.

73. $\dbinom{n}{1} + \dbinom{n}{2} + \cdots + \dbinom{n}{n} = 2^n - 1$

74. Suppose m and n are any positive integers greater than 1. Decide which is larger, $(mn)!$ or $(m!)(n!)$.

11.6 Basics of Probability

The study of probability theory has become increasingly popular because of the wide range of practical applications. In this section, we introduce the basic ideas of probability.

Consider an experiment which has one or more possible *outcomes,* each of which is equally likely to occur. For example, the experiment of tossing a coin has two equally likely possible outcomes: landing heads up (H) or landing tails up (T). Also, the experiment of rolling a die has six equally likely outcomes: landing so the face which is up shows 1, 2, 3, 4, 5, or 6 points.

The set S of all possible outcomes of a given experiment is called the **sample space** for the experiment. (In this text all sample spaces are finite.) A sample space for the experiment of tossing a coin consists of the outcomes H and T. This sample space can be written in set notation as

$$S = \{H, T\}.$$

Similarly, a sample space for the experiment of rolling a single die is

$$S = \{1, 2, 3, 4, 5, 6\}.$$

Any subset of the sample space is called an **event.** In the experiment with the die, for example, "the number showing is a three" is an event, say E_1, such that $E_1 = \{3\}$. "The number showing is greater than three" is also an event, say E_2, such that $E_2 = \{4, 5, 6\}$. To represent the number of outcomes which belong to event E, we use $n(E)$. Then $n(E_1) = 1$ and $n(E_2) = 3$.

The **probability** of an event E, written $P(E)$, is the ratio of *the number of outcomes in sample space S which belong to event E* compared to *the total number of outcomes in sample space S.* Thus,

Definition:
Probability of
Event E

$$P(E) = \frac{n(E)}{n(S)}.$$

To use this definition to find the probability of the event E_1, given above, we note that the sample space for the experiment is $S = \{1, 2, 3, 4, 5, 6\}$ and the desired event is $E_1 = \{3\}$. We already know $n(E_1) = 1$ and we can see there are 6 outcomes in the sample space. Thus,

$$P(E_1) = \frac{n(E_1)}{n(S)} = \frac{1}{6}.$$

EXAMPLE 1   A single die is rolled. Write the following events in set notation and give the probability for each event.
(a) E_3: the number showing is even.
(b) E_4: the number showing is greater than 4.
(c) E_5: the number showing is less than 10.
(d) E_6: the number showing is 8.
Using the definitions above, we have the following.

(a) $E_3 = \{2, 4, 6\}$ and $P(E_3) = \frac{3}{6} = \frac{1}{2}$. (Recall that there are six outcomes in the

sample space and notice that $n(E_3) = 3$.)

(b) $E_4 = \{5, 6\}$ and $P(E_4) = \frac{2}{6} = \frac{1}{3}$.

(c) $E_5 = \{1, 2, 3, 4, 5, 6\}$ and $P(E_5) = \frac{6}{6} = 1$.

(d) $E_6 = \varnothing$ and $P(E_6) = \frac{0}{6} = 0$.   ▢

Note that $E_5 = S$. Therefore the event E_5 is certain to occur every time the experiment is performed. We see that an event which is certain to occur, such as E_5, has a probability of 1. On the other hand, $E_6 = \varnothing$ and $P(E_6)$ is 0. The probability of an impossible event, such as E_6, is always 0, since none of the outcomes in the sample space satisfy the event.

The set of all outcomes in the sample space which do *not* belong to event E is called the **complement** of E, written E'. For example, in the experiment of drawing a single card from a standard deck of 52 cards, let E be the event "the card is an ace." Then E' is the event "the card is not an ace." From the definition of E', we see that for any event E,

$$E \cup E' = S \qquad \text{and} \qquad E \cap E' = \varnothing.*$$

We can illustrate some of these probability concepts with a figure called a **Venn**

*The **union** of two sets A and B is the set $A \cup B$ made up of all the elements from either A or B, or both. The **intersection** of sets A and B, written $A \cap B$, is made up of all the elements that belong to both sets at the same time.

Diagram. The rectangle in Figure 11.3 represents the sample space in an experiment. The area inside the circle represents event E, while the area inside the rectangle, but outside the circle, represents event E'.

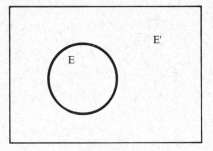

Figure 11.3

EXAMPLE 2 In the experiment of drawing a card from a well-shuffled deck, find the probability of events E, the card is an ace, and E'.

Since there are four aces in the deck of 52 cards, $n(E) = 4$. Also $n(S) = 52$. Therefore, $P(E) = \dfrac{n(E)}{n(S)} = \dfrac{4}{52} = \dfrac{1}{13}$. Of the 52 cards, 48 are not aces, so that we have $P(E') = \dfrac{n(E')}{n(S)} = \dfrac{48}{52} = \dfrac{12}{13}$. $\blacksquare$

In Example 2, note that $P(E) + P(E') = (1/13) + (12/13) = 1$. This is always true for any event E and its complement E', so that in general,

$$P(E) + P(E') = 1.$$

This can be restated as

$$P(E) = 1 - P(E') \qquad \text{or} \qquad P(E') = 1 - P(E).$$

These two equations suggest an alternate way to compute the probability of an event. For example, if it is known that $P(E) = 1/10$, then

$$P(E') = 1 - \frac{1}{10} = \frac{9}{10}.$$

Sometimes probability statements are expressed in terms of odds, a comparison of $P(E)$ with $P(E')$. The **odds** in favor of an event E are expressed as the ratio of $P(E)$ to $P(E')$ or as the fraction $P(E)/P(E')$. For example, if the probability of rain can be established as $1/3$, the odds that it will rain are

$$P(\text{rain}) \text{ to } P(\text{no rain}) = \frac{1}{3} \text{ to } \frac{2}{3}$$

$$= \frac{1/3}{2/3}$$

$$= \frac{1}{2} \qquad \text{or} \qquad 1 \text{ to } 2.$$

On the other hand, the odds that it will not rain are 2 to 1 (or 2/3 to 1/3). If we know that the odds in favor of an event are, say, 3 to 5, then we can see that the probability of the event is 3/8, while the probability of the complement of the event is 5/8. In general, if the odds favoring event E are m to n, then

$$P(E) = \frac{m}{m + n} \quad \text{and} \quad P(E') = \frac{n}{m + n}.$$

We now consider the probability of a **compound event** which involves an *alternative*, such as E or F, where E and F are simple events. For example, in the experiment of rolling a die, suppose H is the event "the result is a 3," and K is the event "the result is an even number." What is the probability of "the result is a 3 or an even number"? We have

$$H = \{3\} \qquad\qquad P(H) = \frac{1}{6}$$

$$K = \{2, 4, 6\} \qquad\qquad P(K) = \frac{3}{6} = \frac{1}{2};$$

therefore $\quad H \text{ or } K = \{2, 3, 4, 6\} \qquad P(H \text{ or } K) = \frac{4}{6} = \frac{2}{3};$

Notice that $P(H) + P(K) = P(H \text{ or } K)$.

Before we assume that this relationship is true in general, let us consider another event for this experiment, "the result is a 2," event G. We have

$$G = \{2\} \qquad\qquad P(G) = \frac{1}{6}$$

$$K = \{2, 4, 6\} \qquad\qquad P(K) = \frac{3}{6} = \frac{1}{2};$$

therefore $\quad K \text{ or } G = \{2, 4, 6\} \qquad P(K \text{ or } G) = \frac{3}{6} = \frac{1}{2}.$

In this case $P(K) + P(G) \neq P(K \text{ or } G)$.

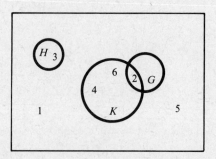

Figure 11.4

As Figure 11.4 shows, the difference in the two examples above comes from the fact that events H and K cannot occur simultaneously. Such events are called **mutually exclusive events.** We see that $H \cap K = \varnothing$, which is true in general for any two mutually exclusive events. Events K and G, however, can occur simultaneously. Both are satisfied if the result of the roll is a 2, the element in their intersection $(K \cap G = \{2\})$. Generalizing the above discussion, we have the following theorem.

Probability of Alternate Events

For any events E and F,

$$P(E \text{ or } F) = P(E \cup F) = P(E) + P(F) - P(E \cap F).$$

EXAMPLE 3   One card is drawn from a well-shuffled deck of 52 cards. What is the probability that it is (a) an ace or a spade? (b) a three or a king?

(a) The events ''drawing an ace'' and ''drawing a spade'' are not mutually exclusive since it is possible to draw the ace of spades, an outcome satisfying both events. Thus,

$$P(\text{ace or spade}) = P(\text{ace}) + P(\text{spade}) - P(\text{ace and spade})$$

$$= \frac{4}{52} + \frac{13}{52} - \frac{1}{52} = \frac{16}{52} = \frac{4}{13}.$$

(b) ''Drawing a 3'' and ''drawing a king'' are mutually exclusive events because it is impossible to draw one card which is both a 3 and a king. We have

$$P(3 \text{ or } K) = P(3) + P(K) - P(3 \text{ and } K) = \frac{4}{52} + \frac{4}{52} - 0 = \frac{8}{52} = \frac{2}{13}. \quad \blacksquare$$

EXAMPLE 4   For the experiment consisting of one roll of a pair of dice, find the probability that the sum of the points showing is at most 4.

We can rewrite ''at most 4'' as ''2 or 3 or 4.'' (A sum of 1 is meaningless here.) Then we have

$$P(\text{at most 4}) = P(2 \text{ or } 3 \text{ or } 4) \tag{1}$$

$$= P(2) + P(3) + P(4),$$

since the events represented by ''2,'' ''3,'' and ''4'' are mutually exclusive. By the definition of probability, $P(2) = 1/36$, $P(3) = 2/36$ (we can get a 2 on the first die and a 1 on the second or a 1 on the first and a 2 on the second), and $P(4) = 3/36$. Substituting into equation (1) above we find

$$P(\text{at most 4}) = \frac{1}{36} + \frac{2}{36} + \frac{3}{36} = \frac{6}{36} = \frac{1}{6}. \quad \blacksquare$$

EXERCISES
11.6

Write a sample space with equally likely outcomes for each of the following experiments.

1. A two-headed coin is tossed once.

2. Two ordinary coins are tossed.

3. Three ordinary coins are tossed.

4. Slips of paper marked with the numbers 1, 2, 3, 4, and 5 are placed in a box. After mixing well, two slips are drawn.

5. An unprepared student takes a three-question true/false quiz in which he guesses the answer to all three questions.

6. A die is rolled and then a coin is tossed.

Write the following events in set notation and give the probability of each event.

7. In the experiment of Exercise 2:
   (a) both coins show the same face;
   (b) at least one coin turns up heads.

8. In Exercise 1:
   (a) the result of the toss is heads;
   (b) the result of the toss is tails.

9. In Exercise 4:
   (a) both slips are marked with even numbers;
   (b) both slips are marked with odd numbers;
   (c) both slips are marked with the same number;
   (d) one slip is marked with an odd number, the other with an even number.

10. In Exercise 5:
    (a) the student gets all three answers correct;
    (b) he gets all three answers wrong;
    (c) he gets exactly two answers correct;
    (d) he gets at least one answer correct.

11. A marble is drawn at random from a box containing 3 yellow, 4 white, and 8 blue marbles. Find the following probabilities.
    (a) A yellow marble is drawn;
    (b) a blue marble is drawn;

(c) a black marble is drawn.

(d) What are the odds in favor of drawing a yellow marble?

(e) What are the odds against drawing a blue marble?

12. A baseball player with a batting average of .300 comes to bat. What are the odds in favor of his getting a hit?

13. In Exercise 4, what are the odds that the sum of the numbers on the two slips of paper is 5?

14. If the odds that it will rain are 4 to 5, what is the probability of rain?

15. If the odds that a candidate will win an election are 3 to 2, what is the probability that the candidate will lose?

16. A card is drawn from a well-shuffled deck of 52 cards. Find the probability that the card is

    (a) a 9

    (b) black

    (c) a black 9

    (d) a heart

    (e) the 9 of hearts

    (f) a face card (K, Q, J of any suit)

17. Mrs. Elliott invites 10 relatives to a party: her mother, two uncles, three brothers, and four cousins. If the chances of any one guest arriving first are equally likely, find the following probabilities.

    (a) The first guest is an uncle or a brother.

    (b) The first guest is a brother or cousin.

    (c) The first guest is a brother or her mother.

18. One card is drawn from a standard deck of 52 cards. What is the probability that the card is

    (a) a 9 or a 10?

    (b) red or a 3?

    (c) a heart or black?

(d) less than a 4? (Consider Aces as 1's.)

19. Two dice are rolled. Find the probability that

(a) the sum of the points is at least 10;

(b) the sum of the points is either 7 or at least 10;

(c) the sum of the points is 2 or the dice both show the same number.

20. A student estimates that his probability of getting an A in a certain course is .4; a B, .3; a C, .2; and a D, .1.

(a) Assuming that only the grades A, B, C, D, and F are possible, what is the probability that he will fail the course?

(b) What is the probability that he will receive a grade of C or better?

(c) What is the probability that he will receive at least a B in the course?

(d) What is the probability that he will get at most a C in the course?

21. If a marble is drawn from a bag containing 3 yellow, 4 white, and 8 blue marbles, what is the probability that

(a) the marble is either yellow or white?

(b) it is either yellow or blue?

(c) it is either red or white?

CHAPTER SUMMARY
11

Key Words

mathematical induction	*n*th term	independent event
Pascal's Triangle	arithmetic sequence	permutations
factorial	common difference	combinations
binomial coefficient	index of summation	outcomes
binomial theorem	geometric sequence	sample space
sequence function	common ratio	event
terms of a sequence	partial sum	probability
sequence	infinite geometric sequence	Venn diagram
general term	successive events property	alternate events

CHAPTER REVIEW EXERCISES
11

Use mathematical induction to prove that each of the following is true for every positive integer n.

1. $1 + 3 + 5 + 7 + \cdots + (2n - 1) = n^2$

2. $2 + 6 + 10 + 14 + \cdots + (4n - 2) = 2n^2$

3. $2^2 + 4^2 + 6^2 + \cdots + (2n)^2 = $

$$\frac{2n(n + 1)(2n + 1)}{3}$$

4. $2 + 2^2 + 2^3 + \cdots + 2^n = 2(2^n - 1)$

5. $1 \cdot 4 + 2 \cdot 9 + 3 \cdot 16 + \cdots + n(n + 1)^2$

$$= \frac{n(n + 1)(n + 2)(3n + 5)}{12}$$

6. $1^3 + 3^3 + 5^3 + \cdots + (2n - 1)^3 = $

$$n^2(2n^2 - 1)$$

Use the binomial theorem to expand each of the following.

7. $(x + 2y)^4$ **8.** $(3z - 5w)^3$

9. $\left(3\sqrt{x} - \dfrac{1}{\sqrt{x}}\right)^5$ **10.** $(m^3 - m^{-2})^4$

Find the indicated term or terms for each of the following expansions.

11. fifth term of $(3x - 2y)^6$

12. eighth term of $(2m + n^2)^{12}$

13. first four terms of $(3 + x)^{16}$

14. last three terms of $(2m - 3n)^{15}$

Write the first five terms for each of the following sequences.

15. $a_n = 2(n + 3)$

16. $a_n = n(n + 1)$

17. $a_1 = 5, a_2 = 3, a_n = a_{n-1} - a_{n-2}$ for $n \geq 3$

18. $b_1 = -2, b_2 = 2, b_3 = -4, b_n = -2 \cdot b_{n-2}$ if n is even, and $b_n = 2 \cdot b_{n-2}$ if n is odd.

19. arithmetic, $a_2 = 6, d = -4$

20. arithmetic, $a_3 = 9, a_4 = 7$

21. arithmetic, $a_1 = 3 - \sqrt{5}, a_2 = 4$

22. arithmetic, $a_3 = \pi, a_4 = 0$

23. geometric, $a_1 = 4, r = 2$

24. geometric, $a_4 = 8, r = 1/2$

25. geometric, $a_1 = -3, a_2 = 4$

26. geometric, $a_3 = 8, a_5 = 72$

27. A certain arithmetic sequence has $a_6 = -4$ and $a_{17} = 51$. Find a_1 and a_{20}.

28. For a given geometric sequence, $a_1 = 4$ and $a_5 = 324$. Find a_6.

Find a_8 for each of the following arithmetic sequences.

29. $a_1 = 6, d = 2$

30. $a_1 = -4, d = 3$

31. $a_1 = 6x - 9, a_2 = 5x + 1$

32. $a_3 = 11m, a_5 = 7m - 4$

Find S_{12} for each of the following arithmetic sequences.

33. $a_1 = 2, d = 3$

34. $a_2 = 6, d = 10$

35. $a_1 = -4k, d = 2k$

Find a_5 for each of the following geometric sequences.

36. $a_1 = 3, r = 2$

37. $a_2 = 3125, r = 1/5$

38. $a_1 = 6, a_3 = 24$

39. $a_1 = 5x, a_2 = x^2$

40. $a_2 = \sqrt{6}, a_4 = 6\sqrt{6}$

Find S_4 for each of the following geometric sequences.

41. $a_1 = 1, r = 2$ **42.** $a_1 = 3, r = 3$

43. $a_1 = 2k, a_2 = -4k$

Evaluate each of the following sums which exist.

44. $18 + 9 + 9/2 + 9/4 + \cdots$

45. $20 + 15 + 45/4 + 135/16 + \cdots$

46. $-5/6 + 5/9 - 10/27 + \cdots$

47. $1/16 + 1/8 + 1/4 + 1/2 + \cdots$

48. $.9 + .09 + .009 + .0009 + \cdots$

Evaluate each of the following sums which exist.

49. $\displaystyle\sum_{i=1}^{4} \frac{2}{i}$ **50.** $\displaystyle\sum_{i=1}^{7} (-1)^{i+1} \cdot 6$

51. $\displaystyle\sum_{i=4}^{8} 3i(2i - 5)$ **52.** $\displaystyle\sum_{i=1}^{6} i(i + 2)$

53. $\displaystyle\sum_{i=1}^{4} \frac{i + 1}{i}$ **54.** $\displaystyle\sum_{i=1}^{12} (8i + 2)$

55. $\displaystyle\sum_{i=1}^{10,000} i$ **56.** $\displaystyle\sum_{i=1}^{6} 4 \cdot 3^i$

57. $\displaystyle\sum_{i=1}^{4} 8 \cdot 2^i$ **58.** $\displaystyle\sum_{i=1}^{\infty} \left(\frac{5}{8}\right)^i$

59. $\displaystyle\sum_{i=1}^{\infty} -10\left(\frac{5}{2}\right)^i$ **60.** $\displaystyle\sum_{i=1}^{\infty} 6(2/3)^i$

Find each of the following.

61. $P(9, 2)$

62. $P(8, 3)$

63. $P(6, 0)$

64. $\binom{8}{3}$

65. $\binom{10}{5}$

66. Four students are to be assigned to 4 different summer jobs. Each student is qualified for all 4 jobs. In how many ways can he jobs be assigned?

67. Nine football teams are competing for 1st, 2nd, and 3rd place titles in a statewide tournament. In how many ways can the winners be determined?

68. How many different license-plate numbers can be formed using 3 letters followed by 3 digits if no repeats are allowed? How many if there are no repeats and either letters or numbers come first?

A card is drawn from a standard deck of 52 cards. Find the probability that the card is

69. a black king

70. a face card or an ace

71. an ace or a diamond

72. not a diamond

73. not a diamond or not black

74. a diamond and black.

75. There is some flexibility in the Σ notation. For instance, $\Sigma_{i=3}^{8}\, i$, $\Sigma_{j=3}^{8}\, j$, $\Sigma_{k=5}^{10}\,(k - 2)$, and $\Sigma_{L=0}^{5}\,(L + 3)$ each equals $3 + 4 + 5 + 6 + 7 + 8$. In each of parts (a), (b), and (c), you are given three expressions; two give the same sum and one is different. Identify the one that does not equal the other two.

(a) $\sum_{k=0}^{7}(2k + 1)$, $\sum_{j=1}^{15} j$, $\sum_{i=2}^{9}(2i - 3)$

(b) $\sum_{k=1}^{7} k^2$, $\sum_{j=0}^{6}(7 - j)^2$, $\sum_{i=1}^{7}(7 - i)^2$

(c) $\left(\sum_{k=7}^{11} k\right)^4$, $\sum_{m=-11}^{-7} m^4$, $\sum_{n=7}^{11} n^4$

76. Prove each of the following statements for real numbers a and b.

(a) $|a| < |b|$ if and only if $a^2 < b^2$

(b) $|a - b| \geq \||a| - |b|\|$

(c) $|a_1 + a_2 + \cdots + a_n| \leq |a_1| + |a_2| + \cdots + |a_n|$

(d) $|a_1 + a_2 + \cdots + a_n| \geq |a_1| - |a_2| - \cdots - |a_n|$

Appendix   Rotation of Axes

If we begin with an xy-coordinate system having origin 0 and rotate the axes about 0 through an angle θ, the new coordinate system is called a **rotation** of the xy-system. We can use trigonometric identities to obtain equations for converting the coordinates of a point from the xy-system to the rotated $x'y'$-system. Let P be any point other than the origin, with coordinates (x, y) in the xy-system and (x', y') in the $x'y'$-system. See Figure 1. Let $OP = r$, and let α represent the angle made by OP and the x'-axis. As shown in Figure 1 on the next page,

Exercise 75: Stanley I. Grossman, *Calculus*, 2nd Edition (Academic Press, New York, 1981), p. 256. Exercise 76: George B. Thomas, Jr. and Ross L. Finney, *Calculus and Analytic Geometry*, 5th Edition (Addison-Wesley, Reading, Massachusetts, 1979), p. 58.

$$\cos (\theta + \alpha) = \frac{OA}{r} = \frac{x}{r}$$

$$\sin (\theta + \alpha) = \frac{AP}{r} = \frac{y}{r}$$

$$\cos \alpha = \frac{OB}{r} = \frac{x'}{r}$$

$$\sin \alpha = \frac{PB}{r} = \frac{y'}{r}.$$

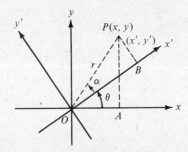

Figure 1

These four statements can be rewritten as

$$x = r \cos (\theta + \alpha), \qquad y = r \sin (\theta + \alpha),$$
$$x' = r \cos \alpha, \qquad y' = r \sin \alpha.$$

Using the trigonometric identity for the cosine of the sum of two angles, we have

$$x = r \cos (\theta + \alpha)$$
$$= r(\cos \theta \cdot \cos \alpha - \sin \theta \cdot \sin \alpha)$$
$$= (r \cos \alpha) \cos \theta - (r \sin \alpha) \sin \theta$$
$$x = x' \cos \theta - y' \sin \theta.$$

In the same way, using the identity for the sine of the sum of two angles, we have $y = x' \sin \theta + y' \cos \theta$. This proves the following result.

Rotation Equations

If the rectangular coordinate axes are rotated about the origin through an angle θ, and if the coordinates of a point P are (x, y) and (x', y') with respect to the xy-system and the $x'y'$-system, then the **rotation equations** are

$$x = x' \cos \theta - y' \sin \theta$$

and $y = x' \sin \theta + y' \cos \theta.$

EXAMPLE 1   The equation of a curve is $x^2 + y^2 + 2xy + 2\sqrt{2}x - 2\sqrt{2}y = 0$. Find the resulting equation if the axes are rotated 45°. Graph the equation.

If we let $\theta = 45°$, then using the rotation equations we have

$$x = \frac{\sqrt{2}}{2}x' - \frac{\sqrt{2}}{2}y'$$

and $y = \frac{\sqrt{2}}{2}x' + \frac{\sqrt{2}}{2}y'.$

Substituting these values in the given equation yields

$$x^2 + y^2 + 2xy + 2\sqrt{2}x - 2\sqrt{2}y = 0$$

$$\left[\frac{\sqrt{2}}{2}x' - \frac{\sqrt{2}}{2}y'\right]^2 + \left[\frac{\sqrt{2}}{2}x' + \frac{\sqrt{2}}{2}y'\right]^2$$

$$+ 2\left[\frac{\sqrt{2}}{2}x' - \frac{\sqrt{2}}{2}y'\right] \cdot \left[\frac{\sqrt{2}}{2}x' + \frac{\sqrt{2}}{2}y'\right]$$

$$+ 2\sqrt{2}\left[\frac{\sqrt{2}}{2}x' - \frac{\sqrt{2}}{2}y'\right] - 2\sqrt{2}\left[\frac{\sqrt{2}}{2}x' + \frac{\sqrt{2}}{2}y'\right] = 0.$$

Expanding these terms,

$$\frac{1}{2}x'^2 - x'y' + \frac{1}{2}y'^2 + \frac{1}{2}x'^2 + x'y' + \frac{1}{2}y'^2 + x'^2 - y'^2$$

$$+ 2x' - 2y' - 2x' - 2y' = 0.$$

Collecting terms gives

$$2x'^2 - 4y' = 0,$$

or finally, $x'^2 - 2y' = 0$

which is the equation of a parabola. The graph of this parabola is shown in Figure 2.   □

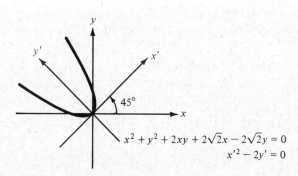

$$x^2 + y^2 + 2xy + 2\sqrt{2}x - 2\sqrt{2}y = 0$$
$$x'^2 - 2y' = 0$$

Figure 2

We have learned to graph equations stated in the general form $Ax^2 + Cy^2 + Dx + Ey + F = 0$. As we saw in the preceding example, the rotation of axes eliminated the xy term. Thus, to graph an equation which has an xy-term, it is necessary to find an appropriate angle of rotation to eliminate the xy-term. The necessary angle of rotation can be determined by using the next result. The proof is quite lengthy and is not presented here.

Angle of Rotation

The xy-term is removed from the general equation

$$Ax^2 + Bxy + Cy^2 + Dx + Ey + F = 0,$$

by a rotation of the axes through an angle θ, $0° < \theta < 90°$, where

$$\cot 2\theta = \frac{A - C}{B}.$$

This result can be used to find the appropriate angle of rotation, θ. To find the rotation equations, we must then find $\sin \theta$ and $\cos \theta$. The following example illustrates a way to obtain $\sin \theta$ and $\cos \theta$ from $\cot 2\theta$ without first identifying the angle θ.

EXAMPLE 2   ☐   Rotate the axes and graph $52x^2 - 72xy + 73y^2 = 200$.
Here we have

$$\cot 2\theta = \frac{52 - 73}{-72} = \frac{-21}{-72} = \frac{7}{24}.$$

To find $\sin \theta$ and $\cos \theta$, we use the trigonometric identities

$$\sin \theta = \sqrt{\frac{1 - \cos 2\theta}{2}} \quad \text{and} \quad \cos \theta = \sqrt{\frac{1 + \cos 2\theta}{2}}.$$

By sketching a right triangle, as in Figure 3, we see that $\cos 2\theta = 7/25$. (Recall: for the two quadrants for which we are concerned, cosine and cotangent have the same sign.)

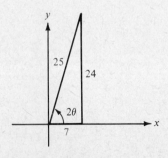

Figure 3

We have $\quad \sin \theta = \sqrt{\dfrac{1 - 7/25}{2}} = \sqrt{\dfrac{9}{25}} = \dfrac{3}{5}$

$\qquad\qquad \cos \theta = \sqrt{\dfrac{1 + 7/25}{2}} = \sqrt{\dfrac{16}{25}} = \dfrac{4}{5}.$

Thus the transformation is given by

$$x = \frac{4}{5}x' - \frac{3}{5}y' \qquad \text{or} \qquad y = \frac{3}{5}x' + \frac{4}{5}y'.$$

Substituting these values of x and y into the original equation yields

$$52\left(\frac{4}{5}x' - \frac{3}{5}y'\right)^2 - 72\left(\frac{4}{5}x' - \frac{3}{5}y'\right)\left(\frac{3}{5}x' + \frac{4}{5}y'\right) +$$

$$73\left(\frac{3}{5}x' + \frac{4}{5}y'\right)^2 = 200.$$

This becomes

$$52\left(\frac{16}{25}x'^2 - \frac{24}{25}x'y' + \frac{9}{25}y'^2\right) - 72\left(\frac{12}{25}x'^2 + \frac{7}{25}x'y' - \frac{12}{25}y'^2\right)$$

$$+ 73\left(\frac{9}{25}x'^2 + \frac{24}{25}x'y' + \frac{16}{25}y'^2\right) = 200.$$

Combining terms, we have

$$25x'^2 + 100y'^2 = 200.$$

Dividing both sides by 200 gives

$$\frac{x'^2}{8} + \frac{y'^2}{2} = 1,$$

which is the equation of an ellipse having intercepts $(0, \sqrt{2})$, $(0, -\sqrt{2})$, $(2\sqrt{2}, 0)$, and $(-2\sqrt{2}, 0)$ with respect to the $x'y'$-system, as shown in Figure 4, next page. $\quad\blacksquare$

The following result enables us to determine, from the general equation, the type of graph to expect.

If the equation

$$Ax^2 + Bxy + Cy^2 + Dx + Ey + F = 0$$

has a graph, then the graph will be

(a) a circle, an ellipse (or a point) if $B^2 - 4AC < 0$;

(b) a parabola (or a line or two parallel lines) if $B^2 - 4AC = 0$;

(c) a hyperbola (or two intersecting lines) if $B^2 - 4AC > 0$;

(d) a straight line if $A = B = C = 0$, and $D \neq 0$ or $E \neq 0$.

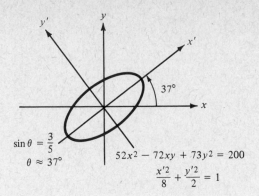

$$52x^2 - 72xy + 73y^2 = 200$$

$$\frac{x'^2}{8} + \frac{y'^2}{2} = 1$$

Figure 4

EXERCISES

Find the angle of rotation θ *which will remove the xy-term in each of the following equations.*

1. $2x^2 + \sqrt{3}xy + y^2 + x = 5$

2. $4\sqrt{3}x^2 + xy + 3\sqrt{3}y^2 = 10$

3. $3x^2 + \sqrt{3}xy + 4y^2 + 2x - 3y = 12$

4. $4x^2 + 2xy + 2y^2 + x - 7 = 0$

5. $x^2 - 4xy + 5y^2 = 18$

6. $3\sqrt{3}x^2 - 2xy + \sqrt{3}y^2 = 25$

Using the given angle of rotation, remove the xy-term and graph each of the following equations.

7. $x^2 - xy + y^2 = 6, \ \theta = 45°$

8. $2x^2 - xy + 2y^2 = 25, \ \theta = 45°$

9. $8x^2 - 4xy + 5y^2 = 36, \ \sin\theta = 2/\sqrt{5}$

10. $5y^2 + 12xy = 10, \ \sin\theta = 3/\sqrt{13}$

Remove the xy-term from each of the following equations by performing a suitable rotation. Graph each equation.

11. $3x^2 - 2xy + 3y^2 = 8$

12. $x^2 + xy + y^2 = 3$

13. $x^2 - 4xy + y^2 = -5$

14. $x^2 - 2xy + y^2 + 4\sqrt{2}x - 4\sqrt{2}y = 0$

15. $7x^2 + 6\sqrt{3}xy + 13y^2 = 64$

16. $7x^2 + 2\sqrt{3}xy + 5y^2 = 24$

17. $3x^2 - 2\sqrt{3}xy + y^2 - 2x - 2\sqrt{3}y = 0$

18. $2x^2 + 2\sqrt{3}xy + 4y^2 = 5$

In each of the following equations, remove the xy-term by rotation. Then sketch the graph.

19. $x^2 + 3xy + y^2 - 5\sqrt{2}y = 15$

20. $x^2 - \sqrt{3}xy + 2\sqrt{3}x - 3y - 3 = 0$

21. $4x^2 + 4xy + y^2 - 24x + 38y - 19 = 0$

22. $12x^2 + 24xy + 19y^2 - 12x - 40y + 31 = 0$

23. $16x^2 + 24xy + 9y^2 - 130x + 90y = 0$

24. $9x^2 - 6xy + y^2 - 12\sqrt{10}x - 36\sqrt{10}y = 0$

Answers to Selected Exercises

Chapter 1

Section 1.1   (page 8)

1. commutative   **3.** commutative   **5.** identity   **7.** associative   **9.** closure   **11.** commutative property of addition and multiplication   **13.** false   **15.** true   **17.** false (9/0 is not a rational number)   **19.** false ($0 - 1$ is not in the set)   **21.** true   **23.** yes   **25.** No [$\sqrt{2} + (-\sqrt{2}) = 0$ is not irrational]   **27.** no [$\sqrt{2}(1/\sqrt{2}) = 1$ is not irrational]   **29.** yes   **31.** no ($1 + 1 = 2$ is not in the set)   **33.** 24   **35.** 17   **37.** -21 **39.** $-6/7$ **41.** not a real number   **43.** 2.22   **45.** natural, whole, integer, rational, real   **47.** integer, rational, real   **49.** rational, real   **51.** irrational, real   **53.** ($-\sqrt{36} = -6$) integer, rational, real   **55.** meaningless   **57.** no; for example, $6 + (5 \cdot 4) \neq (6 + 5)(6 + 4)$ **59.** commutative; distributive; substitution; commutative; substitution

Section 1.2   (page 14)

1. $-9, -4, -2, 3, 8$ **3.** $-|9|, -|-6|, |-8|$ **5.** $-5, -4, -2, -\sqrt{3}, \sqrt{6}, \sqrt{8}, 3$ **7.** $3/4, 7/5, \sqrt{2}, 22/15, 8/5$ **9.** 8   **11.** 6   **13.** 4   **15.** 16   **17.** 6   **19.** -6 **21.** $8 - \sqrt{50}$ **23.** $5 - \sqrt{7}$ **25.** $\pi - 3$ **27.** $x - 4$ **29.** $8 - 2k$ **31.** $56 - 7m$ **33.** $8 + 4m$ or $4m + 8$ **35.** $y - x$ **37.** $3 + x^2$ **39.** (a) 1; (b) 1; (c) 13; (d) 14; (e) 2   **41.** (a) 2; (b) 7; (c) 2; (d) 0; (e) 9   **43.** if $x = y$ or $x = -y$ **45.** if $y = 0$ or if $|x| \geq |y|$ and x and y have opposite signs   **57.** -1 if $x < 0$ and 1 if $x > 0$

Section 1.3   (page 22)

1. 3375   **3.** 1/8   **5.** 31/108   **7.** $-1/64$ **9.** 8   **11.** .001042   **13.** $-72m^5$ **15.** $1/6^4$ **17.** $1/5^4$ **19.** 3^7 **21.** $1/d^8$ **23.** $x^2/(2y)$ **25.** $(4 + s)^6$ **27.** $1/(m^8 n^4)$ **29.** $a^3 b^6$ **31.** $x^6/(4y^4)$ **33.** $9m^5/n^3$ **35.** $1/(r^{10} s^8 t^{12})$ **37.** $k^2/(p + q)^7$ **39.** 1/6   **41.** $-13/66$ **43.** 35/18   **45.** 6.93×10^4 **47.** 6×10^9 **49.** 7.92×10^{-3} **51.** 820,000   **53.** 170,000,000   **55.** .00615   **57.** 4.232   **59.** 1.8×10^{-9} **61.** 2×10^{-3} **63.** 7×10^4 **65.** 4.86×10^7 **67.** 1.29×10^{-4} **69.** -6.04×10^{11} **71.** -2.51×10^5 **73.** $2k$ **75.** x^r **77.** $1/(8b^y)$ **79.** $-2m^{2-n}$

Section 1.4   (page 27)

1. $x^2 - x + 3$ **3.** $3x^3 - 3x^2 + 6x + 2$ **5.** $9y^2 - 4y + 4$ **7.** $p^3 - 7p^2 - p - 7$ **9.** $6a^2 + a - 1$ **11.** $3b^2 - 8b - 1$ **13.** $-14q^2 + 11q - 14$ **15.** $28r^2 + r - 2$ **17.** $18p^2 - 27pq - 35q^2$ **19.** $15x^2 - \frac{7}{3}x - \frac{2}{9}$ **21.** $\frac{6}{25}y^2 + \frac{11}{40}yz + \frac{1}{16}z^2$ **23.** $25r^2 - 4$ **25.** $16x^2 - 9y^2$ **27.** $36k^2 - 36k + 9$ **29.** $16m^2 + 16mn + 4n^2$ **31.** $8z^3 - 12z^2 + 6z - 1$ **33.** $12x^5 + 8x^4 - 20x^3 + 4x^2$ **35.** $15m^4 - 10m^3 + 5m^2 - 5m$ **37.** $-2z^3 + 7z^2 - 11z + 4$ **39.** $27p^3 - 1$ **41.** $8m^3 + 1$ **43.** $m^2 + mn - 2n^2 - 2km + 5kn - 3k^2$ **45.** $a^2 - 2ab + b^2 + 4ac - 4bc + 4c^2$ **47.** $x^{-2} - 4x^{-1} + 4$ **49.** $12m^{-2} - 5m^{-1}n^{-1} - 2n^{-2}$ **51.** $2m^2 - 4m + 8$ **53.** $5x^3 + 10x^2 + 4x - 3/x$ **55.** $5y^2 - 3xy + 8x^2$ **57.** -8 **59.** 1   **61.** -24 **63.** $k^{2m} - 4$ **65.** $b^{2r} + b^r - 6$ **67.** $3p^{2x} - 5p^x - 2$ **69.** $m^{2x} - 4m^x + 4$ **71.** $27k^{3a} - 54k^{2a} + 36k^a - 8$

Section 1.5   (page 32)

1. $4m(3n - 2)$ **3.** $3r(4r^2 + 2r - 1)$ **5.** $2px(3x - 4x^2 - 6)$ **7.** $2(a + b)(1 + 2m)$ **9.** $(x - 3)(x - 8)$ **11.** $(4p - 1)(p + 1)$ **13.** $(2z - 5)(z + 6)$ **15.** $3(2r - 1)(2r + 5)$ **17.** $(6r - 5s)(3r + 2s)$ **19.** $(3x - 4y)(5x + 2y)$ **21.** $x^2(3 - x)^2$ **23.** $z(2r - 3)(r + 1)$ **25.** $(2m + 5)(2m - 5)$ **27.** $9(4r + 3s)(4r - 3s)$ **29.** $(11p^2 + 3q^2)(11p^2 - 3q^2)$ **31.** $(4x^2 + y^2)(2x + y)(2x - y)$ **33.** $(p^4 + 1)(p^2 + 1)(p + 1)(p - 1)$ **35.** $(2m - 3n)(4m^2 + 6mn + 9n^2)$ **37.** $(2 - x)(2 + x)(4 + 2x + x^2)(4 - 2x + x^2)$ **39.** $(4 + m)(x + y)$ **41.** $(q + 3 + p)(q + 3 - p)$ **43.** $(a + b + x + y)(a + b - x - y)$ **45.** $(2x - y)y$ **47.** $(3x - y)(19y^2 - 24xy + 9x^2)$ **49.** $(x + y + 5z)(x + y - 3z)$ **51.** $(3a + 3b + 8c)(2a + 2b - 5c)$ **53.** $4pq$ **55.** $(m^n + 4)(m^n - 4)$

57. $(x^n - y^{2n})(x^{2n} + x^n y^{2n} + y^{4n})$ **59.** $(2x^n + 3y^n)(x^n - 13y^n)$ **61.** $9.44(x^3 + 5.1x^2 - 3.2x + 4)$
63. $7.88(8.2z^4 - 3z^2 + z + 14.5)$ **65.** $p^{-4}(1 - p^2)$ **67.** $4k^{-3}(3 + k - 2k^2)$ **69.** $25p^{-6}(4 - 2p^4 + 3p^8)$

Section 1.6   (page 37)
1. $\{x \mid x \neq 3/5\}$ **3.** all real numbers   **5.** $3p/(2p + 1)$ **7.** $(3k + 2)/(4k - 3)$ **9.** $-(4 + m)/(5m^2)$
11. $a + 2$ **13.** 1 **15.** $(x^2 - 1)/x^2$ **17.** $(x + y)/(x - y)$ **19.** $(x^2 - xy + y^2)/(x^2 + xy + y^2)$
21. $(6x - 1)/(10x^2)$ **23.** $(2y + 1)/[y(y + 1)]$ **25.** $3/[2(a + b)]$ **27.** $(2m^2 + 2)/[(m - 1)(m + 1)]$
29. $4/(a - 2)$ **31.** $(2k^2 + 13k + 18)/(k + 3)^3$ **33.** $(6r^2 + 9r)/[(r + 2)(r + 3)(r - 1)]$
35. $5/[(a - 2)(a - 3)(a + 2)]$ **37.** $(p + 5)/[p(p + 1)]$ **39.** $a(a + 5)/[(a + 6)(a + 1)(a - 1)]$
41. $(6x^2 - 6x - 5)/[(x + 5)(x - 2)(x - 1)]$ **43.** $-1/[x(x + h)]$ **45.** $a + b$ **47.** -1 **49.** $1/(ab)$
51. $y(xy - 9)/(x^2y^2 - 9)$ **53.** $(x + 1)/(x - 1)$ **55.** $(2 - b)(1 + b)/[b(1 - b)]$ **57.** $(2x + 1)/(x + 1)$

Section 1.7   (page 44)
1. $5\sqrt{2}$ **3.** $5\sqrt[3]{2}$ **5.** $-3\sqrt{5}/5$ **7.** $4\sqrt{5}$ **9.** $9\sqrt{3}$ **11.** $-5\sqrt{7}$ **13.** $-\sqrt[3]{12}/2$
15. $\sqrt[4]{24}/2$ **17.** $2\sqrt[3]{2}$ **19.** $7\sqrt[3]{3}$ **21.** $2\sqrt{3}$ **23.** $13\sqrt[3]{4}/6$ **25.** $xyz^2\sqrt{2x}$
27. $2zx^2y\sqrt[3]{2z^2x^2y}$ **29.** $ab\sqrt{ab}(b - 2a^2 + b^3)$ **31.** -7 **33.** 10 **35.** $11 + 4\sqrt{6}$ **37.** $5\sqrt{6}$
39. $2\sqrt[3]{9} - 7\sqrt[3]{3} - 4$ **41.** $\sqrt{6x}/(3x)$ **43.** $x^2y\sqrt{xy}/z$ **45.** $2\sqrt[3]{x}/x$ **47.** $-km\sqrt[3]{k^2}/r^2$
49. $h\sqrt[4]{9g^3hr^2}/(3r^2)$ **51.** $m\sqrt[3]{n^2}/n$ **53.** $2\sqrt[3]{x^3y^3}$ **55.** $\sqrt[6]{4}$ **57.** $\sqrt[18]{x}$ **59.** $-3(1 + \sqrt{2})$
61. $(4\sqrt{3} - 3)/13$ **63.** $4(2 + \sqrt{y})/(4 - y)$ **65.** $(\sqrt{m} + \sqrt{p})/(m - p)$
67. $y(\sqrt{x} - y - z)/[x - (y + z)^2]$ or $y(\sqrt{x} - y - z)/(x - y^2 - 2yz - z^2)$ **69.** $2(3 - \sqrt{1 + k})/(8 - k)$
71. $m\sqrt{p}/p + p\sqrt{m}/m$, or $(m^2\sqrt{p} + p^2\sqrt{m})/(pm)$ **73.** $-2x - 2\sqrt{x(x + 1)} - 1$
75. $5(\sqrt[3]{a^2} - \sqrt[3]{ab} + \sqrt[3]{b^2})/(a + b)$ **77.** $3/(2\sqrt{3})$ **79.** $-1/[2(1 - \sqrt{2})]$ **81.** $x/(\sqrt{x} + x)$
83. $-1/(2x - 2\sqrt{x(x + 1)} + 1)$ **85.** 5.36×10^3 **87.** 1.40×10^{-1} **89.** 2.64×10^{-1} **91.** 9.93 or
9.93×10^0 **95.** approximately 3.146

Section 1.8   (page 48)
1. 2 **3.** 4 **5.** $1/9$ **7.** $27/8$ **9.** $4p^2$ **11.** $9x^4$ **13.** $x^{2/3}$ **15.** $z^{5/3}$ **17.** $y^{9/2}$
19. $m^{5/6}$ **21.** $m^{7/3}$ **23.** $(1 + n)^{5/4}$ **25.** $6yz^{2/3}$ **27.** $r^{3/14}s^{7/20}$ **29.** $a^{2/3}b^2$ **31.** $h^{1/3}t^{1/5}/k^{2/5}$
33. 1 **35.** $x^{11/12}$ **37.** $16/y^{11/12}$ **39.** $4z^2/(x^5y)$ **41.** r^{6+p} **43.** $m^{3/2}$ **45.** $p^{(m+n+m^2)/(mn)}$
47. $(r^2 - 2r + 1)/r$ **49.** $(x + 1)/x^{1/2}$ **51.** $r\sqrt[6]{r^5}$ **53.** $p\sqrt[10]{p^3}$ **55.** $p^2q^2\sqrt[6]{p}$ **57.** $m^2n\sqrt[6]{m^2n}$
59. $mn^2\sqrt[15]{m^{14}n^{13}}$ **61.** $k^{3/4}(4k + 1)$ **63.** $z^{-1/2}(9 + 2z)$ **65.** $p^{-1/4}(p^{-1/2} - 2)$
67. $(p + 4)^{-3/2}(p^2 + 9p + 21)$ **69.** $(2 + 5p)/p^{1/2}$ **71.** $(p + 2)/[2(p + 1)^{3/2}]$ **73.** $(2x^2 + 15)/(2x^2 + 5)^{4/3}$
75. 64 **77.** $\$64,000,000$ **79.** about $\$10,000,000$ **81.** about 86.3 mi   **83.** about 211 mi   **85.** 29
87. 177

Section 1.9   (page 56)
1. $7 - i$ **3.** $3 - 6i$ **5.** $1 - 10i$ **7.** $-10 + 5i$ **9.** $8 - i$ **11.** $31 - 5i$ **13.** $5 - 12i$
15. $-8 - 6i$ **17.** 13 **19.** 29 **21.** 7 **23.** $15i$ **25.** $-198 + 10i$ **27.** $161 + 240i$
29. $3i$ **31.** $-20i$ **33.** $i\sqrt{7}$ **35.** -5 **37.** -4 **39.** 2 **41.** -2 **43.** i **45.** 1
47. $-i$ **49.** 1 **51.** $-i$ **53.** $-i$ **55.** i **57.** $7/25 - (24/25)i$ **59.** $13/20 - (1/20)i$
61. $32/37 - (7/37)i$ **63.** $5/2 + i$ **65.** $-16/65 - (37/65)i$ **67.** $27/10 + (11/10)i$ **69.** $17/10 - (11/10)i$
71. $37/34 + (165/34)i$ **73.** $a = 23; b = 5$ **75.** $a = 18; b = -3$ **77.** $a = -5; b = 1$
79. $a = -3/4; b = 3$ **89.** 25 **91.** $a = 0$ or $b = 0$

Chapter 1   Review Exercises   (page 58)
1. commutative   **3.** inverse   **5.** distributive   **7.** no   **9.** $20/3$ **11.** $9/4$ **13.** 6 **15.** -12,
$-6, 0, 6$ **17.** $-\sqrt{7}, \pi/4, \sqrt{11}$ **19.** natural, whole, integer, rational, real   **21.** $(\sqrt{36} = 6)$ natural, whole,
integer, rational, real   **23.** integer, rational, real   **25.** irrational, real   **27.** irrational, real
29. $-|3 - (-2)|, -|-2|, |6 - 4|, |8 + 1|$ **31.** $3 - \sqrt{7}$ **33.** $m - 3$ **35.** (a) 1   (b) 12
37. if $x \geq 0$ **39.** if $x = 0$ and $y = 0$ **41.** if A and B are the same point   **43.** $-8x^2 - 15x + 16$
45. $24k^2 - 5k - 14$ **47.** $x^2 + 4xy + 4y^2 - 2xz - 4yz + z^2$ **49.** $x^6/2$ **51.** $z(7z - 9z^2 + 1)$
53. $(3m + 1)(2m - 5)$ **55.** $5m^3(3m - 5n)(2m + n)$ **57.** $(13y^2 + 1)(13y^2 - 1)$ **59.** $(x + 1)(x - 3)$
61. $(-r^3 + 2)(7r^6 - 10r^3 + 4)$ **63.** $(m - n)(3 + 4k)$ **65.** $(2x - 1)(b + 3)$ **67.** $(y^k + 3)(y^k - 3)$

69. $(m - 3)(2m + 3)/(5m)$ **71.** $(x + 1)/(x + 4)$ **73.** $(p + 6q)^2(p + q)/(5p)$ **75.** $37/(20y)$
77. $1 - (9/p^2)$ or $(p^2 - 9)/p^2$ **79.** $(q + p)/(pq - 1)$ **81.** $(3m^2 + 2m - 12)/[5(m + 2)]$ **83.** p^8
85. $s/(36r)$ **87.** $2/(27x^{12}y^3)$ **89.** $10\sqrt{2}$ **91.** $\sqrt{21r}/(3r)$ **93.** $-r^2m\sqrt[3]{m^2z}/z$ **95.** 66
97. $m(-9\sqrt{2m} + 5\sqrt{m})$ **99.** $23\sqrt{5}/30$ **101.** $(z\sqrt{z - 1})/(z - 1)$ **103.** $\sqrt{5} - 2$ **105.** $1/216$
107. $16r^{9/4}s^{7/3}$ **109.** $1/p^{3/2}$ **111.** $1/m^{2p}$ **113.** $10z^{7/3} - 4z^{1/3}$ **115.** $3p^2 + 3p^{3/2} - 5p - 5p^{1/2}$
117. $-2 - 3i$ **119.** $5 + 4i$ **121.** $29 + 37i$ **123.** $-32 + 24i$ **125.** $-2 - 2i$ **127.** i
129. $8/5 + (6/5)i$ **131.** $-5/26 + (1/26)i$ **133.** $7/2 - (11/4)i$ **135.** $2i\sqrt{3}$ **137.** 0   **141.** 0
143. $<$ **145.** $1/b < 1/a < 1 < \sqrt{a} < \sqrt{b} < a < b < a^2 < b^2$

Chapter 2

Section 2.1   (page 66)
1. not equivalent   **3.** equivalent   **5.** not equivalent   **7.** not equivalent   **9.** identity
11. conditional   **13.** identity   **15.** identity   **17.** $\{4\}$ **19.** $\{12\}$ **21.** $\{-2/7\}$ **23.** $\{-1\}$
25. $\{3\}$ **27.** $\{4\}$ **29.** $\{3/4\}$ **31.** $\{-12/5\}$ **33.** θ **35.** $\{27/7\}$ **37.** $\{-59/6\}$ **39.** θ
41. $\{-4\}$ **43.** $-3a + b$ **45.** $(3a + b)/(3 - a)$ **47.** $(3 - 3a)/(a^2 - a - 1)$ **49.** $(2a^2)/(a^2 + 3)$
53. 6   **55.** -2 **57.** .72   **59.** 6.53   **61.** -13.26

Section 2.2   (page 74)
1. $V = k/P$ **3.** $l = V/(wh)$ **5.** $g = (V - V_0)/t$ **7.** $g = 2s/t^2$ **9.** $B = 2A/h - b$
11. $r_1 = S/(2\pi h) - r_2$ **13.** $l = gt^2/(4\pi^2)$ **15.** $h = (S - 2\pi r^2)/(2\pi r)$ **17.** $f = AB(p + 1)/24$
19. $R = r_1r_2/(r_2 + r_1)$ **21.** 6 cm   **23.** \$57.65   **25.** \$12,000 at 13% and \$8000 at 16%   **27.** \$54,000
at 8-1/2% and \$6000 at 16%   **29.** \$70,000 for land that makes a profit and \$50,000 for land that produces a loss
31. \$267.57   **33.** no solution—the numbers given are inconsistent   **35.** \$20,000 at 12% and \$40,000 at 10%
37. 2 liters   **39.** 133-1/3 liters   **41.** 12-2/3 kg   **43.** about 840 mi   **45.** 15 min   **47.** 35 kph
49. 3-3/5 hr   **51.** 78 hr   **53.** 13-1/3 hr   **55.** 2 hr   **57.** $m(b - c)/(a - b)$
59. $(100I - nB)/(m - n)$ at $m\%$ and $(Bm - 100I)/(m - n)$ at $n\%$

Section 2.3   (page 83)
1. $\{\pm 4\}$ **3.** $\{\pm 3\sqrt{3}\}$ **5.** $\{3 \pm \sqrt{5}\}$ **7.** $\{(1 \pm 2\sqrt{3})/3\}$ **9.** $\{2, 3\}$ **11.** $\{-5/2, 10/3\}$
13. $\{-3/2, -1/4\}$ **15.** $\{3, 5\}$ **17.** $\{1 \pm \sqrt{5}\}$ **19.** $\{(-1 \pm i)/2\}$ **21.** $\{(1 \pm \sqrt{5})/2\}$
23. $\{(-1 \pm \sqrt{7})/2)$ **25.** $\{3 \pm \sqrt{2}\}$ **27.** $\{(3 \pm i\sqrt{7})/2\}$ **29.** $\{(1 \pm i\sqrt{7})/4\}$ **31.** $\{1 \pm 2i\}$
33. $\{3, -1/4\}$ **35.** $\{3/2, 1\}$ **37.** $\{1.243, -.643\}$ **39.** $\{1.281, -.781\}$ **41.** $\{3.562, -.562\}$
43. $\{(\sqrt{2} + \sqrt{6})/2\}$ **45.** $\{\sqrt{2}, \sqrt{2}/2\}$ **47.** $\{(-i \pm i\sqrt{5})/2\}$ **49.** $\{(1 \pm \sqrt{2})/i\}$ or $\{-i \pm i\sqrt{2}\}$
51. $\{[(1 - i) \pm (\sqrt{7} + i\sqrt{7})]/4\}$ **53.** 0; one real solution   **55.** 1; two real solutions   **57.** 84;
two real solutions   **59.** -23; two complex solutions   **61.** $t = \pm\sqrt{2sg}/g$ **63.** $h = \pm\sqrt{d^4kL/L}$ or $\pm d^2\sqrt{kL}/L$
65. $t = \pm\sqrt{(s - s_0 - k)g}/g$ **67.** $R = (E^2 - 2Pr \pm \sqrt{(2Pr - E^2)^2 - 4P^2r^2})/(2P)$ **69.** 16, 18 or -18, -16
71. 12, -2 **73.** 50 m by 100 m   **75.** 9 ft by 12 ft   **77.** 50 mph   **79.** 16.95 cm, 20.95 cm, 26.95 cm
81. ± 12 **83.** 121/4   **85.** 1, 9   **91.** (a) $x = (y \pm \sqrt{8 - 11y^2})/4$ (b) $y = (x \pm \sqrt{6 - 11x^2})/3$

Section 2.4   (page 89)
1. $\{\pm 3, \pm 2\}$ **3.** $\{\pm\sqrt{10}/2, \pm 1\}$ **5.** $\{4, 6\}$ **7.** $\{(-6 \pm 2\sqrt{3})/3, (-4 \pm \sqrt{2})/2\}$ **9.** $\{7/2, -1/3\}$
11. $\{-63, 28\}$ **13.** $\{0, 1\}$ **15.** $\{\pm i\sqrt{3}, \pm i\sqrt{6}/3\}$ **17.** $\{1/2\}$ **19.** $\{-1\}$ **21.** $\{5\}$
23. $\{9\}$ **25.** $\{9\}$ **27.** $\{5/4\}$ **29.** $\{\pm 2\}$ **31.** $\{2, -2/9\}$ **33.** $\{0, 9\}$ **35.** $\{3/2\}$
37. $\{31\}$ **39.** $\{1, -3\}$ **41.** $\{1/4, 1\}$ **43.** $\{0, 8\}$ **45.** $\{\pm 2.121, \pm 1.528\}$ **47.** $\{\pm .917\}$
49. $h = (d/k)^2$ **51.** $L = P^2g/4$ **53.** $y = (a^{2/3} - x^{2/3})^{3/2}$

Section 2.5   (page 97)
1. $(-1, 4)$ **3.** $(-\infty, 0)$ **5.** $[1, 2)$ **7.** $(-\infty, -9)$ **9.** $-4 < x < 3$ **11.** $x \le -1$
13. $-2 \le x < 6$ **15.** $x \le -4$ **17.** $(-\infty, 4]$ **19.** $[-1, +\infty)$ **21.** $(-\infty, 6]$ **23.** $(-\infty, 4)$
25. $[-11/5, +\infty)$ **27.** $[1, 4]$ **29.** $(-6, -4)$ **31.** $(-16, 19]$ **33.** $(0, 10)$ **35.** $[-2, 3]$
37. $(-\infty, 1/2) \cup (4, +\infty)$ **39.** $((-5 - \sqrt{33})/2, (-5 + \sqrt{33})/2)$ **41.** $(-\infty, -2] \cup [0, 2]$
43. $(-2, 0) \cup (1/4, +\infty)$ **45.** $(-\infty, 1)$ **47.** $(-\infty, -1) \cup (0, +\infty)$ **49.** $(-\infty, 6) \cup [15/2, +\infty)$

51. $(-\infty, 1) \cup (9/5, +\infty)$ **53.** $(-\infty, -3/2) \cup [-13/9, +\infty)$ **55.** $(-\infty, +\infty)$ **57.** $[3/2, +\infty)$
59. $(-\infty, 4)$ **61.** $(5/2, +\infty)$ **63.** $[-8/3, 3/2] \cup (6, +\infty)$ **65.** if k is in $(-\infty, -4\sqrt{2}] \cup [4\sqrt{2}, +\infty)$
67. if k is in $(-\infty, 0] \cup [8, +\infty)$ **69.** $[500, +\infty)$ **71.** $R < C$ for all positive x; this product will never break even
73. $(0, 5/4), (6, +\infty)$ **75.** for any x in $[0, 5/3) \cup (10, +\infty)$ **77.** $32°F$ to $86°F$
85. $(-\infty, 1] \cup [2, 3) \cup (4, +\infty)$

Section 2.6   (page 104)
1. $\{-1/3, 1\}$ **3.** $\{2/3, 8/3\}$ **5.** $\{-6, 14\}$ **7.** $\{7/2, 5/2\}$ **9.** $\{-7/3, -1/7\}$ **11.** $\{-3/5, 11\}$
13. $\{-1/2\}$ **15.** $[-3, 3]$ **17.** $(-\infty, -1) \cup (1, +\infty)$ **19.** $[-10, 10]$ **21.** $(-\infty, -2/3) \cup (2, +\infty)$
23. $[4/3, 2]$ **25.** $(-\infty, -13/4] \cup [1/4, +\infty)$ **27.** $(-\infty, 2) \cup (3, +\infty)$ **29.** $(-3/2, 13/10)$
31. $(-3, -1)$ **33.** $(1, 5) \cup (5, +\infty)$ **35.** $(-\infty, 3/2] \cup [5/2, +\infty)$ **37.** $[7/4, 2) \cup (2, 5/2]$
39. $(-1, 0) \cup (0, 5/11)$ **41.** $(-\infty, 3/8) \cup (7/16, 1/2) \cup (1/2, +\infty)$ **43.** $|x - 2| \le 4$ **45.** $|z - 12| \ge 2$
47. $|k - 1| = 6$ **49.** if $|x - 2| \le .0004$, then $|y - 7| \le .00001$ **51.** $m = 2, n = 20$
53. $(-\infty, -1/5] \cup [1, +\infty)$

Chapter 2   Review Exercises   (page 105)
1. $\{6\}$ **3.** $\{-11/3\}$ **5.** $\{-96/7\}$ **7.** $\{-2\}$ **9.** $\{13\}$ **11.** $x = -6b - a - 6$
13. $x = (6 - 3m)/(1 + 2k - km)$ **15.** $y = 6a/(4 - a)$ **17.** $C = (5/9)(F - 32)$ **19.** $j = n - nA/I$
21. $r_1 = kr_2/(r_2 - k)$ **23.** $L = V/(\pi r^2)$ **25.** $x = -4/(y^2 - 5y - 6p)$ **27.** $500   **29.** 12 hours
31. $\{-7 \pm \sqrt{5}\}$ **33.** $\{5/2, -3\}$ **35.** $\{7, -3/2\}$ **37.** $\{3, -1/2\}$ **39.** $\{-2i \pm i\sqrt{5}\}$
41. -188; two complex solutions   **43.** 484; two real solutions   **45.** 0; one real solution   **47.** 50 m by 225 m,
or 112.5 m by 100 m   **49.** $\{\pm i, \pm 1/2\}$ **51.** $\{-2, 3\}$ **53.** $\{5/2, -15\}$ **55.** $\{63/2\}$ **57.** $\{3\}$
59. $\{-2, -1\}$ **61.** $\{1, -4\}$ **63.** $\{-1\}$ **65.** $\{6\}$ **67.** $\{-1\}$ **69.** $(-7/13, +\infty)$
71. $(-\infty, 1]$ **73.** $(1, +\infty)$ **75.** $[4, 5]$ **77.** $[-5/2, 8)$ **79.** $[-4, 1]$ **81.** $(-2/3, 5/2)$
83. $(-\infty, -4] \cup [0, 4]$ **85.** $(-2, 0)$ **87.** $(-3, 1) \cup [7, +\infty)$ **89.** 0 to 30 (remember, x must start at 0)
91. $\{3, -11\}$ **93.** $\{-1, 5\}$ **95.** $\{11/27, 25/27\}$ **97.** $\{4/3, -2/7\}$ **99.** $[-7, 7]$ **101.** $\varnothing$
103. $(-2/7, 8/7)$ **105.** $(-\infty, -4) \cup (-2/3, +\infty)$ **107.** $[1/2, 1) \cup (1, +\infty)$ **109.** (a) $(-\infty, +\infty)$ (b) $\varnothing$
111. $(x + 2)(36/x + 3)$ or $42 + 3x + 72/x$ **113.** for $x \le 2$, $x = y - 2$ and for $x > 2$, $x = (y + 2)/3$

Cumulative Review Exercises   (page 108)
1. rational, real   **3.** $(\sqrt{49} = 7)$ whole, integer, rational, real   **5.** closed   **7.** not closed   **9.** $-|-7|$,
$-|2|, |-3|$ **11.** $-|-8| - |-6|, -|-2| + (-3), -|3|, -|-2|, |-8 + 2|$ **13.** 0   **15.** $7 - \sqrt{5}$
17. $m - 3y$ **19.** $-1/125$ **21.** 1   **23.** 3^4 or 81   **25.** p^{11}/q^4 **27.** $-5m^2 + 3m + 5$
29. $3k^2 - 29k + 56$ **31.** $27k^3 - 135k^2 + 225k - 125$ **33.** $3y^2 + yz + 11y - 2z^2 + z + 10$
35. $5x^3 + 10x^2 + 4x - 3/x$ **37.** $3m(m^2 + 3 + 5m^4)$ **39.** $(2q - 3)(3q + 4)$
41. $(2a + 5)(4a^2 - 10a + 25)$ **43.** $(r - p)(s + t)$ **45.** $-16z$ **47.** $4/(x - 1)$ **49.** $(3z - 2)/(3z + 2)$
51. $(3m + 11)/[2(m - 7)]$ **53.** $n - m$ **55.** $10\sqrt{10}$ **57.** $-2\sqrt[4]{2}$ **59.** $15\sqrt[3]{75}$
61. $5hyq^2\sqrt[4]{3hy^2q}$ **63.** $-34\sqrt{2}$ **65.** $-(2 + \sqrt{7})/3$ **67.** $1/64$ **69.** $1/(a - b)$ **71.** $24a^{4/3}/b^{5/3}$
73. $rt^2/s^{1/3}$ **75.** $6 + 2i$ **77.** $17 + i$ **79.** $-12 - 5i$ **81.** $7/2 - (1/2)i$ **83.** $2/13 + (3/13)i$
85. $20i$ **87.** $-\sqrt{21}$ **89.** $-i$ **91.** $\{-6/11\}$ **93.** $\{-10/13\}$ **95.** $\{-19/75\}$
97. $t = (v - v_0)/g$ **99.** $B = 2A/h - b$ or $B = (2A - bh)/h$ **101.** 20 cm, 20 cm, 14 cm
103. $60,000   **105.** 10 pounds   **107.** 7 km per hour   **109.** $\{(3 \pm \sqrt{7})/5\}$ **111.** $\{-3/2, -1/4\}$
113. $\{(1 \pm \sqrt{6})/2\}$ **115.** $\{1 \pm i\}$ **117.** $\{(1 \pm 2i\sqrt{2})/6\}$ **119.** $\{i \pm i\sqrt{2}\}$
121. $a = \pm\sqrt{10zyb}/(10yb)$ **123.** $-1, 1$ **125.** 7-1/2 hours   **127.** $\{\pm\sqrt{3}, \pm\sqrt{2}/2\}$ **129.** $\{-124, 9\}$
131. $\{12\}$ **133.** $\{0, 18\}$ **135.** $\{0, \pm\sqrt{7}\}$ **137.** $[6, +\infty)$ **139.** $(2, 9)$ **141.** $(-\infty, -6] \cup [0, +\infty)$
143. $(-\infty, -2] \cup [0, 3]$ **145.** $(-6, 2) \cup [6, +\infty)$ **147.** $(-3, 3)$ **149.** $(-\infty, -4) \cup (-1, +\infty)$
151. $[-3, 1/5]$ **153.** $\{1, 3\}$ **155.** $\{6, -4/5\}$

Chapter 3

Section 3.1   (page 118)

1–5.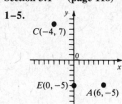

7.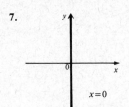

9.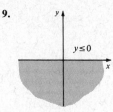

11.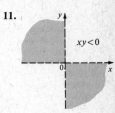

13.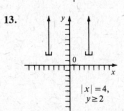

15.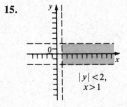

17.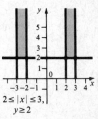

19. $8\sqrt{2}$; $(9, 3)$ **21.** $\sqrt{34}$; $(-11/2, -7/2)$ **23.** $\sqrt{133}$: $(2\sqrt{2}, 3\sqrt{5}/2)$ **25.** 7.616   **27.** 27.203
29. $(13, 10)$ **31.** $(-10, 11)$ **33.** yes   **35.** no   **37.** no   **39.** yes   **41.** no   **43.** $5, -1$
45. $9 + \sqrt{119}, 9 - \sqrt{119}$ **47.** $2\sqrt{2}, -2\sqrt{2}$ **51.**
53. $(2 + \sqrt{7}, 2 + \sqrt{7}), (2 - \sqrt{7}, 2 - \sqrt{7})$
55. $4x + 5y = -41$

Section 3.2   (page 126)

1.  $y = 8x - 3$

3. 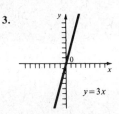 $y = 3x$

5. 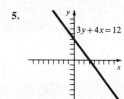 $3y + 4x = 12$

7. 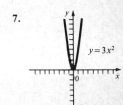 $y = 3x^2$

9. 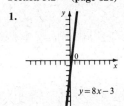 $y = -x^2$

11. 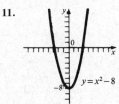 $y = x^2 - 8$

13. 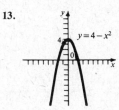 $y = 4 - x^2$

15. 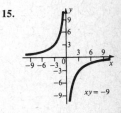 $xy = -9$

17.

$4x = y^2$

19.

$16y^2 = -x$

21.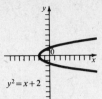

$y^2 = x + 2$

23.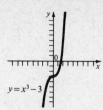

$y = x^3 - 3$

25.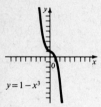

$y = 1 - x^3$

27.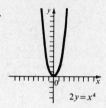

$2y = x^4$

29.

$y = \sqrt{x}$

31.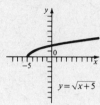

$y = \sqrt{x + 5}$

33.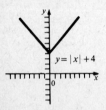

$y = |x| + 4$

35.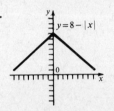

$y = 8 - |x|$

37.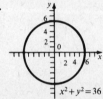

$x^2 + y^2 = 36$

39.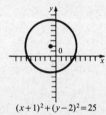

$(x + 1)^2 + (y - 2)^2 = 25$

41.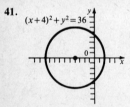

$(x + 4)^2 + y^2 = 36$

43. $(x - 1)^2 + (y - 4)^2 = 9$
45. $(x + 8)^2 + (y - 6)^2 = 25$
47. $(x + 1)^2 + (y - 2)^2 = 25$
49. $(x + 3)^2 + (y + 2)^2 = 4$ **51.** $(-3, -4)$; $r = 4$
53. $(6, -5)$; $r = 6$ **55.** $(-4, 7)$; $r = 0$ (a point)
57. $(0, 1)$; $r = 7$ **59.** $(1.42, -.7)$; $r = .8$
61. $(x + 2)^2 + (y + 3/2)^2 = 149/4$ **63.** yes

Section 3.3   (page 134)

1.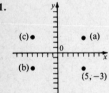

3.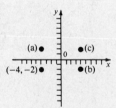

5.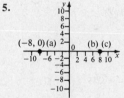

7.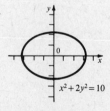

$x^2 + 2y^2 = 10$

9.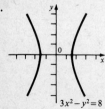

$3x^2 - y^2 = 8$

11.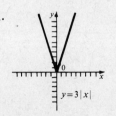

$y = 3|x|$

13.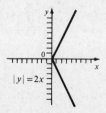

$|y| = 2x$

15.

$|y| = |x + 2|$

17.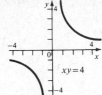

19.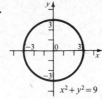

21.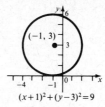

23.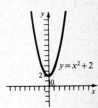

25.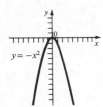

27.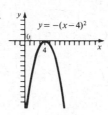

29.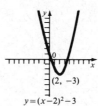

31.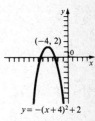

33. reflected about the *x*-axis   **35.** symmetric   **37.** not symmetric   **39.** not symmetric   **41.** symmetric
43. not symmetric

Section 3.4   (page 143)

1. (a) 0   (b) 4   (c) 2   (d) 4   **3.** (a) −3   (b) −2   (c) 0   (d) 2   **5.** −1   **7.** $3a − 1$ **9.** 5
11. $15a − 7$ **13.** $25p^2 − 20p + 4$ **15.** $3m^2 − 1$ **17.** −66.8   **19.** −42.464   **21.** all real
numbers, or $(−\infty, +\infty)$ **23.** all real numbers, or $(−\infty, +\infty)$ **25.** $[−4, 4]$ **27.** $[3, +\infty)$
29. $(−\infty, −2) \cup (−2, 2) \cup (2, +\infty)$ **31.** all real numbers, or $(−\infty, +\infty)$ **33.** all real numbers, or $(−\infty, +\infty)$
35. $(−\infty, −1] \cup [5, +\infty)$ **37.** $(−\infty, −5/3] \cup [1/2, +\infty)$ **39.** domain: $[−5, 4]$; range: $[−2, 6]$ **41.** domain:
$(−\infty, +\infty)$; range: $(−\infty, 12]$ **43.** (a) $x^2 + 2xh + h^2 − 4$ (b) $2xh + h^2$ (c) $2x + h$ **45.** (a) $6x + 6h + 2$
(b) $6h$ (c) 6   **47.** (a) $2x^3 + 6x^2h + 6xh^2 + 2h^3 + x^2 + 2xh + h^2$ (b) $6x^2h + 6xh^2 + 2h^3 + 2xh + h^2$
(c) $6x^2 + 6xh + 2h^2 + 2x + h$ **49.** increasing on $(−\infty, −3]$ and $[3, +\infty)$; decreasing on $[−3, 3]$ **51.** increasing
on $(−\infty, −1]$; decreasing on $[−1, +\infty)$ **53.** increasing on $(−\infty, +\infty)$; never decreasing   **55.** never increasing;
decreasing on $(−\infty, +\infty)$ **57.** increasing on $[0, +\infty)$; decreasing on $(−\infty, 0]$ **59.** increasing on $(−\infty, −2]$;
decreasing on $[−2, +\infty)$ **61.** increasing on $[0, +\infty)$; never decreasing   **63.** even   **65.** even   **67.** neither
69. neither

71.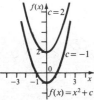

73.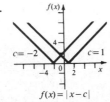

75.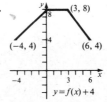

77.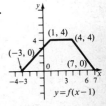

79.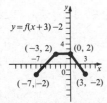

81. domain: $(−\infty, +\infty)$; range: set of all integers
83. (a) 30¢   (b) 57¢   (c) $1.11   (d)

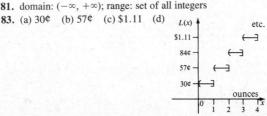

(e) domain: $(0, +\infty)$ (at least in theory); range: $\{.30, .57, .84, 1.11, \cdot\cdot\cdot\}$

85. (a) 90¢   (b) $1.10   (c) $1.60   (d)

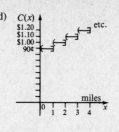

(e) domain: $(0, +\infty)$ (at least in theory);
range: $\{.90, 1.00, 1.10, 1.20, \cdots\}$

87.

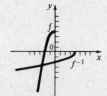

91. no

Section 3.5   (page 149)

1. $10x + 2$; $-2x - 4$; $24x^2 + 6x - 3$; $(4x - 1)/(6x + 3)$; all domains are $(-\infty, +\infty)$, except for f/g, which is $(-\infty, -1/2) \cup (-1/2, +\infty)$ **3.** $4x^2 - 4x + 1$; $2x^2 - 1$; $(3x^2 - 2x)(x^2 - 2x + 1)$; $(3x^2 - 2x)/(x^2 - 2x + 1)$; all domains are $(-\infty, +\infty)$, except for f/g, which is $(-\infty, -1) \cup (-1, +\infty)$ **5.** $\sqrt{2x + 5} + \sqrt{4x - 9}$; $\sqrt{2x + 5} - \sqrt{4x - 9}$; $\sqrt{(2x + 5)(4x - 9)}$; $\sqrt{(2x + 5)/(4x - 9)}$; all domains are $[9/4, +\infty)$, except f/g, which is $(9/4, +\infty)$ **7.** $5x^2 - 11x + 7$; $3x^2 - 11x - 3$; $(4x^2 - 11x + 2)(x^2 + 5)$; $(4x^2 - 11x + 2)/(x^2 + 5)$; all domains (including f/g) are $(-\infty, +\infty)$ **9.** 55   **11.** 1848   **13.** $-6/7$ **15.** $4m^2 + 6m + 1$ **17.** 1122   **19.** 97   **21.** $256k^2 + 48k + 2$ **23.** $24x + 4$; $24x + 35$ **25.** $-5x^2 + 20x + 18$; $-25x^2 - 10x + 6$ **27.** $-64x^3 + 2$; $-4x^3 + 8$ **29.** $1/x^2$; $1/x^2$ **31.** $\sqrt{8x^2 - 4}$; $8x + 10$ **33.** $x/(2 - 5x)$; $2(x - 5)$ **35.** $\sqrt{(x - 1)/x}$; $-1/\sqrt{x + 1}$ **49.** $18a^2 + 24a + 9$ **51.** $16\pi t^2$ **57.** Choose any nonzero real number for a. Then $c = 1/a$. Choose any nonzero real number for d. Then $b = -ad$.

Section 3.6   (page 151)

1. one-to-one   **3.** not one-to-one   **5.** one-to-one   **7.** one-to-one   **9.** not one-to-one   **11.** not one-to-one   **13.** not one-to-one   **15.** one-to-one   **17.** one-to-one   **19.** one-to-one   **21.** not one-to-one   **23.** inverses of each other   **25.** not inverses of each other   **27.** inverses of each other   **29.** not inverses of each other   **31.** inverses of each other   **33.** inverses of each other   **35.** inverses of each other   **37.** not inverses of each other   **39.** $f^{-1}(x) = (x + 4)/3$ **41.** $f^{-1}(x) = 3x$ **43.** $f^{-1}(x) = \sqrt[3]{x - 1}$ **45.** $f^{-1}(x) = 12 - 4x$

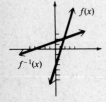

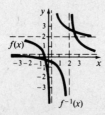

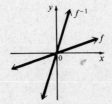

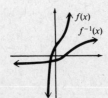

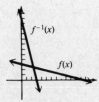

47. not one-to-one
49. $f^{-1}(x) = 1/x$ **51.** $f^{-1}(x) = (x + 3)/(4x - 8)$ **53.** $f^{-1}(x) = -\sqrt{4 - x}$; domain $(-\infty, 4]$

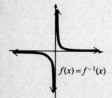

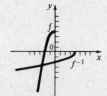

55. $(f \circ g)^{-1}(x) = (\sqrt[3]{19 - x})/2$ **57.** $(g^{-1} \circ f^{-1})(x) = (\sqrt[3]{19 - x})/2$

Section 3.7   (page 162)
1. $a = kb$ **3.** $x = k/y$ **5.** $r = kst$ **7.** $w = kx^2/y$ **9.** 220/7   **11.** 32/15   **13.** 18/125
15. 1075   **17.** 2304 ft   **19.** .0444 ohms   **21.** 140 kg per cm²   **23.** \$1375   **25.** 8/9 metric tons
27. 7500 lbs   **29.** 159 kg per m²   **31.** 4,000,000   **33.** 1600   **35.** about 4.94   **37.** about 7.4 km
39. 263 pounds

Chapter 3   Review Exercises   (page 165)
1.
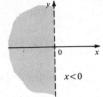
$x < 0$

3.
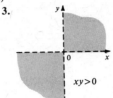
$xy > 0$

5. $d(P, Q) = \sqrt{85}; (-1/2, 2)$
7. $(7, -13)$
9. $-7, -1, 8, 23$
11. no such points exist

15. domain: real numbers   **17.** domain: $(-\infty, +\infty)$ **19.** domain: $(-\infty, 0) \cup (0, +\infty)$ **21.** domain: $[7, +\infty)$

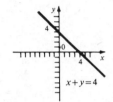
$x + y = 4$

$y = \frac{1}{2}x^2$

$y = -\frac{8}{x}$

$y = \sqrt{x-7}$

23. $(x + 2)^2 + (y - 3)^2 = 25$ **25.** $(x + 8)^2 + (y - 1)^2 = 289$ **27.** $(2, -3); r = 1$ **29.** $(-7/2, -3/2);$
$r = 3\sqrt{6}/2$ **31.** yes; yes; yes; no   **33.** no; no; no; no   **35.** yes; no; no; no   **37.** yes; yes; yes; yes
39.

$y = |x|$

41. reflect graph in x-axis   **43.** translate 1 unit to left; reflect in x-axis; translate 3 units up

$y = -|x|$

$(-1, 3)$
$y = -|x+1| + 3$

45. $(-\infty, +\infty)$ **47.** $(-\infty, +\infty)$ **49.** $(-\infty, 8) \cup (8, +\infty)$ **51.** $[-7, 7]$

53.
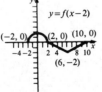
$(8, 3)$
$(-4, 3)$
$(4, 1)$
$y = f(x) + 3$

55.
$y = f(x-2)$
$(-2, 0)$ $(2, 0)$ $(10, 0)$
$(6, -2)$

57.
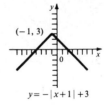
$y = f(x+3) - 2$
$(-5, 0)$
$(-7, -2)$ $(5, -2)$
$(-3, -2)$ $(1, -4)$

59.
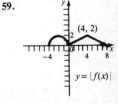
$(4, 2)$
$y = |f(x)|$

61. $-3x^2 - 3xh - h^2 + 4x + 2h$

63.
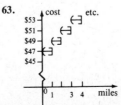
cost   etc.
\$53
\$51
\$49
\$47
\$45
0 1   3 4   miles

65. odd   **67.** odd   **69.** $4x^2 - 3x - 8$
71. 44   **73.** $16k^2 - 6k - 8$
75. $-23/4$ **77.** $(-\infty, +\infty)$
79. $\sqrt{x^2 - 2}$ **81.** $\sqrt{34}$ **83.** 1
85. increasing on $(-\infty, -1]$, decreasing on
$[-1, +\infty)$ **87.** never increasing or
decreasing   **89.** not one-to-one
91. not one-to-one   **93.** not one-to-one
95. not one-to-one

97. $f^{-1}(x) = (x - 3)/12$ **99.** $f^{-1}(x) = \sqrt[3]{x + 3}$ **101.** $f^{-1}(x) = \sqrt{25 - x^2}$; domain: [0, 5]

   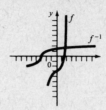   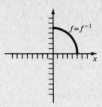

103. $m = kz^2$ **105.** $Y = (kMN^2)/X^3$ **107.** 27/2   **109.** 1372/729   **111.** 36 in
113. The conversion of u U.S. dollars into c Canadian dollars is $c = V(u) = (1 + .12)u = 1.12u$, and the conversion of d Canadian dollars into t U.S. dollars is $t = H(d) = (1 - .12)d = .88d$; therefore $H[V(u)] = .88(1.12)u = .9856u \neq u$.

117.
$$(f \circ g)(x) = \begin{cases} 2 \text{ if } x < 0 \\ x \text{ if } 0 \le x \le 1 \\ 2 \text{ if } x > 1 \end{cases}$$

119.
$$(f \circ g)(x) = \begin{cases} 1 \text{ if } x < 0 \\ 4x^2 \text{ if } 0 \le x \le 1/2 \\ 0 \text{ if } 1/2 < x \le 1 \\ 1 \text{ if } x > 1 \end{cases}$$

121.
$$f^{-1}(x) = \begin{cases} x \text{ if } x < 1 \\ \sqrt{x} \text{ if } 1 \le x \le 81 \\ x^2/729 \text{ if } x > 81 \end{cases}$$

Chapter 4

Section 4.1   (page 177)
1. 3/5   **3.** not defined   **5.** 2   **7.** 5/9   **9.** not defined   **11.** 2   **13.** .5785
15. $2x + y = 5$ **17.** $y = 1$ **19.** $x + 3y = 10$ **21.** $3x + 4y = 12$ **23.** $2x - 3y = 6$
25. $x = -6$
27. $5.081x + y = -4.69$

29.

31.

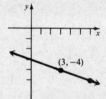

33.

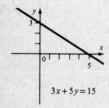

35.

37.

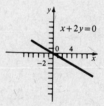

39.

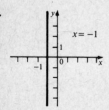

41.

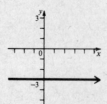

43. $x + 3y = 11$ **45.** $x - y = 7$

47. $-2x + y = 4$ **49.** $x - 11y = 9$ **51.** no

53. (a) $-1/2$ (b) $-7/2$

61.

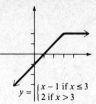

$$y = \begin{cases} x - 1 \text{ if } x \le 3 \\ 2 \text{ if } x > 3 \end{cases}$$

63.

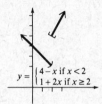

$$y = \begin{cases} 4 - x \text{ if } x < 2 \\ 1 + 2x \text{ if } x \ge 2 \end{cases}$$

65.

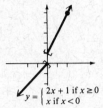

$$y = \begin{cases} 2x + 1 \text{ if } x \ge 0 \\ x \text{ if } x < 0 \end{cases}$$

67.

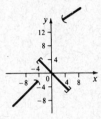

69. (a) 140   (b) 220   (c) 220

(d) 220   (e) 220   (f) 60   (g) 60   (h)

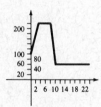

Section 4.2   (page 185)

1.

3.

5.

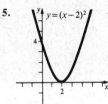

$y = (x - 2)^2$

7.

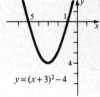

$y = (x + 3)^2 - 4$

9.

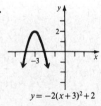

$y = -2(x + 3)^2 + 2$

11.

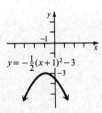

$y = -\frac{1}{2}(x + 1)^2 - 3$

13.

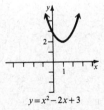

$y = x^2 - 2x + 3$

15.

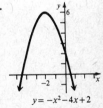

$y = -x^2 - 4x + 2$

17.

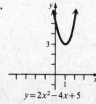

$y = 2x^2 - 4x + 5$

19.

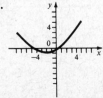

21.

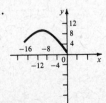

23. $[2, +\infty)$

25. $(-\infty, 0]$

27. $(-\infty, -5/4]$

29. 80 by 160

31. 300 sandwiches; 100

33. 5 in

35. 10, 10

37. (a) $x(500 - x) = 500x - x^2$ (b)

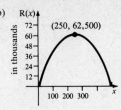

(c) 250   (d) $62,500   **39.**

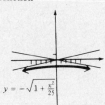

41. $6\sqrt{3}$ cm   **43.** (a) domain: $(-\infty, +\infty)$; range: $[k, +\infty)$ (b) domain: $(-\infty, +\infty)$; range: $(-\infty, k]$ **45.** 25

47. 1/2   **49.** (a) $|y + p|$ (b) The distances from the origin to the line and to the point are the same.   (c) $x^2 = 4py$

Section 4.3   (page 197)

1.

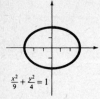

$\frac{x^2}{9} + \frac{y^2}{4} = 1$

3.

$\frac{x^2}{9} + y^2 = 1$

5.

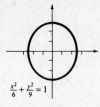

$\frac{x^2}{6} + \frac{y^2}{9} = 1$

7.

$x^2 + 4y^2 = 16$

9.

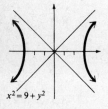

$x^2 = 9 + y^2$

11.

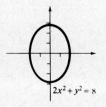

$2x^2 + y^2 = 8$

13.

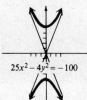

$25x^2 - 4y^2 = -100$

15.

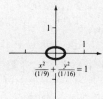

$\frac{x^2}{(1/9)} + \frac{y^2}{(1/16)} = 1$

17.

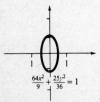

$\frac{64x^2}{9} + \frac{25y^2}{36} = 1$

19.

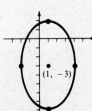

$(1, -3)$

21.

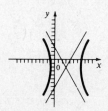

23.

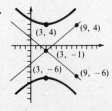

$(3, 4)$ $(9, 4)$
$(3, -1)$
$(3, -6)$ $(9, -6)$

25.

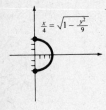

$\frac{x}{4} = \sqrt{1 - \frac{y^2}{9}}$

27. function

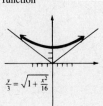

$\frac{y}{3} = \sqrt{1 + \frac{x^2}{16}}$

29.

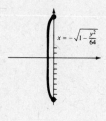

$x = -\sqrt{1 - \frac{y^2}{64}}$

31. function

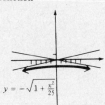

$y = -\sqrt{1 + \frac{x^2}{25}}$

33. $x^2/16 + y^2/12 = 1$ **35.** $x^2/36 + y^2/20 = 1$ **37.** $\dfrac{(x - 3)^2}{16} + \dfrac{(y + 2)^2}{25} = 1$ **39.** $x^2/9 - y^2/7 = 1$

41. $y^2/9 - x^2/25 = 1$ **43.** about 141.6 million miles

Section 4.4   (page 206)

1. hyperbola

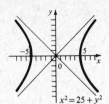

$x^2 = 25 + y^2$

3. ellipse

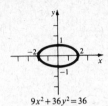

$9x^2 + 36y^2 = 36$

5. circle

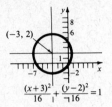

$(-3, 2)$

$\dfrac{(x+3)^2}{16} + \dfrac{(y-2)^2}{\lnot 16} = 1$

7. parabola

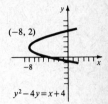

$(-8, 2)$

$y^2 - 4y = x + 4$

9. empty set (no graph)
11. circle

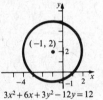

$(-1, 2)$

$3x^2 + 6x + 3y^2 - 12y = 12$

13. parabola

$x^2 - 6x + y = 0$

15. hyperbola

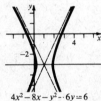

$4x^2 - 8x - y^2 - 6y = 6$

17. ellipse

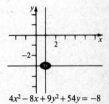

$4x^2 - 8x + 9y^2 + 54y = -8$

19. empty set—no graph   **21.** 1/2   **23.** $\sqrt{2}$ **25.** $\sqrt{21}/7$ **27.** $\sqrt{10}/3$ **29.** 0   **31.** 0; 1

Section 4.5   (page 212)

1. (a)

(b) y-axis   **3.** (a)

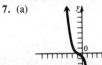

(b) origin

5. (a)

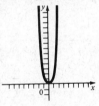

(b) (0, 1)   **7.** (a)

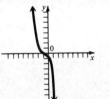

(b) $(-1, 0)$

9. (a)

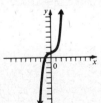

(b) $x = 1$ **11.**

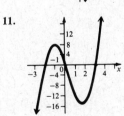

13.

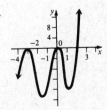

15. 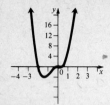 **17.** 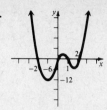 **19.** 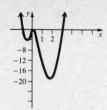 **21.** (a)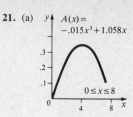

(b) between 4 and 5 hours, closer to 5 hours   (c) from about 1 hour to about 8 hours   **23.** odd   **25.** even
27. neither   **29.** odd   **31.** maximum is 26.136 when $x = -3.4$; minimum is 25 when $x = -3$ **33.** maximum
is 1.048 when $x = -.1$; minimum is -5 when $x = -1$ **35.** maximum is 84 when $x = -2$; minimum is -13
when $x = -1$

Section 4.6   (page 218)

1. 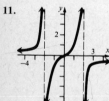   **3.** (a) 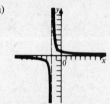   (b)    (c)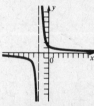

(d) 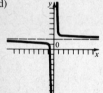   **5.** 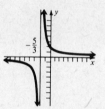   **7.** 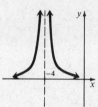   **9.**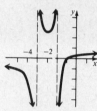

11. 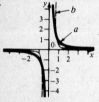   **13.** 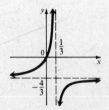   **15.** 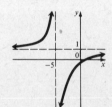   **17.**

19. 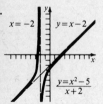   **21.** 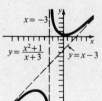   **23.** 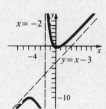   **25.** (a) 12.5, 10, 6.25,
4.76, 3.85

27.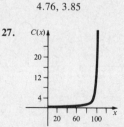

Chapter 4 Review Exercises (page 220)

1. 6/5   **3.** not defined   **5.** 9/4   **7.** 1/5   **9.** 0

11.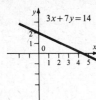

13.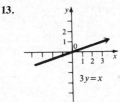

15.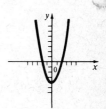

17. $x + 3y = 10$
19. $2x + y = 1$
21. $15x + 30y = 13$
23. $y = 3/4$
25. $5x - 8y = -40$
27. $y = -5$

29.

31.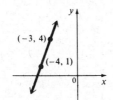

33.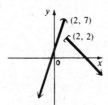

35.

37.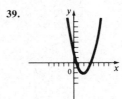

39.

41. $(-3, -9); x = -3$
43. $(7/2, -41/4), x = 7/2$
45. $(-1, 4); x = -1$
47. 5.5 and 5.5
49. a square, 45 m on a side

51.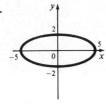

53.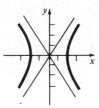

55.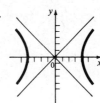

57.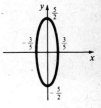

59.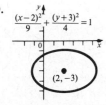

61.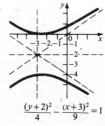

63.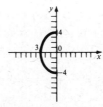

65.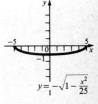

67.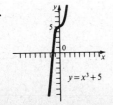

69.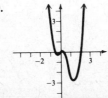

71.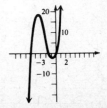

73.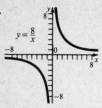

75.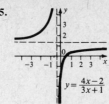

77.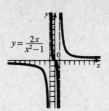

79.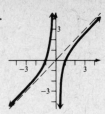

81. $2x - 3y = 14$ **83.** -7 **85.** one square has vertices $(-5, 19)$, $(0, 7)$, $(12, 12)$, $(7, 24)$; another has vertices at $(5, -5)$, $(0, 7)$, $(12, 12)$, $(17, 0)$; a third has vertices at $(0, 7)$, $\left(\dfrac{17}{2}, \dfrac{7}{2}\right)$, $(12, 12)$, $\left(\dfrac{7}{2}, \dfrac{31}{2}\right)$

Chapter 5

Section 5.1   (page 230)

1. (a)

(b)

(c)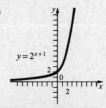

(d)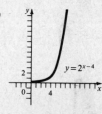

3.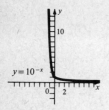

5.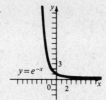

7.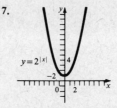

9.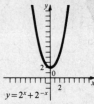

11.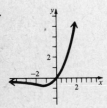

13. $\{1/2\}$ **15.** $\{-2\}$ **17.** $\{0\}$ **19.** $\{3\}$ **21.** $\{8\}$ **23.** $\{1/5\}$

25. $\{0\}$ **27.** $\{3/5\}$ **31.** 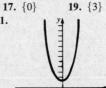   **33.**

35. (a) $500\ g$ (b) $409\ g$ (c) $335\ g$ (d) $184\ g$

(e)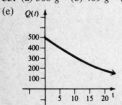

37. (a) \$21,665.74   (b) \$29,836.49   **39.** \$3516.56

41. \$6799.21   **43.** \$8158.66   **45.** (a) about 2,690,000

(b) about 3,020,000   (c) about 9,600,000   (d) about 38,400,000

47. 81.25°C   **49.** 117.5°C   **51.** 11.6 pounds per square inch

55. 0   **57.** 2   **59.** neither   **63.** 2.718

Section 5.2   (page 238)

1. $\log_3 81 = 4$ **3.** $\log_{1/2} 16 = -4$ **5.** $\log_{10} 0.0001 = -4$ **7.** $6^2 = 36$ **9.** $(1/2)^{-4} = 16$ **11.** 2

13. 1   **15.** −3   **17.** 1/2   **19.** −1/6   **21.** 9   **23.** 8   **25.** 9   **27.** {1/5}   **29.** {3/2}
31. {16}   **33.** $\log_3 2 - \log_3 5$ **35.** $\log_2 6 + \log_2 x - \log_2 y$ **37.** $1 + (1/2)\log_5 7 - \log_5 3$ **39.** cannot
be simplified using the properties of logarithms   **41.** $\log_k p + 2 \cdot \log_k q - \log_k m$
43. $(1/2)(\log_m 5 + 3 \cdot \log_m r - 5 \cdot \log_m z)$ **45.** $\log_a (xy)/m$ **47.** $\log_m (a^2/b^6)$ **49.** $\log_x (a^{3/2}b^{-4})$ or
$\log_x (a^{3/2}/b^4)$ **51.** $\log_b [7x(x + 2)/8]$ **53.** $\log_a [(z - 1)^2(3z + 2)]$ **55.** $(1/3) - (1/3) \cdot \log_5 m$

57. (a)   (b)   (c)   **59.**

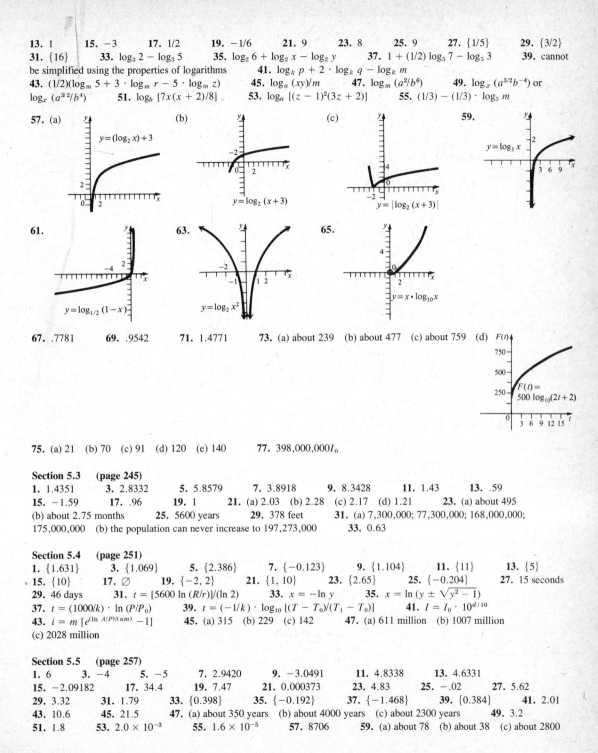

$y = (\log_2 x) + 3$

$y = \log_2 (x+3)$

$y = |\log_2 (x+3)|$

$y = \log_3 x$

61.   **63.**   **65.**

$y = \log_{1/2} (1-x)$

$y = \log_2 x^2$

$y = x \cdot \log_{10} x$

67. .7781   **69.** .9542   **71.** 1.4771   **73.** (a) about 239   (b) about 477   (c) about 759   (d) $F(t)$

750
500
250

$F(t) = 500 \log_{10}(2t + 2)$

0   3 6 9 12 15 t

75. (a) 21   (b) 70   (c) 91   (d) 120   (e) 140   **77.** $398{,}000{,}000 I_0$

Section 5.3   (page 245)
1. 1.4351   **3.** 2.8332   **5.** 5.8579   **7.** 3.8918   **9.** 8.3428   **11.** 1.43   **13.** .59
15. −1.59   **17.** .96   **19.** 1   **21.** (a) 2.03   (b) 2.28   (c) 2.17   (d) 1.21   **23.** (a) about 495
(b) about 2.75 months   **25.** 5600 years   **29.** 378 feet   **31.** (a) 7,300,000; 77,300,000; 168,000,000;
175,000,000   (b) the population can never increase to 197,273,000   **33.** 0.63

Section 5.4   (page 251)
1. {1.631}   **3.** {1.069}   **5.** {2.386}   **7.** {−0.123}   **9.** {1.104}   **11.** {11}   **13.** {5}
15. {10}   **17.** ∅   **19.** {−2, 2}   **21.** {1, 10}   **23.** {2.65}   **25.** {−0.204}   **27.** 15 seconds
29. 46 days   **31.** $t = [5600 \ln (R/r)]/(\ln 2)$ **33.** $x = -\ln y$ **35.** $x = \ln (y \pm \sqrt{y^2 - 1})$
37. $t = (1000/k) \cdot \ln (P/P_0)$ **39.** $t = (-1/k) \cdot \log_{10} [(T - T_0)/(T_1 - T_0)]$ **41.** $I = I_0 \cdot 10^{d/10}$
43. $i = m [e^{(\ln A/P)/(nm)} - 1]$ **45.** (a) 315   (b) 229   (c) 142   **47.** (a) 611 million   (b) 1007 million
(c) 2028 million

Section 5.5   (page 257)
1. 6   **3.** −4   **5.** −5   **7.** 2.9420   **9.** −3.0491   **11.** 4.8338   **13.** 4.6331
15. −2.09182   **17.** 34.4   **19.** 7.47   **21.** 0.000373   **23.** 4.83   **25.** −.02   **27.** 5.62
29. 3.32   **31.** 1.79   **33.** {0.398}   **35.** {−0.192}   **37.** {−1.468}   **39.** {0.384}   **41.** 2.01
43. 10.6   **45.** 21.5   **47.** (a) about 350 years   (b) about 4000 years   (c) about 2300 years   **49.** 3.2
51. 1.8   **53.** 2.0×10^{-3} **55.** 1.6×10^{-5} **57.** 8706   **59.** (a) about 78   (b) about 38   (c) about 2800

Chapter 5   Appendix   Interpolation   (page 261)
1. 3.3701   **3.** 1.6836   **5.** 8.7975 − 10   **7.** 1.4322   **9.** 62.26   **11.** .004534   **13.** 493,200
15. .0001667   **17.** 1.396   **19.** 1.550   **21.** .9308   **23.** 15.57   **25.** .1405

Chapter 5   Review Exercises   (page 262)
1.   **3.**

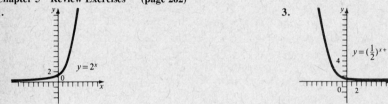

5. 5/3   **7.** 3/2   **9.** 800 g   **11.** $1790.19   **13.** about 12 years   **15.** $\log_2 32 = 5$
17. $10^{-3} = .001$ **19.** $\log_3 m + \log_3 n - \log_3 p$ **21.** $2 \log_5 x + 4 \log_5 y + (1/5)(3 \log_5 m + \log_5 p)$
23. −2.3757   **25.** 5,300,000,000   **27.** 5.97   **29.** 1.44   **31.** $^1/_2 (\log 17/\log 10 + 3) \approx 2.115$
33. {29}   **35.** {3}   **37.** $x = \log_5 [(1 \pm \sqrt{1 + 16y^2})/(4y)]$ **39.** −9.8856   **41.** 2.2553
43. 6.0486   **45.** 1.2 m   **47.** 1.8 m   **49.** $7182.87   **51.** $1241.99   **53.** about 28 quarters or 7 years
55. about 5.3 years   **57.** $16,095.74   **59.** $580,792.63   **61.** (a) − .223; .182   (b) .405 + .288 = .693
(c) .405 + .693 = 1.098; 3(.693) = 2.079   **63.** $100! \approx 9.32 \times 10^{157}$; $200! \approx 7.88 \times 10^{374}$

Chapter 6

Section 6.1   (page 270)
1. $(-\sqrt{2}/2, -\sqrt{2}/2)$ **3.** $(-1, 0)$ **5.** $(1, 0)$ **7.** $(1, 0)$ **9.** $(0, 1)$ **11.** $(-\sqrt{2}/2, -\sqrt{2}/2)$
13. $(\sqrt{2}/2, \sqrt{2}/2)$ **15.** $(-\sqrt{2}/2, \sqrt{2}/2)$ **17.** (a) $(2/3, -\sqrt{5}/3)$;   (b) $(2/3, \sqrt{5}/3)$;   (c) $(-2/3, -\sqrt{5}/3)$;
(d) $(-2/3, \sqrt{5}/3)$ **19.** (a) $(4/5, -3/5)$;   (b) $(4/5, 3/5)$;   (c) $(-4/5, -3/5)$;   (d) $(-4/5, 3/5)$ **21.** (a) $(-1/2, -\sqrt{3}/2)$;
(b) $(-1/2, \sqrt{3}/2)$;   (c) $(1/2, -\sqrt{3}/2)$;   (d) $(1/2, \sqrt{3}/2)$ **23.** (a) $(-2/5, \sqrt{21}/5)$;   (b) $(-2/5, -\sqrt{21}/5)$;
(c) $(2/5, \sqrt{21}/5)$;   (d) $(2/5, -\sqrt{21}/5)$ **25.** 0; 1   **27.** $\sqrt{2}/2$; $\sqrt{2}/2$ **29.** −1; 0   **31.** 0; −1
33. $-\sqrt{2}/2$; $\sqrt{2}/2$ **35.** $-\sqrt{2}/2$; $-\sqrt{2}/2$ **37.** III   **39.** III   **41.** II   **45.** (b) $(1/2, \sqrt{3}/2)$
(c) $(1/2, -\sqrt{3}/2)$; $(-1/2, \sqrt{3}/2)$; $(-1/2, -\sqrt{3}/2)$ **47.** $\sqrt{3}/2$ **49.** −1/2   **51.** $-\sqrt{3}/2$ **53.** 1/2

Section 6.2   Exercises   (page 276)
1. $\tan s = \sqrt{3}/3$; $\cot s = \sqrt{3}$; $\sec s = 2\sqrt{3}/3$; $\csc s = 2$ **3.** $\tan s = -4/3$; $\cot s = -3/4$; $\sec s = -5/3$; $\csc s = 5/4$
5. $\tan s = -\sqrt{3}$; $\cot s = -\sqrt{3}/3$; $\sec s = 2$; $\csc s = -2\sqrt{3}/3$

In Exercises 7–17 and 31–37 answers are given in the order sine, cosine, tangent, cotangent, secant, cosecant. **7.** 0; −1; 0;
undefined; −1; undefined   **9.** $\sqrt{2}/2$; $-\sqrt{2}/2$; −1; −1; $-\sqrt{2}$; $\sqrt{2}$ **11.** $-\sqrt{2}/2$; $\sqrt{2}/2$; −1; −1;
$\sqrt{2}$; $-\sqrt{2}$ **13.** 1/2; $\sqrt{3}/2$; $\sqrt{3}/3$; $\sqrt{3}$; $2\sqrt{3}/3$; 2   **15.** $\sqrt{3}/2$; −1/2; $-\sqrt{3}$; $-\sqrt{3}/3$; −2; $2\sqrt{3}/3$
17. −1/2; $-\sqrt{3}/2$; $\sqrt{3}/3$; $\sqrt{3}$; $-2\sqrt{3}/3$; −2   **19.** +, +   **21.** −, −, +, +, −, −   **23.** II   **25.** III
27. II or III   **29.** III or IV   **31.** $\sin s = \sqrt{65}/9$; $\tan s = -\sqrt{65}/4$; $\cot s = -4\sqrt{65}/65$; $\sec s = -9/4$;
$\csc s = 9\sqrt{65}/65$ **33.** $3\sqrt{13}/13, 2\sqrt{13}/13, 3/2, 2/3, \sqrt{13}/2, \sqrt{13}/3$ **35.** $\cos t = -2\sqrt{5}/5$; $\tan t = -1/2$;
$\cot t = -2$; $\sec t = -\sqrt{5}/2$; $\csc t = \sqrt{5}$ **37.** $a, \sqrt{1 - a^2}, a\sqrt{1 - a^2}/(1 - a^2), \sqrt{1 - a^2}/a, 1/\sqrt{1 - a^2}, 1/a$

In Exercises 39–42, your answers may differ slightly in the last digits, depending on the method you use to get the number.
39. $\sin s = .903687$, $\tan s = -2.11047$, $\cot s = -.473829$, $\sec s = -2.33540$, $\csc s = 1.10658$
41. $\sin t = -.095832$, $\cos t = .995397$, $\tan t = -.096275$, $\cot t = -10.3869$, $\sec t = 1.00462$ **47.** $\cos \theta$
49. $\csc \theta$ **51.** $\tan \theta$ **53.** θ

Section 6.3   Exercises   (page 284)
1. 320°   **3.** 235°   **5.** 90°   **7.** 179°   **9.** $\pi/3$ **11.** $\pi/2$ **13.** $3\pi/4$ **15.** $5\pi/3$
17. $9\pi/4$ **19.** $7\pi/9$ **21.** 60°   **23.** 315°   **25.** 330°   **27.** −30°   **29.** 900°   **31.** 63°
33. .9847   **35.** 42° 07′ 08″   **37.** $\pi/2$ **39.** π **41.** 1800°   **43.** 3/2   **45.** 1.23

47. 3.09309   **49.** 980 mi   **51.** 2200 mi   **53.** 3700 mi   **55.** 477 ft   **57.** We begin the answers with the blank next to 30° and then proceed counterclockwise from there: $\pi/6$; 45°; $\pi/3$; $\pi/2$; 120°; 135°; $5\pi/6$; π; $7\pi/6$; $5\pi/4$; 240°; 300°; $7\pi/4$; $11\pi/6$.   **59.** 2π radians per minute or $\pi/30$ radians per second   **61.** 18π cm   **63.** 4 sec

Section 6.4   (page 292)

In Exercises 1–15, answers are given in the order sine, cosine, tangent, cotangent, secant, and cosecant.
1. $\sqrt{3}/2, -1/2, -\sqrt{3}, -\sqrt{3}/3, -2, 2\sqrt{3}/3$ **3.** $1/2, -\sqrt{3}/2, -\sqrt{3}/3, -\sqrt{3}, -2\sqrt{3}/3, 2$
5. $-\sqrt{3}/2, -1/2, \sqrt{3}, \sqrt{3}/3, -2, -2\sqrt{3}/3$ **7.** $-1/2, \sqrt{3}/2, -\sqrt{3}/3, -\sqrt{3}, 2\sqrt{3}/3, -2$
9. $\sqrt{3}/2, 1/2, \sqrt{3}, \sqrt{3}/3, 2, 2\sqrt{3}/3$ **11.** $1/2, -\sqrt{3}/2, -\sqrt{3}/3, -\sqrt{3}, -2\sqrt{3}/3, 2$ **13.** 0, −1, 0, undefined, −1, undefined   **15.** −1, 0, undefined, 0, undefined, −1   **17.** $\sqrt{3}/3, \sqrt{3}$ **19.** $\sqrt{3}/2, \sqrt{3}/3, 2\sqrt{3}/3$
21. −1, −1   **23.** $-\sqrt{3}/2, -2\sqrt{3}/3$ **25.** 4/5, −3/5, −4/3, −3/4, −5/3, 5/4   **27.** 7/25, 24/25, 7/24, 24/7, 25/24, 25/7   **29.** −5/13, −12/13, 5/12, 12/5, −13/12, −13/5   **31.** −4/5, −3/5, 4/3, 3/4, −5/3, −5/4

In Exercises 33–37, answers are given in the order sine, cosine, tangent, cotangent, secant, and cosecant.
33. 3/5, 4/5, 3/4, 4/3, 5/4, 5/3   **35.** 21/29, 20/29, 21/20, 20/21, 29/20, 29/21   **37.** n/p, m/p, n/m, m/n p/m, p/n **39.** .759260, .650787, 1.16668   **41.** 1   **43.** 23/4   **45.** $1/2 + \sqrt{3}$ **47.** −29/12
49. $-\sqrt{3}/3$ **51.** false   **53.** true   **55.** false   **57.** true   **59.** true

Section 6.5   (page 302)
1. 35°   **3.** 37°   **5.** 69° 50′   **7.** .0684   **9.** .3142   **11.** .8814   **13.** .6338   **15.** −.6248
17. −.5712   **19.** −3.072   **21.** .9636   **23.** −.6361   **25.** .5704   **27.** 1.162   **29.** .1628
31. −.3131   **33.** 58° 00′   **35.** 68° 45′   **37.** 1.4576   **39.** .001   **41.** .01   **43.** .01
45. .10   **47.** 1   **49.** 1   **51.** .8417   **53.** .0100   **55.** 1.0000   **57.** $c = 68.4$ km, $b = 58.2$ km
59. $b = 3330.68$ m, $a = 1311.04$ m   **61.** 2.00 m

Section 6.6   (page 312)
1.

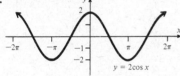

$y = 2\cos x$

3.

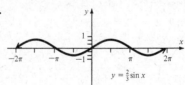

$y = \frac{2}{3}\sin x$

5.

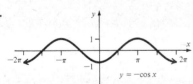

$y = -\cos x$

7.

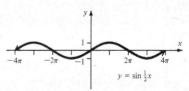

$y = \sin \frac{1}{2}x$

9.

$y = \cos \frac{1}{3}x$

11.

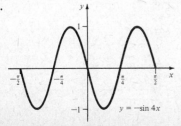

$y = -\sin 4x$

13.

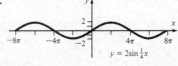

$y = 2\sin \frac{1}{4}x$

15.

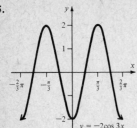

$y = -2\cos 3x$

17.

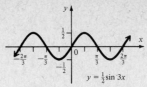

$y = \frac{1}{2}\sin 3x$

19.

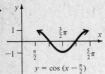

$y = \cos\left(x - \frac{\pi}{2}\right)$

21.

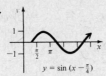

$y = \sin\left(x - \frac{\pi}{4}\right)$

23.

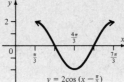

$y = 2\cos\left(x - \frac{\pi}{3}\right)$

25.

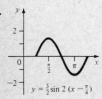

$y = \frac{3}{2}\sin 2\left(x - \frac{\pi}{4}\right)$

27.

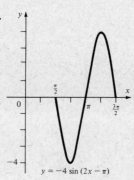

$y = -4\sin(2x - \pi)$

29.

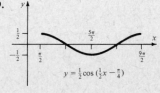

$y = \frac{1}{2}\cos\left(\frac{1}{2}x - \frac{\pi}{4}\right)$

31.

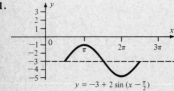

$y = -3 + 2\sin\left(x - \frac{\pi}{2}\right)$

33.

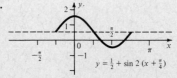

$y = \frac{1}{2} + \sin 2\left(x + \frac{\pi}{4}\right)$

35. odd   **37.** $y = -\cos x$ has the same graph   **39.** $1, 4\pi/3$ **41.** about 35,000 years   **43.** almost 2 hours

45. 20   **47.** (a) 5; 1/60   (b) 60   (c) 5; 1.545; -4.045; -4.045; 1.545   (d)

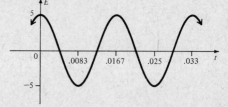

Section 6.7 (page 321)

1.

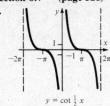

$y = \cot \frac{1}{2} x$

3.

5.

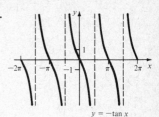

$y = -\tan x$

7.

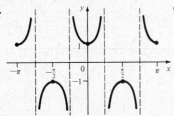

$y = \sec 2x$

9.

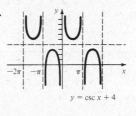

$y = \csc x + 4$

11.

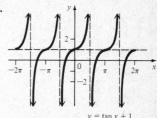

$y = \tan x + 1$

13.

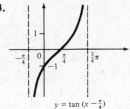

$y = \tan \left(x - \frac{\pi}{4} \right)$

15.

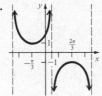

17.

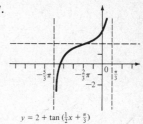

$y = 2 + \tan \left(\frac{1}{2} x + \frac{\pi}{3} \right)$

19.

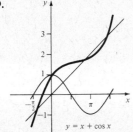

$y = x + \cos x$

21.

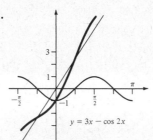

$y = 3x - \cos 2x$

23.

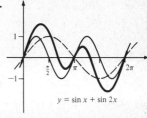

$y = \sin x + \sin 2x$

25.

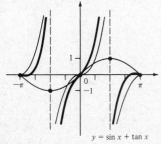

$y = \sin x + \tan x$

27.

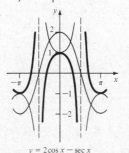

$y = 2\cos x - \sec x$

29.

$y = \cos x + \cot x$

31.

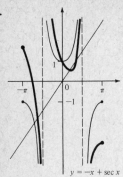

$y = -x + \sec x$

33.

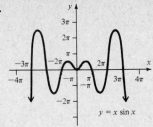

$y = x \sin x$

35.

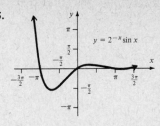

$y = 2^{-x} \sin x$

37.

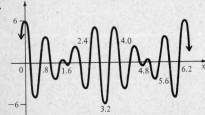

Section 6.8   (page 325)
1. $-\pi/3$ **3.** $\pi/4$ **5.** $-\pi/2$ **7.** $\pi/3$ **9.** $3\pi/4$ **11.** $-\pi/3$ **13.** $-7° \, 40'$ **15.** $113° \, 30'$
17. $22° \, 00'$ **19.** $48° \, 00'$ **21.** $-42° \, 40'$ **23.** $.8058$ **25.** -1.332 **29.** $-26° \, 21' \, 29''$
31. $23° \, 52' \, 41''$ **33.** $\sqrt{5}/2$ **35.** $\sqrt{5}/5$ **37.** $-\sqrt{21}/2$ **39.** $-3\sqrt{10}/10$
41. $.957826$ **43.** $.123430$ **45.** $x = \sin(y/4)$ **47.** $x = (1/2) \tan 2y$ **49.** $x = -2 + \sin y$
51. $\sqrt{1 - x^2}$ **53.** $1/x$ **55.** $\sqrt{1 - x^2}/x$
57. domain: $-1 \le x \le 1$; range: $0 \le y \le 2\pi$ **59.** domain: $-1 \le x \le 1$; range: $(-\pi/2) + 2 \le y \le (\pi/2) + 2$

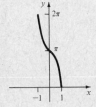

65. false   **67.** false   **69.** false   **71.** false

Chapter 6   Review Exercises   (page 327)
1. $(\sqrt{2}/2, -\sqrt{2}/2)$ **3.** $(-1, 0)$ **5.** $-\sqrt{3}/2, 1/2, -\sqrt{3}$ **7.** $(-2/3, -\sqrt{5}/3)$ **9.** $(2/3, \sqrt{5}/3)$
11. IV   **13.** IV   **15.** $\sin s = -3/5, \tan s = -3/4, \cot s = -4/3, \sec s = 5/4, \csc s = -5/3$ **17.** $\sin s = 4/5$,
$\cos s = -3/5, \cot s = -3/4, \csc s = 5/4$ **19.** $144°$ **21.** $3\pi/2$ **23.** $\sqrt{3}/2$ **25.** $\sqrt{3}$ **27.** -1
29. Sine is $-\sqrt{2}/2$; cosine is $-\sqrt{2}/2$; tangent is 1.   **31.** 1.428 **33.** $-.9216$ **35.** 0.5289 **37.** $79° \, 10'$
39. $62° \, 31'$ **41.** 1.2857 **43.** 2×10^8 m/sec   **45.** $2, 2\pi, \pi$ to the right   **47.** $1, \pi/2,$ none

49.

51.

53.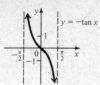

55.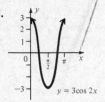

57.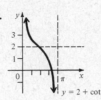

59.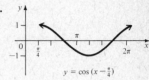

61.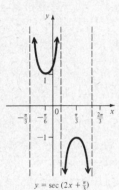

63.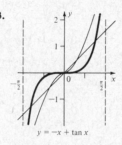

65.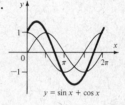

67. $y = (1/3) \sin 4\pi t/3$ **69.** $\pi/4$ **71.** $-\pi/3$ **73.** -0.7214

75.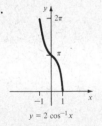

77.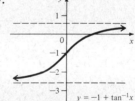

79. (a) $t = [1/(2\pi f)][\arcsin e/E_{max}]$ (b) .0007   **81.** $\sqrt{2}/4$

Chapter 7

Section 7.1   (page 334)
1. $\sqrt{7}/4$ **3.** $-2\sqrt{5}/5$ **5.** $\sqrt{21}/2$ **7.** $\cos \theta = -\sqrt{5}/3$; $\tan \theta = -2\sqrt{5}/5$; $\cot \theta = -\sqrt{5}/2$;
$\sec \theta = -3\sqrt{5}/5$; $\csc \theta = 3/2$ **9.** $\sin \theta = -\sqrt{17}/17$; $\cos \theta = 4\sqrt{17}/17$; $\cot \theta = -4$; $\sec \theta = \sqrt{17}/4$; $\csc \theta = -\sqrt{17}$
11. $\sin \theta = 2\sqrt{2}/3$; $\cos \theta = -1/3$; $\tan \theta = -2\sqrt{2}$; $\cot \theta = -\sqrt{2}/4$; $\csc \theta = 3\sqrt{2}/4$ **13.** $\sin \theta = 3/5$; $\cos \theta = 4/5$;
$\tan \theta = 3/4$; $\sec \theta = 5/4$; $\csc \theta = 5/3$ **15.** $\sin \theta = -\sqrt{7}/4$; $\cos \theta = 3/4$; $\tan \theta = -\sqrt{7}/3$; $\cot \theta = -3\sqrt{7}/7$;
$\csc \theta = -4\sqrt{7}/7$ **17.** (b)   **19.** (e)   **21.** (a)   **23.** (a)   **25.** (d)   **27.** 1   **29.** $-\sin \alpha$
31. 0   **33.** $(1 + \sin \theta)/\cos \theta$ **35.** 1   **37.** -1 **39.** $\sin^2 \theta/\cos^4 \theta$ **41.** 1
43. $(\cos^2 \alpha + 1)/(\sin^2 \alpha \cos^2 \alpha)$ **45.** $(\sin^2 s - \cos^2 s)/\sin^4 s$ **47.** $\cos x = \pm\sqrt{1 - \sin^2 x}$;
$\tan x = \pm\sin x\sqrt{1 - \sin^2 x}/(1 - \sin^2 x)$; $\cot x = \pm\sqrt{1 - \sin^2 x}/\sin x$; $\sec x = \pm\sqrt{1 - \sin^2 x}/(1 - \sin^2 x)$; $\csc x = 1/\sin x$
49. $\tan x = \pm\sqrt{\csc^2 x - 1}/(\csc^2 x - 1)$ **51.** $\cot s = \pm\sqrt{\sec^2 s - 1}/(\sec^2 s - 1)$ **53.** $\sin \theta = \sqrt{2x + 1}/(x + 1)$
57. $(25\sqrt{6} - 60)/12$; $-(25\sqrt{6} + 60)/12$

Section 7.2   (page 338)

1. $1/(\sin\theta\cos\theta)$ **3.** $1 + \cos s$ **5.** 1   **7.** 1   **9.** $2 + 2\sin t$ **11.** $(\sin\gamma + 1)(\sin\gamma - 1)$
13. $4\sin x$ **15.** $(2\sin x + 1)(\sin x + 1)$ **17.** $(4\sec x - 1)(\sec x + 1)$ **19.** $(\cos^2 x + 1)^2$ **21.** $\sin\theta$
23. 1   **25.** $\tan^2\beta$ **27.** $\tan^2 x$ **29.** $\sec^2 x$ **75.** identity   **77.** not an identity   **79.** not an
identity   **81.** not an identity

Section 7.3   (page 344)

1. $\cot 3°$ **3.** $\sin 5\pi/12$ **5.** $\sec 104° 24'$ **7.** $\cos -\pi/8$ **9.** $\csc(-56° 42')$ **11.** true   **13.** false
15. true   **17.** $15°$ **19.** $140/3°$ **21.** $20°$ **23.** $(\sqrt{6} - \sqrt{2})/4$ **25.** $(\sqrt{6} - \sqrt{2})/4$ **27.** $\sqrt{2}/2$
29. $\sqrt{2}/2$ **31.** $(\sqrt{3}\cos\theta - \sin\theta)/2$ **33.** $(\cos\theta - \sqrt{3}\sin\theta)/2$ **35.** $-\sin x$ **37.** $16/65$; $56/65$
39. $(-2\sqrt{10} + 2)/9$; $(-2\sqrt{10} - 2)/9$ **41.** $-36/85$; $84/85$ **43.** $56/65$; $16/65$ **45.** 0; $240/289$
55. $\cos t$; $\sin t$; $-\cos t$; $-\sin t$; $\cos t$; $-\sin t$; $-\cos t$; $\sin t$ **59.** $-.77917$ **61.** $.98209$ **65.** $36/85$

Section 7.4   (page 350)

1. $(\sqrt{6} - \sqrt{2})/4$ **3.** $2 - \sqrt{3}$ **5.** $(-\sqrt{6} - \sqrt{2})/4$ **7.** $(\sqrt{6} + \sqrt{2})/4$ **9.** $\sqrt{2}/2$ **11.** -1
13. 0   **15.** $\sqrt{2}/2$ **17.** $\sqrt{2}(\sin\theta + \cos\theta)/2$ **19.** $(\sqrt{3}\tan\theta + 1)/(\sqrt{3} - \tan\theta)$ **21.** $\sin\theta$
23. $\tan\theta$ **25.** $-\sin\theta$ **27.** $63/65$; $33/65$; $63/16$; $33/56$ **29.** $(4\sqrt{2} + \sqrt{5})/9$; $(4\sqrt{2} - \sqrt{5})/9$;
$(-8\sqrt{5} - 5\sqrt{2})/(20 - 2\sqrt{10})$; $(-8\sqrt{5} + 5\sqrt{2})/(20 + 2\sqrt{10})$ **31.** $77/85$; $13/85$; $-77/36$; $13/84$ **33.** $-33/65$;
$-63/65$; $33/56$; $63/16$ **35.** 1; $-161/289$; undefined; $-161/240$ **49.** $18°$ **51.** $15°$
53. $\sqrt{2}\sin(x + 135°)$ **55.** $13\sin(\theta + 293°)$ **57.** $17\sin(x + 152°)$ **59.** $25\sin(\theta + 254°)$
61. $5\sin(x + 53°)$
63. $2\sin(x + \pi/6)$ **65.** $\sqrt{2}\sin(x + 3\pi/4)$

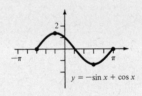

$y = \sqrt{3}\sin x + \cos x$

$y = -\sin x + \cos x$

69. -0.996728 **71.** -0.959882 **73.** $(8\sqrt{17} + \sqrt{85})/51$ **75.** $23/80$

Section 7.5   (page 357)

1. $2\pi/5$ **3.** $\cos 2x$ **5.** $160°, 160°$ **7.** $\cos^2, \sin^2$ **9.** $9°$ **11.** $\tan 170°$ **13.** $-$ **15.** $+$
17. $-\sqrt{42}/12$ **19.** $-\sqrt{6}/4$ **21.** $\sin 22^{1}/_{2}° = \sqrt{2 - \sqrt{2}}/2$, $\cos 22^{1}/_{2}° = \sqrt{2 + \sqrt{2}}/2$, $\tan 22^{1}/_{2}° = \sqrt{3 - 2\sqrt{2}}$
23. $\sin 195° = -\sqrt{2 - \sqrt{3}}/2$, $\cos 195° = -\sqrt{2 + \sqrt{3}}/2$, $\tan 195° = \sqrt{7 - 4\sqrt{3}}$ **25.** $\sin 5\pi/2 = 1$, $\cos 5\pi/2 = 0$,
$\tan 5\pi/2$ does not exist   **27.** $\cos x = -\sqrt{42}/12$, $\sin x = \sqrt{102}/12$, $\tan x = -\sqrt{119}/7$, $\sec x = -2\sqrt{42}/7$,
$\csc x = 2\sqrt{102}/17$, $\cot x = -\sqrt{119}/17$ **29.** $\cos 2\theta = 17/25$, $\sin 2\theta = -4\sqrt{21}/25$, $\tan 2\theta = -4\sqrt{21}/17$,
$\sec 2\theta = 25/17$, $\csc 2\theta = -25\sqrt{21}/84$, $\cot 2\theta = -17\sqrt{21}/84$ **31.** $\tan 2x = -4/3$, $\sec 2x = -5/3$,
$\cos 2x = -3/5$, $\cot 2x = -3/4$, $\sin 2x = 4/5$, $\csc 2x = 5/4$ **33.** $\sin\alpha/2 = \sqrt{3}/3$, $\cos\alpha/2 = -\sqrt{6}/3$,
$\tan\alpha/2 = -\sqrt{2}/2$, $\cot\alpha/2 = -\sqrt{2}$, $\sec\alpha/2 = -\sqrt{6}/2$, $\csc\alpha/2 = \sqrt{3}$ **59.** $4\tan^2 x/(1 - 2\tan^2 x + \tan^4 x)$
61. $\cos 3x = 4\cos^3 x - 3\cos x$ **63.** $\cos 4x = 8\cos^4 x - 8\cos^2 x + 1$ **65.** $84°$ **67.** $60°$ **69.** 3.9
71. -0.843580 **73.** 0.537003 **75.** -1.570905 **77.** $.892230$ **79.** -1.97579 **81.** $\sqrt{3 - 2\sqrt{2}}$
83. 1   **85.** $\sec^2 4x$

Section 7.6   (page 362)

1. $(1/2)[\sin 60° - \sin 10°] = (1/2)[\sqrt{3}/2 - \sin 10°]$ **3.** $(3/2)[\cos 8x + \cos 2x]$
5. $(1/2)[\cos 2\theta - \cos(-4\theta)] = (1/2)[\cos 2\theta - \cos 4\theta]$ **7.** $-4[\cos 9y + \cos(-y)] = -4(\cos 9y + \cos y)$
9. $2\cos 45° \sin 15°$ **11.** $2\cos 95° \cos(-53°) = 2\cos 95° \cos 53°$ **13.** $2\cos 15\beta/2 \sin 9\beta/2$
15. $-6\cos(7x/2)\sin(-3x/2) = 6\cos(7x/2)\sin(3x/2)$

Section 7.7   (page 368)

1. $\{3\pi/4, 7\pi/4\}$ **3.** $\{\pi/3, 5\pi/3\}$ **5.** $\{\pi/6, 7\pi/6, 4\pi/3, 5\pi/3\}$ **7.** $\{\pi/3, 5\pi/3, \pi/6, 11\pi/6\}$ **9.** $\{\pi\}$

11. $\{3\pi/2, 7\pi/6, 11\pi/6\}$ **13.** $\{\pi/4, 3\pi/4, 5\pi/4, 7\pi/4\}$ **15.** $\{0, \pi/2, \pi, 3\pi/2\}$ **17.** $\{\pi/12, 7\pi/12, 13\pi/12,$ $19\pi/12\}$ **19.** $\{3\pi/8, 5\pi/8, 11\pi/8, 13\pi/8\}$ **21.** $\pi/2, 3\pi/2\}$ **23.** $\{71° \ 30', 251° \ 30', 63° \ 30', 243° \ 30'\}$
25. $\{135°, 315°, 71° \ 30', 251° \ 30'\}$ **27.** $\{33° \ 30', 326° \ 30'\}$ **29.** $\{45°, 225°\}$ **31.** $\{0°, 90°, 180°, 270°\}$
33. $\{30°, 60°, 210°, 240°\}$ **35.** $\{0°\}$ **37.** $\{120°, 240°\}$ **39.** $\{270°, 30°, 150°\}$ **41.** $\{90°, 270°, 45°,$ $225°\}$ **43.** $\{70° \ 30', 289° \ 30'\}$ **45.** $\{90°, 270°, 30°, 150°, 210°, 330°\}$ **47.** $\{x \mid x = 90° + 360° \cdot n,$ $180° + 360° \cdot n, n \text{ any integer}\}$ **49.** $\{x \mid x = 30° + 360° \cdot n, 150° + 360° \cdot n, 90° + 360° \cdot n, 270° + 360° \cdot n, n \text{ any}$ $\text{integer}\}$ **51.** $\{x \mid x = 45° + 180° \cdot n, 108° \ 30' + 180° \cdot n, n \text{ any integer}\}$ **53.** $\{x \mid x = 22^1/2° + 180° \cdot n,$ $112^1/2° + 180° \cdot n, n \text{ any integer}\}$ **55.** $\{53° \ 40'; 126° \ 20'; 188° \ 00'; 352° \ 00'\}$ **57.** $\{149° \ 40'; 329° \ 40'; 106° \ 20';$ $286° \ 20'\}$ **59.** no solution   **61.** $\{57° \ 40'; 159° \ 10'\}$ **63.** $\{\pi/12, \pi/2, 5\pi/12, 3\pi/2, 13\pi/12, 17\pi/12\}$
65. $\{0, \pi/4, 3\pi/4, \pi, 5\pi/4, 7\pi/4\}$ **67.** $\{\pi/6, \pi/2, 3\pi/2, 5\pi/6\}$ **69.** $1/4$ sec   **71.** 2 sec   **73.** $\{3/5\}$
75. $\{4/5\}$ **77.** $\{0\}$ **79.** $\{1/2\}$ **81.** $\{-1/2\}$ **83.** $\{0\}$

Chapter 7   Review Exercises   (page 370)
1. $\sin x = -4/5, \tan x = -4/3, \sec x = 5/3, \csc x = -5/4, \cot x = -3/4$ **3.** $\sin (x + y) = (4 + 3\sqrt{15})/20,$
$\cos (x - y) = (4\sqrt{15} + 3)/20$ **5.** $\sin x = \sqrt{2 - \sqrt{2}}/2, \cos x = \sqrt{2 + \sqrt{2}}/2, \tan x = \sqrt{3 - 2\sqrt{2}}$ **7.** j
9. c   **11.** d   **13.** a   **15.** f   **17.** e   **19.** l   **21.** $1/\cos^2 \theta$ or $\sec^2 \theta$ **23.** $1/\sin^2 \theta \cos^2 \theta$
53. $\{\pi/2, 3\pi/2\}$ **55.** $\{\pi/6, 5\pi/6\}$ **57.** $\{\pi/8, 3\pi/8, 5\pi/8, 7\pi/8, 9\pi/8, 11\pi/8, 13\pi/8, 15\pi/8\}$ **59.** $\{\pi/2\}$
61. $\{\pi/3, 5\pi/3, \pi\}$ **63.** $\{60°, 300°\}$ **65.** $\{270°\}$ **67.** $\{0°, 45°, 180°, 225°\}$ **77.** $2 - \sqrt{3}$

Chapter 8

Section 8.1   (page 380)
1. $B = 62°, C = 90°, a = 8.17$ m, $b = 15.4$ m   **3.** $A = 17°, C = 90°, a = 39.1$ in, $c = 134$ in
5. $c = 85.9$ yd, $A = 62° \ 50', B = 27° \ 10', C = 90°$ **7.** $b = 42.3$ cm, $A = 24° \ 10', B = 65° \ 50', C = 90°$
9. $A = 50° \ 50', C = 90°, a = .483$ m, $b = .393$ m   **11.** $B = 30° \ 16', a = 16.19$ feet, $b = 9.445$ feet, $C = 90°$
13. $B = 42.38°, b = 36.01$ cm, $c = 53.42$ cm, $C = 90°$ **15.** $c = 581.9$ cm, $A = 34.24°$ or $34° \ 14',$
$B = 55.76°$ or $55° \ 46', C = 90°$ **17.** 26.6 m   **19.** $52° \ 30'$ **21.** 11 ft   **23.** $35° \ 50'$ **25.** 1581 ft
27. 51.4 m   **29.** 8200 ft   **31.** 446 ft   **33.** 114 ft   **35.** 5.18 m   **37.** $h = k(\tan B - \tan A)$
39. 156 mi   **41.** 120 mi   **43.** 38°   **45.** $m(1 - \cos \alpha)$ **47.** $54° \ 40'$ **49.** 6.993752×10^9 mi
51. $\sqrt{15}/4$ **53.** $\sqrt{5}/3$

Section 8.2   (page 391)
1. $C = 83°, a = 11$ m, $b = 10$ m   **3.** $C = 80° \ 40', a = 79.5$ mm, $c = 108$ mm   **5.** no such triangle
7. $A = 56° \ 00', c = 361$ ft, $a = 308$ ft   **9.** $C = 125° \ 01', b = 576.7$ m, $c = 1224$ m   **11.** $B = 6.50°,$
$A = 148.36°, a = 1171$ cm   **13.** $A_1 = 43° \ 40', C_1 = 106° \ 40', A_2 = 136° \ 20', C_2 = 14° \ 00'$ **15.** no such triangle
17. $B = 27° \ 10', C = 10° \ 40'$ **19.** $C = 21.61°, B = 99.97°, b = 65.64$ m   **21.** $B = 20° \ 40', C = 116° \ 50',$
$c = 20.6$ m   **23.** no such triangle   **25.** $A_1 = 52°10', C_1 = 95° \ 00', c_1 = 9520$ cm, $A_2 = 127° \ 50',$
$C_2 = 19° \ 20', c_2 = 3160$ cm   **27.** $A_1 = 53° \ 20', C_1 = 87° \ 00, c_1 = 37.1$ m, $A_2 = 126° \ 40', C_2 = 13° \ 40', c_2 = 8.77$ m
29. $B_1 = 30°08', C_1 = 120°05', c_1 = 504.5$ cm, $B_2 = 149°52', C_2 = 0°21', c_2 = 3.561$ cm   **31.** 118 m
33. 1.93 mi   **35.** 929 ft   **37.** 111°   **39.** does not exist   **41.** 41.0 ft²   **43.** 356 cm²
45. 722.3 m²

Section 8.3   (page 398)
1. $a = 4.38$ in, $B = 80° \ 10', C = 60° \ 00'$ **3.** $c = 6.47$ m, $A = 53° \ 00', B = 81° \ 20'$ **5.** $a = 156$ cm,
$B = 64° \ 50', C = 34° \ 30'$ **7.** $b = 9.529$ in, $A = 64.6°, C = 40.6°$ **9.** $a = 15.7$ m, $B = 21° \ 30', C = 45° \ 40'$
11. $c = 139.0$ m, $A = 49° \ 20', B = 105° \ 51'$ **13.** $A = 29° \ 00', B = 46° \ 30', C = 104° \ 30'$ **15.** $A = 81° \ 50',$
$B = 37° \ 20', C = 60° \ 50'$ **17.** $A = 42° \ 00', B = 35° \ 50', C = 102° \ 10'$ **19.** $A = 47° \ 40', B = 44° \ 50',$
$C = 87° \ 20'$ **21.** $A = 47° \ 43', B = 72° \ 13', C = 60° \ 04'$ **23.** $A = 28° \ 10', B = 21° \ 56', C = 129° \ 54'$
25. 257 m   **27.** 281 km   **29.** $14° \ 30'$ **31.** 1472 m   **33.** 6.01 km   **35.** 18 ft
37. 25.24983 mi   **39.** 140 in²   **41.** 12,600 cm²   **43.** 3650 ft²   **45.** 1921 ft²   **47.** 33 cans
57. The angle at Mackinac West Base is 16.42821°; the angle at Green Island is 123.13624°; the angle at St. Ignace West Base is 40.43555°.

Section 8.4   (page 410)

1. m and p, n and r **3.** m and p equal $2t$; or t is one-half m or p; also $m = 1p$ and $n = 1r$, $m = p = -q$; $r = -s$

5.

7.

9.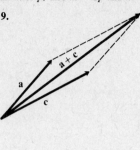

11.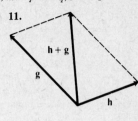

13.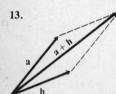

15.

17.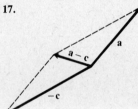

19.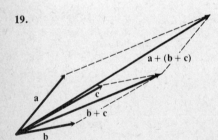

21. $<1, 3>$ **23.** $<22, -22>$ **25.** 14, 14   **27.** 14.2, 24.8   **29.** $-123, 155$ **31.** $-22.3, -65.4$
33. $\sqrt{2}, 45°$ **35.** 16, 315°   **37.** 17, 332°   **39.** 6, 180°   **41.** $-5i + 8j$ **43.** $2i$
45. $4\sqrt{2}i + 4\sqrt{2}j$ **47.** $-.2536i + .5438j$ **49.** 94° 00′   **51.** 17   **53.** 18°   **55.** magnitude 2.86,
equilibrant makes an angle of 124.6° with the 4.72 lb force   **57.** weight 64.8 lb, tension 61.9 lb   **59.** 190, 283
pounds respectively   **61.** 173°   **63.** 39.2   **65.** 237°, 470 mph   **67.** 358° 00′, 169 mph   **75.** -22
77. -50 **81.** 151° 00′

Section 8.5   (page 419)

1.

3.

5.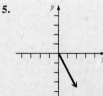

7.

9.

11. $1 + i$ **13.** $-2 + 2i$ **15.** $-2 + 4i$ **17.** $2 + 4i$ **19.** $7 + 9i$ **21.** $(2\sqrt{2}, 315°)$
23. $(3, 90°)$ **25.** $(5, 0°)$ **27.** $(2, 30°)$ **29.** $(2, 60°)$ **31.** $\sqrt{2} + i\sqrt{2}$ **33.** $-\sqrt{2}/2 + i\sqrt{2}/2$
35. $3i$ **37.** $-2 - 2i\sqrt{3}$ **39.** $\sqrt{3}/2 - i/2$ **41.** $\sqrt{2} + i\sqrt{2}$ **43.** $10i$ **45.** $-2 - 2i\sqrt{3}$
47. $\sqrt{3}/2 + i/2$ **49.** $5/2 - 5i\sqrt{3}/2$ **51.** $3\sqrt{2}(\cos 315° + i \sin 315°)$ **53.** $6(\cos 240° + i \sin 240°)$
55. $2(\cos 330° + i \sin 330°)$ **57.** $5\sqrt{2}(\cos 225° + i \sin 225°)$ **59.** $(\sqrt{13}, 56° 20')$,
$\sqrt{13}(\cos 56° 20' + i \sin 56° 20')$ **61.** $-1.0179 - 2.8221i;\ (3, 250° 10')$
63. $-1.8794 + .6840i;\ 2(\cos 160° + i \sin 160°)$ **65.** $6i$ **67.** 4 **69.** $12\sqrt{3} + 12i$
71. $-15\sqrt{2}/2 + 15i\sqrt{2}/2$ **73.** $3\sqrt{3}/2 + 3i/2$ **75.** $-1 - i\sqrt{3}$ **77.** $2\sqrt{3} - 2i$ **79.** $-1/2 - i/2$
81. $\sqrt{3} + i$ **83.** $2.39 + 15.0i$ **85.** $.378 + 3.52i$
87.   **89.**   **91.**

Section 8.6   (page 425)
1. -8 **3.** 1 **5.** $-16\sqrt{3} + 16i$ **7.** $-128 + 128i\sqrt{3}$ **9.** $128 + 128i$ **11.** $-.1892 + .0745i$
13. $5520 + 9550i$
15. $(\cos 120° + i \sin 120°)$,   **17.** $2(\cos 90° + i \sin 90°)$,   **19.** $4(\cos 60° + i \sin 60°)$,   **21.** $\sqrt[3]{2}(\cos 20° + i \sin 20°)$,
$(\cos 240° + i \sin 240°)$, $2(\cos 210° + i \sin 210°)$, $4(\cos 180° + i \sin 180°)$, $\sqrt[3]{2}(\cos 140° + i \sin 140°)$,
$(\cos 0° + i \sin 0°)$ $2(\cos 330° + i \sin 330°)$ $4(\cos 300° + i \sin 300°)$ $\sqrt[3]{2}(\cos 260° + i \sin 260°)$

      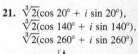

23. $\sqrt[3]{4}(\cos 50° + i \sin 50°)$, $\sqrt[3]{4}(\cos 170° + i \sin 170°)$, $\sqrt[3]{4}(\cos 290° + i \sin 290°)$

25. $1.585 + .6722i,\ -1.375 + 1.036i,\ -.2103 - 1.709i$ **27.** $4.0150,\ -4.0150,\ 4.0150i,\ -4.0150i$
29. $(\cos 0° + i \sin 0°)$ **31.** $(\cos 0° + i \sin 0°)$
$(\cos 180° + i \sin 180°)$ $(\cos 60° + i \sin 60°)$
$(\cos 120° + i \sin 120°)$
$(\cos 180° + i \sin 180°)$
$(\cos 240° + i \sin 240°)$
$(\cos 300° + i \sin 300°)$

33. $(\cos 45° + i \sin 45°)$
$(\cos 225° + i \sin 225°)$

35. $(\cos 0° + i \sin 0°)$, $(\cos 120° + i \sin 120°)$, $(\cos 240° + i \sin 240°)$ **37.** $(\cos 90° + i \sin 90°)$,
$(\cos 210° + i \sin 210°)$, $(\cos 330° + i \sin 330°)$ **39.** $2(\cos 0° + i \sin 0°)$, $2(\cos 120° + i \sin 120°)$,
$2(\cos 240° + i \sin 240°)$ **41.** $(\cos 45° + i \sin 45°)$, $(\cos 135° + i \sin 135°)$, $(\cos 225° + i \sin 225°)$,
$(\cos 315° + i \sin 315°)$ **43.** $(\cos 22\ 1/2° + i \sin 22\ 1/2°)$, $(\cos 112\ 1/2° + i \sin 112\ 1/2°)$,
$(\cos 202\ 1/2° + i \sin 202\ 1/2°)$, $(\cos 292\ 1/2° + i \sin 292\ 1/2°)$ **45.** $2(\cos 20° + i \sin 20°)$, $2(\cos 140° + i \sin 140°)$,
$2(\cos 260° + i \sin 260°)$ **47.** $1.3606 + 1.2637i$; $-1.7747 + .5464i$; $.4141 - 1.8102i$
49. $1.6309 - 2.5259i$, $-1.6309 + 2.5259i$ **51.** $\sqrt[4]{2}(\cos 22\ 1/2° + i \sin 22\ 1/2°)$, $\sqrt[4]{2}(\cos 202\ 1/2° + i \sin 202\ 1/2°)$
53. $\sqrt[4]{18}(\cos 157\ 1/2° + i \sin 157\ 1/2°)$; $\sqrt[4]{18}(\cos 337\ 1/2° + i \sin 337\ 1/2°)$

Section 8.7   (page 430)
1. $x^2 + y^2 - 2y = 0$ or $x^2 + (y - 1)^2 = 1$

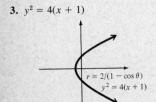

$r = 2 \sin \theta$
$x^2 + (y - 1)^2 = 1$

3. $y^2 = 4(x + 1)$

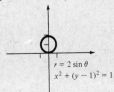

$r = 2/(1 - \cos \theta)$
$y^2 = 4(x + 1)$

5. $x^2 + y^2 + 2x + 2y = 0$ or $(x + 1)^2 + (y + 1)^2 = 2$
    Graph is a circle of radius $\sqrt{2}$ with center at $(-1, -1)$.

7. $x = 2$ **9.** $x + y = 2$ **11.** $y = -2$

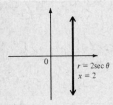

$r = 2\sec \theta$
$x = 2$

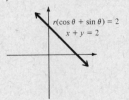

$r(\cos \theta + \sin \theta) = 2$
$x + y = 2$

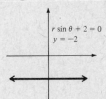

$r \sin \theta + 2 = 0$
$y = -2$

13.   **15.**   **17.**

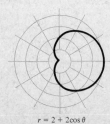

$r = 2 + 2\cos \theta$

$r = 3 + \cos \theta$

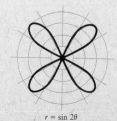

$r = \sin 2\theta$

19.

$r^2 = 4\cos 2\theta$

21.

$r = 4(1 - \cos \theta)$

23.

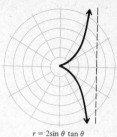

$r = 2\sin \theta \tan \theta$

25.

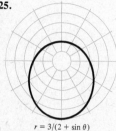

$r = 3/(2 + \sin \theta)$

27.

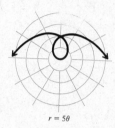

$r = 5\theta$

29.

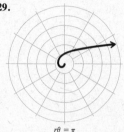

$r\theta = \pi$

31.

$\ln r = \theta$

33. $r \sin \theta = 2$ **35.** $r = 7$ **37.** $r \cos^2 \theta = 4 \sin \theta$ or $r = 4 \sin \theta \sec^2 \theta$ **39.** $r(2 \cos \theta + \sin \theta) = 4$
41. $r^2(\cos^2 \theta + 9 \sin^2 \theta) = 36$

Chapter 8   Review Exercises   (page 431)
1. $B = 42° \, 40'$, $c = 58.4$ cm   **3.** $A = 56° \, 00'$, $B = 34° \, 00'$ **5.** $a = 11.7$ ft, $c = 402$ ft   **7.** 535 ft
9. 70.8 m   **11.** triangle does not exist   **13.** $17° \, 10'$ **15.** 1300 ft   **17.** 52.9 ft   **19.** $25° \, 00'$
21. $19° \, 50'$ **23.** 14.8 m   **25.** 4950 yd^2 **27.** about 2.5 cans; better buy 3
29.

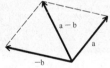

31. $25\sqrt{2}$ or about 35 for each   **33.** horizontal 869; vertical 418   **35.** $-4\sqrt{3}, -4$ **37.** $\sqrt{40}$, $161° \, 30'$
39. 2, 270°   **41.** $2i - j$ **43.** $10i\sqrt{3} + 10j$ **45.** 6   **47.** 10   **49.** $52° \, 10'$ **51.** 30°
53. 270, $56° \, 20'$ **55.** 105 lb   **57.** 306°, 524 mph   **59.** $5 + 4i$ **61.**   **63.**

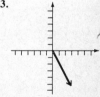

65. $(2\sqrt{2}, 135°)$; $2\sqrt{2}(\cos 135° + i \sin 135°)$ **67.** $\sqrt{2} - i\sqrt{2}$, (2, 315°)   **69.** $-30i$ **71.** $-1/8 + i\sqrt{3}/8$
73. $8i$ **75.** $-1/2 - i\sqrt{3}/2$ **77.** $8^{1/10} (\cos 27° + i \sin 27°)$; $8^{1/10} (\cos 99° + i \sin 99°)$; $8^{1/10} (\cos 171° + i \sin 171°)$;
$8^{1/10} (\cos 243° + i \sin 243°)$; $8^{1/10} (\cos 315° + i \sin 315°)$ **79.** $\cos 0° + i \sin 0°$; $\cos 60° + i \sin 60°$;
$\cos 120° + i \sin 120°$; $\cos 180° + i \sin 180°$; $\cos 240° + i \sin 240°$; $\cos 300° + i \sin 300°$
81. $5 (\cos 60° + i \sin 60°)$; $5(\cos 180° + i \sin 180°)$; $5 (\cos 300° + i \sin 300°)$

83.   **85.**   **87.**   **89.**

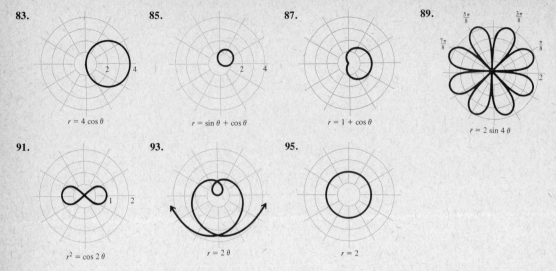

$r = 4 \cos \theta$ $r = \sin \theta + \cos \theta$ $r = 1 + \cos \theta$ $r = 2 \sin 4\theta$

91.   **93.**   **95.**

$r^2 = \cos 2\theta$ $r = 2\theta$ $r = 2$

111. All points of the form $(t, 0)$.

Chapter 9

Section 9.1   (page 442)
1. $\{(-2, 0)\}$ **3.** $\{(1, 3)\}$ **5.** $\{(4, -2)\}$ **7.** $\{(2, -2)\}$ **9.** $\{(12, 6)\}$ **11.** $\{(4, 3)\}$
13. $\{(-4, -1)\}$ **15.** $\{(.38, -.47)\}$ **17.** $\varnothing$ **19.** $\{(2, 2)\}$ **21.** $\{(1/5, 1)\}$ **23.** $\{(1/5, 14/3)\}$
25. $\{(1, 2, -1)\}$ **27.** $\{(2, 0, 3)\}$ **29.** $\varnothing$ **31.** $\{(2, 2, 0)\}$ **33.** $\{(4, 6, 1)\}$ **35.** $\varnothing$
37. $\{(x, 2x - 5, (3x - 4)/3)\}$ **39.** $\{(x, -x, -9x)\}$ **41.** $\{(x, 2x, 6 - 6x)\}$ **43.** goats cost $30 and sheep
cost $25   **45.** 32 days at $14   **47.** 18 liters of 70% and 12 liters of 20%   **49.** $10,000 at 8% and $15,000
at 10%   **51.** $5000 at 8% and $15,000 at 12%   **53.** 200 gallons   **55.** 72 kph and 92 kph   **57.** 15 cm,
12 cm, 6 cm   **59.** 20, 18, 12   **61.** 60 liters of $1, 120 liters of $1.50, and 120 liters of $3   **63.** $(D - 8, D,$
$4(D - 6))$ **65.** $(a, a - 3, 3)$

Section 9.2   (page 449)
1. $\{(-32, -8)\}$ **3.** $\{(1, 3)\}$ **5.** $\{(1, 0, 2)\}$ **7.** $\{(1, 1), (-2, 4)\}$ **9.** $\{(2, 1), (1/3, 4/9)\}$
11. $\{(-3/5, 7/5), (-1, 1)\}$ **13.** $\{(2, 2), (2, -2), (-2, 2), (-2, -2)\}$ **15.** $\{(0, 0)\}$ **17.** $\{(1, 1), (1, -1),$
$(-1, 1), (-1, -1)\}$ **19.** same circles, or $\{(x, y) \mid x^2 + y^2 = 10\}$ **21.** $\{(2, 3), (3, 2)\}$
23. $\{(-3, 5), (15/4, -4)\}$ **25.** $\{(4, -1/8), (-2, 1/4)\}$ **27.** $\{(3, 2), (-3, -2), (4, 3/2), (-4, -3/2)\}$
29. $\{(3, 5), (-3, -5)\}$ **31.** $\{(\sqrt{5}, 0), (-\sqrt{5}, 0), (\sqrt{5}, \sqrt{5}), (-\sqrt{5}, -\sqrt{5})\}$ **33.** $\{(1, 3), (31/14, -9/14)\}$
35. $\{(0, 1)\}$ **37.** $\{(12, 1)\}$ **39.** 6 and 6   **41.** 9 and 15, or -15 and -9 **43.** 5 m and 12 m
45. yes   **47.** $a \neq 5$

Section 9.3   (page 456)
1. $\begin{bmatrix} 2 & 3 & \vdots & 11 \\ 1 & 2 & \vdots & 8 \end{bmatrix}$ **3.** $\begin{bmatrix} 1 & 5 & \vdots & 6 \\ 0 & 1 & \vdots & 1 \end{bmatrix}$ **5.** $\begin{bmatrix} 2 & 1 & 1 & \vdots & 3 \\ 3 & -4 & 2 & \vdots & -7 \\ 1 & 1 & 1 & \vdots & 2 \end{bmatrix}$ **7.** $\begin{bmatrix} 1 & 1 & 0 & \vdots & 2 \\ 0 & 2 & 1 & \vdots & -4 \\ 0 & 0 & 1 & \vdots & 2 \end{bmatrix}$

9. $2x + y = 1, 3x - 2y = -9$ **11.** $x = 2, y = 3, z = -2$ **13.** $3x + 2y + z = 1, 2y + 4z = 22,$
$-x - 2y + 3z = 15$ **15.** $\{(2, 3)\}$ **17.** $\{(-3, 0)\}$ **19.** $\{(7/2, -1)\}$ **21.** $\{(5, 0)\}$
23. $\{(-2, 1, 3)\}$ **25.** $\{(-1, 23, 16)\}$ **27.** $\{(3, 2, -4)\}$ **29.** $\{(2, 1, -1)\}$ **31.** $\varnothing$

33. $\{(3, 3 - z, z)\}$ **35.** $\left\{ \left(\dfrac{-12}{23} - \dfrac{15}{23}z, \dfrac{1}{23}z - \dfrac{13}{23}, z \right) \right\}$ **37.** $\left\{ \left(-6 - \dfrac{1}{2}z, \dfrac{1}{2}z + 2, z \right) \right\}$

39. $\{(-1, 2, 5, 1)\}$ **41.** $\{(1 - 4w, 1 - w, 2 + w, w)\}$ **43.** $\{(1.7, 2.4)\}$ **45.** $\{(.5, -.7, 6)\}$
47. wife 40 days, husband 32 days   **49.** 5 model 201, 8 model 301   **51.** \$10,000 at 8%, \$7000 at 10%,

\$8000 at 9%   **53.** $\begin{bmatrix} 1 & 0 & 0 & 1 & \vdots & 1000 \\ 1 & 1 & 0 & 0 & \vdots & 1100 \\ 0 & 1 & 1 & 0 & \vdots & 700 \\ 0 & 0 & 1 & 1 & \vdots & 600 \end{bmatrix}, \begin{bmatrix} 1 & 0 & 0 & 1 & \vdots & 1000 \\ 0 & 1 & 0 & -1 & \vdots & 100 \\ 0 & 0 & 1 & 1 & \vdots & 600 \\ 0 & 0 & 0 & 0 & \vdots & 0 \end{bmatrix}$ **55.** 1000, 1000   **57.** 600, 600

59. inconsistent system, no solution   **61.** $\{(-15 + z, 5 - 2z, z)\}$

Section 9.4   (page 463)
1. -36 **3.** 7   **5.** 0   **7.** -26 **9.** -16 **11.** $y^2 - 16$ **13.** $x^2 - y^2$ **15.** $0, -5, 0$
17. $-6, 0, -6$ **19.** -1 **21.** 8   **23.** 0   **25.** 0   **27.** $4x$ **29.** $10i + 17j - 6k$ **31.** -88
33. -5.5 **43.** 5/2   **45.** 7   **47.** 8

Section 9.5   (page 469)
1. two rows identical   **3.** one column all zeros   **5.** multiply each element of second row by $-1/4$; then two rows are
identical   **7.** multiply each element of third column by 1/2; then two columns are identical   **9.** rows and columns
exchanged   **11.** two rows exchanged   **13.** multiply each element of second column by 3
15. multiply elements of first row by 1; add products to elements of second row   **17.** multiply elements of second row by 2
19. 0   **21.** 0   **23.** 5   **25.** 32   **27.** -32

Section 9.6   (page 476)
1. $x = 2, y = 4, z = 8$ **3.** $x = -15, y = 5, k = 3$ **5.** $z = 18, r = 3, s = 3, p = 3, a = 3/4$
7. $\begin{bmatrix} 14 & -11 & -3 \\ -2 & 4 & -1 \end{bmatrix}$ **9.** $\begin{bmatrix} -6 & 8 \\ 4 & 2 \end{bmatrix}$ **11.** can't be done   **13.** $\begin{bmatrix} -12x + 8y & -x + y \\ x & 8x - y \end{bmatrix}$

15. $\begin{bmatrix} 7 & 2 \\ 9 & 0 \\ 8 & 6 \end{bmatrix}; \begin{bmatrix} 7 & 9 & 8 \\ 2 & 0 & 6 \end{bmatrix}$ **17.** $\begin{bmatrix} -4 & 8 \\ 0 & 6 \end{bmatrix}$ **19.** $\begin{bmatrix} 2 & 6 \\ -4 & 6 \end{bmatrix}$ **21.** $\begin{bmatrix} -1 & -3 \\ 2 & -3 \end{bmatrix}$ **23.** $\begin{bmatrix} 13 \\ 25 \end{bmatrix}$

25. $\begin{bmatrix} -2 & 5 & 0 \\ 6 & 6 & 1 \\ 12 & 2 & -3 \end{bmatrix}$ **27.** $\begin{bmatrix} 20 & 0 & 8 \\ 8 & 0 & 4 \end{bmatrix}$ **29.** can't be multiplied   **31.** $[-8]$ **33.** $\begin{bmatrix} -6 & 12 & 18 \\ 4 & -8 & -12 \\ -2 & 4 & 6 \end{bmatrix}$

35. (a) $\begin{bmatrix} 17\frac{3}{4} & 23\frac{3}{4} \\ \\ 9 & 12\frac{3}{4} \\ \\ 33 & 42 \\ 6 & 8 \end{bmatrix}$; (b) [220 890 105 125 70]; (c) [4165 5605]   **37.** $\begin{bmatrix} .018 & -.984 \\ 1.002 & -.056 \end{bmatrix}$

39. $\begin{bmatrix} -7.88 & .7 & -.068 \\ -4.87 & -.324 & .259 \\ -1.54 & -.803 & -.192 \end{bmatrix}$ **41.** yes, always true   **43.** yes, always true   **45.** yes, always true
47. always true   **49.** always true   **51.** yes, always true
53. Not true. $(A + B)(A - B) = A^2 + BA - AB - B^2$. Since multiplication is not
commutative, $BA - AB$ is not always 0.

Section 9.7   (page 485)
1. yes   **3.** no   **5.** $\begin{bmatrix} 0 & 1/2 \\ 1/5 & -1/10 \end{bmatrix}$ **7.** does not exist   **9.** does not exist

11. $\begin{bmatrix} -1/12 & 1/6 & -7/12 \\ 7/24 & -1/12 & 25/24 \\ 7/24 & -1/12 & 1/24 \end{bmatrix}$ **13.** $\begin{bmatrix} 1/2 & 0 & 1/2 & -1 \\ 1/10 & -2/5 & 3/10 & -1/5 \\ -7/10 & 4/5 & -11/10 & 12/5 \\ 1/5 & 1/5 & -2/5 & 3/5 \end{bmatrix}$ **15.** $\begin{bmatrix} -2.5 & 5 \\ 12.5 & -15 \end{bmatrix}$

17.
$$\begin{bmatrix} -.4762 & .7937 & 2.6984 \\ .9524 & .0794 & 1.2698 \\ -.7143 & 1.1905 & -.9524 \end{bmatrix}$$
19.
$$\begin{bmatrix} 2 \\ -1 \\ 1 \end{bmatrix}$$
21. $(2, 3)$　**23.** $(-2, 4)$　**25.** $(1, 1, 2)$

27. $(0, 1, 4)$　**29.** $(1, 0, 2, 1)$　**31.**
$$\begin{bmatrix} 10 \\ 4 \\ 12 \end{bmatrix}$$
35. $\dfrac{1}{ad} \begin{bmatrix} d & 0 \\ 0 & a \end{bmatrix}$

Section 9.8　(page 491)

1. $\{(2, 2)\}$　**3.** $\{(2, -5)\}$　**5.** $\{(2, 3)\}$　**7.** $\{(9/7, 2/7)\}$　**9.** no solution, since $D = 0$
11. $\{(-1, 2, 1)\}$　**13.** $\{(-3, 4, 2)\}$　**15.** $\{(0, 0, -1)\}$　**17.** no solution, since $D = 0$
19. $\{(197/91, -118/91, -23/91)\}$　**21.** $\{(2, 3, 1)\}$　**23.** $\{(0, 4, 2)\}$　**25.** $\{(31/5, 19/10, -29/10)\}$
27. $\{(1, 0, -1, 2)\}$　**29.** $\{(-.2, -.8)\}$　**31.** $\{(.2, .5, -.2)\}$　**33.** shirts are \$12, pants are \$18
35. 17.5 g of 12-carat, 7.5 g of 22-carat　**37.** 52/3 lbs　**39.** 164 fives, 86 tens, 40 twenties

Section 9.9　(page 497)

1.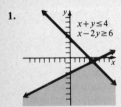
$x + y \le 4$
$x - 2y \ge 6$

3.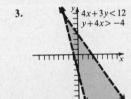
$4x + 3y < 12$
$y + 4x > -4$

5.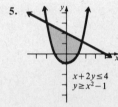
$x + 2y \le 4$
$y \ge x^2 - 1$

7.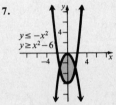
$y \le -x^2$
$y \ge x^2 - 6$

9.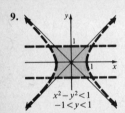
$x^2 - y^2 < 1$
$-1 < y < 1$

11.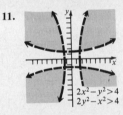
$2x^2 - y^2 > 4$
$2y^2 - x^2 > 4$

13.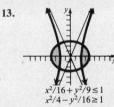
$x^2/16 + y^2/9 \le 1$
$x^2/4 - y^2/16 \ge 1$

15.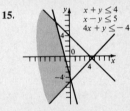
$x + y \le 4$
$x - y \le 5$
$4x + y \le -4$

17.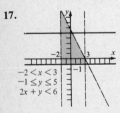
$-2 < x < 3$
$-1 \le y \le 5$
$2x + y < 6$

19.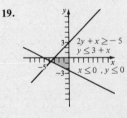
$2y + x \ge -5$
$y \le 3 + x$
$x \le 0, y \le 0$

21.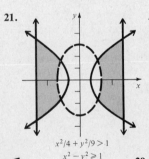
$x^2/4 + y^2/9 > 1$
$x^2 - y^2 \ge 1$
$-4 \le x \le 4$

23.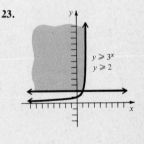
$y \ge 3^x$
$y \ge 2$

25.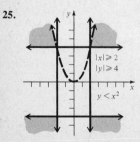
$|x| \ge 2$
$|y| \ge 4$
$y < x^2$

27.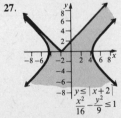
$y \le |x + 2|$
$\dfrac{x^2}{16} - \dfrac{y^2}{9} \le 1$

29.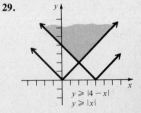
$y \ge |4 - x|$
$y \ge |x|$

31. Let x = number of units of basic, y = number of units of plain; $x \geq 3$, $y \geq 2$, $5x + 4y \leq 50$, $2x + y \leq 16$.

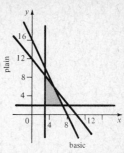

33. Let x = number of red pills, y = number of green pills; $8x + 2y \geq 16$, $x + y \geq 5$, $2x + 7y \geq 20$, $x \geq 0$, $y \geq 0$.

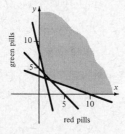

35. Let x = number shipped to warehouse A and y = number shipped to B; $0 \leq x \leq 75$, $0 \leq y \leq 80$, $x + y \geq 100$, $12x + 10y \leq 1200$.

Chapter 9   Review Exercises   (page 499)

1. $\{(-1, 3)\}$ **3.** $\{(6, -12)\}$ **5.** $\{(2, 5)\}$ **7.** $\{(-1, 2, 3)\}$ **9.** 10 at 25¢, 12 at 50¢

11. \$10,000 at 6%, \$20,000 at 7%, \$20,000 at 9%   **13.** $x = \dfrac{15}{11}z + \dfrac{6}{11}, y = \dfrac{14}{11}z - \dfrac{1}{11}$

15. $\{(3\sqrt{3}, \sqrt{2}), (-3\sqrt{3}, \sqrt{2}), (3\sqrt{3}, -\sqrt{2}), (-3\sqrt{3}, -\sqrt{2})\}$ **17.** $\{(-2, 0), (1, 1)\}$

19. $\{((8 - 8\sqrt{41})/5, (16 + 4\sqrt{41})/5), ((8 + 8\sqrt{41})/5, (16 - 4\sqrt{41})/5)\}$ **21.** $a = 5$, $x = 3/2$, $y = 0$, $z = 9$

23. $\begin{bmatrix} 0 & -2 & 7 \\ 6 & -1 & 10 \end{bmatrix}$ **25.** cannot be done   **27.** $\begin{bmatrix} -9 & 3 \\ 10 & 6 \end{bmatrix}$ **29.** $\{(-4, 6)\}$ **31.** $\{(0, 1, 0)\}$

33. $\begin{bmatrix} 3 & -1 \\ -5 & 2 \end{bmatrix}$ **35.** $\begin{bmatrix} 2/3 & 0 & -1/3 \\ 1/3 & 0 & -2/3 \\ -2/3 & 1 & 1/3 \end{bmatrix}$ **37.** $\{(2, 1)\}$ **39.** $\{(-3, 5, 1)\}$ **41.** -25 **43.** -44

45. 138   **47.** rows and columns interchanged   **49.** two identical rows   **51.** exchange columns 2 and 3

53. $\{(-4, 2)\}$ **55.** dependent system   **57.**

59. Let x = number of batches of cakes, let y = number of batches of cookies;

$2x + \dfrac{3}{2}y \le 16$; $3x + \dfrac{2}{3}y \le 12$; $x \ge 0$;

$y \ge 0$

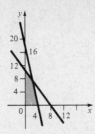

61. The maximum profit is $200, which occurs when 2 batches of cakes and 7 batches of cookies are made.

Chapter 10

Section 10.1   (page 509)

1. $x^2 - 3x + 2$ **3.** $x^2 + 2x + 3$ **5.** $2z^3 - z^2 - (1/2)z - 5/4 + \dfrac{(-13/4)z + 17/4}{2z^2 + z + 1}$ **7.** $p^3 - 5p^2 + 5p - 8$

9. $4m^2 - 4m + 1 + [-3/(m+1)]$ **11.** $x^4 + x^3 + 2x - 1 + 3/(x+2)$ **13.** yes   **15.** yes   **17.** yes

19. no   **21.** yes   **23.** no   **25.** -8 **27.** 2   **29.** $-6 - i$ **31.** $-2 - 3i$ **33.** 42

35. -20 **37.** 4   **39.** $x^3 + 4x^2 - 7x - 10 = 0$ **41.** $x^3 - 3x^2 - 2x + 6 = 0$

43. $2x^4 - 4x^3 - 15x^2 + 17x + 30 = 0$ **45.** -9 **47.** -5.19 **49.** $f(x) = x(x(x+2) - 4) + 7$

51. $f(x) = -2x(x(x - 1.5) + .5) - 1$ **53.** (a) 455.994 (b) 594.25 (c) 714   **59.** -1.41 **61.** 3.32

Section 10.2   (page 514)

1. $f(x) = -\dfrac{1}{6}x^3 + \dfrac{13}{6}x + 2$ **3.** $f(x) = \dfrac{1}{9}x^3 + \dfrac{7}{9}x^2 + \dfrac{16}{9}x + \dfrac{4}{3}$ **5.** $f(x) = -10x^3 + 30x^2 - 10x + 30$

7. $x^2 - 10x + 26$ **9.** $x^3 - 4x^2 + 6x - 4$ **11.** $x^3 - 3x^2 + x + 1$ **13.** $x^4 - 6x^3 + 10x^2 + 2x - 15$

15. $x^3 - 8x^2 + 22x - 20$ **17.** $x^4 - 4x^3 + 5x^2 - 2x - 2$ **19.** $x^4 - 10x^3 + 42x^2 - 82x + 65$

21. $-1 + i, -1 - i$ **23.** $3, -2 - 2i$ **25.** $(-1 + i\sqrt{5})/2, (-1 - i\sqrt{5})/2$ **27.** $3, -2, 1 - 3i$

29. $2 - i, 1 + i, 1 - i$ **31.** $t = 3, t = -2$ **33.** other zeros are $i, -i$; $f(x) = (x - 2)^3(x - i)(x + i)$

Section 10.3   (page 517)

1. $-1, -2, 5$ **3.** $2, -3, -5$ **5.** no rational solutions   **7.** $1, -2, -3, -5$ **9.** $3/2, -1/3, -4$

11. $1/2, -2/3, -3/2$ **13.** $1/2$ **15.** no rational solutions   **17.** $-1, -2, -3, 4$ **19.** $-2, 2/3$ **21.** 1

23. $3, -1, -2/3$ **25.** $1, -5/4$ **27.** 1

Section 10.4   (page 522)

17. (a) positive: 1; negative: 2 or 0   (b) $-3, -1.4, 1.4$
(c)

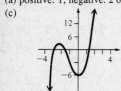

19. (a) positive: 2 or 0; negative: 1   (b) $-2, 1.6, 4.4$
(c)

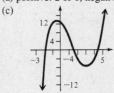

21. (a) positive: 1; negative: 0   (b) 1.5
(c)

23. (a) positive: 3 or 1; negative: 1   (b) $-.7, 1.6, 2.7, 4.4$
(c)

25. $g(x)$

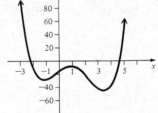

27. .27, 3.73   **29.** $-6.61, .61, 2.76, 7.24$

Section 10.5   (page 529)

1. $-1/x + 4/(x + 1)$ **3.** $2/(3x) - 2/[3(x + 3)]$ **5.** $-1/[2(x + 1)] + 3/[2(x - 1)]$
7. $7/x - 63/[4(3x - 1)] - 7/[4(x + 1)]$ **9.** $-2/(9x) + 2/(3x^2) + 2/[9(x + 3)]$
11. $-3/[25(x + 2)] + 3/[25(x - 3)] + 2/[5(x - 3)^2]$ **13.** $2/(x + 2)^2 - 3/(x + 2)^3$
15. $3/[2(x + 1)] - 3/[2(x + 3)]$ **17.** $-1/[3(2x + 1)] + 2/[3(x + 2)]$ **19.** $1 + 1/x + 5/(x - 2) - 3/(x - 1)$
21. $5/(3x) - 5x/[3(x^2 + 3)]$ **23.** $-1/[3(x + 1)] + (x + 5)/[3(x^2 + 2)]$
25. $3/x - 3/[2(x + 1)] - 3(x + 1)/[2(x^2 + 1)]$ **27.** $1/x - 2x/(x^2 + 1)^2$ **29.** $3/(x - 1) + 1/(x^2 + 1) - 2/(x^2 + 1)^2$
31. $-13/[15(x - 1)] + 46/[15(x + 2)]$ **33.** $1/[2(x^2 + 1)] - 1/[4(x + 1)] + 1/[4(x - 1)]$
35. $3/x - 1/(x + 3) + 2/(x - 1)$ **37.** $a = 1/4; b = -1/4; c = 5/4$ **39.** $a = -1/48; b = 1/48; c = 5/6$

Chapter 10   Review Exercises   (page 530)

1. $q(x) = 2x^2 + 3/2; r(x) = 13/2$ **3.** $q(x) = x^2 - 5; r(x) = 5x + 9$ **5.** $q(x) = 2x^2 + 5x + 1; r = 2$
7. $q(x) = 2x^2 + 4x + 4; r = 14$ **9.** 11   **11.** 28   **13.** $f(x) = x^3 - 10x^2 + 17x + 28$
15. $f(x) = x^4 - x^3 - 9x^2 + 7x + 14$ **17.** no   **19.** no   **21.** $f(x) = -2x^3 + 6x^2 + 12x - 16$
23. $f(x) = x^4 - 3x^2 - 4$ **25.** $f(x) = x^3 + x^2 - 4x + 6$ **27.** $3 + i, 3 - i, 2i, -2i$ **29.** $1/2, -1, 5$
31. $4, -1/2, -2/3$ **33.** none   **43.** $-2.3, 4.6$

45. $3/(x + 2) + 2/(x - 2)$
47. $4/x^2 - 4/(x^2 + 2)$

Chapter 11

Section 11.1   (page 537)

1. $S_1: 2 = 1(1 + 1)$
$S_2: 2 + 4 = 2(2 + 1)$
$S_3: 2 + 4 + 6 = 3(3 + 1)$
$S_4: 2 + 4 + 6 + 8 = 4(4 + 1)$
$S_5: 2 + 4 + 6 + 8 + 10 = 5(5 + 1)$

Section 11.2   (page 544)

1. $x^6 + 6x^5y + 15x^4y^2 + 20x^3y^3 + 15x^2y^4 + 6xy^5 + y^6$ **3.** $p^5 - 5p^4q + 10p^3q^2 - 10p^2q^3 + 5pq^4 - q^5$
5. $r^{10} + 5r^8s + 10r^6s^2 + 10r^4s^3 + 5r^2s^4 + s^5$ **7.** $p^4 + 8p^3q + 24p^2q^2 + 32pq^3 + 16q^4$
9. $m^6/64 - 3m^5/16 + 15m^4/16 - 5m^3/2 + 15m^2/4 - 3m + 1$ **11.** $16p^4 + 32p^3q/3 + 8p^2q^2/3 + 8pq^3/27 + q^4/81$
13. $m^{-8} + 4m^{-4} + 6 + 4m^4 + m^8$ **15.** $p^{10/3} - 25p^4 + 250p^{14/3} - 1250p^{16/3} + 3125p^6 - 3125p^{20/3}$
17. $x^{21} + 126x^{20} + 7560x^{19} + 287,280x^{18}$ **19.** $3^9m^{-18} + 45 \cdot 3^8m^{-20} + 900 \cdot 3^7m^{-22} + 10,500 \cdot 3^6m^{-24}$
21. $7920m^8p^4$ **23.** $180m^2n^8$ **25.** $4845p^8q^{16}$ **27.** $439,296x^{21}y^7$ **29.** $90,720x^{28}y^{12}$ **31.** 1.105
33. 245.937 **35.** 758.650 **37.** $1.002; 5.010$ **39.** $.942$ **41.** $1 - x + x^2 - x^3 + \cdots$ **47.** 6

Section 11.3   (page 552)

1. $10, 16, 22, 28, 34$ **3.** $1, 1/2, 1/4, 1/8, 1/16$ **5.** $1/2, 2/5, 1/3, 2/7, 1/4$ **7.** $-2, 8, -24, 64, -160$
9. $2/3, 5/6, 10/11, 17/18, 26/27$ **11.** $2, 3, 5, 7, 11$ **13.** $4, 9, 14, 19, 24, 29, 34, 39, 44, 49$
15. $2, 4, 8, 16, 32, 64, 128, 256, 512, 1024$ **17.** $1, 1, 2, 3, 5, 8, 13, 21, 34, 55$ **19.** $4, 6, 8, 10, 12$
21. $11, 9, 7, 5$ **23.** $4 - \sqrt{5}, 4, 4 + \sqrt{5}, 4 + 2\sqrt{5}, 4 + 3\sqrt{5}$ **25.** $d = 5, a_n = 7 + 5n$
27. $d = -3, a_n = 21 - 3n$ **29.** $d = \sqrt{2}, a_n = -6 + n\sqrt{2}$ **31.** $d = m, a_n = x + nm - m$
33. $d = -m, a_n = 2z + 2m - mn$ **35.** $a_8 = 19, a_n = 3 + 2n$ **37.** $a_8 = 7, a_n = n - 1$
39. $a_8 = -6, a_n = 10 - 2n$ **41.** $a_8 = -9, a_n = 15 - 3n$ **43.** $a_8 = 5.1, a_n = 2.7 + .3n$
45. $a_8 = x + 21, a_n = x + 3n - 3$ **47.** $a_8 = 4m, a_n = mn - 4m$ **49.** 215 **51.** 125 **53.** 230
55. 16.745 **57.** 32.28 **59.** 18 **61.** 140 **63.** -684 **65.** $500,500$ **67.** 400

69. -78 **71.** $a_1 = 7$ **73.** $a_1 = 0$ **75.** $a_1 = 7$ **77.** 1281   **79.** $n(n+1)/2$ **81.** 4680
83. 100 m, 550 m   **85.** 240 ft, 1024 ft   **87.** 9, 14, 19, 24   **89.** 170 m   **91.** All terms are 0;
or all terms are 1.   **97.** false   **99.** true   **101.** false

Section 11.4   **(page 560)**
1. 2, 6, 18, 54   **3.** 1/2, 2, 8, 32   **5.** 3/2, 3, 6, 12, 24   **7.** $a_5 = 324$; $a_n = 4 \cdot 3^{n-1}$
9. $a_5 = -1875$; $a_n = -3(-5)^{n-1}$ **11.** $a_5 = 24$; $a_n = (3/2) \cdot 2^{n-1}$ or $3 \cdot 2^{n-2}$
13. $a_5 = -256$; $a_n = -1 \cdot (-4)^{n-1}$ or $-(-4)^{n-1}$ **15.** $r = 2$; $a_n = 6 \cdot 2^{n-1}$ **17.** $r = 2$; $a_n = (3/4) \cdot 2^{n-1}$
or $3 \cdot 2^{n-3}$ **19.** $a_n = (9/4) \cdot 2^{n-1}$ or $9 \cdot 2^{n-3}$ **21.** $a_n = (-2/9) \cdot 3^{n-1}$ or $-2 \cdot 3^{n-3}$ **23.** 93   **25.** 33/4
27. 860.95   **29.** 30   **31.** 255/4   **33.** 120   **35.** $r = 2$ **37.** $r = 10$ **39.** $r = 1/2$; sum would
exist   **41.** 64/3   **43.** 1000/9   **45.** 3/2   **47.** 1/5   **49.** 1/3   **51.** 3/7   **53.** 70 m
55. 1600 rotations   **57.** $\$2^{30}$, or $\$1,073,741,824$; $\$2^{31} - \$1$ **59.** .000016%   **61.** 12 m   **63.** $\approx 13.4\%$
65. \$26,214.40   **67.** $.016 \times 2^{11}$ inches, 32.768 inches   **69.** any sequence of the form a, a, a
71. $a_1 = 2$, $r = 3$, or $a_1 = -2$, $r = -3$ **73.** $a_1 = 128$, $r = 1/2$, or $a_1 = -128$, $r = -1/2$

Section 11.5   **(page 567)**
1. 5040   **3.** 720   **5.** 90   **7.** 336   **9.** 7   **11.** 1   **13.** 6   **15.** 56   **17.** 1365   **19.** 120
21. 14   **23.** 1   **25.** 720   **27.** 120, 30,240   **29.** 30   **31.** 120   **33.** 4   **35.** 552
37. 27,405   **39.** 220,220   **41.** 1326   **43.** 10   **45.** 28   **47.** (a) 84;   (b) 10;   (c) 40;   (d) 28
49. $\binom{52}{5} = 2{,}598{,}960$ **51.** 4   **53.** $4 \cdot \binom{13}{5} = 5148$ **55.** $P(5, 5) = 120$ **57.** $\binom{12}{3} = 220$
59. 8   **61.** 4

Section 11.6   **(page 574)**
1. $S = \{H\}$ **3.** $S = \{HHH, HHT, HTH, THH, HTT, THT, TTH, TTT\}$ **5.** Let c = correct, w = wrong
$S = \{ccc, ccw, cwc, wcc, wwc, wcw, cww, www\}$ **7.** (a) $\{HH, TT\}$, 1/2;   (b) $\{HH, HT, TH\}$, 3/4
9. (a) $\{2 \text{ and } 4\}$, 1/10;   (b) $\{1 \text{ and } 3, 1 \text{ and } 5, 3 \text{ and } 5\}$, 3/10;   (c) $\varnothing$, 0;   (d) $\{1 \text{ and } 2, 1 \text{ and } 4, 2 \text{ and } 3, 2 \text{ and } 5, 3 \text{ and } 4,$
$4 \text{ and } 5\}$, 3/5   **11.** (a) 1/5;   (b) 8/15;   (c) 0;   (d) 1 to 4;   (e) 7 to 8   **13.** 1 to 4   **15.** 2/5   **17.** (a) 1/2;
(b) 7/10;   (c) 2/5   **19.** (a) 1/6;   (b) 1/3;   (c) 1/6   **21.** (a) 7/15;   (b) 11/15;   (c) 4/15

Chapter 11   Review Exercises   **(page 575)**
7. $x^4 + 8x^3y + 24x^2y^2 + 32xy^3 + 16y^4$ **9.** $243x^{5/2} - 405x^{3/2} + 270x^{1/2} - 90x^{-1/2} + 15x^{-3/2} - x^{-5/2}$
11. $2160x^2y^4$ **13.** $3^{16} + 16 \cdot 3^{15}x + 120 \cdot 3^{14}x^2 + 560 \cdot 3^{13}x^3$ **15.** 8, 10, 12, 14, 16   **17.** 5, 3, -2,
$-5, -3$ **19.** 6, 2, $-2, -6, -10$ **21.** $3 - \sqrt{5}, 4, 5 + \sqrt{5}, 6 + 2\sqrt{5}, 7 + 3\sqrt{5}$ **23.** 4, 8, 16, 32, 64
25. $-3, 4, -16/3, 64/9, -256/27$ **27.** $a_1 = -29$, $a_{20} = 66$ **29.** 20   **31.** $-x + 61$ **33.** 222
35. $84k$ **37.** 25   **39.** $x^5/125$ **41.** 15   **43.** $-10k$ **45.** 80   **47.** can't be found
49. 25/6   **51.** 690   **53.** 73/12   **55.** 50,005,000   **57.** 240   **59.** $r > 1$ so sum does not exist
61. 72   **63.** 1   **65.** 252   **67.** 504   **69.** 1/26   **71.** 4/13   **73.** 1   **75.** (a) middle sum,
(b) last sum,   (c) first sum

Chapter 11   Appendix   Rotation of Axes   **(page 582)**
1. $30°$ **3.** $60°$ **5.** $22\frac{1}{2}°$

7.

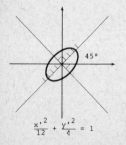

$$\frac{x'^2}{12} + \frac{y'^2}{4} = 1$$

9.

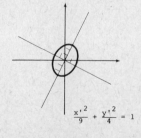

$$\frac{x'^2}{9} + \frac{y'^2}{4} = 1$$

11.

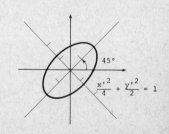

$$\frac{x'^2}{4} + \frac{y'^2}{2} = 1$$

13.

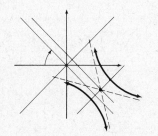

$$x'^2 - 3y'^2 = 5$$

45°

15.

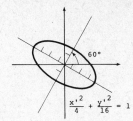

60°

$$\frac{x'^2}{4} + \frac{y'^2}{16} = 1$$

17.

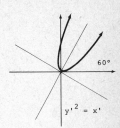

60°

$$y'^2 = x'$$

19.

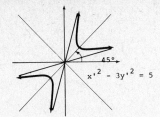

21.

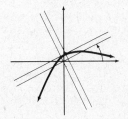

23.

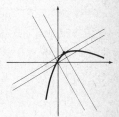

TABLE 1 SQUARE AND SQUARE ROOTS

n	n²	√n	√10n	n	n²	√n	√10n
1	1	1.000	3.162	51	2601	7.141	22.583
2	4	1.414	4.472	52	2704	7.211	22.804
3	9	1.732	5.477	53	2809	7.280	23.022
4	16	2.000	6.325	54	2916	7.348	23.238
5	25	2.236	7.071	55	3025	7.416	23.452
6	36	2.449	7.746	56	3136	7.483	23.664
7	49	2.646	8.367	57	3249	7.550	23.875
8	64	2.828	8.944	58	3364	7.616	24.083
9	81	3.000	9.487	59	3481	7.681	24.290
10	100	3.162	10.000	60	3600	7.746	24.495
11	121	3.317	10.488	61	3721	7.810	24.698
12	144	3.464	10.954	62	3844	7.874	24.900
13	169	3.606	11.402	63	3969	7.937	25.100
14	196	3.742	11.832	64	4096	8.000	25.298
15	225	3.873	12.247	65	4225	8.062	25.495
16	256	4.000	12.649	66	4356	8.124	25.690
17	289	4.123	13.038	67	4489	8.185	25.884
18	324	4.243	13.416	68	4624	8.246	26.077
19	361	4.359	13.784	69	4761	8.307	26.268
20	400	4.472	14.142	70	4900	8.367	26.458
21	441	4.583	14.491	71	5041	8.426	26.646
22	484	4.690	14.832	72	5184	8.485	26.833
23	529	4.796	15.166	73	5329	8.544	27.019
24	576	4.899	15.492	74	5476	8.602	27.203
25	625	5.000	15.811	75	5625	8.660	27.386
26	676	5.099	16.125	76	5776	8.718	27.568
27	729	5.196	16.432	77	5929	8.775	27.749
28	784	5.292	16.733	78	6084	8.832	27.928
29	841	5.385	17.029	79	6241	8.888	28.107
30	900	5.477	17.321	80	6400	8.944	28.284
31	961	5.568	17.607	81	6561	9.000	28.460
32	1024	5.657	17.889	82	6724	9.055	28.636
33	1089	5.745	18.166	83	6889	9.110	28.810
34	1156	5.831	18.439	84	7056	9.165	28.983
35	1225	5.916	18.708	85	7225	9.220	29.155
36	1296	6.000	18.974	86	7396	9.274	29.326
37	1369	6.083	19.235	87	7569	9.327	29.496
38	1444	6.164	19.494	88	7744	9.381	29.665
39	1521	6.245	19.748	89	7921	9.434	29.833
40	1600	6.325	20.000	90	8100	9.487	30.000
41	1681	6.403	20.248	91	8281	9.539	30.166
42	1764	6.481	20.494	92	8464	9.592	30.332
43	1849	6.557	20.736	93	8649	9.644	30.496
44	1936	6.633	20.976	94	8836	9.695	30.659
45	2025	6.708	21.213	95	9025	9.747	30.822
46	2116	6.782	21.448	96	9216	9.798	30.984
47	2209	6.856	21.679	97	9409	9.849	31.145
48	2304	6.928	21.909	98	9604	9.899	31.305
49	2401	7.000	22.136	99	9801	9.950	31.464
50	2500	7.071	22.361	100	10000	10.000	31.623

TABLE 2 COMMON LOGARITHMS

n	0	1	2	3	4	5	6	7	8	9
1.0	.0000	.0043	.0086	.0128	.0170	.0212	.0253	.0294	.0334	.0374
1.1	.0414	.0453	.0492	.0531	.0569	.0607	.0645	.0682	.0719	.0755
1.2	.0792	.0828	.0864	.0899	.0934	.0969	.1004	.1038	.1072	.1106
1.3	.1139	.1173	.1206	.1239	.1271	.1303	.1335	.1367	.1399	.1430
1.4	.1461	.1492	.1523	.1553	.1584	.1614	.1644	.1673	.1703	.1732
1.5	.1761	.1790	.1818	.1847	.1875	.1903	.1931	.1959	.1987	.2014
1.6	.2041	.2068	.2095	.2122	.2148	.2175	.2201	.2227	.2253	.2279
1.7	.2304	.2330	.2355	.2380	.2405	.2430	.2455	.2480	.2504	.2529
1.8	.2553	.2577	.2601	.2625	.2648	.2672	.2695	.2718	.2742	.2765
1.9	.2788	.2810	.2833	.2856	.2878	.2900	.2923	.2945	.2967	.2989
2.0	.3010	.3032	.3054	.3075	.3096	.3118	.3139	.3160	.3181	.3201
2.1	.3222	.3243	.3263	.3284	.3304	.3324	.3345	.3365	.3385	.3404
2.2	.3424	.3444	.3464	.3483	.3502	.3522	.3541	.3560	.3579	.3598
2.3	.3617	.3636	.3655	.3674	.3692	.3711	.3729	.3747	.3766	.3784
2.4	.3802	.3820	.3838	.3856	.3874	.3892	.3909	.3927	.3945	.3962
2.5	.3979	.3997	.4014	.4031	.4048	.4065	.4082	.4099	.4116	.4133
2.6	.4150	.4166	.4183	.4200	.4216	.4232	.4249	.4265	.4281	.4298
2.7	.4314	.4330	.4346	.4362	.4378	.4393	.4409	.4425	.4440	.4456
2.8	.4472	.4487	.4502	.4518	.4533	.4548	.4564	.4579	.4594	.4609
2.9	.4624	.4639	.4654	.4669	.4683	.4698	.4713	.4728	.4742	.4757
3.0	.4771	.4786	.4800	.4814	.4829	.4843	.4857	.4871	.4886	.4900
3.1	.4914	.4928	.4942	.4955	.4969	.4983	.4997	.5011	.5024	.5038
3.2	.5051	.5065	.5079	.5092	.5105	.5119	.5132	.5145	.5159	.5172
3.3	.5185	.5198	.5211	.5224	.5237	.5250	.5263	.5276	.5289	.5302
3.4	.5315	.5328	.5340	.5353	.5366	.5378	.5391	.5403	.5416	.5428
3.5	.5441	.5453	.5465	.5478	.5490	.5502	.5514	.5527	.5539	.5551
3.6	.5563	.5575	.5587	.5599	.5611	.5623	.5635	.5647	.5658	.5670
3.7	.5682	.5694	.5705	.5717	.5729	.5740	.5752	.5763	.5775	.5786
3.8	.5798	.5809	.5821	.5832	.5843	.5855	.5866	.5877	.5888	.5899
3.9	.5911	.5922	.5933	.5944	.5955	.5966	.5977	.5988	.5999	.6010
4.0	.6021	.6031	.6042	.6053	.6064	.6075	.6085	.6096	.6107	.6117
4.1	.6128	.6138	.6149	.6160	.6170	.6180	.6191	.6201	.6212	.6222
4.2	.6232	.6243	.6253	.6263	.6274	.6284	.6294	.6304	.6314	.6325
4.3	.6335	.6345	.6355	.6365	.6375	.6385	.6395	.6405	.6415	.6425
4.4	.6435	.6444	.6454	.6464	.6474	.6484	.6493	.6503	.6513	.6522
4.5	.6532	.6542	.6551	.6561	.6571	.6580	.6590	.6599	.6609	.6618
4.6	.6628	.6637	.6646	.6656	.6665	.6675	.6684	.6693	.6702	.6712
4.7	.6721	.6730	.6739	.6749	.6758	.6767	.6776	.6785	.6794	.6803
4.8	.6812	.6821	.6830	.6839	.6848	.6857	.6866	.6875	.6884	.6893
4.9	.6902	.6911	.6920	.6928	.6937	.6946	.6955	.6964	.6972	.6981
5.0	.6990	.6998	.7007	.7016	.7024	.7033	.7042	.7050	.7059	.7067
5.1	.7076	.7084	.7093	.7101	.7110	.7118	.7126	.7135	.7143	.7152
5.2	.7160	.7168	.7177	.7185	.7193	.7202	.7210	.7218	.7226	.7235
5.3	.7243	.7251	.7259	.7267	.7275	.7284	.7292	.7300	.7308	.7316
5.4	.7324	.7332	.7340	.7348	.7356	.7364	.7372	.7380	.7388	.7396
n	0	1	2	3	4	5	6	7	8	9

TABLE 2   COMMON LOGARITHMS (Continued)

n	0	1	2	3	4	5	6	7	8	9
5.5	.7404	.7412	.7419	.7427	.7435	.7443	.7451	.7459	.7466	.7474
5.6	.7482	.7490	.7497	.7505	.7513	.7520	.7528	.7536	.7543	.7551
5.7	.7559	.7566	.7574	.7582	.7589	.7597	.7604	.7612	.7619	.7627
5.8	.7634	.7642	.7649	.7657	.7664	.7672	.7679	.7686	.7694	.7701
5.9	.7709	.7716	.7723	.7731	.7738	.7745	.7752	.7760	.7767	.7774
6.0	.7782	.7789	.7796	.7803	.7810	.7818	.7825	.7832	.7839	.7846
6.1	.7853	.7860	.7868	.7875	.7882	.7889	.7896	.7903	.7910	.7917
6.2	.7924	.7931	.7938	.7945	.7952	.7959	.7966	.7973	.7980	.7987
6.3	.7993	.8000	.8007	.8014	.8021	.8028	.8035	.8041	.8048	.8055
6.4	.8062	.8069	.8075	.8082	.8089	.8096	.8102	.8109	.8116	.8122
6.5	.8129	.8136	.8142	.8149	.8156	.8162	.8169	.8176	.8182	.8189
6.6	.8195	.8202	.8209	.8215	.8222	.8228	.8235	.8241	.8248	.8254
6.7	.8261	.8267	.8274	.8280	.8287	.8293	.8299	.8306	.8312	.8319
6.8	.8325	.8331	.8338	.8344	.8351	.8357	.8363	.8370	.8376	.8382
6.9	.8388	.8395	.8401	.8407	.8414	.8420	.8426	.8432	.8439	.8445
7.0	.8451	.8457	.8463	.8470	.8476	.8482	.8488	.8494	.8500	.8506
7.1	.8513	.8519	.8525	.8531	.8537	.8543	.8549	.8555	.8561	.8567
7.2	.8573	.8579	.8585	.8591	.8597	.8603	.8609	.8615	.8621	.8627
7.3	.8633	.8639	.8645	.8651	.8657	.8663	.8669	.8675	.8681	.8686
7.4	.8692	.8698	.8704	.8710	.8716	.8722	.8727	.8733	.8739	.8745
7.5	.8751	.8756	.8762	.8768	.8774	.8779	.8785	.8791	.8797	.8802
7.6	.8808	.8814	.8820	.8825	.8831	.8837	.8842	.8848	.8854	.8859
7.7	.8865	.8871	.8876	.8882	.8887	.8893	.8899	.8904	.8910	.8915
7.8	.8921	.8927	.8932	.8938	.8943	.8949	.8954	.8960	.8965	.8971
7.9	.8976	.8982	.8987	.8993	.8998	.9004	.9009	.9015	.9020	.9025
8.0	.9031	.9036	.9042	.9047	.9053	.9058	.9063	.9069	.9074	.9079
8.1	.9085	.9090	.9096	.9101	.9106	.9112	.9117	.9122	.9128	.9133
8.2	.9138	.9143	.9149	.9154	.9159	.9165	.9170	.9175	.9180	.9186
8.3	.9191	.9196	.9201	.9206	.9212	.9217	.9222	.9227	.9232	.9238
8.4	.9243	.9248	.9253	.9258	.9263	.9269	.9274	.9279	.9284	.9289
8.5	.9294	.9299	.9304	.9309	.9315	.9320	.9325	.9330	.9335	.9340
8.6	.9345	.9350	.9355	.9360	.9365	.9370	.9375	.9380	.9385	.9390
8.7	.9395	.9400	.9405	.9410	.9415	.9420	.9425	.9430	.9435	.9440
8.8	.9445	.9450	.9455	.9460	.9465	.9469	.9474	.9479	.9484	.9489
8.9	.9494	.9499	.9504	.9509	.9513	.9518	.9523	.9528	.9533	.9538
9.0	.9542	.9547	.9552	.9557	.9562	.9566	.9571	.9576	.9581	.9586
9.1	.9590	.9595	.9600	.9605	.9609	.9614	.9619	.9624	.9628	.9633
9.2	.9638	.9643	.9647	.9652	.9657	.9661	.9666	.9671	.9675	.9680
9.3	.9685	.9689	.9694	.9699	.9703	.9708	.9713	.9717	.9722	.9727
9.4	.9731	.9736	.9741	.9745	.9750	.9754	.9759	.9763	.9768	.9773
9.5	.9777	.9782	.9786	.9791	.9795	.9800	.9805	.9809	.9814	.9818
9.6	.9823	.9827	.9832	.9836	.9841	.9845	.9850	.9854	.9859	.9863
9.7	.9868	.9872	.9877	.9881	.9886	.9890	.9894	.9899	.9903	.9908
9.8	.9912	.9917	.9921	.9926	.9930	.9934	.9939	.9943	.9948	.9952
9.9	.9956	.9961	.9965	.9969	.9974	.9978	.9983	.9987	.9991	.9996
n	0	1	2	3	4	5	6	7	8	9

TABLE 3   POWERS OF e

x	e^x	e^{-x}	x	e^x	e^{-x}
0.00	1.00000	1.00000	1.60	4.95302	0.20189
0.01	1.01005	0.99004	1.70	5.47394	0.18268
0.02	1.02020	0.98019	1.80	6.04964	0.16529
0.03	1.03045	0.97044	1.90	6.68589	0.14956
0.04	1.04081	0.96078	2.00	7.38905	0.13533
0.05	1.05127	0.95122			
0.06	1.06183	0.94176	2.10	8.16616	0.12245
0.07	1.07250	0.93239	2.20	9.02500	0.11080
0.08	1.08328	0.92311	2.30	9.97417	0.10025
0.09	1.09417	0.91393	2.40	11.02316	0.09071
0.10	1.10517	0.90483	2.50	12.18248	0.08208
			2.60	13.46372	0.07427
0.11	1.11628	0.89583	2.70	14.87971	0.06720
0.12	1.12750	0.88692	2.80	16.44463	0.06081
0.13	1.13883	0.87810	2.90	18.17412	0.05502
0.14	1.15027	0.86936	3.00	20.08551	0.04978
0.15	1.16183	0.86071			
0.16	1.17351	0.85214	3.50	33.11545	0.03020
0.17	1.18530	0.84366	4.00	54.59815	0.01832
0.18	1.19722	0.83527	4.50	90.01713	0.01111
0.19	1.20925	0.82696			
			5.00	148.41316	0.00674
0.20	1.22140	0.81873	5.50	224.69193	0.00409
0.30	1.34985	0.74081			
0.40	1.49182	0.67032	6.00	403.42879	0.00248
0.50	1.64872	0.60653	6.50	665.14163	0.00150
0.60	1.82211	0.54881			
0.70	2.01375	0.49658	7.00	1096.63316	0.00091
0.80	2.22554	0.44932	7.50	1808.04241	0.00055
0.90	2.45960	0.40656	8.00	2980.95799	0.00034
1.00	2.71828	0.36787	8.50	4914.76884	0.00020
1.10	3.00416	0.33287			
1.20	3.32011	0.30119	9.00	8130.08392	0.00012
1.30	3.66929	0.27253	9.50	13359.72683	0.00007
1.40	4.05519	0.24659			
1.50	4.48168	0.22313	10.00	22026.46579	0.00005

TABLE 4 NATURAL LOGARITHMS

x	ln x	x	ln x	x	ln x
0.0		4.5	1.5041	9.0	2.1972
0.1	−2.3026	4.6	1.5261	9.1	2.2083
0.2	−1.6094	4.7	1.5476	9.2	2.2192
0.3	−1.2040	4.8	1.5686	9.3	2.2300
0.4	−0.9163	4.9	1.5892	9.4	2.2407
0.5	−0.6931	5.0	1.6094	9.5	2.2513
0.6	−0.5108	5.1	1.6292	9.6	2.2618
0.7	−0.3567	5.2	1.6487	9.7	2.2721
0.8	−0.2231	5.3	1.6677	9.8	2.2824
0.9	−0.1054	5.4	1.6864	9.9	2.2925
1.0	0.0000	5.5	1.7047	10	2.3026
1.1	0.0953	5.6	1.7228	11	2.3979
1.2	0.1823	5.7	1.7405	12	2.4849
1.3	0.2624	5.8	1.7579	13	2.5649
1.4	0.3365	5.9	1.7750	14	2.6391
1.5	0.4055	6.0	1.7918	15	2.7081
1.6	0.4700	6.1	1.8083	16	2.7726
1.7	0.5306	6.2	1.8245	17	2.8332
1.8	0.5878	6.3	1.8405	18	2.8904
1.9	0.6419	6.4	1.8563	19	2.9444
2.0	0.6931	6.5	1.8718	20	2.9957
2.1	0.7419	6.6	1.8871	25	3.2189
2.2	0.7885	6.7	1.9021	30	3.4012
2.3	0.8329	6.8	1.9169	35	3.5553
2.4	0.8755	6.9	1.9315	40	3.6889
2.5	0.9163	7.0	1.9459	45	3.8067
2.6	0.9555	7.1	1.9601	50	3.9120
2.7	0.9933	7.2	1.9741	55	4.0073
2.8	1.0296	7.3	1.9879	60	4.0943
2.9	1.0647	7.4	2.0015	65	4.1744
3.0	1.0986	7.5	2.0149	70	4.2485
3.1	1.1314	7.6	2.0281	75	4.3175
3.2	1.1632	7.7	2.0412	80	4.3820
3.3	1.1939	7.8	2.0541	85	4.4427
3.4	1.2238	7.9	2.0669	90	4.4998
3.5	1.2528	8.0	2.0794	95	4.5539
3.6	1.2809	8.1	2.0919	100	4.6052
3.7	1.3083	8.2	2.1041		
3.8	1.3350	8.3	2.1163		
3.9	1.3610	8.4	2.1281		
4.0	1.3863	8.5	2.1401		
4.1	1.4110	8.6	2.1518		
4.2	1.4351	8.7	2.1633		
4.3	1.4586	8.8	2.1748		
4.4	1.4816	8.9	2.1861		

TABLE 5 TRIGONOMETRIC FUNCTIONS IN DEGREES AND RADIANS

θ (degrees)	θ (radians)	sin θ	cos θ	tan θ	cot θ	sec θ	csc θ	θ (radians)	(degrees)
7°00'	.1222	.1219	.9925	.1228	8.144	1.008	8.206	1.4486	83°00'
10	.1251	.1248	.9922	.1257	7.953	1.008	8.016	1.4457	50
20	.1280	.1276	.9918	.1287	7.770	1.008	7.834	1.4428	40
30	.1309	.1305	.9914	.1317	7.596	1.009	7.661	1.4399	30
40	.1338	.1334	.9911	.1346	7.429	1.009	7.496	1.4370	20
50	.1367	.1363	.9907	.1376	7.269	1.009	7.337	1.4341	10
8°00'	.1396	.1392	.9903	.1405	7.115	1.010	7.185	1.4312	82°00'
10	.1425	.1421	.9899	.1435	6.968	1.010	7.040	1.4283	50
20	.1454	.1449	.9894	.1465	6.827	1.011	6.900	1.4254	40
30	.1484	.1478	.9890	.1495	6.691	1.011	6.765	1.4224	30
40	.1513	.1507	.9886	.1524	6.561	1.012	6.636	1.4195	20
50	.1542	.1536	.9881	.1554	6.435	1.012	6.512	1.4166	10
9°00'	.1571	.1564	.9877	.1584	6.314	1.012	6.392	1.4137	81°00'
10	.1600	.1593	.9872	.1614	6.197	1.013	6.277	1.4108	50
20	.1629	.1622	.9868	.1644	6.084	1.013	6.166	1.4079	40
30	.1658	.1650	.9863	.1673	5.976	1.014	6.059	1.4050	30
40	.1687	.1679	.9858	.1703	5.871	1.014	5.955	1.4021	20
50	.1716	.1708	.9853	.1733	5.769	1.015	5.855	1.3992	10
10°00'	.1745	.1736	.9848	.1763	5.671	1.015	5.759	1.3963	80°00'
10	.1774	.1765	.9843	.1793	5.576	1.016	5.665	1.3934	50
20	.1804	.1794	.9838	.1823	5.485	1.016	5.575	1.3904	40
30	.1833	.1822	.9833	.1853	5.396	1.017	5.487	1.3875	30
40	.1862	.1851	.9827	.1883	5.309	1.018	5.403	1.3846	20
50	.1891	.1880	.9822	.1914	5.226	1.018	5.320	1.3817	10
11°00'	.1920	.1908	.9816	.1944	5.145	1.019	5.241	1.3788	79°00'
10	.1949	.1937	.9811	.1974	5.066	1.019	5.164	1.3759	50
20	.1978	.1965	.9805	.2004	4.989	1.020	5.089	1.3730	40
30	.2007	.1994	.9799	.2035	4.915	1.020	5.016	1.3701	30
40	.2036	.2022	.9793	.2065	4.843	1.021	4.945	1.3672	20
50	.2065	.2051	.9787	.2095	4.773	1.022	4.876	1.3643	10
12°00'	.2094	.2079	.9781	.2126	4.705	1.022	4.810	1.3614	78°00'
10	.2123	.2108	.9775	.2156	4.638	1.023	4.745	1.3584	50
20	.2153	.2136	.9769	.2186	4.574	1.024	4.682	1.3555	40
30	.2182	.2164	.9763	.2217	4.511	1.024	4.620	1.3526	30
40	.2211	.2193	.9757	.2247	4.449	1.025	4.560	1.3497	20
50	.2240	.2221	.9750	.2278	4.390	1.026	4.502	1.3468	10
13°00'	.2269	.2250	.9744	.2309	4.331	1.026	4.445	1.3439	77°00'
10	.2298	.2278	.9737	.2339	4.275	1.027	4.390	1.3410	50
20	.2327	.2306	.9730	.2370	4.219	1.028	4.336	1.3381	40
30	.2356	.2334	.9724	.2401	4.165	1.028	4.284	1.3352	30
40	.2385	.2363	.9717	.2432	4.113	1.029	4.232	1.3323	20
50	.2414	.2391	.9710	.2462	4.061	1.030	4.182	1.3294	10
		cos θ	sin θ	cot θ	tan θ	csc θ	sec θ	θ (radians)	(degrees)

θ (degrees)	θ (radians)	sin θ	cos θ	tan θ	cot θ	sec θ	csc θ	θ (radians)	(degrees)
0°00'	.0000	.0000	1.0000	.0000		1.000		1.5708	90°00'
10	.0029	.0029	1.0000	.0029	343.8	1.000	343.8	1.5679	50
20	.0058	.0058	1.0000	.0058	171.9	1.000	171.9	1.5650	40
30	.0087	.0087	1.0000	.0087	114.6	1.000	114.6	1.5621	30
40	.0116	.0116	.9999	.0116	85.94	1.000	85.95	1.5592	20
50	.0145	.0145	.9999	.0145	68.75	1.000	68.76	1.5563	10
1°00'	.0175	.0175	.9998	.0175	57.29	1.000	57.30	1.5533	89°00'
10	.0204	.0204	.9998	.0204	49.10	1.000	49.11	1.5504	50
20	.0233	.0233	.9997	.0233	42.96	1.000	42.98	1.5475	40
30	.0262	.0262	.9997	.0262	38.19	1.000	38.20	1.5446	30
40	.0291	.0291	.9996	.0291	34.37	1.000	34.38	1.5417	20
50	.0320	.0320	.9995	.0320	31.24	1.001	31.26	1.5388	10
2°00'	.0349	.0349	.9994	.0349	28.64	1.001	28.65	1.5359	88°00'
10	.0378	.0378	.9993	.0378	26.43	1.001	26.45	1.5330	50
20	.0407	.0407	.9992	.0407	24.54	1.001	24.56	1.5301	40
30	.0436	.0436	.9990	.0437	22.90	1.001	22.93	1.5272	30
40	.0465	.0465	.9989	.0466	21.47	1.001	21.49	1.5243	20
50	.0495	.0494	.9988	.0495	20.21	1.001	20.23	1.5213	10
3°00'	.0524	.0523	.9986	.0524	19.08	1.001	19.11	1.5184	87°00'
10	.0553	.0552	.9985	.0553	18.07	1.002	18.10	1.5155	50
20	.0582	.0581	.9983	.0582	17.17	1.002	17.20	1.5126	40
30	.0611	.0610	.9981	.0612	16.35	1.002	16.38	1.5097	30
40	.0640	.0640	.9980	.0641	15.60	1.002	15.64	1.5068	20
50	.0669	.0669	.9978	.0670	14.92	1.002	14.96	1.5039	10
4°00'	.0698	.0698	.9976	.0699	14.30	1.002	14.34	1.5010	86°00'
10	.0727	.0727	.9974	.0729	13.73	1.003	13.76	1.4981	50
20	.0756	.0756	.9971	.0758	13.20	1.003	13.23	1.4952	40
30	.0785	.0785	.9969	.0787	12.71	1.003	12.75	1.4923	30
40	.0814	.0814	.9967	.0816	12.25	1.003	12.29	1.4893	20
50	.0844	.0843	.9964	.0846	11.83	1.004	11.87	1.4864	10
5°00'	.0873	.0872	.9962	.0875	11.43	1.004	11.47	1.4835	85°00'
10	.0902	.0901	.9959	.0904	11.06	1.004	11.10	1.4806	50
20	.0931	.0929	.9957	.0934	10.71	1.004	10.76	1.4777	40
30	.0960	.0958	.9954	.0963	10.39	1.005	10.43	1.4748	30
40	.0989	.0987	.9951	.0992	10.08	1.005	10.13	1.4719	20
50	.1018	.1016	.9948	.1022	9.788	1.005	9.839	1.4690	10
6°00'	.1047	.1045	.9945	.1051	9.514	1.006	9.567	1.4661	84°00'
10	.1076	.1074	.9942	.1080	9.255	1.006	9.309	1.4632	50
20	.1105	.1103	.9939	.1110	9.010	1.006	9.065	1.4603	40
30	.1134	.1134	.9936	.1139	8.777	1.006	8.834	1.4573	30
40	.1164	.1161	.9932	.1169	8.556	1.007	8.614	1.4544	20
50	.1193	.1190	.9929	.1198	8.345	1.007	8.405	1.4515	10
		cos θ	sin θ	cot θ	tan θ	csc θ	sec θ	θ (radians)	(degrees)

TABLE 5 TRIGONOMETRIC FUNCTIONS IN DEGREES AND RADIANS (Continued)

T7

θ (degrees)	θ (radians)	sin θ	cos θ	tan θ	cot θ	sec θ	csc θ	(radians)	(degrees)
21°00'	.3665	.3584	.9336	.3839	2.605	1.071	2.790	1.2043	69°00'
10	.3694	.3611	.9325	.3872	2.583	1.072	2.769	1.2014	50
20	.3723	.3638	.9315	.3906	2.560	1.074	2.749	1.1985	40
30	.3752	.3665	.9304	.3939	2.539	1.075	2.729	1.1956	30
40	.3782	.3692	.9293	.3973	2.517	1.076	2.709	1.1926	20
50	.3811	.3719	.9283	.4006	2.496	1.077	2.689	1.1897	10
22°00'	.3840	.3746	.9272	.4040	2.475	1.079	2.669	1.1868	68°00'
10	.3869	.3773	.9261	.4074	2.455	1.080	2.650	1.1839	50
20	.3898	.3800	.9250	.4108	2.434	1.081	2.632	1.1810	40
30	.3927	.3827	.9239	.4142	2.414	1.082	2.613	1.1781	30
40	.3956	.3854	.9228	.4176	2.394	1.084	2.595	1.1752	20
50	.3985	.3881	.9216	.4210	2.375	1.085	2.577	1.1723	10
23°00'	.4014	.3907	.9205	.4245	2.356	1.086	2.559	1.1694	67°00'
10	.4043	.3934	.9194	.4279	2.337	1.088	2.542	1.1665	50
20	.4072	.3961	.9182	.4314	2.318	1.089	2.525	1.1636	40
30	.4102	.3987	.9171	.4348	2.300	1.090	2.508	1.1606	30
40	.4131	.4014	.9159	.4383	2.282	1.092	2.491	1.1577	20
50	.4160	.4041	.9147	.4417	2.264	1.093	2.475	1.1548	10
24°00'	.4189	.4067	.9135	.4452	2.246	1.095	2.459	1.1519	66°00'
10	.4218	.4094	.9124	.4487	2.229	1.096	2.443	1.1490	50
20	.4247	.4120	.9112	.4522	2.211	1.097	2.427	1.1461	40
30	.4276	.4147	.9100	.4557	2.194	1.099	2.411	1.1432	30
40	.4305	.4173	.9088	.4592	2.177	1.100	2.396	1.1403	20
50	.4334	.4200	.9075	.4628	2.161	1.102	2.381	1.1374	10
25°00'	.4363	.4226	.9063	.4663	2.145	1.103	2.366	1.1345	65°00'
10	.4392	.4253	.9051	.4699	2.128	1.105	2.352	1.1316	50
20	.4422	.4279	.9038	.4734	2.112	1.106	2.337	1.1286	40
30	.4451	.4305	.9026	.4770	2.097	1.108	2.323	1.1257	30
40	.4480	.4331	.9013	.4806	2.081	1.109	2.309	1.1228	20
50	.4509	.4358	.9001	.4841	2.066	1.111	2.295	1.1199	10
26°00'	.4538	.4384	.8988	.4877	2.050	1.113	2.281	1.1170	64°00'
10	.4567	.4410	.8975	.4913	2.035	1.114	2.268	1.1141	50
20	.4596	.4436	.8962	.4950	2.020	1.116	2.254	1.1112	40
30	.4625	.4462	.8949	.4986	2.006	1.117	2.241	1.1083	30
40	.4654	.4488	.8936	.5022	1.991	1.119	2.228	1.1054	20
50	.4683	.4514	.8923	.5059	1.977	1.121	2.215	1.1025	10
27°00'	.4712	.4540	.8910	.5095	1.963	1.122	2.203	1.0996	63°00'
10	.4741	.4566	.8897	.5132	1.949	1.124	2.190	1.0966	50
20	.4771	.4592	.8884	.5169	1.935	1.126	2.178	1.0937	40
30	.4800	.4617	.8870	.5206	1.921	1.127	2.166	1.0908	30
40	.4829	.4643	.8857	.5243	1.907	1.129	2.154	1.0879	20
50	.4858	.4669	.8843	.5280	1.894	1.131	2.142	1.0850	10
		cos θ	sin θ	cot θ	tan θ	csc θ	sec θ	(radians)	(degrees)
								θ	

θ (degrees)	θ (radians)	sin θ	cos θ	tan θ	cot θ	sec θ	csc θ	(radians)	(degrees)
14°00'	.2443	.2419	.9703	.2493	4.011	1.031	4.134	1.3265	76°00'
10	.2473	.2447	.9696	.2524	3.962	1.031	4.086	1.3235	50
20	.2502	.2476	.9689	.2555	3.914	1.032	4.039	1.3206	40
30	.2531	.2504	.9681	.2586	3.867	1.033	3.994	1.3177	30
40	.2560	.2532	.9674	.2617	3.821	1.034	3.950	1.3148	20
50	.2589	.2560	.9667	.2648	3.776	1.034	3.906	1.3119	10
15°00'	.2618	.2588	.9659	.2679	3.732	1.035	3.864	1.3090	75°00'
10	.2647	.2616	.9652	.2711	3.689	1.036	3.822	1.3061	50
20	.2676	.2644	.9644	.2742	3.647	1.037	3.782	1.3032	40
30	.2705	.2672	.9636	.2773	3.606	1.038	3.742	1.3003	30
40	.2734	.2700	.9628	.2805	3.566	1.039	3.703	1.2974	20
50	.2763	.2728	.9621	.2836	3.526	1.039	3.665	1.2945	10
16°00'	.2793	.2756	.9613	.2867	3.487	1.040	3.628	1.2915	74°00'
10	.2822	.2784	.9605	.2899	3.450	1.041	3.592	1.2886	50
20	.2851	.2812	.9596	.2931	3.412	1.042	3.556	1.2857	40
30	.2880	.2840	.9588	.2962	3.376	1.043	3.521	1.2828	30
40	.2909	.2868	.9580	.2994	3.340	1.044	3.487	1.2799	20
50	.2938	.2896	.9572	.3026	3.305	1.045	3.453	1.2770	10
17°00'	.2967	.2924	.9563	.3057	3.271	1.046	3.420	1.2741	73°00'
10	.2996	.2952	.9555	.3089	3.237	1.047	3.388	1.2712	50
20	.3025	.2979	.9546	.3121	3.204	1.048	3.356	1.2683	40
30	.3054	.3007	.9537	.3153	3.172	1.049	3.326	1.2654	30
40	.3083	.3035	.9528	.3185	3.140	1.049	3.295	1.2625	20
50	.3113	.3062	.9520	.3217	3.108	1.050	3.265	1.2595	10
18°00'	.3142	.3090	.9511	.3249	3.078	1.051	3.236	1.2566	72°00'
10	.3171	.3118	.9502	.3281	3.047	1.052	3.207	1.2537	50
20	.3200	.3145	.9492	.3314	3.018	1.053	3.179	1.2508	40
30	.3229	.3173	.9483	.3346	2.989	1.054	3.152	1.2479	30
40	.3258	.3201	.9474	.3378	2.960	1.056	3.124	1.2450	20
50	.3287	.3228	.9465	.3411	2.932	1.057	3.098	1.2421	10
19°00'	.3316	.3256	.9455	.3443	2.904	1.058	3.072	1.2392	71°00'
10	.3345	.3283	.9446	.3476	2.877	1.059	3.046	1.2363	50
20	.3374	.3311	.9436	.3508	2.850	1.060	3.021	1.2334	40
30	.3403	.3338	.9426	.3541	2.824	1.061	2.996	1.2305	30
40	.3432	.3365	.9417	.3574	2.798	1.062	2.971	1.2275	20
50	.3462	.3393	.9407	.3607	2.773	1.063	2.947	1.2246	10
20°00'	.3491	.3420	.9397	.3640	2.747	1.064	2.924	1.2217	70°00'
10	.3520	.3448	.9387	.3673	2.723	1.065	2.901	1.2188	50
20	.3549	.3475	.9377	.3706	2.699	1.066	2.878	1.2159	40
30	.3578	.3502	.9367	.3739	2.675	1.068	2.855	1.2130	30
40	.3607	.3529	.9356	.3772	2.651	1.069	2.833	1.2101	20
50	.3636	.3557	.9346	.3805	2.628	1.070	2.812	1.2072	10
		cos θ	sin θ	cot θ	tan θ	csc θ	sec θ	(radians)	(degrees)
								θ	

TABLE 5 TRIGONOMETRIC FUNCTIONS IN DEGREES AND RADIANS (Continued)

θ (degrees)	θ (radians)	sin θ	cos θ	tan θ	cot θ	sec θ	csc θ	(radians)	(degrees)
28°00'	.4887	.4695	.8829	.5317	1.881	1.133	2.130	1.0821	62°00'
10	.4916	.4720	.8816	.5354	1.868	1.134	2.118	1.0792	50
20	.4945	.4746	.8802	.5392	1.855	1.136	2.107	1.0763	40
30	.4974	.4772	.8788	.5430	1.842	1.138	2.096	1.0734	30
40	.5003	.4797	.8774	.5467	1.829	1.140	2.085	1.0705	20
50	.5032	.4823	.8760	.5505	1.816	1.142	2.074	1.0676	10
29°00'	.5061	.4848	.8746	.5543	1.804	1.143	2.063	1.0647	61°00'
10	.5091	.4874	.8732	.5581	1.792	1.145	2.052	1.0617	50
20	.5120	.4899	.8718	.5619	1.780	1.147	2.041	1.0588	40
30	.5149	.4924	.8704	.5658	1.767	1.149	2.031	1.0559	30
40	.5178	.4950	.8689	.5696	1.756	1.151	2.020	1.0530	20
50	.5207	.4975	.8675	.5735	1.744	1.153	2.010	1.0501	10
30°00'	.5236	.5000	.8660	.5774	1.732	1.155	2.000	1.0472	60°00'
10	.5265	.5025	.8646	.5812	1.720	1.157	1.990	1.0443	50
20	.5294	.5050	.8631	.5851	1.709	1.159	1.980	1.0414	40
30	.5323	.5075	.8616	.5890	1.698	1.161	1.970	1.0385	30
40	.5352	.5100	.8601	.5930	1.686	1.163	1.961	1.0356	20
50	.5381	.5125	.8587	.5969	1.675	1.165	1.951	1.0327	10
31°00'	.5411	.5150	.8572	.6009	1.664	1.167	1.942	1.0297	59°00'
10	.5440	.5175	.8557	.6048	1.653	1.169	1.932	1.0268	50
20	.5469	.5200	.8542	.6088	1.643	1.171	1.923	1.0239	40
30	.5498	.5225	.8526	.6128	1.632	1.173	1.914	1.0210	30
40	.5527	.5250	.8511	.6168	1.621	1.175	1.905	1.0181	20
50	.5556	.5275	.8496	.6208	1.611	1.177	1.896	1.0152	10
32°00'	.5585	.5299	.8480	.6249	1.600	1.179	1.887	1.0123	58°00'
10	.5614	.5324	.8465	.6289	1.590	1.181	1.878	1.0094	50
20	.5643	.5348	.8450	.6330	1.580	1.184	1.870	1.0065	40
30	.5672	.5373	.8434	.6371	1.570	1.186	1.861	1.0036	30
40	.5701	.5398	.8418	.6412	1.560	1.188	1.853	1.0007	20
50	.5730	.5422	.8403	.6453	1.550	1.190	1.844	.9977	10
33°00'	.5760	.5446	.8387	.6494	1.540	1.192	1.836	.9948	57°00'
10	.5789	.5471	.8371	.6536	1.530	1.195	1.828	.9919	50
20	.5818	.5495	.8355	.6577	1.520	1.197	1.820	.9890	40
30	.5847	.5519	.8339	.6619	1.511	1.199	1.812	.9861	30
40	.5876	.5544	.8323	.6661	1.501	1.202	1.804	.9832	20
50	.5905	.5568	.8307	.6703	1.492	1.204	1.796	.9803	10
34°00'	.5934	.5592	.8290	.6745	1.483	1.206	1.788	.9774	56°00'
10	.5963	.5616	.8274	.6787	1.473	1.209	1.781	.9745	50
20	.5992	.5640	.8258	.6830	1.464	1.211	1.773	.9716	40
30	.6021	.5664	.8241	.6873	1.455	1.213	1.766	.9687	30
40	.6050	.5688	.8225	.6916	1.446	1.216	1.758	.9657	20
50	.6080	.5712	.8208	.6959	1.437	1.218	1.751	.9628	10
		cos θ	sin θ	cot θ	tan θ	csc θ	sec θ	(radians)	(degrees)

(degrees)	(radians)	sin θ	cos θ	tan θ	cot θ	sec θ	csc θ	(radians)	(degrees)
35°00'	.6109	.5736	.8192	.7002	1.428	1.221	1.743	.9599	55°00'
10	.6138	.5760	.8175	.7046	1.419	1.223	1.736	.9570	50
20	.6167	.5783	.8158	.7089	1.411	1.226	1.729	.9541	40
30	.6196	.5807	.8141	.7133	1.402	1.228	1.722	.9512	30
40	.6225	.5831	.8124	.7177	1.393	1.231	1.715	.9483	20
50	.6254	.5854	.8107	.7221	1.385	1.233	1.708	.9454	10
36°00'	.6283	.5878	.8090	.7265	1.376	1.236	1.701	.9425	54°00'
10	.6312	.5901	.8073	.7310	1.368	1.239	1.695	.9396	50
20	.6341	.5925	.8056	.7355	1.360	1.241	1.688	.9367	40
30	.6370	.5948	.8039	.7400	1.351	1.244	1.681	.9338	30
40	.6400	.5972	.8021	.7445	1.343	1.247	1.675	.9308	20
50	.6429	.5995	.8004	.7490	1.335	1.249	1.668	.9279	10
37°00'	.6458	.6018	.7986	.7536	1.327	1.252	1.662	.9250	53°00'
10	.6487	.6041	.7969	.7581	1.319	1.255	1.655	.9221	50
20	.6516	.6065	.7951	.7627	1.311	1.258	1.649	.9192	40
30	.6545	.6088	.7934	.7673	1.303	1.260	1.643	.9163	30
40	.6574	.6111	.7916	.7720	1.295	1.263	1.636	.9134	20
50	.6603	.6134	.7898	.7766	1.288	1.266	1.630	.9105	10
38°00'	.6632	.6157	.7880	.7813	1.280	1.269	1.624	.9076	52°00'
10	.6661	.6180	.7862	.7860	1.272	1.272	1.618	.9047	50
20	.6690	.6202	.7844	.7907	1.265	1.275	1.612	.9018	40
30	.6720	.6225	.7826	.7954	1.257	1.278	1.606	.8988	30
40	.6749	.6248	.7808	.8002	1.250	1.281	1.601	.8959	20
50	.6778	.6271	.7790	.8050	1.242	1.284	1.595	.8930	10
39°00'	.6807	.6293	.7771	.8098	1.235	1.287	1.589	.8901	51°00'
10	.6836	.6316	.7753	.8146	1.228	1.290	1.583	.8872	50
20	.6865	.6338	.7735	.8195	1.220	1.293	1.578	.8843	40
30	.6894	.6361	.7716	.8243	1.213	1.296	1.572	.8814	30
40	.6923	.6383	.7698	.8292	1.206	1.299	1.567	.8785	20
50	.6952	.6406	.7679	.8342	1.199	1.302	1.561	.8756	10
40°00'	.6981	.6428	.7660	.8391	1.192	1.305	1.556	.8727	50°00'
10	.7010	.6450	.7642	.8441	1.185	1.309	1.550	.8698	50
20	.7039	.6472	.7623	.8491	1.178	1.312	1.545	.8668	40
30	.7069	.6494	.7604	.8541	1.171	1.315	1.540	.8639	30
40	.7098	.6517	.7585	.8591	1.164	1.318	1.535	.8610	20
50	.7127	.6539	.7566	.8642	1.157	1.322	1.529	.8581	10
41°00'	.7156	.6561	.7547	.8693	1.150	1.325	1.524	.8552	49°00'
10	.7185	.6583	.7528	.8744	1.144	1.328	1.519	.8523	50
20	.7214	.6604	.7509	.8796	1.137	1.332	1.514	.8494	40
30	.7243	.6626	.7490	.8847	1.130	1.335	1.509	.8465	30
40	.7272	.6648	.7470	.8899	1.124	1.339	1.504	.8436	20
50	.7301	.6670	.7451	.8952	1.117	1.342	1.499	.8407	10
		cos θ	sin θ	cot θ	tan θ	csc θ	sec θ	(radians)	(degrees)

TABLE 5   TRIGONOMETRIC FUNCTIONS IN DEGREES AND RADIANS (Continued)

θ (degrees)	θ (radians)	sin θ	cos θ	tan θ	cot θ	sec θ	csc θ		
42°00′	.7330	.6691	.7431	.9004	1.111	1.346	1.494	.8378	**48°00′**
10	.7359	.6713	.7412	.9057	1.104	1.349	1.490	.8348	50
20	.7389	.6734	.7392	.9110	1.098	1.353	1.485	.8319	40
30	.7418	.6756	.7373	.9163	1.091	1.356	1.480	.8290	30
40	.7447	.6777	.7353	.9217	1.085	1.360	1.476	.8261	20
50	.7476	.6799	.7333	.9271	1.079	1.364	1.471	.8232	10
43°00′	.7505	.6820	.7314	.9325	1.072	1.367	1.466	.8203	**47°00′**
10	.7534	.6841	.7294	.9380	1.066	1.371	1.462	.8174	50
20	.7563	.6862	.7274	.9435	1.060	1.375	1.457	.8145	40
30	.7592	.6884	.7254	.9490	1.054	1.379	1.453	.8116	30
40	.7621	.6905	.7234	.9545	1.048	1.382	1.448	.8087	20
50	.7560	.6926	.7214	.9601	1.042	1.386	1.444	.8058	10
44°00′	.7679	.6947	.7193	.9657	1.036	1.390	1.440	.8029	**46°00′**
10	.7709	.6967	.7173	.9713	1.030	1.394	1.435	.7999	50
20	.7738	.6988	.7153	.9770	1.024	1.398	1.431	.7970	40
30	.7767	.7009	.7133	.9827	1.018	1.402	1.427	.7941	30
40	.7796	.7030	.7112	.9884	1.012	1.406	1.423	.7912	20
50	.7825	.7050	.7092	.9942	1.006	1.410	1.418	.7883	10
		cos θ	sin θ	cot θ	tan θ	csc θ	sec θ	θ (radians)	θ (degrees)

Index